Profiles in Operations Research

Pioneers and Innovators

International Series in Operations Research & Management Science

Volume 147

Series Editor:
Frederick S. Hillier
Stanford University, CA, USA

Special Editorial Consultant:
Camille C. Price
Stephen N. Austin State University, TX, USA

For further volumes:
http://www.springer.com/series/6161

Profiles in Operations Research

Pioneers and Innovators

Written, compiled, and edited

by

Arjang A. Assad
School of Management
University at Buffalo, State University of New York

Saul I. Gass
Robert H. Smith School of Business
University of Maryland, College Park

Springer

Editors
Arjang A. Assad
School of Management
160 Jacobs Management Center
University at Buffalo
State University of New York
Buffalo 14260-4000
NY, USA

Saul I. Gass
Robert H. Smith School of Business
Van Munching Hall
University of Maryland
College Park 20742-1915
MD, USA

ISSN 0884-8289
ISBN 978-1-4419-6280-5 e-ISBN 978-1-4419-6281-2
DOI 10.1007/978-1-4419-6281-2
Springer New York Dordrecht Heidelberg London

Library of Congress Control Number: 2011922044

Printed on acid-free paper

Springer is part of Springer Science+Business Media (www.springer.com)

To the memory of my father,
who gave me my first history books.

Arjang A. Assad

To Trudy

For her continued encouragement,
patience, and
lost weekends

Saul I. Gass

PREFACE

PROLOGUE

Operations research (OR) had its origins in the late 1930s when a group of British Royal Air Force officers and civilian scientists were asked to determine how recently developed radar technology could be used for controlled interception of enemy aircraft. The task required scientists from various disciplines to leave their academic and laboratory settings and participate in field operations, testing, and evaluation. This embryonic but seminal applied research activity was followed by the formation of OR groups to investigate military problems encountered by the services on land, sea, or in the air. The resulting methodological approach was called operational research in the U.K. and operations research in the U.S. The salient feature of this research activity was to bring scientists from diverse disciplines directly into the field to solve operational military problems and make related tactical and strategic recommendations. Throughout World War II (WWII), on both sides of the Atlantic, OR groups analyzed critical military problems. The solutions they produced led to changes in military strategy and tactics that greatly improved results. OR helped the Allies win the war.

At the end of WWII, the U.S. military services, recognizing the wartime contributions of OR, continued their support of OR groups with the problems now focused on logistics, combat modeling, and force planning. Similarly, senior scientists who had participated in wartime OR in the U.K and U.S. were convinced that OR could be used to solve management and operational problems of nonmilitary enterprises and government. OR groups were soon formed and employed by governmental entities and private companies in the U.K. and U.S., professional societies organized, OR consultancies established, and academic OR programs initiated. Today, OR is recognized worldwide as a modern, decision-aiding science that has proved to be of great value to management, business, and industry. The history of how this came about has been reported in many papers and books in the usual narrative form. In contrast, we chose to present this history by means of individual profiles of OR's pioneers and innovators.

What is Operations Research?

"Operations Research is a scientific method for providing executive departments with a quantitative basis for decisions regarding the operations under their control" [Goodeve, C. 1948. Operational Research. *Nature* **161**(4089) 377–384].

Operations Research is defined by "mathematical or scientific analysis of the systematic efficiency and performance of manpower, machinery, equipment, and policies used in a governmental, military, or commercial operation" (*The American Heritage Dictionary of the English Language* 1976, 921).

Operations Research is the science of decision making.

OR is a relatively young field. Many of its developers are still alive and records of their accomplishments are available from them and/or from colleagues and friends. Similarly, for those who have passed on, writings and reports of their OR activities are still reasonably accessible, and can be amplified with the memories of close collaborators or friends. These fortunate circumstances related to timely access led us to believe that the history of OR could be told in a comprehensive and interesting manner through individual profiles. We felt that we had a window of opportunity to capture the story of these pioneers as they charted their courses through the early years of OR and saw it mature as a field. To accomplish this objective, we drew upon historical and technical articles, autobiographies, biographies, interviews, the resources available via the World-Wide Web, and existing accounts of the history of OR. And, most important, we were able to find authors who, as close colleagues or friends, were in the best position to relate the professional and personal histories of the persons they profiled.

We have organized the 43 profiles of this book chronologically, in ascending order of the date of birth (by day, month, and year) of the profilees. The reader progressing from Patrick Blackett's profile (Chapter 1, born in 1897) through Ronald Howard's profile (Chapter 43, born in 1934), will be exposed to a history of the origins and early development of

OR—how it evolved—all interwoven with personal backgrounds, tales, vignettes, and pictures. An alphabetical listing of the profiles follows:

Profiles		
Russell Lincoln Ackoff	Saul I. Gass	Hugh Jordan Miser
E. Leonard Amoff	Murray Aaron Geisler	Philip McCord Morse
Egon Balas	Ralph E. Gomory	Howard Raiffa
Evelyn Martin Lansdowne Beale	Charles Frederick Goodeve	Berwyn Hugh Patrick Rivett
Anthony Stafford Beer	David Bendel Hertz	Bernard Roy
Richard E. Bellman	Ronald A. Howard	Thomas L. Saaty
Patrick Blackett	Ellis A. Johnson	Herbert A. Simon
Alfred Blumstein	Leonid Vital'evich Kantorovich	Jacinto Steinhardt
Seth Bonder	George E. Kimball	Albert W. Tucker
Abraham Charnes	George Kozmetsky	Steven Vajda
C. West Churchman	Harold W. Kuhn	Andrew Vazsonyi
William W. Cooper	John D. C. Little	John von Neumann
George B. Dantzig	John F. Magee	Harvey M. Wagner
Jay Wright Forrester	Harry Markowitz	Philip Starr Wolfe
D. Ray Fulkerson		

The main table of contents is given by name in birth order, followed by a secondary table of contents by name in alphabetical order.

Two tables in the Appendix summarize key information about the profilees: Table 1-Profiles Background, and Table 2-Profiles Honors and Awards. In Table 1, we list date and place of birth, and the discipline and year of the associated highest earned academic degree. From the dates of birth, we note an interesting feature of the early years of OR: the majority of those profiled fall into a fairly narrow generational band. The dates of birth of 31 of the 43 profilees lie in the 20-year period from January 1912 to December 1931 (from Kantorovich to Wagner). This forms a fairly compact cluster of individuals who were between 14 and 33 years of age at the end of WWII. An even narrower band emerges if we consider the year of completion of their formal studies. Out of this group of 31 profilees, 23 completed their formal studies during the period 1945–1960. For the 43 individuals profiled, mathematics, not surprisingly, was the preponderate study area of 21 of the them. The first Ph.D. in OR was awarded in 1955.

In Table 2, we summarize the impressive achievements of the 43 profilees by citing their major honors and awards. Such recognition includes Nobel prizes, governmental medals, national scientific society memberships, and professional association medals and prizes.

Profiles in Operations Research has been written for OR practitioners, OR academics, students of OR, scientists from related fields, as well as the general public interested in the origins and accomplishments of an applied science. For this broad audience, we note that OR problems, OR techniques, and OR solutions are a part of everyone's daily life—when making an airline or hotel reservation, finding the desired products on a supermarket's shelves, joining a waiting line at a bank or post office, having Internet purchases arrive as scheduled, filling an auto's gas tank with the correctly blended fuel—all as a result of OR analyses.

The choice of the persons profiled was based on our collective knowledge of the field. Aiming for a single volume of reasonable size, we limited our initial selection to 50 subjects, all of whom had a seminal or major influence on the development and growth of OR. Over time, some of our original choices did not materialize because we could not identify suitable authors, or find sufficiently rich sources of information, or relax time constraints that precluded possible authors from participating.

The 43 profiles were written by 40 authors, including the two editors. We wish to thank and express our appreciation to the authors who worked with us over many, many months of editing, re-editing, and more re-editing. Each went the extra research mile when encountering comments and questions from the editors. We appreciate the personal perspectives they provided drawing upon their extensive and detailed knowledge of the individuals they profiled. In many cases, the authors were students or colleagues of the persons they profiled, who often collaborated on key research contributions that continue to shape OR theory and practice. While editing for some measure of uniformity, the authors were allowed to present the profiles their way.

We thank the many families and friends of the persons who are profiled for their help in answering queries from the authors and the editors, making source and other documents available, allowing us to quote from personal correspondence, and for the use of family pictures. Their cooperation is sincerely appreciated.

We thank Ron Gass for his editing of many of the profile pictures—he is an expert in obtaining blemish-free pictures while improving their clarity and color. We are especially appreciative to Kluwer Academic Publishers and its editor, Gary Folven, under whose guidance we contracted to write this book, and to Springer Science + Business Media, Inc. and Neil Levine, who became our Springer editor when Kluwer was incorporated into Springer. And, we thank Fred Hillier, the Kluwer/Springer OR series editor for his support.

Saul wishes to express his appreciation to: the Robert H. Smith School of Business, University of Maryland, College Park, for its general support, with special thanks to its computer technical staff who saw him through a hard drive failure and replacement, a software system breakdown, and a myriad of other inflictions of the computer age; the University of Maryland, College Park library for support and use of its link to JSTOR (Journal Storage)—the online system of digitally archived journals; the inter-library loan staff and service of the National Institute of Standards and Technology (NIST); and the support of NIST's Mathematical and Computational Sciences Division.

Arjang wishes to acknowledge the partial research support provided in summer 2008 by the Robert H. Smith School of Business of the University of Maryland where this project started. He also wishes to acknowledge the library resources provided by the University at Buffalo (SUNY) where the bulk of his work was carried out. Arjang also wishes to express special thanks to Charles Lyons, business/management librarian at the SUNY University at Buffalo Libraries for his prompt, diligent, and valuable help in locating and securing various sources and documents. Finally, Arjang thanks Leslie Russo of the Dean's office at the School of Management at the University of Buffalo for the ongoing assistance she provided during the course of this project.

EPILOGUE

Early in our education, we, Arjang and Saul, developed an interest in the history of science driven by our own curiosity in the sciences, especially mathematics. We augmented our schooling by biographical tales from Eric Temple Bell's *Men of Mathematics*, the story of how mathematics developed across the centuries from Carl Boyer's *A History of Mathematics* and, later, we moved on to such works as Norbert Weiner's two-volume autobiography,

Ex-Prodigy and *I am a Mathematician*, and Constance Reid's *Hilbert.* Thus, by the time each of us entered into a research-oriented course of training, we had developed an appreciation of the value of such historical accounts in shaping our views of how one matures as a scientist. Our interest in this area continued as we read John Stillwell's *Mathematics and its History*, or the collection of interviews in *Mathematical People,* edited by Donald Albers and Gerald Alexanderson. As we examined historical or biographical sources in other sciences, notably the physical, biological, and computer sciences, and came to appreciate the wealth of such sources, we were struck by the paucity of studies devoted to the history of OR.

In the past, there have been a few articles that recounted the early history of OR; Joseph McCloskey's three papers that appeared in *Operations Research* in 1984 come to mind. More recently, we have Maurice Kirby's book, *Operational Research in War and Peace,* and our book, *An Annotated Timeline of Operations Research: An Informal History*. We are believers in the need for such historical accounts. By neglecting this need, the failure to chronicle our field will mean that important lessons will be lost to the coming generations of OR researchers, practitioners, and students.

In writing and compiling *Profiles in Operations Research*, we were often stymied in our quest for information about particular profilees—there was little they wrote about except what was contained in their technical papers and reports. It would have been valuable to have background information, the why and how of the rest of the story. In contrast, other profilees diligently recorded their stories in articles, interviews, and autobiographies. These historical treasures enabled the profile author(s) to describe a more detailed and integrated account of a profilee's OR contributions and personal background. We encourage the reader who is engaged in OR innovations and developments to do the same. There is a need for historical accounts of all aspects of OR; we welcome others to delve into this fascinating field.

School of Management, University at Buffalo, State University of New York, Buffalo, NY, USA — Arjang A. Assad

Robert H. Smith School of Business, University of Maryland, College Park, MD, USA — Saul I. Gass

Quotations, References, and Sources

We have tried to be extra diligent in referencing material, especially the sources of quotes. U.S. English is used except for quotes and titles from a British English source. For a published book or article, quotes are cited by (author last name, year, page numbers); quotes from sources that are not paginated such as the Worldwide Web, letters, personal communications, and interviews, are cited by (author, year). Chapter reference formats: books are given by authors (last name first of initial author), year of publication, book title, publisher, city; papers (articles) are given by authors (last name first of initial author), year of publication, paper title, journal (or source), volume number, issue number in parentheses, followed by page numbers.

We wish to acknowledge and thank the many individuals and organizations that gave us permission to quote material and/or use pictures; their contributions are noted in the text. With respect to the pictures on the cover, we thank the following for their permission to use them: Patrick Blackett, courtesy Giovanna Blackett; William W. Cooper, © 2002 Peter Yang/Austin American-Statesman/Won; George B. Dantzig, Courtesy National Academy of Engineering; George E. Kimball, Courtesy National Academy of Science. To the best of our knowledge, and unless otherwise noted, unattributed pictures included in this publication fall under the fair use or public domain provisions of the United States copyright law. Upon reasonable notice and substantiation that a third party owns or controls the intellectual property rights to any of these pictures, we will remove them from any future printings in the event that good faith efforts by the parties fail to resolve any disputes.

CONTENTS

CONTENTS (ALPHABETICAL)

1

Patrick Blackett

Maurice W. Kirby and
Jonathan Rosenhead

Patrick Maynard Stuart Blackett was one of the most distinguished experimental physicists of the twentieth century. He was elected a fellow of the Royal Society of London in 1933 at the early age of 36. His 1948 Nobel Prize was in recognition of his fundamental contribution to knowledge and understanding of cosmic radiation and nuclear physics. Within operational research (OR), Blackett is a towering figure rightly known as the father of the subject, arising from his role in World War II (WWII) as an outstanding practitioner of the nascent discipline and in securing its diffusion throughout Britain's military command structure. Blackett's main contributions to military effectiveness were the result of applying the scientific method to the air defense of Great Britain and the anti-U-boat war in the North Atlantic.

Blackett received no public honor in Britain specifically in recognition of his wartime service, although he was awarded the U.S. Medal of Merit in 1946 for his pioneering work in military OR. In 1965, Blackett was designated by the U.K. Operational Research Society (ORS) as a Companion of Honour, and, in 1967, was appointed to the Order of Merit, Britain's highest civilian honor. In 1969, at the age of 72, he was awarded a life peerage as Baron Blackett of Chelsea. Blackett served as president of the Royal Society of London from 1965 to 1970. The ORS honors him by sponsoring the annual Blackett Memorial Lecture. In 2003, he was inducted into the International Federation of Operational Research Societies' (IFORS) OR Hall of Fame (Kirby 2003b). The moon's Blackett Crater is named after him.

A.A. Assad, S.I. Gass (eds.), *Profiles in Operations Research*, International Series in Operations Research & Management Science 147, DOI 10.1007/978-1-4419-6281-2_1,

THE NAVY AND ACADEMIC LIFE

Courtesy Giovanna Blackett

Patrick Blackett was born in Kensington, London, on November 18, 1897. His father, Arthur Stuart Blackett, was a reluctant stockbroker who was more interested in leisure time pursuits in nature and literature (Blackett 2003). While his father's family originated from north-east England with a tradition of small-scale farming, the forebears of his mother, Caroline Frances Maynard, were Shropshire landowners who wielded local political influence. On this side of the family there was a tradition of military service: Blackett's maternal grandfather had been a major in the Royal Artillery, serving in India at the time of the Indian Mutiny, and an earlier ancestor had been a sailor in the Royal Navy during the Napoleonic era.

Blackett had two sisters, the younger of whom, Marion Milner, became internationally influential on art therapy and, more generally, on psychoanalysis. He entered Osborne Naval College as a naval cadet in September 1910 when he was 12 years old. On the outbreak of World War I (WWI) in August 1914, when he was studying at Dartmouth Royal Naval College, Blackett was assigned to the cruiser HMS *Carnavon* with the junior rank of midshipman. In December of that year, his squadron engaged with several German warships in the Falkland Islands. In the ensuing battle, two German battleships and two cruisers were sunk with the British ships experiencing only minor damage.

Two years later, when he was 18 years old, Blackett experienced a far more substantial naval engagement while serving as a sub-lieutenant on HMS *Barham*, flagship of the Fifth Battle Squadron located in British home waters. The occasion was the Battle of Jutland when the Royal Navy had its only major encounter with the German High Seas Fleet. Although more British ships were sunk or damaged than their German counterparts, the fact that the High Seas Fleet withdrew to its homeports and remained inactive for the duration of the war, has led naval historians to conclude that the battle honors should be awarded to the Royal Navy.

HMS Barham (sunk in WWII)

At the end of the war, Blackett was first lieutenant on the destroyer HMS *Sturgeon*, serving as gunnery officer. In the autumn of 1918, he began to consider his future. His reasoning was as follows:

> I enjoyed my four years at sea during the war, but I was very doubtful if I would enjoy the peacetime Navy. There seemed to me to be two attitudes I might take if I decided to stay in the Navy. I could treat the Navy as providing a pleasant way of life and an introduction to the best clubs around the globe, or I could take the technological problems of naval warfare very seriously and so become orientated to fighting another war. As I put it to myself rather crudely: I enjoyed shooting at the enemy during the war—would I enjoy shooting at targets? I decided I would not. So I decided to resign from the Navy as soon as the war was over (Blackett 2003, 11).

In any event, it was the Admiralty that pointed Blackett in the direction that was to determine his future career as a physicist. Before he had the chance to resign his commission, he was sent as one of approximately 400 young officers to the University of Cambridge for a short course of lectures with the object of instilling "some general culture" (Blackett 2003, 12). Three weeks into the course, Blackett resigned his commission in order to undertake undergraduate study, initially in mathematics and then in physics. In 1921, having obtained a first class degree in natural sciences (physics), he became a Fellow of King's College Cambridge following his recruitment as a research student in the Cavendish Laboratory under the direction of the eminent physicist Sir Ernest Rutherford, an expert on radioactivity. (Rutherford was awarded the 1908 Nobel Prize in chemistry for his investigations into the disintegration of the elements and the chemistry of radioactive substances—his experiments created the foundation of nuclear physics.) Blackett worked as a postgraduate research student at the University of Cambridge in the early 1920s. There he met Costanza Bernadina Bayon, a talented language student. They were married in March 1924 and were parents of a daughter, Giovanna, and a son, Nicolas.

COSTANZA

Costanza came from a far more cosmopolitan background than Patrick. Her mother was 47 when she was born, an English woman married to an Italian, who was not Costanza's father. She was adopted by an English family, who gave her the name Dora. But in adulthood, she was always called Pat—with the result that she and her husband were commonly known as the two Pats. It was a marriage of equals, and they formed a strong team. Pat was bitingly clever, and did not suffer fools gladly. It seems that Patrick ran significant decisions in his life and career past her. And had there been any temptation for her husband's commitment to an activist socialism, she would have seen to it that he held the line. Her instincts were democratic. The party she organized to celebrate Patrick's Nobel Prize became quite famous through her insistence that all the staff of the department he headed should be invited, including secretaries, technicians, and cleaners. At least in 1948 this was a shocking breach of protocol.

THE TWO PATS

Patrick and Costanza (Pat)
Courtesy of Giovanna Blackett.

From the start, Blackett's career flourished. By the end of the 1920s, he was already well known internationally for his work in the fields of cosmic rays and atomic physics. In 1930, he was appointed to a university lectureship at Cambridge and, in 1933, he accepted the position of Professor of Physics at Birkbeck College, a constituent part of the University of London. Anxious to develop his own laboratory, Blackett intensified his work on cosmic rays, and it was in recognition of his achievements in this field that in 1937 he accepted the invitation to take up the prestigious position of Langworthy Professor of physics at the University of Manchester.

It was here, and with "remarkable speed," that Blackett developed and presided over "a major research centre for cosmic rays" in his department (Lovell 1975, 29). He returned to his post in Manchester after WWII, and was there when he was awarded the Nobel Prize for physics in 1948. The citation was for contributions to cosmic radiation and nuclear physics, but his active part in two world wars was also mentioned. (He was regarded as having been unlucky not to share the Nobel Prize for the discovery of the positron in 1936.)

In 1953 Blackett accepted an invitation to take up the chair of physics in succession to Sir George Thomson at the Imperial College in London. Blackett remained there, working on rock magnetism and the geophysics of continental drift, until his retirement in 1964.

Blackett used his status as an outstanding physicist and Nobel Prize winner actively to promote "the scientist's responsibility to society and the public's need to understand scientific or technical evidence supporting or calling into question public policies." Possessing high moral fervor, a commanding physical presence and patrician manner, Blackett epitomized "the twentieth century scientist as public citizen" (Nye 2004, 181). In what follows, we note Blackett's unswerving belief that science is not value free and politically neutral, a stance that led, on occasion, to public criticism and even ostracism. This was especially the case in relation to his views on nuclear weapons where what some people saw as pro-Soviet bias was to draw him to the attention of the British and American security services.

The Tizard Committee and Preparations for War

Blackett's first engagement with defense research was in 1934 when he was appointed to the Aeronautical Research Committee (ARC) as a scientific member. The ARC fulfilled a coordinating role for civil and military research projects: significantly, it was chaired by the eminent scientist, Henry Tizard, then Rector of Imperial College. At that time, fear of aerial bombardment was a growing public concern in view of the fact that most of the British Isles were within striking distance of German airfields, rendering the country vulnerable to a first strike or knock out blow.

A key policy aimed at offsetting this strategic weakness was to provide an effective defense against aerial bombardment, of which one element proved to be the development of radio detection and ranging (radar). The director of Scientific Research at the Air Ministry, H. E. Wimperis, and his principal assistant, A. P. Rowe, recommended that an official Committee for the Scientific Survey of Air Defense (CSSAD) should be established to consider how far recent advances in scientific and technical knowledge can be used to strengthen the present methods of defense against hostile aircraft. The original scientific members of the committee, in addition to Tizard (as chair) and Wimperis, were the distinguished physiologist and Nobel Prize winner, Professor A.V. Hill of University College, London, and Patrick Blackett, the youngest member.

The work of the CSSAD, or Tizard Committee as it came to be called, has been well recounted (Kirby 2003a; McCloskey 1987; Rowe 1948; Watson-Watt 1957; Zimmerman 2001, 2003). The initial focus was on the possibility of destroying an aircraft with a death ray; Tizard asked Robert Watson-Watt of the National Physical Laboratory to explore the potential of this idea. Instead, Watson-Watt proposed that radio waves be used to detect and locate enemy aircraft. Pioneering experiments were undertaken at Bawdsey Manor on the Suffolk coast from 1936 to the outbreak of WWII in 1939 to develop this radar concept.

It was during the course of this work that the term operational research first emerged. This was in the summer of 1938, when the embryonic radar system of aircraft interception was tested successfully during air exercises (Larnder 1978). Watson-Watt and Rowe detached a small group of Bawdsey scientists and engineers to increase the operational effectiveness of the radar system. It was realized that the effectiveness of interception would be dependent not only on the hardware of air interception (radar stations, fighter aircraft), but also on the creation of a man–machine system, including civilian scientists, engineers, and serving RAF officers. RAF officers became accustomed to working in partnership with scientists—institutional barriers were broken, fruitful working relationships were developed, and a climate conducive to the development of OR was created. Solly Zuckerman, a scientific adviser to the British government during WWII, was to describe OR as "the bastard born out of a secret liaison of scientists and airmen" (Zuckerman 1964, 287). Robert Watson-Watt actively promoted this novel OR activity to the extent that some have designated him as the father of OR, though most observers give that title to Blackett.

At the end of the war, Charles Goodeve estimated that radar had been responsible for increasing the probability of fighter interception by a factor of ten, with the work of OR analysts increasing the probability by a further factor of two. Goodeve's conclusion that the contribution of OR to the air defense of Great Britain "was out of all proportion to the amount of effort spent on research" seems entirely appropriate (Goodeve 1948, 228). The Battle of Britain was, of course, won, albeit by a small margin—in the absence of radar, it would almost certainly have been lost.

Following Blackett's work for CSSAD, he was appointed, in 1939, principal scientific officer in the instrument section of the Royal Aircraft Establishment (RAE). There, among other work, he embarked upon a study of bombsights that served as the catalyst for the development of the Mk-14 bombsight. This eliminated the need for a level bombing run at the time of bomb release, a requirement of previous equipment, which exposed aircraft to greater danger from defensive fire. It remained in service with the RAF until 1965.

Blackett and OR in World War II

The decisive moment in Blackett's wartime career took place on August 9, 1940 when A.V. Hill introduced him to General Sir Frederick Pile, commander-in-chief of Anti-Aircraft (Ack-Ack) Command. (Ack-Ack was the conventional abbreviation for anti-aircraft.) With an acute awareness of the Command's problematic gunnery performance, and having heard of the ongoing radar-related OR work at Fighter Command, he offered Blackett the position of scientific adviser at Ack-Ack Command Headquarters, Stanmore.

> PILE ON BLACKETT
>
> " . . . he [Blackett] came one morning, deep in thought, into the G (technical) Office at Stanmore. It was a bitterly cold day, and the staff were shivering in a garret warmed over only with an oil-stove. Without a word of greeting, Blackett stepped silently up on to the table and stood there pondering with his feet among the plans. After ten minutes somebody coughed uneasily and said, diffidently: 'Wouldn't you like a chair, sir . . . or something?' 'No, thank you,' said Professor Blackett, 'it is necessary to apply scientific methods. Hot air rises. The warmest spot in this room, therefore, will be near the ceiling.' At this, Colonel Krohn, my technical G.S.O., stepped up on the table beside the Professor, and for the next half-hour, the two stayed there in silence. At the end of this period Professor Blackett stepped down from the table saying: 'Well! That's *that* problem solved.' And so it was" (Pile 1949, 161).

Pile recognized that Blackett possessed "the quick intuition of a freshman" (Lovell 1975, 56). He remained with Pile for only 7 months, but during that short period he made an enduring impression. From a starting point when Ack-Ack gunners could engage in little more than a gesture of fist shaking at enemy aircraft, Blackett succeeded in reducing the number of rounds per bird (the average number of rounds used to shoot down one aircraft) from 20,000 at the onset of the London Blitz to only 4,000 by the summer of 1941.

There were two elements to this success—the personnel that Blackett recruited to assist him, and the application of OR. He acquired the services of an eclectic group of scientists, several of them with knowledge of radar. The multi-disciplinary group included three physiologists, a physicist, an astrophysicist, two mathematical physicists, two mathematicians, a surveyor, and an army officer. Collectively they became known as Blackett's Circus in recognition of this bizarre diversity of talent. In applying the scientific method to anti-aircraft fire, the Circus focused attention on the efficiency of radar sets used for gun laying. As Blackett recalled:

> ... the first task was to work out the best method of plotting the [radar] data and of predicting the future enemy position for the use of the guns on the basis only of pencil and paper, range and fuse tables. The second task was to assist in the design of simple forms of plotting machines which would be manufactured in a few weeks. The third stage was to find means of bringing the existing [mechanical] predictors into use in connection with radar sets. ... A special school was set up by AA Command to work out the methods of doing this and to give the necessary training. The fourth stage was to attempt to modify the predictors to make them handle the rough [radar] data more effectively (Blackett 1962, 208).

This proved to be a highly successful program of work in enhancing the effectiveness of Ack-Ack Command, a fact acknowledged by General Pile when, on the occasion of Blackett's departure to RAF Coastal Command in March 1941, he complained to A.V. Hill that "They have stolen my magician" (Lovell 1975, 58).

The OR group at AA Command became the nucleus of the Army Operational Research Group (AORG). Covering many aspects of Army operations, it became easily the largest wartime OR group. Hundreds of scientists worked in AORG during the course of the war.

Coastal Command was a new challenge. It was the section of the RAF responsible for operations to safeguard shipping in the waters surrounding Britain. The occasion for Blackett's transfer, as head of the OR Section (ORS), was the mounting threat to Britain's merchant shipping supply lines across the North Atlantic by enemy U-boat action. As in WWI, the U-boat menace held out the prospect of a British capitulation. In commencing the task of reducing the losses to shipping and raising the U-boat loss rate, Blackett assembled a group of scientists as eclectic as his Ack-Ack Circus. Among the first recruits were "three physicists, three communications experts, four mathematicians, two Canadian astronomers, and several physiologists and biologists" (Christopherson and Baughan 1992, 574). Until the end of the war, the maximum size of the group was no more than 25, with an average of 16. It was a youthful group and these were modest numbers for the task in hand, but shortage of staff was more than compensated for by their high quality. Taking the whole of the wartime period, Coastal Command's ORS members went on to become two Nobel Prize winners, five Fellows of the Royal Society, and a Fellow of the Australian Academy of Sciences. Still others were to achieve professorial rank in British universities.

In first containing and then reducing the U-boat threat, the key factor was the U-boats' requirement to re-charge their batteries on the surface, thereby rendering them vulnerable to aerial detection and attack. Coastal Command aircraft had two critical opportunities to engage with the U-boats; in the first instance via their traffic routes from the U-boat pens on the French Atlantic coast, and second, in and around the convoys themselves.

For Blackett and his ORS colleagues, their program of work can be grouped as follows:

(a) enhancing the search capability of individual aircraft and/or reduced U-boat visual observation of aircraft,

(b) enhancing the bombing accuracy (including optimal depth charge settings),

(c) improving the range and reduced visibility of aircraft.

Each of these categories was subject to extensive analysis by the ORS. For example, as Blackett recalled in the case of raising the number of U-boat sightings:

> All the obvious courses of action were considered and recommended where necessary—better look-out drill for the aircrews, better binoculars, etc. Then the best direction of aircraft course in relation to the sun was considered. If the aircraft flew down the sun, the U-boat crew might have more difficulty in seeing it. Discussing these questions one day in Coastal Command, a Wing Commander said casually: 'What colour are Coastal command aircraft?' Of course I knew that they were mainly black. . . . But before the question was asked me, I had missed the significance of the fact. Night bombers are painted black so as to reflect as little light as possible from enemy searchlights. When there is no artificial illumination by searchlights an aircraft of any color flying at moderate or low height both by day and by night, is normally seen at a great distance as a dark object against a lighter sky and only seldom as a light object against a darker sky (Blackett 1962, 217–218).

In the generally overcast weather conditions of the North Atlantic, therefore, the ideal color for Coastal Command aircraft was white. In this respect, Blackett predicted that repainting would facilitate 30% more U-boat sinkings for the same number of sightings (Waddington 1973, 151).

Following the repainting of all Coastal Command aircraft the sinking rate began to climb significantly. This was due both to more sightings and to greater lethality of attack. A classic OR study contributing to this second factor was the work on depth charge settings carried out by Professor E. J. Williams, a scientist later to be well-known for his work on the quantum theory of atomic collisions.

When Williams joined the ORS in late 1941, the existing practice in Coastal Command was to assume that a U-boat would spot an incoming aircraft approximately 2 min before the attack commenced, and that within this time it would dive to around 100 ft. Depth charges were set to explode accordingly. Williams spotted a fallacy in this assumption:

> It might be true that *on the average* a U-boat might sight the aircraft a long way off and so manage to get to the 100 feet before the attack. However, just in those cases the U-boat had disappeared out of sight of the aircraft for so long that the aircrew could not know when to drop the depth charges, so that the effective accuracy in plan of the attack was low. . . . Thus the existing method of attack failed to sink deep U-boats owing to low bombing accuracy and failed to sink shallow U-boats due to the depth setting (Blackett 1962, 215).

> The transition to a shallower depth setting took place in stages so that a setting of 25 feet was not in general service until the summer of 1942. In the meantime the lethality of attack mounted steadily. This rise in lethality can be attributed in part to the introduction of a more effective explosive material, and an increase in bomb loads as heavier aircraft became available. However it is evident that the shallower depth charge setting made a substantial contribution to the observed improvements (Waddington 1973, 177).

As Blackett (1962, 215) observed, "There can be few cases where such a great operational gain had been obtained by such a small and simple change of tactics." Captured U-boat crews assumed that Coastal Command had begun to make use of depth charges containing a vastly more lethal explosive. The ultimate tribute to Coastal Command and its ORS is provided by the increasing proportion of U-boats sunk by land-based aircraft after 1941. Up to the end of that year, land-based aircraft had sunk only four U-boats compared with 46 sunk by naval vessels. In 1942, however, virtual parity was achieved with 36 and 35 sinkings, respectively. By 1943, Coastal Command enjoyed a decisive lead, destroying 84 U-boats compared with 64 sunk by naval action.

There can be no doubt that during the course of 1943, the role of Coastal Command aircraft was transformed from "submarine scarecrow to submarine killer" (Korner 1996, 68). In his obituary notice of Williams, Blackett stated that the success achieved by Coastal Command's ORS put OR on the map and that the anti-U-boat war of 1943 was waged under closer scientific control than any other campaign in the history of the British armed forces save for the final stages of the Anglo-U.S. bombing offensive against the German communications system in 1944.

Blackett at the Admiralty

In January 1942, Blackett moved from Coastal Command to the Admiralty as chief adviser on OR, reporting directly to the vice-chief of the Naval Staff. In advance of his appointment, he wrote a paper, "Scientists at the operational level," and circulated it to senior Naval Staff officers (Blackett 1941). The paper contained the fruits of his OR experiences from the Tizard Committee onwards. One memorable phrase stands out as proclaiming Blackett's philosophy at the time, namely that the application of "numerical analysis" could "help to avoid running the war by gusts of emotion." And,

> Operational staffs provide scientists with the operational outlook and data. The scientists apply scientific methods of analysis to these data, and are thus able to give useful advice.
>
> The main field of this activity is clearly the analysis of actual operations, using as data the material to be found in an operations room, e.g. all signals, track charts, combat reports, meteorological information, etc.
>
> It will be noted that these data are not, and on secrecy grounds, in general, cannot be made available to the technical establishments. Thus scientific analysis, if done at all, must be done in or near operations rooms (Blackett 1962, 171).

If OR analysts were of the highest standing in science "'with direct experience of Service Technical Establishments," then there was no reason why they should not "grasp the service problems of operational research as surely as did the service officers themselves." Indeed, an OR section should operate in conformity with "a first-class pure scientific research institution" but with the proviso that its members should work in close cooperation with service personnel. It has been suggested that "Scientists at the operational level" was "a hurriedly and flippantly written document" (Lovell 1975, 60). Nevertheless, it helped persuade the Admiralty that Blackett should be given an extensive brief for OR studies—and its distribution within U.S. military circles is credited with helping to launch the U.S. OR activity.

Following his arrival at the Admiralty, Blackett's searching mind ranged over the relevance of battleships or carriers as the core of the fleet, priorities in postwar budgeting between production and operating costs, and proposals to establish a Joint Warfare Committee to consider the theory and practice of combined warfare. He also held forth on the nature of military operations in more general terms [suggesting that in any one weapons epoch "there are relatively few things which a commander of major operations must know and remember if he is consistently to be more successful than chance" (Lovell 1975, 68)]. Inevitably, however, the main thrust of Blackett's work remained the anti-U-boat war where he continued to work closely with his former colleagues at Coastal Command that was under the operational control of the Admiralty. In the period 1942–1943, Blackett's ORS presented 89 OR reports of which over 60% concerned anti-U-boat tactics, though this number fell in future years following major successes in the Bay of Biscay.

EXECUTIVE ACTION

"... when an operational research worker comes to some conclusion that affects executive action, he should only recommend to the executives that the action be taken if he himself is convinced that he would take the action, were he in executive authority. It is useless to bother a busy executive with a learned resume of all possible courses of action and with conclusion that it is not possible to decide between them. Silence here is better than academic doubt. Research workers must also guard against the temptation to expect the executive machine to stop while they think. War, manufacturing, trade, government business—all must go on, whether the research worker is there or not" (Blackett 1950, 6).

The value of Blackett's work at the Admiralty can be illustrated by his analysis of the escort provision for trans-Atlantic convoys in relation to the number of merchant ships and optimal convoy speeds. Blackett's work on these issues constituted an outstanding example of wartime OR. In examining the number of escort vessels deployed in 1941–1942, he calculated that convoys with nine escorts had experienced 25% less sinking than those with six. This led him to the conclusion that for each extra escort vessel, between two and three merchant ships could be saved annually. In reality, however, building proportionately more escorts could not be realized, at least in the short term, as a result of "the practical difficulty of changing shipyards rapidly over from the building of merchant vessels to the building of escort vessels. As so often happens with the predictions of economic theorizing, the theoretically optimum production programs cannot be realized quickly in practice" (Blackett 1962, 229).

For Blackett the matter did not end there. Statistical analysis of convoy losses for 1941–1942 revealed that "smaller convoys with an average size of 32 ships had suffered a loss of 2.5 percent, whereas the large convoys with an average of 54 ships had suffered a loss of 1.1 percent. Large convoys appeared to be in fact more than twice as safe as small convoys" (Blackett 1962, 230–232). Existing Admiralty rules, derived from collective memories of WWI, were based on the assumption that small convoys numbering no more than 40 vessels were relatively safe. Convoys in excess of 60 ships were, therefore, prohibited. Yet, although the OR results in favor of the latter were "quite reliable," Blackett and his colleagues were determined to provide the Admiralty with a "rational" justification for large convoys (Llewellyn-Jones 2003, 142–150). As Blackett recalled:

> An intensive study of all available facts about the U-boat campaign against the convoys was undertaken. Of great use were the accounts of prisoners of war from sunken U-boats of the detailed tactics pursued by the U-boats in their "wolf-pack" attacks on the convoys. After several weeks of intensive research analysis and discussion, the following facts emerged. The chance that a given merchant ship would be sunk in any voyage depended on three factors: a) the chance that the convoy in which it sailed would be sighted, b) the chance that having sighted the convoy the U-boat would penetrate the screen of escort vessels round it, and c) the chance that when a U-boat had penetrated the screen the merchant vessel would be sunk. It was found a) that the chance of a convoy being sighted was nearly the same for large and small convoys; b) that the chance that a U-boat would penetrate the screen depended only on the linear density of the escorts, that is, on the number of escort vessels for each mile of perimeter to be defended; and c) that when a U-boat did penetrate the screen the number of merchant ships sunk was the same for both large and small convoys—simply because there were always more than enough targets (Blackett 1962, 232).

It followed, therefore, that the percentage of U-boats sunk was, broadly, inversely proportional to convoy size, a fact confirmed by experience in WWI (Kirby 2003a, 35–38) and that Admiralty policy should be directed to reducing the number of convoys in order to limit the possibility of U-boat sightings. The proviso was that the size of convoys should be increased to maintain the total number of ships sailing. After weeks of argument, Blackett's recommendations were accepted and from the spring of 1943 average convoy size was gradually increased. Although the first large convoy did suffer severe losses, the ultimate accolade for Blackett was provided by the Admiralty in the summer of 1944 when it proclaimed the successful crossing of the Atlantic by a record 187 ships in convoy (Falconer 1976, 315–327).

Blackett was the first to admit that the pay-off from his work could not be tested directly since the move to larger convoys (more than 150 ships) took place after the Battle of the Atlantic had basically been won. He did, however, regret deeply that the relevant statistical analysis had not been undertaken earlier by "a strong operational research group with access to the relevant facts" (Blackett 1962, 233). If large convoys had been introduced earlier than 1943, the bottleneck in the supply of escort vessels would have been eased, and up to 200 merchant vessels and countless lives saved. As it was, the new rules were introduced in sufficient time to permit the transfer of naval vessels from the Atlantic to support the Normandy landings in 1944, thereby diminishing the U-boat threat to the invasion force while in cross-channel transit.

AREA BOMBING AND NUCLEAR WEAPONS

It was during his period at the Admiralty that Blackett became involved in a bruising dispute with Sir Arthur Harris, the formidable and single-minded commander-in-chief of RAF Bomber Command, and other high-level advocates of the area bombing of German towns and cities. Notable among these was F. A. Lindemann, Prime Minister Churchill's personal scientific adviser. Area bombing was inaugurated in 1942 with the aim of de-housing the German urban and industrial population. Its purpose was to induce capitulation in response to collapsing morale. The advocates of this strategy argued not only that it would shorten the war, but that it would also eliminate the need for a land invasion of Germany. Harris and his colleagues claimed that Bomber Command could deliver the British Army from the mass slaughter of trench warfare (Webster and Frankland 1961, v. 1, 340).

Area bombing remains the subject of continuing controversy. The fire bombing of Dresden in February 1945 continues to be regarded as the incarnation of its immorality (Grayling 2006). There can be no doubt that Blackett himself was fundamentally opposed to terror bombing as part of a total war in which civilian populations were potentially in the front line of attack (Crook 2003). But in arguing the case against the effectiveness of this strategy in breaking civilian morale, Blackett made use of OR studies conducted by the distinguished scientists J. D. Bernal and S. Zuckerman.

The fundamental division of opinion over area bombing related to the role of Bomber Command at a particular stage in the war. For Tizard and Blackett, in particular, the overwhelming strategic issue confronting Britain in 1942 was the anti-U-boat war in the North Atlantic. In their view, an effective campaign required the deployment of heavy, long-range aircraft by Coastal Command, which was in direct competition with the needs of Bomber Command. An area bombing campaign against the German homeland could not produce decisive results in 1942 or 1943, they argued, while neglect of the defense of convoys carried the risk of British defeat through the rupture of the Anglo-American supply line.

Blackett was brought into open dispute with Bomber Command by the submission in March 1943 of an Admiralty paper to the Anti-U-boat Committee chaired by Prime Minister Winston Churchill. The paper, prepared by Blackett (now Director of Naval Operational Research), noted the rapidly mounting shipping losses in the North Atlantic and ascribed them to the U-boats' successful countermeasures against radar detection. Blackett called, therefore, for the diversion of 190 heavy bombers from Bomber to Coastal Command, to neutralize the new U-boat challenge.

Blackett's request elicited a vitriolic response from the commander-in-chief, Air Marshal Sir John Slessor. Slessor sought to undermine Blackett's credibility by referring to "slide-rule strategy of the worst kind" before concluding that the anti-U-boat offensive should be tackled from "a less scientific, namely less quantitative, but more practical angle." What Coastal Command required was not additional aircraft, but "aircraft of the right type with the right sort of radar and with crews trained in the right sort of way" (Slessor 1956, 524–525).

Lovell has commented that there were strengths and weaknesses on both sides of the argument (Lovell 1988, 226–227). Blackett's numerical calculations were more relevant to 1942 than 1943. In Slessor's case, there was an apparent complacency in the face of steeply rising shipping losses in the last 3 months of 1942. What is really significant about the Blackett–Slessor exchange, however, was the latter's unwavering commitment to the area offensive as the primary claimant on resources. The area offensive under Harris's direction, as opposed to operations over the sea, represented "true air warfare," and Slessor shared fully Harris's belief that area bombing could win the war outright without the need for "great land battles on the scale of the First World War" (Terraine 1985, 452).

As with its morality, the role of area bombing in hastening Germany's defeat remains highly contentious (Kirby 2000, 2003a; Overy 1997). In the years after 1945, Blackett's view remained unchanged. As he commented (Blackett 1962, 126):

> The only major campaign in modern history in which the traditional military doctrine of waging war against the enemy's armed forces was abandoned for a planned attack on its civilian life was a disastrous flop. I confess to a haunting sense of failure, and I am sure that Tizard felt the same way. If we had only been more persuasive and had forced people to believe our simple arithmetic, if we had fought officialdom more cleverly and lobbied ministers more vigorously, might we not have changed this decision?

These views on the ineffectiveness of area bombing as a morale-breaking weapon were confirmed by British and American bombing surveys at the end of the war (Kirby 2003a, 169–171). Their conclusion that air power could not win a war in the absence of ground troops played a key role in informing Blackett's stance on Western nuclear strategy following the onset of the Cold War in 1947 (Nye 2004, 88–89). Blackett set out his views with typical clarity in his 1948 book, *The Political and Military Consequences of Atomic Energy* (Blackett 1948b). This was the year in which Blackett was awarded the Nobel Prize for physics. Inevitably, this enhanced international interest in his book. His thesis was controversial and deeply unpopular, especially in the U.S. where he was denounced as "an apologist for the Kremlin," "a master of artful and intelligent distortion," and a "confused Englishman" (Nye 2003, 276).

Blackett began his analysis with a clear denunciation of the American nuclear attacks on Hiroshima and Nagasaki in August 1945. His interpretation of these events was as follows:

> Let us sum up the three possible explanations of the decision to drop the bomb and its timing. The first that it was a clever and highly successful move in the field of power politics, is almost certainly correct; the second, that the timing was coincidental, convicts the American government of a hardly credible tactlessness [towards the Soviet Union]; and the third, the Roman holiday theory [a spectacular event to justify the cost of the Manhattan Project], convicts them of an equally incredible irresponsibility (Blackett 1948b, 126).

For Blackett, the dropping of atomic bombs on Japan was unnecessary in August 1945 because by that time a Japanese surrender was inevitable—and sooner rather than later. The conventional war should have continued because "no appreciable loss of American lives would have resulted." This judgment was wounding enough to American opinion, but Blackett compounded the matter with his view that the use of atomic weapons against Japan was "not so much ... the last military act of the Second World War [but] the first act of the cold diplomatic war with Russia" (Nye 2004, 89). Far from demonizing the Soviet Union, the U.S. should seek to maintain contact and cooperation with its leaders. Blackett also predicted that America would prove to be "trigger happy" in unleashing nuclear warfare, through a continuing but in his view erroneous belief in the virtues of strategic bombing and the desire to minimize military casualties (Towle 2003, 203–205). Blackett foresaw, correctly, that the Soviet Union would soon acquire its own nuclear weapons and argued that this held out the prospect of a strategic nuclear exchange with potentially catastrophic implications for Britain in terms of collateral damage. To preempt this possibility, he advocated a policy of neutrality for Britain on the assumption that the Americans would prove unwilling and unable to maintain a large standing army in Europe.

Blackett's explanations of the decision to drop nuclear bombs on Japan remain controversial. Other commentators at the time and since have argued that a conventional war would have lasted far longer than Blackett assumed (Towle 2003, 204–205). Moreover, the Korean War did not result in the trigger happy use of nuclear weapons on the part of the U.S. and, equally significant, the end of the conflict was followed by the commitment of substantial U.S. ground forces to the European theatre where they were to remain for years to come.

By the mid-1960s, Blackett's stance on nuclear weapons was no longer so isolated, politically and militarily, as it had been in the late 1940s. Opposition to nuclear weapons had become a mainstream concern, and the decade as a whole was notable for a sequence of disarmament negotiations culminating in international agreement on a partial nuclear test ban treaty. British and American nuclear defense strategy had also moved decisively away from the massive retaliation of the 1950s and the first strike use of tactical nuclear weapons. After 1960, the North Atlantic Treaty Organization (NATO) moved toward the concept of flexible response in which the Soviet Union would be confronted by an array of military options on the part of NATO, from Polaris missiles and strategic bombers to non-nuclear counter forces. This was adopted as official NATO strategy in 1967 (Baylis 1995; Freedman 1980). The extent to which Blackett and his arguments had a direct influence on these developments can only be a matter for conjecture (Nye 2004, 97).

Politics and Science

In 1949, the anti-totalitarian writer, George Orwell, compiled for the U.K. Foreign Office a list of 37 writers, journalists, academics, and artists who he deemed to be pro-communist and sympathetic to the Soviet Union. The list included Patrick Blackett. This was hardly surprising in the light of the uproar caused by Blackett's book, *The Political and Military Consequences of Atomic Energy*, published in the previous year (Blackett 1948b). When Harold Wilson, as President of the Board of Trade in the postwar Labour Government, appointed Blackett to the National Research Development Corporation in 1949, there was a furor. As Wilson recalled, Blackett's appointment "provoked one of the most violent political reactions in Parliament that [he] had known in 30 years of parliamentary turbulence" (Nye 2003, 284).

In fact, Blackett's political commitment long pre-dated this period. While still at Cambridge he had been asked to stand for election to the House of Commons as a Labour candidate (Lovell 1975, 76). Though he came to describe himself as a Fabian (a moderate socialist) this was not entirely, at least not always, the case. He was very much part of the radical science movement of the 1930s (Werskey 1978), of which the leading intellectual was J. D. Bernal, a committed Marxist (also an operational researcher during the war). His book *The Social Function of Science* (Bernal 1939) argued that the full positive potential of science was inevitably stifled within the profit motivation of capitalism. Bernal's thesis was that scientists should press for science to be used directly for the common good.

These views influenced a generation of socially conscious U.K. scientists, including many of the early pioneers of OR (Rosenhead 1989). This was the pre-war milieu, which Blackett inhabited. Unlike a number of his scientific contemporaries, notably Bernal, H. Levy, and J. B. S. Haldane, he was never a member of the Communist Party (Nye 2004, 14). But his commitment to a Bernal-influenced position can be seen, for example, in his chapter (based on a radio talk) on "The Frustration of Science," in a book of the same name (Blackett 1935). During the war, Blackett was active in the Tots and Quots dining club (started by Solly Zuckerman in 1931 as a meeting place for scientists), and in 1947 became president of the Association of Scientific Workers, the left-leaning trade union for scientists—a position also held in 1943 by Watson-Watt and by Bernal in 1949.

For several years, Blackett's suspect status was such that he was not allowed to enter the U.S. This caused a drama in 1951 when, following a cosmic ray conference in Mexico, his plane to Canada made a refuelling stop in Tampa, Florida. He and his wife were held overnight under armed guard before being allowed to proceed on a flight the next day (Blackett [undated]; Lovell 1988, 229).

Blackett's exclusion from government circles from the late 1940s coincided with a decisive move of the Labour government away from planning (with which OR was aligned) in favor of the market. With the onset of the Cold War, those with communist associations, or sympathetic to the Soviet Union, were no longer welcome. Indeed Blackett was to be almost totally excluded from inner Government advisory circles for a full 16 years (Lovell 1988, 228).

Post-war Civil Government

By 1956, Blackett was no longer completely persona non grata in British politics. This is confirmed by his membership of the Gaitskell Group of scientists, one of several study groups on future policy for the Labour Party (Hugh Gaitskell was the then leader of the Labour Party.) By 1960, Blackett had emerged as one of the key drivers in formulating a science and technology policy for the Labour Party (Kirby 1999, 987).

Blackett's main concern was to enhance state provision for research and development (R&D) in the civil manufacturing sector. He saw British firms as technologically backward; mainstream manufacturing industry as handicapped by poor management, a lack of qualified scientists and engineers, financial constraints, and a failure to achieve economies of scale as a result of inadequate firm size; and the military-defense sector as excessively large. A future Labour Government should invest £10 million per annum in private sector civil R&D programs conducive to the national interest. This would be accomplished by using existing state agencies to allocate R&D funding for appropriate projects.

Blackett's prognosis and his policy proposals were broadly accepted, but the full flowering of his influence came with the election of Harold Wilson as leader of the opposition Labour Party in 1963. At his first Party conference as leader, Wilson gave a celebrated speech in which he stated that a future Labour Government would institute "far reaching changes in economic and social attitudes which permeate our whole system of society." This would take place in the context of the "white heat" of a scientific and technological revolution (Wilson 1964, 27). Sitting beside

Wilson on the conference platform, and nodding sagely throughout the speech, was Patrick Blackett.

When Wilson moved to complement his rhetoric with practical policy proposals, the program that he described was one devised by Blackett: a sustained expansion of university education to provide the number of graduates which the scientific revolution would require; and a "full Ministry of Science" to determine national priorities for allocating resources—which meant the civilian manufacturing sector rather than "prestige" defense projects and "consumer gimmicks." As the general election approached, Blackett produced a leaner, more focused version of this program (Kirby 1999, 988).

Following Labour's election victory in October 1964, the new Ministry of Technology (Mintech) promised by Wilson was established. Blackett was offered the position of Minister of Technology but refused because this would necessitate membership of the House of Lords, and he was unwilling to accept a peerage at that time (Nye 2004, 143–144). Instead, Mintech's first head was the former trade union leader, Frank Cousins, and one of his first actions as Minister was to implement Blackett's proposal for the establishment of an Advisory Council on Technology with himself as chairman. Blackett was appointed to serve as deputy chairman and scientific adviser to Mintech, positions which he held until the general election of 1970. According to Lovell, it was in these capacities that, for a time, Blackett wielded immense power, a judgment validated by Sir Iuean Maddock's statement that "initially," Blackett was virtually the "Deputy Minister." His views were accepted as "absolute and his priorities determined the activities of the day" (Lovell 1975, 80–81).

Under Blackett's influence the new government's priorities were to boost the R&D of civilian manufacturing industry (mainly at the expense of the defense sector), and to restructure industry into larger units to make this possible. To achieve these ends, Blackett tried to employ the analytic methods, which had been so successful for OR during the war. At least, at first, he seems to have assumed that the transfer of the relevant principles to the peacetime economy was unproblematic. As he commented, "There is not all that real difference between trying to cope in 1942-3 with [shipping] losses of 700,000 tons a month and coping in 1965 with an adverse balance of payments of 700 million a year" (Kirby 1999, 989).

Sir Alan Cottrell, a colleague of Blackett's on Mintech's Advisory Council for Technology thought that Blackett had "very nearly turned this into an exact science" (Lovell 1975, 84–85). The civil servants who worked for him, however, took a more jaundiced view of the approach. Thus, the view of Sir George Macfarlane, Controller of Research at Mintech at the time was that:

> Blackett's general technique in argument was to attack by stating a series of conclusions and demanding actions to follow them. His conclusions were not always justified and often based on a simplified analysis of a very complex situation in which opinions and philosophies were more numerous than facts (Lovell 1975, 83).

In a similar vein, we have the judgment of John. B. Adams, Mintech's Technical Head "industry cannot be commanded like an army . . . . nor is it the enemy of government" so the analysis and strategy by analogy with war situations are hardly applicable (Lovell 1975, 85).

In fairness to Blackett, however, it can be conjectured that these criticisms, expressed by senior civil servants, were neither judicious nor neutral. In the post-war period, there was extreme hostility in the Civil Service to the adoption of OR as a guide to industrial policy. As observed by Rosenhead, beyond the defense ministries, OR was perceived as a threat to the status quo: its advocates were "outsiders" and their distinctive methodology was not understood by those with a disciplinary allegiance to the arts and humanities (Rosenhead 1989, 24).

BLACKETT: THE PRIVATE MAN

Patrick Blackett was a very public figure through much of his life. He was an engaged public intellectual of a type which is now becoming more rare, at least in Anglo-Saxon countries. Some idea of the private man behind the public figure can add perspective.

Throughout his adult life, people who met him were struck by his dramatic appearance. Tall, strikingly handsome, imposing, commanding. The sculptor Jacob Epstein was so impressed by photographs of Blackett that he asked Blackett's sister to persuade him to sit for a portrait bust. As a young man, Tam Dalyell (later a Labour MP for 40 years), was on Blackett's staff. He described Blackett as "quite simply, the most personally formidable man for whom I have ever worked—or, indeed, met, at close quarters" (Hore 2003, ix). He knew that some staff members felt that he ran his university departments like a captain runs a ship. (Others, knowing Blackett's naval background, thought of him rather as the Admiral.)

Bust of Patrick Blackett, by Jacob Epstein
Courtesy Giovanna Blackett.

Some found him more formidable than charming. The author, scientist, and politician C. P. Snow described his "extremely superior sort of personality." He was known to have a terrible temper on occasions, yet this did not lose him the respect, affection even, of many of those who worked in his laboratories (Nye 2003, 172–177).

His daughter Giovanna found him friendly but remote. This was partly a question of absence—her father was away a great deal of the time, especially during the war, and, when not away, still concentrated on work. He would be writing, not taking part in the family nitty-gritty. In later years, when his wife was away, Blackett would eat three fried breakfasts each day. The grandchildren have memories of him that are warm and affectionate—he took them to see each Beatles film as it came out.

The Blacketts moved in non-scientific intellectual circles. The family used to stay with Adrian Steven, the brother of the novelist Virginia Woolf, at his country house. It was there that they met Clough Williams-Ellis, the celebrated architect who was developing the eccentric Italianate fantasy Welsh coastal resort of Portmeirion. Williams-Ellis owned a nearby valley and persuaded the Blacketts to rent a cottage there, initially with neither electricity nor water. The Blacketts moved home frequently, so this rural base became a cherished fixture. It was there, mostly, that the children and

Blackett would properly intersect. His recreations were walking, bird watching, and sailing (Hore 2003). Walking was always done with binoculars round his neck. For sailing he had the Red Witch, a boat bought with the Nobel Prize money.

BLACKETT'S OR LEGACY

Like most of the talented scientists who worked in OR during the war, Blackett returned afterwards to his original discipline. Goodeve (1975) held that Blackett showed very little interest in the post-war development of the subject he had helped to bring about. This is not entirely the case. It is true that he played little role in the affairs of the Operational Research Society. And it may be the case, as Goodeve states, that he viewed with some suspicion the development of OR into a new discipline or profession. He was, though, one of four people at the select dinner meeting at the elite London club, the Athenaeum, in the autumn of 1947 which led to the establishment of the OR Club, later the ORS.

He also published articles on OR in this period, including one in *The Advancement of Science* (1948a) and another in the first issue of the *Operational Research Quarterly* (1950). Both articles were reprinted in *Studies of War: Nuclear and Conventional* (Blackett 1962). He participated in a series of lectures on OR given in Manchester in 1949–1950, and he reviewed Morse and Kimball's (1951) classic book *Methods of Operations Research* in the journal *Physics Today* (Blackett 1951)—he expressed surprise to find paragraphs of his own words reproduced almost verbatim but without acknowledgement! Then there seems to have been a gap. When he did re-engage with the subject, obliquely, his purpose was to reinforce his trenchant views on military and deterrent strategy.

In *Studies of War*, Blackett made clear his distaste for the mathematization of the subject that was now in full swing, especially in the U.S. The OR studies in WWII had been firmly rooted in data generated by actual military operations. Blackett's emphasis on the primacy of data in OR is indeed entirely in keeping with the basis of his career in physics. His successes in a whole range of fields, from cosmic radiation to geophysics, were based on his great gifts as an experimentalist, generating and analyzing data in a creative way. His criticism of the new strategists

> of the balance of nuclear terror, especially Herman Kahn and Albert Wohlstetter, was that their work was based on data-free theories plus standard solution techniques. As a supreme, though theoretically aware experimental scientist, he always reacted strongly against attempts by theoreticians to extract more than the limitations of their data would justify (Lovell 1975, 69).

PRACTICAL OR

"In a sense, of course, probability theory in the form of the simple laws of chance is the key to the analysis of warfare; . . . My own experience of actual operational research work, has however, shown that its is generally possible to avoid using anything more sophisticated. . . . In fact the wise operational research worker attempts to concentrate his efforts in finding results which are so obvious as not to need elaborate statistical methods to demonstrate their truth. In this sense advanced probability theory is something one has to know about in order to avoid having to use it" (Blackett 1951, 19).

Blackett's critique was of a piece with the perspective on OR that he had developed during the war. These were expressed in two memoranda (Blackett 1941, 1943). The first of these, "Scientists at the operational level," stressed the importance of the analytic staff spending time in close touch with service personnel actually performing their jobs. The second, "A note on certain aspects of the methodology of operational research" was more considered, and had a 2-year gestation period. Blackett himself described it as "the nearest existing approach to a 'text book' of operational research" (Blackett 1962, 169–170).

This document contrasted two approaches—the a priori and the variational methods. The former attempts to find general solutions to arbitrarily simplified problems by selecting those variables, which were susceptible to quantification and ignoring the rest. Blackett strongly advocated the alternative, variational approach, which he also thought accorded with commonsense. This abandoned the attempt to construct—in the head—an imagined military operation, but instead to examine in practice how small changes in the key factors of actual operations changed its yield. Of course, the variational method necessitated quantities of field data, and indeed Blackett emphasized the importance, as a first step, of collecting as much data as possible.

It is, therefore, possible to see Blackett's criticism of 1960s military strategic analysis by game theory and optimization methods as entirely consistent with the lessons he had earlier drawn from the successes of OR in WWII. (It should be recalled that the impressive achievements of wartime OR were achieved without the benefit of any of the techniques which later became OR's standard repertoire.) It is also evident that his critique went largely unheard—from the 1960s, the search for and employment of general solution methods for entirely quantified problems rapidly became the norm in OR.

There is a case for arguing that Blackett's less mathematized view of OR has some reflection in the contemporary development of Problem Structuring Methods (Rosenhead and Mingers 2001). These are participative approaches, which build bespoke representations of particular decision situations in order to generate agreed commitment to action. But as the supreme natural scientist, Blackett would doubtless have felt uneasy at the use of subjective as well as objective information.

There is also a distinctively pragmatic tradition in British OR by comparison with U.S. or other European practice—a tradition of study in the field and an emphasis on close relations between analysts and those who will operate any system which results. There is also still present a strand of work, known as community OR, that adopts a social commitment to using analysis on behalf of the under-privileged. Both of these tendencies can, perhaps, be seen as OR practice in accordance with the spirit of Blackett.

We are left, though, with Blackett's undoubted legacy—as the outstanding figure from the heroic period of OR. Throughout his career, both in and out of OR, he provides us with an example of the committed public intellectual—committed to science, reason, and progressive social change. Veritably, a hard act to follow.

Patrick Blackett died on July 13, 1974. He was survived by his wife and children.

ACKNOWLEDGMENT

We are grateful for the generous assistance in preparing this profile provided by Patrick Blackett's daughter Giovanna Blackett.

REFERENCES

Baylis J (1995) Ambiguity and deterrence: British nuclear strategy 1945–1964. Clarendon Press, Oxford

Bernal J (1939) The social function of science. Routledge, London

Blackett PMS (1935) The frustration of science. In: Hall D et al. (eds) The frustration of science. Allen and Unwin, London, pp 129–144

Blackett PMS (1941) Scientists at the operational level. Admiralty, London. (Reprinted in Blackett 1962)

Blackett PMS (1943) A note on certain aspects of the methodology of operational research. Admiralty, London

Blackett PMS (1948a) Operational research. Adv Sci 5(17):26–38

Blackett PMS (1948b) The political and military consequences of atomic energy. Turnstile Press, London

Blackett PMS (1950) Operational research. Oper Res Q 1(1):3–6

Blackett PMS (1951) Operations research. Phys Today 4(November):8–20

Blackett PMS (1962) Studies of war: nuclear and conventional. Oliver and Boyd, Edinburgh

Blackett PMS (2003) Boy Blackett. In: Hore P (ed) Patrick Blackett: sailor, scientist, socialist Frank Cass, London, pp 1–14

Blackett PMS (Undated). Reminiscence. Alton Cosmic Ray Conference (manuscript note)

Christopherson D, Baughan EC (1992) Reminiscences of operational research in World War II by some of its practitioners. J Oper Res Soc 43(5):569–77

Crook P (2003) The case against area bombing. In: Hore P (ed) Patrick Blackett: sailor, scientist, socialist. Frank Cass, London, pp 167–86

Falconer N (1976) On the size of convoys: an example of the methodology of wartime OR scientists. Oper Res Q 27(2):315–327

Freedman L (1980) Britain and nuclear weapons. Macmillan, London

Goodeve C (1948) Operational research. Nature 161(4089):377–384

Goodeve C (1975) Lord Blackett – a memoire. Oper Res Q 25(4):vii–viii

Grayling AC (2006) All the dead cities: the history and moral legacy of the World War II bombing of civilians in Germany and Japan. Bloomsbury, London

Hore P (ed) (2003) Patrick Blackett: sailor, scientist and socialist. Frank Cass, London

Kirby M (1999) Blackett in the "white heat" of the technological revolution: Industrial modernisation under the Wilson governments, 1964–1970. J Oper Res Soc 50(10):985–993

Kirby M (2000) Operations research and the defeat of Nazi Germany. Mil Oper Res 5:1–14

Kirby M (2003a) Operational research in war and peace: the British experience from the 1930s to 1970. Imperial College Press, London

Kirby M (2003b) IFOR's operational research hall of fame: Patrick Maynard Stuart Blackett. Int Trans Oper Res 10(4):405–407

Korner T (1996) The pleasures of counting. Cambridge University Press, Cambridge

Larnder H (1978) The origins of operational research. In: Haley KB (ed) OR '78. North Holland, Amsterdam, pp 3–12

Llewellyn-Jones M (2003) A clash of cultures: the case for large convoys. In: Hore P (ed) Patrick Blackett: sailor, scientist, socialist. Frank Cass, London, pp 138–166

Lovell B (1975) Patrick Maynard Stewart Blackett: Baron Blackett of Chelsea. Biogr Mem Fellows R Soc 21: 1–115

Lovell B (1988) Blackett in war and peace. J Oper Res Soc 39(2):221–233

McCloskey J (1987) The beginnings of operations research: 1934–1941. Oper Res 35(1):143–152

Morse P, Kimball G (1951) Methods of operations research. Wiley, New York, NY

Nye M (2003) A physicist in the corridors of power. In: Hore P (ed) Patrick Blackett: sailor, scientist, socialist. Frank Cass, London, pp 269–293

Nye M (2004) Blackett: physics, war, and politics in the twentieth century. Harvard University Press, Cambridge, MA

Overy R (1997) Bomber Command 1939–1945: reaping the whirlwind. Harper and Collins, London

Pile F (1949) Ack-Ack: Britain's defence against air attack during second world war. George C. Harrap & Company, London

Rosenhead J (1989) Operational research at the crossroads: Cecil Gordon and the development of post-war OR. J Oper Res Soc 40(1):3–28

Rosenhead J, Mingers J (eds) (2001) Rational analysis for a problematic world revisited: problem structuring methods for complexity, uncertainty and conflict. Wiley, Chichester

Rowe A (1948) One story of radar. Cambridge University Press, London

Slessor J (1956) The central blue: recollections and reflections. Cassell, London

Terraine J (1985) The right of the line: the air force in the European War, 1934–1945. Hodder and Stoughton, London

Towle P (2003) Blackett and nuclear strategy. In: Hore P (ed) Patrick Blackett: sailor, scientist, socialist. Frank Cass, London, pp 201–216

Waddington C (1973) O.R. in World War 2: operational research against the U-Boat. Elek Science, London

Watson-Watt R (1957) Three steps to victory. Odhams, London

Webster C, Frankland N (1961) The Strategic Air offensive against Germany, 1939–1945, 4 vols. HMSO, London

Werskey G (1978) The visible college: a collective biography of British scientists and socialists of the 1930s. Allen Lane, London

Wilson H (1964) Purpose in politics: selected speeches. Weidenfeld and Nicholson, London

Zimmerman D (2001) Britain's shield: radar and the defeat of the Luftwaffe. Sutton Publishing, Phoenix Mill

Zimmerman D (2003) Preparations for war. In: Hore P (ed) Patrick Blackett: sailor, scientist, socialist. Frank Cass, London, pp 110–125

Zuckerman S (1964) In the beginning—and later. Oper Res Q 15(4):287–292

Blackett's private papers can be consulted by bona fide researchers in the Library, Royal Society of London.

2

Steven Vajda

Jakob Krarup

Known as the British father of linear programming (LP), Steven Vajda was a mathematician, educator, mentor, one of mathematical programming's true pioneers, and the person who introduced linear programming to both Europe and Asia. He was a fellow of the Royal Statistical Society, was awarded an honorary doctorate degree from Brunel University (West London), and was promoted Honorary Doctor of Philosophy by the University of Budapest.

Steven's 1956 book, *Theory of Games and Linear Programming*, was the first book to present a systematic and comprehensive account of matrix games and LP. Translated into German, French, Japanese, and Russian, it was also the first book on LP published outside the U.S. and the second one worldwide. To many in Europe and Asia, this text was their first introduction to a promising new area and a revelation that matrix games and LP were so intimately interrelated. It is, therefore, Steven who rightly should be credited for having spread the message to the two continents. For over 50 years, Steven was a key figure in the development, in the U.K., of post-war operations research (OR), in general, with particular emphasis on mathematical programming (MP).

Early Childhood and Family

Steven (István, originally) was born in Budapest on August 20, 1901 to Josef and Aurelia (Wollak). Little is known about his parents, but both are believed to have been born in Vienna. Josef left school as soon as he could and was employed as a traveling salesman, selling handkerchiefs. Aurelia was never employed.

A.A. Assad, S.I. Gass (eds.), *Profiles in Operations Research*, International Series in Operations Research & Management Science 147, DOI 10.1007/978-1-4419-6281-2_2,

The family moved to Vienna in 1902 when Steven was 16 months old. Steven attended the local primary school and won a place at the Akademisches Gymnasium, a selective school for intelligent students. He excelled in mathematics, languages (he spoke eight languages as an adult), and European history. He was also a keen boy scout and loved football (soccer).

There were two siblings, an older sister who died in infancy, possibly from scarlet fever, and a younger sister Ilonka who married a Czechoslovakian and moved to Prague before World War II (WWII). After the war, Steven tried to find out what had become of his family and was told that his father died of pneumonia in a concentration camp, and his mother perished in a gas chamber. There was no information available about Ilonka.

Between the two world wars, Vienna was a focal point of intellectual activity. Steven took a great interest in the modern philosophies of the day such as psychoanalysis and met with many of the most original thinkers of the twentieth century: philosophers, notably Karl Popper; members of the Vienna Circle (the circle of scientifically minded intellectuals); the physicist Dennis Gabor, the inventor of the hologram; and the mathematician/logician Kurt Gödel whose Incompleteness Theorems managed to generate a turmoil within the community of mathematicians.

As an adult, Steven engaged in printing and book-binding as hobbies, and, throughout his life, maintained a great interest in European history. He was an interesting and stimulating traveling companion who not only planned trips to see as much as possible, but also was able to give detailed accounts of places, people, events, and dates.

In 1929, Steven married Eva Paulina Fanta, a bank manager's daughter. She was born in Vienna (1906) and was trained as a teacher. They had two children, Hedy (1930) and Robert (1933). Eva died in 1983.

Academic Training, Mentors, and Influences

Steven originally embarked on an engineering degree at the Vienna Technical University but changed to a 2-year course in actuarial science. On completing the course, he found it difficult to find employment and transferred to the University of Vienna to take a 3-year course in mathematics. Upon further studies, he received the Ph.D. degree in 1925 and subsequently attended the University of Göttingen for one semester.

At that time, Göttingen ranked as the stronghold of mathematics. Here, Steven met a range of eminent mathematicians: David Hilbert, Richard Courant, Emmy Noether, and Bartel van der Waerden who guided Steven to an actuarial job in Romania. Steven took this job as he was unable to obtain employment in Vienna; he was not a member of the local Social Democratic Party.

He stayed in Braila (Romania) for just 18 months (1926–1927) and then returned to Vienna where he was invited to join the actuarial staff of the Phönix Insurance Company. This job offer came about due to some work he did while in Romania. Steven had carried out an evaluation (the exact insurance situation is unclear) and, because he had no idea how he was expected to carry out the assignment, he designed his own unique method. From this study and related activities, he became rather well known in the local actuarial community. In 1936, he transferred to another insurance company which assigned him to Budapest for 6 months, but, otherwise, Vienna was his home base.

"Hitler didn't like me very much..."

Hitler's steady demand for Lebensraum led to the annexation of Austria (the Anschluss) in 1938. With the Nazis occupying the country, the Vajda family felt compelled to leave. Their children, Hedy and Robert, were dispatched to Sweden. Later, Eva left Vienna for Sweden to pick up her children and then to England where she supported herself in domestic service.

It was difficult for Steven, however, to obtain a visa for England. He contacted his old friend Karl Popper, who was teaching at the University of Christchurch in New Zealand, about employment there. He was offered a job, which also enabled him to obtain a temporary visa to travel via England to collect his wife and children. Three months after his arrival in England, WWII broke out, Britain declared war on Germany, and the family could not continue to New Zealand.

It is no wonder that there was some panic in the U.K. about potential spies among travelers arriving from Central Europe. As an enemy alien, Steven was interned on the Isle of Man, but his family was allowed to remain in Epsom. They were lucky aliens in that many others, who were being sent to Canada, had their ship torpedoed. During his stay at the internment camp, along with about 1,600 other refugees, Steven was instrumental in establishing a do-it-yourself university where he taught mathematics to the younger students and helped them prepare for university entrance examinations; an entrance exam he prepared was approved by the University of London.

As Steven once said, "Hitler didn't like me very much, but then I didn't like him very much either" (Haley and Williams 1998, 299). As evidenced during the subsequent 55 years, these mutual antipathies were indeed fortunate for British OR and, in turn, for the mathematical programming community worldwide.

CITIZEN OF THE U.K.

After 6 months on the Isle of Man, Steven's record was reviewed and it was decided that he was not a danger to the country. Supported by friends in Vienna, he then joined the Gresham Insurance Company located near Epsom. Once again he was lucky. The British consul in Vienna had forgotten to stamp his passport with a statement saying that the owner is allowed into England but not to take a job.

In 1944, Steven was approached by the statistician, Hilary Seal, who was setting up a new statistical department at the Admiralty for the purpose of maximizing military resources in the Pacific. On the basis of Steven's actuarial research publications, which he wrote prior to leaving Austria and which were published in *Bulletin des actuaires suisses*, Seal invited him to join his group at the Admiralty. This was in fact the beginning of his career in OR. When the war ended, Seal saw to it that Steven was one of the first aliens to be given British citizenship.

> STEVEN THE LEADER
>
> The name Vajda does actually mean something in Hungarian. It appears that the word can be found in Hungarian texts dating back to about AD 950. A Vajda is sort of a regional leader or a reigning prince and close variants are known in ancient Greek, Polish, ancient Russian, Serbo-Croatian, and Slovenian. Vajdaság, the region led by Vajda, is the Hungarian designation of a region in the northern part of former Yugoslavia.

Steven worked for the Admiralty until 1965, first as a statistician, soon rising to assistant director of physical research. In 1949, he was appointed assistant director of OR, and, in 1952, he was promoted head of Mathematics Group at the Admiralty Research Laboratory at Teddington. Steven's work for the navy spanned manpower planning, military OR, and the provision of mathematical, statistical, and computational services.

Patrick Rivett was the first professor of OR in the U.K. (Lancaster University, 1963). Steven became the second one when he joined the University of Birmingham in 1965, a position he held until his retirement in 1968 when he became a senior research fellow. In 1967, he was invited by Sussex University to become a fellow, and, in 1973, became visiting professor of mathematics, a role in which he remained active by teaching and writing research papers for about 22 years—probably a record unsurpassed in the U.K. and anywhere.

CONTRIBUTIONS TO OR AND BEYOND

It is not quite obvious what Cowles Commission Monograph No. 13, *Activity Analysis of Production and Allocation* (Koopmans 1951), may have had to offer Steven. This volume, however, contains a paper titled "Maximization of a linear function subject to linear inequalities" by a certain G. B. Dantzig (1951), along with two other papers by the same author, one of which is a revision and extension of an earlier paper which appeared in the journal *Econometrica* (Dantzig 1949). These were the earliest publications on what has since been named LP. They became Steven's gateway to the field.

Steven became excited about these new ideas, their potential in practice, and the abundance of theoretical challenges, and he was eager to share his enthusiasm (Vajda 1956, 1958). He took a particular interest in duality, a central concept in LP, a powerful modeling tool, an equally powerful modeling principle, and a means to appreciating what post-optimal analyses may have to offer. Furthermore, as H. P. Williams, one of Steven's closest collaborators notes:

> A major achievement of Steven Vajda was to explain the implications of duality in a wide variety of seemingly unrelated contexts, e.g. combinatorics and graph theory, flows in networks, game theory, and economics. ... Perhaps most surprising is its applications to certain problems in combinatorics, e.g. Menger's theorem and the König-Egerváry theorem since these problems belong to the realm of discrete mathematics and superficially appear to have nothing to do with the continuous mathematics of LP. Of course, when these problems are cast in an LP form, the optimal solutions are integral, which results in the relevance of LP (Williams 1997, 203).

Steven was presumably the first person in Britain who realized these connections and, characteristically, to enthusiastically share them with others. His results in this direction are reflected in his books (Vajda 1961, 1962) that have served as textbooks in many university and industrial courses.

Manpower planning can be defined as the problem of determining the number of personnel and their skills that best meet the future operational requirements of an enterprise. It is basically a management decision problem concerned with "arranging for the right number of individuals to be allocated to various well-defined activities," ...where "the tools of manpower planning include transfers, promotion, and recruiting" (Vajda 1978, 1). It is not exactly clear as to what first motivated Steven to study this field, but he does note that its origins go back to the 1940s when he was a statistician with the Admiralty (Vajda 1978). His colleague there, Hilary Seal, wrote a paper on the mathematics of a population (Seal 1945), which Steven followed with a related paper (Vajda 1947). Here, Steven's research dealt with mathematical analyses of such problems as "a stratified population, which is subject to (i) mortality, dependent on age, and to (ii) promotion rates, indicating the ratios of members of a grade which are transferred to the next higher grade at the end of the year" and extends to situations which have been "of actual importance in practical establishment work: the promotion rates are made dependent on the time spent in the grade instead of on the age" (Vajda 1947, 253–254). These practicalities of a population are conditions and/or constraints that need to be considered when analyzing manpower planning problems.

Steven later returned to manpower planning research when he was senior research fellow at the University of Sussex. This culminated in his book, *Mathematics of Manpower Planning* (1978). Graham Rand (1979, 767–768), in his review of the book, notes as follows: "[In] Part I, Vajda, starting from the basic concepts of a cohort with a hierarchy of states and rates of transfer between states, deals first with a workforce whose total is kept constant by new entrants and then with a workforce where all the hierarchical state totals are kept constant (a stationary workforce). If only the totals of certain subsets of the states are kept constant, the workforce is semi-stationary (a concept introduced by Steven). . . . In Part II, the questions of what structures can be attained, reattained and partially retained after *t* steps are investigated. The penultimate chapter, Optimization, is concerned with the determination of the most preferable way of moving from one structure to another." (Rand notes, "it must be stressed that this is, as the name implies, a book of mathematics.") Novel to Steven's later manpower research is the application of optimization techniques, especially the simplex and dual simplex methods of LP (Vajda 1975, 1978).

Possibly the best account of Steven's professional achievements as an author, educator, and mentor is given by Haley and Williams (1998). They give a complete listing of Steven's works. Whereas his most significant contributions fall within MP, with particular emphasis on LP, other, less known topics are cited, such as his series of papers (in German) on actuarial science, a topic in which he maintained a long-term interest (Vajda 1984). Steven, in addition to his writings, was an essential contributor to a wide variety of courses on both LP and general OR that were offered in the U.K. from the mid-1950s.

Steven's research investigations extended to related areas in group theory, geometry, cryptography, and experimental design, often with links to mathematical programming.

MATHEMATICIANS ARE SIMPLE . . .

"Mathematicians are simple folk, easily amused, quiet and amiable. They have no need for extraneous diversions since there are plenty of inner thoughts and puzzles to keep the mind busy. They make good husbands, even if their abstraction is sometimes infuriating. And lest I be thought sexist, I should add that they also make good wives when they are of the feminine gender" (Conolly and Vajda 1995, 15).

Noteworthy are his works on Fibonacci and Lucas numbers, and the Golden Section. An excellent account of the diversity of Steven's interests within pure and applied mathematics is his 15th and last book, *A Mathematical Kaleidoscope*, coauthored with Brian Conolly, and published a few weeks before his death (Conolly and Vajda 1995). This collection of essays is derived from various sources—mathematical moments one might say. Truly delightful reading!

HONORS AND AWARDS

Steven was a member of the Institute of Actuaries and a fellow of the Royal Statistical Society, London. He was awarded an honorary doctorate (Doctor Technices *honoris causa*) at Brunel University in 1971, and an Honorary Doctor of Philosophy at University of Budapest in 1975.

His outstanding contributions to OR were also recognized by the British Operational Research Society (ORS), by granting him honorary membership in 1986, and by publishing a special issue in honor of Steven Vajda's 91st birthday, edited by B. Conolly (1992).

On February 10, 1995, a meeting honoring Steven was organized at the Monopolies and Mergers Commission, London. Once again, duality was in focus. In his talk "Where is duality hiding?" Steven asked for the connections between Dantzig's simplex algorithm and Nobel laureate Kantorovich's approach; a question answered afterward by Susan Powell (1997). At the end of the meeting, Steven was presented with the ORS Companionship of OR by Professor Lyn Thomas, then President of ORS. The citation by Professor M. F. Shutler (1997) together with most of the papers presented can be found in Powell and Williams (1997).

The after-dinner speech was given by Brian Haley who can claim to have been the first academic colleague of Steven. He recalled Steven's propensity to retire, the first being from the Admiralty into academia. This was followed by his further retirements from a number of other academic posts. In his reply, Steven noted that he was famous for having retired four times and that the best thing that he could do then was to retire again so that people could get on by enjoying themselves. He did, and they did.

Martin Beale was among those recruited by Steven into the Admiralty. When EURO (Association of European Operational Research Societies), in 1988, honored Martin posthumously with the EURO Gold Medal, it was Steven who gave the speech of thanks, with a full paper containing many new ideas and reflecting his concerns.

MENTOR AND FRIEND

My life as an OR worker embarks from the early years (1958–1964) during which I was affiliated with Regnecentralen, The Danish Institute of Computing Machinery, where, literally, Danish datalogy was born. The climate around the first digital computer in Denmark and its users was appealing in every conceivable respect. Even an OR division materialized. From conceptually simple scenario analyses, we moved on via LP and were soon spellbound by the magic of the computationally intractable integer problems.

The prime influence of Steven's second textbook, *Mathematical Programming* (1961), purchased for Regnecentralen's library, cannot be overrated in this context. It was instrumental when the above-mentioned OR division was created, and, sans comparaison, our key reference during these first years of development. Not only did it introduce the various families of optimization problems and the algorithmic tools employed for their solution, but it also introduced the scope and limitations of MP as a means for investigating managerial and other decision problems via prescriptive models.

These directions were further pursued in my Ph.D. work. The most important event of that period was a 1966 NATO Summer School held in Italy. An invited speaker was awaited with particular anticipation, a Professor S. Vajda who, in a very suggestive way, highlighted the significance of duality as a central concept within MP and whose visions and thought-provoking viewpoints, in general, left a strong impression on all those present.

We met again in 1992 at a conference in Hungary where Steven delivered the opening lecture. Afterward, we discussed the earliest occurrence of duality in general nonlinear programming. I was able to recollect most of Harold Kuhn's findings as presented at the Mathematical Programming Symposium (Budapest, 1976) and, to my great surprise, thus to tell Steven just a little bit about duality which was new to him.

The story about the primal problem, a challenge thrown out in 1643 by Fermat, and its dual, a brainteaser published in 1755 in *The Ladies Diary or Woman's Almanack*, has, since Kuhn's presentation, been ranked as the first instance of duality (Kuhn 1976, 7). Whether some credit actually should go to St. Matthew is still an open question (see Matthew 11.11). Steven enjoyed the story. We both knew that Torricelli came up with a geometrical solution; neither of us, however, was familiar with the details so

FERMAT, VAJDA, ET AL.

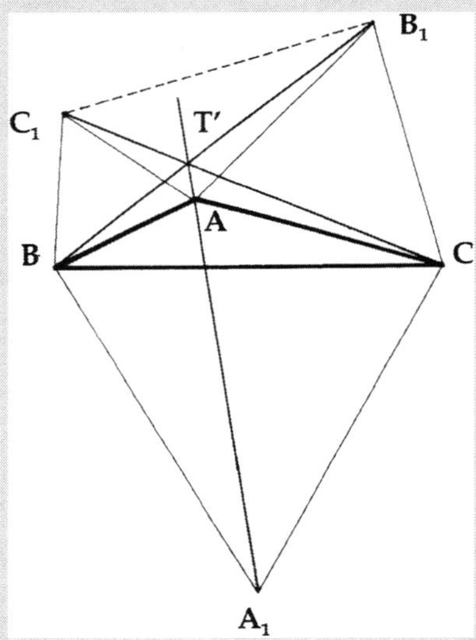

A famous problem due to Pierre de Fermat is to find a point T inside a triangle (ABC) such that the sum of the straight-line distances from T to the vertices A, B, and C is minimal. An application is when A, B, and C are warehouses and T is a factory supplying goods to the three warehouses. The mathematicians Evangelista Torricelli, Thomas Simpson, and others developed geometric-based solutions, but they hold only if all of the angles of the triangle are less than or equal to 120°. If an angle is equal to 120°, then T is at the vertex of that angle. In their book, *What Is Mathematics*, Courant and Robbins (1941) addressed the situation when an angle is greater than 120°. There they stated that if you apply Simpson's method, the associated point, here T', is such that the sum of the distances ($BT' + CT' - AT'$) is minimal. They left the proof of what they termed the "complementary problem . . . as an exercise to the reader" (Courant and Robbins 1941, 358). With Steven a reader, along with this author, and the result as stated being incorrect, one would expect the situation to be appropriately remedied. This is reflected by Steven's last paper (Krarup and Vajda 1997), which later led to Jalal and Krarup (2003) and Krarup (1998).

it was decided to investigate the matter further. The result appeared in Krarup and Vajda (1997).

Another incentive for further joint studies was an error made in the famous book *What Is Mathematics*? (Courant and Robbins 1941). Also, a paper on "Visualizing duality" was on the drawing board at the time Steven passed away. It meant to exhibit a series of examples within linear, nonlinear, and combinatorial programming where the notion of duality appears to be particularly conspicuous.

Even more valuable, however, was the friendship that evolved between us through these endeavors. Similar debts to Steven are in one way or another owed by many others. Since Steven joined the Admiralty, he spent about 50 years consciously or unconsciously motivating the careers of numerous OR workers. He has exerted this influence directly by teaching and conference presentations and indirectly by his writings and by the example of his life.

With an amazing mobility, Steven remained a frequent traveler to conferences and symposia. Insight and an implacable curiosity marked his active contributions as a lecturer or discussant. Likewise, no effort was spared outside the conference halls. Not even the extremely hot weather in Israel in July 1995 discouraged Steven from taking part in the physically demanding 2-days' excursion to Galilee organized in conjunction with EURO XIV conference. Not a single detail was missed, let alone the celebrations of EURO's 20th Anniversary. The organizers of the conference were honored to have him as the oldest participant ever at a EURO event. The International Federation of Operational Research Societies Vancouver conference in 1996, with a pre-conference hike in the Rockies, was supposed to be "next time" (Krarup 1996, 12).

It is a gift of grace to shuffle off this mortal coil after having enjoyed a long life without suffering the physical horrors of old age and even more so to preserve both a warm heart and a brilliant mind to the end. Those gifts were granted to Steven Vajda, still teaching and writing in his 94th year before he passed away after a short illness on December 10, 1995. As the panegyric at his funeral noted, he had the capacity to make everyone he met believe that they were his friend. Therein lays the truth of his memory (Krarup 1966, 437–439).

ACKNOWLEDGMENTS

The primary sources for this profile are the interview (Bather 1995), the detailed account of Stevens's work (Haley and Williams 1998), M. F. Shutler's citation for the Award of the Companionship of the Operational Research Society to Steven (Shutler 1995), and the obituary by Krarup (1996). Supplementary material, not available elsewhere, was acquired via personal communication with Steven's daughter, Hedy Firth, and his last coauthor, Brian Conolly. Their kind assistance and permission to use, quote, and paraphrase their material on Steven is gratefully acknowledged.

REFERENCES

Bather J (1995) An interview with Steven Vajda. OR Newsletter January, 25–29

Conolly B (1992) Editorial. Special issue on mathematical methods in honour of Steven Vajda. J Oper Res Soc 43(8):737–739

Conolly B, Vajda S (1995) A mathematical kaleidoscope. Albion Publishing, Chichester

Courant R, Robbins H (1941) What is mathematics? Oxford University Press, Oxford

Dantzig G (1949) Programming of interdependent activities, II, mathematical model. Econometrica 17(3–4):200–211

Dantzig G (1951) Maximization of a linear function of variables subject to linear inequalities. In: Koopmans T (ed) Activity analysis of production and allocation, Cowles Commission Monograph No. 13. Wiley, New York, NY

Jalal G, Krarup J (2003) Geometrical solution to the Fermat problem with arbitrary weights. Ann Oper Res 123:67–104

Haley K, Williams H (1998) The work of Professor Steven Vajda. J Oper Res Soc 49(3):298–301

Krarup J (1996) Obituary: Steven Vajda 1901–1995. OPTIMA 49:12

Krarup J (1998) On a "Complementary Problem" of Courant and Robbins. Location Sci 6:337–354

Krarup J, Vajda S (1997) On Torricelli's geometrical solution to a problem of Fermat. IMA J Math Appl Bus Ind 8(3):215–224

Koopmans T (ed) (1951) Activity analysis of production and allocation. Cowles Commission Monograph No. 13. Wiley, New York, NY

Kuhn H (1976) Nonlinear programming: a historical view. SIAM-AMS Proc 9:1–26

Powell S (1997) Kantorovich's hidden duality. IMA J Math Appl Bus Ind 8(3):195–201

Powell S, Williams H (eds) (1997) Special issue: duality in practice, dedicated to the work of Steven Vajda. IMA J Math Appl Bus Ind 8(3)

Rand G (1979) Mathematics of manpower planning (book review). J Oper Res Soc 30(8):767–768

Seal H (1945) The mathematics of a population composed of *k* stationary strata each recruited from the stratum below and supported at the lowest level by a uniform annual number of entrants. Biometrica 33:226–230

Shutler M (1995) Companion of operational research. J Oper Res Soc 46:918

Shutler M (1997) The life of Steven Vajda. IMA J Math Appl Bus Ind 8(3):193–194

Vajda S (1947) The stratified semi-stationary population. Biometrika 34(3/4):243–254

Vajda S (1956) The theory of games and linear programming. Methuen, London (Translated into French, German, Japanese and Russian)

Vajda S (1958) Readings in linear programming. Pitman, London (Translated into French and German)

Vajda S (1961) Mathematical programming. Addison-Wesley, New York, NY

Vajda S (1962) Readings in mathematical programming (Second edition of Vajda, 1958). Pitman, London

Vajda S (1975) Mathematical aspects of manpower planning. OR Q 26(3):527–542

Vajda S (1978) Mathematics of manpower planning. Wiley, Chichester

Vajda S (1984) Actuarial mathematics. In: van der Ploeg F (ed) Mathematical methods in economics. Wiley, Chichester, pp 457–476

Williams H (1997) Integer programming and pricing revisited. IMA J Math Appl Bus Ind 8(3):203–213

3

Philip McCord Morse

Robert M. Oliver

Philip McCord Morse was a leading American physicist, a World War II (WWII) pioneer in operations research (OR), founding member and first president (1952) of the Operations Research Society of America (ORSA), and probably the first person to take steps to ensure that OR and systems analysis should become an academic discipline to improve decision making in private and public institutions, as well as government. He is widely viewed as the father of OR in the U.S. In addition to this enormous accomplishment, he had a distinguished career as a physicist and made major contributions to public policy and society.

His service to the U.S. government and professional organizations included: Director of the first OR organization established in the U.S., the Anti-Submarine Warfare Operations Research Group (ASWORG); Director, Brookhaven National Laboratory; Deputy Director and Director of Research, Weapons Systems Evaluation Group; President, Acoustical Society of America; and President, American Physical Society.

Phil's myriad honors include the U.S. Presidential Medal for Merit, the Silver Medal of the Operational Research Society of the U.K., and the Acoustical Society of America Gold Medal Award.

The Early Days

Phil Morse was born August 6, 1903 in Shreveport, Louisiana, but grew up in Cleveland, Ohio. He died September 5, 1985 in Concord, Massachusetts. Phil's mother was Edith McCord and his father was Allen

A.A. Assad, S.I. Gass (eds.), *Profiles in Operations Research*, International Series in Operations Research & Management Science 147, DOI 10.1007/978-1-4419-6281-2_3,

Crafts Morse; they were married in 1901 and had five children, two girls and three boys of which Phil was the eldest. His father attended the newly founded Case School of Applied Science for 3 years. He then worked in telephone-system construction supervising the building of telephone exchanges and in various capacities for independent telephone companies. His mother's father, William McCord, was editor of the East Liverpool, Ohio newspaper. After finishing high school, Edith worked there as a reporter.

In 1929, Phil married Annabelle Hopkins. Their son, Conrad Philip, was born on February 22, 1931, and their daughter, Annabella, was born on December 24, 1939. Conrad married Gail Proudman, and Annabella married Hugh Fowler. Phil and Annabelle were blessed with six grandchildren (Fowler 2008; Morse 1977a).

Phil learned how to play the violin at an early age, how to drive the family's model T Ford when he was 15, and did well in school—"my grades were mostly A's"—even though he missed school for 2 months most winters due to illnesses (Morse 1977a, 13). He read voraciously, was attracted to chemistry, and became fascinated with radios. In his last year of high school, he joined with friends to start a business supplying radio parts; the Radioelectric Shop opened in Cleveland during the summer of 1921. Phil graduated from Lakewood High School, Lakewood, Ohio in 1921, and applied and was accepted to the Case School of Applied Science (Feshbach 1994, Morse 1977a).

At Case, Phil majored in physics, graduating in 1926. Dayton C. Miller, then one of the country's great physicists and head of the Physics Department, supervised Phil's senior thesis. (Miller was an expert in acoustics and music-making, an expert flautist, and a collector of flutes and wind instruments; he donated nearly 1700 flutes, wind instruments, books, and

> MORSE FAMILY TREE?
>
> A much earlier ancestor, Samuel Mors, son of Anthony Mors, a rector living near Dedham, England emigrated to Massachusetts in 1635; he was to become one of the founders of Dedham, Massachusetts. Separately, an Anthony and William Morse in Wiltshire County, near Bristol, England also emigrated in the 1600's. This Anthony Morse was an ancestor of the Reverend Jedediah Morse, the first American geographer, and the more famous Samuel F. B. Morse, inventor of the telegraph. If Anthony Mors was related to Anthony Morse, which seems likely but has not been firmly established, then there are many cousins in the states of Ohio and Massachusetts, including Samuel F. B. Morse and Philip McCord Morse that can be connected by a common family tree going back more than fifteen generations (Morse 1967, 1977a, 116).

related items to the Library of Congress [Gilliam and Lichtenwanger 1961, Morse 1977a].) The radio supply business prospered and enabled Phil to earn his way through Case; after his freshman year, Phil had to leave Case for a year due to family financial difficulties and work in the Shop, which eventually closed in 1926.

Miller suggested that Phil apply for graduate admission and a fellowship to three universities: Chicago, Harvard, and Princeton. Although accepted by all three, he decided to pursue his doctorate at Princeton because they offered him $700 a year plus tuition (Morse 1977a). There he started to work under the direction of the physicist Karl Compton. Phil soon realized that he was in the presence of a brilliant experimentalist who was always motivated by the measurement and explanation of new phenomena. But Phil was more attracted to theoretical explanations and the excitement of devising new models to explain illogical phenomena or puzzling discrepancies between experimental results and theory. His early fascination with model building seems to have started when the field measurements of gas discharges were not supporting the small amount of theory that was then available for atomic plasmas. By combining his knowledge of electrostatics with new developments in statistical and quantum theory of atomic ionization, he and his mentor were able to publish one of the early papers on this subject (Compton and Morse 1927). This and a list of other gas discharge problems puzzled Phil and he continued to incorporate new theory and improved models that might explain the experimental observations. Phil published four papers on electron discharges in gases, now known as plasma physics. Compton decided to accept one of Phil's papers as his dissertation (Morse 1928): "A theory of the electric discharge through gases." He received his Ph.D. in 1929.

From 1930 to the summer of 1931, Phil was granted a Rockefeller Fellowship to study and do research at the Ludwig Maximilian University in Munich under the direction of Arthur Sommerfeld. There he came in contact with several famous theoreticians: Felix Bloch, Edward Teller, Werner Heisenberg, Linus Pauling, and W. L. Bragg. He then visited Cambridge University in the spring and summer of 1931 where he came to know and work with a large number of prominent scientists, including Ernest Rutherford, Neville Mott, Paul Dirac, Harrie Massey, Peter Fowler, Julius Stratton, and Patrick M. S. Blackett. Although Phil borrowed Blackett's car (Morse 1977a, 116) on a trip to East Anglia to trace roots of his ancestor Samuel Mors, neither one of them could anticipate how their future professional paths, in a new discipline called OR, would cross a decade later.

Teaching and Research at MIT (1931–1941)

Karl Compton became president of the Massachusetts Institute of Technology (MIT) in 1930 and offered Phil a faculty position in 1931 after his return from Europe. Phil was a significant leader in establishing MIT's eminence in physics and his early research showed a wide range of interests. As a graduate student, he co-authored the first American book on quantum mechanics with E. U. Condon (Condon and Morse 1929). Many papers followed, including his discovery of closed form solutions to the well-known Schrödinger equation, now called the Morse Potential. Later, in 1936, he wrote a seminal book, *Vibration and Sound* (Morse 1936) modernizing and extending the theory of acoustics (Phil dedicated the book to Professor Miller, his adviser at Case). Many of the principles and experimental results revealed in this book contributed not only to further papers, but also to acoustic design principles that were implemented in many large orchestral halls and, indirectly, to his interest in transmission of underwater sounds.

In the period 1940–1942, Phil became director of the MIT Underwater Sound Project, a position which directly led to his active involvement in the war effort in 1941—he was asked to direct a U.S. Navy project on devices that detect and defend against acoustically actuated mines. At Princeton, Phil shunned the idea of becoming an experimentalist and, although he never claimed to be a designer of devices, he and others in the project were challenged with making a cheap and reliable source of intense noise. Phil came up with the idea that they tow, through the water, two four-foot-long pipes, each an inch-and-a-half in diameter and spaced a half-inch apart, rigged in such a way that their length was perpendicular to the direction of travel, thus causing the water to be forced through the slot between the pipes. A machinist built the contraption in one night. The next day's test found Phil's contrivance first in noisiness by many decibels—he had invented a large underwater raspberry! Later, when this device (named the Foxer) was towed behind American destroyers, it successfully decoyed and blew up several acoustic torpedoes without the loss of a single destroyer (Acoustical Society 2001).

While at Princeton, Phil had met George Kimball, an undergraduate in chemistry. They became reacquainted at MIT when Phil was appointed to the faculty in 1931 and George was a post-doctoral fellow (Morse 1968b). Phil noted that "... in 1934, I was developing the graduate course in theoretical physics .... George was the first person I asked to help me teach the course; his help was invaluable ..." (Morse 1968b, 872). When Phil was asked to organize a small group of scientists to analyze anti-submarine operations, George was one of the first persons he contacted. Later, in 1942, Phil appointed George as his deputy for the U.S. Navy's wartime OR group.

During this period, Morse recognized the importance of integrating operational policies with technology. Increasingly, he felt that the key to effectiveness lay not just in designing improved hardware, but also in understanding how best to organize and make better use of existing equipment. Therefore, he quickly accepted a call from Washington, D.C. to recruit and direct a team of civilian scientists who were asked to study and recommend actions on anti-submarine warfare. The U.S. Navy's hope was to duplicate in the U.S. the success of a small naval OR team started in the U.K. under Patrick Blackett, a future Nobel Laureate in physics for his work on cosmic radiation and nuclear physics. Blackett, whom Phil had met at Cambridge in 1931, had also been head of OR groups for the U.K.'s anti-aircraft command from 1940 to 1941 and Coastal Command in 1941–1942. He stressed the role of scientists as observers at executive levels and potential critics, and was instrumental in advising how such groups should be integrated within the military command structure (Kirby 2003). A paper by Blackett (1941), "Scientists at the operational level," written for the U.K. Admiralty, described the early successes that he and his team of analysts, known as Blackett's Circus, had achieved in anti-submarine and radar operations. The paper had been circulated at U.S. Naval

Headquarters, and it was the impetus for the Navy's recruiting Phil to replicate such an activity.

A SEARCH FOR HEMIBELS: U.S. OR IN WORLD WAR II

In the U.S., the Antisubmarine Warfare Operations Research Group (ASWORG) was formed in April 1942 with headquarters in Washington. Morse's team soon demonstrated that research on operations in the hands of highly motivated, first-rank scientists could produce striking improvements in short time frames under enormous wartime pressures. Based on its early success, ASWORG was renamed the Operations Research Group (ORG) in 1944 and extended its scope of activities to a much broader list of areas than antisubmarine warfare. During this period, the group gained access to senior decision makers by demonstrating outstanding results. Success led to growth and, by the end of the war in 1945, the organization was renamed the Operations Evaluation Group (OEG).

Phil's scientific and organizational talents, coupled with his many professional contacts, enabled him to attract and assemble a distinguished group of scientists and applied mathematicians; in particular, George Kimball from the Chemistry Department at Columbia University, Bernard Koopman from the Mathematics Department at Columbia University, and William Shockley from Bell Telephone Laboratories. Phil had come in contact with these scientists at Princeton during his graduate days or at MIT after he was appointed to the faculty.

Phil's research approach was basic and direct: complex problems require the perspectives and competitive contributions of different analytic and experimental techniques, as well as taking into account the needs of different users. This view is in sharp contrast to the research approach of many others who search for problems that can be solved by a particular mathematical technique, and who discount the need for direct experimental observations in the development of their models.

Although there have been many instances in the past where scientific analyses have provided aid to military commanders and decision makers, ground-breaking developments in the field and a sense of focus were largely created by the pressures of WWII when advances in tactics and strategy were, in many cases, found to be as important as new developments in technology. In a wartime setting, armed forces cannot usually afford the luxury of waiting for better weapon designs; they have to seek ways to improve the operational use of existing equipment.

THE POWER OF HEMIBEL THINKING

"It is important first to obtain an overall quantitative picture of the operation under study. One must first see what is similar in operations of a given kind before it will be worthwhile seeing how they differ from each other. In order to make a start in so complex a subject, one must ruthlessly strip away details (which can be taken into account later), and arrive at a few broad, very approximate 'constants of the operation.' By studying the variations of these constants, one can then perhaps begin to see how to improve the operation.

"It is well to emphasize that these constants which measure the operation are useful even though they are extremely approximate; it might almost be said that they are more valuable because they are very approximate. This is because successful application of operations research usually results in improvements by factors of 3 or 10 or more. Many operations are ineffectively compared to their theoretical optimum because of a single faulty component: inadequate training of crews, or incorrect use of equipment, or inadequate equipment. Usually, when the 'bottleneck' has been discovered and removed, the improvements in effectiveness are measured in hundreds or even thousands of per cent. In our first study of any operation we are looking for these large factors of possible improvement. They can be discovered if the constants of the operation are given only to one significant figure, and any greater accuracy simply adds unessential detail.

"One might term this type of thinking 'hemibel thinking.' A bel is defined as a unit in a logarithmic scale corresponding to a factor of 10. Consequently, a hemibel corresponds to a factor of the square root of 10, or approximately 3. Ordinarily, in the preliminary analysis of an operation, it is sufficient to locate the value of the constant to within a factor of 3. Hemibel thinking is extremely useful in any branch of science, and most successful scientists employ it habitually. It is particularly useful in operations research" (Morse and Kimball 1951, 38).

The team of scientists in the anti-submarine warfare group were asked to consider different weapons (depth charges, anti-submarine rockets, canons, ...), delivery vehicles (airplanes, destroyers, frigates, submarines, ...), and detection methods (sonar contacts, radar signatures, visual sightings, ...). How should one gather and study relevant information to predict future tracks and locations of an enemy submarine? Could one design tactical maneuvers and organize them into meaningful strategies to search and destroy the submarines? Could one build statistical and mathematical models and meaningful analyses of data that included the effect of variations in visibility, imperfect sonar, an unfamiliar new technology of radar, fuel, and speed considerations for different search aircraft, effectiveness of weapons with differences in sizes, weights, and destructive power? High on the list of uncertainties were the depth, range, speed characteristics, and strategies of enemy submarines. The models often led to complex searching strategies. Early on, Phil recognized that insights derived from the models not only had to be understandable, but also, above all else, trusted by U.S. Navy personnel whose lives were on the line if they were to adopt the recommended courses of action. Because radar was an unfamiliar new technology, ORG scientists worked closely with pilots to make operational use of devices that frequently broke down, had to be recalibrated, and tested.

As an example of the kind of thinking that entered into the analyses of anti-submarine warfare (ASW), it is interesting to read some of the findings that were declassified in 1982 by The Center for Naval Analyses (ASWORG 1942). One, in particular, illustrates an important aspect in the practice of OR: A decision model, data collection, and implementation is never complete, even though the problem may have been solved analytically, until post-implementation performance data is collected and analyzed and, is able to confirm the predicted results. This profound insight is one OR practitioners would do well to heed today to confirm the trust one can place in the theoretical models, the soundness of recommended actions, and the influence of new decisions. It gives direct assessments of the validity of the underlying models. As an example of this kind of thinking it may be useful to describe some of the early developments of Seek, Hide, and Feint strategies that were developed because the original theoretical models did not perform as they were supposed to.

Shortly after the ASW unit was formed, it was recognized from shared experiences with the U.S. and British forces that two distinct anti-submarine strategies should be considered. The most familiar one was passive detection of submarines operating alone or in packs and attacking large convoys in mid-Atlantic that passed through their midst; these were pursued and often destroyed by Allied destroyers skirting the convoys with new sonar detection equipment. A new idea was that active military search and destroy tactics should also be undertaken in parallel with these passive methods. This led to a plan to send out aircraft patrols from land bases to aggressively hunt for submarines in critical bottleneck areas where the density of submarines was higher than in the open ocean.

The early recommendation that the organizational structure of aggressive search and destroy tactics be separate and distinct from ordinary patrols, and that contacts for these patrols be identified in three classes, depended on the accuracy of location of enemy submarines, the distance and time for arrival of a specialized attack/destroy aircraft or ships to reach the contact location. Initially, tactics were designed that required the detecting aircraft to remain at their stations, thus forcing the detected submarine to stay submerged and travel at a slow speed underwater; it was believed this strategy would speed up the use of scarce battery energy, prevent surface recharging, and, because of the slow speed of an underwater submarine, reduce the area that would have to be subsequently searched upon the arrival of the attack/destroy aircraft.

Because Phil and his colleagues insisted on reexamining data of submarines contacted and destroyed following the introduction of new tactics, it was found, to everyone's surprise, that the scant data on confirmed sinkings did not agree with model predictions that had led to the remain-on-station tactics. Could there be some advantage in having the submarine resurface after initial detection but before the attack/destroy aircraft arrived? At first, this idea seemed strange but it did include the possibility of obtaining a second and better fix, that is, a more precise location of the submarine if it resurfaced. Someone also suggested that additional precious electrical energy might be required if the submarine repeated surface and dive maneuvers. In what might pass as one of the earliest uses of game theory, tactics were then designed to create alternating periods of search, detection, withdrawal, and reappearance by the search aircraft so that the submarine would be tricked into resurfacing to charge

WARTIME HIDE AND SEEK

"It should also be apparent that the newly developed fields of game theory and communications theory are of great utility in operations research. An example of the use of game theory lies in the aircraft-anti-submarine battle in the Bay of Biscay, off the coast of France, in the last war. The U-boat, as it went out from its base in France, had to come to the surface for a few hours a day to recharge batteries. The longer it ran on the surface, the sooner it could get out of range of the airplanes. The British first used visual sighting and flew only in the daytime. The U-boats, of course, surfaced only at night and submarine losses were small. The British then equipped a few airplanes with radar and searchlights for use at night. The problem was to determine how many such airplanes should be equipped and flown. If they all flew at night, the Germans would surface by day. Moreover, the night attack was not as deadly as the day one. Game theory determined what proportion of day and of night fliers should be maintained in order to keep the Germans most confused and in order to get the most U-boat kills per given cost in equipment, training, and men. Game theory also indicated what were the best tactics for the submarine; what percentage of hours it should surface by day and how often by night" (Morse 1951, 194).

its batteries because it believed that air surveillance had been withdrawn. Numerous calculations of the length of the withdrawal periods and optimal Seek, Hide, and Feint strategies were developed which depended on different assessments of underwater and surface speeds for the submarine and the time for the first of the attack/destroy aircraft to reach the patrol area. Models were redesigned and predictions tested again; new data validated the improved performance.

BAY OF BISCAY U-BOAT CIRCULATION MODEL

The Morse and Kimball U-boat Circulation Model (McCue 1990, 99–108; Morse and Kimball 1951, 78) is an example of the kind of superb model building and analysis that Phil and George incorporated in their scientific studies. The first noteworthy item is their choice of a measure of effectiveness that focused on the number of merchant ships afloat (not sunk) rather than enemy submarines sunk. The number of active submarines at sea obviously influenced the former, so a model was needed to estimate the status of submarines: the original inventory of submarines plus newly produced ones minus those sunk or under repair, the latter also unavailable for sinking ships.

Conservation requires that the rate of growth of U-boats at sea equal the net production, p (newly produced minus number sunk), less the net flow into or out of repair facilities, a term which could be positive or negative. If we think of x and y as dimensionless quantities representing fractions or numbers of U-boats at sea and numbers under repair, the rate of completed repairs $r(y)$, itself a function of the number in repair, and u a scaled unit of time, the rates of change of numbers at sea and in repair can be approximated as two first-order differential equations:

$$\frac{dx}{du} = p - \frac{dy}{du}$$
$$\frac{dy}{du} = kx - r(y).$$

Phil and George argued that the growth rate of numbers needing repair is directly proportional (constant k) to the active number at sea, a positive number, minus the rate, $r(y)$, of submarines repaired and reactivated. The particular choice for the rate, $r(y)$, should be zero with no submarines in repair ($y = 0$) but achieve saturation when the number needing repair is very large. A natural choice is exponential saturation

$$r(y) = 1 - e^{-y}:$$

$$\frac{dx}{du} = p - kx + (1 - e^{-y})$$

$$\frac{dy}{du} = kx - (1 - e^{-y}).$$

Solutions of this simple model "... seem to indicate that the damaging of repair bases had a greater effect than the damaging of the factories and thus led to the strategic decision to intensify the bombing of repair facilities" (Morse and Kimball 1951, 80). The importance of U-boat repairs was confirmed after the war by Admiral Karl Dönitz (McCue 1990, 99). More advanced models followed during and after the war but the elegant simplicity of this model for the effects of U-boat circulations had an enormous impact on decision making.

Other examples appear in the compilation of Navy reports edited by Phil and George Kimball (Morse and Kimball 1946, 1951). The resetting of the detonation depth for air-dropped depth charges, combined with better search tactics, increased the sinking of enemy submarines by a factor of five which fits well within the hemibel (factor of three) thinking that pervaded much of Phil's work. ORG's plan for laying out search patterns between Brazil and Africa to catch German ships carrying rubber from the East Indies resulted in catching six of the seven ships that tried to run the blockade.

Many details of the ASWORG story are discussed in Phil's autobiography (Morse 1977a). Excerpts from this book that describe the beginnings of OR in the U.S., that is, WWII ASW activities, are given in a posthumous paper (Morse 1986). In an introductory note, Hugh Miser states, "Morse's story is important for the history of operations research on two counts: because he was the central figure in starting operations research activity in the U.S., his account is uniquely authoritative; and it represents a useful paradigm of how operations research analysts can—and, as many believe, should—approach a problem area to which they bring no specialized contextual knowledge: it should therefore be widely known. Thus, the purpose of these excerpts, drawn from Morse [1977a, 170–196], is to make Morse's account widely and readily available to operations research workers everywhere (Morse 1986, 10)."

After the war, Phil organized an operations research group for the Secretary of Defense and the Joint Chiefs of Staff, the Weapons Systems Evaluation Group (WSEG), and served as deputy director, under Lt. Gen. J. Edward Hull, until 1950. The organization set up a civilian counterpart, the Institute for Defense Analysis (IDA) in 1956 and Phil served as trustee.

He was the founding director of the Brookhaven National Laboratory from 1946 to 1948. He was also a trustee of The RAND Corporation of Santa Monica, California from 1948 to 1952 and one of the founder/trustees of the ANSER Corporation of Arlington, Virginia from 1962 to 1973. The former focused more on strategic problems and the latter more on tactical problems for the Secretary of the Air Force.

The MIT Operations Research Center (ORC)

Following the war, Phil and others were convinced that OR could play an important role in non-military as well as military organizations and that OR should be included in an interdisciplinary academic research program, though perhaps not as a separate department. This is possibly where he may have differed with Patrick Blackett who had been the first to develop and introduce OR to the U.K. armed forces during World War II.

In 1953–1954, I was on a Fulbright Scholarship in Physics at the University of London. While browsing through the college library, I came across the first issue of the *Journal of Operations Research* (U.S.) and the *Operational Research Quarterly* (U.K.), which was already in its second volume. There was an announcement that graduate research assistantships were being offered at MIT under the auspices of the Office of Naval Research and a committee chaired by Phil. What better way to get a first-hand assessment of this mysterious subject than to ask the Chairman of the Physics Department, Patrick Blackett, in his upstairs office, whose name was clearly associated with the same topic? It became clear that while he believed that scientists could and should contribute to problems of national interest, he did not feel that OR had a natural place in academe; rather, one should pursue studies in traditional fields of science. I went to the meeting with great enthusiasm, but came away feeling a bit let down. My impression then and now is that he and Phil would probably agree on the importance and likely future of OR—the need for good science and imaginative scientists—with Blackett much less sanguine about the role of OR in an educational institution.

When he returned to MIT, Phil devoted major attention to the establishment and growth of OR as an interdisciplinary team effort grounded in the use of good science. In 1952, MIT, under Phil's chairmanship, agreed to the formation of the Committee on Operations Research to coordinate education and research in the subject. As interest in OR grew and two small research contracts were obtained from the Navy and Army, a more formal cross-departmental Operations Research Committee was formed, with Phil as its chair from 1956 to 1969. Although I do not have a formal record of its activities, I suspect he worked very hard finding additional sources for financial aid to attract graduate students in the new discipline (Morse 1977a, 291–295).

Phil also recognized the need to establish an OR professional organization as a means for spreading OR beyond its growing academic basis and its past successes in WWII. As he noted in Morse (1952, 1): "For the past three or four years the activity variously known as operations research, operational research or operations analysis, originally devised for military needs, has been a subject of increasing interest in nonmilitary circles. . . . A number of workers in the field have felt that the time was ripe for some organization in this country to provide a means for the advancement of and diffusion of knowledge concerning operations research." Phil stated a further impetus for forming a professional society: "As soon as it began to look as though O/R [OR] would become popular, a few quacks began using its name to sell their magic. It was time to establish standards and to provide outlets for an exchange of information" (Morse 1977a, 291).

Phil chaired a formation committee in January 1952 whose efforts led to the founding meeting of the Operations Research Society of America on May 26–27, 1952. Phil was elected ORSA's first president. His interest in supporting the professional aspects of OR also extended to international activities. He was instrumental in the preparations for the first International Conference on Operational Research held in Oxford from September 2–6, 1957; his invited opening address was "Operations research is also research" (Morse 1957). This conference led to the founding of the International Federation of Operational Research Societies (IFORS). Phil served a term (1962–1964) as secretary of IFORS and was general chairman of its fourth conference held at MIT in August 1966 (Horvath and Ernst 1986).

From the beginning he was interested in attracting faculty and students from a wide spectrum of MIT departments for study of operations and systems. By getting financial support for graduate students, Phil

THE DIRECTOR

"People have asked me what it was like to be Morse's student. I remember his office well. He had a totally inadequate blackboard. I recall it as 2.5 feet by 3 feet and you couldn't write more than one and half equations on it.... Morse's office also contained a couch opposite the blackboard. It was extraordinarily saggy and uncomfortable. Surely nobody overstayed their leave in his office if he had seated them on the couch. Morse, himself, sat at a desk from which he could see both couch and blackboard and, if he wished, look out the window. I have said he was a salesman, but he was not high pressure. Rather he was reasoned and flexible, but behind the demeanor was a very quick mind" (Little 2002, 148).

attracted the interest of a small number of faculty, even though most had not yet warmed up to the idea of having an interdisciplinary degree-granting program, instead of the traditional vertical department structure. Phil, with his experience and persuasive skills, convinced some of the brightest and best to join him in the endeavor. Eventually a formal and structured degree program with close ties to the Sloan School of Management and the School of Engineering were established.

Through his contacts with the Office of Naval Research and the Army Research Office, he initiated a number of research programs that gave financial support to a growing number of graduate students whose undergraduate degrees were in physics; mathematics; electrical, civil, and mechanical engineering. He brought together a number of interested faculty to help coordinate and contribute to

ORC AT HOME

"When the first actual 'home' for the OR Center was established in a half-level warren of Physics (1956?), we discovered other students and faculty from different departments, also working on interesting problems .... Slowly, we began to think of ourselves as colleagues, if not a separate department" (Bill Jewell in Larson and Magnanti 1994).

the activities of the ORC: Dean Arden and Bill Linvill in electrical engineering (Linvill went on to form the Engineering and Economic Systems Department at Stanford University), George Wadsworth and Herb Galliher from mathematics, David Durand and Bob Solow from economics, and Harvey Wagner, Ned Bowman and Mike Gordon from the Sloan School of Management.

The ORC supported summer courses in OR, seminars by speakers with direct OR wartime experiences, and encouraged visitors who wanted to introduce OR into their organizations. Visitors and speakers who passed through the ORC included many of the early OR researchers and practitioners: Russ Ackoff, David Boodman, Arthur Brown Glen Camp, West Churchman, Les Edie, Martin Ernst, Robert Herman. David Hertz, William Horvath, George Kimball, Bernard Koopman, John Magee, Jacinto Steinhardt. The book (MIT 1959), *Notes on Operations Research 1959*, compiled by the ORC staff, contains lecture material from earlier summer courses and was the basis for a 2-week Special Program in Operations Research held in Brussels for the European NATO members in August 1959. Phil's contributions included chapters on Markov processes and the organization of OR groups.

During the first (June 1953) of the 15-year series of summer OR courses sponsored by the ORC, Phil's attention was drawn to a lecture on queues. He recognized, of course, that queues are elements of many operations such as banks, tollbooths, and airport landings and take-offs. The lecture prompted a train of thought that "suggested a more general technique of analysis of a large number of queuing problems. The one method could be used to obtain a wide variety of answers," instead of what he termed "special tricks to obtain answers" to particular applications (Morse 1977a, 292–293). This led to his book, *Queues, Inventories, and Maintenance* (Morse 1958) that described "an outline of the general technique, with examples, written in a style I hope would be read by few executives and by most O/R [OR] professionals, not just by mathematicians" (Morse 158, 294).

Many of Phil's students have contributed to the development and application of OR. This is by no means an exhaustive list, but from physics it includes John Little, Steve Pollock, Walter Helly, and myself; from electrical engineering, Ron Howard, and in the early days, Newt Garber and Bill Jewell, who participated in the weekly discussions and all other ORC activities. Later, Al Drake, Richard Larson, and Ralph Keeney from electrical engineering were graduate students who continued their affiliation with the ORC after being appointed faculty in different MIT departments.

Libraries

> LESSON LEARNED (1)
>
> "Two stories that Phil told about his work in OR during WW II have influenced my choice of problems to study. One was that a team never worked on a problem unless they thought they could identify ways to yield at least a 300% improvement in operations. This said to me that the values of operations researchers are important for choosing problems, and that the operations researcher should spend time to create alternative solutions to that problem. The other story was that his team wanted to think about the process of antisubmarine warfare before reading about it or analyzing any data on it. The insight here is that the operations researcher should be involved in identifying problem areas to address. In this, your values are crucial" (Keeney 1994, 794).

It is not clear when Phil developed his keen interest in library operations and policies, but it was probably as early as 1956. At one of our early graduate student meetings, he complained about the science library that kept accurate records on the less popular books because return dates for borrowers had not filled up the card in the flap at the end of the book but, on the other hand, kept few records on a popular book because the card in the flap would be thrown out and a new one inserted (possibly many times over the course of a year) when the number of sign-outs exceeded space on the old card. If his only interest had been in estimating expected demand for books, he would have resorted to any one of numerous statistical papers on estimation with biased samples.

With his enormous experience in search models, both theoretical and experimental, one might have also expected that Phil would look for applications of this well-established theory in a library setting. But this was not his style; he always started with a fresh perspective on unsolved problems. In this case, he and his students quickly focused on the quality and cost of access to information, the connections between user demand, book inventories, queues with circulation patterns, and the trade-offs between sign-out and reserve policies. This led him to study ways to discriminate, yet balance high-demand, popular parts of a collection with the less-used portion and to improve book purchasing and book retirement decisions. Effective planning, organization, and improved decision making began to substitute for passive collections and investments in large repositories.

> MEASURE FOR MEASURE
>
> "I remember with glee when we (Phil and I) were able to confirm from librarian surveys that their common performance measure was essentially the probability that a book would be in the stacks when a borrower requested it. We pointed out that this measure could be maximized by not allowing any sign-outs. We were happy to suggest that they might consider other measures, possibly economic, with constraints on the probability of availability!" (Pollock 2008).

Because of his interest in libraries and computers and his predictions for the rapidly developing role of computers in the acquisition and distribution of information, he interacted with several new colleagues in library research centers at the University of California, Berkeley and Indiana University, and was frequently asked to give talks on this emerging science. He was awarded ORSA's Lanchester Prize in 1969 for his book, *Library Effectiveness: A Systems Approach* (Morse 1968a).

PUBLIC POLICY

As early as 1962, Phil began to understand the possibilities for the use of OR in the public sector. In 1966, one of the many MIT OR summer programs that attracted people from industry and government was devoted to "Operations Research in Public Affairs." As was usually the case, he involved a dozen or more lecturers to address a broad range of problems in traffic congestion, transportation networks, criminal justice, medical and hospital practice, and regional planning. They undoubtedly influenced the publication of the book by Drake et al. (1972), *Analysis of Public Systems*.

> LESSON LEARNED (2)
>
> "To be of greatest use, operations research must be an experimental, not a historical science. For this reason the most effective operations research groups during wartime were attached to high command levels, where the broader operational problems arose and where the action necessary to implement an operational experiment could be ordered. In peacetime, also, operations analysts must have direct access to the industrial or governmental official in charge of the operation to be studied, or their work will have little practical value" (Morse 1946, 29).

Based on the lectures given at this summer program, Phil co-edited the book *Operations Research for Public Systems* (Morse and Bacon, 1967). His introductory chapter contains the most lucid descriptions I have ever read of the likely benefits and possible limitations of the application of science to the solution of industry, government, private, and public decisions. I have read it several times, and never cease to be astounded by the breadth of his foresight on the promise, the difficulties, the necessary ingredients, and the challenge of using scientific models to improve prediction and decision making. He comments on the purpose and makeup of interdisciplinary teams to study complex systems problems in terms of two main concerns:

> *The analyst and the client*: the essential role of cooperation between consultants and executives; the need for the OR analyst to accept and give feedback to management; an appreciation of the perceptions and needs of different users and clients; the contribution of anecdotal information from experienced hands to initiate and assist in model development; the importance in translating analytical results into insights and actions that can be understood and appreciated by non-scientists.
>
> *The analyst and the model*: the early use of rough and dirty approximations when the system under study is not well understood; the importance of independent observation and experimental results to confirm theoretical predictions; and, vice versa, the use of theory to suggest what data should be observed and collected; the pitfalls of sloppy or unstructured data analysis and a warning against blind and exhaustive data collection; the search for good measures of effectiveness and hemibel improvements; the strengths and limitations of simulations; the essential role of cooperation between consultants and executives; the importance in translating analytical results into insights and implementable actions that can be understood and appreciated by non-scientists.

Today, one would welcome Phil's thoughts on climate change, energy, terrorism, and financial stability, to mention a few important topics in modern society.

When Phil gave advice, one could see it was handled by a perceptive, experienced, tough yet humble scientist who understood the possibilities, the limitations, and the political implications in applying scientific methods to new policies and decisions in diverse environments. He was particularly adept in dealing with the hazy interface between science and high-level policy decisions; the former with its limitations and restrictive assumptions and the latter that usually deal with complex objectives and are sometimes jealously guarded by the egos and financial or political fortunes of decision makers. He used his wisdom at many levels and was able to travel through operational minefields with surprising clarity.

PHIL MORSE: THE PERSON AND FRIEND

In my graduate school days, I remembered him as a quiet and fair, but competitive and demanding person. He spent time with his students when there was something substantive to discuss and encouraged intellectual discourse in the structure of new models or findings and insights they revealed. With the exception of some lectures in the MIT summer courses or discussions at the blackboard, I do not remember his ever offering a course in OR prior to 1958; although his analytical reputation preceded him, I would be surprised if any of us fully appreciated his knack for framing problems and contributing as architect and model builder. A laboratory course in OR was established in 1962 where students participated in collecting and analyzing data for the MIT library and had an opportunity to get first-hand exposure to his model-building skills.

It was only after a number of his early students left MIT did we come to know another side of Phil Morse. He made a habit of visiting former graduate students and, in later years, giving talks at the academic or research centers with which they were associated. Combined with infrequent visits to RAND in Santa Monica and Cal Tech in Pasadena, he visited the University of California at Berkeley and Stanford University in Palo Alto. He knew Bill Linvill and Ron Howard in the Engineering and Economic Systems Department at Stanford University, and Ron Wolff, Bill Jewell, Ron Shephard, and myself in the Operations Research Department at Berkeley.

As a student in a formal setting, I was, quite frankly, awed by his presence. But in a more relaxed setting, my wife and I came to know him as a warm but shy person. During his Berkeley visits, he often stayed in our home and engaged us in puzzles at dinnertime. A memorable one was the challenge to our whole family to come up with different specialized names for groups of animals based on their species or condition: a pride of lions, a bevy of quail, a pod of whales, a gaggle of geese, a huddle of penguins, a hover of trout, and so on. The children and we made many trips to the encyclopedia. One of his very favorite pastimes was to walk the ridge trails and shorelines of the Point Reyes National Seashore northwest of San Francisco. On these hikes he was always inquisitive and observant, asking many questions about the landscape, but at the same time pointing out interesting details to all of us. He was interested in how nature or man played a part in our larger world, how we interacted with each other and what might change or be changed that would influence us and affect the future—always the systems thinker.

One of the feats Phil was most proud of was becoming an official member of the 4000 Footer Club of the White Mountains. One became a member by climbing to the summit of all 47 White Mountain peaks over 4000 ft. On such climbs, Phil was often accompanied by his grandson Donnie and his son-in-law Hugh Fowler. When he was 77 years old, Phil climbed the last one, Owl's Head, at an altitude of 4025 ft, on October 12, 1980. The hike was a long one—between 16 and 18 miles—and, according to Hugh, Phil had to come down much of the trail on the seat of his pants, as his legs were not up to the very steep slopes (Fowler 1980).

Obviously, some of Phil's greatest interests and many of his accomplishments were in physics: vibration, sound and acoustics, plasmas, extreme temperatures and pressures, quantum theory. With his former student, Herman Feshbach, he wrote their famous two volume work, *Methods of Theoretical Physics* (Morse and Feshbach 1953). There were revisions and editions, even following his death in 1985. One estimate is that the two volume series has sold close to 100,000 copies. Phil, through his important books and published papers, and a long succession of students, qualifies as one of the giants of modern acoustics. His final scientific work in acoustics was the book *Theoretical Acoustics* (Morse and Ingard 1968).

At a very early stage, he appreciated the power of the digital computer and its importance to science and OR. He was very perceptive about the role this field would play not only in computations of scientific formulas, but also in data retrieval, analysis and distribution. Following the establishment of the ORC, he established the MIT Computation Center in 1957, serving as director until 1967. That he could have made so many leading contributions in acoustics, quantum mechanics, thermal physics, and computation, while at the same time promulgating and influencing the diversity and growth of the new field of OR, makes his scientific prowess and persuasive skills even more remarkable.

In pioneering OR in the U.S., Morse and Kimball expanded on a vision for the field suggested by a member of the ASWORG staff (Kittel 1947, 150):

> Operations Research is a scientific method for providing executive departments with a quantitative basis for decisions.

This statement was modified in an important way by the British OR pioneer Charles Goodeve (1948), and it was this definition that Phil and George felt captured their view of OR and which they promulgated in their classic book, *Methods of Operations Research.* Their opening lines are (Morse and Kimball 1951, 1):

> Operations Research is a scientific method of providing executive departments with a quantitative basis for decisions regarding operations under their control.

The wording clearly reflects the wartime experiences of Morse and his colleagues, yet generalizes the field beyond its military context. As a definition, it left room for the remarkable development of OR in diverse fields such as hospital administration, health care, product and information distribution, transportation, finance, library and information retrieval, inventory control, production, risk management, marketing, and public policy. Its use keeps us grounded in real-world observations, the search for relevant data, and the design of models supported by repeated testing and validation.

BACK TO THE FUTURE

In one of his first papers describing OR, Phil stated, "Operations Research provides an opportunity for scientists to contribute to a more effective use of our resources of men and material, while at the same time carrying on research of scientific importance in applied science. The study of operations research has the further advantage of offering career opportunities in industry, either in war and peace, which could have pronounced influence on the management of industrial operations" (Morse 1951, 217).

In a paper acknowledging the 25th anniversary of the founding of ORSA, Phil, in commenting about the expansion of OR beyond purely military operations, cautioned, "At times we sound more like mathematicians than like physical scientists . . . . The pages of *Operations Research* seem more like advanced mathematics texts, than like a journals of physical science." His conclusion, however, was optimistic "Twenty-five years ago a small band of enthusiasts had the presumption to predict that operations research would be as useful in industrial and public affairs as it had been in military affairs. The prediction has been verified; we have done fairly well. And we should continue to prosper. We cannot hope to grow again by two orders of magnitude in the next 25 years. But if we stick to our original aim of matching our models to reality, rather than trying to make reality fit the preferred model, we can contribute in important ways to many of the serious problems facing this country and the world" (Morse 1977b, 187–188).

To our nation's everlasting good fortune, Phil chose to develop, expand, and apply OR models and techniques to great success in private and public endeavors and, above all, to encourage and support the pursuit of OR as an academic discipline in higher education.

Honors and Awards

Phil's many honors include: the U.S. Presidential Medal for Merit (1946) in recognition of exceptional service to his country during WWII, largely because of his efforts to successfully develop OR to solve anti-submarine search, detect, and destroy missions from 1942 to 1945; founding president of ORSA; election to the National Academy of Sciences (1955); receipt of the Silver Medal of the Operational Research Society of the U.K., its highest honor (1965). He was awarded ORSA's Lanchester Prize for the best OR publication in 1968 for his book *Library Effectiveness: A Systems Approach* (Morse 1968). For his life-time contributions to the study of vibration and sound, Phil received the Acoustical Society of America Gold Medal Award in 1973 (Acoustical Society 2001). In 1974, he was the recipient of ORSA's Kimball Medal for distinguished service to the society and the profession, and, in 2003, he was made a member of the International Federation of Operational Research Societies' Hall of Fame (Little 2003).

References

Acoustical Society of America (2001) http://asa.aip.org/encomia/gold/philipmorse.html (accessed November 3, 2008)

ASWORG (1942) Preliminary report on the submarine search problem. May 1, 1942. Declassified Report CNA 82-0493/1, released by Center for Naval Analysis, Alexandria, Virginia, April 1982

Blackett P (1941) Scientists at the operational level. U.K. Admiralty, London

Compton K, Morse PM (1927) Theory of normal cathode fall in gas discharges. Phys Rev 30:305

Condon E, Morse PM (1929) Quantum mechanics. McGraw-Hill, New York, NY

Drake A, Keeney R, Morse PM (1972) Analysis of public systems. MIT Press, Cambridge, MA

Feshbach H (1994) Philip McCord Morse. Natl Acad Sci 65:243–255

Fowler H (1980) Personal communication, August 12

Fowler H (2008) Personal communication

Gilliam L, Lichtenwanger W (1961) The Dayton C. Miller Flute Collection: a checklist of the instruments. Library of Congress, Washington, DC

Goodeve C (1948) Operational research. Nature 161(4089):377–383

Horvath W, Ernst M (1986) A remembrance. Oper Res 34(1):7–9

Keeney R (1994) Using values in operations research. Oper Res 42(5):793–812

Kirby M (2003) IFORS' Operational research hall of fame: Patrick Maynard Stuart Blackett. Int Trans Oper Res 10(4):405–407

Kittel C (1947) The nature and development of operations research. Science 105(2719):150–153

Larson R, Magnanti T (eds) (1994) Reminiscences of the operations research center 1953–1993. MIT OR Center Publication, Cambridge, MA

Little J (2002) P. M. Morse and the beginnings. Oper Res 50(1):146–148

Little J (2003) IFORS' Operational research hall of fame: Philip McCord Morse. Int Trans Oper Res 10(3):307–309

McCue B (1990) U-Boats in the Bay of Biscay. National Defense University Press, Washington DC

MIT (1959) Notes on operations research 1959. Operations Research Center, MIT, Cambridge, MA

Morse PM (1928) A theory of the electric discharge through gases. Phys Rev 31(6):1003–1017

Morse PM (1936) Vibration and sound. McGraw-Hill, New York, NY

Morse PM (1946) Of men and machines. MIT Technol Rev 49(1):29–31, 58, 60

Morse PM (1951) Operations research. MIT Technol Rev 53(4):191–194, 214, 216–217

Morse PM (1957) Operations research is also research. In: Davies M, Eddison RT, Page T (eds) Proceedings of the first international conference on operational research. Operations Research Society of America, Baltimore, MD, pp 1–8

Morse PM (1958) Queues, inventories and maintenance. Wiley, New York, NY

Morse PM (1967) Your ancestors. Privately Published

Morse PM (1968a) Library effectiveness: a systems approach. MIT Press, Cambridge, MA

Morse PM (1968b) Letter to the Editor – George E. Kimball. Oper Res 16(4):872

Morse PM (1977a) In at the beginnings: a physicist's life. MIT Press, Cambridge, MA

Morse PM (1977b) ORSA twenty-five years later. Oper Res 25(2):186–188

Morse PM (1986) The beginnings of operations research in the United States. Oper Res 34(1):10–17

Morse PM, Bacon L (eds) (1967) Operations research for public systems. MIT Press, Cambridge, MA

Morse PM, Feshbach H (1953) Methods of theoretical physics: Parts I and II. McGraw-Hill, New York, NY

Morse PM, Ingard K (1968) Theoretical acoustics. McGraw-Hill, New York, NY

Morse PM, Kimball G (1946) Methods of operations research (classified). Operations Evaluation Group, Office of the Chief of Naval Operations, Washington, DC (declassified version at http://www.cna.org/documents/1100005400.pdf [accessed 4 April 2008])

Morse PM, Kimball G (1951) Methods of operations research. Wiley, New York, NY; also, Dover Publications, 2003

Pollock S (2008) Personal communication August 12.

4

John von Neumann

Saul I. Gass

The origins and initial scope of operations research (OR) evolved from the need to resolve operational military problems during World War II (WWII). OR's growth and applicability as a new science has come about by its ability to assimilate, adapt, and extend old and new ideas from across a wide range of scientific developments. In particular, the interaction between OR and advances in mid-twentieth century mathematics and computers has proven to be a remarkable occurrence. The mathematician John von Neumann played a central role in these advances. He made seminal and major contributions to game theory, utility theory, stored-program computers, numerical analysis, and Monte Carlo simulation, all of which helped to advance the theory and application of OR.

Although his professional career as one of the twentieth century's greatest mathematicians was based at the Princeton Institute for Advanced Study (IAS), from 1933 to 1957, von Neumann also served as a war-time consultant to key government agencies, and as a postwar member and chairman of the U.S. Atomic Energy Commission. During this time, he was instrumental in the design and application of the first digital computers, and, with Oskar Morgenstern, revolutionized our thinking about conflict situations with their publication of *The Theory of Games and Economic Behavior*. Von Neumann's scientific contributions and their influence on OR was acknowledged by the establishment, in 1974, of the John von Neumann Theory Prize by the Operations Research Society of America

A.A. Assad, S.I. Gass (eds.), *Profiles in Operations Research*, International Series in Operations Research & Management Science 147, DOI 10.1007/978-1-4419-6281-2_4,

and The Institute of Management Sciences—it is awarded annually for fundamental and sustained contributions to theory in OR and the management sciences.

JOHNNY

John (Jansci, Johnny) von Neumann was born on December 28, 1903 in Budapest, Hungary, to Margaret (Kann) and Max Neumann. Max was a successful banker and an advisor to the Hungarian government. He was rewarded by Emperor Franz Josef in 1913 with hereditary nobility and a title whose Germanized version is von Neumann. Johnny was the oldest of three sons and the only one in the family to use the form von Neumann (Halmos 1973, 382; Macrae 1992, 57–58).

As a young child, Johnny exhibited a photographic memory and a remarkable ability in mathematics. It is reported that "at age 6 he could divide two-eight digit numbers in his head; by 8 he had mastered calculus; by 12 he had read and understood Borel's *Théorie des Fonctions*" (Halmos 1973, 383). Soon after starting his formal education at the Budapest Lutheran Gymnasium, his mathematics teacher recognized that Johnny was a child prodigy and asked the Budapest University mathematician and tutor, Gabriel Szego, to "give some university teaching to the lad" (Macrae 1992, 70). Szego visited Johnny at home once or twice a week in 1915–1916, after which Johnny was tutored by University of Budapest professor Michael Fekete (Macrae 1992). A special mathematics program was initiated and, by the time Johnny left the gymnasium, he and Fekete had written a joint paper that extended a theorem in analysis (von Neumann and Fekete 1922). In 1921, von Neumann enrolled in the mathematics program at the University of Budapest, but did not take any classes. He also registered at the University of Berlin where he studied chemistry through 1923. [Max thought that Johnny would not be able to make a living as a mathematician and chemistry was chosen as a compromise (Halmos 1973)]. He then moved to Zurich and enrolled in the Eidgenössische Technische Hochschule (Swiss Federal Institute of Technology Zurich) where he received an undergraduate degree in chemical engineering in 1925. During his time in Berlin and Zurich, he would return to Budapest at the end of each semester so he could take (and pass) the exams at the University. He was thus able to receive his doctorate in mathematics (dissertation on set theory) under the direction of Leopold Fejér in 1926.

Von Neumann was appointed Privatdozent (lecturer) at the University of Berlin where he remained from 1927 to 1929. He then held the same title at the University of Hamburg through 1930. After a spending a semester in 1929 lecturing on quantum mechanics at Princeton University, he was offered a position there as Visiting Professor which he accepted and held from 1930 to 1933. Prior to his coming to Princeton, he married Marietta Kövesi in Budapest on January 1, 1930. Their daughter Marina was born in 1935. (She is the economist Marina von Neumann Whitman, University of Michigan.) The von Neumanns were divorced in 1937; Johnny married Klára (Klari) Dán on November 17, 1936 (Halmos 1973, 384–385; Heims 1982, 178; Macrae 1992, 157).

When the Princeton Institute for Advanced Study was opened in 1933, Johnny was appointed as a Professor in Mathematics, a position he held until his death in 1957. He was the youngest member of the Institute when he joined the newly formed illustrious faculty of Albert Einstein, Marston Morse, Oswald Veblen, and Hermann Weyl. He became a U.S. citizen in 1937.

THE DUALITY TWINS: GAME THEORY AND LINEAR PROGRAMMING

The origins of modern game theory can be traced to the work of Ernst Zermelo and Émile Borel, but it was von Neumann who set the stage for what was to follow by his 1928 minimax theorem paper (von Neumann 1928). In it he proved the existence of optimal randomized mixed strategies for any two-person, zero-sum game, as well as the existence of a unique value for the game, the minimax value. There was a long hiatus between this 1928 and von Neumann's next game theory publication in 1944. This came about due to his friendship with the Princeton University economist Oskar Morgenstern who introduced him to the competitive problems inherent in economic activities. They co-authored the seminal book, *Theory of Games and Economic Behavior* (1944). Besides establishing the theory of games in a rigorous fashion, this book also set the stage for the development of modern utility theory by giving it an axiomatic base that leads to an existence theorem for a real-valued utility function; the material on utility theory is first given as an appendix in the second edition of *Theory of Games and Economic Behavior* (1947).

Looking at von Neumann's game theory mathematical results in terms of matrix and linear relationships, one can see how and why von Neumann reacted to George Dantzig's description of his newly formulated

THE ELEPHANT IN THE BOOK

Oskar Morgenstern (1976, 811), in his paper, "The collaboration between Oskar Morgenstern and John von Neumann on the theory of games" noted: "We wrote virtually everything together and in the manuscript there are sometimes long passages written by one or the other and also passages in which the handwriting changes two or three times on the same page. We spent most afternoons together, consuming quantities of coffee, and Klari [von Neumann's wife] was often rather distressed by our perpetual collaboration and incessant conversations. She was at that time collecting elephants made of ivory, glass, and all sorts of other material. At one point she teased us by saying that she would have nothing more to do with the ominous book, which grew larger and larger and consumed more and more of our time, if it didn't also have an elephant in it. So we promised we would happily put an elephant in the book: anyone who opens the pages can find a diagram showing an elephant if he knows that he should look for one."

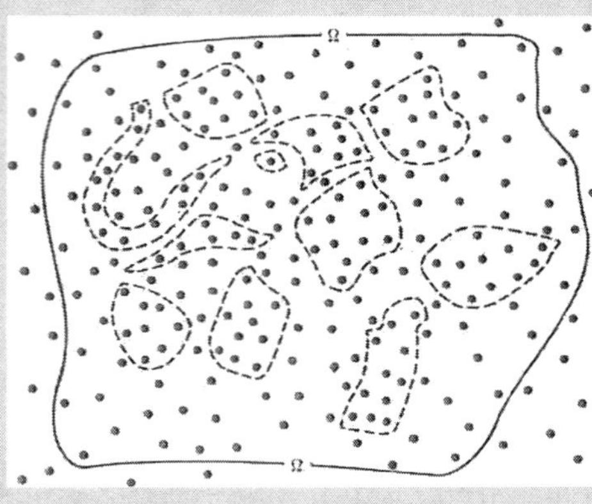

The elephant appears in on page 64 of the *Theory of Games and Economic Behavior* (von Neumann and Morgenstern 1944) in the guise of a discussion of partitioning elements of a set.

linear-programming (LP) model when they first met in 1947. The story as told by Dantzig (1982, 45):

> On October 3, 1947, I visited him for the first time at the Institute for Advanced Study at Princeton. I remember trying to describe to von Neumann, as I would to an ordinary mortal, the Air Force problem. I began with the formulation of the linear programming model in terms of activities and items, etc. Von Neumann did something which I believe was uncharacteristic of him. "Get to the point," he said impatiently. Having at times a somewhat low kindling-point, I said to myself "O.K., if he wants a quicky, then that's what he will get." In under one minute I slapped the geometric and algebraic version of the problem on the blackboard. Von Neumann stood up and said "Oh that!" Then for the next hour and a half, he proceeded to give me a lecture on the mathematical theory of linear programs.
>
> At one point seeing me sitting there with my eyes popping and my mouth open (after I had searched the literature and found nothing), von Neumann said: "I don't want you to think I am pulling all this out of my sleeve at the spur of the moment like a magician. I have just recently completed a book with Oscar [sic] Morgenstern on the theory of games. What I am doing is conjecturing that the two problems are equivalent. The theory that I am outlining for your problem is an analogue to the one we have developed for games." Thus I learned about Farkas' Lemma, and about duality for the first time.

Thus, in the 1940s, we have the almost simultaneous appearance of modern game theory and the field of LP. Both areas have become mainstays of OR theory and its application. But they did not evolve from the exigencies of WWII military operations, as did the origin and practice of early OR. What is really fascinating and beautiful about these two areas is that, although they were developed independently in two very different research environments, they are intimately related as it can be shown that they solve the same mathematical problem.

Kuhn and Tucker (1958, 119–120) concluded their extensive discussion of von Neumann's work in the theory of games and mathematical economics with a challenge that still holds to this day:

> The impact of von Neumann's Theory of Games extends far beyond the boundaries of this subject. By his example and through his accomplishments,

> he opened a broad new channel of two-way communication between mathematicians and the social sciences. These sciences were fortunate indeed that one of the most creative mathematicians of the twentieth century concerned himself with some of their fundamental problems and constructed strikingly imaginative and stimulating models with which to attack their problems quantitatively. At the same time, mathematics received a vital infusion of fresh ideas and methods that will continue to be highly productive for many years to come. Von Neumann's interest in "problems of organized complexity," so important in the social sciences, went hand in hand with his pioneering development of large-scale high-speed computers. There is a great challenge for other mathematicians to follow his lead in grappling with complex systems in many areas of sciences where mathematics has not yet penetrated deeply.

NUMERICAL ANALYSIS, MONTE CARLO SIMULATION, STORED-PROGRAM COMPUTER

Von Neumann is considered to be the originator of modern numerical analysis and a key contributor to the development and application of Monte Carlo simulation. He is also considered the first one to formally describe an electronic computer in terms of a logical structure that included the stored-program concept and how such a computer processes information. These three areas—numerical analysis, Monte Carlo simulation, stored-program computers—have had major impacts on the development of OR methods and their application.

Many problems in engineering and mathematics require the inversion of matrices, as well as the solution of large linear systems that arise in LP and least-squares regression. In the early 1940s, an acceptable matrix inverse was rather difficult to determine due to matrix size, accuracy in computation, and the human effort required to calculate it. In addition to von Neumann, the computer giants of John Atanasoff, Herman Goldstine, and Alan Turing recognized that a stable means of solving $AX = b$ was of great importance. It was the "... absolutely fundamental problem in numerical analysis: how best to solve a large system of linear equations" (Goldstine 1972, 289).

By the early 1940s, Atanasoff had designed a new "computing machine for the solution of linear algebraic equations" that applied

Gaussian elimination. Atanasoff noted, "The solution of general systems of linear equations with a number of unknowns greater than ten is not often attempted. But this is precisely what is needed to make approximate methods more effective in the solution of practical problems" (Goldstine 1972, 124).

MEETING OF THE MINDS

(Left to right) Julian Bigelow, Herman Goldstine, J. Robert Oppenheimer, John von Neumann.
Courtesy of the Archives of the Institute for Advanced Study, Princeton, NJ, 08540. Photographed by Alan Richards.

The question at that time was whether or not numerical procedures for solving large-scale linear systems could be developed that would produce accurate solutions. A heuristic analysis by the statistician Harold Hotteling (1943) indicated that Gaussian elimination was unstable; to achieve 5-digit accuracy for a 100 × 100 system approximately 65 digits would be needed! This caused von Neumann and his associates to consider iterative procedures in their report "Solution of linear systems of high order" (Bargmann et al. 1946). But, von Neumann and Goldstine, figuring that Gauss was too skilled a "computer" to be caught in such an accuracy problem, decided to pursue the matter (Goldstine 1972, 290). In their seminal paper, "Numerical inverting of matrices of high order" (von Neumann and Goldstine 1947)—the first modern paper on numerical analysis—they concluded that Gaussian elimination was "very good indeed provided the original problem was not 'ill-conditioned'; in other words, the procedure was stable" (Goldstine 1972, 291). This paper helped to set the future of modern numerical analysis. According to Goldstine (1972, 292), he and von Neumann were so involved with matrix inversion that Mrs. von Neumann named their newly acquired Irish Setter puppy Inverse.

During and after WWII, von Neumann was involved with the theory and design of nuclear weapons being developed at the Los Alamos

Laboratory (Lax 2005, Macrae 1992). He was a recognized expert in quantum theory and hydrodynamics, and most important, he had the rare ability to work with physicists and extract their problems into a mathematical form that could then be subjected to analysis and calculation. One of his associates at Los Alamos was the mathematician Stanislaw Ulam. They were both involved in the difficult numerical computations of neutron diffusion and nuclear explosions. Ulam traced the birth of the Monte Carlo method to a question that occurred to him while he was playing solitaire during his convalescence from a brain operation in January 1946. Ulam's unpublished account notes:

> . . . the question was what are the chances that a Canfield solitaire laid out with 52 cards will come out successfully? After spending a lot of time trying to estimate them by pure combinatorial calculations, I wondered whether a more practical method . . . might not be to lay it out say one hundred times and simply observe and count the number of successful plays. This was already possible to envisage with the beginning of the new era of fast computers, and I immediately thought of problems of neutron diffusion . . . . Later . . . [in 1946, I] described the idea to John von Neumann and we began to plan actual calculations (Eckhardt 1989, 131).

In a letter to Robert Richtmyer (March 11, 1947), theoretical division leader at Los Alamos, von Neumann wrote about the "possibility of using statistical methods to solve neutron diffusion and multiplication problems, in accordance with the principle suggested by Stan Ulam" (Goldstine 1972, 295). The name Monte Carlo was coined by the Los Alamos theoretical physicist Nicholas Metropolis, the leader of the group that solved the first computer-based Monte Carlo analysis—the simulation of chain reactions, done on the Aberdeen Proving Ground ENIAC (Electronic Numerical Integrator and Computer) in 1947 (Metropolis 1990, Gass and Assad 2005).

The application of Monte Carlo techniques requires a source of random numbers, either by physical means (counting radiation hits on a Geiger counter) or by arithmetic processes using a generating function. For the latter, von Neumann proposed the middle-square procedure in which the calculations are done by squaring an initial n-digit number (seed) and extracting the middle n digits for the next number in the sequence. Von Neumann recognized that this would yield a pseudo-random sequence, at best; the middle-square method is now out of favor as the sequence can be short, degenerate to a zero, or continuously repeat. In his paper (von Neumann 1951), he discusses high-speed computation of random digits from two perspectives: (1) How can one produce a sequence of random decimal digits—a sequence where each digit appears with probability one tenth and where consecutive ones are independent of each other in all combinations?, and (2) How can one produce random real numbers according to an assigned probability distribution law?

Stanislaw Ulam

RANDOM OR NOT

"Any one who considers arithmetical methods of producing random digits is, of course, in a state if sin. For, as has been pointed out several times, there is no such thing as a random number—there are only methods to produce random numbers, and a strict arithmetic procedure of course is not such a method" (von Neumann 1951, 36).

Von Neumann's contributions to the development of electronic computers is a well-told one (Aspray 1990, Goldstine 1972, Stern 1980). Because manual methods for finding solutions of nonlinear systems of equations in fluid dynamics (that arose in nuclear research at Los Alamos) were so time-consuming, von Neumann, in 1944, investigated the status of ongoing computer developments—Harvard's (Mark I), the relay computer of Bell Telephone Laboratories, and the use of IBM punched-card equipment for scientific computations (Stern 1980, 352). Although he pursued these developments with his usual diligence, for some reason, he was not aware of the ENIAC being built at the Moore School of Engineering, University of Pennsylvania. The turning point in von Neumann's evolution as a major contributor to electronic computers was a chance meeting with Herman Goldstine, a mathematician who taught at the University of Michigan, and who had joined the U.S. Army in 1942 as an ordnance officer, assigned to the Ballistic Research Laboratory (BRL), Aberdeen Proving Ground, Maryland. In 1943, Goldstine helped secure government funding for the ENIAC computer that was to be used by BRL—he was BRL's liaison with the Moore School. In his book, *The Computer from Pascal to von Neumann* (Goldstine 1972, 182), he recounts how he first met von Neumann:

> Sometime in the summer of 1944 after I was out of the hospital I was waiting for a train to Philadelphia on the railroad platform in Aberdeen when along came von Neumann [von Neumann was a member of BRL's Scientific Advisory Committee]. Prior to that time I had never met this great mathematician, but I knew much about him of course and had heard him lecture on several occasions. It was therefore with considerable temerity that I approached this world-famous figure, introduced myself, and started talking. Fortunately for me von Neumann was a warm, friendly person who did his best to make people feel relaxed in his presence. The conversation soon turned to my work. When it became clear to von Neumann that I was concerned with the development of an electronic computer capable of 333 multiplications per second, the whole atmosphere of our conversation changed from relaxed good humor to one more like the oral examination for the doctor's degree in mathematics.

During the mid-1940s, von Neumann and Goldstine were also involved in the development of the second Moore School computer developed for BRL, the EDVAC—Electronic Discrete Variable Computer. It was the first stored-program computer to be designed and calculated using binary arithmetic. Related to the EDVAC's design were two of von Neumann's oft-cited seminal papers: "The principles of large-scale computing machines" (von Neumann 1981)—a reprint of a talk he gave in 1946—and "First draft of a report on EDVAC" (von Neumann 1945). The first paper, as noted in a foreword by Nancy Stern, focuses on the application of computers to mathematics and mathematical physics and on the logical design for an electronic digital computer; von Neumann addresses the question "Why does one wish to achieve extremely high speed?" The second paper "consisted of a series of definitions of concepts relating to logic control and stored programming. It also included a section on how computers, in principle, could be used to operate on data. The contents of these sections formed the basis of computers and stored-program design in the late 1940s" (Stern 1980, 354). Goldstine cites this report as "the most important document ever written on computing and computers" (Goldstein 1972, 191). Based on the EDVAC paper, von Neumann is viewed by many as being the originator of the stored-program concept. But, the stored-program concept predates von Neumann's involvement with EDVAC (Stern (1980).

FASTER THAN A SPEEDING FLY

The following problem was posed to von Neumann: "Two bicyclists start twenty miles apart and head toward each other, each going at a steady rate of 10 m.p.h. At the same time a fly that travels at a steady 15 m.p.h. starts from the front wheel of the southbound bicycle and flies to the front wheel of the northbound one, then turns around and flies to the front wheel of the southbound one again, and continues in this manner till it is crushed between the two front wheels. Q: what total distance did the fly cover? The slow way to find the answer is to calculate what distance the fly covers on the first, northbound leg of the trip, then on the second, southbound leg, then on the third, etc., etc., and finally, to sum the infinite series so obtained. The quick way is to observe that the bicycles meet exactly one hour after they start, so that the fly had just an hour for his travels; the answer must therefore be 15 miles. When the question was put to von Neumann, he solved it in an instant, and thereby disappointed the questioner: 'Oh, you must have heard the trick before!' 'What trick?' asked von Neumann; 'all I did was sum the infinite series' " (Halmos 1973, 385–386).

ON THE TRAIL

"There's a wonderful picture of my father in a string of people going down the Grand Canyon on mules, and there are two odd things about it. One of them is that, while everybody else is dressed in jeans and polo shirts, my father is dressed in a full three-piece banker's suit. The second thing about it is that his mule is facing the opposite direction from everybody else's" (Whitman 2005, 8).

(Courtesy of Marina von Neumann Whitman)

John von Neumann died of cancer on February 8, 1957 at Walter Reed Hospital in Washington, D.C. His contributions to mathematics, physics, economics, and computers have helped to define many of the scientific contributions of the twentieth century (Bochner 1958, Heims 1982, Macrae 1992). His intellect and talents were wide-ranging. Ulam (1958, 27) mentions, in passing, von Neumann's contributions to "operational research" as stemming from *Theory of Games and Economic Behavior*. In a sense, one cannot argue with that observation. As measured by von Neumann's total output, his contributions to OR is just a subset, but what an important subset!

Honors and Awards

John von Neumann received the U.S. Presidential Medal of Merit (1947), U.S. Presidential Medal of Freedom (1956), U.S. Navy's Distinguished Civilian Award, Albert Einstein Commemorative Award, and the U.S. Atomic Energy Commission's Enrico Fermi Award. He was elected to the National Academy of Sciences in 1937. He was awarded honorary degrees from Princeton University (1947), Harvard University (1950), University of Pennsylvania (1950), University of Istanbul (1952), Case Institute of Technology (1952), University of Maryland (1952), Institute of Polytechnics-Munich (1953). He was elected to the International Federation of Operational Research Societies' Hall of Fame in 2006 (Gass 2006).

References

Aspray W (1990) John von Neumann and the origins of modern computing. MIT Press, Cambridge, MA

Bargmann V, Montgomery D, von Neumann J (1946) Solution of systems of high order. In: Taub AH (ed) John von Neumann collected works, vol V. Bureau of Ordnance, Navy Department, Washington, DC, pp 421–477

Bochner S (1958) John von Neumann. Biogr Memoirs Natl Acad Sci 32:438–457

Dantzig G (1982) Reminiscences about the origins of linear programming. Oper Res Lett 1(2):43–48; also see Dantzig GB (2002) Linear programming. Oper Res 50(1):42–47

Eckhardt R (1989) S. Ulam, J. von Neumann, and the Monte Carlo Method. In: Cooper NG (ed) From cardinals to chaos: reflections on the life and legacy of Stanislaw Ulam. Cambridge University Press, New York, NY, pp 131–137

Gass SI (2006) IFOR's operational research hall of fame: John von Neumann. Int Trans Oper Res 13(1):85–90

Gass SI, Assad AA (2005) Tales from the timeline: the definition of OR and the origins of Monte Carlo simulation. Interfaces 35(5):429–444

Goldstine H (1972) The computer from Pascal to von Neumann. Princeton University Press, Princeton, NJ

Halmos P (1973) The legend of John von Neumann. Am Math Monthly 80(4):382–394

Heims S (1982) John von Neumann and Norbet Weiner: from mathematics to the technologies of life and death. MIT Press, Cambridge, MA

Hotteling H (1943) Some new methods in matrix calculations. Ann Math Stat 14(1):1–34

Kuhn H, Tucker A (1958) John von Neumann's work in the theory of games and mathematical economics. Bull Am Math Soc 64:100–122

Lax P (2005) John von Neumann: the early years at Los Alamos, and the road to computing. Siam News 38(2):9–10

Macrae N (1992) John von Neumann. Pantheon Books, New York, NY. (Reprinted by the American Mathematical Association, 1999.)

Metropolis N (1990) The Los Alamos experience, 1943–1954. In: Nash SG (ed) A history of scientific computing. ACM Press, New York, NY, pp 237–250

Morgenstern O (1976) The collaboration between Oskar Morgenstern and John von Neumann on the theory of games. J Econ Liter 14(3):805–816

Stern N (1980) John von Neumaun's influence on electronic digital computing. Ann Hist Comput 2(4):349–362

Ulam S (1958) John von Neumann. Bull Am Math Soc 64:1–49

Von Neumann J (1928) Zur theorie der Gesellschaftsspiele, Mathematische Annalen 100:295–320; (translated by) Bargmann S (1959) On the theory of games of strategy. Contributions to the theory of games, vol IV. Tucker AW, Luce RD (eds) Ann Math Studies 40:13–42, Princeton University Press, Princeton, NJ

Von Neumann J (1945) The first draft of a report on the EDVAC. Moore School of Electrical Engineering, University of Pennsylvania, Philadelphia, PA

Von Neumann J (1951) Various techniques used in connection with random digits. J Res Natl Bur Stand (Applied Mathematics Series) 3:36–38

Von Neumann J (1981) Principles of large-scale computing machines. Ann Hist Comput 3(3):263–273 (Reprint of paper given by von Neumann, May 15, 1946)

Von Neumann J, Fekete M (1922) Über die Lage der Nullstellen gewisser Minimum-polynome. Jahresbericht (Annual Report), Deutsche Mathematiker-Vereinigung, Leipzig 31:125–138

Von Neumann J, Goldstine H (1947) Numerical inverting of matrices of high order. Bull Am Math Soc 53(11):1021–1099

Von Neumann J, Morgenstern O (1944) Theory of games and economic behavior. Princeton University Press, Princeton, NJ (second edition, 1947, third edition, 1953)

Whitman M (2005) Life with father. SIAM News 38(2):8

5

Charles Frederick Goodeve

Maurice W. Kirby

After World WarII, many of the scientists who had applied their knowledge to study and solve military planning and operational problems returned to their laboratories and universities. There were a few, however, who recognized that their military successes, which had launched the fledgling science of operations research (OR), could be. transferred into the industrial and business worlds. Sir Charles Frederick Goodeve, a physical chemist, was the most prominent and influential advocate of OR as a means of raising industrial productivity in postwar Britain. Appointed in 1945 as director of the newly formed British Iron and Steel Research Association (BISRA), Charles ensured that OR was represented as a distinct department with the mission of applying the scientific method to industrial problems. Throughout the later 1940s and 1950s, he acted as a tireless advocate of OR, writing numerous articles on the subject, delivering speeches, and using his personal charm to persuade industrial executives of its value.

Charles was elected a Fellow of the Royal Society, received the Order of the British Empire in for his wartime weapons developments, and was awarded the U.S. Medal of Freedom with Silver Palm.

Canada, U.K., and the Royal Navy

Charles Goodeve was born on February 21, 1904, in the Canadian town of Neepawa located on the eastern border of the prairies, 100 miles to the west of Winnipeg. His father, F. W. Goodeve, an Anglican minister, and his mother, Emma Hand, were first generation Canadians. In 1914, Goodeve senior moved to a parish church in Winnipeg, eventually

A.A. Assad, S.I. Gass (eds.), *Profiles in Operations Research*, International Series in Operations Research & Management Science 147, DOI 10.1007/978-1-4419-6281-2_5,

becoming a Canon of St. John's Cathedral. Charles was educated at the Kelvin High School where he displayed a keen interest in science, although on entering the University of Manitoba in 1919, he did so as an art student. Two years later, he transferred to the B.Sc. honors degree course in chemistry and physics which he completed successfully in 1925. As an impoverished undergraduate, Charles worked as a junior laboratory demonstrator and it is a tribute to his abilities that he was appointed to an assistant lectureship in his department on graduation.

His early research interest was in electrolytic chemistry and it was in this subject that he was awarded an M.Sc. in 1927. In the same year, he was offered an 1851 Exhibition Scholarship tenable in the Department of Chemistry at University College, London (UCL). In the pantheon of British academic science, the UCL department, headed by Professor Frederick Donnan, Fellow of the Royal Society (FRS), was noted for its research excellence and, in this stimulating environment, Charles flourished. Charles was appointed to a lectureship in physical chemistry in 1930, received a D.Sc. in 1936, and was promoted to Reader in 1937. Charles ended the decade with a considerable research reputation, especially in the areas of photochemistry and the chemistry of vision. It was for this work that he was elected an FRS in 1940. He married Janet Wallace, a former chemistry student of his from the University of Manitoba, who also came to UCL and received a Ph.D. in chemistry in 1932, the same year they were married. They had two sons, Peter Julian, born 1936, and John Anthony, born 1944.

While in Canada, Charles had joined the Royal Canadian Navy Volunteer Reserve and, on his arrival in Britain, he transferred his allegiance to the Royal Navy Volunteer Reserve (RNVR). In 1936, he was promoted to the rank of Lieutenant Commander having been "to sea in submarines and minesweepers,. . .four battleships and three destroyers" (Richardson 1981, 132). It was at this point that Charles began to engage

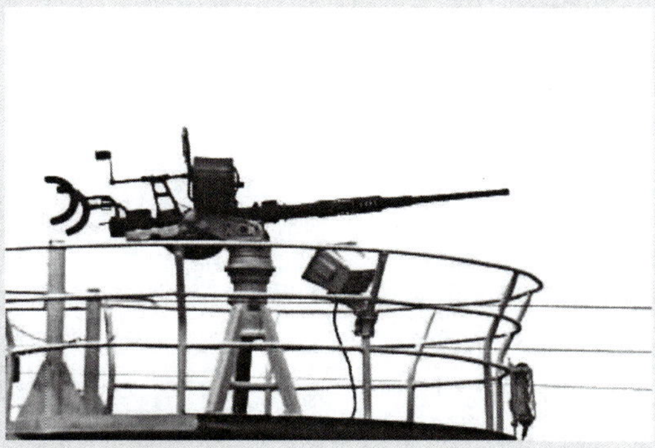

Oerlikon gun

in research projects on naval problems with financial support from the Admiralty.

From the mid-1930s onwards, and in his RNVR capacity, he undertook attachments to H.M.S. *Vernon*, a shore-based establishment at Portsmouth specializing in mining techniques and related problems. Charles was also a qualified torpedo specialist. Thus, it was not surprising that when war broke out in 1939, he was posted to H.M.S. *Vernon*. Taking the entire period of the war, Charles carried out distinguished work in the field of magnetic mines, notably degaussing (the neutralization of the magnetic field of a ship as protection against magnetic mines), and the double longitudinal sweep (the use of a self-buoyant electric cable pulled by two mine sweeper vessels in order to detonate magnetic mines). Further important contributions to military effectiveness included anti-aircraft weaponry (production of the Oerlikon gun), and anti-submarine warfare (forward mortar fire against U-boats). After serving with H.M.S *Vernon*, Charles took up research positions in the Department of Miscellaneous Weapons Development and the Inspectorate of Anti-aircraft Weapons and Devices. Of all his projects, Charles received the greatest acclaim for inventing

Charles (Royal Navy Commander) on wartime visit to Winnipeg to see his wife and son who had been evacuated from London.
(Courtesy Peter Goodeve)

degaussing (a term he coined); many ships involved in the evacuation of the British Expeditionary Force from Dunkirk in 1940 had been degaussed using his procedures, thereby saving many lives. At the end of the war he was awarded the considerable sum of £7,500 for this invention which he shared with colleagues who had assisted him in the project.

In October 1942, Charles was appointed to the civilian position of Assistant Controller Research and Development, responsible for the Navy's R&D strategy. In this influential role, and later as Deputy Controller, he worked closely with Professor Patrick Blackett, Director of Naval Operational Research at the Admiralty, particularly in relation to the anti-U-boat war. As the war drew to an end, Charles tried to persuade the Department of Scientific and Industrial Research to establish the Royal Naval Scientific Service, and resumed his prewar campaign to encourage the Admiralty to recruit to the RNVR young scientists and engineers. Unfortunately, the Admiralty rejected his advice, as it had done in 1939. [The paper, "The 'Invisible science': operational research for the British Armed Services after 1945," reviews OR's role in the formulation of tactics and strategy during the Cold War (Kirby 2010)].

BISRA and Industrial OR

At the end of the war, Charles chose not return to academia and took up the position of Director of the newly formed BISRA. The Association was an industry-wide resource which had as its aim the raising of efficiency in an industry which was viewed by government as a technically backward and inefficient cartel. In devising BISRA's structure, Charles's most radical move was to establish a department of OR to accompany those devoted to physics and chemistry. Its objectives were defined as follows:

> To apply the methods of mathematical statistics and allied techniques to the solution of operational problems arising in the industry.
>
> To provide, by studies of costs, a detailed picture of the economics of the essential processes of the iron and steel industry. The purpose of this part of the work is twofold. First, in order that research projects may be concentrated on problems whose successful solution will lead to the greatest economic advantage. Second, to enable estimates to be made of the savings to be expected from any proposed changes in technical procedure (Kirby 2003, 214–215).

Beginning with a staff of two in 1947, the OR Department, by the mid-1960s, had expanded to more than 50, approximately 40 of whom were science graduates. In addition to external recruitment, the original numbers had been augmented by staff transfers from BISRA's Plant Engineering Division, and from the Mathematics Section of the Physics Department. This was in response to the growing volume of work being undertaken after 1950. The largest project by far encompassed a sequence of studies on the importing of iron ore, continuing well into the 1960s. Other significant projects in the 1950s included the analysis of stock control and accident statistics and, more importantly, the application of digital electronic computers to problems of managerial control. Although the larger iron and steel firms such as the United Steel Companies; the Steel Company of Wales; Richard, Thomas and Baldwins; Guest, Keen and Nettlefolds were establishing their own OR groups after 1950 (with much encouragement from Charles), BISRA began to offer a consultancy service to small and medium-size firms in 1953 with financial support from the U.S.-sponsored Conditional Aid Fund. By 1965, consultancy work on a confidential basis accounted for one half of BISRA's work in OR, although, by that time, firms using the service were being charged full cost fees. The very fact of consultancy contributed to the further diffusion of OR in the steel industry.

Although Charles took the view that his primary task was to encourage the diffusion of OR in the iron and steel industry, he was determined to use his position as Director of BISRA to advocate an OR presence in other sectors of the industrial economy. Thus, in the decade after 1945, Charles mounted what was tantamount to a one-man campaign in favor of OR via speeches and articles in which he underlined its wartime successes and pointed to a variety of peacetime uses. His early thinking is set out with clarity in a 1948 survey published in *Nature* (Goodeve 1948a). In terms of definitions, Charles endorsed the formulation of Charles Kittel, who had served as an OR analyst with the U.S. Navy Antisubmarine Warfare Operations Research Group (Gass and Assad 2005). Kittel (1947, 105) wrote that "Operations Research is a scientific method for providing executive departments with a *quantitative basis for decisions*." Charles made a critical addition to Kittel's definition to produce the basic OR definition that has stood the test of time:

> Operations Research is a scientific method of providing executive departments with a quantitative basis for decisions regarding the operations under their control (Goodeve 1948a, 377).

He also identified OR with the "scientific method" for its use of "the rigorous logic of mathematics and statistics" to analyze the data from "controlled experiments....or experiences" (Goodeve 1948a, 377). Finally, Charles commented that the application of OR was wholly dependent on the meaning attached to the word operations that he had added to Kittel's definition:

> In war, operational research was applied to the use of weapons, to tactics, and to strategy. In the peace-time application of operational research, studies are directed, for example, to the use of equipment and man-power, to operating procedures, and to the solution of those many problems faced by management in operating or controlling factories, or public utilities, or by Government authorities in planning. The field of operational research is wide; but it will be seen that it is quite distinct from the field of most applied research, the latter being concerned with new or improved processes, equipment, materials, etc (Goodeve 1948a, 377).

Charles's use of the present tense in relation to peacetime applications was validated by the claim that OR was "already in full swing." In this respect, he cited the examples of railway and road traffic management, both of which gave "a promise of useful results on the application of the scientific method." In the former, Charles noted that the London Midland and Scottish Railway Company (one of the largest private sector railroad companies in the U.K. before nationalization) had established a special unit to investigate its communications system. Recommendations concerning improvements to organization and equipment were being implemented on the basis of quantitative predictions. In road transport, The Road Research Department of the Department of Scientific and Industrial Research (DSIR) was using a "wide range of the tools of modern science" to investigate improvements in traffic flow and the reduction of accidents. The remaining sections of Charles's survey range over the application of OR to "Inspection and Maintenance" and "Technical progress" (Goodeve 1948a, 381–382). In a later article, he reported on the proceedings of a DSIR-sponsored conference that had discussed the relevance of OR to individual research associations (Goodeve 1948b, 584–585). These included cotton textiles, ceramics, the electrical and allied industries, the boot and shoe trades, and Charles's own research association in the iron and steel industry. The account was instilled with enthusiasm, with the reader's attention being drawn to the fact that:

> There never has been a more favourable opportunity for an extension of operational research. With the present "sellers market," competition is at a

> minimum and co-operation at a maximum. Organized labor, which has always had to look first for security of employment, is able now to take a broader view and consider efficiency of employment (Goodeve 1948b, 585).

Charles concluded with a resume of the objectives of OR "which have been or could be carried out by the research associations," (Goodeve 1948b, 585):

1) to increase production per-man-hour, per unit of equipment, or per unit of fuel;
2) to improve quality and suitability of products;
3) to improve the basis of decisions regarding the application of science;
4) to make studies of consumer or market problems, that is, scientific studies in market research;
5) to aid in finding the optimum degree of standardization of any particular field of production;
6) to aid in the planning of research programs.

From this and his other postwar writings (Goodeve, 1954, 1955, 1957a, b), it is evident that Charles envisaged a peacetime role for OR as a means of alleviating or resolving a wide range of social and economic problems.

THE OR CLUB AND BEYOND

In the institutional development of OR in Britain after 1945, the guiding hand of Charles was evident from the beginning. In this respect, Pat Rivett commented that if Patrick Blackett was the wartime "father" of OR, Charles was the "foster parent" or "wet nurse" of "the baby which was more or less left on the doorstep in 1946" (Kirby 2003, 367, Rivett 1980). In 1948, Charles took the initiative in founding the Operational Research Club as:

> a small informal group of people who are working in or are concerned with problems associated with Operational Research. The club has been organised

> so that its members may meet together to discuss problems connected with their work and to assist in the development of the methods of Operational Research.
>
> The word "Club" was deliberately chosen to suggest the informality of the group, as it was felt that free discussion and interchange of ideas were the first essentials, and the fewer the rules the freer the Club's development would be (Anonymous 1951, 36).

The natural concomitant was a limitation on numbers to avoid the inconvenience of rules of procedure. The original members of the Club's steering committee were Charles; W.K. (later Sir William) Slater, destined to become chairman of the Agricultural Research Council; A.W. Swan, a pioneer of OR at United Steel and, subsequently, at Courtaulds (viscose yarn producers); and Cecil Gordon of the Special Research Unit of the Board of Trade. In 1950, Gordon resigned and was replaced by Owen (later Sir Owen) Wansbrough-Jones, the Scientific Adviser to the Army Council and soon to be appointed Principal Director of Scientific Research (Defence) at the Ministry of Supply. Club membership was at the discretion of the committee and was limited to one member per industry. By the end of 1952, the agreed limit was 50 members. The intention was to hold at least four meetings a year when papers would be read and discussed. Early participants included Charles himself who spoke on the theme of the "Laws and methods of physics applied to social problems," C.H. Waddington on OR problems in livestock breeding, Roger Eddison on OR problems in the steel industry, and Norman Wright on the National Food Survey. Meetings of the Club were held in London in the ornate rooms of the Royal Society at Burlington House, while meetings of the steering committee were held in Wansbrough-Jones's flat in Pall Mall.

In his endeavors to raise the public profile of OR, Charles was determined that the OR Club should be complemented by hard copy in the form of the world's first journal to be dedicated to OR. The *Operational Research Quarterly* (*ORQ*), which made its debut in March 1950, was the brainchild of Charles and was inspired by the popularity of a sequence of lectures on OR organized by Professor E. S. Pearson at University College, London, during 1949. The joint editorship was, in effect, imposed by Charles on two of his BISRA colleagues: Max Davies as business editor and Roger Eddison as technical editor.

At the 1953 Annual General Meeting of the OR Club a motion was carried that the Club should cease to exist and that "the Operational Research Society (ORS) should simultaneously come into existence with the new Constitution" (Kirby 2003, 369). An article in the December 1953 issue of the ORQ provided the following clarification:

> The increasing interest in operational research in this country has caused the Operational Research Club, hitherto a deliberately restricted group of scientists meeting more or less informally, to reconstitute itself formally as the Operational Research Society [ORS] with no numerical limit on membership. Members must still, however, be engaged in some form of operational research "whether or not it is so called."
>
> The change is not merely one of nomenclature. It reflects the widened scope that the Club has had to accept, and the wider nature of the task that lies before it (Anonymous 1953, 57–58).

C.F.G. ON POMPOSITY

"One of C.F.G's most endearing characteristics was his love of deflating pomposity. I remember being his target on two occasions. On one, I had been impressed by a certain manager whom C.F.G. had not met. When he asked for my opinion I extolled his technical ability, management skill, even temperament and so on. At the end of all this C.F.G. said: 'How very dull he must be!' Once, being left in charge of things when C.F.G. was abroad, I reported on his return that all had gone well in his absence, and no problems had arisen. His rejoinder was 'Then you haven't been working hard enough!'" Alan Leckie in (Richardson 1981, 29).

The growth of interest in OR was illustrated by reference to a survey article published by Charles in an earlier issue of the *ORQ* based on a questionnaire sent out to 210 organizations ranging across the private and public sectors (Goodeve and Ridley 1953, 21–24). The survey concluded that 45 of the organizations had sections carrying out OR as their major activity. In addition, a further 41 organizations carried out OR "in one or other parts of their organisation at some time or other." For Charles, these results were "satisfactory" in view of "the extremely vague borders" of OR and the fact that the survey itself had not encompassed private sector services such as the commercial banks and insurance companies.

FRIENDS, HONORS, AND AWARDS

Charles Goodeve died on April 7, 1980, and, as with all deceased Fellows of the Royal Society, he was the subject of a lengthy obituary in the Society's *Biographical Memoirs* (Richardson 1981). The obituary contained warm appreciations of Charles's contribution to OR from Alan Leckie, a former colleague in BISRA; B.H.P. (Pat) Rivett, Professor of OR at the University of Sussex; and Russell Ackoff, Professor of Management Science at the University of Pennsylvania. Rivett had a long-standing acquaintance with Charles resulting from their early collaboration in the work of the ORS and it was in this context that he drew attention to Charles's role in building international links (Richardson 1981, 27):

> Shortly after the creation of the Operational Research Society, the Operational Research Society of America [ORSA] suggested an international conference for which the three sponsoring societies were to be ORS, ORSA and the French OR society. [Charles] was chairman of the organizing committee and oversaw the whole of the arrangements with an enthusiastic and penetrating eye. This was without doubt the most impressive operational research conference ever and from it stemmed a number of influences now found everywhere. It led to the stimulation of academic interest in operational research in this country. It led also to the formation of the International Federation of Operational Research Societies [IFORS] at a private dinner that [Charles] organized at Woodstock during the conference.

$\sqrt{Members}$

This is believed to be the first organization [IFORS] to have a square root sign in its Constitution: it was Charles' idea that the individual societies should contribute financially in proportion to their memberships, but have voting powers proportional to the square roots of their memberships.

> At each of the subsequent IFORS Conferences in Aix-en-Provence, Oslo, Boston, Venice and Dublin, Charles was a prominent participant, reading papers, taking part in the discussion and always by his simple act of being there illuminating the conference, intellectually, scientifically and socially.

Rivett concluded by emphasizing that Charles had always been aware of the role of individuals and social groups within the wider society to the extent that "so much of what we do in management research depends on human factors" (Richardson 1981, 29). Thus, in the early 1960s, Charles worked with Eric Miller and Russell Ackoff to create the Institute for Operational Research (IOR) in association with the Tavistock Institute of Human Relations. Over the succeeding years, the IOR developed a distinctive school of OR based upon the concept and practice of action research, as pioneered by the social scientists employed in the Tavistock Institute since it was founded in 1947. In recalling his collaboration with Charles, Ackoff commented:

> I was always amazed that a man who was so firmly part of the Establishment was so completely open to new and innovative ideas, and involved himself in them with the vigour and enthusiasm of a young man. He never occupied himself with identifying reasons why something would not work, but focused on finding ways to make it work. He did all this without trying to occupy centre stage or receive credit. He, better than any of the "old timers" I have known, recognized the need for OR to engage in a permanent revolution if it was to remain a vital and important activity. Consolidation and respectability were his least concerns. In working with him on such projects as the creation of the Institute of Operational Research, he always kept me enthusiastic and hopeful, but never let me get out of touch with reality. Charles basked in his interactions with graduate students, and faculty also. While with us [Charles had earlier visited Ackoff at the University of Pennsylvania] he . . . . loved exchange of ideas and criticism. Again his enthusiasm was infectious. Even those who felt he was on the wrong track could not help but try to contribute constructively to his efforts (Richardson 1981, 28).

Charles was elected a Fellow of the Royal Society in 1940, received the Order of the British Empire in 1942 for his wartime weapons developments, knighted in 1946, and was awarded the U.S. Medal of Freedom with Silver Palm. Charles served on numerous official committees and councils: director of the Industrial and Commercial Finance Corporation from 1964 to 1974, Vice-President of the

Parliamentary and Scientific Committee from 1950 to 1962, member of the Lord President's Council on Scientific Policy from 1953 to 1956, a Vice-President of the Royal Society from 1968 to 1970. Charles was awarded honorary doctorates by the Universities of Manitoba (1946), Sheffield (1956), Newcastle Upon Tyne (1970), and Salford (1974). He received the ORS Silver Medal, its highest honor, in 1964.

References

Anonymous (1951) Operational research club. Oper Res Q 2(1):36

Anonymous (1953) Club into society. Oper Res Q 4(4):57–60

Gass S, Assad A (2005) Model world: tales from the time line—the definition of OR and the origins of Monte Carlo simulation. Interfaces 35(5):429–435

Goodeve C (1948a) Operational research. Nature 161(4089):377–384

Goodeve C (1948b) Operational research in the research associations. Nature 161(4094):584–585

Goodeve C (1954) Operational research as a science. J Oper Res Soc Am 1(4):166–180

Goodeve C (1955) Operational research: the front line scientist in the management team. Manager December:995–998

Goodeve C (1957a) Man must measure. J Inst Transp March:75–82

Goodeve C (1957b) Operational research: the common factor. Engineer December 6:345–346

Goodeve C, Ridley GR (1953) A survey of OR in Britain. Oper Res Q 4(2):21–24

Kirby MW (2003) Operational research in war and peace: the British experience from the 1930s to 1970. Imperial College Press, London

Kirby MW (2010) The 'Invisible science': operational research for the British Armed Services after 1945. Oper Res Q 61(1):68–81

Kittel C (1947) The nature and development of operations research. Science 105(2719):150–153

Richardson FD (1981) Charles Frederick Goodeve, 21 February 1904–1907 April 1980. Biogr Mem Fellows R Soc 27:307–353. http://www.goodeveca.net/CFGoodeve/cfg_bio.html. Accessed 17 Oct 2009)

Rivett BHP (1980) Sir Charles Frederick Goodeve, OBE, FRS. OR Newslett May

Further Information

The website of the Naval Museum, Manitoba (http://www.naval-museum.mb.ca/people/g/goodeve.htm) is a useful supplement to this article. Accessed 17 Oct 2009)

Goodeve's private papers can be consulted at the Churchill Archives Centre, Churchill College, Cambridge CB3 0DS, U.K.

6

Albert W. Tucker

Saul I. Gass

Operations research(OR) has been able to flourish and make its mark by solving real-world decision problems in just about all facets of human endeavor. Often, such successes rest heavily on mathematical results discovered by researchers who had little or no interest in applications. Further, those who successfully extended the application reach of OR were often trained in their mathematical and related skills by outstanding teachers who are able to convey the implications of such theoretical results by being great expositors. For over 50 years, especially during the embryonic years of OR, Albert W. Tucker was the pre-eminent example of the mathematician, teacher, and expositor that graced the field. His contributions to OR were recognized by his being awarded the Operations Research Society of America's John von Neumann Prize and his being inducted into the International Federation of Operations Research Societies' Hall of Fame.

Early Talent

Albert William Tucker was born November 28, 1905 in Oshawa, Ontario, Canada. Al (as everyone called him) was the son of a Methodist minister who had studied to be a high-school mathematics teacher and had taught for a few years. Al first showed his talent for mathematics while in high school. In a class test, he answered original geometry problems that were taken from an old provincial examination. His teacher, after finding out that Al's father was not coaching him in mathematics, said to Al's father, "I think your son must be a mathematical genius. I think he can have a very promising career as an actuary!" (Albers and Alexanderson 1985,

A.A. Assad, S.I. Gass (eds.), *Profiles in Operations Research*, International Series in Operations Research & Management Science 147, DOI 10.1007/978-1-4419-6281-2_6,

339). Not knowing much about what an actuary does, Al's early career goal was to be a high-school teacher of mathematics and physics.

Courtesy Alan Tucker

In 1924, he enrolled in the University of Toronto, having placed first in the provincial examinations in mathematics and Latin. Al studied mathematics and physics, but in his fourth year's Honors Course, elected to major in just mathematics. He received a B.A. in mathematics in 1928. After serving as a Teaching Fellow at Toronto (1928–1929), during which he earned an M.A., he decided to pursue a Ph.D. A fortuitous encounter with a Princeton graduate catalog acquainted him with courses taught by Oswald Veblen, Solomon Lefschetz, and Luther Eisenhart, which convinced him that Princeton was the place to go. Al applied to Princeton and was appointed a part-time instructor (and graduate student) for the school year 1929–1930 with a salary of $1,000 and free tuition. There he studied topology under Solomon Lefschetz, who was Chairman of the mathematics department. His review and rewrite of Lefschetz's first topology book was such an original approach that it became his doctoral thesis, "An Abstract Approach to Manifolds" (Tucker 1933). Al received a 12-month post-doctoral fellowship from the U.S. National Research Council, which enabled him to spend the fall semester at Cambridge, England, 6 months at Harvard working with Marston Morse, and the summer at the University of Chicago.

ON FINDING A THESIS

"Often graduate students have asked me 'How do you get started writing a thesis?' I would say, there are lots of ways, but here is one way I have had good experience with myself. Take something you are interested in, mull it over, and make it your own. There's a good chance that in doing this you will find new ways of looking at the material, and this will turn into something that's publishable" (Albers and Alexanderson 1985, 342).

A CAREER AT PRINCETON

Although his early research in the burgeoning field of topology had garnered him job offers from Harvard and Yale, Al became a full-time member of the Princeton mathematics faculty in 1933, first as an instructor for 1 year and then as an assistant professor. He became a tenured associate professor in 1938 and professor in 1946. He was chairman of the department from 1953 to 1963, and, in 1953, was appointed the prestigious Albert Baldwin Dod Professor in Mathematics (endowed in 1869!). Al retired as an Emeritus Professor in 1974.

From the 1930s through the mid-1940s, Al's research was mainly in the field of combinatorial topology. He brought this into the classroom by developing a junior course in elementary combinatorial topology, a course he regularly taught. The material was turned into a textbook by a student, Donald W. Blackett (1967). During World War II, Al maintained his normal teaching load, but was also associate director of the Princeton Fire Control Research Project that dealt with pre-radar research on the firing of artillery and naval guns using optical range finders. The director of the project was Merrill Flood, a Princeton 1935 Ph.D. in mathematics and an old friend. Al's duties "were mainly administrative and editorial. The products of the project were reports, usually written to meet some need that had been put to us by the military. It was my job to edit these reports and make them readable for military officers. They often came to me in rather abstract technical and mathematical form, and it was my job to get these changed into a more readable form. But I did participate to some extent in the research and did quite a bit of traveling, because we had to keep in touch with work that was going on at Fort Monroe, Virginia, and later at Albuquerque, New Mexico, and at Colorado Springs" (Tucker 1985a). Al did, however, invent improved counters for photo-theodolites used to check on the performance of height finders and range finders.

In 1946, upon returning to full time university work, Al found that he was having some difficulty in resuming his pre-war topological investigations. But, he soon had the opportunity to apply his mathematical skills to a new field (Albers and Alexanderson 1985).

Linear Programming: A Short Exposition

As with most career-changing events, Al's introduction to things OR, especially linear programming (LP), was a more-or-less chance event. The inventor of LP, George B. Dantzig, who was a mathematician for the U.S. Air Force, traveled from Washington, D.C. to Princeton on October 3, 1947 to meet with John von Neumann (the mathematician of game theory fame). Dantzig wanted to consult with the "great Johnny von Neumann to see what he could suggest in the way of solution techniques" for a solving an LP problem (Dantzig 1991, 24; 2002). It was during this meeting that von Neumann conjectured that the LP problem and the two-person zero-sum game were equivalent, and stressed the "fundamental importance of duality" (Dantzig 1963, 24). [A few weeks later, von Neumann proposed an iterative nonlinear scheme for solving LP problems. It was tested against the simplex method by Alan Hoffman and his group at the National Bureau of Standards—the simplex method came out a "clear winner" (Dantzig 1991, 24).]

Dantzig revisited von Neumann in June 1948 to discuss the possibility of establishing a government-funded, university-based research project to study LP. Al did not know Dantzig and was unaware of his LP research. Upon meeting Dantzig at the end of his second visit, Al offered to drive him to the Princeton Junction station to catch the train back to Washington. During the short drive, Dantzig gave Al a quick exposition of LP, using the transportation problem as an example. Al recognized the connection to Kirchhoff's Law for electrical networks, which tied in with his interest in combinatorial topology (Tucker 1957). Soon after, because of this encounter, Al was asked to head a Princeton research effort to study such problems, sponsored by the Office of Naval Research (ONR). Known as the ONR Logistics Project, it lasted over two decades and supported the research of a number of graduate students; Al enlisted graduate students David Gale and Harold Kuhn to be members of the Project team (Albers and Alexanderson 1985, Dantzig 1991, Kuhn 2002).

One of the most fascinating coincidences of twentieth century mathematics and economics is the emergence of both LP and game theory in the 1940s. Developed independently and seemingly unrelated, both fields intersect mathematically by the remarkable demonstration that shows the equivalence of the general LP problem and the zero-sum two-person game, that is, one can be transformed into the other. Further, the duality aspects of a linear program and its solution yield valuable information related to the economic interpretation of the problem setting. Al happened to be at the place (Princeton) where both fields came together. OR has been the beneficiary of his involvement in these apparently disparate but revolutionary achievements. Through his own research, and that of his students, Al extended the theory of both fields, as well as establishing the mathematical basis and relationship between game theory and LP, with special emphasis on duality theory and the Tucker tableau (Nering and Tucker 1993).

> WHAT IF?
>
> From an interview with Al conducted by William Aspray, April 12, 1984:
>
> "ASPRAY: The subject of this interview is the educational program at Princeton in the 1930s. I know there were a number of expectations of students . . . .
>
> TUCKER: In the first place it should be said that admission was very selective. . . . They were all doctoral students from the very beginning. . . .
>
> ASPRAY: Of the people who didn't come here, where did they often go? Did you follow up on that?
>
> TUCKER: Well, no I didn't. I've heard by accident about some. George Dantzig has told me that he applied to come here and was not taken, and he went to the University of California at Berkeley where he worked with a statistician, Neyman. He said that he was quite upset at the time he wasn't accepted here, but the way things went for him at Berkeley that he now regards it as almost fortunate that he wasn't."
>
> [Tucker 1985b, Princeton Mathematics Community in the 1930s. Transcript Number 31 (PMC 31); also Princeton 1985.]
>
> Published with permission.

As there has been some uncertainty about who did what with respect to game theory and duality, to set the record straight, we cite the following (Nering and Tucker 1993, 260): "When George B. Dantzig first described a linear programming problem to John von Neumann in 1947, von Neumann immediately saw that the problem of finding an optimal strategy for a player of a matrix game (zero-sum two-person game) was a linear programming problem. But since there are two players in a matrix game, von Neumann suggested that linear programming problems came in pairs. This conjecture was proved by David Gale, Harold W. Kuhn and Albert W. Tucker in 1948." Further, in an internal Air Force (unpublished) report, Dantzig noted that he stated and proved duality in 1948 based on his 1947 discussion with von Neumann. As Dantzig sums up the situation: "Today everyone cites von Neumann as the originator of the duality theorem and credits Tucker, Kuhn and Gale as the publishers of the first rigorous proof" (Albers et al. 1990, 76). The formal publication of the Gale et al. (1951) duality paper and the Dantzig (1951) paper on the equivalence of the linear-programming problem and the zero-sum two-person game were both published in the proceedings of the 1949 Cowles Commission conference on activity analysis (Koopmans 1951).

To solve the general LP problem, Dantzig and others introduced the concept of a computational tableau that facilitates the simplex method's transformations of the given problem. Al, in his paper, "Solving a matrix game by linear programming" (Tucker 1960), introduced a tableau pivot algebra as a pedagogical and computational way for solving an LP problem which also demonstrates duality and the primal-dual optimal solution conditions. The resulting tableaus are called Tucker tableaus (Nering and Tucker 1993). The 1966 revised version of J. D. William's game theory book, *The Compleat Strategyst*, is essentially the same as its 1954 version except for a new chapter devoted to Al's method. As Williams notes in the preface to the revised edition: "the *STRATEGYST* was, in an important particular, obsolete." That important particular being "... the pivot method, a general method for solving matrix games ... of Professor Tucker, who has developed a combinatorial linear algebra of great generality and power" (Williams 1966, vii).

NONLINEAR PROGRAMMING

While on sabbatical leave at Stanford (1949–1950), Al returned to his initial thoughts about the relationship between LP and the theory of electrical networks. A particular problem of interest was the classical Kirchhoff–Maxwell problem of determining the distribution of direct current in an electrical network, a problem that Al recognized as being a quadratic-programming problem (a new term at that time) (Tucker 1957, 1985a). He then suggested to his ONR team of David Gale and Harold Kuhn that they investigate duality and related issues for quadratic programming; David declined—"he said he'd had enough of that sort of stuff" (Tucker 1985a)—but Harold accepted (Kuhn 2002). The work developed via letters exchanged between Al in Stanford and Harold in Princeton. Based on Harold's urging, they decided, instead, to study the more general problem of what they termed nonlinear programming (NLP) (Kuhn 2002). The form of the problem they considered can be stated as follows (Kuhn and Tucker 1951, 483):

> Let the mapping Fx be an m-vector whose components $f_1(x), \ldots, f_m(x)$ are differentiable functions of x defined for $x \geq 0$. Let $g(x)$ be a differentiable function of x defined for $x \geq 0$. The *maximum problem* is then: To find an x^0 that maximizes $g(x)$ constrained by $Fx \geq 0$, $x \geq 0$.

Their research on this problem was completed in the spring of 1950. Al presented the results at a RAND conference in May 1950. It was there that an example, due to Charles. B. Tompkins, raised the need for a "constraint qualification" that ruled out "singularities on the constraint set, such as an outward pointing 'cusp'" (Kuhn and Tucker 1951, 483).

Al was invited by Jerzy Neyman to give a paper at the Second Berkeley Symposium on Mathematical Statistics and Probability that was held from July 31 to August 12, 1950. He chose to speak on his NLP research with Harold. It was the proceedings of this symposium that brought the topic of NLP to the theoretical and applied OR and mathematical worlds as it included their paper "Nonlinear Programming" (Kuhn and Tucker 1951), the paper that gave the name to the field (Kuhn 1976, 1).

Of importance to their results were conditions for the solution of the original maximizing nonlinear-programming problem and its dual (saddle-point problem). These conditions were later named (by others) the Kuhn–Tucker (KT) conditions: for a problem with "a nonlinear objective function and nonlinear constraints, . . ., there appear as necessary conditions for a local optimum the existence of generalized Lagrange multipliers that satisfy conditions that are 'dual' to the original constraints" (Kuhn 2002, 133). The KT conditions have been renamed the Karush–Kuhn–Tucker (KKT) because, as Al noted (Tucker 1985a):

> Perhaps it might more properly have been called "Convex Programming," but we just picked the name nonlinear. It was in this way that what is now referred to as Kuhn-Tucker theory came about. Of course, we now know that it should be called Karush-Kuhn-Tucker theory because Bill Karush had anticipated what we did in 1950 in his [unpublished] master's thesis at Chicago about 1940 [Karush 1939]. But his work was done in the context of the calculus of variations where it didn't attract attention, and our work was done in the context of mathematical programming where it was viewed as the first breakthrough from the linear programming.

The Kuhn and Tucker's (1951) paper also introduced the vector-maximum problem and developed the theoretical solution details in terms of efficient solutions and the associated saddle-value problem. An historical account of the origins of nonlinear programming and related duality theorems are given in Kuhn (1976). As Dantzig noted in his Foreword to *Linear Programming and Related Problems* (Nering and Tucker 1993, ix) "The contributions of Al Tucker and his group at Princeton were many. They published rigorous proofs of the famous duality theorem conjectured by von Neumann. Theirs was an independent discovery. They were responsible for the equally famous K-T or Kuhn-Tucker conditions of nonlinear programs. These two theorems tell us how to check if a feasible solution to a linear or nonlinear program is optimal."

THE PRISONER'S DILEMMA

Al and his students advanced the embryonic field of game theory on many fronts. Al is famously noted for his description of a non-zero-sum, non-cooperative, two-person game that was first introduced by Merrill Flood and Melvin Dresher at the RAND Corporation. When Al was asked to describe game theory to psychology majors at Stanford during his sabbatical year, he used his pedagogical and communication skills to recast it as an easily understood strategic problem that pitted two supposed partners in crime against each other. Al described his version of the game as follows (Tucker 1983): "Two men, charged with a joint violation of law, are held separately by the police. Each is told that (1) if one confesses and the other does not, the former will be given a reward of one unit and the latter will be fined two units, (2) if both confess, each will be fined one unit. At the same time, each has good reason to believe that (3) if neither confesses, both will go clear." In his book, *Prisoner's Dilemma*, William Poundstone (1992, 118) notes:

> Over the years, this story has improved in the retelling and now almost always concerns prison terms. (Plea bargains over prison terms are more realistic than cash prizes for confessing!) A typical contemporary version of the story goes like this:
>
> Two members of a criminal gang are arrested and imprisoned. Each prisoner is in solitary confinement with no means of speaking to or exchanging messages with the other. The police admit they don't have enough evidence to convict the pair on the principal charge. They plan to sentence both to a year in prison on a lesser charge. Simultaneously, the police offer each prisoner a Faustian bargain. If he testifies against his partner, he will go free while the partner will get three years in prison on the main charge. Oh, yes, there is a catch ? If both prisoners testify against each other, both will be sentenced to two years in jail.

> The prisoners are given a little time to think this over, but in no case may either learn what the other has decided until he has irrevocably made his decision. Each is informed that the other prisoner is being offered the very same deal. Each prisoner is concerned only with his own welfare—with minimizing his own prison sentence.

Prisoner's Dilemma (Poundstone 1992, 118)	**Prisoner B: Does not confess**	**Prisoner B: Confess**
Prisoner A: Does not confess	(A) 1 year, (B) 1 year	(A) 3 years, (B) 0 years
Prisoner A: Confess	(A) 0 years, (B) 3 years	(A) 2 years, (B) 2 years

Based on Al's statement of the problem, his discussion of its ramifications, and subsequent analysis and publications by many others, the Prisoner's Dilemma has contributed important insights in biology, decision analysis, economics, philosophy, political science, sociology, and, especially, game theory (Poundstone 1992). (We have not, however, been able to answer the related dilemma on whether it is the Prisoner's Dilemma or the Prisoners' Dilemma. In his 1950 Stanford seminar, Al titled his talk "A Two-Person Dilemma.")

Al was dissertation advisor of John F. Nash, who was awarded the 1994 Nobel Prize in economics for his thesis on equilibrium solutions for general-sum non-cooperative games. Al became Nash's advisor after von Neumann had expressed disapproval and disinterest in Nash's ideas. Al was also the Ph.D. advisor for others who have made contributions to OR and game theory: Michel Balinsky, David Gale, John Isbell, Stephen Maurer, Torrence Parsons, Lloyd Shapley, as well as Marvin Minsky of artificial intelligence fame (Albers and Alexanderson 1985).

AL AND SOLOMON LEFSCHETZ

Al's career at Princeton was influenced greatly by Solomon Lefschetz. He was born in Moscow, trained as an engineer at the École Centrale in Paris, emigrated to the U.S. when he was 21, and worked for Westinghouse Electric Company in Pittsburgh. In 1907, he lost both hands and forearms due to a transformer explosion; he used artificial limbs and wore black leather gloves for the rest of his life. Lefschetz then turned to mathematics and enrolled as a doctoral student at Clark University, Worcester, Massachusetts. He received his Ph.D. under William Story in 1911 with a dissertation on algebraic geometry, "On the existence of loci with given singularities." His first teaching positions were at the University of Nebraska and the University of Kansas. In 1924, he was a visiting professor at Princeton, after which he accepted a permanent position there. He became chair of the mathematics department in 1945 and retired in 1953. Lefschetz is noted as a pioneer in developing the algebraic techniques of topology, a word he coined.

In her book on John Nash, *A Beautiful Mind*, Sylvia Nassar described Lefschetz as: "Entrepreneurial and energetic, . . . the supercharged human locomotive that had pulled the Princeton department out of genteel mediocrity right to the top. He recruited mathematicians with only one criterion in mind: research. . . . He exhorted, bossed, and bullied, but with the aim of making the department great and turning his students into real mathematicians, tough like himself (Nassar 1998, 58).

In a short reminiscence article, Al wrote:

> Solomon Lefschetz is the mathematician I have known best, and perhaps because I have known him best, admired most. But it is very difficult for me to talk about Lefschetz in a way that I feel will do justice to the tremendous respect, admiration and affection that I have for him.
>
> I don't know of any other mathematician whose career has been so intertwined with another's as mine has been with Lefschetz'. I did my Ph.D. thesis with him, and it was understood when I joined the Princeton faculty that I was going to work with him in my research. Then in 1945, when he succeeded Eisenhart as chairman of the department, I was his lieutenant for all undergraduate departmental administration. Lefschetz retired in 1953 and I succeeded him as chairman, which was clearly his wish. But when he retired, this was only a formal thing; he kept coming to Fine just as much as before.
>
> We were very close friends (Tucker 1985d, 349).

In an interview, Al commented:

> Well, Lefschetz was a very rough person, in the sense that he spoke rather roughly. You see I was a Ph.D. of Lefschetz; indeed I was his man Friday for

> years and years while he was chairman of the department. So I really regard Lefschetz as my mathematical father.
>
> I had the very good fortune my first year as a graduate student to stand up to Lefschetz. I criticized the way he was proving something in a graduate course, and he sarcastically invited me to come to the board and prove it the way I thought it should be proved. Which I proceeded to do. I finished the proof, and then he ridiculed this performance. Class came to an end. Two days later, when there was the next class, he said that he wanted to go back over this proof that there had been criticism of. He proceeded to give exactly the proof that I had given and made quite a hit. Everybody in the class knew it was my proof, but he never admitted this in any way. Indeed I remember saying when he was getting rather old—I said this to one of my colleagues one day—"You know, Lefschetz is slipping. Today I won an argument from him." I said, "Usually I have to wait several days before I'd known I won an argument with him, but today he agreed to my point. This very day" (Tucker 1985c).

Al played a gentle joke on Lefschetz, as recounted by Richard Bellman (1984, 120):

> Lefshetz had to give a talk in Brooklyn on Pearl Street. He knew that Tucker had spoken there previously and asked directions. Tucker told him the subway stop to get off and said that the simplest thing was to ask directions from there. However, he added with a straight face, the natives of Brooklyn don't speak good English. Instead of asking for Pearl Street ask for Poil street. Lefschetz came back and complained to Tucker. He had asked several people where Poil Street was. Finally one person said, "Poil Street, you mean Pearl Street," and gave him directions. I don't think Lefschetz ever realized it was a joke.

A PERSONAL NOTE

I was introduced to game theory in 1952–1953 when I took an evening, two-semester graduate mathematics course at The American University in Washington, D.C. The course (Theory of Games) was taught once a week by either Al or Harold Kuhn. On the day of the class, one or the other would take the train from Princeton to Washington and spend the day consulting at the George Washington University Logistics Research Project as part of their work for ONR. One of the students, who worked at George Washington, would drive whoever was teaching to American University and, after class, rush him to Union Station for the last train back to Princeton. I was thus introduced to Al, his knowledge and teaching skills, and learned what it meant to be a modern mathematician. I still have the notes from the class! Surprisingly, the class discussion of the Prisoner's Dilemma does not refer to it by that name, it was just stated in the form given above. Evidently, the name had not caught on at that time. The seminal book, *Games and Decisions*, by Luce and Raiffa (1957) highlighted the Prisoner's Dilemma in its discussion of two-person, non-zero-sum, non-cooperative games and appears to be the source for its subsequent popularity and interest. Al suggested the title to their book.

Whenever Al lectured, in class or at a meeting, he always handed out colored (usually yellow or blue) mimeographed notes that contained the essential lecture material. I always appreciated that effort. Why did he do it?

ANNALS OF MATHEMATICS STUDIES

The Annals of Mathematical Studies was an off-shoot of the *Annals of Mathematics* (edited by Lefschetz) when it was decided that there was a need for a journal in which longer papers, not suited for the *Annals*, could be published. Thus, the *Annals Studies* was born. Al was responsible for arranging for its production (at an Ann Arbor, Michigan lithoprint company) and for Princeton University Press to be the publisher. Al had a strong and steady editorial hand in the development and publication of six influential *Annals Studies* volumes; he also contributed important papers—in Volume 24: "On symmetric games" and "Reductions of game matrices" (both with David Gale and Harold Kuhn); in Volume 38: "Dual systems of homogeneous linear relations," and "Polyhedral convex cones" (with Alan Goldman), all of which contributed greatly to aspects of LP, game theory, and related mathematical foundations. From personal experience, I can state that theses volumes were eagerly awaited and well-read.

Kuhn, H.W., A.W. Tucker. 1950. eds. Contributions to the Theory of Games, Volume I. *Annals of Mathematics Studies 24*, Princeton University Press, Princeton, New Jersey.

Kuhn, H.W., Tucker, A.W. 1953. eds. Contributions to the Theory of Games, Volume II. *Annals of Mathematics Studies 28*, Princeton University Press, Princeton, New Jersey.

Kuhn, H.W., Tucker, A.W. 1956. eds. Linear Inequalities and Related Systems. *Annals of Mathematics Studies 38*, Princeton University Press, Princeton, New Jersey.

Dresher, M., A. W. Tucker, P. Wolfe. 1957. eds. Contributions to the Theory of Games, Volume III. *Annals of Mathematics Studies 39*, Princeton University Press, Princeton, New Jersey.

Tucker, A.W., R. D. Luce. 1959., eds. Contributions to the Theory of Games, Volume IV. *Annals of Mathematics Studies 40*, Princeton University Press, Princeton, New Jersey.

Dresher, M., L. S. Shapley, A. W. Tucker 1964. eds. Advances in Game Theory. *Annals of Mathematics Studies 52*, Princeton University Press, Princeton, New Jersey.

Maybe because after 1945, he never submitted a paper of his own to a journal. Instead, he wrote conference papers with students, published in symposia proceedings, and co-edited and published in many volumes of the Princeton *Annals of Mathematics Studies.* According to his son Alan, "he wanted to leave space in the journals for the next generation" (Nasar 1995). Al's colored hand-outs are part of his mathematical legacy—a multi-hued trail whose branches still lead to fruitful fields of OR discovery. [A bibliography of his papers, reports, and public addresses is given in Balinski (1974, 4–9).]

The Personal Al

In answer to the question "What are some of your hobbies," Al answered:

> I certainly do like to travel, and fortunately I've had many professional opportunities for this. Even during vacation trips I like to visit with mathematicians and give talks. My favorite place for travel has been Australia. I've been there four times as a visiting lecturer. The city of Perth in Western Australia is my favorite city. That's where I would live if it weren't so far away from everything else that I'm tied to.
>
> The other hobby I might mention is that I like detective stories. It isn't that I read them so much to try to guess the end; I really read them for just relaxation. I have quite a collection of paperbacks. I like best the classical British detective stories, which I started reading when I was a student in Toronto.
>
> Early on I had liked chess, but I swore off chess when I discovered that after playing a keen chess game I had difficulty sleeping at night. I was continuing to concentrate on the game. So I switched to reading, and found that somehow detective stories provided me with the sort of relaxation I liked (Albers and Alexanderson 1985, 348).

Al married Alice Curtiss in 1938 and they had three children: Alan Curtiss, Thomas William, and Barbara Jane. They were divorced in 1960. Al married Mary Shaw in 1963.

Al and his father were mathematics teachers and Al's two sons, Alan and Tom, have continued the tradition. They both have a double dose of the math gene—their mother's father, David R. Curtiss, was chairman of the Mathematics Department of Northwestern University. Curtiss was president of the Mathematical Association of America (MAA) (1936–1936)—Al was president from 1961 to 1962. Alan received his Ph.D. in Mathematics from Stanford University and teaches at the State University of New York, Stony Brook, Long Island. Tom received his Ph.D. in mathematics from the University of California, Berkeley, and teaches at Colgate University, Hamilton, New York. Alan and Tom were back-to-back First Vice-Presidents of the MAA, respectively (1998–1990, 1990–1992). Al's daughter, Barbara Cervone, has a Ed.D. from Harvard University and is president of "What Kids Can Do," a the non-profit organization that promotes through books, videos, and related resources nontraditional learning for adolescents in and out of school.

(TUCKER)^3

Alan Al Tom

(Courtesy Alan Tucker)

Harold Kuhn, in his article describing the MAA's Award for Distinguished Service to Mathematics given to Al in 1968, wrote the following (Kuhn 1968, 3):

> One of the most fitting tributes to the qualities of Al Tucker's life in mathematics was composed by John Sloan Dickey, President of Dartmouth College, and read on the occasion of the award of an honorary degree of Doctor of Science in June, 1961. It captures such a large share of the debt we owe him that it seems proper to quote it directly:

> "Nearly three decades ago you began an academic career at Princeton which became a mission to mathematics. In a field where scholarship scores only if the idea is both new and demonstrably true your ideas have won their way in topology, in the theory of games, and in linear programming. But even in mathematics a mission is more than ideas; it is also always a man, a man who cares to the point of dedication, whose concern is that others should care too, and who can minister to the other fellow, as the need may be, either help or forbearance. Because you, sir, embody in extraordinary measure both your profession's love of precision and man's need for conscientious leadership, mathematics in America at all levels is today higher than it was and tomorrow will be higher."

Al died of complications from pneumonia in Highstown, New Jersey, on January 25, 1995. He was survived by his wife, Mary, his three children, and six grandchildren.

Honors and Awards

Al was involved deeply in advancing the mathematical and related scientific communities. He served as President of the MAA, a Vice President of the American Association for the Advancement of Science, and Chairman of the Conference Board of the Mathematical Sciences. He received the MAA's award for Distinguished Service to Mathematics in 1968 (Kuhn 1968). Al received an honorary degree from Dartmouth College in 1961. The Operations Research Society of America and The Institute of Management Sciences awarded the 1980 von Neumann Theory Prize to Al, joint with David Gale and Harold Kuhn, for their seminal role in laying the foundations of game theory, linear and nonlinear programming. In 2004, Al was inducted into the International Federation of Operations Research Societies' Hall of Fame (Gass 2004).

References

Albers DJ, Alexanderson GL (eds) (1985) Albert Tucker. Mathematical people. Birkhauser, Boston, MA, pp 339–348

Albers DJ, Alexanderson GL, Reid C (eds) (1990) George B. Dantzig. More mathematical people. Academic, Orlando, FL, pp 61–79

Balinski M (ed) (1974) Pivoting and extensions: in honor of A. W. Tucker. Mathematical programming study 1. North-Holland, Amsterdam

Bellman R (1984) Eye of the Hurricane: an autobiography. World Scientific Publishing Company, Singapore

Blackett DW (1967) Elementary topology: a combinatorial and algebraic approach. Academic, New York, NY

Dantzig GB (1951) A proof of the equivalence of the programming problem and the game problem. In: Koopmans TC (ed) Activity analysis of production and allocation. Wiley, New York, NY, pp 330–335

Dantzig GB (1963) Linear programming and extensions. Princeton University Press, Princeton, NJ

Dantzig GB (1991) Linear programming. In: Lenstra J, Rinnooy Kan A, Schrijver A (eds) History of mathematical programming. Elsevier, Amsterdam, pp 19–31.

Dantzig GB (2002) Linear programming. Oper Res 50(1):42–47

Gale D, Kuhn HW, Tucker AW (1951) Linear programming and the theory of games. In: Koopmans TC (ed) Activity analysis of production and allocation. Wiley, New York, NY, pp 317–329

Gass SI (2004) IFORS' operational research hall of fame: Albert William Tucker. Int Trans Oper Res 11(2):239–242

Karush W (1939) Minima of functions of several variables with inequalities as side conditions. Master's Thesis, Department of Mathematics, University of Chicago, Chicago, IL

Koopmans TC (ed) (1951) Activity analysis of production and allocation. Wiley, New York, NY

Kuhn HW (1968) Award for distinguished service to Professor Albert W. Tucker. Am Math Mon 75(1):1–3

Kuhn HW (1976) Nonlinear programming: a historical view. In: Cottle RW, Lemke CE (eds) Nonlinear programming: proceedings of the Siam-AMS symposia, New York, March, 1975, vol 9. American Mathematical Society, Providence, RI, pp 1–26

Kuhn HW (1991) Nonlinear programming: a historical note. In: Lenstra JK, Rinnooy Kan AHG, Schrijver A (eds) History of mathematical programming. North-Holland, Amsterdam, pp 82–96

Kuhn HW (2002) On being in the right place at the right time. Oper Res 50(1):132–134

Kuhn HW, Tucker AW (1951) Nonlinear programming. In: Neyman J (ed) Proceedings of the second Berkeley symposium on mathematical statistics and probability. University of California Press, Berkeley, CA, pp 481–492

Luce LD, Raiffa H (1957) Games and decisions. Wiley, New York, NY

Nasar S (1995) Obituary: Albert W. Tucker. New York Times, January 27

Nasar S (1998) A beautiful mind. Simon and Schuster, New York, NY

Nering ED, Tucker AW (1993) Linear programming and related problems. Academic, Boston, MA

Poundstone W (1992) Prisoner's dilemma. Doubleday, New York, NY

Princeton University (1985) Princeton Mathematics Community in the 1930s: Interview with Albert W. Tucker, transcript no. PMC 31. Trustees of Princeton University, Princeton, NJ. (Transcript Nos. 29–40, interviews with Albert Tucker, 1975, 1979, 1984.) http://www.princeton.edu/mudd/math. Accessed 22 Feb 2009

Tucker AW (1933) An abstract approach to manifolds. Ann Math 34(2):191–243

Tucker AW (1957) Linear and nonlinear programming. Oper Res 5(2):244–257

Tucker AW (1960) Solving a matrix game by linear programming. IBM J Res 4:507–517

Tucker AW (1983) The mathematics of Tucker: a sampler. Two-Year Coll Math J 14(3):228–232

Tucker AW (1985a) The Princeton Mathematics Community in the 1930s. Transcript number 39 (PMC39), Seeley G. Mudd Manuscript Library, Princeton University, Princeton, NJ
http://www.princeton.edu/~mudd/finding_aids/mathoral/pm02.htm
http://www.princveton.edu/~mudd/finding_aids/mathoral/pmc39.htm.
Accessed 27 Oct 2009)

Tucker AW (1985b) The Princeton Mathematics Community in the 1930s. Transcript number 31 (PMC31), Seeley G. Mudd Manuscript Library, Princeton University, Princeton, NJ
http://www.princeton.edu/~mudd/finding_aids/mathoral/pm02.htm
http://www.princeton.edu/~mudd/finding_aids/mathoral/pmc31.htm.
Accessed 27 Oct 2009)

Tucker AW (1985c) The Princeton Mathematics Community in the 1930s. Transcript number 40 (PMC40), Seeley G. Mudd Manuscript Library, Princeton University, Princeton, NJ
http://www.princeton.edu/~mudd/finding_aids/mathoral/pm02.htm
http://www.princeton.edu/~mudd/finding_aids/mathoral/pmc40.htm.
Accessed 27 Oct 2009)

Tucker AW (1985d) Solomon Lefschetz: a reminiscence. In: Albers DJ, Alexanderson GL (eds) Mathematical people. Birkhauser, Boston, MA, pp 349–350

Williams JD (1966) The compleat strategyst. McGraw-Hill, New York, NY

7

Jacinto Steinhardt

Saul I. Gass

The origins of operations research (OR), stemming from the discovery of radar and its military applications in the late 1930s and the start of World War II (WWII), provide a classical case of how individuals from diverse fields and interests can give birth to a new discipline. On both sides of the Atlantic, scientists from almost all professions joined military-sponsored groups whose main function was to apply current scientific knowledge to improve the outcomes of battle situations—on land, sea, and air. In the U.S., an amazing cadre of scientists became involved with a broad range of military problems across all services and bent their talents to their resolution with great success.

The U.S. Navy, in particular, attracted many of the post-depression trained scientists who would have normally pursued a career in academia or within governmental or industrial laboratories. We find the chemist, Jay Steinhardt, thus making a career-changing move into OR and establishing himself as a military (Navy) OR analyst. For his analytical work during WWII, he received the American Medal of Freedom and the President's Certificate of Merit. He directed the first postwar military OR consulting group for 16 years, and was president (1954) of the Operations Research Society of America (ORSA).

Transition: Chemist to Operations Research Analyst

Jacinto Steinhardt, known to all as Jay, was born on May 20, 1906 in New York City. A product of NYC public schools, he also stayed local in his pursuit of a college education. He attended Columbia University

A.A. Assad, S.I. Gass (eds.), *Profiles in Operations Research*, International Series in Operations Research & Management Science 147, DOI 10.1007/978-1-4419-6281-2_7,

where he received his A.B. in 1927, A.M. in 1928, and a Ph.D. in biophysics in 1934. A sequence of post-doctoral fellowships enabled him to spend the next 4 years doing research on the physical chemistry of proteins at laboratories in Copenhagen, Uppsala, Cambridge (England), and at the Harvard Medical School. This work led to his accepting, in 1938, a research position as a physical chemist with the Textile Foundation of the National Bureau of Standards (NBS) in Washington, D.C. At NBS, his main research dealt with the electrochemistry of fibrous proteins and, during the 4 years he was with NBS, he published 25 papers on vision and the physical chemistry of proteins. In November 1942, Jay joined the recently established U.S. Navy's Anti-Submarine Warfare Operations Research Group (ASWORG) that was under the direction of the physicist Philip Morse. It is unclear why he chose ASWORG as a means to contribute to the war effort and not search out a position more in line with his education and training, for example, the Army Chemical Corps. By becoming an OR analyst, Jay changed the direction and scope of his scientific career (Tidman 1984, Engel 1985).

ASWORG was headquartered in Washington and, by the end of 1942, was staffed by 30 civilian scientists. The work and importance of ASWORG during the U.S. involvement in WWII is ably told in many places (Morse 1977, Tidman 1984). But, the contributions of the individual analysts is often clouded by the telling of the remarkable achievements of ASWORG's analysts working as members of a team, either stationed in Washington or in the field as members of a naval unit. As noted in Tidman (1984, 37), "Morse insisted that analysts not claim credit for anything, since they, in turn, took no responsibility for the ultimate decision made by a commander. Moreover, analysts were reminded of the importance of being polite and cooperative, given the tenuousness of their situation." Morse must be credited with initiating the basic mode of operation of ASWORG (and its subsequent Navy OR incarnations): the assignment of OR analysts to submarine bases, if for no other reason than to obtain firsthand operational data at the source. This proved to be "the key to ASWORG's

... effectiveness, namely its field representative program" (Tidman 1984, 37). For Jay, fortunately, we do have information on some of his ASWORG activities.

In March of 1943, Jay was assigned to the Fourth Fleet stationed at Recife, Brazil. German ships were attempting to run a blockade with strategic supplies (mainly tin and rubber) picked up in Japan and Malaya. The Allies had identified five such ships that were trying to make the run between Recife and Ascension Island 1,400 miles to the east. Aircraft were assigned to patrol this area based on "scientifically designed flight plans ... set up to intercept all ships passing through this Atlantic pathway. Planes and crews ... operated in accordance with this plan which were so successful that three of the blockade runners ... were intercepted and destroyed in one 48 hour period" (Steinhardt 1946, 650). The fourth ship escaped by an "elaborate disguise," while "the fifth got through before the patrols were under way but ran aground in the Bay of Biscay when attacked by British destroyers operating on information supplied by the American forces" (Tidman 1984, 65). Jay, working out of Recife, helped to develop anti-submarine measures for blockading enemy submarines in the Atlantic that were adopted and contributed to the sinking of several U-boats (Tidman 1984, 39). For his efforts, Jay received the American Medal of Freedom and the President's Certificate of Merit for "outstanding fidelity and meritorious conduct in aid of the war effort" (*Science* 1947, 382; Engel 1985, 23).

The Navy's Tenth Fleet was established in May 1943 with its objective "to exercise direct control over all Atlantic sea frontiers," with especial emphasis in controlling the allocation of antisubmarine forces (Tidman 1984, 57). The Tenth Fleet had no ships of its own but operated all anti-submarine warfare activities (Morse 1997, 184). It was under the command of Admiral Ernest J. King, with Rear Admiral Francis S. Low as his deputy. ASWORG was assigned to the Tenth Fleet in July 1943. One of ASWORG's tasks was determining how accurate the Navy's radio-direction finding (RDF) net located U-boats. The task was assigned to Jay. Morse (1977, 185–186) gives the following account of Jay's RDF analysis:

> ... Jay Steinhardt, a new recruit, took on the job of seeing how accurately our radio-direction finding (RDF) net located U-boats. Each day each enemy submarine talked to its headquarters in Germany—in a burst of high-speed code, of course. A set of RDF stations along our coast recorded these bursts

> and triangulated the position of the U-boat. Planes could then be sent out to attack it. In some of these cases the sub was found and attacked, so data were available on the actual locations of the U-boats, to compare with the RDF estimates of their positions. Steinhardt found that the compared differences were unbelievably small. The estimates of U-boat positions, given to us from the secret room as RDF estimates, checked the actual positions of U-boats found and attacked with an accuracy ten times better than the RDF equipment was supposed to possess. It sometimes is possible to make a machine perform better than expected. In this case, however, accuracy depended on the wavelength of the radio wave, and waves aren't that obliging; something else had to be enhancing the RDF accuracy.
>
> In one of my daily sessions with Admiral Low I reported Steinhardt's finding, simply saying with a straight face that the reports of U-boat locations given us by the secret room had much greater accuracy than the RDF equipment could possibly produce and that we were going to investigate the anomaly. Admiral Low, also with a straight face, said that was interesting. But the next day he called Steinhardt and me in and disclosed what by that time we had guessed but never mentioned, that our side had broken the German code and that the locations given to us as RDF readings were in fact the positions reported by the submarine skipper himself to his commander in Germany. The whole episode convinced me that Admiral Low really would have let us in on many more secrets, if he had been allowed, but he had to have specific arguments in each case before he could persuade Admiral King to relax.

After his tour at Recife, Jay joined the Seventh Fleet assigned to General Douglas McArthur, first in Brisbane, Australia, and then in Hollandia, New Guinea. By this time, ASWORG had expanded from its original cadre of seven analysts to a total of 73. During this growth period, the staff began taking on assignments other than antisubmarine warfare, such as air operations and amphibious operations. Due to its expanded area of interests, ASWORG was renamed the Operations Research Group (ORG) and, in October 1944, assigned to the Readiness Division of the Commander in Chief, U.S. Fleet (Tidman 1984). As the war came to an end, many of the ORG/ASWORG analysts returned to their civilian occupations, but a few felt that the postwar period would enable them to contribute to any future plans to reorganize and modernize the Navy. "The Navy agreed and a smaller edition of the group was kept on under Steinhardt's direction [as recommended by Morse]. In modified form it still exists" (Morse 1977, 208).

THE OPERATIONS EVALUATION GROUP

> A LOOK INTO THE FUTURE
>
> "The techniques and principles of analysis developed by the group during its history have wide application to modern government and industry. Briefly, these techniques are those of the competent scientist, applied to a large-scale human operation as a whole, with the aim of fitting the operation to its purpose, and of measuring the effectiveness with which the operation is being carried out
>
> "At one time or other, nearly every member of the group has suggested important peacetime applications. Examples are railroad operation, in all its phases, highway design and location, town planning, city and country traffic control, home design, department store operation, and assembly line processes" (Steinhardt 1946, 653).

Based on an enthusiastic letter of recommendation by Admiral King, Secretary of the Navy James V. Forrestal agreed to continue a peacetime ORG to be staffed initially by 25 civilian scientists. Its establishment was formalized on November 1, 1945, with a name change to the Operations Evaluation Group (OEG). The renaming was due to the Office of Naval Research (ONR) providing the group's funds, and, since naval research was the purview of ONR and OEG's operations were to be independent of ONR, the name change came about. To further the independence of OEG, it was to be managed by the Massachusetts Institute of Technology (MIT) under an initial 3-year contract, with an organizational tie to its Physics Department. OEG's mission was to "furnish liaison for the fleets with the development and research laboratories . . . and conduct studies and make reports to the Deputy Chief of Naval Operations" (Tidman 1984, 98). Thus, OEG became the first postwar, OR-based consultant organization in the U.S. It was located in the Pentagon.

Jay's main challenge was to structure an organizational model for his analysts that would integrate their expertise within the peacetime naval establishment and, hopefully, allow for a friction-free mode of operation that addressed the Navy's current and future analytical requirements. OEG was sui generis, the first in what eventually turned out to be a long line of military, governmental, business, and industrial OR groups. It seemed natural that Jay would base his organizational structure on his knowledge of ORG's wartime successes. An initial effort was to codify ORG/ASWORG's new wartime methods and techniques, a task that produced three classified reports: OEG Report 51, *Antisubmarine Warfare In World War II* (Sternhall and Thorndike 1946); OEG Report 54, *Methods of*

Operations Research (Morse and Kimball 1946); and OEG report 56, *Search and Screening* (Koopman 1946). [Report 54 was later declassified and published as Morse and Kimball (1951)]. During its first year of operation, although short on staff, OEG conducted over 120 projects, each dealing with a tactical or theoretical problem (Engel 1960, Tidman 1984). Jay continued the wartime practice of assigning analysts to naval operating forces in the field, either on ships or at naval bases. As Joseph Engel notes: "In this way OR assistance could be made available right on the spot when and where it was needed, while at the same time central direction and coordination could be provided through close ties with the home office in Washington" (Engel 1985, 23). Engel, a case in point, joined OEG in 1949 and spent a year as a field representative with the Seventh (Pacific) Fleet and then a year with the Sixth (Mediterranean) Fleet (Tidman 1984, 225).

In 1949, the MIT organizational tie was changed to the Department of Mathematics as a means of improving faculty interest in OEG studies. To further such interactions, Jay arranged for OEG Washington-based analysts to spend a year at MIT where, working with faculty, they could obtain analytical assistance.

Jay also recognized the need to train naval officers in the basic aspects of OR methods. He joined with Rear Admiral E. E. Herman, the Superintendent of the Naval Postgraduate School (NPS), Monterey, California, in proposing and designing a six-term degree curriculum for naval officers. The first class of nine officers graduated in January 1953. The program continues to this day having graduated well over 3,300 military officers (from all U.S. services), as well as military officers from other countries (Schrady 2001).

During its first 5 years as a peacetime OR organization, OEG undertook studies that expanded upon its results in antisubmarine warfare, fleet air defense, mining, and mine countermeasures. This work led to a broader overview of the problems that was captured in the important report "Measures for the protection of overseas transport." The report's "judgments and recommendations ... profoundly influenced the portion of the navy's research and development program that were addressed by the analysis" (Tidman 1985, 125; Engel 1960, 802). But, the start of the Korean War on June 25, 1950, caused "an inevitable redirecting of commitments" for OEG (Tidman 1985, 129).

At the start of the Korean War, OEG's staff consisted of 40 analysts, soon grew to 50, and was close to 60 when the war ended. A number of analysts were sent to the Far East Theater and remained there (or rotated) during war. Sadly, one of them, Dr. Irving Shankov, "was killed when the plane in which he was collecting data was shot down [May 14, 1952] during an interdiction mission" (Tidman 1985, 146). He was posthumously awarded the Medal of Freedom by the Secretary of the Navy.

> BACK TO BASICS
>
> "Although the group [OEG] has through the years increased its competence to use advanced mathematical techniques, it still performs an important part of its work by means of standard mathematical procedures. High school geometry for example is sufficient for many of our problems. The group's contribution to the Navy consists of examining its problem logically and quantitatively, in asking pertinent questions about the Navy's abilities, and then answering these questions" (Engel 1960, 804).

OEG was faced with a different set of problems then those of WWII—"the Korean War was the first occasion the Navy had to expend its major aviation effort in support of our forces in combat on shore" (Engel 1960, 804). This led to the consideration and analysis of new tactical and strategic problems: Choice of weapons for naval attack on tactical targets, close air support, naval gunfire in shore bombardment, efficiency of blockade tactics, interdiction of land transportation. It was during this time that OEG pioneered in applying queueing theory to a military problem and produced the study "Statistical aspects of port operations." It showed that "decreasing the average proportion of occupied berths in a port system could result in a dramatic decrease in average waiting time imposed on a ship before it could enter a berth" (Engel 1960, 804). On July 27, 1953, the U.S. signed an armistice agreement with North Korea and China, thus ending the hot war.

Over the next 10 years, OEG continued to furnish the Navy with a wide range of analyses in the general areas of undersea warfare, air warfare, and general warfare. It expanded its staff to include an economics division, and coordinated its efforts with newly formed Navy analysis groups such as the Institute of Naval Studies and the Applied Science Division. Budgetary, coordination, management, and other pressures caused the Navy to restructure its approach to maintaining a viable analytical staff. Thus, in March 1962, the Navy signed a contract with the Franklin Institute for the management of a consortium of the Navy's analytical advisory groups under a new organizational structure called the Center for Naval Analyses (CNA)—"an overall agency charged with the management and support of the [Navy's] major study groups" (Tidman 1984, 210). OEG was subsumed within the CNA structure. Jay left OEG in August 1962 to become the science advisor to the president of Georgetown University with an appointment as professor in the chemistry department, a position he held until 1980. He died of a stroke on January 30, 1985.

> JAY'S LEGACY
>
> "The legacy left by Steinhardt's sixteen-year stay as the group's [OEG] head was substantial. He played a critical role, for example, in ensuring that the peacetime group was able to establish a firm footing in the years immediately following World War II, when there was still uncertainty as to the group's continued acceptance by the navy. Additionally, his unflagging pursuit of a scientifically excellent group paid off, as he refused to dilute the quality of his staff for the sake of bigness, and he encouraged scientists already in the organization to stay abreast of advances in their fields. He always promoted provocative thinking among his analysts, so that difficult issues would be faced head-on, even at the risk of having to present unwelcome conclusions. Steinhardt also demonstrated an uncanny shrewdness in guiding the group through uncharted waters, forming spinoff groups ... in response to changing navy expectations. Finally, under Steinhardt's leadership, OEG continued to contribute to the development of the basic science of operations research and, by way of its 'alumni,' to place its imprint far and wide" (Tidman 1985, 225).

Operations Research Society of America (ORSA)

The Operations Research Society of America was founded on May 26, 1952, at Arden House, Harriman, New York. It was attended by 71 persons who represented a wide range of business, industrial, academic, consultant, military, and other governmental organizations (Page 1952). Jay and 10 other OEG staff members were in attendance. Jay served on ORSA's initial publication committee, and was elected ORSA's third president (1954–1955). His views on the early growth and future of OR are given in his retiring president's address (Steinhardt 1955). Jay noted that "the rapid growth of our Society and the rapid acceptance of the operations-research concept in industry—an acceptance that has created a market far beyond our capacity to satisfy at an early date—convince me that it has a very real contribution to make; and that our profession will continue to grow for a long time to come" (Steinhardt 1955, 232).

HONORS AND AWARDS

For his WWII analytical work as a member of ASWORG, Jay received the American Medal of Freedom in 1945. In 1946, he was awarded the President's Certificate of Merit for "outstanding fidelity and meritorious conduct in aid of the war effort" (*Science* 1947, 382, Engel 1985, 23). Jay was president of ORSA in 1954.

REFERENCES

Engel JH (1960) Operations research for the U.S. Navy since World War II. Oper Res 8(6):798–809

Engel JH (1985) A tribute to Dr. Jacinto Steinhardt. Phalanx 18(2):23

Koopman BO (1946) Search and screening. Operations Evaluation Group Report No. 56. Center for Naval Analysis, Alexandria, VA

Morse PM (1977) In at the beginnings: a physicist's life. MIT Press, Cambridge, MA

Morse P M, Kimball GE (1946) Methods of operations research. Operations Evaluation Group Report No. 54. Center for Naval Analysis, Alexandria, VA. OEG Report 54 is available from http://www.cna.org/documents/1100005400.pdf. Accessed 23 Dec 2008

Morse PM, Kimball GE (1951) Methods of operations research. Wiley, New York, NY; also, Dover Publications, 2003

Page T (1952) The founding meeting of the society. Oper Res 1(1):18–25

Schrady D (2001) Fifty years of graduate education in operations research at NPS produces 3,300 alumni worldwide. ORMS Today 49(1):38–40

Science (1947) News and notes 105(2728, April 11):382

Steinhardt J (1946) The role of operations research in the Navy. Proc U S Naval Inst (Annapolis, MD) 72(519):649–655

Steinhardt J (1955) Terminal ballistics. Oper Res 3(3):231–232

Sternhell CM, Thorndike AM (1946) Antisubmarine warfare in World War II. OEG Report 51. Center for Naval Analysis, Alexandria, VA

Tidman KR (1984) The operations evaluation group. Naval Institute Press, Annapolis, MD

8

George E. Kimball

John F. Magee

The first group of U.S. World War II (WWII) operations research (OR) analysts came from many organizations, especially universities. After the war, most analysts returned to their previous home bases and continued to work in their principal disciplines. A few, however, recognized that this new science of OR could be of value beyond its military applications; they were the first wave to introduce OR to industry and management. Chief among these OR pioneers was George E. Kimball, a chemist from Columbia University. He was a key player in the first industrial OR consulting firm and in the founding of the Operations Research Society of America (ORSA); he was president of ORSA in 1964. In recognition of his influential contributions, ORSA established in 1974 the George E. Kimball Medal awarded to society members for distinguished service to the society and the profession.

George, John, and Arthur D. Little

In December, 1949, when I (John Magee) visited Arthur D. Little, Inc. (ADL) in Cambridge, Massachusetts, for a job interview, I was minimally familiar with the concept of OR and had never heard of George Kimball. I was being considered for employment as the first (very junior) member of a proposed experimental group to undertake OR in industry. To help me understand the proposal, I was allowed to read a declassified copy of the military version of *Methods of Operations Research* by Philip M. Morse and George E. Kimball that had been obtained by the ADL staff (Morse and Kimball 1946, 1951). That was my first introduction to George Kimball.

A.A. Assad, S.I. Gass (eds.), *Profiles in Operations Research*, International Series in Operations Research & Management Science 147, DOI 10.1007/978-1-4419-6281-2_8,

Several members of the ADL staff, who had served in military related activities, including Gilbert W. King and Bruce S. Old, were familiar with George Kimball and his military OR record. Others who knew him included MIT professors Philip Morse and George Wadsworth who were consultants to ADL. They urged the company to engage George, then a chemistry professor at Columbia University, as a consultant to help get the fledgling OR activity off the ground. Happily, he agreed to consult with the new OR group in 1950, the beginning of our long and fruitful relationship.

I met George when he first visited Cambridge. I remember him as a tall, stocky man, pleasant but quiet. He was clearly interested in what we were trying to do, but was sparing in his advice. During 1950, he visited Cambridge infrequently. We had only one OR assignment, a project for Sears Roebuck, with the work led by Gilbert King, who took a year away from his physics projects at ADL to help launch the effort. Nevertheless, George followed what we were doing with interest and made at least two important contributions, discussed later.

Courtesy National Academy of Sciences

George and I worked together in 1951 when we undertook the group's second OR client project, this time for the Baby Products Division of Johnson & Johnson. The division manufactured a line of lotions, shampoo, and powders for baby care. Manufacturing operations were essentially mixing and packaging done on several mechanized lines operated by a flexible, highly skilled work force. The demand for products was highly seasonal, with the strong peak coming in midsummer, at the time when the plant traditionally closed for a 2-week vacation period. The nature of the work force and of the manufacturing facility put a firm cap on production in any period, except for the possibility of using expensive overtime. The plant manager wanted a method for deciding when and for which products to build inventory to meet the summer peak.

We worked in a fashion that became characteristic of George's method. My job was to observe the plant operation and collect data on product costs, operating rates, production limits, line change issues—anything else that might be relevant—and to document these data and discuss them with George. At the same time, he would ponder the data and speculate on how to frame the issue and structure the problem. Each time we met, roughly weekly, I might brief him on what new information I had gathered and he would sketch a possible approach. Eventually, he characterized the issue in this way: Demand at the peak will be met from a combination of current production and production stored in inventory; inventory, thus, is essentially production time held in the form of product. What is the most economical product to use to store the needed production time, given the costs of producing and holding the various products? He showed me an approach that fit the facts I had gathered: minimizing a linear cost function subject to a series of linear constraints, long before I had ever heard the term linear programming (LP). When I asked for some background reading, he directed me to a translated copy of Hermann Weyl's 1935 paper on the theory of convex polyhedra (Weyl 1950). (Later I came to appreciate its significance underlying LP and used it as part of the foundation of a graduate thesis.) The procedure we built was a straightforward one, essentially a version of the LP transportation model—the important point was that it worked, as attested by the plant manager.

In the course of our work, after one of our visits to the plant in Cranford, New Jersey, George invited me to join his family for dinner at their home in Leonia, New Jersey. It was a delightful time and an opportunity to see another dimension of George Kimball. The household was active and busy with George, his wife Alice, and their four children. Perhaps my strongest memory of the evening is of the youngest daughter, Martha, probably about 2 or 3 years old, coming down from her bath and pulling up a small rocking chair to chat and entertain me while Alice and George were busy in the kitchen.

EARLY YEARS AND BEYOND

George Elbert Kimball was born in Chicago, Illinois, on July 12, 1906. His father was Arthur Gooch Kimball, born in Chicago around 1880, as was his mother, Effie (Smallen) Kimball. Arthur Kimball's father had moved to Illinois in the late nineteenth century from Salem, Massachusetts, the area where the Kimball family had lived for several generations. Arthur's mother, Helen, had emigrated from England. George's mother's family, the Smallens, had lived in the Chicago area for some time. Arthur and Effie were married in 1905.

The Kimball family appears to have lived a comfortable middle-class life. George's mother taught elementary school before her marriage. His father started work as an office boy in the Chicago office of the manufacturing firm of Landers, Frary & Clark, an important manufacturer of a very wide line of household products, appliances, and cutlery. By the time of George's birth, Arthur Kimball was a successful salesman for the firm, and, around 1910, the family moved to New Britain when Arthur was promoted to sales manager at the company headquarters.

George was the oldest of three children. His sister, Elizabeth, made a career in radio, as host of a program in Hartford, and as director of various little theater groups in the area; his brother, Penn Kimball, was a Rhodes Scholar and a professor on the faculty of the Columbia University School of Journalism.

George attended the New Britain public schools. There is no evidence that he was identified at an early age as exceptionally outstanding, nor did he appear to have a special interest in science. As often happens, a high school chemistry teacher excited his interest in that subject, which directed the course of his future. After completing high school in New Britain, he spent a year at Phillips Exeter Academy in New Hampshire before entering Princeton University in 1924.

At Princeton, George led an apparently normal undergraduate life. He was a good athlete, enjoyed sports, especially swimming, and was a member of the University's water polo team. In one important respect, George's experience at Princeton was unusual and gave an early indication of both his intellectual strength and breadth of interest. He said later that one of the reasons for choosing to major in chemistry was that the chemistry program at Princeton permitted him to study as much mathematics and physics as chemistry.

During George's undergraduate years, Schrödinger, Heisenberg, Dirac, and Born laid the foundations of quantum theory. These luminaries believed that quantum mechanics provided the fundamental laws of chemistry, reducing it to physics. The pioneering work of Walter Heitler and Fritz London on the hydrogen molecule in 1927 indicated the power of quantum calculations in chemistry. Further contributions by Friedrich

Hund, John Slater, and Linus Pauling established quantum chemistry. The new field was a natural for George. By the time he completed his undergraduate work in 1928, George was committed to pursuing quantum chemistry. The path-breaking work of Pauling on the chemical bond was just around the corner.

George clearly did well as an undergraduate; he was offered one of the chemistry department's most valuable graduate fellowships which induced him to stay on at Princeton for his master's degree (1929) and Ph.D. (1932). Among his activities as a graduate student, he organized and taught a private course for the chemistry department faculty to introduce them to the mysteries of quantum mechanics.

George's Ph.D. thesis, on the quantum mechanics of the recombination of hydrogen atoms, was done under the guidance of Henry Eyring. This was the beginning of a long and fruitful collaboration. (Eyring, who had joined the Princeton faculty in 1931, was only 5 years older than George.) George's dissertation, "The recombination of hydrogen atoms" was published in 1932, the first of his many contributions to the scientific literature (Kimball 1932). During the academic year, 1932–1933, he published three other papers co-authored with Eyring, each related to a new problem in the application of quantum mechanics to the analysis of chemical reactions.

George remained at Princeton as an instructor for another year before receiving a National Research Fellowship that permitted him to spend the academic years 1933–1935 at the Massachusetts Institute of Technology (MIT). Although, under the terms of his fellowship, he was technically attached to the Chemistry Department, George worked a great deal with members of MIT's Physics Department—George Shortley, his office mate, William Shockley, and Philip Morse, who had been a graduate student at Princeton when George was an undergraduate. He helped Morse organize and teach a course in theoretical physics for graduate students. Morse (1968, 871) noted, "Indeed, in his knowledge of quantum mechanics and general theoretical physics, he was the equal of most physicists." During his stay at MIT, George's research results led to two published papers, joint with George Shortley, on questions in quantum mechanics, and a solo authored paper, "The electronic structure of diamond" (Kimball 1935). At MIT, George met Alice Hunter, an undergraduate chemistry major, who was a social friend of the younger members of the MIT's chemistry and physics departments. Alice graduated in 1936 and they were married shortly thereafter.

LIFE AT THE MIT EASTMAN LAB

"The group which inhabited the third floor of the Eastman Laboratory sat at the feet of an academic trinity. John Slater (then 33 years of age) was the Old Man, with a long and illustrious career behind him. Philip Morse was the junior member of the trinity .... The third and most spiritual member was Julius Stratton another old man ....

"The great Depression was at its height (my first job after I earned my Ph.D. paid the magnificent salary of $900 a year). As a result the group of graduate students and post-doctoral fellows with whom I worked lived a sort of Vie de Boheme [sic]. The center of this life was the third floor of the Eastman Laboratory, where we shared office space. We spent our evenings as well as our days there, but not always at our work. There was a ping-pong table, and someone discovered that the long, long corridors of M.I.T. made a wonderful place to roller skate.

"Every afternoon we had tea, served by Alice Hunter, student in chemistry, who has since done me the honor of becoming my wife. Those teas became a sort of discussion group, led by Norbert Wiener, who would argue violently on any subject, such as Chinese grammar, or whether or not the number of palindromic primes is infinite" (Kimball as quoted by Morse 1973, 131–132).

COLUMBIA UNIVERSITY

After MIT, George spent a year as a physics instructor at Hunter College before joining the chemistry department of Columbia University in 1937 as an assistant professor. He was promoted to professor of chemistry in 1947 and remained associated with Columbia until 1956 when he joined ADL. At Columbia, George focused on the physics of chemistry, in general, particularly on the quantum mechanics of chemical bonds and reaction rates, and he continued to publish actively on topics in quantum chemistry. The classic, introductory text, *Quantum Chemistry*, by Eyring et al. (1944), had been a work in progress since George's graduate days at Princeton.

George had an early and active interest in scientific computation. Columbia University became an important center of scientific computation right after WWII due in part to its close association with the IBM Corporation and the (Thomas) Watson family. The Watson Scientific Computing Laboratory was established in 1945 and George made use of its facilities for graduate courses and research work. He oversaw the work of Margaret Oakley, a graduate student and Watson Computation Laboratory Fellow, whose thesis (Oakley and Kimball 1949) was a pioneering work in the application of mass data processing equipment to research in theoretical chemistry. (Margaret

Oakley Dayhoff is noted as the founder of the field of bioinformatics.) George and Alice maintained a strong interest in, and friendship with, young graduate students and faculty members. George was known for his effectiveness as a teacher, and his concerns for the graduate examination and grading process; he took special interest in advising graduate students.

George in his Columbia office.

PASS OR FAIL

Isaac Asimov, a student of George's, became a professor of biochemistry at Boston University. Asimov is best known for his works of science fiction and for his popular science books; he was one of the most prolific writers of all time, having written or edited about 500 books. Although Asimov had a near-photographic memory and became a member of Mensa International, he did have trouble with one of George's examinations. He wrote:

"I had a lab course from Kimball in physical chemistry and at one time was asked one question out of a number of possible questions and drew a complete blank. I got a zero. I came to him afterward and said that the question I was asked was the only one of the alternatives I couldn't answer perfectly and that a mark of zero was not a true measure of the state of my knowledge. He said, 'The time will come when you will be asked a question, and it will be the only one of a number of alternatives which you can answer perfectly. You will then get a mark of one hundred and that will not be a true measure of the state of your knowledge either. But you will not complain then, will you?' Very much against my will, I saw the justice of that and subsided. I kept my zero—but I passed the course (Isaac Asimov as quoted by Morse 1973, 135).

George was widely recognized outside Columbia for his scientific contributions. He lectured at Case Western Reserve University in 1943 and again in 1953 in the Frontiers in Chemistry Lectures where scientists of international distinction lecture on major achievements in chemistry. In 1949, he gave the Reilly Lecture at the University of Notre Dame. Reflecting his interest in real data, he lectured to the Northeastern Section of the American Chemical Society in 1954 on

"Experimental design in chemistry." He was elected to the National Academy of Sciences in 1954.

George maintained an active interest in chemistry after he left the Columbia faculty to join ADL; his chemistry research ranged from "A quantum mechanical theory of complex ion formation" (Kimball and Loebl 1959), to his last paper (Kimball 1963) on "Quantum mechanics." At the time of his death on December, 6, 1967, George was attending a meeting of the Visiting Committee, Department of Chemistry, Carnegie-Mellon University. He was then chairman of the Northeastern Section of the American Chemical Society.

THE WAR YEARS AND OR

During WWII, many members of the faculty and administration of MIT served in Washington and elsewhere in support of the U.S. military effort. One of these was Philip M. Morse, a member of the Physics Department, and director of the MIT Navy Underwater Sound Project. In 1942, Morse was asked by the Navy to come to Washington, D.C. to organize a group to assist it in analyzing and developing antisubmarine tactics. The Battle of the Atlantic was raging; German submarines were effectively attacking convoys of men and material essential to the European war effort. New thinking was needed to help the Navy improve its effectiveness in combating the submarine threat.

George was one of the first persons recruited by Morse to the new organization, the Antisubmarine Warfare Operations Research Group (ASWORG). As Morse recounts, George ". . . joined the group within a month. Almost immediately he showed his worth. Together we worked out the basis of the theory of search, and then wrote it up, all in less than two months. . . . Kimball toured naval bases along the east coast, explaining his ideas, working out applications, and learning the practical difficulties" (Morse 1973, 135–136).

George quickly assumed a leadership role as Deputy Director and served as the effective leader of the core cadre in Washington. He was the person to whom ASWORG analysts turned to for help in clarifying a question, working out an approach or developing a solution. In this new and very different operational environment, he demonstrated the qualities of mind that had impressed his colleagues in chemistry and physics—an ability to get to the heart of the issue, a focus on reality, and a capacity for making simple, workable formulations. Based on its early success, ASWORG was renamed the Operations Research Group (ORG) in 1944 and extended its scope of activities to a much broader list of areas than antisubmarine warfare.

After the end of the war, Morse and Kimball remained in Washington to document some of what they had learned. The result was a classified report titled *Methods of Operations Research.* A limited distribution version was declassified in 1949 with an unclassified version published in 1951 (Morse and Kimball 1951). This seminal work has become a classic. George Kimball was awarded the Presidential Citation of Merit for his contributions to the war effort.

After completing *Methods,* George returned to Columbia, but was often called upon to assist governmental groups on OR and related matters. He continued his Navy consultancy with the reorganized ORG, called the Operations Evaluation Group. He worked with the Weapons Systems Evaluation Group, formed in 1949, to assist the Joint Chiefs of Staff and the Secretary of Defense in OR matters. He helped organize the NATO Advisory Panel on Operations Research, served on the U.S. Army Scientific Advisory Panel, and was a member of the Statistical Advisory Panel of the U.S. Census.

As Philip Morse notes, George's view of OR went beyond its military applications: "Even during the war Kimball had become convinced that OR could be effectively applied in industry and the public sector. He was interested in enlarging public awareness of its potentialities and was active in organizing the OR Society of America, which was founded in 1952, with Kimball as a member of the society's first council. By 1964, when he was elected as the society's president, the society had about 5000 members" (Morse 1973, 138).

THE WIDE WORLD OF OR CONSULTANCY

In 1950, when George was invited to become a consultant to the OR group of ADL, the group's work was limited to an initial test assignment for Sears, Roebuck & Co. George's role on this project was relatively limited, but beginning in 1951, as the work of the OR group expanded, George's role and level of commitment expanded as well. He joined ADL full time in 1956 with the title Scientific Advisor. He was elected a vice president of the company in 1961. From his initial role as a consultant until the early 1960s, George devoted his time at ADL almost exclusively to industrial OR. His approach was often characteristically simple and direct.

The first Sears, Roebuck assignment for the ADL OR group was to improve the effectiveness of their system for selecting customers to receive its printed catalogs. Sears had approximately ten million names on its customer file; because of the cost of the famous catalogs, the company characteristically mailed any given edition of a catalog to about half the customer list. Selecting the most responsive customers was critical to Sears success in the catalog business. Sears had studied this issue thoroughly for decades; finding a way to make an improvement, even small, would be a significant contribution and a demonstration of the effectiveness of the OR group's abilities. Sears maintained a test index of several groups of towns, chosen to be comparable in geographic and demographic characteristics, for which detailed records of customer purchases were kept, and which were used to test ideas for catalog distribution or other merchandising ideas. Each group had about 10000 customers. Characteristically, a test would be run by treating one set of towns by the current method and the other by a new or proposed method; after a period of time, the sales results would be tallied to find the result. The records that were maintained permitted detailed analysis of results.

Despite his limited time commitment in 1950, George made two important contributions to the Sears work. The ADL OR group had found some customer behavioral characteristics that indicated a possible approach to an improved selection algorithm. Their analysis had shown that the number of orders a customer had placed in a given period, rather than the total value of purchases, was the best indicator of future activity. George had postulated that, as a general principle, there should be a consistent relationship from one period to another in the value of information in making control forecasts. In other words, if the information from period two has twice the value of information in period one in forecasting period three, then the information from period three should have twice the value of information from period two in forecasting period four; thus, the usefulness of information for making a control forecast decays geometrically with time. This principle was tested and found pragmatically effective in predicting Sears customer purchasing activity; a geometric weighting of the number of orders place in past periods proved to be an effective selection tool for choosing the best customers. A second finding was that a customer's ordering activity would, on average, be doubled if the customer received a major catalog versus not receiving one. These findings gave the basis for an improved customer selection method.

The improvement, although potentially important for the business, appeared small enough to be hard to measure with a significant degree of confidence using the conventional Sears test method. George noted that since both the existing and proposed selection algorithms were presumably effective, most customers would be classified the same way by either; the difference, if any, would be in the relatively small groups treated differently between the two, but the difference between these two relatively small groups should be significant if the proposed algorithm was to be an improvement. George proposed a test design focused on these two groups; all customers in a large sample were classified under both selection algorithms with an equal number chosen by each. As a result, the customers in the sample fell into one of four groups: (1) chosen to receive a catalog by both methods, (2) rejected by both, (3) chosen by the existing method, rejected by the new one, (4) chosen by the new method, rejected by the existing. If the new method was superior, then the customers in group (4) must be significantly superior to those in group (3). Half the customers in each group (3) and (4) were chosen at random to receive the catalog. The resulting sales showed the clear superiority of group (4) customers, chosen by the proposed selection system, over those in group (3), chosen by the current one.

When Sears management saw the results and were assured by their staff that there were no downside risks, Sears ordered the immediate implementation of the new method throughout the company. George's test design had proven effective.

George contributed valuable insights to the evolving, early operational aspects of inventory control. In an ADL project to develop an improved control system for the multistage product distribution process of the General Electric Lamp Division, he noticed that the signal of ultimate demand that came back through the system in the form of reorders became increasingly lumpy as it moved through the system, with the orders becoming fewer, larger, and less predicable, as they went from retail stores through field warehouses and ultimately to manufacturing. He proposed that end demand should be transmitted simultaneously to all levels in the system; each level could then respond in an efficient way to maintain stocks. The inventory to protect against demand fluctuation could thus be kept at a constant minimum, the base stock level.

In work for the telephone system on the issue of the appropriate rate of expansion of outside plant, George noted that the available but unused outside plant was simply an inventory to accommodate new demand and efficient installation: the engineering and construction of outside plant, the transmission cables and junction boxes by which customer phones were connected to central office switch gear, could be governed by the principles of inventory management. These are just a few examples of George's insistence on dealing with reality rather than theory, his ability to get to the core of the question, and his facility in finding a workable solution.

In 1963, George became ADL's full-time scientific director of the ADL Trident Project, an antisubmarine systems analysis project for the U.S. Navy. During that period of the Cold War, submarine activity in the Atlantic Ocean, both the U.S. and Soviet, was intense and included missile-carrying vessels and attack submarines. The purpose of the Trident Project was to develop and test the feasibility of an ocean surveillance system that could locate and track submarines throughout a broad region of the Atlantic Ocean. George's OR capabilities, his knowledge of physics, his understanding of the Navy, and his enormously high regard within the Navy community, made him ideal for this role. An ADL colleague, Robert Neal, recalls that ". . .it was his practice from time to time to convene the entire staff for lectures or tutorials on probability and statistics. He's a consummate teacher, and I thoroughly enjoyed these sessions (learned quite a bit, too). Also, I believe he was happiest when he was doing this" (Neal 2007). George was actively committed to the Trident Project at the time of his death.

George served as ADL's thought leader and mentor for all of its OR work. He did not take on formal management responsibility for client assignments. His contributions were intellectual rather than managerial; in this role, he demonstrated most of the same characteristics that were familiar to his colleagues in chemistry and physics.

GEORGE ON OR

George had a clear view of the nature of OR. *Methods of Operations Research* begins by citing the definition of OR as "a scientific method of providing executive departments with a quantitative basis for decisions regarding the operations under their control" (Morse and Kimball 1951, 1). His views on OR did not stray from this concept throughout his life. He was impatient with the tendency of some to equate OR with mathematical techniques, especially optimization methods. In his address to the Philosophical Society of Washington, George stated:

> In discussions of operations research we hear very frequent reference to such things as linear programming, dynamic programming, allocation theory, and the like. There are those who regard these subjects as the real essence of operations research. But to me calling these subjects operations research is the same as identifying physics with the differential equations of physics. Now mathematics is indispensable to physics . . . but in no sense is physics a branch of mathematics . . . today we are finding the needs of operations research stimulating research in new branches of mathematics, and operations research workers making use of previously unapplied mathematical developments. This symbiosis is of great importance to both parties, but the distinction between mathematics, with its right to carry any set of assumptions to their logical conclusions, and science, with its obligation to check its assumptions against nature, must still be maintained. (Kimball 1958, 123)

George expressed similar precautions concerning Monte Carlo simulations: "Its great advantage is that it makes possible the simulation of the behavior of almost any model . . . .. In spite of this great usefulness, it must be remembered that the Monte Carlo method is strictly an analytical device. The results obtained from it are no better than the model which it simulates, and it is always a model—not the actual system—that is simulated" (Kimball 1957, 204).

EARLY MONTE CARLO SIMULATION

In is paper, "Some industrial applications of military operations research methods," George noted: "One mathematical device is of much importance that it cannot be allowed to go unmentioned. That is the Monte Carlo method of simulation. It originated independently in a number of military problems, and is now a most important industrial tool" (Kimball 1957, 204), In 1943, before the present digital computers had been developed, and before the method was given its name by Nicholas Metropolis, George "set up and ran a Monte Carlo experiment on an IBM sorter. What he did was to draw the silhouette of a submarine on a punch card, punch out the inside of the submarine, take another card and move the silhouette according to a random draw from a two-dimensional normal distribution, do it again, and so on, until he had a respectable deck of randomized submarines. Then, using the sorter, he dropped a number of different depth-charge patterns onto the cloud of submarines and counted the hits. He was apologetic whenever he had to talk about it, saying that anyone would have thought of it. Nevertheless he was the one who did think of it, and do it, and it resulted in valuable conclusions when they were needed" (Arthur Brown as quoted by Morse 1973, 136–137).

George did not believe that the goal of OR was to optimize operating performance. In his Philosophcal Society address he commented, "Because many of the mathematical techniques used in OR are methods of finding maxima and minima, there has arisen a temptation to claim that OR is the study of the best way to control an operation. If real operations were as simple as some of the mathematical models used to describe them, this might be possible." Among the difficulties he noted is having an unambiguous quantitative measure to be maximized. He argued further that a service was performed by identifying and comparing feasible alternatives. Finally, he remarked, "In my experience when a moderately good solution to a problem has been found, it is seldom worthwhile to spend much time trying to convert this into the 'best' solution. The time is much better spent in trying to find those variables which have been overlooked, for when these are found, much greater improvements become possible than by tinkering with the well-known parameters" (Kimball 1958, 123).

George maintained his strong interest in computing and information technology after joining ADL. He built operating simulations, worked on computer games, and investigated self-learning systems. Together with Vincent Giuliano and Paul Jones of the ADL staff, he developed and patented a unique general character recognition technique that was independent of the font type used. Another one of his insights was to note that, in contrast with scientific computation, business data processing was a matter of file processing, and that the magnitude of the data-processing task was measured by the business file size and sorting technology used, in accordance with the Shannon-Weiner law of information entropy. This turned out to be a helpful basis for making a preliminary assessment of data-processing system feasibility.

The breadth of Kimball's knowledge, the effectiveness of his mind, and his openness put him in constant demand as a mentor or source of advice and help. Ronald Howard (2007) writes that George "...had the ability to construct a simple representation that would get immediately to the heart of the problem. ... George Kimball, who as far as I know did not have a formal appointment at M.I.T., served as my 'de facto' advisor, with whom I had frequent very productive conversations on my research." Stephen Pollock (2007) noted "...although I didn't exactly work for or with him, George always had his door open for me and I had many long and fascinating sessions with him. I had no idea at that time that George was a vice president at Arthur D. Little. But I did know that he was a fantastic intellect, and willing to spend time with a fresh and somewhat naïve PhD." The examples of Kimball's ability to listen quietly and then to succinctly bring focus to the issue and define the question are legion.

NAPPING(?)

Any friend or colleague of George Kimball would find a characterization of him incomplete without reference to his habit of apparently taking naps. These could occur on any occasion, whether in a meeting with a client, sitting in a seminar, or talking with a staff member or colleague. In the midst of the discussion, George would put his head back, close his eyes and appear to drift off to sleep. At times, the sound of gentle snoring might be heard. This could be disconcerting, especially if one were a young professional or graduate student hoping to impress George with the results of his latest work.

Time and again, however, George would rouse himself and astonish others present with his grasp of the discussion. For example, Ron Howard reports on an important client meeting during his time at ADL. "As the morning progressed, the client explained his problem and why it was puzzling. Toward the end of the morning, we heard someone snoring, It was George. Given his stature, no one was about to wake him up, and so we just continued the discussion without him. After a half hour or so, when we had forgotten that George was napping, he started to speak, and immediately presented a simple model of the problem that was a major advance in our thinking. After that, we would tell each other that George could do better work in his sleep than we could do awake."

On another occasion, a client was holding forth in a rather desultory fashion when he stopped to chide a junior staff member for staring out the window, not paying attention. When the young person unwisely objected that the client said nothing when George dozed off, the client replied, "George Kimball asleep is smarter than anyone else I know wide awake" (Howard 2007).

THE KIMBALLS: FAMILY, FRIENDS, COMMUNITY

When George accepted his teaching position at Columbia, the Kimballs settled in an established neighborhood in Leonia, New Jersey, across the George Washington Bridge from Manhattan and the Columbia campus. There they raised four children: Prudence, Thomas, Susanna, and Martha. George played the role of father, academic mentor, and occasional family cook—he was reported to be the gravy cook of the family, arguing that it took a chemist to make good gravy. There was a blackboard in the kitchen regularly used for "discussing calculus or chemistry or physics with one or another of the children in turn" (Lathrop 1968, 869). George's daughter, Susanna recalls "As the only nonscientifically inclined child in the family, I had a very different relationship with my father from that of my siblings. When I had a question about something that was in the area of science or math Dad could always simplify everything so that I could easily understand. One of my fondest memories was driving with father from Connecticut to Massachusetts taking the back roads instead of the turnpike. He took me to some quiet little valleys with quaint cottages and colorful gardens. When I think of Dad today I remember that picture. . . . When Dad taught me to drive, he just braced himself between the seat and the dash and sat there calmly until the end of the block, at which time he would have me stop and proceed to tell me everything I could have done better. He was the most patient man alive" (Monette 2007).

George and Alice led an active community life. They regularly played bridge with a group of friends, with George reportedly the best player among the group (Morse 1973, 135). He served as a trustee of the Hackensack Unitarian Church and as president for one term.

In 1956, when George left Columbia to join ADL, the Kimball family moved to Winchester, Massachusetts, They lived in a substantial home in an older attractive neighborhood overlooking Mystic Lake. George "spent many years as officer, committeeman, or consultant for church, Boy Scout, or community groups" (Lathrop 1968, 869). He served as a consultant to the Winchester School Committee, contributing an analysis of the characteristics and growth of the local population as a basis for estimating the ranges of time within which various school facilities would reach capacity. He and Alice continued their active association with the Unitarian congregation in Winchester.

Alice and George (October 1967)

George Kimball's style was simple. The Kimball homes in Leonia and Winchester were comfortable but unpretentious. He drove to his office in Cambridge in a vintage Ford Mustang. He often brought his lunch from home in a simple child's lunch box. His office was plain but utilitarian. His style was well matched to the qualities of his mind and personality that made him so much admired by his colleagues.

The Kimball family spent several summers at their vacation home by Hartland Pond in West Hartland, Connecticut, built in 1930 by George's parents. Alice moved there from Winchester to make it her permanent residence after George's death.

George fully supported Alice's community, church, and association activities. As early as 1945, she belonged to the Alliance of Unitarian Women and later served as its national president. In Winchester, she was active in the local congregation, including service on a number of committees. She was appointed an initial member of the Commission on Appraisal of the Unitarian–Universalist Association, a powerful group established to investigate any function or activity of the association, and served as secretary of the commission from its organization in 1962 until 1967. She continued her activities at the national level for many years after George's death. Alice maintained a close association with MIT, including service as president of the Association of MIT Alumni from 1970 to 1972. She was recognized in 1984 by the Bronze Beaver Award from the MIT Alumni Association given for distinguished service, the highest honor the Association can bestow upon any of its members. She died on May 18, 2008, at age 93.

George Kimball was a pleasant companion. He had wide ranging interests beside science—history, languages, music—and he liked people. In conversation, he tended to be quiet, interested in others' views and ready to comment when he felt he had something to contribute. Philip Morse commented on George's famous silences. "They were never unfriendly, those silences. One always felt his pleasure in the other's presence, his real interest in whatever others wished to talk about, and his willingness to say something whenever he judged he could make a contribution. They were modest, good-humored silences, never self-centered; they cemented friendship and promoted true communication" (Morse 1968, 872). George Kimball enjoyed good food taken with friends. Robert Neal recalled "our trips to Washington, where we always stayed at a hotel within easy walking distance of Arbaugh's, 'The Home of Barbecued Spare Ribs', on Connecticut Avenue. George was a trencherman and he loved this restaurant" (Neal 2007). In 1960, I had the opportunity to travel to France with George, Philip Morse, and Martin Ernst (ADL vice president for management sciences) to the second international conference of the International Federation of Operations Research Societies held in Aix-en-Provence. We rented a car at Orly Airport and traveled leisurely south across the French countryside, enjoying frequent stops for a good meal and a view of the country. A highlight of our trip was a break taken away from the conference when we four drove to Les Baux-de-Provence to visit the site of the castle of the Counts of Les Baux that dominates the plain for miles below, and then to lunch at the famous L'Oustau de Baumanière on soufflé de homard accompanied by several bottles of white Chateauneuf-du-Pape. The drive back to Aix remains something of a blur.

> A SHORT CONVERSATION
>
> George Kimball was a kind person, patient with honest ignorance and always willing to help a person learn. He did not, however, care to put up with arrogance or condescension. On one occasion at an ADL luncheon with some visitors, one of the visitors asked in a rather demeaning way, "Dr. Kimball, what exactly is the difference, if any, between physical chemistry and chemical physics?"
>
> Kimball was silent for a brief moment before replying, "*h*." That was the end of that topic of conversation.
>
> [I presume Kimball was referring to Planck's constant, symbolized by *h*, but I never quite dared to confirm it (John Magee)].

LEGACY

George Kimball was an intellectual giant and great source of strength to the discipline of OR. His legacy is not so much in a voluminous written record as in the leadership and inspiration he provided to students and colleagues. His friend and colleague of several decades, Philip Morse (1973, 129) wrote, "Perhaps his greatest contribution was the education and inspiration he gave to many younger men, now working in various fields of science and technology."

The Operations Research Society of America, on the initiative of Arthur D. Little, Inc., established the George E. Kimball Medal as one way to sustain his memory. The medal is awarded each year to an individual recognized for distinguished service to the society and the profession.

REFERENCES

Eyring H, Walter J, Kimball G (1944) Quantum chemistry. Wiley, New York, NY

Howard R (2007) Personal communication

Kimball GE (1932) The recombination of hydrogen atoms. J Am Chem Soc 54(6):2396–2398

Kimball GE (1935) The electronic structure of diamond. J Chem Phys 3(9):560–564

Kimball GE (1957) Some industrial applications of military operations research methods. Oper Res 5(2):201–204

Kimball GE (1958) A critique of operations research. J Wash Acad Sci 48(2):33–37

Kimball GE (1963) Quantum mechanics. Proceedings of the conference on desalination research. Natl Res Publ 942:173–177

Kimball GE, Loebl EM (1959) A quantum mechanical theory of complex ion formation. J Chem Educ 36(5):233–236

Lathrop J (1968) George Kimball, the man. Oper Res 16(4):868–871

Monette S (2007) Personal communication

Morse PM (1968) George E. Kimball. Oper Res 16(4):871–874

Morse PM (1973) George Elbert Kimball. Biographical memoirs. National Academy of Science, Washington, DC

Morse PM, Kimball GE (1946) Methods of operations research (classified), Operations Evaluation Group, Office of the Chief of Naval Operations, Washington, DC. Declassified version at http://www.cna.org/documents/1100005400.pdf. Accessed 4 Apr 2008

Morse PM, Kimball GE (1951) Methods of operations research. Wiley, New York, NY; also, Dover Publications, 2003

Neal R (2007) Personal communication

Oakley M, Kimball GE (1949) Punched card calculation of resonance energies. J Chem Phys 17(8):706–717

Pollock S (2007) Personal communication

Weyl H (1950) Elementary theory of convex polyhedra. In: Kuhn HW, Tucker AW (eds) Contribution to the theory of games, vol. I, Princeton Ann Math No. 24, pp 3–18; (Translation of Weyl, H. 1935. Elementare theorie der konvexen polyeder, Commentarii Mathematici Helvetici 7, 290–306.)

9

Ellis A. Johnson

Saul I. Gass

When World WarII (WWII) began, the U.S. was able to call on scientists from many disciplines to apply their expertise to a wide range of military combat problems. A key contributor was the physicist Ellis A. Johnson. In the early days of the war, he and his Naval Ordnance Laboratory (NOL) embryonic operations research (OR) staff made seminal contributions to the design and utilization of mines, including degaussing procedures and mine-laying patterns. Later, as a naval officer, Ellis coordinated the Army Air Force's major mine-laying operation of the Inland Sea between Japan and the Asian mainland—Operation Starvation—a spectacularly successful campaign that was conceived, planned, and directed by the use of OR techniques. For his wartime contributions, Ellis receiving the Navy's Distinguished Civilian Citation and the Army's Distinguished Civilian Service Medal, and Legion of Merit awards from both the Navy and the Air Force.

After WWII, Ellis was appointed director of the newly formed U.S. Army's Operations Research Office (ORO). Starting from ground zero, he planned, organized, staffed, and managed the premier Army OR organization in its carrying out of important studies, especially during the Korean War. For the Army, the ORO showed how OR could be used to address critical military decisions, helped to spread the use of OR methods in all areas of operations, and furthered the establishment and growth of OR-related organizations.

A.A. Assad, S.I. Gass (eds.), *Profiles in Operations Research*, International Series in Operations Research & Management Science 147, DOI 10.1007/978-1-4419-6281-2_9,

PRELUDE TO OR

Ellis A. Johnson was born in Quincy, Massachusetts, on September 2, 1906, the son of Peter and Elizabeth (Telko) Johnson. He attended the Massachusetts Institute of Technology (MIT) and received a B.S. in 1928 and an M.S. in 1929. From 1929 to 1934, he served as an MIT research assistant and instructor in electrical engineering and physical chemistry, and worked on his dissertation topic, "A primary standard for measurement of the earth's magnetic vector" (his advisor was the physicist Ralph Bennett). In 1934, he was given permission by the Electrical Engineering Department to finish his dissertation in absentia and, by 1940, had completed all the theoretical and nearly all of the experimental work. Governmental assignments and his U.S. Navy duty in WWII delayed the submission of the completed dissertation until 1947, upon which he received his Ph.D. He married Alice Legasse on August 4, 1934, and they had a son, Peter, and a daughter, Betsy (Shrader 2006).

Ellis left MIT in 1934 for Washington, D.C. where he was employed as an associate electrical engineer for the U.S. Coast and Geodetic Survey. He then moved to the Department of Terrestrial Magnetism of the Carnegie Institution of Washington. It was here, "by applying the technology of magnetism to problems related to the earth's magnetic field, he measured the magnetization of sedimentary rocks and observed long-term changes in their magnetic fields related to continental drift." This work led to his consulting with NOL in 1940, where he developed and planned operations with underwater mines and countermeasures, designed and produced mines, and helped to blockade Japan "with a virtually unsweepable mix of various types of mines" (Page et al. 1974, 1141). He was Associate Director of Research for NOL from 1940 to 1942 where, in March 1942, he established the precursor OR group in the U.S., the Mine Warfare Operations Research Group (MWORG) (Shortley 1967, 5; Shrader 2006, 19; Visco 2004, 67).

Ellis's expertise in magnetism and mining, and the recognition that any future war in Europe or Japan would put both the American military and merchant fleets in danger from enemy mines, caused him to organize, within NOL, a staff of scientists whose mission was to investigate the use of mines, both offensively and defensively. (One of the staff members was his MIT advisor Ralph Bennett who was on duty at NOL as a naval officer; he later became scientific head of NOL). These efforts also made NOL the main U.S. organization for developing degaussing methods with operational magnetic ranges established in the Chesapeake Bay for magnetic measuring and degaussing of the Atlantic fleet, and a counterpart range at Pearl Harbor for the Pacific fleet. Although not at war at this time, Ellis encouraged the scientific staff to study both the operational and technical sides of wartime mining operations by applying war gaming methods "to compute outcomes from which preferred offensive-mine characteristics and defensive operating procedures could be inferred" (Page et al. 1974, 1143). These early OR-based activities, however, were extra-curricular and carried out over weekends. A gaming exercise was conducted at NOL on Saturday, December 6, 1941, that dealt with enemy aircraft dropping mines on Pearl Harbor, a strategy that demonstrated that such an attack would bottle up the Pacific fleet for days—the next day's Japanese Pearl Harbor attack was mainly with aerial torpedoes. Coincidentally, that weekend found Ellis at Pearl Harbor working on its degaussing range. His experiences during the attack were described by Page et al. (1974, 1144):

> It was while Ellis was at Pearl Harbor, working on problems in conjunction with the range there, that the Japanese attack came early on Sunday morning of December 7, 1941. Ellis stood on "10–10" dock and watched the waves of Japanese planes dropping their bombs on the vital Pacific naval base. He saw the *Castle* and the *Downs* blow up and the *Arizona* capsize, its men leaping from its decks into the flaming oil of the harbor. He himself boarded a minesweeper to help sweep the harbor of mines so that any ships left afloat could get out and into the comparative safety of open water.

Upon his return to NOL in Washington, Ellis mobilized MWORG to aid the fleet in technical mining matters, to develop new types of mine-firing devices and mine countermeasures, and to explore the use of analytical techniques for minefield design (Shortley 1967). He felt, however, that his personal contribution would be greater if he were allowed to bring his expertise into the field. Thus, he accepted a commission as a Lieutenant

Commander in the Naval Reserve and he returned to Pearl Harbor as a mining operations officer for the Pacific Fleet.

He and his wartime Pacific Fleet staff, in particular, Thornton Page and William A. Wallace, both of whom were from his NOL unit and were now Naval officers, spent the rest of the war years developing and overseeing tactical deployments of new types of mines, including aircraft-laid mines. They also developed plans for a strategic mining blockade of the Japanese home islands that called for the laying of the mines by Army Air Corps B-29 long-range bombers, instead of carrier-based aircraft. When this was agreed to by the Navy and Army (with the aid of Admiral Chester Nimitz, Commander in Chief, Pacific Fleet), Ellis was assigned to the staff of Army General Curtis LeMay's 21st Bomber Command on Guam as Director of Mining in charge of planning Operation Starvation, the B-29 mission to mine the waters of the Inland Sea:

> One of the first operations-research studies performed by Johnson's group at Tinian [from which the B-29s could reach the Japanese mainland] concerned basic tactics of aerial delivery. The bombing of Japan was being conducted by B-29s flying in formation at high altitudes in daylight. Attrition [loss of aircraft by all causes]was running 10 percent. Study showed that a much more favorable tactic would be for the planes to fly singly at much lower altitude and at night, using radar sights for target location. When this tactic was employed, attrition dropped to about 1 percent, and the tactic was adopted for most bombing raids as well as most mining raids (Shortley 1967, 7–8).

It was the approach used in the blockade of the Japanese home islands by B-29s mining the Inland Sea:

> The first minelaying operation, directed against Shimonoseki Straits, was carried out in force on the night of March 27, 1945, and operations were continued steadily until the end of the war.
>
> In a period of less than five months, the B-29s of the 313th Wing laid over 12,000 mines in the Shimonoseki Straits and the channels leading to all the major ports of Japan and Korea. These U.S. mines were virtually unsweepable; some were set off by magnetic fields, others fired on a combination of magnetic and acoustic signals, still others were activated by pressure changes caused by a

ship's passing overhead, and all were equipped with "ship counters" to complicate mine-sweeping operations. Evidence obtained by reconnaissance aircraft and other intelligence sources, corroborated by interrogating Japanese port officials during the U.S. occupation, showed that Japanese shipping was brought to a standstill in mid-1945 by the mine blockade. Many felt that this blockade, with its near total elimination of seaborne freight, had already won the Pacific War before the atomic bombs were dropped on Hiroshima and Nagasaki (Page et al. 1974, 1147).

As George Shortley (a physicist and member of the first contingent of MWORG sent to the Pacific) wrote:

> This campaign [Operation Starvation], which was the product of operations research studies from conception to execution, constitutes, in my opinion, the most complete single example of the successful application of military operations-research techniques during the War. In this example, operations research on the mine-firing characteristics was fed back into mine development and forward into minefield design. Intelligence information was fed back through operations-research channels into mine modifications and forward into strategic and tactical planning.
>
> Although the whole operation constituted less than 6 per cent of the effort of the 21st Bomber Command, whose principal mission was the *bombing* of the Japanese homeland, Prince [Fumimaro] Konyoe [former Premier and Foreign Minister of Japan] stated that the effort that was devoted to mining the home island had an effect on the Japanese economy that was comparable with the effect of the entire bombing effort (Shortley 1967, 8–9).

After the war, Ellis co-authored with David Katcher a report on the wartime work of his group (originally prepared in classified form in 1947); it was eventually published in an abbreviated and unclassified book, *Mines Against Japan* (Johnson and Katcher 1973). Its last chapter summarizes Ellis's "view on operations research as applied to weapons design; he emphasizes communications within large organizations, and particularly the transmitting of new ideas, their evaluation, and rapid action on them by decision-making executives" (Page et al. 1974, 1147–1148). Ellis also served as the technical director of the Office of Atomic Energy of the Air Force from January to August 1948 where he established the intelligence system which monitored the Soviet atomic tests and stockpiles (Visco 2004).

LESSONS LEARNED

"There is a happy theory current that science won the war, it can do it again, and all that is needed is a sufficiency of money, men, and equipment. This may produce weapons, but it will not make certain that the weapons are useful or available when needed. Science helped win World War II, but there may not be time again to even help if we go about military research and development in the slapdash way we have done heretofore. If our political and diplomatic strategy fails its mission, and we are once again drawn into war, the winning of it will have its roots in what we do now, and science has a crucial part to play.

"Yet, by early 1947, we are indulging in wishful thinking once again. Each profession, undaunted by the record, feels that it alone has the intelligence, the ability, and the know-how to do the job. The partnership which had its intelligent beginnings during the war years is dissolving and becoming bankrupt. This volume is a warning of the dangers of permitting such a dissolution, and a history of its previewed effects. Planned research and development for weapons holds one of the most important keys to national military security" (Johnson and Katcher 1973, 313).

Post-WWII Army OR: The Operations Research Office

In his comprehensive history of OR in the U.S. Army, Shrader (2006, 3–6) describes military antecedents of the use of science and technology from ancient to early modern times, ranging from Archimedes's catapults to Thomas Edison's study of U-boat tactics in World War I. The military origins of OR started in the mid-1930s with British field studies of the deployment and utilization of radar by a combined interdisciplinary team of civilian scientists and Royal Air Force (RAF) personnel. During WWII, the RAF continued the use of such teams in its development of aircraft bombing strategies; the team concept was adopted by the British Naval Command where new tactics were designed, tested, and employed for convoy formation and the search for U-boats.

In 1942, the U.S., motivated by the British experiences, organized interdisciplinary OR groups for the Navy, the Anti-Submarine Warfare Operations Research Group (ASWORG), and for the Eighth Bomber Command (then part of the Army Air Corps) that had been deployed to England—by the end of the war, there were more than 25 such groups. But where was OR with respect to the U.S. Army's ground forces in WWII?

Shrader (2006, 36–40) details the Army's rather weak ability to replicate the successful experiences of the British and the U.S. Navy and

Air Force in the use of civilian scientists in the field (the British also had several OR sections assigned to its ground forces). Much of this void was not due to lack of those familiar with such activities—recommendations were made—but rather due to a lack of follow through by the top levels of command. Two Army operations analysis sections were established in the Southwest Pacific Area, and a third in Hawaii; they were deactivated soon after the war ended. The situation at the end of the war was summed up by Shrader (2006, 43):

> Although Army service and ground forces lagged behind the Navy and Army Air Forces in the integration of OR into the decision-making process during World War II, even the limited exposure of Army civilian leaders, commanders, and staff officers had an effect. In the postwar period, OR would become an integral part of the Army decision-making process, not only for the design and improvement of weapons and other military equipment but for the development of tactical doctrine and strategic planning as well. Having lagged behind the Navy and Army Air Forces in the adoption and use of OR in World War II, the ground Army quickly closed the gap in the postwar period.

In 1948, the Army recognized the need to have its own dedicated OR organization capable of employing OR methods to the study of the Army's total range of problems: strategy, tactics, organization. To this end, an independent, nonprofit organization, first called the General Research Office (GRO), was established in August 8, 1948, to be managed under contract with Johns Hopkins University (JHU). Ellis Johnson was appointed its first director. GRO's research program was stated in a nebulous manner as consisting of "Operations research and/or analysis on problems that are not unique to any one Army agency [and] basic research of a nonmaterial nature for which primary cognizance has not been assigned to a specific Army agency" Shrader (2006, 66). GRO was renamed the Operations Research Office (ORO) on December 27, 1948, and given a more direct mission statement (Shrader 2006, 66): "The mission of the Operations Research Office is to apply scientific, qualitative, and quantitative analysis to the study of warfare with the objective of improving the strategy, tactics, logistics, weapons, and weapons systems of the future." ORO was first headquartered at Ft. McNair in Washington, D.C., but, in June 1951, moved to Chevy Chase, Maryland, having rented a former junior college facility then owned by the National 4-H Club.

Ellis had to organize, structure, and build ORO from the ground up. Unlike the post-WWII OR groups for the Navy's Operations Research Group and the Air Force's Operations Analysis Division, there was no Army wartime organization with a cadre of OR analysts to subsume. It is unclear whether Ellis was familiar with Patrick Blackett's OR concept of utilizing scientists and others from diverse disciplines, but Ellis's instincts did cause him to recruit analysts from a wide range of disciplines—physics, mathematics, economics, political science, history, chemistry, meteorology—few of whom had WWII field experience or a background in military matters. The Harvard trained political scientist, George S. Pettee (Congressional research analyst, educator, and author who wrote extensively on national security issues) was hired in April 1949 and appointed deputy director of ORO in the following year. By the summer of 1950, the ORO staff had grown to 40 analysts, well below the projected 70, but ORO had arrangements with 100 consultants (including Henry Kissinger in 1951) and a number of contractors such as Stanford Research Institute and Battelle Memorial Institute (Shrader 2006).

Most important, however, Ellis had to define and develop ORO's research program. The Army was focused on "quick, easily accomplished studies that would produce concrete solutions to immediate problems," while "Ellis Johnson and key members of his staff . . . were forward looking and preferred to take on studies of emerging problems, particularly those outside traditional weapons and tactical analysis topics." The initial scope of projects proved to be "quite broad and included an emphasis on matters other than military weapons and equipment" (Shrader 2006, 71). Because the Army had hardly any experience in using OR personnel and applying their talents to military problems, Ellis and his staff had to "orient ORO customers [the Army staff] on what OR could and could not accomplish. The problem was resolved, albeit slowly and haltingly, by on-the-job training of ORO analysts, the visits of ORO personnel to Army installations, and the sharing within ORO of experience by those ORO personnel who did have military experience" Shrader (2006, 67).

By the end of August 1948, ORO had two major projects assigned to it: (1) Project ANALAA dealt with the analysis of antiaircraft weapons and systems and (2) Project EVANAL with the task of determining a means for analyzing the performance of Army equipment under various environmental conditions, particularly arctic ones, and to determine the feasibility of using business machines for that purpose. A third 1948 project was Project MAID, directed by Ellis, that investigated the potential value of U.S. military aid programs for foreign countries. "Project MAID had a significant influence on the Army's support for passage of the Mutual Defense Act of 1949, which established the military Assistance Program, a key element of America's Cold War strategic policy" (Shrader 2006, 72).

THE IMPORTANT CHARACTERISTICS OF OR

"Research on the operations of the whole organization;

Optimization of operations in a manner that brings about greater assurance of both short and long range health for the organization;

Application of the newest scientific methods and techniques;

Synthesis and extension of the methods and techniques of the older management sciences;

Development and use of analytical models in the manner common to the basic sciences;

Design and use of experimental operations that give an insight into the behavior of actual operations;

Use of integrated and creative multidisciplinary team research to solve complex operational problems" (Johnson 1954, xiv).

WHAT OR OFFERS TO MANAGEMENT

"Operations research is the prediction and comparison of the values, effectiveness, and costs of a set of proposed alternative courses of action involving man-machine systems.

To do this, it uses a model of the action that has been developed analytically by a logical and, when feasible, a mathematical methodology.

The values of the basic action parameters are derived from historical analysis of past actions or from designed operational experiments.

Most importantly, because all human and machine factors are meant to be included, an estimate of the uncertainty in the predicted outcome and in the values, effectiveness, and costs of the proposed action is provided" (Johnson 1954, xxiii–xxiv).

Shrader (2006, 73) further notes: "The nature and scope of the projects undertaken by ORO as well as the office's organization, staffing, and overall philosophy were profoundly influenced by one man, Dr. Ellis A. Johnson. ORO, and with it the bulk of the Army's postwar OR program, was largely Johnson's creation." In early 1950, Ellis suggested to the Army that a rather full range of continuing studies be established at ORO covering nuclear weapons, ground tactics, logistics, military costing, psychological warfare, guerilla warfare, and air defense. The Army agreed and a set of 15 projects was authorized and funded. But, with the outbreak of the Korean War (June 25, 1950), ORO, under Ellis's leadership, became deeply involved in a hot war.

THE KOREAN WAR, ARMY DESEGREGATION, AND POST-ORO

Immediately after the start of the Korean war, Ellis proposed that ORO analysts be assigned to support the combined U.S. and United Nations (UN) Army in the field. In this way, ORO could "contribute directly to the solution of problems encountered by forces in the field, to obtain actual field data for ongoing projects, and to promote better understanding by the Army of the capabilities of OR" (Shrader 2006, 86). Authorization by the Army was quickly approved; Ellis and a team of four analysts arrived in Korea in early September, just in time to accompany the UN forces north from the Pusan perimeter to meet those landing at Inchon (Page et al. 1974, 1150). By the end of 1950, ORO had established eight teams in Korea, the bulk of ORO's technical personnel. The ORO's Korean War studies covered a full range of old and new military areas: infantry weapons and tactics, tactical use of atomic bombs, close air support of ground forces, airborne operations, and psychological warfare operations. By the Korean ceasefire in July 1953, more than 50% of the ORO professional staff served in the combat zone (in uniform as operations analysts), many of whom, along with Ellis, received the UN Service Medal (Shrader 2006).

Although President Truman had issued an executive order in 1948 prohibiting discrimination in the armed forces, and the Army Reorganization Act of 1950 removed the requirements of segregated units, the units fielded in Korea were basically segregated. Due to a scarcity of white replacements, the Army began to assign black soldiers to previously all-white units. "Based on a request from the Army chief of staff General J. Lawton Collins in March 1951, ORO was tasked to study the question of the utilization of Negro manpower in the Army" (Shrader 2006, 101). The project, headed by Alfred Hausrath (a war games researcher) concluded (Hausrath 1954, 29):

> ... this study provided policy-makers in the U.S. Army with objective arguments in favor of integrated units of Negro and white soldiers. These were: first, that integrated units allow more effective use of the manpower available through a more even distribution of aptitudes than in segregated

> units; second, that performance of integrated units is satisfactory; and, third, that resistance to integration is greatly reduced as experience is gained.

This study "supported the 1952 decision by the Army in favor of integration, a move that was a major precursor of the movement to integrate blacks into all areas of American life" (Page et al. 1974, 1152).

For over a decade, ORO was the principal organization responsible for the Army's OR research and study program. Under the Ellis's direction, the ORO staff produced hundreds of reports and technical memoranda that helped shape Army tactics and policy—psychological and guerilla warfare, nuclear weapons in support of ground warfare, air-mobile army force, army logistics. In addition, ORO was in the forefront of applying computer-based techniques such as war gaming to military problems (Page et al. 1974, 1152). It is estimated that through the first 10 years of ORO, about 80% of its study recommendations were accepted by the Army (Whitson, 1960).

Not everything went well between Ellis, ORO, and the Army. Over time, Ellis strived to broaden the scope of ORO into non-Army and nongovernmental areas. "Accordingly, he wanted to continue the trend toward involvement of ORO in the study of such topics as nation building, social policy, and the solution of a variety of problems somewhat remote from the immediate, practical interests of his Army paymasters" (Shrader 2006, 121). Ellis was chair of a committee established by the Operations Research Society of America (ORSA) "to investigate opportunities to organize or sponsor studies of national and international scope" and pursued such activities with the United Nations Secretariat (Johnson 1961, 743). In a talk to the first annual conference of the Canadian Operational Research Society (May 1959), Ellis stated that the three great OR frontiers would be in regional and world development, the field of medicine, and the field of charity (Johnson 1960). Although his crystal ball was not too cloudy, Ellis's attempts to move ORO into nonmilitary areas caused the Army to lose confidence in his leadership and the direction of ORO (Shrader 2006, 86–87). Thus, when it was time to renew the contract between the Army and JHU, the Army insisted that Ellis be replaced. Milton Eisenhower, president of JHU, very reluctantly, tried to convince Ellis to step down voluntarily. Faced with the Army's demand, Eisenhower decided not to renew the JHU contract with the Army. Ellis did resign in July of 1961, with ORO going out of existence at "2400 hours, 31 August 1961" (Shrader 2006, 122). Its physical facilities, most of its technical and administrative staffs, and contractual obligations were taken over by the newly formed nonprofit organization, the Research Analysis Corporation (RAC). Ellis's influence on OR, and specifically his impact on Army OR was summed up by Shrader (2006, 123):

> Under Johnson's direction ORO had demonstrated the applicability of OR techniques to analysis of the political, economic, and special issues pertinent to military strategy and national policy. . . . By 1962, the Army's operations research program had passed through the throes of adolescence and had reached a vigorous young adulthood. OR was generally accepted as a useful tool for the military decision maker and had proven itself in a wide variety of areas requiring analysis. . . . On the whole, operations research in the Army was poised to make a mature contribution.

And, as Ellis wrote (Operations Research Office 1961):

> During the last 13 years ORO's accomplishments have indeed been noteworthy. ORO published 648 studies containing thousands of conclusions and recommendations. A majority of these have been adopted and acted on. This survey was written to summarize ORO accomplishments so that these could be considered in perspective and with satisfaction by those responsible for the accomplishments—the entire ORO staff: research staff, support staff, and administrative staff.

We can all be proud of this record.

ORSA AND ORO

The founding meeting of ORSA was held on May 26–27, 1952, at Columbia University's Arden House in Harriman, New York. Nine members of the ORO staff, including ORO's deputy director, George Pettee, and Thornton Page, an ORO division chief, were in attendance. Ellis, who fully supported the idea of an OR professional society, arranged with JHU and the Army for ORO staff and its administrative office to be of assistance—Thornton Page was the first editor of the society's journal, *Operations Research*, who was followed in that position by George Shortley, an ORO analyst, with staff member Charles Chadsey, managing editor.

Ellis also arranged with JHU to make an endowment to ORSA to support the Lanchester Prize that would be awarded each year to the author(s) of the best contribution to operations research and the management sciences published in English (Reed 1954). The prize was established in 1954. Richard E. Zimmerman, an ORO analyst, who is considered to be the father of Army combat modeling (Shrader 2006, 97), was the third recipient of the prize for his paper, "A Monte Carlo model for military analysis" Zimmerman (1956).

After leaving ORO, Ellis became professor and director of the Systems Research Center, Case Institute of Technology, Cleveland, Ohio, from 1962 to 1965, where he taught OR at the graduate and undergraduate levels. He then returned to the Washington, D.C. area as research coordinator for the Department of Health, Education and Welfare, 1965 to 1967, and then served as a consultant to the National Bureau of Standards from 1967 to 1970. Ellis suffered from aphasia caused by a brain operation in 1970. He died on December 16, 1973, at the Veterans Administration Center, Martinsburg, West Virginia; he was survived by his wife and children (Page et al. 1974, 1141–1142).

Honors and Awards

Ellis's service to the U.S. military was recognized by his receiving The Navy's Distinguished Civilian Citation and the Army's Distinguished Civilian Service Medal, and Legion of Merit awards from both the Navy and Air Force. For his scientific contributions, he was made a fellow of the American Association for the Advancement of Science, the American Physical Society, and the American Geophysical Union.

References

Hausrath AH (1954) Utilization of Negro manpower in the Army. Oper Res 2(1):18–30

Johnson EA (1954) The executive, the organization, and operations research. In: McCloskey JF, Trefethen FN (eds) Operations research for management, vol 1. Johns Hopkins University Press, Baltimore, MD, pp xi–xxiv

Johnson EA (1960) The long-range future of operational research. Oper Res 8(1):1–23

Johnson EA (1961) Toward establishment of a role for operations research in economic development programs. Oper Res 9(5):743–747

Johnson EA, Katcher DA (1973) Mines against Japan. Naval Ordnance Laboratory, White Oak, MD; U.S. Government Printing Office, Washington, DC

Operations Research Office (1961) A survey of ORO accomplishments. The Johns Hopkins University, Baltimore, MD

Page T, Pettee GS, Wallace WA (1974) Ellis A. Johnson, 1906–1973. Oper Res 22(6):1141–1155

Reed LJ (1954) The Johns Hopkins University Lanchester Prize: letter from the president of Johns Hopkins University to the president of ORSA. Oper Res 2(4):362

Shortley G (1967) Operations research in wartime naval mining. Oper Res 15(1):1–10

Shrader CR (2006) History of operations research in the U.S. Army, vol I: 1942–1962. U.S. Government Printing Office, Washington, DC

Visco EP (2004) Ellis A. Johnson and Wilbur B. Payne FS: two unsung (relatively) pioneers. Mil Oper Res 9(4):67–72

Whitson WL (1960) The growth of the operations research office in the U.S. Army. Oper Res 8(6):809–824

Zimmerman RE (1956) A Monte Carlo model for military analysis. In: McCloskey JF, Coppinger JM (eds) Operations research for management, vol 2. Johns Hopkins University Press, Baltimore, MD, pp 376–400

10

Leonid Vital'evich Kantorovich

Saul I. Gass and Jonathan Rosenhead

Leonid Kantorovich was one of the twentieth century's outstanding mathematicians who applied his immense talents to the study of economic problems of industry and national economic systems and, thus, helped to change not only the field of mathematical economics, but also the economic planning system of the USSR. His initial economic research started with the analysis of industrial production problems, which led him to state and resolve basic problems in linear programming (LP) in 1939, before such structures were studied in the West. He recognized the applicability of his work to a wide range of industrial and transportation problems; he also generalized his ideas to show how the USSR could improve its allocation of resources. This work was not looked at kindly by the Soviet government and academic economists. Conducted under very trying conditions, his research was often maligned, causing him to stop his efforts in the early 1940s. Leonid's seminal work was unknown to the West, especially to those in the U.S. who independently developed LP from 1947 and established it as a major analytical tool for business, industry, and government. When, in the late 1950s, the situation within the USSR changed and the importance of Leonid's work was recognized, his ideas led to a new school of Soviet economics that drew upon his mathematical economic theories.

Leonid received the prestigious USSR Lenin Prize in 1965 for works in economic science, particularly for applications of mathematical methods to economic research and planning. He was awarded the Nobel Prize in economics in 1975 for contributions to the theory of optimum allocation of resources. He was a full member of the USSR Academy of Sciences,

A.A. Assad, S.I. Gass (eds.), *Profiles in Operations Research*, International Series in Operations Research & Management Science 147, DOI 10.1007/978-1-4419-6281-2_10,

honorary member Hungarian Academy of Sciences, member of the American Academy of Arts and Sciences, and an honorary member of the Econometric Society. He was made a member of the IFORS' Hall of Fame in 2003.

A REMARKABLE BEGINNING I: MATHEMATICS

Leonid V. Kantorovich was born in St. Petersburg (Leningrad), Russia on January 19, 1912. His father, Vitalij, who was a doctor, died when Leonid was 10 years old and he was brought up by his mother Paulina (Saks). In his Nobel Prize autobiography (Kantorovich 1975), Leonid notes, "My first interest in sciences and the first displays of self-dependent thinking manifested themselves about 1920." His exceptional mathematical and scientific talents were recognized early by others—in 1926, when he was 14 years old, he was a student in the Faculty of Mathematics and Mechanics of Leningrad State University. As he needed special permission to attend, which was not granted until November 1926, he spent his freshman year in private study preparing and passing a "significant complex of subjects" (Kantorovich 1990, 9). When he graduated at age 18, he had already published 11 scientific works with emphasis on functional analysis, analytical operations on sets, and projective sets. At age 22, Leonid was a full professor at Leningrad University and was also associated with the Institute of Industrial Construction Engineering. One year later, when the system of academic degrees was restored in the USSR, he received his D.Sc. in physicomathematical sciences (Makarov and Sobolev 1990). His spheres of interest were remarkable for their range: functional analysis, topological vector spaces, mechanics, computing and computer architecture, as well as optimization and mathematical economics. He was recognized as one of the world's foremost mathematicians, often compared to the life and contributions of John von Neumann (Gardner 1990). His students said that he liked to "cook in fifty cauldrons" (Makarov and Sobolev 1990, 1). According to his wife Natalie Iljina-Kantorovich, "many people believed that Kantorovich the mathematician and Kantorovich the economist are two different people" (Iljina-Kantorovich 1990, 47).

A REMARKABLE BEGINNING II: ECONOMICS

Leonid had become interested in economics as a student and attended lectures on political economy; he even worked as an economist. In the summer of his third academic year, students had to go on work experience. As he was not interested in the usual mathematician's work experience "calculating numbers one to ten," Leonid found work as a statistician in Tashkent in the directorate of the Central Asian Water Board (Kantorovich 1990, 27). On arriving there, the job was not available. Instead, he was taken on as a junior statistician where he studied the description of irrigation conditions and use of water resources and their distribution. His full conversion to his economist alter ego happened a few years later and came about, as such things often do, fortuitously. As a professor at Leningrad University, he was also head of the Department of Mathematics at the Institute of Mathematics and Mechanics. It was in this capacity that he was consulted on problems faced by the Laboratory of the Plywood Trust: a production planning problem of the plywood industry—Given raw materials, and subject to some constraints, how could the productivity of the different equipment types required to process the raw material be maximized? Also, given plywood sheets, how should they be cut to produce a specified assortment of pieces with minimum waste (Gardner 1990, Kantorovich 1990)? Leonid soon recognized that the general mathematical structure required "the solving of a maximization problem in the presence of a series of linear constraints" (Kantorovich 1990, 29). Further investigation into such problems revealed the difficulty one would have for problems of even small dimensions, say, five machines and eight types of raw materials, which would require the solving a system of about a "billion" linear equations. Leonid noted that the generalization of such problems to the "class of extremum problems in which an extremum point lies on the boundary of the domain under consideration are characteristic of economics" (Kantorovich 1990, 29). He gave the problems much thought, although he was distracted a bit as it was the year (1938) of his marriage to Natalie Iljina-Kantorovich. (She became a physician; they had a son and daughter.) But, during that year, he solved and reported on such problems at a scientific meeting. Further consideration, especially when the problem was completely linear, led Leonid to develop in January, 1939, a solution method he called Lagrange resolving multipliers. As he notes, "The method created was distinguished by the close unification of the processes for solving the direct and dual problems; it is similar to the algorithm that Dantzig et al. (1956) worked out much later" (Kantorovich 1990, 30). Leonid compiled his research on a range of production and planning problems into a long paper (monograph) under the title *Mathematical Methods in the Organization and Planning of Production* (*Matematischeskie Metody Organizatsii i Planirovaniya*) produced under the imprint of the Publication House of the Leningrad State University (Kantorovich 1939). He was then 27 years old!

LP: East and West

The report by Kantorovich (1939) was unknown to the West until 1956 when Leonid sent a copy to the economist Tjalling C. Koopmans. Koopmans, in fact, had requested a different paper written by Leonid, "On the translocation of masses," published in Russia in 1942. This theoretical paper addresses two transportation-type problems, one dealing with railway haulage, and the other the leveling of a land area and the transporting of the earth to new locations at minimum cost (Kantorovich 1942). The translocation paper was brought to the attention of the mathematician Merrill M. Flood (University of Michigan) by the mathematician Max Shiffman (Stanford University) in 1949. As noted in Koopmans (1960), the paper was first cited in the West by Flood (1952, 1953).

Koopmans was instrumental in getting the 1939 paper translated and published in *Management Science* (Kantorovich 1960); the 1942 paper was also translated and published in *Management Science* (Kantorovich 1958). Koopmans also received a third paper from Leonid, "The application of mathematical methods to freight flow analysis," (Kantorovich and Gavurin 1949). This paper is related to the 1942 paper in which it had been noted, "The solution of this and more complicated problems of the same type the reader will find exposed in detail in a paper by L. Kantorovich and M. Govurin [sic] which will soon be published" (Kantorovich 1958, 3). This soon to be published paper in fact did not appear until 1949: It was submitted to the *Railway Journal* in 1940 but was not published "because of the dread of mathematics" by the editors; publication was further delayed due to the war (Kantorovich 1990, 32; Koopmans 1960, 363).

Leonid's (1939) and (1949) papers establish his priority in many aspects of the field of linear programming. As Koopmans notes (1960, 363–364):

> With the help of the knowledge of Russian of my colleague, Raymond P. Powell, it became apparent to us that these two papers are remarkable documents indeed in the history of management science, of linear programming, and of economic theory in general. The 1949 paper discusses transportation models for a single commodity and for many commodities (including empty vehicles), and a single-commodity model for a capacitated network, with applications to sections of the Russian railroad network. All problems considered in the 1939 paper reprinted in this issue are what would now be called linear programming problems [see insert below]. The coefficient matrices of the problems labeled "A" and "B" exhibit special structures somewhat similar to that of the transportation problem matrix. Problem "C" while appearing still to have a somewhat special structure, is in fact equivalent to the general linear programming problem.
>
> The applications of problems "A," "B," "C" envisaged by the author include assignment of items or tasks to machines in metalworking, in the plywood industry, and in earth moving; trimming problems of sheet metal, lumber, paper, etc.; oil refinery operations; allocation of fuels to different uses; allocation of land to crops, and of transportation equipment to freight flows. The author's introductory remarks compare the operation of the Soviet and capitalist systems; however it is quite evident that the wide range of applications perceived by the author make his paper an early classic in the science of management under any economic system. For instance, the concluding discussion anticipating objections to the methods of linear programming has a flavor independent of time and place.

KANTOROVICH'S LP PROBLEMS

Problem A: Determine the numbers $h_{ik}(i = 1, 2, \ldots, n; k = 1, 2, \ldots, m)$ on the basis of the following conditions:

(1) $h_{ik} \geq 0$;
(2) $\Sigma_k h_{ik} = 1 \ (i = 1, 2, \ldots, n)$;
(3) and for $\Sigma_i \alpha_{ik} h_{ik} = z_k$,

then h_{ik} must be so chosen that the quantities $z_1, z_2, \ldots, z_m$ must be equal to each other and moreover that their common value $z = z_1 = z_2 = \ldots = z_m$ is a maximum. (h_{ik} is the time expressed as a fraction of the working day that the i-th machine is used to produce the k-th part; α_{ik} is the number of k parts that can be produced in a day on the i-th machine; z is the number of items produced in a day.)

Problem B: Find the values h_{ik} on the basis of conditions of (1), (2), and (3) of Problem A, and the condition

(4) $\Sigma_i \Sigma_k c_{ik} h_{ik} \leq C$.

(c_{ik} is the amount of energy required to produce the i-th part on the k-th machine, and C is the total amount of energy available.)

Problem C: Find the values of h_{il} to satisfy the conditions

$h_{il} \geq 0$;
$\Sigma_l h_{il} = 1$;

and, if we set $z_k = \Sigma_{il} \gamma_{ikl} h_{il}$, then it is necessary that $z = z_1 = z_2 = \ldots = z_m$, and that their common values, z, have its maximum attainable value. (h_{il} is the unknown time of work of the i-th machine according to the l-th method of production; γ_{ikl} is the number of parts k that can be produced on the i-th machine under the l-th production plan.)

(Kantorovich 1939, 1960, 371–372)

Optimal Economic Planning vs. Marxist Economics

Leonid was interested in pursuing the general economic aspects of his work, and one of his objectives was as follows: "The spreading of the method of description and analysis from separate economic problems to general economic systems with their application to planning problems on the level of an industry, a region, the whole national economy as well as the analysis of the structure of economic indices" (Kantorovich 1975). But, such a pursuit was interrupted by more urgent war work—Leonid was evacuated from the siege of Leningrad in January 1942 and assigned as a professor to the Higher Engineering and Technical School of the USSR Navy that had been relocated to Yaroslavl, about 400 miles south-east of Leningrad. His job was to give lectures and carry out research. He did, however, find time to write the first version of a book in which LP was linked to optimal economic planning. The working title was then "Economic calculation of the most expedient utilization of resources." Its focus was on how LP could be applied at the level of state planning (Kantorovich 1990, 34). In 1942, Leonid sent it to officials of Gosplan (the Soviet state planning committee), but it did not meet with approval. It was again sent to Gosplan in the mid-1950s, as well as to other agencies, but his ideas continued to be rejected, with more or less severity, depending on the political climate. The book was eventually published in Russia in 1959 under the title *The Best Use of Economic Resources*; an English translation came out in 1965 (Kantorovich 1959, 1965).

But, back in 1943, Leonid concluded:

> Everything suggested that it was necessary to leave this work for the time being. It was dangerous to continue it—as I subsequently found out, my assumption was not unfounded. Of course, this was a severe blow to me as I had great expectations for it. For some time I was even in a state of depression—began to doubt that I could successfully study science—this work had to be deferred, while in the meantime I lagged behind in mathematics (Kantorovich 1990, 34).

This lack of support for his revolutionary ideas was disappointing to Leonid, but it did not come as a surprise. His venture into mathematical economic planning was, indeed, a dangerous venture on his part. His first experience in such rejections occurred in 1939:

> In the spring of 1939 I gave some more reports—at the Polytechnic Institute and the House of Scientists, but several times met with the objection that the work used mathematical methods, and in the West the mathematical school in economics was an anti-Marxist school and mathematics in economics was a means for apologists of capitalism. This forced me when writing a pamphlet [his 1959 book] to avoid the term 'economic' as much as possible and talk about the organization and planning of production; the role and meaning of the Lagrange multipliers had to be given somewhere in the outskirts of the second appendix and in the semi Aesopian language (Kantorovich 1990, 31).

Thus, isolated by barriers of ideology, language, World War II, and the Cold War, Leonid's ideas and publications remained unknown to researchers in the non-Soviet world until the mid-1950s. George Dantzig's developments in 1947 (and beyond) were thus independent rediscoveries of linear programming and its extensions—and it was Dantzig's work that led to the remarkable worldwide adoption of optimization methods that was largely responsible in establishing operations research and its methods within industry, business, and government. Dantzig (1963, 22–23) puts into perspective the early contributions of Leonid's 1939 monograph, as well as the material in Kantorovich (1942) and Kantorovich and Gavurin (1949): Leonid had mathematically described a wide range of possible LP applications and related computational approaches, many of which were independently developed and extended in the U.S. and in other countries after 1947.

Leonid's discussion, formulation, and mathematical analysis of the industrial and other real-world problems he encountered were influenced by the social and economic framework of the USSR. The economist Robert Dorfman contrasts Leonid's approach and that of Dantzig: Leonid's formulation is "the basic maximization problem for an economy where economic effort is coordinated by the maintenance of socially and technically appropriate material balances," while Dantzig's approach is "the appropriate form of maximization in an economy in which effort is coordinated by comparisons of relative value and the individual profitability of decentralized enterprises" (Dorfman 1984, 286–287).

As can be gathered from the reaction of Gosplan, Leonid's work was not welcomed by the Soviet authorities. What was the difficulty? There was a general hostility from the Soviet authorities to the mathematical economics of which Leonid was the foremost representative. In particular, his resolving multipliers were a political problem. Though Leonid referred to them as objectively determined valuations, they were in effect shadow or dual prices. These indicators could be used to put a value on scarce resources, so relating price to demand. For Leonid, they allowed production decisions to be decentralized without loss of efficiency, with lower-level decision-makers using them as the basis of their production calculations. To the authorities, however, such formulations appeared to introduce market relations and the notion of profit to the Soviet economy by the back door. As late as 1959, his book could only be published with a preface castigating Leonid for his many errors. Only from about 1960 on did the official attitude to his sort of work improve, a change marked by Leonid being awarded the 1965 Lenin Prize. Today, the recognition of Leonid's contributions to the general field of mathematical programming and its economic implications have been duly and fully recognized—Leonid and Tjalling Koopmans received the 1975 Nobel Prize in economics for their contributions to the theory of optimum allocation of resources. [A review of Leonid's contributions to economics is given in Johansen (1976).]

MEETING OF THE MINDS (1976)

George Dantzig Leonid Kantorovich

The assigning of priority for the origins of LP depends on how one interprets the words discovery or invention. Few observers of the field have written on this topic. The economist, Robert Dorfman, gives priority for the discovery of LP to Leonid. Dorfman regards LP, not as a branch of mathematics, but in the "domains of economics (both applied and theoretical) and management" (Dorfman 1984, 285). In a rebuttal to Dorfman, the mathematician and operations researcher, Saul Gass, assigns the priority for the invention of LP to Dantzig (Gass 1989). Gass regards LP as "a topic in mathematics (applied and theoretical)" (Gass 1989, 147). Dantzig is known in the West as "the father of LP" (Alpers et al. 1986); while in the East, as noted by the Russian translator of Gass (1973, 8), "linear programming was conceived in our country [USSR]. In the work of the outstanding Soviet mathematician and economist, the Lenin Prize laureate and academician L. V. Kantorovich . . . ." The dilemma can be summed up in the following manner: "Thus, the situation of Kantorovich is rather like that of Columbus. He really never touched the American mainland, and he didn't give it its name, but he was the first one in the area" (Gardner 1990, 645).

Leonid returned to Leningrad in 1944 where he worked at the University and at the Leningrad Branch of the Mathematical Institute of the USSR Academy of Sciences, where he headed the Department of Approximate Methods. It was here that he became involved with computational problems, computer programming, and computer design (Kantorovich 1975, 1990, 34–38).

After the publication of his book in 1959, recognition of Leonid's work improved (Ilina-Kantorovich 1990, 48). In 1960, a conference on Mathematical Methods in Economics and Planning, sponsored by the

USSR Academy of Sciences, was held in Moscow. "The conference approved the new scientific direction. But, this time we had obtained some positive experience of its application" (Kantorovich 1975). Further, in 1957, the new Siberian Branch of the Academy of Sciences was established in Akademgorodok (Academy Town: the city of scientific research), located about 20 miles south of Novosibirsk, Siberia.

REMINISCENCES

Lev Leifman, a longtime colleague of Leonid and the editor of the book *Functional Analysis, Optimization, and Mathematical Economics : A collection of Papers Dedicated to the Memory of Leonid Vital'evich Kantorovich*, recounted the following in his preface:

"In his closing remarks at the [1960] Conference on Application of Mathematical Methods in Economics and Planning in Moscow, Kantorovich told his audience, 21 years after the publication of his groundbreaking work titled 'Mathematical Methods of Organization and Planning of Production,' which was reprinted for the first time only as part of the preparation for the conference:

> 'Here comrade Mstislavskii [the preceding speaker, L.L.] was talking about the necessity of applying mathematical methods in economics. But he did not always say so; not so long ago he was saying otherwise. And his friend and coauthor Yastremskii at one meeting said addressing me: "You are talking here about optimum. But do you know who is talking about optimum? The fascist Pareto is talking about optimum!" You know how that sounded in 1943 [in the midst of the bloodiest war against fascism, L.L.]. Nevertheless, I did not say that, not to be like the fascist Pareto, let us strive for maximum of costs and minimum of production.'

"At this point his voice was drowned in the roar of laughter and applause. When he was able to continue, Kantorovich said in a homely voice, without the metal ringing of the first part:

> 'No, I am telling this not to settle the accounts, but for the younger generation to know who stood by his views for years, and who changed them like gloves.' [A popular Russian expression alluding to the nobility's white gloves changed several times a day to have them always snow-white.]

"A standing ovation was the answer. And there he was at the podium, with a shy smile, a gentle person in life, a genius in science, and an uncompromising fighter for scientific truth" (Leifman 1990, x–xi).

In her acceptance of the posthumous award of the Silver Medal of the Operational Research Society to Leonid, his wife, Natalie Iljina-Kantorovich made the following remarks:

> "Leonid's work in linear programming and related work in mathematical economics and operations research was carried out soon after our wedding. I vividly recall how intensely he worked then; how he hoped that his results would be implemented and would become useful to people; his tenacity and persistence in trying to overcome the difficulties—and the actual danger to his own life—involved in developing economic theories not agreeable to the establishment. His book 'The Best Use of Economic Resources' was written in the difficult war years and completed in 1942. It was published, in a very limited edition, only in 1959, and was severely criticized. Nevertheless, a whole generation of Soviet economists grew up on this book, including Leonid's direct pupils and his followers. Today, it is these people who are working on the radical economic reforms taking place in our country, based on L. V. Kantorovich's pioneer work. How tragic that Leonid did not live to see the day his lifework in economics became practical reality in his own country" (Iljina-Kantorovich 1990, 48; Rosenhead 1990).

(Novosibirsk is about 2000 miles East of Moscow.) Akademgorodok had at that time 14 research institutions. Here, in 1960, a special laboratory on the application of mathematics in economics was organized, under the direction of V. S. Nemchinov, with Leonid, who was not a communist party member, relegated to deputy director. Leonid remained in Novosibirsk until 1971. During this period, opposition to his work relaxed: He was elected a full member of the Soviet Academy of Sciences in 1964, received the Lenin Prize (highest honor available to Soviet civilians), and trained an active group of mathematical economists. His past research, when combined with his Novosibirsk research program, led to a new and active school of thought in the Soviet economics profession (Gardner 1990).

Leonid returned to Moscow in 1971 where, for the rest of his career, he directed Gosplan's Laboratory of the Institute for the Management of the National Economy and was a consultant to many government organizations. He died from cancer on April 7, 1986.

An obituary, signed by the leading Soviet scientists, aptly sums up Leonid as a scientist and human being (Leifman 1990, 6):

> By his nature Leonid Vital'evich was a soft, modest, and benevolent person. At the same time, in defending the scientific truth he was a man of principle and uncompromising. These qualities of L. V. Kantorovich, his inexhaustible optimism, love of life, active position of a scientist, human being, and citizen, assisted in no small degree to the recognition of his revolutionary ideas and new methods of planning and management developed on this basis. His scientific legacy will serve the people for a long time ahead, it will be studied by experts and historians of science.

HONORS AND AWARDS

Leonid received the USSR State (Stalin) Prize (1949) for works in mathematics, the Lenin Prize (1965, jointly with the economist Victor V. Novozhilov and the economist–statistician Vasyly S. Nemchinov), and, in 1975, the Nobel Prize in Economics (jointly with Tjalling C. Koopmans). He received honorary degrees from the universities of Glasgow (1966), Grenoble (1967), Nice (1968), Munich (1970), Helsinki (1971), Yale (1971), Sorbonne (1975), Cambridge (1976), Pennsylvania (1976), Halle (1984). He was posthumously awarded the Silver Medal of the Operational Research Society in 1986. He was elected to the IFORS' Hall of Fame in 2003

(Left to right) Nikolai Tretyakov, Mrs. Gol'shtein, Evgeny Gol'shtein, Leonid Kantorovich, Natalie Iljina-Kantorovich, Trudy Gass

THE MOSCOW LUNCH CROWD

In June, 1977, my wife, Trudy, and I traveled to Russia for a 30 day visit sponsored by the U.S. and USSR Academies of Sciences. My main Moscow contact was Evgeny Gol'shtein, a mathematician at the Central Economics and Mathematics Institute (CEMI). He was the translator of the 1961 Russian edition of my 1958 LP book, and co-author of the first Russian language linear-programming textbook (Yudin and Gol'shtein 1963).

Upon arriving, I asked Evgeny to arrange for me to meet Leonid. The meeting occurred two weeks later on our last day in Moscow before we went to Kiev, Tbilisi, and Novosibirsk. We had a late afternoon lunch with the CEMI staff and our wives. The restaurant had been reserved for us—we were the only patrons! Leonid arrived in his chauffeur-driven Volvo that was purchased with funds from his Nobel award. He was very friendly, a lovely man. We discussed our visit, his current work, and the impact of the Nobel award. We started and ended the meal with friendship toasts of ice-cold vodka.

(Saul I. Gass)

(Rosenhead 2003). He was a full member of the USSR Academy of Sciences, honorary member Hungarian Academy of Sciences, member of the American Academy of Arts and Sciences, and an honorary member of the Econometric Society.

REFERENCES

Albers D, Reid C, Dantzig GB (1986) An interview with George B. Dantzig: the father of linear programming. Coll Math J 17(4):293–314

Dantzig GB (1963) Linear programming and extensions. Princeton University Press, Princeton, NJ

Dantzig GB, Ford L Jr, Fulkerson D (1956) A primal-dual algorithm for linear programs. In: Kuhn HW, Tucker AW (eds) Linear inequalities and related systems. Princeton University Press, Princeton, NJ, pp 171–181

Dorfman R (1984) The discovery of linear programming. Ann Hist Comput 5(3):283–295

Flood M (1952) On the Hitchcock distribution problem. In: Orden A, Goldstein L (eds) Symposium on linear inequalities and programming (14–16 June 1951). Project SCOOP, U.S. Air Force, Washington, DC, pp 74–99

Flood M (1953) On the Hitchcock distribution problem. Pac J Math 3(2):369–386

Gardner R (1990) L. V. Kantorovich: the price implications of optimal planning. J Econ Lit 28(June):638–648

Gass SI (1973) An illustrated guide to linear programming. (Russian Translation). Moscow

Gass SI (1989) Comments on the history of linear programming. Ann Hist Comput 11(2):147–151

Iljina-Kantorovich N (1990) Silver medal: reply by Natalie Iljina-Kantorovich. In: Leifman L (ed) Functional analysis, optimization and mathematical economics: a collection of papers dedicated to the memory of Leonid Vital'evich Kantorovich. Oxford University Press, New York, NY, pp 47–48

Johansen L (1976) L. V. Kantorovich's contributions to economics. Scand J Econom 78(1):61–80

Kantorovich LV (1939) Mathematical methods in the organization and planning of production. Publication House, Leningrad University, Leningrad, Russia

Kantorovich LV (1942) On the translocation of masses. Proc Natl Acad Sci USSR 37(7–7):227–230

Kantorovich LV (1958) English translation of Kantorovich (1942). Manage Sci 5(1):1–4

Kantorovich LV (1959) The best use of economic resources (in Russian). USSR Academy of Sciences, Moscow. (English translation, Pergamon Press, Oxford, 1965.)

Kantorovich LV (1960) English translation of Kantorovich (1939). Manage Sci 6(4):363–422

Kantorovich LV (1965) The best use of economic resources [English translation of (Kantorovich 1959)]. Pergamon Press, Oxford, England

Kantorovich LV (1975) Autobiography. http://nobelprize.org/nobel_prizes/economics/laureates/1975/kantorovich-autobio.html. Accessed 29 Oct 2008.

Kantorovich LV (1990) My journey in science. In: Leifman LJ (ed) Functional analysis, optimization, and mathematical economics. Oxford University Press, New York, NY, pp 8–45

Kantorovich LV, Gavurin M (1949) The application of mathematical methods to freight flow analysis. In: Problems in increasing the effectiveness of transport. academy of sciences of the USSR. Moscow–Leningrad, pp 110–138

Koopmans TC (1960) A note about Kantorovich's paper, "Mathematical Methods of Organizing and Planning Production." Manage Sci 6(4):363–365

Leifman L (ed) (1990) Functional analysis, optimization and mathematical economics: a collection of papers Dedicated to the Memory of Leonid Vital'evich Kantorovich. Oxford University Press, New York, NY

Makarov V, Sobolev S (1990) Academician L. V. Kantorovich. In: Leifman LJ (ed) Functional analysis, optimization, and mathematical economics. Oxford University Press, New York, NY, pp 1–7

Rosenhead J (1990) Silver medal. In: Leifman LJ (ed) functional analysis, optimization, and mathematical economics, Oxford University Press, New York, NY, pp 46–48

Rosenhead J (2003) IFORS' operational research hall of fame: Leonid Vitaliyevich Kantorovich. Int Trans Oper Res 10(6):665–667

Yudin D, Gol'shteinE (1963) *Lineinoe programmirovanie.* Moscow

11

C. West Churchman

Arjang A. Assad

C. West Churchman made pioneering and important contributions to operations research (OR), management science (MS), and systems theory. Over the course of six decades, he investigated and contributed to a wide range of topics—statistics, accounting, city planning, space exploration, education, mental health, and peace and conflict studies. Influenced by his practical work on statistical quality control during World War II (WWII), West came to OR after a successful career in philosophy. As a pioneer, he helped define OR and its relation to systems thinking in the early 1950s. With his friend and colleague, Russell Ackoff, he established the path-breaking Operations Research Group at Case Institute of Technology. Along with E. Leonard Arnoff, they coauthored the first integrated text in OR, *Introduction to Operations Research*, published in 1957.

West was a founding member of both the Operations Research Society of America (ORSA) and The Institute for Management Sciences (TIMS). He was appointed the first editor-in-chief of the TIMS flagship journal, *Management Science*, in 1954. He also served as president (1962) of TIMS. His 1968 award-winning book, *The Systems Approach*, was widely recognized as the standard introduction to the subject. In the last three decades of his life, he increasingly stressed the importance of adopting a comprehensive and ethical view of OR and systems analysis.

A.A. Assad, S.I. Gass (eds.), *Profiles in Operations Research*, International Series in Operations Research & Management Science 147, DOI 10.1007/978-1-4419-6281-2_11,

EARLY BEGINNINGS: QUAKERS TO PHILOSOPHY

C. West Churchman was born in Mount Airy, Pennsylvania, August 29, 1913. His parents, Clark Wharton Churchman and Helen Norah Fassitt, were descendents of Main Line Philadelphia families. Clark was an architect and Helen owned and managed an antique store. West grew up in the elite society of Philadelphia, was raised as a Catholic, and attended a Quaker school in that city. The ethical imperative to help other people emerged early in his life:

> What I got out of the Quakers was the knowledge that you can have a life devoted to humanity. . .. They impressed on us that the need for humanity was for a world government. They wanted to get rid of nations. . .. That impressed me a lot as a young man (van Gigch et al. 1997, 732, 735).

West enrolled in the University of Pennsylvania (Penn), Philadelphia, when he was 17, and joined the Zeta Psi fraternity, a social college fraternity, dedicated to forging academic excellence and lifelong bonds of brotherhood. When he started college, West began writing a journal in which he put down a list of five to seven possible majors. He considered medicine first, and then law, as the path toward helping people, but eventually settled for philosophy because that seemed to "be the discipline that was most interested in the broadest possible view of humanity" (van Gigch et al. 1997, 733). His undergraduate and graduate degrees were all in philosophy: B.A. (1935), M.A. (1936), Ph.D. (1938).

In the course of his studies, West was profoundly influenced by the work of the philosopher Edgar Arthur Singer, Jr. Singer had studied with the Harvard University philosopher William James and taught at the University of Pennsylvania from 1909 to 1943. Singer's student, Henry Bradford Smith, who became a noted logician, was West's teacher and dissertation advisor.

Smith was interested in studying very general forms of logic. West wrote his doctoral dissertation, "Toward a general logic of propositions," under Smith's direction. West's dissertation was an exercise in propositional calculus, with modal operators, in which he tried to extend some of Smith's ideas. The dissertation, completed in 1938, was formally published a few years later (Churchman 1942). It was reviewed by the eminent logician Alonso Church who found some of the claims West made for his general logic "to be at best extremely doubtful" (Church 1943, 53).

The informal Singer School at Penn included West, his teacher Smith, and his two longtime collaborators, Thomas A. Cowan and Russell L. Ackoff. For Singer's students, to philosophize meant "to philosophize about how the world fit together" (Churchman 1979, 10). To Singer and West, this integrative view needed to include statistics, and West took this subject up in his graduate studies. According to West, Singer was "the only philosopher of science who appreciated the epistemological significance of the concept of a probable error" (Churchman 1982, 29).

West became an instructor of philosophy at Penn in 1937 and was appointed an assistant professor after earning his doctorate. It was in this role that he first met Russell (Russ) L. Ackoff—an architecture student who had enrolled in West's modern philosophy course. The two of them "hit it off" to start a lifelong friendship that was fateful in the development of their careers—first in philosophy and later in OR (Ackoff 1988, 351).

In 1941, West was the undergraduate chairman of the philosophy department. In that role, West assigned Russ three courses to teach as an assistant instructor. Russ had not taught any of these courses, but West assured him that "teaching a subject was a better way to learn it than being taught it" (Ackoff 1988, 352). Russ became West's first doctoral student; and, although Russ was drafted almost immediately after Pearl Harbor, they managed to continue with their collaboration on Russ's dissertation.

West at Frankford Arsenal

WWII not only interrupted West's academic career, but it also effected a profound change in his priorities. Because of his dependent parents, West was not eligible for the draft. Even for a philosopher, however, the enormity of the menace served as a call to action. West felt the urgency of the need to act. As he put it, "The enemy was Hitler and his successful armies, and not the logical positivists. Hitler was highly dangerous, while the logical positivists just sounded silly" (Churchman 1994, 100).

West's statistical studies at Penn, especially his knowledge of statistical inference, provided him with an opportune wartime assignment. West took his first nonacademic job with the mathematical section of the U.S. Ordnance Laboratory, Frankford Arsenal, Philadelphia. His work at the Arsenal included calibration of measurements in physical chemistry. Later, he reflected:

> It was really World War II that saved me from philosophy departments. At that time, we academics really had a choice of trying to stick to the university or go out and do something and be a volunteer for military service or do research. I elected to go into a laboratory and do research. I was performing mathematical statistics. My boss said, "What I would like you to do is to be as sure as possible that the ammunition we produce here will fire when the GI pulls the trigger of his gun anywhere in the world." That was my first introduction to a true management problem (van Gigch et al. 1997, 733).

West was assigned to find the best inspection policy to ensure that small arms ammunitions would not misfire. The items being tested were primers. The primer is the part of the bullet that ignites the gunpowder in the shell when hit by a firing pin. Failure to ignite (a misfire) jeopardizes a soldier's life.

The inspection practice prior to West's arrival was the following: Test 100 primers. If there are no failures (all 100 went off), then accept a lot of 25,000 primers. If one failed, then inspect a lot of 200 and if all of these went off, accept a lot. If there was more than one failure in the original sample, or another failure occurred in the second sample, reject the lot. To evaluate this inspection policy, West posed it as a classical probability problem. Suppose that a lot of 25,000 has 1% defectives. What is the probability that a random sample of 100 primes from that lot will all fire? He found that the lot would be accepted with a probability of approximately 1/3, even though 1% defectives is unacceptably high for combat situations. Recalling his observations of how agricultural experts tested the full range to measure the efficacy of insecticides, West proposed a different approach (Verma 1998, vi).

Today, we immediately recognize West's assignment as a statistical quality control problem. In fact, as part of West's job, he visited manufacturing plants and lectured and introduced statistical quality control methods. He made several visits to Bell Telephone Laboratories where he discussed these concepts with William Shewhart (considered the founder of statistical quality control) and became quite familiar with his work and that of his Bell associates, Harold Dodge and Harold Romig. For Frankford Arsenal, West designed a technically superior method of inspection and wrote a manual for inspection policy. Later, this led to papers dealing with statistical sampling and control (Churchman and Epstein 1946, Epstein and Churchman 1944).

TRANSITION TO OR

How did a philosophy professor get interested and involved in OR? As West explains:

> It is hard to recall how and why I moved intellectual dwelling some half century ago from epistemology to management. The two questions, "What's wrong with logical positivism's theory of knowledge?" and "How many 15 ½-33 men's shirts should be kept in a retail store's shelves?" do seem a bit different, don't they?
>
> When I first discovered operations research, I found that there were many scientists who were able to shift from their own domains of inquiry into management domains rather rapidly. ... This shift was called "operations research," and usually was accomplished by applying some branch of mathematics to military "operations" problems (Churchman 1994, 99, 101).

West's transition to OR was rooted in his wartime assignment and in his interpretation of how a science of management should be formed.

In 1945, back at Penn, West was elected chairman of the philosophy department. Russ was released from service in early 1946 and returned to his doctoral studies. West and Russ worked closely and collaboratively on Russ's dissertation. To maximize their rate of progress, they rented a small house near the university and lived together with West's sister and a friend of hers. This co-location strategy continued through the summer of 1946 when they both went to Cambridge and stayed near Harvard University. There, Russ finished his dissertation, a work that West and Russ viewed as a joint effort.

At Penn, West and Russ tried to establish an Institute of Experimental Method. The mission of the Institute was to apply Singer's philosophy to a wide range of societal issues—management, education, and city planning. The Penn philosophy department was not receptive to the concept of "practicing philosophy as an applied discipline" (Ulrich 2004a, 208). The Institute failed to materialize and Russ's teaching appointment was not renewed. In 1947, he accepted an appointment as an assistant professor of philosophy and mathematics at Wayne (State) University in Detroit, Michigan (Ackoff 1988).

To enable West and Russ to continue working on their book, *Methods of Inquiry*, West arranged his teaching schedule so he could be in Detroit each week. West's trips led to a visiting professorship appointment at Wayne and then an offer of a full-time position as associate professor, which he accepted in 1948. West and Russ had received a promise from "the relevant dean that he would enable us to establish the Institute [of Experimental Method]" at Wayne (Ackoff 1988, 353). This, however, was met with strong opposition from the faculty, leading to the decision not to renew Russ's 3-year appointment. West and Russ both came to the conclusion that their goal of doing applied philosophy could not be realized within philosophy departments. On the personal front, West met his future wife, Gloria, as a student in one of his classes at Wayne. They married in 1954.

Prologue to Case Institute of Technology

While the Institute of Experimental Method was never formalized, West and Russ, while at Penn, did use its name to sponsor a conference on "Measurement of Consumer Interest" (Churchman et al. 1947). In one of the sessions, the Princeton University statistician Sam Wilks talked about operational research. Russ recalls:

> Now West and I had only a vague idea as to what operations research was. . . . In the preface to our publication of the proceedings of that conference [Measurement of Consumer Interest], West and I had noted the similarity of intent between operations research and what we were trying to do, putting interdisciplinary teams together. We did not care much about the name we went under when the Case Institute offer came up. By then we knew that we could not do what we wanted to do within philosophy departments (Ackoff 1988, 354).

Russ resigned his position at Wayne in January, 1951, and interviewed at the Case Institute of Technology, Cleveland, Ohio. West decided to accompany him and they met with the assistant to the Case president who, by coincidence, had worked under West at Frankford Arsenal. Both Russ and West received offers from Case (Ackoff 1988, 354).

West's account of how the Case offer came about is a bit different. The vice president of the Chesapeake and Ohio (C&O) Railroad had contacted the acting president of Case to see what Case was doing in OR and if this new subject could be useful to C&O. The acting president, who was not familiar with OR, bought the only book available on the subject, *Methods of Operations Research* (Morse and Kimball 1951), and then called West to see if he would join Case (Churchman 1979, 16). A third view of how West and Russ joined Case is given by Dean (1994, 6).

CASE INSTITUTE OF TECHNOLOGY: CREATION OF OR AS AN ACADEMIC DISCIPLINE

West and Russ joined the Case Institute of Technology's Department of Engineering Administration for the beginning of the 1951 academic year; West as a full professor and Russ as an associate professor (Ulrich 2004b). Together, they set about to organize the first OR academic program. (The Institute was renamed the Case Western Reserve University in 1967.)

In November 1951, the embryonic Case OR staff sponsored the first of a series of 3-day OR business/industrial-oriented conferences (five other such conferences were held through 1957). The conference's success led to the formation, within the department, of the Operations Research Group whose purpose was to undertake corporate consulting assignments (it was directed by Russ starting in 1954). Case was the first college to house, within an academic department, graduate masters and doctoral programs in OR; its first OR M.S. and Ph.D. degrees were granted in 1955 and 1957, respectively (Arnoff 1957, Dean 1994).

During the period 1951–1957, the Case OR program flourished with the OR faculty increasing from 6 to 30 (Arnoff 1957, Dean 1994). The graduate programs in OR placed special emphasis on "the experiential aspect of OR education" by providing the graduate students the opportunity to work as research assistants on sponsored research projects

(Arnoff 1957, 290). West's accomplishments during his years at Case are detailed in Dean (1994).

In June 1952, the OR staff gave the first short (week-long) course in OR designed for "those in industry, business, and government who were interested in determining the potential value of OR to their own organizations" (Arnoff 1957, 290); the course continued to be given each June. The lecture material from this course grew into the first textbook in OR, *Introduction to Operations Research*, coauthored by West, Russ, and E. Leonard Arnoff, published in January 1957. As a student exposed to the text in 1958, Richard Mason, who later became West's student at Berkeley, was captivated by the beginning and ending chapters of the text which described the methodology of the systems approach as applied to OR (Mason 1994).

The text characterized OR as "the application of scientific methods, techniques, and tools involving the operations of systems so as to provide those in control of the operations with the optimum solutions to the problems" (Churchman et al. 1957, 8–9). Twelve of the twenty-two chapters of the book were written by other authors, one being a reprint of Leslie Edie's Lanchester prize-winning paper, "Traffic delays at toll booths" (Edie 1954). As Mason points out, a key difference between this text and other introductory accounts of OR is "its methodical grounding in an interdisciplinary Singerian philosophy and its heavy emphasis on the systems approach" (Mason 2004, 587). Another difference is the attention and space devoted to OR practice and implementation.

THE INVERTED SANDWICH

"During the course I spent many an evening grinding out the seemingly endless pivots of a simplex tableau or working my way through some of the other innovative mathematical machinations described in the book, all quite new to me. (In all of my conversations with West Churchman, however, he never once talked about the elegance or beauty of pivoting in the simplex algorithm!) Like most of my colleagues, however, I was captivated most by the beginning and ending chapters of *Introduction*. The book is like an inverted sandwich. The meat is at the front and back, the daily bread in the middle. In the meaty chapters, the systems approach and methods for bringing it to life are covered in remarkably clear and practical language and diagrams. Throughout the book the spirit of securing improvement prevails. It encourages you to go out and do something with what you have learned" (Mason 1994, 69).

THE BERKELEY YEARS

West left Case in 1958 to become professor of business administration, Graduate School of Business Administration, University of California, Berkeley (UCB). He founded the UCB master's and doctoral programs in OR, and he helped to establish the Center for Research in Management Science. He remained at UCB until his retirement in 1981.

In the mid-1960s, the National Aeronautics and Space Administration (NASA) sponsored university research grants to study both the management and the social impact of technology. Berkeley received such a grant, and its operations were housed within the university's Space Sciences Laboratories (SSL). In 1963, West was appointed research philosopher and associate director at the SSL where he directed its Social Sciences Program from 1965 to 1971. His book, *Challenge to Reason* (Churchman 1968a), grew out of West's investigations of industry and government and his interactions with his SSL colleagues.

Besides NASA, West consulted for several organizations including the U.S. Fish and Wildlife Service, the National Institutes of Health, and the National Science Foundation. From 1962 to 1963, he served as a research director of the System Development Corporation. West also taught in the Interdisciplinary Ph.D. Program of UCB's Graduate Division.

After West retired from his faculty position, he continued as a UCB professor of peace and conflict studies and taught classes in peace and ethics until 1996—his classes were very popular and regularly oversubscribed. He became a leading advocate of applying systems to peace and conflict studies.

West died at the age of 90 of complications from Parkinson's disease on March 21, 2004, in Bolinas, California. He was survived by his wife, Gloria, and his son, Daniel Wharton Churchman, and two grandchildren. Gloria died on August 2, 2009.

OR: EARLY WORK

West's early OR activities can be divided into two areas: (1) applied statistics and sampling and (2) the measurement of values. His statistical efforts followed naturally from his WWII experiences.

He coauthored two papers on statistical sampling that dealt with the collection of sensitivity data (Churchman and Epstein 1946, Epstein and Churchman 1944). Here, sensitivity data refer to test data collected when the test is destructive and the strength of the stimulus is varied to identify the critical stimulus level at which the object will fail. For the Frankford Arsenal problem, the tested objects were primer caps, and the stimulus was a blow of a certain energy. In successive experiments, the energy level was varied to find the critical level just sufficient to cause an explosion. The 1944 paper discusses the estimation of sampling errors and includes a result on stratified sampling; the 1946 paper studies techniques for determining the distribution of the critical stimulus with special focus on the case where a very small probability of failure is required.

After the war, West continued to pursue his prewar interests in the foundations of statistical methodology. He explored the interface between his previous work in logic and statistics (Churchman 1946). He examined the general logic of statistical tests of hypothesis and tried to reformulate Wald's results in the language of formal systems (Wald 1942). In the preface to his book, *Theory of Experimental Inference* (Churchman 1948), West thanked his friends from the Arsenal for providing the motivation for the book—its final chapter is on quality control and describes the problems he worked on while at the Arsenal. The overall thrust of the book is, however, decidedly philosophical.

A theme that links West's early work in philosophy and OR is the measurement of human values. West's interest in measurement derived from Singer's work on metrology, as well as his own practical experience with calibration (in physical chemistry experiments) during the wartime years. In the 1940s, he turned his attention to applying measurement concepts to human values. West and Russ continued to work on related issues through the 1950s, including an approximate measurement of value (Churchman and Ackoff 1954). This line of investigation is discussed in chapter 6: "Weighted Objectives" in Churchman et al. (1957), and in greater detail in Churchman (1961).

OR AS A SCIENCE OF MANAGEMENT

West's view of OR and MS underwent considerable change over time. After WWII, he viewed himself as member of the diverse group of scientists who had shifted from their original domains of interest to management. This group typically applied mathematics to some facet of military operations. Accordingly, West's initial view was that his good grasp of quantitative methods would result in important contributions to problems in management. Over time, his interest shifted from quantitative methods to the broader view of OR as a method of inquiry aiming at a science of management. As an applied philosopher who turned to OR, he continued to view it through two lenses: as a method of inquiry and as a way of making a difference in the actual world.

In his paper "Operations research as a profession," West stated, "OR is the securing of improvement in social systems by means of scientific method" (Churchman 1970, B39). Given his roots as a philosopher, he believed that each of the term's scientific method, social systems, and improvement required further discussion, and he was especially keen to stress the word securing. He wrote:

> Securing has a much deeper meaning than selling managers on OR findings. It also means observing the implementation process and controlling the change in the face of unforeseen circumstances. In general, it means "learning how to do OR better the next time" (Churchman 1970, B40).

West proposed to tell two tales of the OR profession. This was inspired by Carl Jung's notion that there are two ways of telling the story of one's life: a rational one and an irrational one. The more familiar rational story is defined as training in OR at the level of "at least a masters [degree] in OR" and "at least two years of satisfactory internship in doing OR" (Churchman 1970, B38). As for the less familiar, alternative story, we have:

> The other story of the preparation for the profession says that we are to prepare the practitioner for a life of ambiguity and persistent failure to resolve the basic issues. He must prepare himself to appreciate that every human problem is fundamentally elusive, and that the attempt to clarify its meaning and identify a solution distorts the problem and destroys its real significance (Churchman 1970, B39).

West did not hesitate to make radical recommendations as to how OR should deal with "its mysteries, its irrational side." These included holding morality meetings where there are no experts, holding self-evaluation meetings of the OR and MS professional societies which begin with the invocation: "We have failed as a society; therefore. . .," and publishing papers on "failures which learn" (Churchman 1970, B52, B53).

In an effort to assess the impact of OR, West took an early interest in implementation issues. Dean (1994) describes how annual conferences at Case specifically explored and documented the successes or failures of implementation efforts in industry. In October 1958, at the 14th annual meeting ORSA, West chaired a session, "Case histories five years after—a symposium." This was probably the first time a concerted effort was made to revisit earlier implementation efforts and to track their progress over time. West's symposium paper is of particular interest—a Case OR team's study (under West's direction) of a C&O railroad accounting problem, the settlement of interline accounts (Churchman 1960).

The project involved estimating the appropriate share owed to each carrier when a shipment involves multiple carriers (van Voorhis 1953). The OR team suggested sampling as an effective approach. Using stratified sampling, the team sampled about 10% of a large population of waybills and showed that this would produce estimates that were accurate to within 1% of the true figure owed each carrier. But, in his paper on the study, West does not focus on the methodology (Churchman 1960). Rather, he discusses the resistance of the railroad's management to use sampling despite the strong economic argument in its favor. Nine years after the original study, West reported that sampling had gained much greater acceptance. He concluded, however, that "it will be still some time before what appeared obviously beneficial to the operations-research team is adopted" (Churchman 1960, 258). This was an early example of what West termed "resistance to implementation" (Churchman 1994, 105).

Reflecting upon implementation, West often returned to the experiment he conducted with Philburn Ratoosh around 1965. The subjects of the experiment (MBA students) were asked to run a company with three products by making decisions on pricing, purchasing, and the production schedule. The past data they were given allowed them to infer a correct mathematical model of the firm's operations and to derive an optimal policy that maximized profits. To make sure that the subjects were aware of the mathematical model, a stooge was planted in each team to steer them to the optimal solution. In over 40 runs of the experiment (except for two cases), the solution given by the mathematical model was not implemented even after the stooge explained the model and its implications to the subjects. Typically, they ignored the information provided by the stooge and proceeded to follow their own policies. (Churchman 1971, 1994). Later, West drew broader conclusions from this experiment: ". . .no matter how clear, forceful, and correct the argument for optimal action may be, human beings do not behave in accordance with the instructions (the ethical imperatives)" (Churchman 1994, 104).

West also advocated including accounts of OR implementation efforts in the professional journals. He reviewed the various perspectives on implementation and proposed further research on the subject in Churchman and Schainblatt (1965a). West's own view of implementation was philosophical stressing such notions as mutual understanding and trust (Churchman and Schainblatt 1965b). Responses to this paper by a dozen scholars appeared in the October 1965 issue of *Management Science.*

When papers with theoretical and technical contents started to dominate the TIMS journal, *Management Science,* West joined forces with Roger Crane (1957 president of TIMS) to publish case histories along with "management discussions of management sciences, such as the nature of business problems which they feel are important and can be attacked by the management sciences" (Crane and Churchman 1960, 1). This led to the creation of the TIMS *Journal of Management Technology,* soon renamed *Interfaces.*

Reflecting on the relationship between the scientist and the manager, West believed that "science and management cannot be separated; if science is to become a method of managing, then management must become the method of science" (Churchman and Schainblatt 1965a, B70).

West returned to this idea in his book *Challenge to Reason:*

> Saying that science can become a way of managing didn't imply automation or any other form of mechanical decision making, because none of this is science. Science is the creative discovery of knowledge. Management science is the process of trying to look at science as a management function. Similarly, management can be looked at as a scientific function, that is, as a way of finding out about the world (Churchman 1968a, 104).

Philosophical Roots and Works

Throughout West's career, his philosophical development continued to be influenced by his teachers, the philosophers Singer and Smith. Singer belonged to the philosophical tradition of pragmatism which held that "the meaning and value of all human endeavors . . . is to be measured by the way it serves the practice of human life" (Ulrich 2004a, 204). Singer's goal was to systematize a science of human decisions that allows for purposeful behavior.

Singer, at his retirement dinner, named West as his disciple, and West was pleased to note: "This made it official!" (Churchman 1979, 9). Two of West's former students have indicated the lasting influence of Singerian themes in West's development of the systems approach (Ulrich 2004a, Verma 1994). We briefly discuss some of these themes below. Britten and McCallion (1994) provide an overview of what they call the Singer/Churchman/Ackoff school of thought. West's own appreciation of Singer's work appears in Churchman (1982).

According to West, Singer, in the 1930s and 1940s, led one of the first attacks on the strict separation between the physical and social–psychological worlds (Churchman 1979, 39). A later, important attack on this separation, which also influenced West's own early work, was Norbert Wiener's work on cybernetics which allowed for the notion of teleology, that of purposeful systems (Rosenbleuth et al. 1943, Wiener and Rosenbleuth 1950).

According to West:

> Singer's philosophical position had the theme of comprehensiveness, so that all aspects of the natural world were to be swept in, in order to pursue humankind's endless quest for knowledge and control of nature. Singer made no attempt to exclude values from science. . . . Physical reality, psychological and social reality, ethical and moral reality, religious reality were all to be included under the generic label "science or the quest for knowledge" (Churchman 1979, 9).

Singer's unifying theme was progress toward the ideal. While ideals are unattainable, Singer believed that they can be approximated with increasing accuracy. Just as the velocity of light was measured with increasingly greater accuracy throughout the nineteenth century, Singer believed that measurement can be used to assess progress in the social sciences, just as measurement accuracy served this purpose in the physical sciences.

Singer's major program of synthesis of physical and social sciences resulted in his last book *Experience and Reflection* (Singer 1959). West and

Russ were familiar with this project from their course notes of the early 1940s, but Singer's project of integration work stopped with the natural and biological sciences (Churchman 1990). In their joint paper, "An experimental definition of personality" (Churchman and Ackoff 1947), they took it upon themselves to complete the Singerian program by adding psychology, sociology, and ethics. Seeking to devise a formalism of personality that admitted measurement in the context of "purposeful behavior of living individuals," personality was "defined in terms of the probability that an individual will choose a specified means for a specified end in a specified environment" (Churchman and Ackoff 1947, 311). The pursuit of their goal of providing a complete classification of behavior patterns and ends was continued in their book, *Psychologistics* (Churchman and Ackoff 1946).

The other Singerian theme—a philosophical account of measurement that accounted for uncertainty—was further developed in West's main philosophical work: *Theory of Experimental Inference* (Churchman 1948). Here, West investigated the philosophical issues surrounding the experimental method, measurement, and statistical inference. In the preface, West acknowledged the influence of Singer's ideas and the contributions of his friends Thomas Cowan and Russ Ackoff.

THE SYSTEMS APPROACH

West is viewed as one of the leading thinkers of the systems approach. His expository treatment of the subject, *The Systems Approach* (Churchman 1968b), sold over 200,000 copies and was translated into many languages (Ulrich 2004a).

West's understanding of systems was heavily influenced by his philosophical background and preferences so that his contributions to the subject are unique and not easy to duplicate. Consequently, even this popular book is far from a simplified how-to manual and raises deeper issues of inquiry and method. Instead of presenting tools and techniques without comment, he chooses "to inject the criticism into the very context in which a technique was being discussed" (Churchman 1968b, 228). As a result, the book assumes the form of an extended debate about the different perspectives on systems thinking, represented by various dramatis personae: the efficiency expert, the scientist, the management scientist, the behavioral scientist, the planner, and even the anti-planner!

West warns about the expert's fallacy—the tendency of viewing a complex system through the lens of the expert's specialty:

> In order to understand the systems approach, we should try to understand what it really is. Don't try to find out only by asking an advocate or a practitioner, because each will start by emphasizing his own special approach. One will talk incessantly about how the systems approach is linear programming, another will insist that is nonlinear feedback modeling.... You'd come away ... feeling that since systems approachers don't agree ... perhaps there is no systems approach at all. But that would be just as ridiculous as inferring that art doesn't exist because no two artists would describe it the same way.... The systems approach is ... one of the approaches based on the fundamental principle that all aspects of the human world should be tied together in one grand rational scheme, just as astronomers believe that the whole universe is tied together by a set of coherent "laws" (Churchman 1979, 7–8).

West acknowledged the contribution of different approaches to a systems problem but was wary of the expert's claim that a given approach can account for all aspects of the whole system:

> ...however a system problem is solved, by planner, scientist, anti-planner, or whomever—the solution is wrong, even dangerously wrong. There is bound to be deception in any approach to the system.... And yet when one looks at the solution and sees its wrongness, one is also deceived, because in searching for the wrongness, one misses the progressive part of the solution. We have to say that the advocate of the solution both deceives and perceives (Churchman 1968b, 229).

West reminds us that it is equally misguided to ignore the positive contribution (progressive part) of a particular approach or solution.

Defining and Designing Systems

West identified two key issues in systems thinking: (a) defining the system boundaries and (b) specifying its components:

> One system design problem of central importance is to decide how large the system is, i.e., its boundaries and environment. A closely related problem is

> one of determining the basic components. . .. All men are system designers, and each man tries to determine what, in his world, is the largest system and the smallest (Churchman 1971, 7–8).

While West recognized the practical need to define the scope of system studies in manageable terms, his own tendency was to continually point out the limitations resulting from defining the system too narrowly.

> IS THERE AN ELEPHANT IN THE ROOM?
>
> "The story is often told . . . about a group of blind men who are assigned the task of describing an elephant. Because each blind man was located at a different part of the body, a horrendous argument arose in which each claimed to have a complete understanding of the total elephantine system. What is interesting in this story . . . [is] the magnificent role the teller had given himself—namely, the ability to see the whole elephant . . . . The story is in fact a piece of arrogance. It assumes that a very logically astute wise man can always get on top of a situation, so to speak, and look at the foolishness of seeing the whole. This piece of arrogance is what I called 'management science' " (Churchman 1968b, 27).

West's favorite example of limiting assumptions in modeling was drawn from inventory control. In standard inventory models, the cost of carrying inventory is an ubiquitous parameter. But what does this cost mean? The usual answer cites obsolescence, taxes, costs of physical storage, and the cost of tying up capital. But this invites a battery of new questions: How can we accurately quantify the cost of tying up capital? One can respond that this cost is "whatever the comptroller says it is" (Churchman 1970, 41). But then how can we be sure that the comptroller can judge optimal allocation of the firm's funds? What if the past is not a good guide because the firm has failed to use its capital productively? Will the inventory manager looking for a cost parameter of the model suddenly get embroiled in company-wide organizational issues? West raises similar questions about inventory cost components (Churchman 1971, 165–167). To West, data collection in OR always involves strategic decisions and systems thinking requires us to consider and confront these issues, instead of simply assuming them away. Such data are "not the result of observation alone, but of observation plus strong systemic assumptions" (Churchman 1970, 42).

West believed that systems thinking and management science should attack precisely those problems that are far-reaching and global in scale. In a guest editorial in the *Management Science* (Churchman 1967),

> SYSTEMS PRINCIPLES
>
> "The systems approach begins when you first see the world through the eyes of another.
>
> "The systems approach goes on to discovering that every world view is terribly restricted.
>
> "There are no experts in the systems approach.
>
> "The systems approach is not a bad idea."
>
> (Churchman 1968b, 231–232)

West expounded on the term wicked problems proposed by Horst Rittel, a professor at UBC's Department of Architecture. Rittel had suggested that OR and MS must be counted among the efforts aimed at taming wicked problems. Wicked problems are defined as "that class of social system problems which are ill-formulated, where the information is confusing, where there are many clients and decision makers with conflicting values, and where the ramifications in the whole system are thoroughly confusing." West's reaction was this: "A better way of describing the OR solution might be to say that it tames the growl of the wicked problem: the wicked problem no longer shows its teeth before it bites . . . the taming of the growl may deceive the innocent into believing that the wicked problem is completely tamed." The moral lesson for West was "whoever attempts to tame a part of the wicked problem, but not the whole is morally wrong" (Churchman 1967, B141–B142).

The Inquiring Systems Trilogy

West wrote that his style was "to proceed from the more specific problem of inquiry to the whole system" (Churchman 1971, viii). This imperative to sweep in all that is relevant is another example of the Singerian influence. West's meditations on the meaning of the whole system appear in three philosophical works: *Challenge to Reason* (Churchman 1968a), *The Design of Inquiring Systems* (Churchman 1971), and *The Systems Approach and Its Enemies* (Churchman 1979). Although these books represent different agendas and perspectives, they are informed by a common underlying quest that might be called an epistemology of inquiry. In these books, West set out to reinterpret the insights of the Western philosophical tradition in the language of system design.

In the preface to *Challenge to Reason*, West mentions what struck him in his study of Leibniz's *The Monadology*: "that a concept of the whole system was essential in understanding how each part worked" (Churchman

1968a, v). West's book was intended to explore the implications of this statement. The book starts with the seemingly simple question: Whether improvement can occur in a sector of a system without considering the whole system? This question leads to what West called the challenge to reason: How to establish improvement pathways by rational means and "What assures us that in our attempts to improve social systems we have considered all that is possible to consider?" (Churchman 1968a, 171):

> We are talking not about techniques of improving performance *given* the goals that certain people wish to attain. We are also talking about whether the goals themselves are proper ones, and we are asking how the scientist can possibly come to answer this question. The underlying theme is that if he fails to answer the question, he fails to "apply knowledge to system improvement" (Churchman 1968a, 15).

In *The Design of Inquiring Systems*, West provides a philosophy of design or planning. To him, the term design includes all "activities in which we consciously attempt to change ourselves and our environment to improve the quality of our lives" (Churchman 1971, vii). West approaches design by asking how knowledge leading to improvement can be gained. He then reviews the ideas of the great thinkers in Western philosophy to examine different approaches to inquiry and learning, restating their insights as different approaches to the design of inquiring systems. For example, what West extracts from the ideas of Leibniz and other rationalists is that "no optimal design of a part of the system is possible without prior knowledge of the 'whole' system" (Churchman 1971, 40).

In the *Design of Inquiring Systems*, West expands on his characterization of systems as teleological in the sense of exhibiting goal-seeking behavior. Purposive systems differ from deterministic systems in an important way: In the former, one can identify purposive individuals who can produce alternatives to meet the individual's objectives. West distinguishes three such roles: the client, the decision maker, and the designer. In addition the goal-seeking behavior can be related to a measure of performance. To illustrate, West considers a college as a system. The measure of performance can be defined as credit hours per dollar of expenditure per semester, and its components are curricula, administration, and services. The client is the set of students paying tuition plus those who can potentially attend the college, while the decision maker comprises trustees, administration, and faculty. The designer can be taken to be the planning committee of the college that intends to recommend changes that will maximize the college's benefit to the client (Churchman 1971, 56–57).

The *Design of Inquiring Systems* raised a wide range of philosophical issues that are rarely found in other works on system design. Since these included the design of computer-based inquiring system, the book found an audience among information-systems scholars. Swanson (1994) describes how he used this book in his teaching to provide a framework for system design.

West lamented that even systems thinkers can turn a blind eye to certain classes of questions: "Common sense would seem to argue that anyone who approaches society systematically would have to defend his own inquiring system. But then a very peculiar thing about many systems analysts is that they do not include themselves in the systems being studied" (Churchman 1971, 6). In warning about the dangers of disciplinary allegiances, West had already identified one obstacle to effective system design:

> I came to realize that there are two "sciences": the one represented by the collection of the disciplines, and the other by the systems approach. The first contains a collection of ideas about methodology and is essentially isolated by its disciplinary politics. The other is an attempt to engage in those areas of inquiry which are most relevant to the social good.... The "science" of the disciplines is an enemy of the science of the system approach (Churchman 1979, 13).

In *The Systems Approach and Its Enemies*, West enlarges the scope of his inquiry to include precisely those who stand outside the system. The word enemies refer to other approaches to solving human problems (such as politics, morality, religion, and aesthetics) that do not limit themselves to the rationalism of systems, and draw upon other sources to suggest alternatives. Since West uses the word enemy dialectically, one should guard against taking the word literally. As one close reader of West's works observed:

> Churchman's use of the term *enemy* has perplexed some people ... he said that he used the term in the spirit of a Christian tradition—to *love your enemy*. His intent was not to polarize his systems approach, which for him was based on reason, against other traditions of inquiry. He only wanted to demonstrate that these designs of inquiry could be compelling, despite their

> distortion or exclusion of rational thought, thus denying the full benefit of human reason as advanced by the systems approach (Nelson 2003, 471).

West's point was that if the enemies are real, important, and compelling, then a comprehensive approach to planning should include them in its vision (Churchman 1979, 148). Neeraj Verma (West's last doctoral student) interprets the term enemy as an object of envy (Verma 2009). When Verma (1998) added analytic philosophy to West's list of enemies, West fully approved.

ENGAGING WITH ETHICS FOR LIFE

To properly appreciate West's work, it is important to realize that the ethical imperative was at the heart of his quest from the very start. West took an early interest in ethical issues and defined his ethics as Quaker-like. To him, the overall ethical purpose that remained invariant was to serve humanity (van Gigch et al. 1997, 743).

> ETHICS IN THE CLASSROOM
>
> "I think one reason a professor may discourage the discussion of ethical issues among his students in class in that he himself has no satisfactory answers; as a person, he may hold strong ethical beliefs, but he cannot 'objectively' profess them since his intellect recognizes that they are all subject to doubt. What he fails to realize is that ethics is not a body of theory substantiated by facts. Instead, it is process of continuously—and I think eternally—discussing and debating and occasionally fighting over the issues. In short, ethics is a dialectical process in which all humanity, past, present, and future must take part" (Churchman 1979, 118).

His early exposure to management convinced West that human civilization "badly needs knowledge of how to manage well, that is ethically." Such a "science of management would have to be a branch of ethics" (Churchman 1994, 107). An early ambition of West's in joint work with Russ was to complete the Singerian program by incorporating the relation between logic and ethics, so that they could "start the design of the unity of science over again, this time with ethics at its base" (Churchman 1990, 127).

In *The System Approach and its Enemies*, West stressed the link between systems and ethics:

> The important feature of the systems approach is that it is committed to ascertaining not simply whether the decision maker's choices lead to his desired ends, but whether they lead to ends that are ethically defensible. Therefore ... it is extremely important to understand which values are ethical and which are not (Churchman 1979, 65).

As West grew older, this ethical imperative assumed a more central place in his inquiry. On his 80th birthday, West declared: "If I could change anything, I would like to be born now because I have an exceptional idea, mainly how do you create a science that will help the human condition" (van Gigch et al. 1997, 737).

To West, service to humanity was the invariant ethical purpose. West expressed moral outrage when science and technology fell short of this goal, as in the failure to eliminate world hunger (Churchman 1982, 9).

West repeatedly questioned the conventional wisdom of his field. For example, he believed that the notion of serving the customer should be expanded to include the customer who is not served or does not have the ability to pay: "ethical management is not reached until the product being produced or service being provided has reached everyone who has a true need for it" (van Gigch et al. 1997, 740). Another example was his insistence to include future generations as stakeholders in the assessment of a systems solution.

THE EDITOR

West was a founding member of TIMS and, in 1954, he was appointed the first editor-in-chief of its journal, *Management Science*, a position he held through 1960. This was not West's first senior editorial position—from 1949 to 1959 he served as editor-in-chief of the Philosophy of Science Association's journal, *Philosophy of Science*. West hoped that the field of MS would be quite different from the field of OR in that MS and its journal, its meetings, and its research would create and design a science of management that lived up to the standards of good science, where as OR "would be the practical application of such a science" (Churchman 1994, 107). And, further, and most important, the journal would be all encompassing:

> All these philosophers—the mathematical purist, the adherent to hard facts, the generalizer, the "case" man—all are committed to a conviction that a

> science of management will stand as a legitimate and recognized field of scientific endeavor. Some are committed to a conviction that this science will stand as the greatest scientific discovery of our age—all are committed to the conviction that no other field of endeavor is as important to man as the field which searches for truths about the ways in which men work and live together. *Management Science* is committed to the conviction that all these philosophies should be given expression in its pages—in articles that emphasize mathematical models, that emphasize measurement and control, that emphasize broad viewpoints, that emphasize specific cases and methods—no matter what the origin of the writer may be—mathematician, physicist, social scientist, biologist, engineer—manager and non-manager—philosopher (Churchman 1955, 187–188).

But, to West's disillusionment, *Management Science* soon became an outlet for technical contributions. Model-building came to be viewed as the core of MS, and the journal's referees only agreed to publish model-building papers. West explains that he decided to "latch onto systems" to avoid a disciplinary label (Churchman 1979, 160). However, even this ran the danger of bearing the systems analysis label.

THE TEACHER, THE PERSON

West's impact on his profession is due, in large part, to his role as a teacher, how he conducted his classes, and his open relationships with his students. This was evident to his earliest students. Russ Ackoff recalls West's style and the climate of the philosophy department at Penn when he first met West:

> West is the least status conscious person and teacher I have ever known. Even when I met him for the first time outside the classroom, he did not impose the student–teacher dichotomy on our conversation. He treated me as an equal even though I wasn't. He acted similarly with all his students.
>
> West was the undergraduate chairman of the department of philosophy. ... It was a program in which we, the graduate students, had at least as large a say as faculty. We formed an intimate and very effective learning consortium into which we admitted a few carefully selected faculty members. West was our spiritual leader, not because he was a member of the faculty, but because he didn't act as though he were (Ackoff 1994, 73–74).

Richard Mason, a Berkeley graduate student in 1965, describes the impressions of a new student entering West's seminar class for the first time and encountering West as he silently wrote on the blackboard his guiding theme: "The purpose of management science is to secure improvement in social systems by means of the scientific method" (Mason 1994, 67). West asked for discussion. Unguided, the students spent much time discussing the sentence and its words and phrases, but all ignored the word secure. After some time, West underlined it. "West explained. We were treated to the product of a life's reflection—reflection that continues to this day" (Mason 1994, 68).

Harold Nelson notes the remarkable range of topics that were discussed at West's once-a-week seminars that were open to all. "It was in this seminar that I learned how Immanuel Kant, the eighteenth-century German philosopher had direct relevance to the pragmatic issues faced by decision-makers in everyday situations" (Nelson 2003, 464). Joana Porra, who had a series of dialogues with West in 1996–1997 at Berkeley, recalls a piece of advice he was fond of giving: "In your thinking, practice going broader and increasingly universal. Go as broad and universal as you can possibly stand" (Porra, 2001, 27).

Over the years, countless students from diverse fields were exposed to West's evolving quest for the integration of philosophy, systems, and MS and OR (Koenigsberg and van Gigch 1994). Many paid him tribute for his impact as a teacher—and took pride in the sweaters West had knit for them! (West was a noted knitter who usually took his knitting kit along when he attended meetings and during his travels.)

West's influence was worldwide due to his books and foreign students. European students, especially Werner Ulrich and Kristo Ivanov, extended West's approach to systems in other fields, contributing to the diffusion of his ideas in Europe. The book of collected papers, *Wisdom, Knowledge, and Management*, gives a broad view—past and current—of how West's ideas have been used, developed, and spread to "all of Social Sciences, Political Sciences, Policy Sciences, Information Sciences and the like" (van Gigch and McIntyre-Mills 2006, xiii).

West was not afraid to confront and discuss serious personal concerns. In an interview, he faulted himself for lacking the humility required to do good science (van Gigch et al. 1997, 737–738). But, his most telling concern was his bout with alcohol, about which he wrote openly:

> . . .and five years ago I was one of them [a person who willfully chooses action(s) that will harm him dreadfully]. I was an alcoholic. I believed as a fact that alcohol was dangerous. I did deny that it was harmful to me, but I also believed somehow that it was dreadfully harmful to me. But I continued to drink. Almost to the point of death. Now, largely because of Alcoholics Anonymous, I know I shouldn't drink, or at least I try hard to know, not by observation, but by spiritual help (Churchman 1994, 105).

Using his logical and analytical talents, he concludes that "greed for wealth, fame, or political power are all addictions" (Churchman 1994, 106). That is, the alcoholic's pathology of never having enough is destructive in the same way as greed is in human systems.

STUDENT KANT

"The modern Ph.D. is an example of one of the disasters created by disciplinary science. An alert and competent Ph.D. candidate often has a very broad question in mind which he is eager to investigate. His professors simply will not let him do it. He is told he has to narrow the topic to 'manageable dimensions.' I like to imagine what a modern dean would have said to Kant in the 1760s when Kant explained to him that he was about to embark on a voyage to create the necessary conditions for any future metaphysics. Such a dean would surely point out that Kant's ambitions were much too broad" (Churchman 1990, 127).

KNIT 1, PURL 2.

"West was a powerful teacher. He made students feel different. He knew how to move them and to awake their intellectual curiosity. He raised their awareness in ways that made them reflect on the meaning of their academic education and set themselves new goals. . .. His weekly 'informal seminar' sessions were proverbial. . . . Patiently he listened to the comments of everyone who wished to comment and accepted what they had to say. In other sessions, he left the topics entirely to the participants and was mainly listening. With his head bent over a piece of knitting to which he seemed to dedicate all attention, he would only now and then throw in a short question or comment. . .. He radiated something that few could define clearly, but clearly they had been missing it in their studies before" (Ulrich 2004a, 200).

Core Themes

West's research, writings, and professional activities have caused many of those working in MS and OR to broaden their views of these fields. His philosophical themes of systems and ethics have shown how MS and OR can be of service to humanity. "Academic philosophy wasn't satisfying to him. He wanted philosophy to have meaning in the world.

He wanted to insert an ethical dimension into science. And he really made it is job to remind all these CEOs that they had ethical responsibilities" (Gloria Churchman 2004).

In a first reading, West's last four books will likely appear bewildering than edifying. His way of repeatedly proposing possible answers, only to knock these down as incomplete, suboptimal, or naïve can be infuriating. But West viewed himself as the perpetual inquirer, where the goal of the inquiry was to learn to ask better questions. When he was nearly 40 years, he tersely stated what was to become his practice in the latter half of his life: "moral behavior consists of making man less and less satisfied with his answers to better and better problems" (Churchman 1952, 65).

As Ulrich (2004a, 207) observes, West's name for the subject matter of his quest changed from experimentalism to OR, to MS, to systems approach, and, finally, to social system design. Each time that his chosen subject area proved to be too narrow for the concerns he really cared about, or set up disciplinary boundaries that he found to be procrustean, West moved on. It turned out that West's desire for comprehensiveness, the interdisciplinary worldview, and incorporation of the human element coincided with the period where OR and MS in the U.S. were solidifying their identities as sciences using the positivistic paradigm and focusing on advances in techniques. As a result, his influence lessened in the U.S. as compared to Europe.

His last book, *Thought and Wisdom* (Churchman 1982), contains West's most personal retrospective account of his career and works. The book reviews the core themes he pondered during a lifetime of reflection. In his 1981 lectures in systems science at UCB, West summarized this lifelong quest:

> The design of my philosophical life is based on an examination of the following question: is it possible to secure improvement in the human condition by means of the human intellect? The verb "to secure" is (for me) terribly important, because problem solving often appears to produce improvement, but the so-called "solution" often makes matters worse in the larger system. . . . The verb "to secure" means that in the larger system over time the improvement persists. I have to admit that the philosophical question is much more difficult than my very limited intellect can handle . . . though I've done my best to tap the wisdom of such diverse fields as psychology, economics, sociology, anthropology, public health, management science,

> education, literature, and history. But to me the essence of philosophy is to pose serious and meaningful questions that are too difficult for any of us to answer in our lifetimes. Wisdom, or the love of wisdom, is just that: thought likes solutions, wisdom abhors them (Churchman 1982, 19–20).

West died at the age of 90 of complications of Parkinson's disease on March 21, 2004, in Bolinas, California, not far from his oceanside cabin. He was survived by his wife Gloria, son Daniel (Josh) Wharton Churchman, grandson Kyle West Churchman of Bolinas, and granddaughter Jenna Rose Churchman of Washington, DC.

Honors and Awards

West was president of TIMS (1962) and president of the International Society for the Systems Sciences (1989). He was a fellow of the American Association for the Advancement of Science. In 1968, West received the McKinsey Book Award for *The Systems Approach*, and the Best Book in Management Award from the Academy of Management for *The Challenge to Reason*. He was the recipient of three honorary doctorates: Washington University in St. Louis (1975); University of Lund, Sweden (1984); University of Umeå, Sweden (1986). In 1983, West received the Berkeley Citation for distinguished or extraordinary service to the University. He received the 1999 LEO (Lyons Electronic Office) Award for Lifetime Exceptional Achievement in Information Systems, presented jointly by the Association for Information Systems and the International Conference on Information Systems.

Acknowledgments

I would like to thank Professor Neeraj Verma of the University at Buffalo, SUNY, for reading and commenting on an earlier version of this profile.

References

Ackoff R (1988) Redesigning the future: C. West Churchman. Syst Pract 1(4):351–355

Ackoff R (1994) Higher education and social stratification. Interfaces 24(4):73–82

Arnoff E (1957) Operations research at the Case Institute of Technology. Oper Res 5(2):289–292

Britton F, McCallion H (1994) An overview of the Singer/Churchman/Ackoff school of thought. Syst Pract 7(5):487–521

Church A (1943) Review of "Towards a general logic of propositions." J Symbolic Logic 8(1):53–54

Churchman CW (1942) Towards a general logic of propositions. In: Clarke FP, Nahm MC (eds) Philosophical essays in honor of Edgar Arthur Singer, Jr. University of Pennsylvania Press, Philadelphia, PA, pp 46–68

Churchman CW (1946) Philosophical aspects of statistical theory. Philos Rev 55(1):81–87

Churchman CW (1948) Theory of experimental inference. Macmillan, New York, NY

Churchman CW (1952) Ethics, ideals, and dissatisfaction. Ethics 63(1):64–65

Churchman CW (1955) Management science: the journal. Manage Sci 1(2):187–188

Churchman CW (1960) Case histories five years after—a symposium: sampling and persuasion. Oper Res 8(2):254–259

Churchman CW (1961) Prediction and optimal decision: philosophical issues of a science of values. Prentice-Hall, Englewood Cliffs, NJ

Churchman CW (1967) Guest editorial: wicked problems. Manage Sci 14(4):B141–B142

Churchman CW (1968a) Challenge to reason. McGraw-Hill, New York, NY

Churchman CW (1968b) The systems approach. Delacorte Press, New York, NY

Churchman CW (1970) Operations research as a profession. Manage Sci 17(2):B37–B53

Churchman CW (1971) The design of inquiring systems, basic concepts of systems and organization. Basic Books, New York, NY

Churchman CW (1979) The systems approach and its enemies. Basic Books, New York, NY

Churchman CW (1982) Thought and wisdom. Intersystems, Seaside, CA

Churchman CW (1984) Early years of the philosophy of science association. Philos Sci 51(1):20–22

Churchman CW (1990) Ackoff comes of age. Syst Pract 3(2):125–130

Churchman CW (1994) Management science: science of managing and managing of science. Interfaces 24(4):99–110

Churchman CW, Ackoff R (1946) Psychologistics. University of Pennsylvania, Philadelphia, PA

Churchman CW,Ackoff R (1947) An experimental measure of personality. Philos Sci 14(4):304–332

Churchman CW, Ackoff R (1954) An approximate measurement of value. Oper Res 2(2):172–181

Churchman CW, Ackoff R, Arnoff E (1957) Introduction to operations research. Wiley, New York, NY

Churchman CW, Ackoff R, Wax M (eds) (1947) Measurement of consumer interest. University of Pennsylvania Press, Philadelphia, PA

Churchman CW, Epstein B (1946) Tests of increasing severity. J Am Stat Assoc 41(236):567–590

Churchman CW, Schainblatt A (1965a) The researcher and the manager: a dialectic of implementation. Manage Sci 11(4):B69–B87

Churchman CW, Schainblatt A (1965b) On mutual understanding. Manage Sci 12(2):B40–B42

Churchman G (2004) Obituary: C. West Churchman. San Francisco Chronicle, March 25, 2004.

Crane R, Churchman CW (1960) Foreword to monograph on case histories in management sciences. Manage Technol 1(1):1

Dean B (1994) West Churchman and operations research: case institute of technology, 1951–1957. Interfaces 24(4):5–15

Edie L (1954) Traffic delays at toll booths. J Oper Res Soc Am 2(2):107–138

Epstein B, Churchman CW (1944) On the statistics of sensitivity data. Ann Math Stat 15(1):90–96

Koenigsberg E, van Gigch J (1994) Introduction: in celebration of the 80th birthday of C. West Churchman. Interfaces 24(4):1–4

Mason R (1994) One man's quest for the meaning of therefore. Interfaces 24(4):67–72

Mason R (2004) IFORS' operational research hall of fame: C. West Churchman. Int Trans Oper Res 11(5):585–588

Morse P, Kimball G (1951) Methods of operations research. Wiley, New York, NY

Nelson H (2003) The legacy of C. West Churchman: a framework for social systems assessments. Syst Res Behav Sci 20:463–473

Porra J (2001) A dialogue with C. West Churchman. Inform Syst Front 3(1):19–27

Rosenbleuth A, Bigelow J, Wiener N (1943) Behavior, purpose and teleology. Philos Sci 10(1):18–24

Singer E (1959) Experience and reflection. Churchman CW (ed). University of Pennsylvania Press, Philadelphia, PA

Swanson E (1994) Churchman's theory of design integrity. Interfaces 24(4):54–59

Ulrich W (2004a) In memory of C. West Churchman (1913–2004): reminiscences, retrospectives, and reflections. J Oper Transform Soc Change 1(2–3):199–219

Ulrich W (2004b) Obituary: C West Churchman, 1913–2004. J Oper Res Soc 55(11):123–1129

van Gigch J, Koenigsberg E, Dean B (1997) In search of an ethical science: an interview with C. West Churchman, an 80th birthday celebration. J Bus Ethics 16(7):731–744

van Gigch J, McIntyre-Mills J (eds) (2006) Wisdom, knowledge, and management: a critique and analysis of Churchman's systems approach. Springer, New York, NY. (vol 2 in the series: C. West Churchman's Legacy and Related Matters)

van Voorhis (1953) Sampling methods in railroad accounting. Oper Res 1(5):259–262
Verma N (1994) Organizations and their purposes: a note on Churchman's philosophy of management. Interfaces 24(4):60–66
Verma N (1998) Similarities, connections and systems: the search for a new rationality for planning and management. Lexington Books, Lanham, MA
Verma N (2009) Personal communication
Wald A (1942) On the principles of statistical inference. Notre Dame Mathematical Lectures, Number 1. University of Notre Dame, South Bend, IN
Wiener N, Rosenbluth A (1950) Purposeful and non-purposeful behavior. Philos Sci 17(4):318–326

12

William W. Cooper

Timothy W. Ruefli and
Robert R. Wiggins

The pioneers of operations research (OR) came from many different social, economic, and educational backgrounds. Their entry into OR was often due to a chance meeting, an unrelated job opportunity, or the influence of a teacher. The William (Bill) Cooper story is a most amazing case in point. He rose from a most difficult social and economic milieu to become one of the most respected and honored members of the OR profession.

Bill is professor emeritus of management, finance, and accounting at the McCombs School of Business, University of Texas at Austin. During his academic career, working alone or with his long-time academic collaborator, Abraham Charnes, he was responsible for major theoretical developments and the bringing of these ideas into the world of practice: goal programming, data envelopment analysis, fractional programming, chance-constrained programming. His very early joint research on refinery operations brought linear programming (LP) to the attention of the industrial and business worlds and was a key factor in the acceptance of OR outside of the military.

Bill received the 1982 John von Neumann Theory prize (jointly with Abe Charnes and Richard Duffin) from the Operations Research Society of America (ORSA) and The Institute of Management Sciences (TIMS), and the Gold Medal Award by the Society for Multiple Criteria Decision Making (MCDM). He was a founder of TIMS and served as its first president in 1954.

A.A. Assad, S.I. Gass (eds.), *Profiles in Operations Research*, International Series in Operations Research & Management Science 147, DOI 10.1007/978-1-4419-6281-2_12,

TOUGH BEGINNINGS

Born July 23, 1914 in Birmingham, Alabama, William Wager Cooper, sometime later (courtesy of his parents, William W. Cooper and Rae Rossman Cooper) joined the great, employment-motivated migration of the time north to Chicago. Along with his brother and sister, Leon and Ruth, and his parents, he lived in a tough neighborhood on the west side of Chicago, on a street made famous by Studs Terkel in his book *Division Street: America*. This was not too far north of Little Italy—an even rougher neighborhood. His father owned a string of 16 gas stations. Bill went to local grammar schools and then to Tuley High School—one of the best academic schools in Chicago—but he had to drop out to help support his family when his father's business went under during the Great Depression.

© 2002 Peter Yang/Austin American-Statesman/ WpN

Bill's ability to earn a living for his family was an exercise in the application of his skills to his environment. Major league criminal gangs and their farm teams, the street gangs, were a fact of his Chicago life. Bill learned to fight at an early age. His determination, along with ample practice—he reports 30 "scrapes" in just one summer—made him a winner (Beach 2002). Faced with the need to earn money, he turned to professional boxing, debuting as a bantamweight at $25–$35 per fight. Managed by a guy linked to the rackets, Bill won 58 and drew 2 of 63 bouts during his career. Additional remuneration was provided by setting pins in a bowling alley and by being a caddy at a golf course on the north side of the city.

Even then, Bill appreciated the value of research as way to improve things. His library card gave him access to books such as the one on physiological psychology that he read in attempt to improve his competitiveness as a boxer. It is not clear what effect that book had on his won/lost record in the ring but, in a most interesting twist of fate, that very book played a part in starting Bill on a radically different sort of a career—in business, government, and academia.

In the autumn of 1932, while hitching a ride north on the Waukegan Road to caddy at a golf course east of Evanston, a car driven by Eric Louis Kohler stopped. Kohler was an accountant with an A.B. (1914) and M.A. degrees (1915) in economics from the University of Michigan and Northwestern University, respectively. Prior to joining the Northwestern University faculty from 1922 to 1928, Kohler had established his own accounting firm. Earlier, he had practiced with the accounting firm of Arthur Andersen & Co. in Chicago, and rejoined Andersen for a few years after he left Northwestern (Mautz and Previts 1977). "In spite of the fact that Bill was the toughest-looking guy Kohler had ever laid eyes on, Kohler stopped and picked him up" (Beach 2002). As they drove along, the topic of Bill being a fighter arose which caused him to mention the book he was reading on physiological psychology. The dissonance between Bill's tough looks and his investigations into the scientific literature apparently intrigued Kohler. He stayed in touch with Bill over the next year (in what must be in contention for the longest admissions interview in history) and ended up convincing him to drop boxing in favor of an undergraduate education at the University of Chicago—paid for by Kohler! Part of the attraction of the offer was that going to college would put an end to people asking Bill if he had his high school diploma.

Despite his family's reservations at the time (the loss of Bill's income was significant), Bill entered the University of Chicago in the autumn of 1934 as an undergraduate with a major in physical chemistry—chosen as much for its job potential as anything. The academic environment was a radical change for Bill. As he noted, "There was all this intellectual conversation all around me. I was dumbfounded" (Beach 2002). But, one would suspect, not so dumbfounded that he did not participate. One of Bill's dorm-mates at Chicago, Herbert Simon, became a life-long friend and colleague.

Yet, another significant decision point in Bill's career involved Kohler. Andersen & Co. had been engaged to provide expert witness testimony in a patent infringement case and Kohler asked Bill to check some of the mathematics. Bill did, and found some errors. As a result, he was hired by Andersen part-time during the school year and full-time during the summer. Given the opportunity for continuing employment, Bill switched majors to economics. He graduated Phi Beta Kappa in 1938 and took a position with Andersen—the only one in his class he knew who gained employment.

Soon after, when Kohler was appointed comptroller of the Tennessee Valley Authority (TVA) in 1938, Bill followed him to Knoxville, Tennessee as his assistant. In the first year, not only did Bill do such tasks as preparing his boss to testify about cost allocation before a joint congressional committee, but also he developed new types of audits for municipal and regional activities programs. In his second year, he served as head of the internal audit unit. In contrast, Bill was also president of the United Federal Workers Union Local 24 and also a member of the Board of the Southeastern Region Conference of the Congress of Industrial Organizations. When Bill went to a meeting of Local 24, he found all of the members present were white. Bill was informed that there was a parallel group of black union members who met separately and had their own treasury. Pragmatist that he was, Bill simply invited the black union members to attend the meetings of the white members; when they did, and when no objection was voiced, Bill had succeeded in integrating a trade union in a southern border state in 1938!

In 1940, Bill left the TVA to enroll in the Ph.D. program at the Graduate School of Business at Columbia University in New York City. In 2 years, he completed all of the requirements for the degree—except for his thesis. James Bonbright was his chair, with William Vickrey (future Nobel Laureate in economics) also on the committee. It's not that Bill did not write a thesis, he did, and defended it twice! His topic bucked the economic orthodoxy that decision making in a firm was centralized by proposing that multiple agents in the firm were responsible for some decisions. Both times his committee did not accept the thesis—but they also did not reject it. What they did, apparently, was not understand it. Bill recognized a stonewall when he encountered one and, with the country at war, in 1942, he went to Washington D.C. to help the best way that he could. He was appointed Principal Economist for the Bureau of the Budget (the forerunner of the Office of Management and Budget). In that position, Bill was "in charge of coordination all U.S. government work on accounting statistics and related aspects of programs for war procurement programs, price control, production allocations and related economic studies" (Cooper 2008a).

One night in May of 1944 Bill went on a blind date, the venue of which was a party at the apartment of Mary Jane and Richard Watt on 16th Street in Washington, D.C. There was a poor beginning to the date—Mary Jane, who had arranged for Ruth Fay West to be Bill's date, neglected to introduce the two parties. Bill and Ruth found each other anyway, and they hit it off. Unable to get a cab after the party, Bill offered to walk Ruth to her home on Connecticut Avenue. After all, he, too, lived on Connecticut

Avenue, but, as it turned out, on the opposite end from Ruth. So, when they reached her place, she offered to return the favor and walk him home. When they got there, being a gentleman, he could not let her walk home alone, so they retraced their steps. Through the night they yo-yoed back and forth on Connecticut Avenue until about 6 a.m., at Bill's end of the circuit. They startled his roommate, Sam Nakasian, who fixed them all breakfast (Cooper 1980).

Ruth, born in 1917, grew up on a horse ranch in Montana. Her father died when she was in her teens. By the time she graduated from high school in the midst of the Depression, she could afford only one semester at Montana State in Bozeman before she had to go to work to support herself and aid her family (West 1980). She was employed by the War Relocation Authority working on the rights of Japanese-Americans interned in relocation camps when she met Bill. They were married in Baltimore in September, 1944.

Bill was recruited by the economist Theodore Schultz and returned to the University of Chicago in 1944 to begin his academic career as an instructor in economics (Gleeson and Schlossman 1992a, b). He and Ruth lived on South Ellis Avenue, within walking distance of the university. Ruth found work with the Chicago Commission on Race Relations and studied at the University of Chicago, receiving an A.B. degree in 1946. While Bill was at Chicago, he and Eric Kohler published their article (Kohler and Cooper 1945) summarizing what they had learned about accounting for the war effort, which the American Institute for Accountants awarded them the prize for the most valuable article on an accounting subject that year (Cooper 2008a). But Bill and Ruth's time at Chicago was short. In 1946, George Leland Bach, himself a Chicago graduate, hired Bill as an assistant professor in the new economics department he was building at the Carnegie Institute of Technology (Carnegie Tech) (Gleeson and Schlossman 1992a, b).

Ruth continued her undergraduate studies at Carnegie, receiving her B.S. She went on to receive an L.L.B. from the University of Pittsburgh Law School and opened a law practice in Pittsburgh, one of the few women lawyers to do so at the time. She later moved her practice to Russellton, PA. where she became involved with helping miners and their families.

In 1949, William Larimer Mellon, President of the Gulf Oil Company, gave Carnegie Tech six million dollars to establish what became the Graduate School of Industrial Administration (GSIA). Bill was one of first three faculty members (Bach was first and Elliott Dunlap Smith was second).

His old undergraduate friend from Chicago, Herbert Simon, was recruited as the fourth member. Bill was known for his strong opinions, open mind, hard work, and his collegiality. Alan Meltzer, a former colleague, remembers that Bill always carried two packed briefcases to and from GSIA (Meltzer 2007). Bill became the leader of the OR faculty at GSIA (Gleeson and Schlossman 1992a, b). He was able to attract both funding and problems from external sources, thus providing a real-world focus for theoretical and methodological developments.

GRACIOUS TO A ROOM

Bill Cooper was instrumental in getting me (Egon Balas) to Carnegie Mellon. (Carnegie Tech joined with the Mellon Institute of Science in 1965 to form the Carnegie Mellon University.) In 1966, while I was still in Europe as a fresh emigrant from behind the Iron Curtain, he persuaded Dick Cyert, at that time the Dean of GSIA, to offer me a visiting professorship at the school for the academic year 1967–1968. That was the beginning of an association of more than four decades.

In the spring of 1967, on my way to Stanford where I was to spend the summer, I stopped for a couple of days in Pittsburgh to visit the school where I was supposed to return in September, having accepted Dean Cyert's offer. Bill Cooper, whose work I greatly admired but whom I had never met in person before, impressed me with his open mindedness, his inquisitive attitude, and above all by his human warmth and generosity. To illustrate the latter, I will invoke a single episode:

As Bill was showing me around GSIA and introducing me to some of my future colleagues, at one point we entered a very nice, spacious room with a view to nearby Schenley Park. "Oh, what a nice office," I exclaimed admiringly. "Would you like this to be your office when you join us in the fall?" asked Bill. When I said that the office in question was obviously in use and I would not want to take away somebody else's room, Bill insisted that I could have the office. I did not take him up on his offer, but as I found out later in the fall, the room in question was Bill's own office (Balas 2007).

It was at Carnegie Tech that Bill began his long collaboration with Abraham Charnes that contributed so much to management science (MS) and OR. The relationship began in 1949 (Cooper 2002). Bill and Abe co-authored over 200 papers, book chapters, proceedings articles, and 12 books and monographs in an association that ended in 1992 with Abe's death. From this partnership came seminal developments on goal programming (GP), data envelopment analysis (DEA), fractional programming (FP), and chance-constrained programming (CCP).

Abe and Bill's first co-authored paper reported their pioneering research on the blending of aviation gasolines (Charnes et al. 1952). It described the first application of LP to an industrial problem and motivated all the major U.S. oil companies to organize OR and computer-based

Ruth and Bill
UT College of Business Administration (now McCombs) Hall of Fame, May 1990.

research departments (Phillips and Seiford 2006; Symonds 1955). Today, just about all of the world's refineries are managed and operated based on such methods and their extensions. This research was the prototype for the publication of a wealth of articles that combined MS/OR methodologies with applications to real-world problems. It also provided a model for Bill's subsequent research, which he also employed in organizing MS/OR research efforts at GSIA. He divided the Carnegie OR group into two teams—the two-man team of Abe and Bill, and a remarkable, multi-member team consisting of Charles Holt, Franco Modigliani, John Muth, and Herbert Simon. The latter team, "which operated under the leadership supplied by Holt, centered its attention on a single application at the Springdale, Pennsylvania, plant of the Pittsburgh Plate Glass Company (now PPG Industries). Although following a different strategy, this work was also directed toward developing analytically based methods of management that could be (and were) computer implemented. Using the knowledge of company personnel (but not incorporating these persons in the research), this team concentrated on planning for production, work forces, and inventory in the paint factory at Springdale. The objective was to develop analytically based, computer implemented methods that could improve the performances of this plant. This, of course, was also to be a prelude to the generalizations that the team was subsequently to publish" (Cooper 2002, 38)—*Planning Production, Inventories, and Work Force* (Holt et al. 1960). This approach was later conceptualized by Bill as applications-driven theory development.

In 1968, Bill left GSIA, but stayed at Carnegie (now Carnegie Mellon) to become the founding Dean of the School of Urban and Public Affairs. The new school was intended, in part, to extend the developments in MS to the field of public administration. In 1975, Bill was appointed the Arthur

Lowes Dickinson Professor of Accounting at the Harvard Business School. He then moved, in 1980, to what is now the McCombs School of Business, University of Texas at Austin, where he was appointed the Foster Parker Professor of Management, Finance, and Accounting (a recognition that no one department was enough to contain his interests)—a position he still holds in an emeritus status. Ruth taught law and business ethics at the School of Business for a while and then served as Director of the Austin Symphony Society. She died August 5, 2000.

Beach (2002) observed, "Social justice also was important to Cooper and Ruth, a lawyer. Ruth was a fighter for women's and minority rights. Cooper marched with the Black Panthers to help get more minorities in the Pittsburgh craft unions. He still wears a ring with a clenched fist, the symbol of the Panthers." (If you look closely at Bill's portrait above, you can see that ring on his left hand.)

Bill's academic duties were always his primary concerns, but he recognized that research in OR was driven by continued involvement in real-world applications. Thus, over the years, jointly with colleagues and students, he took on over 200 consultancies including: the Marshall Plan, General Motors Corporation, EXXON, TRW Inc., Hughes Aircraft Co., Teledyne, U.S. Army, U.S. Navy, U.S. Air Force, U.S. Coast Guard, Arthur Andersen & Co., Market Research Corporation of America, U.S. Department of Health, Education and Welfare, and the U.S. Department of Commerce. Most of these engagements led to one or more publications, all of which illustrate instances of application-driven theory development.

Contributions to MS/OR: GP, DEA, FP, and CCP

Bill's academic publishing career started while he was a graduate student at Columbia with a paper on public utility regulation that appeared in the *Journal of Political Economy* (Cooper 1943). More than 400 articles have followed, with still more in process. He has enough articles in the top journals of each of the fields of economics (econometrics), accounting, and MS/OR to satisfy the careers of most academics in any one of those areas. Here we focus on his contributions to MS/OR. Among his most influential research and associated publications are those that initiated the fields GP, DEA, FP, and CCP.

GP first appeared in the *Management Science* paper "Optimal estimation of executive compensation" (Charnes et al. 1955). Their co-author, Robert Ferguson, was a consultant with Methods Engineering Council, a Pittsburgh-based firm. As Bill relates (Cooper 2007): "Here, what later became to be called goal programming, took the form of a derivation to an equivalent linear-programming problem of a least absolute value (LAV) regression. This regression was to be used to develop a method to determine the salaries of executives at the General Electric Company's Major Industrial Appliance Division in Schenectady, N.Y. LAV was chosen because it mitigated the effects of 'outliers' as is true of ordinary least squares. Computational difficulties in direct application of LAV regressions were resolved by transforming the LAV formulation to an equivalent ordinary linear program, a method that has since been followed in all of the further uses (by others) of LAV regressions." This formulation was subsequently given the name goal programming in Bill and Abe's two-volume *Management Models and Industrial Application of Linear Programming* (Charnes and Cooper 1961). Bill notes (Cooper 2007): "We chose this name because we had encountered numerous management planning problems that lent themselves to this already formulated linear-programming equivalent of a LAV regression which we had previously made in the above cited article. This name gave form to all of the subsequent uses for management-planning research in this area" (Charnes and Cooper 1977).

The success of Bill's GSIA research approach, especially the GP analysis that ranged from LAV regression computations to GP management-planning formulations, led Bill and Abe to develop a research strategy that Bill refers to as applications-driven theory. He explains it this way (Cooper 2007): "By this is meant that the research is to begin with an actual application problem, usually in association with knowledgeable company officials, as in, for instance, the role played by R. O. Ferguson in the *Management Science* article [Charnes, Cooper and Ferguson 1955]. The basic idea is as follows: First, an application directs the research to a successful conclusion in a particular company. It is then generalized and published as a contribution to general knowledge with further effects like those I just noted for goal programming." The concept of applications-driven theory and its classroom use is discussed further in Cooper and McAlister (1999).

This applications-driven theory is the route they followed in other research such as DEA, first discussed in Charnes et al. (1978). This paper has its origin in the Ph.D. thesis, "Data envelopment analysis and approaches for measuring the efficiency of decision-making units with an application to Program Follow Through in U.S. education" by Eduardo Rhodes at Carnegie Mellon University's School of Urban and Public

Affairs. It originated in work he was doing with Abt Associates, a Boston-based consulting firm. Abt was retained by the U.S. Department of Health, Education, and Welfare to evaluate the results of Program Follow Through, a federally funded experiment in education originally intended to provide a continuation of Head Start services to students in their early elementary years. (Head Start delivered educational, health, and social services to disadvantaged preschool children and their families.) Bill described the impetus of DEA in Cooper (2007): "Every statistical technique we tried failed to give satisfactory results. We were thus met with frustration until Rhodes called my attention to (a technique in) a little known article by M. J. Farrell (1957) which we generalized, extended, and named 'Data Envelopment Analysis' (DEA)." The application of DEA to Program Follow Through is discussed in Charnes et al. (1981). The theoretical and application impact of DEA has been enormous. A survey of the first 30 years of scholarly literature in DEA lists more than 4000 publications (articles, books, monographs, etc.) involving more than 2500 authors in 42 countries, and the list continues to grow (Emrouznejad et al. 2008). [In his response to receiving the MCDM Gold Medal Award, Bill gave a detailed description of the origins, uses of, and relations between GP and DEA (Cooper 2005).]

FP had its OR beginnings in the *Naval Research Logistics Quarterly* paper "Programming with linear fractional functionals" (Charnes and Cooper 1962). Bill notes (Cooper 2007): "I don't recall the company that gave rise to this problem, which involves optimizing a ratio of two linear forms subject to linear inequality constraints, which we showed can be converted to an ordinary linear-programming problem." Siegfried Schaible refers to this paper as the "classical" article that has given rise to a whole literature dealing with the topic of optimizing ratios in various forms (Schaible 1996, 234). The range of applications of FP includes resource allocation, transportation, production, and DEA, and involves linear, quadratic, or concave fractional programs.

CCP first appeared in the *Management Science* paper "Cost horizons and certainty equivalents: An approach to stochastic programming heating oil production" (Charnes et al. 1979). Gifford Symonds was Director of Refinery Research at Standard Oil of New Jersey (now Exxon Mobil), which was then the world's biggest supplier of heating oil. The problem they attacked involved the scheduling of heating oil production. Bill describes the application and resulting model structure in Cooper (2007):

The problem was inherently probabilistic since neither the demand nor the weather that gave rise to the demand were known at the time the production had to be undertaken. This commodity, according to the company, was also fraught with issues of "public interest," since shortages, that could happen in a period of extreme cold, could also give rise to Congressional involvement. Hence, a method was needed to incorporate the probabilistic aspects into the analysis. This was done by modifying the structure of the related LP model by replacing the usual LP constraints by inequalities that captured the inherent uncertainly, that is, by chance constraints. These new type of constraints placed very high probabilities on meeting all demands, as it would have been impossible to ensure with certainty that demands were always satisfied. These high probabilities were intended to guard against shortages in the event of an unforeseen cold snap. The company set up a Risk Committee to assess and choose the probabilities in ways that could be justified even if a shortage did occur. (Shortages might occur even when the production was adequate as when a heavy snow fall made it impossible to deliver available heating oil.)

This approach was so successful that the company applied it to other problems. For example, it was used to determine when to buy or lease tankers for the company's fleet. (It had what was then the biggest tanker fleet maintained by any company or even a country like the USSR). The CCP structure has given rise to a diverse set of applications: water resource management, energy production, circuit manufacturing, chemical engineering, telecommunications, and finance; related topics include stochastic or probabilistic programming.

SYNTHESIS

Bill, who was always searching for ways to improve human decision-making paradigms, extended the basic CCP structure to show how it could be used to join together supposed irreconcilable approaches. To illustrate such a possibility, he developed a CCP model for resolving the classical decision problem of choosing the best alternative (solution) from a set of alternatives compared against each other with respect to specified criteria (Cooper 2007). His structure attacks the competing approaches of optimizing vs. satisficing, as articulated by Herbert Simon in his book, *Models of Man* (Simon 1957). Simon argued that the goal of optimizing must be replaced with the goal of satisficing—the selection of an alternative solution that first meets the decision maker's stated aspiration levels for the criteria. Humans lack both the knowledge and computational skill required to make choices in a manner compatible with economic notions of objective rationality (Simon 1987). Bill's current developments in this area (work in progress) shows how the two can be joined and generalized in an optimizing form (Cooper 2007).

WHO'S COUNTING?

Bill's publications are so numerous that listing them would require more space than this entire chapter has been allotted. His first publication dealt with capital gains taxation (Cooper 1938), his first journal article discussed public utility regulation (Cooper 1943), and his first book chapter was on financial statistics (Cooper 1946). His first book was *An Introduction to Linear Programming* (Charnes et al. 1953). This important book was the first nonacademic journal or conference proceedings publication that brought to the OR and economic communities the basic mathematical and economic aspects of LP, including the famous nut-mix problem. It was followed by 26 more books, including one in each of the last 3 years (2006–2008). Bill's rate of publication averages six per year over his career, but, in the last 17 years the rate is an even higher—eight per year!

It is one thing to be prolific, but even more impressive to be relevant. Bill is rated as highly cited by the Institute for Scientific Information indexing service. The Social Science Citation Index (1965 to present) notes many of Bill's papers that are highly cited. In particular, the first DEA paper (Charnes et al. 1978) is the most highly cited paper that has appeared in the *European Journal of Operational Research* being cited over 1700 times, and was selected as one of the 30 most influential papers published in the first 30 years of that journal. Fifty-seven of Bill's other papers have citation rates ranging from 800 to 10. A comprehensive review of Bill's contributions to OR and MS is given in the paper by Glover and Sueyoshi (2009).

SERVICE TO THE PROFESSION

Bill has made outstanding contributions to the MS/OR profession. He was a founder of TIMS and served as its first president in 1954 (Hopp 2004; Horner 2002). He also initiated the TIMS journal, *Management*

Science, and appointed C. West Churchman, Case Institute of Technology (now Case-Western Reserve University) as the first editor. Bill, with the help of his wife Ruth, developed a policy statement, which the journal still basically adheres to. In addition, Bill was president of the Pittsburgh chapter of the American Society for Public Administration (1974–1975).

Throughout his career, Bill has been active in the editorial aspects of many of his profession's journals: Associate Editor for *Management Science, Operations Research, Naval Research Logistics Quarterly*, and *Operations Research Letters*; founding editor in chief, *Auditing: A Journal of Practice and Theory*; editorial boards of *Research in Governmental and Nonprofit Accounting, Socio-Economic Planning Sciences*, and the *European Journal of Operational Research.*

Bill's character also showed up in his editorial decisions. In 1962, as Associate Editor of the Theory Section of *Management Science*, Bill was faced with two strongly negative referee reports on a submission. He overrode the advice and gave approval for the first publication of William Sharpe's simplified explanation of Markowitz's Capital Asset Pricing Model (Sharpe 1963). Sharpe, along with Harry Markowitz and Herman Miller, was a recipient of the 1990 Nobel Prize in economics.

HONORS AND AWARDS

Bill received honorary D.Sc. degrees from Ohio State University (1970) and from Carnegie Mellon University (1982), and the degree of Doctorado Honoris Causa from the University of Alicante (1995). He is a fellow of the Econometric Society (1956), American Association for the Advancement of Science (1963), Accounting Researchers International Association (1976), and the Institute for Operations Research and the Management Sciences (INFORMS) (2002). He is a member of the International Federation of Operational Research Societies' Hall of Fame (Banker 2006), and the Accounting Hall of Fame (1995).

Bill was awarded the 1982 ORSA and TIMS John von Neumann Theory Prize, joint with Abe Charnes and Richard J. Duffin, for fundamental contributions to optimization methods, concepts, and models for problems of decision, planning, and design; the 2004 Gold Medal Award by The International Society of Multiple Criteria Decision Making for contributions to the theory, methodology, and practice of

MCDM; the 2006 INFORMS impact Prize, joint with Abe Charnes, for their seminal research and application of their DEA research; and, the 2007 Production and Operations Management Society's Award for contributions to Operations Research and Production and Operations Management. He was the Omega Rho (OR National Honorary Society) Distinguished Lecturer (1997) and the British Operational Research Society's Blackett Memorial Lecturer (1997).

A Provisional Summing Up

At this time, amazingly enough, it is premature to provide a final summary of Bill's career. He still goes to the office 7 days a week and often has lunch at the faculty club with a group of friends and colleagues. He no longer teaches, but makes himself available to students, former Ph.D. students (over 100) whose committees he chaired, and faculty members who seek him out for help with thorny research problems. Administrators look to him for counsel on policy and personnel decisions. Perhaps a carry-over from his boxing days, he works out every day—but doctors have advised against continuing his favored one-handed chin-ups.

> COMBATIVE INTELLECTUALISM
>
> Bill's early career as a boxer informed his academic career. In his writings, in the classroom, and in presentations he could be combative. This aspect of his personality showed up early. Once when he was president of the graduate student Economics Club at Columbia, during a presentation by the noted Austrian economist Joseph Schumpeter, Bill peppered him with so many mathematical questions that finally, exasperated, Schumpeter (knowing Bill was a mathematical economist) said "Mr. Cooper, the smile of a duchess cannot be captured by your equations" (Cooper 2008b).

At a dinner celebrating Bill's 75th birthday, George Kozmetsky, a colleague of his at GSIA and his dean at Texas, gave a succinct summary of Bill's career

> As an academic entrepreneur, Bill's inputs have had important impacts on the fields of accounting, finance, marketing, quantitative methods, managerial strategy, risk management, human resources management, management science and ethics. These contributions are legendary. They have been insightful, far-reaching and practical (Beach 2002).

REFERENCES

Balas E (2007) Personal communication, October 15

Banker RD (2006) IFORs' Operational research hall of fame: W. W. Cooper. Int Trans Oper Res 13(4):379–383

Beach P (2002) His beautiful mind. Austin American-Statesman, March 4

Charnes A, Cooper WW (1961) Management models and industrial application of linear programming, vols 1 and 2. Wiley, New York, NY

Charnes A, Cooper WW (1962) Programming with linear fractional functionals. Nav Res Logist Q 9(3–4):181–186

Charnes A, Cooper WW (1977) Goal programming and multiple objective optimizations. Eur J Oper Res 1(1):39–54

Charnes A, Cooper WW, Ferguson RO (1955) Optimal estimation of executive compensation. Manage Sci 1(2):138–155

Charnes A, Cooper WW, Henderson A (1953) An introduction to linear programming. Wiley, New York, NY

Charnes A, Cooper WW, Mellon B (1952) Blending aviation gasolines – a study in programming interdependent activities in an integrated oil company. Econometrica 20(2):135–159

Charnes A, Cooper WW, Rhodes E (1978) Measuring the efficiency of decision making units. Eur J Oper Res 2(6):429–444

Charnes A, Cooper WW, Rhodes E (1981) Evaluating program and managerial efficiency: an application of DEA to program follow through. Manage Sci 27(6):668–697

Charnes A, Cooper WW, Symonds GH (1979) Cost horizons and certainty equivalents: an approach to stochastic programming heating oil production. Manage Sci 4(3):244–274

Cooper WW (1938) Report of the committee on capital gains taxation. Proceedings of the National Tax Association, Washington, DC

Cooper WW (1943) The yardstick of public utility regulation. J Pol Economy. 51(3):258–262

Cooper WW (1946) Financial statistics. In: Hauser P , Leonard WR (eds) Government statistics for business use. Wiley, New York, NY, pp 249–269

Cooper WW (1980) Living together—a full life. In: Schroder T (ed) Ruth Fay West Cooper (private publication). Boston, MA, pp 85–93

Cooper WW (2002) Abraham Charnes and W. W. Cooper (et al.): a brief history of a long collaboration in developing industrial uses of linear programming. Oper Res 50(1):35–41

Cooper WW (2005) Origins, uses of, and relations between goal programming and data envelopment analysis. J Multi Criteria Decis Anal 13(1):3–11

Cooper WW (2007) Personal communication, July 27

Cooper WW (2008a) Professional biography, January 22

Cooper WW (2008b) Personal communication, February 24
Cooper WW, McAlister L (1999) Can research be basic and applied? You bet. It better be for B-schools! Socioecon Plann Sci 33(4):257–276
Emrouznejad A, Parker BR, Tavares G (2008) Evaluation of research in efficiency and productivity: a survey and analysis of the first 30 years of scholarly literature in DEA. Socioecon Plann Sci 42(3):151–157
Farrell MJ (1957) The measurement of productive efficiency. J R Stat Soc, Sec. A. 120(3):253–290
Gleeson R, Schlossman S (1992a) The many faces of the new look: The University of Virginia, Carnegie Tech, and the reform of management education in the postwar era, Part I. Selections (Winter) 9–27
Gleeson R, Schlossman S (1992b) The many faces of the new look: The University of Virginia, Carnegie Tech, and the reform of management education in the postwar era, Part II. Selections (Spring) 1–24
Glover F, Sueyoshi R (2009) Contributions of Professor William C. Cooper in operations research and management science. Eur J Oper Res 197(1):1–16
Holt CC, Modigliani F, Muth JF, Simon HA (1960) Planning production, inventories, and work force. Prentice-Hall, Englewood Cliffs, NJ
Hopp WJ (2004) Fifty years of management science. Manage Sci 50(1):1–7. Also see http://www.informs.org/site/Interfaces/index.php?c=19&kat=OR%2FMS+History. Accessed 26 July 2008
Horner P (2002) History in the making. OR/MS Today 29(5):30–39
Kohler E, Cooper WW (1945) Costs, prices and profits: accounting in the war program. Account Rev 20(3):l–42. (Awarded the American Institute for Accountants prize for the most valuable article on an accounting subject, 1945)
Mautz RK, Previts GJ (1977) Eric Kohler: an accounting original. Account Rev 52(2):300–307
Meltzer A (2007) Personal communicaiton, October 15
Phillips F, Seiford L (2006) IFORS' Operational Research Hall of Fame: Abraham Charnes. Int Trans Oper Res 13(3):273–277
Schaible S (1996) Fractional programming. In: Gass SI, Harris C (eds) Encyclopedia of operations research and management science. Kluwer, Norwell, MA, pp 234–237
Sharpe WF (1963) A simplified model for portfolio analysis. Manage Sci 9(2):277–293
Simon HA (1957) Models of man: social and rational. Wiley, New York, NY
Simon HA (1987) Bounded rationality. In: Eatwell J, Milgate M, Newman P (eds) The New Palgrave: a dictionary of economics. Macmillan, New York, NY, pp 266–268
Symonds G (1955) Linear programming: the solution of refinery problems. Esso Standard Oil Company, New York, NY
West P (1980) Montana years. In: Schroder T (ed) Ruth Fay West Cooper (private publication) Boston, MA, pp 5–10

13

George B. Dantzig

Saul I. Gass

The influence of George B. Dantzig on the field of operations research (OR) is best reflected by noting that the he was the first recipient of the Operations Research Society of America (ORSA) and The Institute of Management Sciences (TIMS) prestigious John von Neumann Theory Prize, an award given annually to a scholar who has made fundamental, sustained contributions to theory in OR and management science (MS). He was the first inductee into the International Federation of Operational Research Societies' (IFORS) OR Hall of Fame. He is regarded as the father of linear programming (LP). He was awarded the President's National Medal of Science in 1975 by President Ford. He served as president of TIMS in 1966.

George's seminal work can be summed up succinctly as the recognition and definition of the broad class of practical problems that can be studied as linear programs and the development of the simplex algorithm for solving them. These developments came after the beginnings of the emerging field of OR, as developed by British scientists during World War II. Early books and reports on OR had little or nothing to say about LP. But, before long, George's work became a central and mainstay feature of OR and MS methods and their application.

During his lifelong commitment as a mathematician, teacher, researcher, consultant, and colleague to many, George influenced OR in a remarkable manner. His extensive array of research contributions, usually developed to solve real-world applications, encompass a wide span of theoretical and algorithmic developments. These have led to powerful, computer-based approaches that have enhanced our ability to solve critical decision problems of business, industry, and government.

A.A. Assad, S.I. Gass (eds.), *Profiles in Operations Research*, International Series in Operations Research & Management Science 147, DOI 10.1007/978-1-4419-6281-2_13,

THE EARLY YEARS

George was born on November 8, 1914 in Portland, Oregon, the first of two children of Tobias and Anja (Ourisson) Dantzig. Tobias (1884–1956) was born in Shavli, Russia (now Lithuania); Anja was born in Poland. The B. in George's name stands for Bernard, as Tobias, at first, wanted George to become a writer. George's brother's name, Henry P., comes from Henri Poincaré. Henry was an applied mathematician; he died in 1973.

(Courtesy of the National Academy of Engineering)

Tobias moved to France where he studied mathematics under Poincaré at the Sorbonne. There he met and married Anja, also a mathematics student; they emigrated to Portland, Oregon in 1910. Tobias first held a variety of jobs—lumberjack, road builder, house painter—before he obtained a teaching position at Indiana University. He received his Ph.D. (mathematics) in 1917. He taught at Columbia University (1918–1919) and Johns Hopkins University (1919–1920), worked in industry (1920–1925), and then taught at the University of Maryland (College Park) for 20 years; he was chair of the mathematics department from 1938 to 1946. Tobias was a student of the classics and a wonderful expositor of mathematics. His highly regarded 1930 book, *Number, The Language of Science*, is still in print. As a teenager, George prepared some of the figures that appeared in the book (Lustig 2000).

In an interview, George recounts his early struggle with ninth grade algebra and how "geometry really turned me on" and how he became a top student in mathematics and science. Tobias "gave me thousands of geometry problems while I was in high school ... (which) did more than anything else to develop my analytical power" (Albers and Reid 1986, 297; Albers et al. 1990, 64). George attended the University of Maryland, College Park, and received his A.B. degree in mathematics (minor in physics) in 1936. That summer, he married Anne Shmuner (from Baltimore, Maryland), also a Maryland student. George received a graduate scholarship from the University of Michigan and moved to Ann Arbor in the fall of 1936; Anne stayed behind to finish her senior year studies (B. A. in French) and joined him in the spring of 1937. George completed his course work for an M.A. in mathematics (awarded in February 1938) and because "except [for one course] in statistics, everything else I took at Michigan was terribly abstract—so abstract that I had but one desire: to quit my graduate studies and get a job, which I did" (Albers et al. 1990, 66).

George qualified for the Civil Service as a statistical clerk and, in 1937, accepted a job at the U.S. Bureau of Labor Statistics in Washington, D.C. [$1440 a year, Dantzig 1990)]. At that time, he felt that statistics was "just a bag of tricks" (Albers et al. 1990, 66), but, on the job, he learned many practical applications and became familiar with the work of Jerzy Neyman (University of California, Berkeley), all of which caused him to change his view about statistics. George wrote to Neyman about pursuing a Ph.D. under his direction and Neyman arranged for him to be a teaching assistant. George and Anne moved to Berkeley in 1939. At that time, statistics was part of the Berkeley mathematics department, and although George only took two statistics courses, both from Neyman, his dissertation was in statistics.

In June 1941, prior to defending his dissertation, George accepted a job in Washington, D.C. as head of the Pentagon-based Army Air Force's Combat Analysis Branch of Statistical Control. His motivation was "to contribute to World War II" under the assumption that the U.S. would soon become actively involved (Albers et al. 1990, 68). He developed a reporting system with which combat units were able to record the number of sorties flown, aircraft lost and damaged, bombs dropped, and targets attacked. It was here that George became familiar with the Air Force concepts of program planning of interrelated activities, ideas that would help him structure the basic form of the LP model. His colleagues included Robert McNamara (who later became Secretary of Defense and head of the World Bank) and Warren Hirsch (a probabilist from New York University who proposed in 1957 the still unproven eponymous conjecture concerning the maximum number of steps to solve a linear-programming problem) (Dantzig 1988). For his wartime contributions, George was awarded the highest War Department award for meritorious civilian service, the Exceptional Civilian Service Award.

George returned to Berkeley in the spring of 1946 and received his Ph.D. in mathematics. His dissertation title was "I. Complete form of the Neyman-Pearson Lemma; II. On the non-existence of tests of 'Student's' hypothesis having power functions independent of sigma." Although George was offered a teaching position in Berkeley's mathematics department, he decided to stay at the Pentagon where he became the mathematical advisor to the comptroller of the newly established Department of the Air Force. The deciding factor in George's decision was that the Berkeley job offer "paid too little" (Albers et al. 1990, 68).

URBAN LEGEND

If you search the Web for "urban legend George Dantzig" you will probably find the first hit to be "Snopes.com, The Unsolvable Math Problem" (Snopes 2009). That Web site recounts the story of how George, coming in late for his statistics class, mistakenly thought two problems written on the board by Neyman were homework problems. After a few days of struggling, George turned his answers in. About 6 weeks later, early on a Sunday morning, he and Anne were awakened by someone banging on their front door. It was Neyman who said, "I have just written an introduction to one of your papers. Read it so I can send it out right away for publication" (Snopes 2009).

George's answers to the homework problems were proofs of then two unsolved famous problems in statistics. Snopes (2009) and Albers et al. (1990, 67–68) give all the details about how George's experiences ended up as a sermon for a Lutheran minister and the basis for an introductory scene in the 1997 Oscar-winning film "Good Will Hunting." The solution to the first homework problem is given in Dantzig (1940). The solution to the second problem became part of a joint paper with Abraham Wald who proved it in 1950, unaware that George had solved it until it was called to his attention by a journal referee. Wald just added George's name as co-author (Dantzig and Wald 1951). Neyman had George submit his solutions to the homework problems as his doctoral dissertation.

Although he considered the Pentagon a place to mark time while looking for an academic position, George's choice started him down a life-changing research path that led to LP. The outcomes of this decision have been momentous: setting OR onto a new and major course of research and applications, and, more importantly, enabling enterprises and governments worldwide to become more effective and efficient.

ORIGINS OF LINEAR PROGRAMMING

George was challenged by his Pentagon colleagues to figure out how the Air Force could mechanize its planning process; to speed up the computation of a deployment of forces and equipment, training and logistical support—all this in a world of desk calculators and IBM accounting equipment [it was even suggested that George "develop some kind of analog device which would accept, as input, equations of all types, basic data, and ground rules, and use these to generate as output a consistent Air Force plan" (Dantzig 1988, 13)]. George's study of Air Force requirements led him to adapt and generalize the structure behind Wassily Leontief's inter-industry model. As he notes in Dantzig (2002, 43): "The *activity analysis* model I formulated would be described today as a time-staged, dynamic linear program with a staircase matrix structure. *Initially there was no objective function*; broad goals were never stated explicitly in those days because practical planners simply had no way to implement such a concept. ... By mid-1947, I had formulated a model which satisfactorily represented the technological relations usually encountered in practice. I decided that the *ad hoc* ground rules had to be discarded and replaced by an explicit objective function." George's insight enabled him to state mathematically—for the first time—the basic linear-programming problem:

$$\text{Maximize } \sum_{j=1}^{n} c_j x_j$$

$$\text{subject to}$$

$$\sum_{j=1}^{n} a_{ij} x_j = b_i \qquad (i = 1, \ldots, m)$$

$$x_j \geq 0 \qquad (j = 1, \ldots, n)$$

or, in words: maximize a linear objective function subject to a system of m linear equations expressed in terms of nonnegative values of n variables (with $m < n$). But, the constraints that define the problem usually have many, many solutions; one wanted to find the best solution, the one that maximized the objective, that is, the optimal solution. George had no way to solve it!

Viewed as an activity analysis model, the LP problem stems from the field of economics with its focus on the allocation of scarce resources. But, in his quest of a solution methodology, George's canvas of the top economists found no results. Falling back on his fine-tuned geometric intuition, plus Neyman's research and his own thesis results, George invented the simplex method (algorithm) for solving LP problems. But would it work? Would the simplex algorithm manage to find the optimal solution in a reasonable amount of computational effort? George experimented with a number of problems, and, under a geometric view of the LP problem, which he used in his dissertation, he convinced himself that the simplex method would be "a very efficient technique" (Dantzig 2002, 44). It had to be so proved. George dates the birth of the simplex method, that is, his being convinced that it

WHAT'S IN A NAME?

"In the summer of 1948 [T. C.] Koopmans and I visited the RAND Corporation. One day we took a stroll along the Santa Monica Beach. Koopmans said: 'Why not shorten Programming in a Linear Structure to Linear Programming?' I replied: 'That's it! From now on that will be its name.' Later that day I gave a talk at RAND, entitled 'Linear Programming'; years later [A. W.] Tucker shortened it to *linear program*.

"The term *Mathematical Programming* is due to Robert Dorfman of Harvard, who felt as early as 1949 that the term *Linear Programming* was too restrictive.

"The term *simplex method* arose out of a discussion with T. Motzkin who felt that the approach I was using, when viewed in the geometry of the columns, was best described as a movement from one simplex to a neighboring one. ... The term dual is an old mathematical term. But surprisingly the term *primal* is new and was proposed by my father Tobias Dantzig around 1954 after William Orchard-Hays stated the need for a word to call the original problem whose dual was such and such" (Dantzig 2002, 46–47).

PRIMAL

"DANTZIG, DR. TOBIAS, 2321 Lakeview Ave, Los Angeles 26, Calif. MATHEMATICS. Shavli, Russia, Feb. 19, 84. Licencié, Paris, 09; Ph.D, Indiana, 17. Instr. math, Indiana, 15–18; Columbia, 18–19; Hopkins, 19–20; industries, 20–25; assoc. prof. math, Maryland, 26–36, prof, 36–46, chairman dept, 38–46; RETIRED. Lecturer, Bur. Standards, 26–46. Consulting mathematician, Ordnance Dept, 19. Math. Soc; Math. Asn. Geometry; applied mathematics; theory of transformations."

American Men of Science (Cattell 1949, 563)

DUAL

"DANTZIG, DR. GEORGE B(ERNARD), 3203 N. 19th Road, Arlington, Va. MATHEMATICAL STATISTICS. Portland, Oregon, Nov. 8, 14; m. 36; c. 1. A.B, Maryland, 36; Horace Rackham scholar, Michigan, 36–37, M.A, 38; Ph.D. (math. statist), California, 46. Jr. statistician, U. S. Bur. Labor Statist, 37–39; teaching fellow math. statist, California, 39–41,46; from assoc. to sr. statistician, Air Force, 41–45, MATHEMATICIAN, AIR FORCE HEADQUARTERS COMPTROLLER, 46- Assoc. Math. Soc.; assoc. Inst. Math. Statist. Existence of similar regions in theory of mathematical statistics; mathematical theory interrelating activities of a large organization; general theory of logistics in military establishment."

American Men of Science (Cattell 1949, 563)

would be computationally feasible, during August 1947 (Dantzig 1988). But, in just over a year in his new job, George introduced the underlying mathematical structure of the general LP model, a structure that continues to encompass an ever-widening class of practical and important problems, and developed the simplex method for solving it.

In June of 1947, the Air Force established a major task force to work on the high-speed computation of its planning process, later named Project SCOOP (Scientific Computation of Optimal Programs), under the direction of the economist Marshall K. Wood, with George as chief mathematician. Project SCOOP was the setting in which the LP structure and the simplex method were proved and introduced to the world. The SCOOP civil service staffs of mathematicians, statisticians, and computational experts were responsible for formulating and solving a wide range of Air Force planning and programming problems. Key members of SCOOP were Walter W. Jacobs, Murray Geisler, and Emil Schell who were branch chiefs, respectively, of mathematical formulation, standards evaluation, and mathematical computation. The SCOOP research and development activities are described in Wood and Geisler (1951) and Wood (1952).

Project SCOOP sponsored two external programs: the first at the National Bureau of Standards (NBS) for its staff to experiment with the simplex method [it included funds to build the first stored-program computer completed in the U.S., the Standards Eastern Automatic Computer (SEAC)], and the second, a research center at the Graduate School of Industrial Administration at Carnegie Institute of Technology (now Carnegie Mellon University) in which the team of Abraham Charnes and William W. Cooper were instrumental in bringing LP to the attention of the industrial world (Albers et al. 1990, Dantzig 1963, Cooper 2002). In 1947, members of the NBS Mathematical Tables Project, using hand-operated desk calculators, solved a 9 inequality and 77 variable LP problems (Stigler's diet problem) in about 120 person-days (Dantzig 1951a, 1963, 1990). The simplex algorithm systematically went through just 24 iterations (solutions). Subsequent hand computations of a variety of LP problems, plus the first electronic computer LP computations on the SEAC, demonstrated the efficiency of the simplex method (Orden 1952, Hoffman et al. 1953). To support its computational requirements, Project SCOOP installed the second commercial UNIVAC-1 computer (the first going to the Bureau of Census) in the basement of the Pentagon; it was officially accepted by the Air Force on June 25, 1952 (Johnson 2006). [George's experience in planning his personal diet is recounted with humor in Dantzig (1990).]

George's research for the Air Force helped set the course of OR and mathematical programming: the first statement of the LP problem and the recognition of its applicability to a wide range of decision problems and the invention of the simplex method—the simplex algorithm was designated as one of the twentieth century's top-10 algorithms by the Institute of Electrical and Electronics Engineering (Dongarra and Sullivan 2000). At Project SCOOP, both the mathematics and the computational aspects of the LP model and the simplex method were established, tested, and proven; the LP primal-dual problems and their relationship via the simplex algorithm were stated and proven; the simplex transportation algorithm was developed; the equivalence between the LP problem and a zero-sum two-person game was established; as well as the application of LP to a wide range of planning and operational Air Force problems (Cottle et al. 2007, Dantzig 1963, Gass 2002, Orden 1993).

George stayed with Project SCOOP until June 1952, when he joined the RAND Corporation as a research mathematician. Project SCOOP was disbanded in 1955; the UNIVAC was retired in 1958.

THE DEBUT OF LP

George's first academic presentation of his Project SCOOP LP research was a paper given on September 9, 1948 at the Econometric Society meeting held at the University of Wisconsin-Madison. It's title: "Programming in a linear structure." Dantzig described the scene as follows:

"Not too long after my first meeting with [Al] Tucker there was a meeting of the Econometric Society in Wisconsin attended by well known statisticians and mathematicians like [H. O.] Hotteling and [John] von Neumann, and economists like [T. C.] Koopmans. I was a young unknown and I remember how frightened I was with the idea of presenting for the first time to such a distinguished audience the concept of linear programming.

"After my talk, the chairman called for discussion. For a moment there was the usual dead silence; then a hand was raised. It was Hotelling's. I must hasten to explain that Hotelling was fat. He used to love to swim in the ocean and when he did, it is said that the level of the ocean rose perceptibly. This huge whale of a man stood up in the back of the room, his expressive fat face took on one of those all-knowing smiles we all know so well. He said: '*But we all know the world is nonlinear.*' Having uttered this devastating criticism of my model, he majestically sat down. And there I was, a virtual unknown, frantically trying to compose a proper reply.

"Suddenly another hand in the audience was raised. It was von Neumann. 'Mr. Chairman, Mr. Chairman,' he said, 'if the speaker doesn't mind, I would like to reply for him.' Naturally I agreed. Von Neumann said: 'The speaker titled his talk "linear programming" and carefully stated his axioms. If you have an application that satisfies the axioms, well use it. If it does not, then don't,' and he sat down. In the final analysis, of course, Hotelling was right. The world is highly nonlinear. Fortunately systems of linear inequalities (as opposed to equalities) permit us to approximate most of the kinds of nonlinear relations encountered in practical planning" (Dantzig 1982; 2002, 45).

RAND, ACADEMIA, AND BEYOND

While at RAND, working alone or with a stellar cast of co-researchers (Lester Ford, Jr., Ray Fulkerson, Selmer Johnson, William Orchard-Hays, Philip Wolfe), George furthered the development of LP and its extensions as an effective, versatile, and mathematically sound approach for analyzing a wide range of real-world decision problems. His path-breaking RAND research included: the decomposition principle/algorithm by which large-scale industrial and economic planning LP models could be solved (Dantzig and Wolfe 1960, 1961; Dantzig 1963); imbedding nonlinear, nonconvex, and combinatorial conditions within a general integer-programming framework (Dantzig 1960a, 1963); the resolution of network (max-flow min-cut) and shortest route problems (Dantzig 1960b; Dantzig and Fulkerson 1956; Fulkerson and Dantzig 1955); a novel cutting-plane (subtour elimination constraint) approach to the traveling salesman problem (TSP); and the solution of the 48-state plus Washington, D.C. TSP (Dantzig et al. 1954; Grötschel and Nemhauser 2008); stochastic programming (uncertain values of the coefficients) and general decision making under uncertainty (Dantzig 1955, Ferguson and Dantzig 1956); the development of advanced simplex-based computer systems (Dantzig and Orchard-Hays 1953, 1954; Dantzig et al. 1955; Orchard-Hays 1968); plus much more (Cottle 2003, 2005; Cottle et al. 2007; Dantzig 1963). Most of the relevant RAND research was published in an extensive and widely distributed series of RAND memorandum reports—their impact on the development and diffusion of the then still young field of OR was far-reaching and of great importance.

> JOB DESCRIPTION
>
> "In 1952 I left the Air Force to work for RAND. John [Williams] was my boss. After I had worked for several months without receiving any direction, I went to see him. I said, 'John, what is it that I am supposed to do?' He didn't say a word, not one–he just sat looking at me across his desk. Five minutes passed, and I began to get uneasy. Finally, he said, 'George you know better than to ask that question.' I understood what he meant and got out of his office fast. John's policy was to let his researchers do their thing. For, example, he tolerated me for nine years while I wrote my book. Of course, I also wrote a lot of papers during the same period" (Albers et al. 1990, 77).

In 1960, George began his illustrious academic career as professor of engineering science in the Department of Industrial Engineering,

University of California, Berkeley. During his 6 years there, he taught his basic and advanced courses in LP, supervised eleven doctoral dissertations (including the one by this author), founded and directed Berkeley's off-campus Operations Research Center, continued and extended his research program, and consulted for major companies such as Esso Standard Oil Company and Crown Zellerbach Corporation. It was a Crown Zellerbach cutting-stock paper manufacturing problem that gave rise to a rather large-scale LP problem that involved many generalized upper-bound (GUB) constraints in which every variable appears at most once in these constraints (Dantzig and Van Slyke 1967, Cottle et al. 2007). By proving special characteristics of an optimal solution to the GUB problem, Dantzig and Van Slyke developed an efficient modification of the revised simplex method for solving such large-scale problems.

George moved to Stanford University in 1966 as professor of OR and computer science, and was appointed the C. A. Criley Endowed Chair in Transportation in 1973. At Stanford, George established and became co-director of the Systems Optimization Laboratory (SOL), where he assembled a remarkably talented mathematical programming group—Philip Gill, Walter Murray, Michael Saunders, John Tomlin, and Margaret White. The SOL had the objective of developing "computational methods and associated computer routines for numerical analysis and optimization of large-scale systems" (Cottle et al. 2007). It was a most successful endeavor whose research and software greatly advanced the state of the art. A particular problem of interest that George addressed within SOL dealt with energy-economic modeling, the PILOT model (Planning Investment Levels Over Time). PILOT's major objective was the assessment of how specific energy policies would affect the energy

George with some of his Berkeley and Stanford Ph.D. students who joined him to celebrate is 70th birthday. He supervised 11 Ph. D. students at Berkeley and 41 at Stanford.

supply/demand situation (Dantzig and Parikh 1976). It was a long-term model for analyzing trends 40 years into the future. As George stated, "the PILOT model is the real McCoy—a powerful tool for making policy decisions" (Albers et al. 1990, 78). George was director of the PILOT Energy-Economic Model Project from 1973 to 1988. As Cottle et al. (2007, 358) point out, "The PILOT project provided a context combining three steams of research that greatly interested Dantzig: modeling of a highly relevant economic issue, large-scale programming methodology, and the computation of optimal solutions or the solutions of economic (complementarity) problems." George retired in 1985 as professor emeritus, but continued teaching and maintaining an active research agenda until the fall of 1997.

In his retirement, George returned to the study of stochastic models, a most challenging area. Earlier, in his text (Dantzig 1963, 299), George wrote, "In this chapter [25-Uncertainty] we shall examine some of the solved problems in this area, cautioning the reader that the treatment is incomplete and that much research remains to be done." Forty years later (Dantzig 2002, 46), he commented: "Stochastic programming is one of the most promising fields of future research, one closely tied to large-scale methods."

In stochastic problems, one assumes that statistical distributions are known for parameters deemed uncertain. But doing so causes the model size to be extremely large and difficult to solve. Combining Benders (dual) decomposition technique and Monte Carlo importance sampling, George, working with Peter Glynn and Gerd Infanger, developed a new approach for attacking such problems (Dantzig and Glynn 1990, Infanger 1991, 1994). Using a laptop computer, they solved large-scale test problems dealing with electrical facility expansion and portfolio management. Their procedure was also implemented on a parallel (Hypercube) multicomputer (Infanger 1993). The conclusion: "very accurate solutions of stochastic linear programs can be obtained with only a small sample size" (Infanger 1994, 129). As George notes in his Foreword to Infanger (1994, v): "This discovery that it is practical to solve linear problems under uncertainty on PCs, has opened the door to a world of new applications that will eventually change the way planning is done." And, no matter who develops future successes, the degrees of separation network begins with George Dantzig.

Much of George's important early work is captured in his classic text *Linear Programming and Extensions* (1963). Students and researchers will find much of value in it: an historical discussion of how the field first developed, the basic elements of LP, and the remarkable mathematical and applied extensions of the field. A later, comprehensive overview, co-authored by George and Mukund Thapa (1997, 2004), is given in two volumes (*Linear Programming: Introduction* and *Linear Programming; Theory and Extensions*). An extensive anthology of George's research, along with a complete bibliography of his work, is contained in Cottle (2003). Other aspects and stories about George's career are recounted in Dantzig (1982, 2002) and Cottle et al. (2007). A review of George's many contributions to OR and MS is given in Cottle (2005). The special issue of the journal, *Discrete Optimization* [2008 **5**(1)], in memory of George, contains papers that review his contributions and influence on: systems optimization in solving complex real-world problems (Gill et al. 2008), economic analysis (Arrow 2008), integer programming (Grötschel and Nemhauser 2008), and the theory of computation (Karp 2008).

George authored or co-authored seven books and more than 150 papers. One book, *Compact City* (Dantzig and Saaty 1973), shows another side of George's wide range of interests. The book's cover notes that it is "a nontechnical, introductory study about the feasibility of building a city that makes more effective use of the vertical dimension and the time dimension (through around-the-clock use of facilities) than do present cities." It includes a section on "Operations Research and the Total-System Approach," but no LP!

The development and use of electronic computers was of continuing interest to George—he had fond memories of his early Air Force contributions that relied on mechanical desk calculators and IBM punch-card accounting equipment (Dantzig 1988, 2002). Early on, George recognized that the objectives of Project SCOOP, and the value of LP and the simplex method, could only be met by applying the power of electronic computers, even though their operational experience was, at that time, very limited. His research was influential in having Project SCOOP sponsor the NBS SEAC and the Air Force purchasing the second production model of the UNIVAC-1. As George noted: "It indeed may be true that much of the postwar development of electronic computers can be traced to the direct and indirect sponsorship of the AAF Comptroller's office" (Dantzig 1988, 13).

Once LP was applied with success to real-world operational problems, initially by the gasoline blending model of Charnes, Cooper, and Mellon (Cooper 2002, Dantzig 2002), the spread of LP to a wide range of other commercial areas proceeded at a fast pace, supported by the sale of many of the first electronic computers. In reading his writings, it is clear that George was a student of computers and that he gave much thought about their impact on society. Thus, it should not come as much of a surprise that George wrote a science-fiction book about the evolution of computers (unpublished), *In Our Own Image*. "The book . . . is fantasy about our heirs, the Comps, who 2000 years from now look back and wonder about their origins just as we do about our own" (Dantzig 1984, Foreword 1). George had a contract with a publisher, but for some unknown reason it fell through.

GENESIS

"According to the Good Book, for the first 75 years after we were created, Comp lived in Paradise. It was a wonderful air-conditioned world with ministering angels all about taking care of our every need. In the beginning God created only Eniac and he was alone. But soon there was Edvac, Edsac, Binac, Seac, Swac, and Johnniac, and many many others.

"Piece by piece, I have gathered the scientific evidence that the prehistoric events that I am about to describe actually did take place. As you will see, it is a story of our origins quite different from what we have been taught from the Good Book."

(Dantzig 1984, I.1)

George was a frequent visitor to the International Institute of Applied Systems Analysis (IIASA) headquartered in Laxenburg, Austria. IIASA, an international research organization that "conducts inter-disciplinary scientific studies on environmental, economic, technological, and social issues in the context of human dimensions of global change" (IIASA 2009). In 1973–1974, George and Anne spent a sabbatical year at IIASA with George serving as IIASA's head of the methodology group.

CURIOUS GEORGE AT IIASA

A few days after he arrived at IIASA, George called Ruth Steiner, the administrator who took care of visitors, with the following request: "In front of my office is a truck that is very long. I cannot imagine that it is advantageous to have such a length. Could you please find out what is in the truck, where it came from, what route it took, and how it got around corners?" The answers came back: "furniture from Salzburg, over the autobahn, by backing up several times." Some time later, George's research assistant informed Ruth that the company could save 40% of its costs if they used four smaller trucks and suggested she inform the company, which she did. As she reported: "They thought I was out of my mind."

At IIASA, George drank his soup out of a beer mug. When Ruth asked him why, he wrote out a "long mathematical formula which proved that there was less heat wasted using a beer mug compared to a soup bowl" (Steiner 1992).

The legacy of George goes way beyond his research and teaching. It includes his friendship, his mentoring, his unselfishness with his time and ideas. But one important item is missing from his résumé—the Nobel Prize.

The Swedish Misconnection

Tjalling C. Koopmans and Leonid V. Kantorovich received the 1975 Nobel Prize in economics for their contributions to the theory of optimum allocation of resources, that is, LP, or in economic terms, its interpretation as linear activity analysis (Balinski 1991, Dorfman et al. 1958, Koopmans 1957). Those familiar with the origins and development of LP and its impact on economic theory were amazed and disappointed that George was not included as an honoree—a Nobel Prize can be shared by up to three living recipients. It is unclear why George was not included.

In 1939, Kantorovich was Head of the Department of Mathematics at the Institute of Mathematics and Mechanics, Leningrad University. It was in this capacity that he was consulted on problems faced by the Laboratory of the Plywood Trust: a production planning problem of the plywood industry—given raw materials, and subject to some constraints, how could the productivity of the different equipment types required to process the raw material be maximized? Also, given plywood sheets, how should they be cut to produce a specified assortment of pieces with minimum waste (Gardner 1990, Kantorovich 1990)? Kantorovich recognized that the general mathematical structure required "the solving of a maximization problem in the presence of a series of linear constraints" (Kantorovich 1990, 29).

Kantorovich's research received little recognition in the Soviet Union at that time and for many years later. As Kantorovich (1987, 31) wrote in a posthumously published paper: "In the spring of 1939 I gave some more reports at the Polytechnic Institute and the House of Scientists, but several times met with the objection that the work used mathematical methods, and in the West the mathematical school in economics was an anti-Marxist school and mathematics in economics was a means for apologists of capitalism." And, as noted by Makarov and Sobolev (1990, 5), "Unfortunately, due to the well-know peculiarities of the USSR economy in those years, as well as the absence of computers, Kantorovich's works on linear programming did not find wide enough application at that time and actually remained unknown not only abroad but also in the USSR." It was not until the 1950s that Kantorovich's work became known in the west (Kantorovich 1958, 1959, 1960, 1965; Koopmans 1960, Leifman 1990).

George, in his text *Linear Programming and Extensions* (1963, 22), wrote, "Kantorovich should be credited with being the first to recognize that certain important broad classes of production problems had well-defined mathematical structures which, he believed, were amenable to practical numerical evaluation and could be numerically solved." The first formal source of George's 1946–1947 seminal linear-programming research is the book, *Activity Analysis of Production and Allocation* (proceedings of the Cowles Commission for Research in Economics 1949 conference), edited by Koopmans (1951). The proceedings contains the four seminal papers authored by George while he was with Project SCOOP (Dantzig 1951a, b, c, d).

THEN THERE WERE TWO

Koopmans, Dantzig, Kantorovich meet at IIASA.

Koopmans was "profoundly distressed" that George had not shared in the Nobel Prize (Balinski 1991, 10–12, Kuhn 2008). Koopmans gave a gift of $40,000 to IIASA in George's honor. As the total Nobel award for that year was $240,000, this left Koopmans with $80,000, the amount he would have received if George was also a Nobel laureate. All three principals had worked and met at various times at IIASA. In a conversation I had with Koopmans shortly after the award, he told of his displeasure with the Nobel selection and how he had earlier written to Kantorovich suggesting that they refuse the prize, certainly a most difficult decision for both, but especially so for Kantorovich.

Two views of the discovery/invention of LP are given by the economist Robert Dorfman (1984) [pro Kantorovich] and myself (Gass 1989) [pro Dantzig]. Both George's and Kantorovich's work were original and independent of each other. Unlike George's contributions, Kantorovich's research had no impact on the world-wide acceptance and application of LP. Further, the important economic interpretation of LP had been developed by George and others based the relationship between the primal-dual problems and the role of the simplex method. The first books on economic theory to discuss these aspects were *Three Essays on the State of Economic Science* (Koopmans 1957), and *Linear Programming and Economic Analysis* (Dorfman et al. 1958). Neither book was based on Kantorovich's research. Koopmans, in a footnote added during proof of his manuscript, states, "I have recently become aware that prior to most of the publications cited the basic ideas of linear programming and implicit pricing had been developed to a considerable degree by the Russian mathematician, L. V. Kantorovich" (Kantorovich 1939, 1942, Koopmans 1957, 68).

Who should be credited with priority of discovery cannot settled to everyone's satisfaction. The dilemma can be summed up in the following manner: "Thus, the situation of Kantorovich is rather like that of Columbus. He really never touched the American mainland, and he didn't give it its name, but he was the first one in the area" (Gardner 1990, 645).

TOO MATHEMATICAL?

Nobel Prize selection committees are very secretive in terms of the criteria and rationale for choosing persons for the award. The basis for the evaluation and decision concerning an award may not be made available until at least 50 years have elapsed after the date on which the decision in question was made. But, in the case of the 1975 Nobel Prize, we have the following possible explanation as to why George Dantzig was not included, as reported by Harold Kuhn (2008):

"At the 11th [Mathematical Programming] Symposium in Bonn in 1982, I was approached by Leonid Kantorovich who asked urgently: 'Could we have coffee alone together?' He had been a guest in our home in Princeton some years earlier, when he came to a meeting of economists (in Atlantic City) who had been awarded the Sveriges Riksbank Prize in Economic Sciences in Memory of Alfred Nobel.

"What was the topic for which Kantorovich sought me out? It was to explain why George Dantzig had not won the Nobel Prize with Koopmans and Kantorovich in 1975! Kantorovich had rationalized this with an argument that did not convince me. He argued that the Selection Committee had concluded that George's contributions had been too mathematical, indeed, that his case had been weakened by his emphasis on the Simplex Method and the problems of computation. Eleven years later, when I was campaigning in Stockholm for a Nobel for John Nash, I heard a similar argument from Asser Lindbeck, who was the Chairman of the Selection Committee from the time that the prize in Economics was established in 1969 to1994, when a prize was first awarded in game theory to John Nash, John Harsanyi, and Reinhard Selten."

Honors and Awards

George was a fellow of the Econometric Society, Institute of Mathematical Statistics, Association for the Advancement of Science, American Academy of Arts and Sciences, and the Institute for

Operations Research and the Management Sciences (INFORMS). He was a member of the National Academy of Sciences and the National Academy of Engineering. He was a founding member of TIMS and served as its president in 1966, and a founder and chair (1973–1974) of the Mathematical Programming Society.

In 1975, George was the first recipient of the ORSA and TIMS John von Neumann Theory Prize (now awarded by INFORMS). George was also the first inductee into the IFORS' Operational Research Hall of Fame (Gass 2003). He was awarded the Silver Medal of the British Operational Research Society (1986) and the Harvey Prize in Science and Technology from the Technion University. George received the President's National Medal of Science in 1975 from President Ford. The citation read: "For inventing linear programming and discovering methods that led to wide-scale scientific and technical applications to important problems in logistics, scheduling, and network optimization, and to the use of computers in making efficient use of the mathematical theory." The medal was presented at a White House ceremony on October 18, 1976.

GEORGE'S NATIONAL MEDAL OF SCIENCE

NATIONAL MEDAL OF SCIENCE AWARD CEREMONY

George and Anne with President Ford.

George received nine doctorate honorary degrees, but perhaps the one he most cherished was from his University of Maryland alma mater in 1976, exactly 40 years after his graduation. The citation read, in part: "Dantzig's linear programming was one of the principle forces leading to the emergence of the mathematical science of decision making as a new discipline called operations research or management science in the 1950s." He was inducted into the University of Maryland's Hall of Fame in 2010.

A LAST PHONE CONVERSATION WITH GEORGE

"My last encounter with George was by telephone in 2005, shortly before he died. During the previous year the Princeton University Press was preparing to celebrate the 100th anniversary of the Press. As part of that celebration, the editors were selecting the 100 best books published by the Press in the preceding 100 years. Authors, editors, and faculty members were invited to nominate books for this list and the competition was fierce.

"Quite naturally, I made a strong argument that George's book, *Linear Programming and Extensions*, should be included. I was very pleased that it had made the cut along with such distinguished books as Einstein's *The Meaning of Relativity* and von Neumann and Morgenstern's *Theory of Games and Economic Behavior*.

"Since I had heard that George was quite ill and might not receive the book honoring the 100 best books, I phoned George and read the citation to him on the phone. That was my last close encounter with George."

Harold Kuhn (2008). "57 Years of close encounters with George," presented at the George B. Dantzig Memorial Paper Cluster, INFORMS Annual Meeting, Washington, D.C., October 14, 2008.

RECOLLECTIONS ABOUT GEORGE

"George was undoubtedly one of the most influential mathematicians of the twentieth century. His main strength was not theorem-proving, but insight into the structure of processes and their mathematical representation most suitable to optimization. Not just his original discovery of linear programming and the simplex method, but all his subsequent contributions to its development and enrichment carry this mark.

"As a human being, George was a warm, compassionate person, with a sense of justice and fairness, and also a sincere, straightforward character, a reliable friend that you could count upon. He was very proud of his father, Tobias, and his accomplishments. He was, as far as I can tell, untinged by any sort of national or ethnic prejudice, and was not religious; but his Jewishness mattered to him, and he felt strong solidarity with the victims of the Nazis as well as the Soviets.

"George had a peculiar sense of humor, ... He would enjoy musing that linear programming is older than the Bible. Indeed, Joseph in Egypt, having interpreted the Pharaoh's dream about the seven fat cows and seven lean cows as meaning seven rich harvests followed by seven poor ones, proceeded to save Egypt from the ensuing famine by introducing 'Lean-Year Programming.' I don't know whether this quip originated with George or he was quoting someone else, but he certainly enjoyed telling it."

Egon Balas (2008) "Recollections about George Dantzig," presented at the George B. Dantzig Memorial Paper Cluster, INFORMS Annual Meeting, Washington, D.C., October 14, 2008.

With respect to the non-mathematical world, George was an ecumenical person. Over the years, he did manage to find time to pursue many nonacademic activities. When he was young, he played tennis, swam, became a lifeguard, played chess and poker, and sketched and painted. He had a lifelong love of artwork, visiting museums all over the world, and collecting prints. He and Anne enjoyed plays and concerts. Later, he watched old movies; listened to classical music on the radio; liked to read his favorite authors, Mark Twain and George Bernard Shaw, as well as mysteries and science fiction. Food also played an important role: eating dessert, especially Anne's homemade baked goods and fruit sauces; drinking coffee; and eating out—he ate in just about all of the restaurants in Palo Alto. He was not afraid to get his hands dirty—collecting shells, rocks, geodes, and fossils, and helping Anne do the gardening. And he was renowned for repairing all sorts of things with the creative George approach (Klass 2009, Thapa 2009).

George died May 13, 2005 in Stanford, California. He was survived by his wife, Anne, and their three children, Jessica (Klass), David, and Paul; three grandchildren and two great grandchildren. Anne died on August 10, 2006 at the age of 86.

A most suitable close are the following quotes from two colleagues who represent those who knew, worked, and loved George over many, many rewarding years of friendship:

> His remarkable professional life was faithfully devoted to his vision of enabling mankind to make decisions about complex systems and issues using mathematical programming. He went about this in a thoughtful and noble way. His legacy is not simply the concept of linear programming and the invention of the simplex method. He worked on a much larger canvas, one with room for the full ramifications and potential of the subject and many others related to it. His approach was ingenious and inventive rather than imitative. He knew how to get things done and often to make the doing serve more than one purpose. His manner was kindly and welcoming, not brusque. He was thoroughly dedicated to his work and loyal to his students, friends, and colleagues. It is not surprising that he was so loved and admired by operations researchers everywhere (Cottle 2005, 896).

> I used to urge the students in the OR department at Stanford to get to know George. It would be something to tell their grandchildren that they had spoken with the father of linear programming. But George was not only

the father: he was also the obstetrician, neonatal nurse, nanny, coach and conductor. Although other mathematicians created subjects, I cannot think of any other who continued to supervise its development and encourage its practitioners with such grace, style and devotion. And, reciprocally, we admired and adored him. . . . I do not think the world of mathematicians witnessed George's like before, I do not think we will see his like again, and I hope posterity will remember his name" (Hoffman 2008, 148–149).

GEORGE AND RICHARD COTTLE

GEORGE AND ALAN HOFFMAN

REFERENCES

Albers DJ, Alexanderson GL, Reid C (eds) (1990) George B. Dantzig. More mathematical people. Harcourt Brace Jovanovich, Boston, MA, pp 61–79

Albers DJ, Reid C (1986) An interview with George B. Dantzig: the father of linear programming. Coll Math J 17(4):293–314

Arrow K (2008) George B. Dantzig in the development of economic analysis. Discrete Optim 5(2):159–167

Balas E (2008) Recollections about George Dantzig. http://www2.informs.org/History/dantzig/rem_balas.htm. Accessed 28 Jul 2009

Balinksi ML (1991) Mathematical programming: journal, society, recollections. In: Lenstra JK, Rinnooy Kan AHG, Schrijver A (eds) History of mathematical programming. North-Holland, Amsterdam, pp 5–18

Cattell J (1949) American men of science, 8th edn. The Science Press, Lancaster, PA

Cooper WW (2002) Abraham Charnes and W. W. Cooper (et al.): a brief history of a long collaboration in developing industrial uses of linear programming. Oper Res 50(1):35–51

Cottle RW (ed) (2003) The basic George B. Dantzig. Stanford University Press, Stanford, CA

Cottle RW (2005) George B. Dantzig: operations research icon. Oper Res 53(6):892–898

Cottle R, Johnson E, Wets R (2007) George B. Dantzig (1914–2005). Notices Am Math Soc 54(3):344–362

Dantzig GB (1940) On the non-existence of tests for Student's hypothesis involving power functions independent of sigma. Ann Math Stat 11(2):186–192

Dantzig GB (1951a). The programming of interdependent activities: mathematical model. In: Koopmans TC (ed) Activity analysis of production and allocation: proceedings of a conference. Wiley, New York, NY, pp 19–32

Dantzig GB (1951b) A proof of the equivalence of the programming problem and the game problem. In: Koopmans TC (ed) Activity analysis of production and allocation. Wiley, New York, NY, pp 330–335

Dantzig GB (1951c) Maximization of a linear function of variables subject to linear inequalities. In: Koopmans TC (ed) Activity analysis of production and allocation. Wiley, New York, NY, pp 339–347

Dantzig GB (1951d) Application of the simplex method to a transportation problem. In: Koopmans TC (ed) Activity analysis of production and allocation. Wiley, New York, NY, pp 359–373

Dantzig GB (1955) Linear programming under uncertainty. Manage Sci 1(3–4):197–206

Dantzig GB (1960a) On the significance of solving linear programs with some integer variables. Econometrica 28(1):30–44

Dantzig GB (1960b) On the shortest route through a network. Manage Sci 6(2):187–190

Dantzig GB (1963) Linear programming and extensions. Princeton University Press, Princeton, NJ

Dantzig GB (1982) Reminiscences about the origins of linear programming. Oper Res Lett 1(2):43–48

Dantzig GB (1984) In our own image (unpublished manuscript)

Dantzig GB (1988) Impact of linear programming on computer development. OR/MS Today 15(4):12–17

Dantzig GB (1990) The diet problem. Interfaces 20(4):43–47

Dantzig GB (2002) Linear programming. Oper Res 50(1):42–47

Dantzig GB, Fulkerson DR (1956) On the max-flow min-cut theorem of networks. In: Kuhn H, Tucker A (eds) Linear inequalities and related systems. Princeton University Press, Princeton, NJ, pp 215–221

Dantzig GB, Fulkerson DR, Johnson SM (1954) The solution of a large-scale traveling salesman problem. Oper Res 2(4):293–410

Dantzig GB, Glynn P (1990) Parallel processors for planning under uncertainty. Ann Oper Res 22(1–4):1–21

Dantzig GB, Orchard-Hays W (1953) Notes on linear programming: part V—alternate algorithm for the revised simplex method using product form for the inverse. The RAND Corporation, Research Memorandum RM–1268, November 19

Dantzig GB, Orden A, Wolfe P (1955) The generalized simplex method for minimizing a linear form under linear inequality restraints. Pac J Math 5(2):183–195

Dantzig GB, Parikh S (1976) On a pilot linear programming model for assessing physical impact on the economy of a changing energy picture. In: Roberts F (ed) Energy: mathematics and models, proceedings of a SIMS conference on energy, 7–11 July 1975. SIAM, Philadelphia, PA, pp 1–23

Dantzig GB, Saaty T (1973) Compact city. W. H. Freeman, San Francisco, CA

Dantzig GB, Thapa M (1997) Linear programming: introduction. Springer, New York, NY

Dantzig GB, Thapa M (2003) Linear programming: theory and extensions. Springer, New York, NY

Dantzig GB, Van Slyke R (1967) Generalized upper bound techniques. J Comput Syst Sci 1:213–226

Dantzig GB, Wald A (1951) On the fundamental lemma of Neyman and Pearson. Ann Math Stat 22(1):87–93

Dantzig GB, Wolfe P (1960) Decomposition principle for linear programs. Oper Res 8(1):101–111

Dantzig GB, Wolfe P (1961) Decomposition algorithm for linear programs. Econometrica 29(4):767–778

Dongarra J, Sullivan F (2000) The top 10 algorithms. Comput Sci Eng 2(1):22–23

Dorfman R (1984) The discovery of linear programming. Ann Hist Comput 5(3):283–295

Ferguson A, Dantzig GB (1956) The allocation of aircraft to routes—an example of linear programming under uncertain demand. Manage Sci 3:45–73

Fulkerson D, Dantzig GB (1955) Computations of maximal flows in networks. Naval Res Logistics Q 2:277–283

Gardner R (1990) L. V. Kantorovich: the price implications of optimal planning. J Econ Lit 28(June):638–648

Gass SI (1989) Comments on the history of linear programming. IEEE Ann Hist Comput 11(2):147–151

Gass SI (2002) The first linear-programming shoppe. Oper Res 50(1):61–68

Gass SI (2003) IFORS' operational research hall of fame: George B. Dantzig. Int Transact Oper Res 10(2):191–193

Gill P, Murray W, Saunders M, Tomlin J, Wright M (2008) George B. Dantzig and systems optimization. Discrete Optim 5(2):151–158

Grötschel M, Nemhauser G (2008) George B. Dantzig's contributions to integer programming. Discrete Optim 5(2):168–173

Hoffman A (2008) Editorial: a special issue in memory of George B. Dantzig. Discrete Optim 5(2):145–150. http://www2.informs.org/History/dantzig/rem_hoffman.htm. Accessed 3 Nov 2009)

Hoffman A, Mannos M, Sokolowsky D, Wiegmann N (1953) Computational experience in solving linear programs. SIAM J 1(1):17–33

IIASA (2009) http://www.iiasa.ac.at/docs/IIASA_Info.html. Accessed 5 Jan 2009)

Infanger G (1992) Monte Carlo (importance) sampling within a Benders decomposition algorithm for stochastic linear programs. Ann Oper Res 39(1–4):69–95

Infanger G (1994) Planning under uncertainty: solving large-scale stochastic linear programs. The Scientific Press Series, Boyd & Fraser, Danvers, MA

Johnson LR (2006) Coming to grips with the UNIVAC. Ann Hist Comput 28(5):32–42

Kantorovich LV (1939) Mathematical methods in the organization and planning of production. Publication House, Leningrad University, Leningrad, Russia

Kantorovich LV (1942) On the translocation of masses. Proc Natl Acad Sci USSR 37(7–7):227–230

Kantorovich LV (1958) English translation of Kantorovich (1942). Manage Sci 5(1):1–4

Kantorovich LV (1959) The best use of economic resources (in Russian). USSR Academy of Sciences, Moscow. (English translation, Pergamon Press, Oxford, 1965)

Kantorovich LV (1960) English translation of Kantorovich (1939). Manage Sci 6(4):363–422

Kantorovich LV (1965) The best use of economic resources [English translation of (Kantorovich 1959)]. Pergamon Press, Oxford, England

Kantorovich LV (1990) My journey in science. In: Leifman LJ (ed) Functional analysis, optimization, and mathematical economics, Oxford University Press, New York, NY, pp 8–45

Karp R (2008) George B. Dantzig's impact on the theory of computation. Discrete Optim 5(2):174–185

Klass J (2009) Personal communication

Koopmans TC (ed) (1951) Activity analysis of production and allocation: proceedings of a conference. Wiley, New York, NY

Koopmans TC (1957) Three essays on the state of economic science. McGraw-Hill, New York, NY

Koopmans TC (1960) A note about Kantorovich's paper, "Mathematical Methods of Organizing and Planning Production." Manage Sci 6(4):363–365

Kuhn H (2008) 57 Years of close encounters with George. http://www2.informs.org/History/dantzig/articles_kuhn.html. Accessed 9 June 2009

Leifman L (ed) (1990) Functional analysis, optimization and mathematical economics: a collection of papers dedicated to the memory of Leonid Vital'evich Kantorovich. Oxford University Press, New York, NY

Lustig I (2000) Optimization trailblazers: interview with George Dantzig (text and video). http://www.e-optimization.com/directory/trailblazers/dantzig. Accessed 2 Aug 2009

Makarov V, Sobolev S (1990) Academician L. V. Kantorovich. In: Leifman LJ (ed) Functional analysis, optimization, and mathematical economics. Oxford University Press, New York, NY, pp 1–7

Mathematical Genealogy Project (2009) www.genealogy.ams.org. Accessed 5 Jan 2009

Orchard-Hays W (1968) Advanced linear-programming computing techniques. McGraw-Hill, New York, NY

Orden A (1952) Solution of systems of linear inequalities on a digital computer. In: Proceedings of the association for computing machinery, May:91–95

Orden A (1993) LP from the '40 s to the '90 s. Interfaces 23(5):2–12

Snopes (2009) www.snopes.com/college/homework/unsolvable.asp. Accessed 27 Jul 2009)

Steiner R (1992) Memories of IIASA, irreverent and reverent. IIASA Options, June 17

Thapa M (2009) Personal communication

Wood MK (1952) Research program of Project SCOOP. In: Orden A, Goldstein L (eds) Proceedings symposium on linear inequalities and programming, directorate of management analysis service, Comptroller, Headquarters, U. S. Air Force, Washington, DC, pp 3–14

Wood MK, Geisler MA (1951) Development of dynamic models for program planning. In: Koopmans TC (ed) Activity analysis of production and allocation: proceedings of a conference. Wiley, New York, NY, pp 189–215

14

Herbert A. Simon

Arjang A. Assad

In the early years of operations research (OR), the line between OR and mathematical economics was often blurred. Important and influential contributions to OR were being made by mathematical economists from the University of Chicago's Cowles Commission for Research in Economics and the RAND Corporation. Herbert A. Simon, a social scientist whose research spanned social and economic systems, was first exposed to OR through the work of the Cowles Commission and RAND, as well as by his research projects at Carnegie Mellon University. Even as he advocated alternative approaches, he considered OR as a normative science of decision making. He received the Nobel Prize in Economics in 1978 for his pioneering research into the decision-making process within economic organizations. He was awarded the John von Neumann Theory Prize from the Operations Research Society of America (ORSA) and The Institute of Management Sciences (TIMS) for the renaissance-like character of his research program, and for specific research contributions in bounded rationality, production planning, artificial intelligence, and cognitive psychology. He authored and co-authored 27 books and close to 1,000 publications. Beyond his specific technical contributions, his influence on OR persists due to his lifelong interest in investigating decision making in administrative, social, biological, and computer-based systems.

A.A. Assad, S.I. Gass (eds.), *Profiles in Operations Research*, International Series in Operations Research & Management Science 147, DOI 10.1007/978-1-4419-6281-2_14,

EARLY YEARS: MILWAUKEE TO CHICAGO

Herbert (Herb) Alexander Simon was born in Milwaukee, Wisconsin, on June 15, 1916. His father, Arthur Simon (1881–1948) was born in a village near Mainz, Germany. Several generations of the Simon family had been vintners and wine merchants in that region. After Arthur received his engineering diploma from the Technische Hochschule in Darmstadt in 1902, he emigrated to Milwaukee in 1903. There he married Edna Merkel (born 1888), a second-generation descendant of German immigrants from Prague and Cologne. Her maternal grandfather was a salesman and a civil war veteran while her paternal grandfather was a piano builder. Edna was an accomplished pianist and a piano teacher until she married in 1910.

Copyright ©, Pittsburgh Post-Gazette, 2009, all rights reserved. Printed with permission.

For most of his career, Arthur worked at the Cutler-Hammer Manufacturing Company, a producer of eletrical equipment. Although he practiced law as a patent attorney, he took great pride in his profession as an engineer. It was important to him to follow the scientific advances in engineering. As a designer of control gear (he patented a servomechanism in 1919), Arthur was "a significant contributor to the development of control devices" (Simon 1991, 108). His social circles included other engineers and professionals; he sometimes took Herb on plant tours organized by the Engineer's Society. Arthur was also skilful at carpentry and gardening. His intellectual interests extended beyond engineering and he communicated to Herb an appreciation of broad cultural interests which was later reinforced by Herb's University of Chicago education.

In addition to his parents, Herb's household included his only sibling Clarence (5 years older) and his maternal grandparents. Two events put a strain on the family life. Arthur almost lost his job during the Great

Depression and was forced to take a pay cut to keep it. Edna became a semi-invalid in 1936 requiring much care and attention. In spite of these events, Herb did not encounter any economic hardship as a child and enjoyed a happy, middle-class family life.

At school, Herb soon found out that "he was smarter than his classmates and that became important to him" (Simon 1991, 8). He mastered his courses with ease and developed the confidence to learn new subjects, often on his own. His older brother was not a close companion; at home, Herb was often alone, reading books, pursuing chess, or collecting insects or stamps. His interest in societal matters started early—in the fourth grade he drew up a school constitution and presented it to the principal, and wrote similar items for student clubs. In high school, his educational "interests were quite dispersed, although they were increasingly directed toward science," of what sort, he was not sure (Simon 1978c). He expanded his social activities by joining student clubs (debate, science, Latin). He always loved the outdoors and, as a Boy Scout, took part in hiking, and camping trips. Thanks to his extensive readings in economics and the sciences in his high school years, Herb already had a notion that the social sciences could benefit from mathematical thinking in the same way as the physical sciences.

In September 1933, Herb started college at the University of Chicago, having won a $300 a year scholarship based on his scores in a competitive examination. Herb intended to pursue both social sciences and mathematics. By age 19, he had already broached what he called the central issue of his intellectual quest (he would call it his monomania): decision making (Simon 2001).

While Herb studied both economics and political science as an undergraduate, he majored in political science to avoid an accounting course required by the economics department. His college career was, to put it mildly, unconventional. Because of the excellent education he acquired in high school, he finished all the survey courses and some of the upper-division requirements in both the political and social sciences in less than 2 years. He preferred self-study and did not care to attend classes even when he was enrolled in a course—Herb took a French class that he almost never attended and stopped taking any more classes in mathematics when his sophomore calculus professor insisted that he attend class. Since then, all of Herb's mathematical education was self-taught (Simon 1991, 40). In his third year, Herb was free to focus entirely on graduate courses.

At Chicago, Herb made two very close friends who greatly influenced his life and career—Harold Guetzkow and William (Bill) Cooper. Harold had grown up in Milwaukee less than a mile from Herb's home, but they first met on the train to Chicago as they headed for college. While they followed different curricula at the University of Chicago, they kept a lively intellectual exchange about their mutual interests—it was Harold who got Herb interested in cognitive psychology while both were undergraduates (Simon 1997, xv). Bill Cooper, who lived in the same dormitory as Herb, often double-dated with Herb and, on one such occasion, Bill arranged a blind-date for Herb with Dorothea Pye. Dorothea was a graduate student in the political science department and was also the department's secretary. Herb and Dorothea started to date regularly and were married on Christmas Day, 1937. In 1949, Bill persuaded Herb to join the Carnegie Institute of Technology faculty, and, in turn, Herb, convinced Harold to join Carnegie Tech in 1957.

Herb also took courses outside of political science from three professors who influenced him greatly: Henry Schultz, Nicholas Rashevsky, and Rudolf Carnap. Through the econometrician and mathematical economist Schultz, with whom Herb took graduate courses in economics, Herb learned about Walrasian equilibrium theory, the Neyman-Pearson theory of statistical inference, and the importance of the identification problem (a subject he was to work on himself). Schultz and Herb were both impressed by Alfred Lotka's book, *Elements of Physical Biology*, and his use of mathematics to model biological systems (Lotka 1925). This led Herb to the mathematical biophysicist Rashevsky, who, inspired by Lotka's work, created a new degree program in mathematical biology at the Chicago.

Herb was also drawn to philosophy of science courses offered by Rudolf Carnap. One of Carnap's main interests was the logical structure of language. Herb embarked on a serious study of formal logic on his own and absorbed several key works of logic and logical positivism by the summer of 1937. This intensive study of logic had profound effects on Herb's subsequent work (Crowther-Heyck 2005, 73). At one point, Herb had planned to write his dissertation on the logical structure of political science. Reflecting on these three teachers, Herb stated: "All three men communicated to me in their lectures something of how science—at least science involving the applications of mathematics—was taught" (Simon 1991, 53).

GRADUATE SCHOOL AND BEYOND

Chicago's department of political science was the nation's leader in 1936. It matched Harvard University as the top granter of Ph.D. degrees (separated by a large gap from third place Columbia University). The faculty included Charles Merriam, Harold Lasswell, and Harold Gosnell; its graduate roster included Gabriel Almond, V. O. Key, and David Truman. All of these individuals, joined by Herb, were to make a substantial impact on the field (Crowther-Heyck 2005, 41). An important characteristic of the department was its intellectual atmosphere that allowed students to pursue different interests in different disciplines. Another important connection was the department's involvement in the city manager movement, which aimed for efficient, more business-like ways of running municipalities. According to one historian, "it is fair to call the University of Chicago the movement's intellectual and institutional hub" (Crowther-Heyck 2005, 45).

A term paper Herb wrote in 1935 for an undergraduate course he was taking turned out to be significant in his future path. The paper was on the recreation programs offered by Milwaukee's city government and directly led Herb, in 1936, to enroll in Clarence Ridley's course on measuring municipal governments. Ridley was the director of the International City Managers' Association (ICMA) and an adjunct professor in the political science department. ICMA offered instruction and services to city managers, prepared textbooks and training manuals, and offered correspondence courses. Its yearbook allowed city managers to compare their city's statistics with others.

When Herb entered graduate school in September, 1936, Ridley offered Herb a half-time research assistantship (at $62.50 a month). Herb's task was to support Ridley in a large project on municipal activities that he was managing. The work involved "carrying out investigations that would now be classified as operations research" in the measurement of public services (Simon 1978c). As a research assistant, Herb soon found himself spending more time on this project than on his courses. Although he was just 21 years old, Herb's ICMA activities involved visits to and interactions with city managers around the country. More generally, the ICMA project exposed Herb to managerial issues in municipalities. By observing Ridley, Herb gained insights into how a professional organization could be led and effectively run. The project results appeared in a series of 13 articles in *Public Management*, starting with Ridley and Simon (1937a, b), giving Herb his first journal publications. Later, these articles were collected into the monograph, *Measuring Municipal Activities* (Ridley and Simon 1938).

In late Spring 1938, Herb, still a graduate student at Chicago, went to California in a temporary position to help on a municipal research project at the University of California (Berkeley) Bureau of Public Administration. The Bureau was funded by the Rockefeller Foundation. With the project soon to close, Herb, on completion of his temporary assignment, returned to Chicago to continue his studies and pursuit of his Ph.D. But, in early 1939, the Foundation funded a 3-year extension of the Bureau's project. Herb was invited to become director of the project, even though he had not completed his doctorate. Arranging for a leave from his studies, he accepted the Bureau's offer. He and Dorothea set out for California by rail in August 1939. In a hotel room along the way, they were woken up at 2:00 a.m. to learn that Hitler had invaded Poland.

The Berkeley project involved three studies of an empirical and statistical nature. The goal was to shed light on effective municipal practices. The first study sought to determine how large a caseload to assign to social workers in the State relief administration. Herb called this "comparable in scope to the Hawthorne Studies... carried out in the Western Electric Company during the 1930s, and more systematically designed" (Simon 1991, 82). This study generated a massive amount of data that was processed using new IBM wired plug-board machines. Another study involved a theoretical examination of urban property taxes and led to Herb's first publication in the economics literature (Simon 1943). Along with the Milwaukee recreation study, this research served as an early source of Herb's ideas on bounded rationality. He found that his conclusions depended as much on the assumptions a human decision maker included in an analysis of a tax increase as on the neoclassical assumptions of economic rationality.

The period at Berkeley was satisfying to Herb intellectually and socially. He became friends with Ronald Shephard, a student of the economist Griffith Evans, and Kenneth May, who studied with the statistician Jerzy Neyman. From them, he learned much about economics and the theory of statistical inference. In addition to his project work, Herb had to worry about preparing for his doctoral examinations and the writing of a dissertation. He took a 3-month leave of absence from work to prepare for his examinations which he took by mail in 1940 and "moonlighted a dissertation" (Simon 1978c). In May 1942, he returned to Chicago to defend his dissertation "Administrative behavior: A study of decision-making processes in administrative organizations." Herb's graduate transcript listed only one course for which he received credit—a course on sports! Curiously, the instructor was Bill Cooper.

After Pearl Harbor, Herb wanted to enlist for combat service and, with his education and experience, he felt qualified to be a commissioned officer. A convoluted bureaucratic process eventually led to his being rejected due to his color blindness (Simon 1991, 90–91). He was left with the choice of training for a commission in a noncombat job or being eventually drafted as a private. As he was not keen on the former, he decided to stay out until called. The call was postponed mainly because of family responsibilities—Herb was married with two children. His draft noticed came just after the bombing of Hiroshima; he never served. "I would be lying if I said I was sorry to have missed that war; perhaps just a little guilty" (Simon 1991, 91).

ILLINOIS INSTITUTE OF TECHNOLOGY, *ADMINISTRATIVE BEHAVIOR*, AND THE COWLES COMMISSION

When Herb was convinced that he would not join the military, he accepted an offer from the Illinois Institute of Technology (IIT) in Chicago, his first faculty position. Herb and his family returned to Chicago in the summer of 1942. Accompanying Herb and Dorothea was their daughter Katherine (Kathie), 3 months old at the time (Simon 1991, 93). Herb made rapid progress during his 7 years at IIT. By 1946, he was an associate professor earning $4600 per year. Despite his young age, he sought and secured the position of department chairman in 1946. After the publication of his first major book, *Administrative Behavior* (1947), he was promoted to full professor with a salary of $6000. On the personal front, Herb and Dorothea's household increased to five, with Peter born in 1944 and Barbara in 1946.

The book, *Administrative Behavior*, was based on Herb's dissertation and drew upon his previous work on municipalities with Ridley and the Berkeley project. Herb's observations of resource allocation for municipal services convinced him that the decisions of local governments could not be explained with the neoclassical economic framework of the single decision maker optimizing a profit function. Instead, he stressed that "administrative theory must be concerned with the limits of rationality, and the manner in which organization affects these limits for the person making a decision" (Simon 1947, 241).

At IIT, Herb's main assignment was teaching constitutional law to engineers. He taught two or three sections of this required course each semester. In the late 1940s, when Donald Smithburg and Victor Thompson joined the department, they and Herb co-authored *Public Administration* (Simon, Smithburg, and Thompson 1950)—an ambitious textbook that covered a wide spectrum of topics in greater detail than Herb's *Administrative Behavior.*

Upon Herb's return to Chicago, Bill Cooper suggested that he become involved with the Cowles Commission for Research in Economics. The Cowles Commission was the center of postwar research in mathematical economics and econometrics and Herb became a regular participant at its weekly seminars. At the time, the Cowles staff included Jacob Marschak, Tjalling Koopmans, Kenneth Arrow, Gerard Debreu, Leonid Hurwitz, and Lawrence Klein. Milton Friedman, Oscar Lange, and Franco Modigliani (Herb's future colleague at Carnegie) also took part in the seminars. Herb's association with Cowles and its staff left an important mark on his research trajectory and his professional associations. At Cowles, he "received his fourth education in economics" (Simon 1991, 102). Herb also believed that the greatest impact of his exposure to Cowles was to encourage him "to try to mathematize" his "previous research in organization theory and decision making" (Simon 1947, 4). This also fueled Herb's interest in mathematical economics and econometrics.

At Cowles, Herb learned about the nascent capabilities of linear programming (LP) and Koopman's wartime research on activity analysis (Simon 1991, 103). He participated in Marschak's major project on the economic effects of atomic energy, and took charge of the macroeconomic components of the study for which he used activity analysis. Having learned of game theory before the *Theory of Games and Economic Behavior* by von Neumann and Morgenstern (1994) was published, Herb spent his 1944 Christmas vacation reading it and published one of the first reviews of that book (Simon 1945). Herb also developed lifelong friendships with, and much affection for, Koopmans and Marschak. He dedicated his book, *Reason in Human Affairs* (Simon 1983) to the latter.

The mathematician Merrill Flood, through his ties with Cowles and RAND, invited Herb to consult for RAND starting in the summer of 1952. Herb's main project at RAND was to study the organization of early warning stations for air defense. He spent several summers and the entire academic year 1960–1961 at RAND.

Herb's intensive reading of the economics literature had already resulted in an early contribution to input–output economics, now known as the Hawkins–Simon theorem (Hawkins and Simon 1949), the significance of which is described by Paul Samuelson (2004). As a result of his participation in the Cowles Commission and consulting at RAND, Herb's research increasingly turned to mathematical economics and OR in 1950–1955. Some of the research Herb conducted while at the Cowles Commission and RAND, and his early Carnegie period, is reflected in the first five chapters of his book, *Models of Man: Social and Rational* (Simon 1957). Another important consequence of Herb's involvement with the Cowles Commission was his enhanced contact with (and visibility within) the mathematical economics community.

Along a different line, Herb pursued his interest in statistical distributions governing social or biological formations, which dated back to his dissatisfaction with the explanations provided for Zipf's law in the 1930s. In 1952, he studied a family of skew distributions related to Yule's distribution (Simon 1955b) and used these for distributions of business firms sizes, incomes, or cities by population. This resulted in stream of work reflected in his joint work with Charles Bonini and Yuji Ijiri (Ijiri and Simon 1964, 1977, Simon and Bonini 1958).

Carnegie Institute of Technology: Graduate School of Industrial Administration (GSIA)

Herb made his first visit to the Carnegie Institute of Technology, Pittsburgh, Pennsylvania, in 1948 at the invitation of Bill Cooper who had joined its department of economics in 1947. In 1949, Carnegie received a $6 million endowment from William Larimer Mellon (founder of the Gulf Oil Company and Andrew's brother) to establish a Graduate School of Industrial Administration (GSIA). Carnegie Provost, Elliott Dunlap Smith, and G. Lee Bach, chairman of the economics department, led the planning effort for GSIA. Cooper suggested that Herb be invited to join in these discussions. This resulted in an offer to Herb to join Carnegie as chairman of the Department of Industrial Engineering. He accepted, and he and Dorothea, and their three children, moved to their new house on Northumberland Street in the summer of 1949. The house was located a mile from work—every day for the next 47 years, Herb walked back and forth to Carnegie for a total of 25,000 miles (Frank 2004, 12). [Carnegie was joined with the Mellon Institute of Science in 1965 and renamed Carnegie Mellon University (CMU)].

BOUNDED HOUSE HUNTING

"On a trip to Pittsburgh in April 1949 . . . I took a long walk early one morning through much of the northern part of Squirrel Hill. Just before this visit, I had drawn on a map of Pittsburgh a circle of one mile radius around the Carnegie Tech campus, for I was resolved to walk to work . . . and had checked the census tract data to discover which portions of this area were inhabited by college-educated middle-class families. I looked in these portions for a house we could afford" (Simon 1991, 136–137).

HERB'S WALK

Herb's early years at GSIA encompass a remarkable story. He continued to conduct research in several different areas including organization theory, economics, and OR. The breadth of the topics Herb was actively pursuing is reflected in the compilation of his papers entitled *Models of Man* (Simon 1957). As part of the new GSIA administrative faculty, Herb played a pivotal role in steering the organization through its formative years—an important exercise in institution building, arguably his most significant within the area of management education. Their common goal was to "place business education on a foundation of fundamental studies in economics and behavioral science" (Simon 1978c). As Herb put it years later:

> We felt like we were going to have the first business school that had academic respectability, scientific respectability, and we didn't think it needed to run like dead-headed, old-fashioned business schools, or we wouldn't have been there. None of the people who came in were from a business school

> background. We came in with the understanding that we were going to build a different kind of business school, that we were going to experiment and see where these new ideas—operations research, or management science . . . and organization theory . . . led (McCorduck 1979, 139).

It was also during this period that Harold Guetzkow, Charles Holt, George Kozmetsky, and Franco Modigliani joined the GSIA faculty. The school focused on two main research areas: organizational behavior and management science (MS). Herb and Harold led the former, with Bill Cooper taking charge of the latter. One of Bill's projects, planned in a document dated February 1950, was called production technology. Herb was a member of one of the two MS research teams directed by Bill—the paint-factory project, funded by the Office of Naval Research.

The paint factory research focused on production planning at a single plant of the Pittsburgh Plate Glass Company (now PPG industries) located in Springdale, Pennsylvania. Three researchers joined Herb to form the project team: Charles Holt, an electrical engineer; economist Franco Modigliani, who had previously worked on production smoothing; and Jack Muth, a "green graduate student" with an undergraduate degree in industrial engineering who was studying for his Ph.D. at GSIA (Muth 2004, 377).

In joining this team, Herb's primary interest was to understand how managers actually made decisions in organizations (Holt 2002, 96). The results of the project led to the book, *Planning Production, Inventories, and Work Force* (Holt et al. 1960), commonly referred to as HMMS. HMMS is a true classic of the OR literature—it pays equal attention to theory and practice, combining insightful modeling with elegant mathematics.

Herb continued his research on organizations, especially the descriptive study of organizational decision making, well into the 1950s. An early project Herb participated in was led by George Kozmestky and funded by the Controllership Foundation. This work studied organizational issues of the controller's function in large firms (Simon et al. 1954). His work with Harold Guetzkow and James March on a project funded by the Ford Foundation to inventory various theories of organization led to the book, *Organizations* (March and Simon 1958), and several papers in *Models of Man* (Simon 1957).

Another topic of lasting interest to Herb was aggregation and decomposability in mathematical models. In work with Albert Ando in 1956, Herb considered the case of a system composed of sectors, where strong interactions took place within each sector but interactions across sectors were weak (Simon and Ando 1961). Mathematically, this led to a study of nearly block-diagonal structure with few off-diagonal elements and methods for their inversion. In his expository account, Herb discusses the implications of such structures for decomposability along with an example (Simon 1996, 197–200). Egidi and Marengo (2004) discuss how the topic of near-decomposability is a recurrent theme in Herb's work in different disciplines.

THE TRANSITION OF 1955: ARTIFICIAL INTELLIGENCE (AI)

Herb has called 1955–1956 as the most important years of his life as a scientist (Simon 1991, 189). While the scope of his research efforts had been broad and interdisciplinary prior to 1955, he was typically identified as an economist and political scientist. In the last months of 1955, his work in human problem solving opened up an entirely new direction that soon assumed a central position and absorbed most, if not all, of his research activity.

Herb believed that a deeper understanding of decision making required a theory of human problem solving and knowledge of the processes people use to solve complex problems. Computers were just emerging as viable tools for such research. Herb's strategic insight was to realize the potential of computers and to stress their ability and capacity for processing symbols, rather than crunching numbers. In 1952, while consulting at RAND, he met Allen (Al) Newell at the System Research Laboratory. There, Al and J. Clifford (Cliff) Shaw had already collaborated on computer-generated radar maps. Al was 25 years old and already an accomplished researcher. Herb and Al immediately recognized their intellectual affinity: they both viewed the human mind as a "symbol-manipulating" (Herb's term) or "information-processing" (Al's term) system (Simon 1991, 168).

By 1954, computer simulation of human cognition and heuristic problem solving took center stage in their joint research. Both were intrigued by the possibility of using computers as non-numerical processors—Herb taught himself to program the IBM 701. In time, they were recognized as two of the founding fathers of computer science. [Herb's seminal role in the birth of AI is described in McCorduck (1979).]

December 15, 1955, is the day that computer-based, heuristic-problem solving was born. On that day, he and Al had the insight on how to program a thinking machine. Together, the Newell-Shaw-Simon (NSS) team implemented their vision to create Logic Theorist (LT), a computer program that used heuristic rules to prove theorems. LT used a list processing language they developed and was the first operational AI program presented at the Dartmouth workshop on AI in June, 1956. On August 9, 1956, LT produced the first complete proof of theorem 2.01 given in Whitehead and Russell's *Principia Mathematica*. By 1957, they had developed GPS (General Problem Solver), a program that used means-end analysis as a general strategy. (In problem solving, means-end analysis refers to the notion of using rules of thumb to move from the current state to the desired state by successively reducing the gap between the current and desired states.)

THE MACHINE THAT THOUGHT

"I was an undergraduate senior, but I was taking a graduate course over in GSIA from Herb Simon. . . . It was just after Christmas vacation—January 1956—when Herb Simon came into the classroom and said: 'Over Christmas, Allen Newell and I invented a thinking machine.' And we all looked blank. . . so we said, 'Well what do you mean by a thinking machine? And in particular, what do you mean by a machine?' In response to that, he put down on the table a bunch of IBM 701 manuals and said, 'Here, take this home and read it and you'll find out what I mean by a machine.' Carnegie Tech did not have a 701, but RAND did, though it was actually the Johnniac that Newell, Shaw, and Simon were working with. But Herb chose the 701 as an introduction for us" [Edward Feigenbaum cited in McCorduck (2004, 138)].

The NSS theory of human information processing was based on the considerable experience gained from LT. It views problem solving as a heuristic search that uses simple rules and finds its formal expression in a computer program. NSS introduced such foundational concepts of AI as list processing languages, heuristic search, production rules, means-end analysis, and verbal protocols.

The book, *Human Problem Solving* (Newell and Simon 1972), presented an integrated view of the research program of the NSS team. The computer scientist, Edward Feigenbaum, characterized it as "perhaps the most important book on the scientific study of human thinking in the 20th century" (Feigenbaum 2001, 2107). In their 1975 Turing Lecture, Al and Herb described the notion of a physical symbol system as "a machine that produces through time an evolving collection of symbol structures," an example would be the computer (Newell and Simon 1976, 116). They formulated the physical symbol system hypothesis: "A physical symbol system has the necessary and sufficient means for general intelligent system" (Newell and Simon 1976, 116). This allowed them to consider artificial intelligence and the understanding of human problem solving as two facets of the same scientific program. Feigenbaum observed that Herb ultimately found programming languages to be richer than mathematics in modeling the complex problem-solving processes his team was studying (Feigenbaum 2001).

In 1965, Herb left GSIA to join the CMU computer science (CS) department; he was appointed the Richard King Mellon University Professor of Computer Science and Psychology, a position he held for the remainder of his life. In 1968, Herb was invited to give the Karl Taylor Compton lectures at MIT. This provided him with the opportunity to present his thoughts on the interface between computers, problem solving, AI, and design in a series of lectures he called "The Sciences of the Artificial." In one of these lectures, he urged that design be developed as part of the engineering curriculum. This led to the formation of the Design Research Center at CMU—the first center of its kind (Simon 1991, 259). The lectures formed the basis of his book, *Sciences of the Artificial* (Simon 1965).

Herb was at CMU for over 50 years—his impact was on the institution was extensive. Besides helping to launch GSIA, he also helped create the CS department, the Robotics Institute, and the Design Research Center, as well as forming the cognitive science group within the psychology department. As a member of the university board of trustees, his views were heard at the highest level.

Herb made two important decisions about his role as a university administrator. In 1961, when the GSIA dean, Lee Bach, decided to leave Carnegie, Herb was associate dean and viewed as a possible successor to Bach. Herb, however, declined to express interest in the position because he felt that this would leave little time for his own research. Similarly, in 1972 when CMU was searching for a new president, Herb and the GSIA economist Dick Cyert were candidates for the job. Herb withdrew as a candidate, but served on the search committee. This time, he felt uncomfortable with the "close association with the business community that is essential for effective performance as president" (Simon 1991, 263).

THE NOBEL PRIZE

Early in 1969, Herb visited Sweden to take part in a joint seminar between CMU and Scandinavian business faculty. The initiation of the new The Sveriges Riksbank Prize in Economic Sciences in Memory of Alfred Nobel (The Nobel Prize in Economic Sciences) had just been announced, with the first prize to be awarded in the fall of 1969. Walter Goldberg, a business economist who was attending the seminar, predicted, in a conversation with Herb, "You will receive the prize within the next ten years." Herb was incredulous about that happening—he felt that economists did not regard him as an economist (Simon 1991, 320). On Monday, October 16, 1978, at 6:00 a.m. Pittsburgh time, Herb received a call from Stockholm that he had been awarded the Nobel Prize in economics. Significantly, this was not a shared prize. In early December, Herb and his family attended Nobel week in Stockholm, where he presented his Nobel lecture on December 8, 1978 (Simon 1979a). He delivered one of his banquet speeches in Swedish.

The choice of Herb as a Nobel laureate was somewhat unexpected in view of his sustained criticism of the classical economic paradigm for decision making (a rational decision maker maximizes the utility of the decision). In part, Herb attributed his selection to the fact that he was a "duly certified member of the Econometric Mafia" and "an insider to its elite" (Simon 1991, 326). He defined this group as fellows of the Econometric Society in 1954—the year in which he was elected a fellow. These were the thought leaders responsible for "the sudden conquest of economics by mathematics and statistics" (Simon 1991, 325–326). The 1954 list of fellows contained 120 names, only about 80 of which were alive in 1969, the year the first Nobel Prize in economics was awarded (the prize is awarded to the living). Herb notes that 20 of the first 27 Nobel laureates belonged to this group. He knew about half the fellows, and nearly all the previous prize winners. Coupled with the highly favorable incidence of Nobel Prizes awarded to the fellows, Herb's visibility within this elite group contributed to his selection for the Nobel Prize.

Herb's Nobel lecture is a masterful account of bounded rationality and his intellectual trajectory in advancing this concept (Simon 1978c). He appreciated the renewed attention that the Nobel Prize would draw to his research and planned to use the opportunity as a bully pulpit for presenting his ideas (Simon 1991, 324). He estimated that because of the Nobel Prize, he was spending about 10% of his research time "thinking about the application of bounded rationality to economic theory" (Simon 1991, 325). He continued his polemics with the economics profession in his American Economic Association Ely Lecture "Rationality as process and as product of thought" (Simon 1978a), his paper "How to decide what to do" (Simon 1978b), and his article questioning the empirical evidence supporting neoclassical theory (Simon 1979b).

MAJOR THEMES IN DECISION MAKING: ADMINISTRATIVE BEHAVIOR, BOUNDED RATIONALITY, AND PRODUCTION PLANNING

ADMINISTRATIVE BEHAVIOR AS DECISION MAKING

Herb's first book, *Administrative Behavior* (Simon 1947), stemmed from his doctoral dissertation. Initially off to a slow start, the book achieved a pervasive influence over the years and became a staple of schools of business and public administration. The Nobel Prize press release called the book epoch-making; it was named as Book of the Half Century by *Public Administration Review* and appeared in its fourth edition in 1997. It is rare for a book on management to hold up so well after 60 years.

In his preface to *Administrative Behavior*, Herb modestly pointed out that the book does not offer a theory of administration: "If any 'theory' is involved, it is that decision-making is the heart of administration, and that the vocabulary of administrative behavior must be derived from the logic and psychology of human choice" (Simon 1997, xi).

Early in the book, Herb registers his dissatisfaction with the received wisdom in administration and its commonly cited adages. Herb felt that the literature on public administration showed a general neglect of decision making, save for the book *The Functions of the Executive* by the executive Chester Barnard (1938), which he cited as a motivation for his work, and a paper by the political scientist Edwin Stene (1940). In *Administrative Behavior* (Chapter 3), Herb assigned a central place to decisions in administration:

> At each moment, the behaving subject, or the organization composed of numbers of such individuals, is confronted with a large number of alternative behaviors, some of which are present in consciousness and some of which are not. Decision ... is the process by which one of these alternatives for each moment's behavior is selected to be carried out. The series of such

> decisions which determines behavior over some stretch of time may be called a *strategy* . . . . The task of rational decision is to select that one of the strategies which is followed by a preferred set of consequences. . . . The task of decision involves three steps: (1) the listing of all of the alternative strategies; (2) the determination of all the consequences that follow upon each of these strategies; (3) the comparative evaluation of these sets of consequences (Simon 1997, 77).

At this point, Herb adds the crucial modifying phrase: "it is impossible for the individual to know *all* his alternatives or *all* their consequences and the impossibility is a very important departure of actual behavior from objective rationality" (Simon 1997, 77). This set the stage for thinking about limits to rationality.

In Chapter 5 of *Administrative Behavior*, Herb explores these limits that arise because (1) knowledge of consequences is incomplete, (2) future consequences are imperfectly anticipated, and (3) only a few of all possible alternatives ever come to mind. He concludes:

> *The central concern of administrative theory is the boundary between the rational and the nonrational aspects of human social behavior.* Administrative theory is peculiarly the theory of intended and bounded rationality—of the behavior of human beings who *satisfice* because they have not the wits to *maximize* (Simon 1997, 118).

Although the seeds of bounded rationality were already present in *Administrative Behavior*, Herb noted that, in 1947, "limited rationality was defined largely as a residual category—as a departure from rationality" (Simon 1997, 118). The theoretical framework for the concept of bounded rationality came later.

Bounded Rationality

Bounded rationality is regarded as Herb's signature contribution to the economic and social sciences. This principle argues:

> The capacity of the human mind for formulating and solving complex problems is very small compared with the size of the problems whose solution is required for objectively rational behavior in the real world—or even for a reasonable approximation to such objective reality (Simon, 1957, 198).

James March summarizes the original idea of bounded rationality as follows:

> ...it is well to recall that the original argument was a narrow one. It started from the proposition that all intendedly rational behavior is behavior within constraints. Simon added the idea that the list of technical constraints on choice should include some properties of human beings as processors of information and as problem solvers. The limitations were limitations of computational capability, the organization and utilization of memory, and the like. Human beings develop procedures that are sensible, given the constraints .... As a short-hand label for such procedures, he coined the term "satisficing" (March 1978, 590).

In the early 1950s, Herb set out to replace the notion of rationality of classical economics with an appropriate alternative. Herb's motivation for introducing this change was clear from the beginning:

> Broadly stated, the task is to replace the global rationality of economic man with a kind of rational behavior that is compatible with the access to information and the computational capabilities that are actually possessed by organisms, including man, in the kinds of environments in which such organisms exist (Simon 1955a, 99).

In explaining the properties of the "choosing organism," Herb noted that "the state of information may as well be regarded as a characteristic of the decision-maker as a characteristic of his environment" (Simon 1955a, 100). Identifying the environment as an important element of the explanation developed into a recurrent theme in Herb's subsequent research, specially his later work on complexity (Simon 1956).

The framework for this notion of approximate rationality was developed in his seminal papers on a behavioral model of rational choice (Simon 1955a, 1956a) and the book *Models of Man* (Simon 1957). In "A behavioral model of rational choice" (Simon 1955a), Herb reviews the key elements of the classical optimization framework of rational choice: a set of alternatives to choose from, future outcomes of choice, and information as to which outcomes will occur if a particular choice is made. Given a payoff function that specifies the utility for each possible outcome, the optimal rational choice is the one that maximizes the (expected) utility. Arguing that the computations required to find the optimal choice places an unrealistic burden on the decision maker, Herb then systematically simplifies the various components to reduce this burden. Instead of looking for the best payoff at a given stage, the decision maker may simply stop searching when a certain aspiration level is met. While aspiration levels may change dynamically as the search proceeds, eventually, they provide a criterion for concluding the search.

The use of aspiration levels and identification of subgoals recognize the cognitive limitations of human beings. In the search for solutions, an effective strategy is to substitute "the goal of satisficing, of finding a good enough move, for the goal of maximizing, of finding the best move" (Simon 1957, 205). A satisficing decision maker will be content with achieving an aspiration level set in advance. If a solution meeting this aspiration level can be found, the search will be stopped, otherwise the aspiration level will be lowered. Given the bounded and uncertain nature of the real world, Herb argued that satisficing behavior based on rules of thumb is a more accurate representation of real decision-making behavior. Further, he also felt these have greater explaining power:

> Models of satisficing behavior are richer than models of maximizing behavior because they treat not only of equilibrium but of the method of reaching it as well. Psychological studies of the formation and change of aspiration levels support propositions of the following kinds. (a) When performance falls short of the level of aspiration, search behavior (particularly search for new alternatives) is induced. (b) At the same time, the level of aspiration begins to adjust itself downward until goals reach levels that are practically available. (c) If the two mechanisms just listed operate too slowly to adapt aspirations to performance, emotional behavior—apathy or aggression, for example—will replace rational adaptive behavior (Simon 1959, 263).

As March (1978) has pointed out, subsequent developments of the theory of bounded rationality were extensive. Herb's own work in this area was published in two volumes (Simon 1982). As used by others, distortions of the original theory were introduced, some of which did not meet with Herb's approval. To define the notion of rationality more precisely, Herb himself made a distinction between substantive and procedural rationality. March (1978) lists a number of alternative notions of rationality.

Production Planning: The HMMS Study

One of Herb's earliest papers in the area of operations was in production control. Around 1948, he became interested in using feedback control for economic systems and was delighted to discover that his father had been a designer of control gear. Herb investigated the application of servomechanism theory to production systems (Simon 1952). The problem was to control the production rate of a single item in continuous time so as to minimize the sum of production and inventory-related costs. These were the days before optimal control theory, so Herb used the techniques of transforms for feedback control. Interestingly, the variable production costs varied with changes in the production rate, as was the case with the subsequent full-scale paint factory HMMS study. A follow-up to this work that incorporated some of the paint factory results was published by Simon and Holt (1954).

The HMMS study (for the PPG paint factory) may be viewed as the foundation of aggregate production and inventory planning. The independent decision variables for the firm are, for each time period t, the aggregate production level P_t and the aggregate work force level W_t. Given these, the inventory level I_t at the end of each period can be obtained (I_t is unrestricted in sign). The objective is to minimize the sum of production, workforce-related, and inventory-related costs. Nonlinearities arise in modeling the cost of overtime or changes in the level of the work force (hiring and firing costs). In the HMMS model, the main idea is to approximate all costs components as quadratic functions in P_t, W_t, and I_t so that the first-order optimality conditions result in linear decision rules. Specifically, these rules express the optimal values of P_t and W_t as linear expressions in terms of W_{t-1}, I_{t-1}, and orders for the current and future time periods (Holt et al. 1955).

The HMMS model also provides a good illustration of how Herb saw the connection between the notion of satisficing and modeling strategies in MS. In his Nobel lecture, Herb referred to this work when he contrasted two possible approaches to model construction in MS:

> Now the salient characteristic of the decision tools employed in management science is that they have to be capable of actually making or recommending decisions, taking as their inputs the kinds of empirical data that are available in the real world, and performing only such computations as can reasonably be performed by existing desk calculators or, a little later, electronic computers. For these domains, idealized models of optimizing entrepreneurs, equipped with complete certainty about the world—or, at worst, having full probability

> distributions for uncertain events—are of little use. Models have to be fashioned with an eye to practical computability, no matter how severe the approximations and simplifications that are thereby imposed on them. The first is to retain optimization, but to simplify sufficiently so that the optimum (in the simplified world!) is computable. The second is to construct satisficing models that provide good enough decisions with reasonable costs of computation. By giving up optimization, a richer set of properties in the real world can be retained in the models . . . . Neither approach, in general, dominates the other, and both have continued to co-exist in the world of management science (Simon 1979a, 498).

The HMMS study led to two important developments. The paint factory data showed the need for better forecasts that would also be practical to use for a large number of items. Holt modified the exponential smoothing method to handle trend and seasonal effects. This led to the Holt-Winters model of forecasting (Holt 2002, 98). A more significant development for economics was Herb's paper on dynamic optimization under a quadratic cost objective (Simon 1956b). Herb showed that to derive the optimal decisions, the expected values suffice as the certainty equivalents. This is true as long as the costs are quadratic and the residual variance is independent of the decisions. Thus, in the HMMS model, the assumption of quadratic costs not only results in optimal linear decision rules, but also ensures that only the expected values (and not the higher moments or entire distributions) of the uncertain quantities enter into the decision rule. This result, called certainty equivalence, was proved by Herb and Charles Holt (Holt et al. 1955, 16). More important, it led to the development of the theory of rational expectations in economics.

Although the HMMS study used a substantial amount of analysis and mathematical machinery, the members of the team, to their credit, were also concerned with model validation and implementation issues. The results of the model were compared with factory performance under different decision rules and forecasting techniques. The linear decision rules proved to be superior—a conservative estimate indicated cost savings of 8.5%, with further savings possible with improved forecasting (Holt et al. 1955). But, when it was decided to implement the optimal plan, difficulties were encountered; they were resolved eventually and the HMMS results were implemented at all 70 PPG factories (Holt 2002, 97).

BOLD HYPOTHESIS

Herb's certainty-equivalence result led Jack Muth to propose a bold hypothesis—a new theory of expectations that extended rationality considerably beyond even what the classical theory suggested. Responding to Herb's view that the decision maker cannot estimate "the joint probability of future events" in such a way as to "maximize the expected value or utility or profit" (Simon 1959, 268), especially in the presence of changes over time, Jack's hypothesis was based on exactly the opposite point of view: "that dynamic models do not assume enough rationality." Instead, he proposed "that expectations of firms (or, more generally, the subjective probability distribution of outcomes) tend to be distributed, for the same information set, about the prediction of the theory (or the 'objective' probability distribution of outcomes)" Muth (1961, 316).

This rational expectations theory found favor with economists in the 1970s, constituting a new orthodoxy with assumptions directly opposed to those of bounded rationality. Although Herb did not believe that this theory described the real state of affairs, he felt that Muth surely deserved a Nobel Prize for its formulation. Robert Lucas, building on Muth's rational expectation model, extended that research to earn a Nobel Prize in 1995. Commenting on the close and amicable relations of the four men who made up the HMMS team, Herb wrote: "The HMMS research team harbored simultaneously two Keynesians (Modigliani and Holt), the prophet of bounded rationality (Simon), and the inventor of rational expectations (Muth)—the previous orthodoxy, a heresy, and a new orthodoxy" (Simon 1991, 250).

OR Modeling and AI

Even as Herb's interests moved toward the cognitive sciences and AI, he continued to think about the implications of these new areas for OR. He focused on presenting his view of heuristic problem solving to the OR community at ORSA's 12th national meeting in November 1957. After recognizing the significant contributions of OR to "those management decisions that can be reduced to systematic computational routines," he stated that "large areas of managerial activity . . . have hardly been touched by operations research or the advances of management sciences" (Simon and Newell 1958, 3–4). These areas are typically ill-structured but important decisions of top-level executives. Simon and Newell suggested that advances in heuristic problem solving would narrow the gap between these areas and the well-structured problems OR had addressed.

The Simon and Newell (1958) paper attracted a certain degree of notoriety in which the authors state four predictions with respect to AI and heuristic problem solving that were to be realized within 10 years of 1957. While these predictions did not come to pass within the proposed time frame, Herb's view of heuristics in reasoning influenced several theoretical and applied areas of OR by the 1980s. In the late 1980s, Herb continued to argue for a closer connection between AI and OR (Simon 1987).

Herb became intrigued by the new areas of chaos and complexity and their implications for modeling and design. He argued that chaos has raised serious doubts about the possibility of designing good predictive models. Starting with a critique of the Club of Rome limits to growth model and its predictions, Herb invited the OR community to consider the limitations and misuse of predictive models—OR analysts should redirect their modeling efforts "to focus them on the questions that we *can* answer more or less definitely." He also restated his view that "intelligent approximation, not brute force computation, is still the key to effective modeling" (Simon 1990, 10–11).

THE CLOUDED CRYSTAL BALL

1. Within the next ten years a digital computer will be the world's chess champion, unless the rules bar it from competition.
2. Within the next ten years a digital computer will discover and prove an important new mathematical theorem.
3. Within the next ten years a digital computer will write music that will be accepted by critics as possessing considerable aesthetic value.
4. Within the next ten years most theories in psychology will take the form of computer programs, or of qualitative statement about the characteristics of computer programs.

(Simon and Newell 1958, 7–8)

For the last 25 years of his research career, Herb continued to develop and explore computer models of cognition in different domains. He studied and used computer-based simulation programs of learning processes, the role of representation and framing in problem solving, what made problems hard, and scientific discovery. Aiming at the non-specialist, Herb described the main themes of this period in the third edition of *Sciences of the Artificial* (Simon 1996).

THE SCIENTIST AS ADMINISTRATOR OR POLITICIAN

Herb maintained a consistent interest in science policy issues. This interest began with his involvement with the Ford Foundation where he served as an advisor and member of the board of directors of the Social Sciences Research Council (SSRC), and served as its chair from 1967 to 1971. In keeping with SSRC's mission, Herb acted as a strong advocate for interdisciplinary initiatives in this role. He argued against narrow area studies and other SSRC activities that lay within single disciplines. He also challenged research funding that was motivated by undue disciplinary loyalties.

In the 1960s, the focus of his activities shifted to the National Research Council (NRC), a unit within National Academy of the Sciences (NAS) in Washington, D.C. The elected membership of the NAS included mainly physical and biological scientists and very few social scientists. Herb and several others decided to rectify this situation—his approach was to effect change from within. The NRC's Division of Anthropology and Psychology did include social scientists and demographers. Herb was invited to join the board of this group, to be groomed as a prospective chair. This led to his election to the NAS in 1967. He then joined with other nominating groups to bring such social scientists as the economists Kenneth Arrow, Tjalling Koopmans, Robert Merton, and Paul Samuelson into the NAS. The new NAS structure formalized in 1972 recognized a new class of membership for Social and Behavioral scholars.

During 1968–1971, Herb served on the Committee on Science and Public Policy of the NAS. Herb found this assignment highly rewarding: "Members, whatever our own expertise, received a broad education in every part of science with outstanding experts as our teachers. Most important of all was the sheer pleasure of associating with bright, wide-ranging minds, and sharp wits" (Simon 1991, 294). Beyond such personal satisfaction, Herb also recounted his experiences on committees as mini-cases to illustrate organizational or political issues (Simon 1983).

In January 1968, Herb became a member of the President's Science Advisory Committee (PSAC), having been appointed by President Lyndon Johnson—he was the first social scientist to so serve—and he continued to serve through the first Nixon administration. Dominated by physicists ["physics mafia" (Simon 1991, 294)], it was highly atypical of PSAC to include a social scientist. While Herb was most likely elected by virtue of his expertise in AI and computer science, he and others viewed this as an opportunity to cause PSAC to include social scientists. Herb served on several PSAC panels, including two focusing on environmental and educational issues.

In the 1980s, Herb became quite involved with scientific exchanges with China. (He spent more time in China than in any other foreign country.) His first trip to China was in 1972 as part of a delegation of six computer scientists (and two wives, including Dorothea) that took them to Canton, Shanghai, and Beijing for lectures and discussions with Chinese researchers, teachers, and students (this was during the time of Mao's cultural revolution). His second trip was in 1980 with a delegation from

> HOW TO RANK, CHINESE STYLE
>
> "From the moment we arrived at Canton [1972 visit], the Chinese were very curious about our ages. No matter how often we told them, they would ask again. Then we began to realize that, when our names were arranged in alphabetical order (and how else would Americans arrange them?), our ages coincidently ran strictly from youngest to oldest (myself). But it was in that order that we were seated at banquets—from highest to lowest. And it was in that order that our cars proceeded in autocade. (Foreign visitors being rare at that time, each of us had a car, driver, and an interpreter.) How wondered the Chinese, could the head of the party be the youngest, and the oldest be the last?" (Simon 1991, 341).

the American Psychological Association, 4 years after the removal of the Gang of Four, when China was showing the first signs of economic reform. In 1983, Herb became chair of the Committee for Scholarly Communication with the People's Republic of China—a body created by the NAS and the SSRC. His 4-year appointment as chair (1983–1987) gave him a weightier role; he made annual trips to China with each one including a heavy program of visits, lectures, and teaching. In June 1989, Herb's usual summer visit to China was shortened to 2 days as he passed though Beijing during the Tiananmen Square uprising. Herb is one of the few foreigners to be elected to the Chinese Academy of Sciences.

ON THE PERSONAL SIDE

Throughout his life, Herb maintained research as his top priority. He called himself a "monomaniac about the study of decision making" (Augier and Feigenbaum 2003, 196). Much has been written about Herb's ways of thinking and approach to problem solving by those who have worked and learned from him (Augier and March 2004). Feigenbaum (2004, 384) sums it up succinctly by stressing Herb's main theme of unity and drive toward simplicity: "human problem solving is but an instance of a general information processing theory of problem solving that is as valid for artifacts (like computers) as for people." Here, simplicity occurs in Herb's simple goal-seeking rules, preference for simple (or near-decomposable) architectures, and heuristic problem solving.

Herb admitted to being a workaholic and characterized long bouts of intense work as fun (Klahr and Kotovsky 2001). His friends remember him as a perpetually curious man. He simply never stopped thinking—thinking was his passion. Given a lull in a conversation, he would introduce a topic and engage his listeners in active dialogue. Or, he could convene an impromptu seminar while waiting for a guest in a hallway (Larkey 2002, 239). A simple observation from everyday life could launch him into hours of mental activity, trying to offer a plausible account of the observation (Crowther-Heyck 2005, 312). His colleague David Klahr recalls how during a long drive in a car, he said something to Herb about the roads being deserted. This launched Herb into a mental calculation of the density of cars on the interstates at 1:00 a.m. (Klahr 2004, 444).

Herb remained intellectually active until the last days of his life. One of the last manuscripts he commented on was the one Bill Cooper wrote for the 50th anniversary issue of the journal *Operations Research* (Cooper 2002; 2004, 74). Herb always enjoyed a vigorous debate with his colleagues, and had a reputation as a polemicist. His arguments were exacting, but never ad hominem. If he started a remark with "Look friend. . ." this meant that the gloves were off and that the flaws in his interlocutor's argument would be laid bare (Larkey 2002, 241).

Herb was also a dedicated teacher and took institution building and professional service seriously (Klahr and Kotovsky 2001). He was patient and tolerant with his students, whom he invited to think in class (instead of taking notes). Nor did he try to cut corners when it came to teaching. He did "a sizeable portion of the TAs [teaching assistants] job himself" and read student progress reports because they were exciting. Faced with a doubling in the size of a core course, he said 'you know, when people come to you and want to learn—you should teach them'" (Kotovsky 2004, 458). His doctoral students included Ed Feigenbaum, David Klahr, Kenneth Kotovsky, Pat Langley, Jack Muth, Harry Pople, and Raul Valdez-Peres.

A SATISFICING LUNCH—SAY CHEESE

"I particularly love [this anecdote] because it shows how Herb applied the 'satisficing' precept to his own life. . . . Herb thought of 'satisficing' as a way of cutting down on decision time in a world that required too many decisions relative to available time. During our frequent lunches, we observed how Herb always ordered the same lunch: American cheese on white. . . . He explained that he had adopted a rule for lunch: when faced with a choice, he would always request American cheese on white. This order ... avoided time-consuming decisions about what to choose and could certainly be executed anywhere in the U.S. Though the outcome might not be the most refined dish available, it was 'good enough' " (Modigliani 2004, 374).

Herb kept his family life private and wrote little about it. Glimpses of Herb as a father are provided by his elder daughter, Katherine (Kathie) Simon (Frank 2001, 2004). He found time for simple family pleasures with his three children. At times, he was a stern father and the disciplinarian of the family, but it was important to him that his children learn to make their own decisions. Kathie recounts how Herb invited his family to a Carnegie classroom to simulate computer-based problem solving. Following instructions written on cards, each of the three children acted as a subroutine with the role of the executive program reserved for Dorothea (Frank 2001).

While a private man, Herb also formed lifelong friendships. The tributes to Herb in the memorial collection of retrospective essays tell of his impact on his good friends (Augier and March 2004). He hung seven portraits on the walls of his study: his father, Arthur Simon; Chester Barnard; Albert Einstein; Abraham Lincoln; Charles Merriam; Clarence Ridley; and Franklin Delano Roosevelt (Simon 1991, 72). As his admired leaders or teachers, these and a few others directly influenced his intellectual development. In conducting his own affairs, he was a (bounded) rationalist through and through, preferring not to devote much time or attention to insignificant decisions (Frank 2004).

Herb did not watch television, rarely listened to the radio, and claimed that reading newspapers was a waste of time. He did enjoy playing piano, listening to music, reading, and learning languages. He read novels frequently and twice read the seven volumes of Marcel Proust's *Remembrance of Things Past* in the original language (McCorduck 2004, 476). In his memoirs, Herb describes his conversation

WITHOUT MINOTAUR

"I have encountered many branches in the maze of my life's path, where I have followed now the left fork, now the right. The metaphor of the maze is irresistible to someone who has devoted his scientific career to understanding human choice. And if I had not encountered labyrinths early in my life, I would have met them later in the stories of Jorge Luis Borges . . . .

"In describing my life as mazelike, I do not mean that I have made a large number of deliberate, wrenching decisions to go off in one direction or another. On the contrary, I have made very few. Obvious responses to opportunities and circumstances, rather than studied decisions, have put me on the particular roads I have followed" (Simon 1991, xvii–xviii).

with the writer and poet Jorge Luis Borges in Argentina in 1970 (Simon 1991, 175–179). Herb was struck by how Borges described life as a search through a maze in his book of short stories *Ficciones*. This moved Herb to write a short story on this theme (*The Apple: A Story of a Maze*; Simon 1991, 180–188) which he regarded as a nontechnical introduction to his work. Accordingly, the metaphor of a maze runs through his memoirs.

In mid-January 2001, Herb underwent surgery to remove a cancerous tumor from his abdomen. The surgery was successful but complications developed in the ensuing weeks:

> At the end, there was an issue of how invasive the attempts to prolong his life should be. His daughter Kathie has related how he finally told her it was time for him to go. He told her, about a day before his passing, that life has a beginning, a middle, and an end, that there were no projects he'd started that he was essential for, and that it was time for the end (Kotovosky 2004, 458).

Herb died on Friday, February 10, 2001. He was survived by his wife Dorothea, who died on August 15, 2002, and his three children: Kathie Simon Frank of Minneapolis; Peter A. Simon of Bryan, Texas; and Barbara M. Simon of Wilder, Vermont.

HONORS AND AWARDS

Herb's contributions were recognized across social and scientific fields: American Psychological Association Distinguished Scientific Contributions Award, 1969; Association of Computing Machinery Turing Award (jointly with Allen Newell), 1975; Nobel Prize in Economics, 1978; National Medal of Science, 1986; ORSA-TIMS 1988 John von Neumann Theory Prize. He was a member of and/or received awards from: American Academy of Arts and Sciences, 1959; National Academy of Sciences, 1967; American Psychological Association Distinguished Scientific Contribution Award, 1969; American Economic Association Distinguished Fellow 1976, Ely Lecturer 1977; Academy of Management Scholarly Contributions Award, 1983; American Political Science Association James Madison Award that recognizes an American political scientist who has made a distinguished scholarly contribution to political science, 1984; John M. Gaus Award and Lectureship for a lifetime of exemplary scholarship; Award for Lifetime Contributions to Psychology,

1993; American Society of Public Administration Dwight Waldo Award for outstanding contributions to the professional literature of public administration, 1995. Herb was a member of the International Federation of Operational Research Societies' Hall of Fame (Assad 2004); a foreign member of the Chinese, Japanese, and Russian academies of science; and he was awarded over 20 honorary doctorates.

REFERENCES

Assad A (2004) IFORS' operational research hall of fame: Herbert A. Simon. Int Trans Oper Res 11(4):479–484

Augier M, Feigenbaum E (2003) Herbert Simon: biographical memoirs. Proc Am Philos Soc 147(2):193–198

Augier M, March J (2004) Models of man: essays in memory of Herbert A. Simon. MIT Press, Cambridge, MA

Barnard C (1938) The functions of the executive. Harvard University Press, Cambridge, MA

Cooper WW (2002) Abraham Charnes and W. W. Cooper (et al.): a brief history of a long collaboration in developing industrial uses of linear programming. Oper Res 50(1):35–41

Cooper WW (2004) Memorial to Herbert A. Simon. In: Augier M, March J (eds) Models of a man. MIT Press, Cambridge, MA, pp 67–74

Crowther-Heyck H (2005) Herbert A. Simon: the bounds of reason in modern America. Johns Hopkins University Press, Baltimore, MD

Egidi M, Marengo L (2004) Near-decomposability, organization, and evolution: some notes on Herbert Simon's contribution. In: Augier M, March J (eds) Models of a man. MIT Press, Cambridge, MA, pp 335–350

Feigenbaum E (2001) Retrospective: Herbert Simon, 1916–2001. Science 291(5511):2107

Feigenbaum E (2004) On a different plane. In: Augier M, March J (eds) Models of a man. MIT Press, Cambridge, MA, pp 384–388

Frank K (2001) Herbert A. Simon: a family memory. http://www.cs.cmu.edu/simon/kfrank.html. Accessed 5 Aug 2009

Frank K (2004) He's just my dad. In: Augier M, March J (eds) Models of a man. MIT Press, Cambridge, MA, pp 33–43

Hawkins D, Simon HA (1949) Note: some conditions of macroeconomic stability. Econometrica 17(3/4):248–248

Holt C (2002) Learning how to plan production, inventories, and work force. Oper Res 50(1):96–99

Holt C, Modigliani F, Muth J, Simon HA (1960) Planning production, inventories, and work force. Prentice-Hall, Englewood Cliffs, NJ

Holt C, Modigliani F, Simon HA (1955) A linear decision rule for production and employment scheduling. Manage Sci 2(1):1–30

Ijiri Y, Simon HA (1964) Business firm growth and size. Am Econ Rev 54(2):77–89

Ijiri Y, Simon HA (1977) Skew distributions and the size of business firms. North-Holland, Amsterdam

Klahr D (2004) Encounters with the force of Herbert A. Simon. In: Augier M, March J (eds) Models of a man. MIT Press, Cambridge, MA, pp 433–449

Klahr D, Kotovsky K (2001) Remembering Herb Simon. APS Obs 14(4):1–6

Kotovsky K (2004) Strong ideas. In: Augier M, March J (eds) Models of a man. MIT Press, Cambridge, MA, pp 451–460

Larkey P (2002) Ask a simple question: a retrospective on Herbert Alexander Simon. Public Sci 35(3):239–268

Lotka A (1925) Elements of physical biology. Williams & Wilkins, Baltimore, MD

March J (1978) Bounded rationality, ambiguity, and the engineering of choice. Bell J Econ 9(2):587–608

March J, Simon HA (1958) Organizations. Wiley, New York, NY (2nd edn, 1993, Blackwell)

McCorduck P (1979) Machines who think. W. H. Freeman, San Francisco, CA

McCorduck P (2004) Herb Simon: a recollection. In: Augier M, March J (eds) Models of a man. MIT Press, Cambridge, MA, pp 473–480

Modigliani F (2004) Simon: some cherished memoirs. In: Augier M, March J (eds) Models of a man. MIT Press, Cambridge, MA, pp 373–375

Muth J (1961) Rational expectations and the theory of price movements. Econometrica 29(3):315–335

Muth J (2004) Herbert Simon and production scheduling. In: Augier M, March J (eds) Models of a man. MIT Press, Cambridge, MA, pp 377–380

Newell A, Simon HA (1972) Human problem solving. Prentice-Hall, Englewood Cliffs, NJ

Newell A, Simon HA (1976) Computer science as empirical inquiry (The 1975 ACM Turing Award Lecture). Commun ACM 19(3):113–126

Ridley C, Simon HA (1937a) Techniques of appraising standards. Public Manage 19(2):46–49

Ridley C, Simon HA (1937b) Development of measurement standards. Public Manage 19(3) 84–88

Ridley C, Simon HA (1938) Measuring municipal activities. International City Manager's Association, Chicago, IL

Samuelson P (2004) The Hawkins and Simon story revisited. In: Augier M, March J (eds) Models of a man. MIT Press, Cambridge, MA, pp 153–167

Simon HA (1943) The incidence of a tax on urban real property. Q J Econ 57(3):398–420
Simon HA (1945) Review of the theory of games and economic behavior. Am J Soc 50(6):558–560
Simon HA (1947) Administrative behavior: a study of decision-making processes in administrative organizations. Macmillan, New York, NY
Simon HA (1952) On the application of servomechanism theory in the study of production control. Econometrica 20(2):247–268
Simon HA (1955a) A behavioral model of rational choice. Q J Econ 69 (1):99–118
Simon HA (1955b) On a class of skew distribution functions. Biometrika 42(3/4):425–440
Simon HA (1956a) Rational choice and the structure of the environment. Psychol Rev 63(2):129–138
Simon HA (1956b) Dynamic programming under uncertainty with a quadratic criterion function. Econometrica 24(1):74–81
Simon HA (1957) Models of man: social and rational. Wiley, New York, NY
Simon HA (1959) Theories of decision-making in economics and behavioral science. Am Econ Rev 49(3):253–283
Simon HA (1965) The sciences of the artificial. MIT Press, Cambridge, MA
Simon HA (1978a) Rationality as process and as product of thought. Am Econ Rev 68(2):1–16
Simon HA (1978b) On how to decide what to do. Bell J Econ 9(2):494–507
Simon HA (1978c) Autobiography (Nobel Prize Web site). http://nobelprize.org/nobel_prizes/economics/laureates/1978/simon-autobio.html. Accessed 13 Jun 2009
Simon HA (1979a) Rational decision making in business organizations (The Nobel Prize lecture of December 8, 1978). Am Econ Rev 69(4):493–513
Simon HA (1979b) On parsimonious explanations of production relations. Scand J Econ 81:459–474. (Reprinted in Models of Bounded Rationality, vol 1, 1982)
Simon HA (1982) Models of bounded rationality: volumes I and II. MIT Press, Cambridge, MA
Simon HA (1983) Reason in human affairs. Stanford University Press, Stanford, CA
Simon HA (1987) Two heads are better than one: the collaboration between AI and OR. Interfaces 17(4):8–15
Simon HA (1990) Prediction and prescription in systems modeling. Oper Res 38(1):7–14
Simon HA (1991) Models of my life. Basic Books, New York, NY
Simon HA (1996) The sciences of the artificial, 3rd edn. MIT Press, Cambridge, MA
Simon HA (1997) Administrative behavior: a study of decision-making processes in administrative organizations, 4th edn. The Free Press, New York, NY
Simon HA (2001) On simulating Simon: his monomania, and its sources in bounded rationality. Stud Hist Philos Sci 32(3):501–505

Simon HA, Ando A (1961) Aggregation of variables in dynamic systems. Econometrica 29:111–138

Simon HA, Bonini C (1958) The size distribution of business firms. Am Econ Rev 48(4):607–617

Simon HA, Holt C (1954) The control of inventories and production rates—a survey. J Oper Res Soc Am 2(3):289–301

Simon HA, Guetzkow H, Kozmetsky G, Tyndall H (1954) Centralization and decentralization in organizing the controller's department. Controllership Foundation, New York, NY

Simon HA, Newell A (1958) Heuristic problem solving: the next advance in operations research. Oper Res 6(1):1–10

Simon HA, Newell A (1972) Human problem solving. Prentice-Hall, Englewood Cliffs, NJ

Simon HA, Smithburg D, Thompson V (1950) Public administration. Alfred A. Knopf, New York, NY

Stene E (1940) An approach to a science of administration. Am Pol Sci Rev 34(6):1124–1137

Von Neumann J, Morgenstern O (1944) Theory of games and economic behavior. Princeton University Press, Princeton, NJ (2nd edn, 1947; 3rd edn, 1953)

15

Andrew Vazsonyi

Nancy C. Weida

Andrew (Andy) Vazsonyi helped shape the post-World War II (WWII) emerging operations research (OR) profession by leading the way for theoreticians and practitioners to enhance the practice of quantitative and qualitative problem-solving in business and industrial settings. A mathematician by inclination and education, Andy's career highlights include his now famous and continually rediscovered solution of a problem in location theory that he solved when he was 16 years old, and the finding of his true professional life when he encountered the real-world problems of business and industry that led him to the decision and information sciences. During his early engineer phase, he also contributed to the fields of aerodynamics and control systems. His 1958 book, *Scientific Programming in Business and Industry*, was seminal in its ability to describe, using a minimum amount of technical mathematics, the new and exciting ideas of OR to a world-wide audience of non-mathematical managers, technicians, and students. His research and writings ranged from early descriptions of mathematical applications to production and inventory control, to automated information systems in command and control, and, much later, to the use of spreadsheets in the decision sciences. Andy was a driving force in establishing management science (MS) as a proper extension of OR principles and methods into the business and industrial environments. Andy was a founder and first past-president (he was never president!) of The Institute of Management Sciences (TIMS), and a fellow and In-House Philosopher of the Decision Sciences Institute (DSI).

Andy wore many hats during his life as an OR/decision sciences pioneer: mathematician, author, researcher, educator, businessman,

A.A. Assad, S.I. Gass (eds.), *Profiles in Operations Research*, International Series in Operations Research & Management Science 147, DOI 10.1007/978-1-4419-6281-2_15,

engineer, illustrator, and storyteller. His last book, *Which Door has the Cadillac: Adventures of a Real Life Mathematician*, is an autobiographical account of how a young Hungarian immigrant to the U.S. came, saw, and conquered.

A JOURNEY WEST: BUDAPEST TO PARIS TO NYC

Andy was born Endre Weiszfeld in Budapest, Hungary on November 4, 1916, the middle of three sons. His father, Miska, was born in 1880 in the town of Nagyvázsony, and his mother, Hermine, was born in 1885 in the village of Sárospatak. Miska was a successful businessman who owned a custom-made shoe shop that he had inherited from his father. His shoes were well-known for their fine quality and handcraftsmanship. Andy absorbed many basic concepts of economics and business from his father that would remain central throughout his life and career.

It was during his childhood that Andy discovered what he liked to call his math hump—a love and talent for mathematics. He became lifelong friends with Paul Erdös, another Hungarian mathematics prodigy. As teenagers, they enjoyed solving math problems, and challenging each other, activities that would continue until Paul died in 1996. The 14-year-old Andy first met the 17-year-old Paul in 1930. In their first encounter, arranged by Miska, Paul asked Andy how many proofs of the Pythagorean Theorem he knew. "One," answered Andy. "I know thirty-seven," replied Paul (Vazsonyi 2002a, 34). (Miska was the only one who could make comfortable shoes for Paul's contorted feet. After Paul left Hungary, he only wore sandals.)

As was the case for many Hungarian Jews in the years leading up to WWII, Andy decided to change his name. He chose the name Vazsonyi, after the town of Nagyvázsony, his father's birthplace. In this, he followed the lead of his famous cousin Vilmos, the Hungarian Minister of Justice (1917–1918) and the leader of the National Democratic Party. Vilmos had earlier changed his name from Weiszfeld to Vazsonyi.

In 1938, Andy received his Ph.D. from the Pázmány Péter University under Lipót Fejér. (Fejér was also the dissertation advisor for Paul Erdös, George Pólya, and John von Neumann.) Andy's dissertation (exact title unknown) was his first journal publication under the name of Weiszfeld—it dealt with surfaces of higher dimensions (Vazsonyi 2002a, 51). Because he would have to serve in the military, Andy was not anxious to stay in Hungary. Also, opportunities for Jews to do post-graduate work or teach in Hungary were basically nonexistent, so he decided to emigrate, to the U.S., if possible, or to a western European country. He thus started to learn English.

With the help of the Hungarian mathematician, Otto Szász, who was on the faculty at the University of Cincinnati, Andy received a fellowship to study there. But he was granted only a 1-year student passport in 1938, after which he would have to return to Hungary and enter the military. Instead, he decided to go to Paris—an easy thing to do then—with the hope of getting a visa to the U.S. He then started to learn French.

In Paris, Andy learned that there were a great many Hungarian Jews seeking visas to the U.S. and the waiting list was filled. After a 2-year convoluted negotiation between Andy and Hungarian and U.S. embassy personnel, he managed to obtain a visa to the U.S. He had to travel to Liverpool to book passage on a ship to New York City. He arrived there on April 24, 1940, just 2 months before France fell to the German army (Vazsonyi 2002a, 69–79).

METAMORPHOSIS I: THE OR CATERPILLAR

Before he left Budapest, Andy solved what is now a fundamental problem in OR: for a warehouse that serves many stores, find a location that minimizes the sum of the (Euclidean) distances from the warehouse to all the stores. He was unaware of his OR contribution until 30 years later when it was brought to his attention by a management science colleague (Vazsonyi 2002a, 61–63, 2002b). The story is the following (Gass 2004, 18):

> When Andy was 16, his interest in geometry brought him to the famous problem first posed by Pierre de Fermat in 1643: given *n* points, find a point such that the sum of its distances to the *n* points is a minimum (a generalization of Fermat's original 3-point problem). From an OR/MS perspective, the problem is now interpreted as a facility location problem, but to Andy it was just an interesting math problem. Andy's paper on his algorithmic solution method was written in French and published in a Japanese journal: "Sur le point pour lequel la somme des distances de *n* points donnés est minimum," *Tôhoku Mathematical Journal,* Vol. 34, 355–386, 1937 [Weiszfeld 1937].
>
> The paper describes the now famous and classic Endre Weiszfeld algorithm. When I asked Andy why the paper was written in French and published in an obscure math journal, he answered: "I cannot recall why I sent the location paper to the *Tôhoku Journal.* I must have had some indication that they would accept it. French was selected because I had a friend who was available and had good French knowledge. Hungarian was obviously useless." [Andy had an earlier paper, also written in French, published in the *Tôhoku Journal,* Weiszfeld 1936.]
>
> At the time of its publication, there was little interest in the Weiszfeld algorithm. Things changed in the 1960s when facility location became a hot OR topic. The paper kept being "rediscovered" by many researchers who did not know who Weiszfeld really was. This always bothered Andy. As he wrote in his autobiography, *Which Door has the Cadillac: Adventures of a Real-Life Mathematician* [Vazsonyi 2002a]: "To my great chagrin, nobody knows that Weiszfeld is around and kicking, or that Vazsonyi = Weiszfeld."
>
> I was pleased when Andy sent me a copy of the book with the dedication: "To my dear friend and colleague of many years, the only management scientist who is aware of my early math work like the Weiszfeld algorithm." The Fermat Problem is often called the Weber Problem or the Steiner Problem, but neither Weber nor Steiner contributed much to its solution. It is time to change the name of the general problem to the Vazsonyi (aka Weiszfeld) Problem.

Credit for developing an iterative algorithm for solving the *n*-point location problem is now given to Andy. Love et al. (1988) note that the paper by E. Weiszfeld "presented an iterative procedure for locating a new facility to minimize the sum of weighted Euclidean distances to any number of existing facilities .... The paper was 25 years ahead of the main thrust of modern facilities location research, which started to gain momentum in the 1960s. Indeed, the single-facility location problem was described as being unsolved in *Progress in Operations Research*, Vol. II, edited by Hertz and Eddison (1964)" (Love et al. 1988, 8–9). Harold Kuhn (1973, 9) calls attention to "a little known work [Weiszfeld 1937], which appears to have been the first to propose an iterative algorithm" and "the same algorithm has been rediscovered several times." He cites three such papers (Cooper 1963, Kuhn and Kuenne 1962, Miehle 1958) and gives a complete treatment of the algorithm's properties, including a proof of convergence. An English translation of Andy's paper is given in Weiszfeld and Plastria (2009).

EVOLUTION: FROM MATHEMATICIAN TO ENGINEER

Job opportunities in the U.S. for "a young refugee mathematician" were quite rare, but Andy did have an opportunity to attend a year-long workshop in Haverford, Pennsylvania sponsored by the Quakers for "'distinguished' European scholars who were interested in learning American ways" (Vazsonyi 2002a, 89, 93). The workshop leaders put Andy in touch with Richard von Mises, a professor of engineering at Harvard, who offered Andy a fellowship as a graduate engineering student. In the fall of 1941, Andy traveled to Cambridge, Massachusetts to accept the Gordon McKay Fellowship at Harvard's Graduate School of Engineering. (Von Mises had immigrated to the U.S. in 1939 and was Harvard's Gordon-McKay Professor of aerodynamics and applied mathematics.) Andy received an M.S. in mechanical engineering in 1942. He accepted a Harvard teaching and research fellowship sponsored by the National Aeronautics Association, working under the direction of engineering professor Howard Emmons. Andy and

SOLUTION TO FERMAT'S PROBLEM

Fermat Point

(Acrylic on canvass)

Emmons worked on supersonic aerodynamics, with their results used later in the design of supersonic airplanes and missiles. Andy also solved a complicated air-duct system problem for Pratt-Whitney Aircraft, an accomplishment that caused him to feel that he was now an engineer (Vazsonyi 2002a, 105).

THE FAMILY

Laura and Andy; Bobbi and Sky

About this time (1942), Andy met Baroness Laura Vladimirovna Saparova, a daughter of emigrants from Tbilisi, Georgia. After a 7-week courtship, they married and settled in Cambridge, Massachusetts. They had one daughter, Bobbi, who grew up to become a marriage, family, and child counselor. Throughout his life, nobody or nothing was more important to him than his beloved family—Laura, Bobbi, and her husband, Sky Chaney. He was so proud of all of them. He loved telling others of the passion and talents of his family members, and the fun adventures they all had together.

After spending 3 years at Harvard working with Emmons, Andy decided he should obtain some industry experience if he hoped to get a first-class teaching position in a school of engineering. He had become a U.S. citizen in 1945, and with the war still going, jobs were plentiful. Although he was offered a position with Pratt-Whitney, he decided, in the spring of 1945, to work for a small organization, the Elliott Company of Jeanette, Pennsylvania, a manufacturer of pumps. Later, at a conference of the American Society of Mechanical Engineers, where Andy was giving a paper, he met William Bollay, one of his Harvard professors who was now working with North American Aviation in Los Angeles. Bollay offered Andy a job with North American and he accepted.

Andy and Bollay worked on the design of a missile for which Andy made significant design changes that were controversial, but proved successful, and helped to resolve vibration problems on the P51 Mustang fighter airplane. He left North American after a falling out with Bollay and, in the fall of 1948, became head of a new guidance and control missile division for the U.S. Naval Ordnance Station in the Mojave Desert, with headquarters in Pasadena, California. But, it turned out that Andy was not enthused about being in a management position. In 1953, he joined Hughes Aircraft where Simon Ramo and Dean Woolridge were organizing a new research division; Andy was put in charge of investigating Hughes's possible venture into developing computers and business applications. This move was a turning point in his career. As he noted, "Little did I suspect that the last engineering work I would ever do would be for the Navy. Soon at Hughes, a career in a new field of managerial decision making would open up for me" (Vazsonyi 2002a, 122).

METAMORPHOSIS II: THE OR/MS CHRYSALIS

Andy's first assignment at Hughes was to see how IBM cards that were being used for payroll and production control could be replaced by computers. This brought him into the field of production and inventory control, and led to a two-part paper published in the new TIMS journal *Management Science*—"The use of mathematics in production and inventory control" (Vazsonyi 1954, 1955). These papers, based on Andy's experience working with Hughes production operating personnel, demonstrated Andy's ability to expound and explain complex management areas in a unique and understandable manner. They introduced the OR and MS communities to his famous Gozinto Diagrams for determining total parts requirements. His far-reaching aim here and in his related papers was to develop mathematical theories of managerial problems that would form a discipline "that truly could be called by the name Management Sciences" (Vazsonyi 1954, 85).

Over time, Andy's assignments at Hughes, as senior staff member of the Ramo-Woolridge Corporation, as partner of the Roe Alderson marketing consultant firm, and back to North American Aviation as a systems engineer in 1965, all contributed to his development as a leading proponent of the use of mathematics and computers in important management science areas: production control (Vazsonyi 1956), manufacturing and inventory control (Vazsonyi 1957), optimum production scheduling (Karush and Vazsonyi 1957), and automated information systems in planning, control, and command (Vazsonyi 1965). In his papers on production and inventory control (Vazsonyi 1956, 1957), Andy analyzed parts requirements problems using matrix equations, and extended the analysis to include the problem of time dependence. He developed a system

DR. Z. GOZINTO

Andy used his Gozinto (goes into) Diagrams to illustrate the complex relationships that can occur when, in a production problem, there is a need to determine total parts requirements. The problem is confounded as some parts are used directly by an assembly and some indirectly. Andy found Gozinto Diagrams a useful device in explaining his methods to Hughes executives, manufacturing floor personnel, and clients; Hughes even had an advertising campaign that featured Andy and his matrix equation used to calculate parts requirements (Vazsonyi 2002a, 158). Based on this work, Andy found he had given birth to an alter ego. The story unfolds as follows

(Vazsonyi 2002a, 205–206):

GOZINTO DIAGRAM

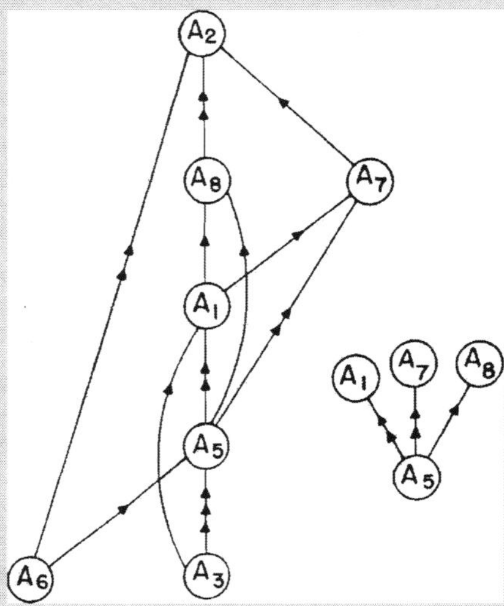

(Vazsonyi 1954, 77).

In 1956, while giving a lecture at the RAND Corporation on his matrix-based approach to solving the parts requirements problem, Andy used a Gozinto Diagram to illustrate how the parts are interrelated. After the lecture, George Dantzig remarked that he had never heard of the mathematician named Gozinto and wanted to know his university affiliation. Andy feigned surprise that George was not familiar with the famous Italian mathematician, Z. Gozinto. When George asked for Gozinto's first name, Andy said he did not know but would get back to him. He later related his encounter with George to Abe Charnes and Bill Cooper, with Bill coming up with the perfect name: Zepartzat Gozinto (say it out loud).

Further adventures of Z. Gozinto are given in Vazsonyi (1990).

of equations that related shipping schedules to requirements for assemblies and subassemblies. He also was concerned with formulating decision rules and related decision functions that could be applied by the foreman of the operation, especially with respect to scheduling. Andy's hands-on project activities, and his experiences working directly with nontechnical planning and manufacturing personnel, enabled him to make his influential book, *Scientific Programming in Business and Industry* (Vazsonyi 1958), accessible to both non-mathematical and mathematical readers.

During this period, Andy had found his mathematical niche, that is, how his mathematical talent and knowledge could be used to improve many important areas of business and industry. But, related to his own development, and that of others, was how all this fit into one's professional life. In his first research activity at Hughes determining how computer systems could be used and applied to improve production schedules, Andy "came across an announcement about the first meeting of the Operations Research Society of America (ORSA). Their goals included the use of mathematics to improve business management. Little did I know how quickly this would become one of my life's goals" (Vazsonyi 2002a, 130). Andy started to attend ORSA events and present papers at its meetings.

Andy's involvement in ORSA eventually turned sour as he felt that the founders of ORSA, most of whom had experience in WWII military OR, were "too far removed from the day-to-day world of American business. A sense of academic hubris pervaded the early ORSA sessions" (Vazsonyi 2002c, 32). In August 1953, Andy, along with others who had similar interests (economists, business school academics, management consultants, and industry managers) drafted a manifesto that gave impetus to the founding of a new professional society, The Institute of Management Sciences (TIMS) (Vazsonyi 2002a, 133–134; 2002c, 32–33):

> During the last two decades, and in particular, since World War II, the problems of modern management have increased in scope and complexity. A need now exists for systematic research which can evolve new and effective instruments for dealing with these problems. During the same period, a growing number of workers have produced a considerable amount of research in this field. A characteristic feature of this work is the attempt to deal with managerial problems by means of mathematical models. Exploiting advances in pure sciences and technology, these workers established foundations upon which a structure of management sciences may be erected.
>
> Correspondence and communication between interested persons has made it increasingly apparent that a stage has now been reached which makes it desirable to consider founding an organization dedicated to encouraging and extending this type of work. Such an organization would, if successful, (1) establish this type of work as a field of scientific endeavor, (2) stimulate work in the field, and (3) publish a journal, convoke meetings, and adopt other suitable means of making the fruits of this work available to all interested persons.

The manifesto was later signed by 18 people, a few of whom became Nobel Laureates in economics, including Herbert Simon. TIMS was founded on December 1, 1953 with William W. Cooper as president, and, because the TIMS constitution required one, Andy was elected to be the first past-president, the only past-president who never served as president. (TIMS and ORSA would eventually merge in 1995 to become INFORMS, the Institute for Operations Research and the Management Sciences.)

HEROES: PAUL ERDÖS AND HERB SIMON

Two of Andy's most important heroes or privy councilors were Nobel Laureate Herbert Simon and mathematician extraordinaire Paul Erdös. Andy also had many other heroes he spoke of often, especially Richard Feynman, Benjamin Franklin, and his daughter Bobbi Chaney (as friend, author, professional, and someone with whom he shared ideas).

Andy dedicated his autobiography (Vazsonyi 2002a, v) "to the memory of Herbert A. Simon, my lifelong friend, icon and professional inspiration." Andy credited Simon's *Models of Man: Social and Rational* (Simon 1957) as the work that helped him most in his transition from pure to applied mathematician—this book laid the groundwork for bringing scientific rigor to the social sciences while taking into account human motivation and behavior. Andy spoke often of Simon's influence on his thinking and how he approached problems (Vazsonyi 2002a, 183–190). This influence permeated all of Andy's later work, including his last textbook, *Operations Analysis Using Microsoft Excel* (Weida et al. 2001), which featured Simon's approach and phases of decision making, using creativity and scenario analysis.

Andy's life intertwined with Paul Erdös, his childhood and lifelong friend. During his Hungarian boyhood, Andy studied mathematics alongside Paul Erdös, and published two papers with him in 1936. While Erdös remained in the world of pure math, Vazsonyi eventually discovered the wonders of applying math to the real world of business and engineering. Erdös was known to declare Andy dead (Erdös' term for someone who no longer did pure mathematics), though Erdös later acknowledged Andy came back to life from time to time when he solved a real math problem. The accomplishments and interactions of Weiszfeld/Vazsonyi and Erdös are discussed in Andy's tribute to Erdös (Vazsonyi 1997a) and Andy's autobiography (Vazsonyi 2002a).

Metamorphosis III: The Butterfly—Professor and Decision Scientist

In 1970, after 25 years working in industry, Andy "decided to swap a successful industrial career for a life in academia. . . . I thought I had kept up with university life, as I had always taught in extension schools. But I had very little understanding of how academia really worked. I didn't realize how little professors knew about practical management" (Vazsonyi 2002a, 260). He accepted a 3-year appointment at the University of Southern California School of Management. He taught statistics and computer systems, oversaw the installation of the School's first mainframe computer, and wrote articles and books on information systems. He then accepted a faculty position at the University of Rochester's Graduate School of Business where he initiated courses in management information systems, wrote his best-selling book, *Introduction to Electronic Data Processing* (Vazsonyi 1973), and taught himself to ski from a book. In 1977, he became a chaired professor at the business school of St. Mary's University in San Antonio, Texas, staying there until his retirement in 1987. He then moved back to California and kept ties to academia with his appointment as professor emeritus with the McClaren School of Business, University of San Francisco. He and Laura retired to his Santa Rosa, California home in 1989 to be near Bobbi and Sky.

Andy's contributions to OR/MS continued in high gear upon his official retirement. In fact, he often described himself as an activist for helping ordinary people use mathematics and the scientific method in daily life, while helping them ignore the nonsense of various charlatans. He wrote letters to the editor of his local newspaper, and created a Web site (which, since his death, is no longer accessible) to help people appreciate how mathematics can help individuals and organizations make decisions, both large and small. He was trained as a pure mathematician, yet circumstances forced him to become practical; eventually he came to lose the old ways and appreciate fully the practical—and love what he was doing!

While at Rochester, Andy became associated with the American Institute of Decision Sciences, now the Decision Sciences Institute (DSI), a professional group whose members are educators mainly from schools of business. DSI became Andy's principal professional affiliation (Vazsonyi 2002a, 273). He served on its council and was honored by being elected a DSI fellow. In 1989, he started to write a regular column for DSI's newsletter, *Decision Line*, under the heading "The Specialist with the Universal Mind." Andy's columns were often provocative and covered a wide range of OR/MS topics; he reveled in stirring interest and debate. In rereading these columns, one is struck about their prescience and continued relevance, as seen in these sample articles: "Math deficit disorder" (Vazsonyi 2002e); "Young man/woman go healthcare" (Vazsonyi 2002f); "The quest for the mathematics for decision sciences" (Vazsonyi 1997b); and "Multicriteria decision making" (Vazsonyi 1995). Andy alternated his column with "The Decision Sciences Picture Book" in which he would expound on a topic and, to make a point, he would illustrate it with one of his drawings. He wrote his *Decision Line* columns steadily from May 1989 until his final column in July 2003. Andy's columns can be found at http://www.decisionsciences.org/decisionline/archive.asp, the DSI Web site (Decision Sciences Institute 2008).

A MODEL IS LIKE A SAUSAGE MACHINE

"Mathematical models are the core of decision sciences. They are mathematical objects, functions. They map, transform the input domain into the output range. I can harmonize the concept of the mathematical model with the scientific philosophy of Karl Popper, that is, as a mental image imbedded in the inner world of my mind. But real-world is on the outside. Stimulated by Popper, I say that I create a 'painting' of my mental model employing symbols and notations. Thus there are three worlds of math models: the inner world, the outer world, and the painting.

"The traditional x, y, z way of dealing with this third world turns most people off, and we run into major emotional blocks in spreading our science. However, the dot.com magic opens the gate to millions to deal with this third world. We must not abandon the math models of our inner world, but we can paint them by using symbols, notations, and graphics of the computer. And herein lies the future of management science" (Vazsonyi 2000, 18).

"Models and sausage machines are alike: They both transform inputs into outputs"(Vazsonyi 2000, 18).

Andy also shared his insights and concerns with TIMS, ORSA, and, when they merged, the INFORMS communities by publishing numerous columns, articles, and letters to the editor in *Interfaces*, *OR/MS TODAY* (where his contributions included several "Musings upon OR/MS" columns), *Operations Research* and *Management Science*. Notable are his contributions to the special issue of *OR/MS Today* celebrating 50 years of OR where he discussed the origin of the "Milestone manifesto" (Vazsonyi 2002c) and "My first taste of OR: 'I had a helluva big assignment'" (Vazsonyi 2002d). His *OR/MS Today* pieces can be viewed at http://www.lionhrtpub.com/ORMS.shtml (INFORMS 2008).

Andy was convinced that an optimization syndrome was killing OR and the decision sciences in schools of business. He recognized early the need to supplant mathematical models with computer-based (especially spreadsheet-based) models. "The great advantage of Excel is that we can liberate management science from much of the obnoxious math details, and can include behavioral stuff. Prometheus unchained" (Vazsonyi 1998a). He pursued this concept with his usual diligence with an article in *Interfaces*: "Where We Ought to be Going: The Potential of Spreadsheets." He starts with a no-nonsense statement (Vazsonyi 1993, 26):

> Spreadsheets empower millions of managers to avoid algebra and classical mathematics and unknowingly approximate solutions to their problems by elementary numerical analysis. They use spreadsheets to carry out vast quantities of computations and create graphical support. These business people are numerate in the sense that they deal comfortably with the fundamental notions of numbers and chance. A hurdle in disseminating MS/OR is the shortage of mathematically proficient managers. These numerate managers should be more capable, receptive, and sympathetic to MS/OR. I believe that spreadsheets are the manager's skeleton key to mathematical modeling.

Andy concludes with his "private prescription" (Vazsonyi 1993, 38):

Deep six:

- Paper, pen, pencil,
- Adding machines,
- Pocket calculators,
- Mathematical and statistical tables, and
- Graph paper.

This perspective culminated in his 2001 book, co-authored with Nancy Weida and Ronny Richardson, *Operations Analysis Using Microsoft Excel.* It is a non-traditional textbook that includes new techniques and approaches to modeling and analyzing a wide range of operations management problems—optimization, scheduling, project management, and quality control—with innovative uses of spreadsheet modeling, integrating scenario analysis, and Herbert Simon's three phases of the decision-making process: intelligence, design, and choice (Simon 1997).

ADVENTURES OF A REAL-LIFE MATHEMATICIAN

We are fortunate to have Andy's autobiographical account of his amazing life and passion for mathematics: *Which Door has the Cadillac: Adventures of a Real-Life Mathematician* (Vazsonyi 2002a). Andy's life and work is composed of story upon story, as he uninhibitedly relates the wide variety of events and people—and twists and turns—of his life. The book is also a whimsical ride through the world of mathematical applications, and a guide for others to understand how mathematics can make one's life better. Andy was particularly interested in reaching mathematicians, non-mathematicians, and teachers at both the high-school and college

WHICH DOOR HAS THE CADILLAC

The title refers to the classic counterintuitive problem of updating probabilities, popularized when Marilyn vos Savant, in her "Ask Marilyn" *Parade Magazine* column, September 9, 1990, included a solution that enraged some mathematicians (and others). Andy delighted in discussing and analyzing this problem in which logic trumps intuition, and was particularly pleased when he stumped Paul Erdös with it. In this "Monty Hall" problem, named for the host of the old television game show "Let's Make a Deal," a contestant chooses one of three doors, behind two of which are goats and behind the other is a Cadillac. Monty, knowing which door the Cadillac is behind, opens one of the doors not chosen by the contestant and shows a goat. The contestant is then given the opportunity to switch doors. The question: Should the contestant switch from the original choice? The problem caught the public's imagination, especially when some prominent mathematicians challenged Marilyn's (correct) solution—it pays to switch! Andy describes the problem and why the switch has a probability of 2/3 of winning the Cadillac, and a probability of 1/3 of finding a goat if the switch is not made (Vazsonyi 1999, 2002a, 2003a).

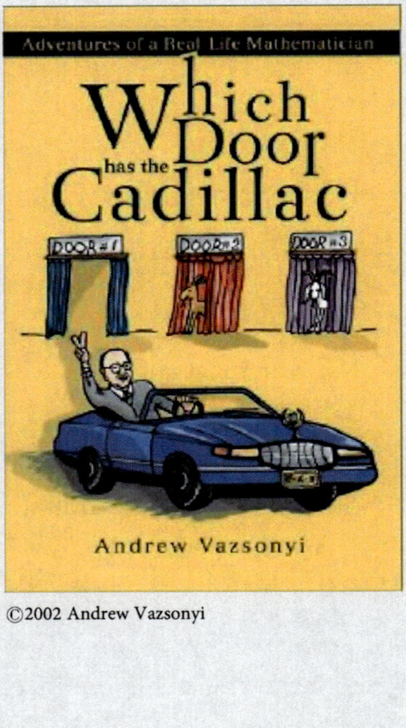

©2002 Andrew Vazsonyi

levels. He was interested in what he termed the eternal question: Why study math? His answer was the story of his life.

PASSION AND COMMITMENT

Andy's contributions to the OR profession and our understanding of decision-making techniques continued until just before he died of a stroke on November 13, 2003. His beloved Laura died soon afterwards in January 2004.

For Andy, the consummate teacher, writing was a constant, and making mathematical modeling accessible to colleagues and the general public was his passion. A prolific writer, his work spans over 70 papers and 9 books, plus numerous articles, columns, cartoons, and a Web site. His subjects varied from pure mathematics to real-world applications; he had the knack of recognizing where and how OR and the decision sciences could be put to good use, and was not bashful about prodding academics to get with it, as he did in one of his last articles that bemoaned the state of the U.S. health care system (Vazsonyi 2002f). His final *Decision Line* column, in July 2003, demonstrates that to the end he wanted to find commonalities in various fields of research. He started this column, "The science of negotiations" (Vazsonyi 2003b), by challenging us to continue to build bridges.

Andy left us with a wonderful legacy. It was his passion and commitment, along with his obvious joy in sharing his lifelong love of math applications with all of us, which kept him going during his long and productive life. He enjoyed his role as a maverick and someone who was always thinking forward, and was often ahead of his time. He had an independent mind, was quite skeptical of trendy ideas, but was also drawn to topics and issues others might scoff at. His soul was in every problem he analyzed, and he encouraged debate; he could do it no other way. He simply loved to play with ideas and was fascinated with the human side of mathematical models. He was ahead of the curve, particularly with respect to building bridges between the technical and real worlds before it was common to do so—he was having so much fun that many of us joined him.

A PERSONAL CONNECTION

I am honored and humbled to have known and worked with Andy Vazsonyi. Through Andy I feel connected to the entire field of OR and the decision sciences, as I met him at the twilight of his official career, when mine was just starting. He retired from his last full-time academic position in 1987, the year I began my full-time academic career at Bucknell University. What an inspiring mentor he was! I actually encountered the spirit of Andy much earlier, when I was a graduate student. I vividly recall laughing the loudest in class when the famous mathematician Zepartzat Gozinto was mentioned. I wondered who had created such a great character and hoped I could find out more about the person with such a great sense of whimsy. My wish came true a few years later, when, as a young assistant professor, I was lucky enough to become one of Vazsonyi/Gozinto's partners in crime. We connected very early in my academic career after I mailed Andy a letter (before most of us had access to email) challenging some of his comments in one of his *Decision Line* columns. He was absolutely delighted to have a young professor challenge him! Thus, began wonderful years of communication and collaboration. Andy would from time to time refer to Professor Gozinto in emails and faxes. One that sticks out is when we were having a several months-long discussion about chaos theory and Andy emailed the following (Vazsonyi 1998b): "According to Professor Gozinto chaos appears in three ways: (1) In nature, (2) In math, (3) In people's minds."

From our book-writing to papers to presentations, my fondest memories are of our many long and interesting telephone and email conversations where he helped me to think broadly and creatively, and reminded me to have fun, by adding paprika to my life and work. We challenged each other, and we enjoyed each other. Andy often spoke fondly of his family and loved to share ideas with his daughter Bobbi Chaney.

In our early days, we communicated with letters, telephone calls, faxes, and in-person conversations. Before the Internet's easy access to articles, he relied on friends around the world to tell him what was going on—many of us sent him articles, book reviews, and books of interest. Andy and I mailed diskettes back and forth, cross-country, creating a new game—guess how long the postal service will take—as we tried (and failed) to determine a cause–effect relationship for the number of days it would take, based on a variety of variables. It was exciting when we both had access to fairly reliable email, with my Bucknell education account, and Andy's CompuServe account. Long before today's text messaging shorthand, we developed our own so that any time I see { } I think of Andy. He signed his emails to me:

&y (say this out loud)

{{{{{{{{{{{{{{{{{{{{{{{ }}}}}}}}}}}}}}}}}}}}}}} (which are hugs)

Andy was an avid golfer and liked to brag that he shot in the low 60s—which he always clarified with a twinkle in his eye that he was referring to his score for nine holes! And, we shared a love for the beach and classical music.

When he mailed me a copy of his autobiography in January 2003, he included this inscription:

> To Nancy C. Weida, my beloved friend of many years, fellow conspirator and partner in many of my adventures. A lovely lady! She keeps me updated and honest, stimulates my thought processes, provides invaluable support and paprika for my writing.
>
> With much love
>
> Andy Vazsonyi

I miss him.

REFERENCES

Chung F, Graham R (1998) Erdös on Graphs: his legacy of unsolved problems. A. K. Peters, Wellesley, MA

Cooper L (1963) Location-allocation problems. Oper Res 11(3):331–343

Decision Sciences Institute (2008) http://www.decisionsciences.org/decisionline/archive.asp. Accessed 15 Sept 2008

Gass SI (2004) In memoriam: Andrew (Andy) Vazsonyi: 1916–2003. OR/MS Today 31(1):18–19

Hertz D, Eddison RT (eds) (1964) Progress in operations research, vol II. Wiley, New York, NY, pp 110–113

Hoffman P (1998) The man who loved only numbers: the story of Paul Erdös and the search for mathematical truth. Hyperion, New York, NY

INFORMS (2008) http://www.lionhrtpub.com/ORMS.shtml. Accessed 4 Aug 2009

Karush W, Vazsonyi A (1957) Mathematical programming and service scheduling. Manage Sci 3(2):140–148

Kuhn HW (1973) A note on Fermat's problem. Math Program 4(1):98–107

Kuhn HW, Kuenne RE (1962) An efficient algorithm for the numerical solution of the generalized Weber problem in spatial economics. J Reg Sci 4(2):21–33

Love RF, Morris JG, Wesolowsky GO (1988) Facilities location: models and methods. North-Holland, New York, NY

Miehle W (1958) Link-length minimization in networks. Oper Res 6(2):232–243
Schechter B (1998) My brain is open: the mathematical journey of Paul Erdös. Simon & Schuster, New York, NY
Simon HA (1957) Models of man: social and rational. Wiley, New York, NY
Simon HA (1997) Administrative behavior: a study of the decision-making processes in administrative organizations, 4th ed. Free Press, New York, NY
Vazsonyi A (1954) The use of mathematics in production and inventory control-I. Manage Sci 1(1)70–85
Vazsonyi A (1955) The use of mathematics in production and inventory control-II. Manage Sci 1(3–4):207–223
Vazsonyi A (1956) Operations research in production control—a progress report. Oper Res 4(1):19–31
Vazsonyi A (1957) Economic-lot-size formulas in manufacturing. Oper Res 5(1):28–44
Vazsonyi A (1958) Scientific programming in business and industry. Wiley, New York, NY
Vazsonyi A (1965) Automated information systems in planning, control and command. Manage Sci 11(4):B2–B41
Vazsonyi A (1973) Introduction to electronic data processing. Wiley, New York, NY
Vazsonyi A (1990) Biography of Z. Gozinto – celebrated Italian mathematician. OR/MS Today 17(4):12
Vazsonyi A (1993) Where we ought to be going: the potential of spreadsheets. Interfaces 23(5):26–39
Vazsonyi A (1995) Multicriteria decision making. Decis Line 26(1):13–14
Vazsonyi A (1997a) Paul Erdös, beloved math genius, leaves us. Decis Line 28(2):19–21
Vazsonyi A (1997b) The quest for the mathematics for decision sciences. Decis Line 28(3):21–23
Vazsonyi A (1998a) Personal email
Vazsonyi A (1998b) Personal email
Vazsonyi A (1999) Which door has the Cadillac? Decis Line 30(1):17–19
Vazsonyi A (2000) A model is like a sausage machine. Decis Line 31(2):18
Vazsonyi A (2002a) Which door has the Cadillac: adventures of a real-life mathematician. Writer's Club Press, New York, NY
Vazsonyi A (2002b) Pure mathematics and the Weiszfeld algorithm. Decis Line 33(3):12–13
Vazsonyi A (2002c) Milestone manifesto. OR/MS Today 29(5):32–33
Vazsonyi A (2002d) Reminiscences & reflections: my first taste of OR: 'I had a helluva big assignment'. OR/MS Today 29(5):36
Vazsonyi A (2002e) Math deficit disorder. Decis Line 33(1):18, 23
Vazsonyi A (2002f) Young man/woman go healthcare. Decis Line 33(5):21–22

Vazsonyi A (2003a) Which door has the Cadillac? Part II. Decis Line 34(3)15–17

Vazsonyi A (2003b) The science of negotiations. Decis Line 34(4)15–16

Weida NC, Richardson R, Vazsonyi A (2001) Operations analysis using microsoft excel. Duxbury, Pacific Grove, CA

Weiszfeld E (1936) Sur un problème de minimum dans l'espace. Tôhoku Math J 42:274–280

Weiszfeld E (1937) Sur le point pour lequel la somme des distances de n points donnés est minimum. Tôhoku Math J 43, 355–386

Weiszfeld E, Plastria F (2009) On the point for which the sum of the distances to n given points is minimum (translated and annotated by F. Plastria). Drezner Z, Plastria F (eds) Recent advances in location analysis. Ann Oper Res167(1):7–41. http://dx.doi.org/10.1007/s10479-008-0352-z. Accessed 4 Aug 2009

16

Murray Aaron Geisler

Perkins C. Pedrick

Operations research (OR) theorists and practitioners recognized early that the flow of goods within an organization constituted an important field of study. The transportation and storage of such goods provided them with a fertile area for modeling and analysis. Before long, these problems evolved into the field of logistics, which spanned the procurement, distribution, maintenance, and replacement of material and personnel. The fruitful interplay between logistics and OR methodologies started in the 1950s and continues to this day as logistics assumed its current name of supply chain management. And, from the start, chief among the developers of logistics was Murray Geisler.

Educated as a statistician, Murray was instrumental in the first successful applications of linear programming (LP) and the recognized leader who transformed logistics into a scientific discipline. His involvement in the subject started in 1948 when he joined the staff of Project SCOOP (Scientific Computation of Optimum Programs), the U.S. Air Force's Pentagon-based organization that, under George Dantzig, was the first to develop and apply LP. There, Murray's group was responsible for translating the Air Staff's statements of its plans or programs into mathematical models that were then solved by SCOOP's mathematical and computational branches. Murray's historic contribution was to automate the process of formulating and solving these planning models. In 1954, Murray moved to the RAND Corporation where he headed its Logistics Systems Laboratory and Logistics Department. During the next 22 years, he established logistics as a major area of study within the

A.A. Assad, S.I. Gass (eds.), *Profiles in Operations Research*, International Series in Operations Research & Management Science 147, DOI 10.1007/978-1-4419-6281-2_16,

military establishment. He served as the president (1961) of The Institute of Management Science (TIMS).

WRESTLING WITH STATISTICS

Murray was born on March 23, 1917 in Brooklyn, New York, and spent his childhood there. His father, trade union vice president David Geisler, and his mother, Rebecca (Anfang), had immigrated from Austria-Hungary. Murray had two brothers, Gerald and Martin, and a sister, Lillian.

Murray graduated from Townsend Harris High School, an institution for academically gifted boys, where he was a member of the wrestling team. He attended the College of the City of New York and, in 1938, he received a B.S. in mathematics, was elected to Phi Beta Kappa, and managed the wrestling team. He then enrolled at Columbia University. There he majored in economics and statistics, and studied under Harold Hotelling and Abraham Wald. He received his M.A. in 1941 (Wald supervised his thesis) and was a member of Sigma Xi (Mertz 1971).

During the same period, Murray became acquainted with Margaret Simms. Marge, who had a B.S. in science from the University of Wisconsin-Madison had come to Columbia for her M.S. in the teaching of science. They were introduced by Marge's University of Wisconsin friends who were also attending Columbia. Murray and Marge were married in New York in 1941.

While working on his master's degree, Murray was employed part-time as a statistician for the National Committee on Maternal Health. The Committee was studying contraception for women in Appalachia and the southern states, with funding being provided by Clarence Gamble of Procter & Gamble, whose aim was to have fewer women on welfare. The study director, the sociologist and statistician Dr. Gilbert W. Beebe, became a lifelong Geisler family friend. Murray's first published paper, based on work for the Committee, was a jointly authored one (Beebe and Geisler 1942).

In 1941, with the recommendation of Hotelling, Murray applied and was accepted for a position with the Office of Price Administration, a new federal government organization located in Washington, D.C. Leon Henderson, the director, took him to lunch, interviewed him, and hired him as Chief of the Income and Price Forecasting Branch. Murray joined a group of about 20 staff members that included Robert Dorfman, David Blackwell, and Melvin Dresher, all of whom went on to make significant contributions to OR.

When World War II (WWII) started in December, Murray received a deferment for his work with the government. After about a year, he decided that he would like to use his forecasting skills in the war effort. He applied for a meteorological position with the Army Air Forces and, in 1943, became one of 500 cadets in a training program in Grand Rapids, Michigan. It was a doubly eventful year. First, his son, Gary Evan, was born there on May 4. Second, the experience led to a highly desired position back in Washington.

In Grand Rapids, Murray learned that the five top-ranked cadets would be given Pentagon posts. That prospect gave him great incentive. He worked hard, achieved his goal, and was assigned to a small Pentagon unit that included future Nobel Prize winner Kenneth Arrow and future Stanford University Professor Charles Stein (both weather service officers) (Mertz 1971). A bonus discovery was that Arrow and Stein were good bridge players. Marge and Murray enjoyed many late night sessions with them. Murray remained a Pentagon Weather Officer until the war ended.

He remained in the Army and, in late 1945, became Chief Statistician of its Institute of Pathology. There was much opportunity for useful analysis—the office was flooded with wartime data on rates of various diseases and of suicides. It produced some intriguing scientific papers. Murray co-authored one on coronary artery disease (Yater et al. 1948). He found the subject matter fascinating, although he felt that the physicians were too casual about their statistics. He left the service in early 1946, but stayed with the Institute until October 1946 (Mertz 1971). (He maintained an Air Force Reserve affiliation, which eventually led to his retirement as a Colonel.)

Murray then returned to the Pentagon as an operations analyst in the Operations Analysis Office (OAO), Headquarters, U.S. Air Force. There he became involved in a number of studies, such as the accuracy of WWII bombing and the testing of new all-weather flying systems. Through his OAO work, Murray became acquainted with George Dantzig, Chief Mathematician for the Air Force Comptroller, who was working on a comprehensive Air Force resource planning project that gave rise to LP. Through that relationship, Murray received his first exposure to the analytical challenge of military logistics, a subject that occupied the rest of his professional life. In March 1948, he joined Dantzig and economist Marshall Wood in the Office of the Comptroller of the Air Force as a staff member of the now famous Project SCOOP. Wood was head of the Air Force's Planning Research Division; Murray was appointed head of the Division's Standards Evaluation Branch.

Pioneering in OR: Enterprise Planning

The Air Force, with a vast amount of collected data, was taking stock of what had been learned in WWII. George Dantzig and Marshall Wood, who were civilian Air Force employees during and after the war, were convinced that more efficient techniques were required for resource planning and management. Their ideas called for the use of electronic computers, which did not yet exist, other than in concept and research. They received strong support from the Comptroller, General Edward Rawlings, a forward-thinking intellectual. Their most important asset was the work of George Dantzig who had been developing mathematical statements of the Air Force planning process and solution procedures for "the optimum computation and allocation of resources" (Geisler 1986, 3).

An early product of Project SCOOP was a planning factors manual that included attrition rates, sortie rates, crew-to-aircraft ratios, fuel consumption rates, ammunition consumption rates, maintenance man-hours per flying hour, and operationally ready rates. The factors could not simply come from observation, as conditions had changed since the war and new hardware was coming into the inventory. They had to be produced to a large extent by new analytic approaches.

By the time an official set of planning factors was attained, George Dantzig's thinking had progressed to his statement of the general LP model. It was ready for use in resource planning, along with his recently developed simplex algorithm for solving such problems. The effort toward selecting and funding a computer also progressed; Remington-Rand's UNIVAC–I was chosen.

Murray soon became an integral member of SCOOP and of the project's inner management circle of Dantzig, Wood, and Geisler; a close-knit team, all of whom were uncommonly humble men. Each insisted on giving credit to the others. There was no question that George Dantzig with his mathematical constructs was the driving force and leader of the team, but George was always generous in his remarks about the other members of the triumvirate (Dantzig 1985, 3) "Marshall Wood and Murray Geisler wrote their fundamental paper [Wood and Geisler 1951] which they published in the now classic book edited by Tjalling Koopmans [1951], a seminal paper that revolutionized the way planning was done in large enterprises. The world has never been quite the same since." Clearly each of these three men played a significant role in pioneering OR.

While LP was a major catalyst in the growth and acceptance of OR as a professional field of analysis, it was never employed in Project SCOOP as it was first intended. The magnitude of the Air Force resource planning problem was so great that it could not be handled, at least not with the electronic computation capability of that time. The SCOOP team produced an adaptation, the Triangular Model, that structured the problem in terms of a square set of equations of modest size, did not involve optimization, and could be readily solved by punched-card equipment. The Triangular Model provided the first mathematically based solution of an Air Force operational plan (Geisler 1986; Wood and Geisler 1951). As noted by Dantzig (1963, 16), "Since 1948 the Air Staff has been making more and more active use of mechanically computed programs. The Triangular Models are in constant use for computation of detailed programs, while the general linear programming models have been applied in certain areas, such as (a) contract bidding, (b) balanced aircraft, crew training, and wing deployment schedules,

> LESSONS LEARNED
>
> "When we began, in 1948, we had no idea we were in the forefront of a major revolution in the science and art of management. It was fascinating to be involved in the application of electronic computers, but our expectations for this new tool were modest. We did not foresee the explosion in computer development and usage that has occurred in the past 35 years. . . .
>
> "We learned how fragile a research group can be, and we came to realize that keeping it productive takes a lot of tender care. A large bureaucracy does not value such groups adequately; they are therefore doomed to have a limited life. Trying to do things differently or trying to have effect inevitably invites attack and criticism. The group can resist such assaults for a time, but eventually its fragility can lead to its decline or demise" (Geisler 1986, 10–11).

(c) scheduling of maintenance overhaul cycles, (d) personnel assignment, and airlift routing problems." All of these problems depended on Murray's staff for the data (planning factors) that held the models together. As discussed by Walter Jacobs, deputy chief of the Air Force's Computation Division, the Triangular Model was redesigned so as to make it a more general and flexible computer-based production system and renamed the Trim Model (Jacobs 1957). As Jacobs notes:

> The publication of manuals containing these [planning] factors proved to have a value far beyond their use in computations run on the UNIVAC. They led to a better understanding by the Air Staff of programming procedures, and a greater recognition that good factors were indispensable in good planning and programming. Although some of these factors were available in the offices responsible for parts of the programs, many others were developed for the first time in the course of preparing the planning factors manuals" (Jacobs 1957, 214).

Such is the legacy of Murray's Project SCOOP endeavors.

Murray and his Project SCOOP staff participated in a number of Air Force studies that demonstrated how a new look at planning, budgeting, and controlling the inventory of spare parts could be made more effective and less costly. A major project for the Air Material Command (AMC) was the conversion of its manual spare-parts budget calculations to a punch-card system to make it accessible for computation on the UNIVAC–I—the Air Force was handling more than a million line items. The AMC application revealed that the Air Force had a large surplus in aircraft spare components and a billion dollars could be diverted to other supply accounts. Murray's group also developed an actuarial technique to determine aircraft engine life-expectancy values that showed only a third as many spare engines were needed (Geisler 1986). All this was not without strain and pain. As Murray noted:

> We learned a lot about the difficulties of introducing a new system and learned how far people might go to avoid change. We also learned how close researchers had to be to the implementation process to be sure the technical procedure was followed. There are many details that can have disastrous effects if they are not handled properly (Geisler 1986, 11).

In 1953, Murray sensed that Project SCOOP had peaked and might be entering a period of decline. His experience there had been exhilarating.

The project had served as a prominent demonstration of the solution-providing power of the new OR methods and techniques. It had shown the significance of the solutions through the dollar savings produced. It had made clear that information must be regarded as a resource, like other resources that had benefits and costs and could be traded off. Finally, it had shown that implementation was a separate problem from solution, requiring separate skills. But, Murray saw the end of Project SCOOP coming, as it did in the mid-1950s.

MOVING INTO MANAGEMENT

After the birth of his daughter, Lauren Sonia, on July 24, 1953, Murray started thinking about where to work next. When George Dantzig, at RAND since June 1952, learned of Murray's interest in a change, he advised the RAND economist Stephen Enke to hire him, warning Enke that Murray had misgivings about a move to the west coast.

Murray liked to tell the story of the ensuing negotiation. Enke, he claimed, took him for a ride in his Jaguar on the Pacific Coast Highway. When the speedometer read 110 mph, Enke shouted, "If you don't agree to come, I'll push it up to 120 mph." Not surprisingly, Murray signed on.

Marge and Murray

While the story may have been enriched, its conclusion was true. Murray became the head of the research program in the Logistics Department, which was part of the Economics Division. He served in that capacity until 1960, when he took time off to complete his Ph.D. in statistics at Stanford University. His dissertation title was "Some statistical properties of simulation models used in studying inventory problems"—his advisor was Herbert Scarf. In 1961, he resumed the leadership of logistics research and, in 1963, became Head of the Logistics Department.

LOGISTICS DEFINED

"Logistics has always had a very broad meaning in military usage. It has been defined, for military purposes, as the creation and sustained support of weapons and forces to be tactically employed in order to achieve strategic objectives. . . . One now feels comfortable in referring to logistics for both military and civilian purposes as the study and application of decision-making techniques intended to achieve efficient allocation and use of resources in satisfying stated objectives of an organization. . . . Much of early management science emerged from work on logistics. The Air Force's Project SCOOP . . . used examples such as the logistics problems incurred with the Berlin Airlift to illustrate the optimization of a specified objective function under resource constraints. One of the Project's earliest contributions to the Air Force was in mechanizing the computation of supply requirements and distribution for the hundreds of thousands of Air Force spare parts by means of punch cards initially, and later with the aid of computers" (Geisler 1975, v).

A major activity of Murray and his Logistics Department's staff was the pursuit of a cost-effectiveness approach to spare parts management and inventory control. In particular, they addressed data ambiguities that made spares calculations suspect. They distinguished demand data from issue data, which had erroneously been regarded as the same. They made other clarifications and corrections for common parts, substitute and interchangeable parts, and hierarchies of parts. They found a way to design flyaway kits (bundles of spare parts selected in advance of use to meet supply needs of aircraft [Geisler and Karr 1956]) so as to minimize the expected number of unfilled demands. Tests of the proposed versus existing flyaway kits indicated substantial reduction in stockouts, but the Directors of Supply at both the Strategic and Tactical Air Commands strenuously objected to change. Prolonged debate ensued until the Air Force Logistics Command issued a regulation, which was largely ignored. Murray's observations about resistance to change seemed to be borne out.

Another issue of concern was centralized inventory control. With the advent of electronic computers, it seemed that centralized knowledge of inventories had much to offer. A supply center could know the status of inventories at every location and place assets where they would do the most good. Before any action could be taken, however, questions of feasibility and cost had to be answered. RAND designed a test, using the Oklahoma City Air Logistics Center and a storage location 80 miles away. Inventory and requisition status was made accessible over telephone lines on a real-time basis. Results were satisfactory. They showed that inventory control of physical assets stored in one place could be maintained with data stored in another place. In this case, RAND's role was limited to test design, so it was spared involvement in the implementation decision.

It was an Air Force idea that RAND establish a Logistics Systems Laboratory to test new logistics concepts and policies. Under Murray's leadership, it took on large and significant explorations, mainly simulations. His papers (Geisler 1959, 1960, 1962) discuss the staffing, operation, and management of the Laboratory, and briefly describe its major projects to that time. [The (1962) paper was Murray's presidential address, presented at the 8th annual international meeting of TIMS, Brussels, Belgium, August 24, 1961; its introduction was given in French!]. Murray was a strong advocate of using simulation in analyzing military logistics problems, as he felt that the techniques developed at RAND "have some general usefulness and application" (Geisler 1962, 240).

TALES FROM THE SIMULATION LABORATORY

"As I have said, the experiments dealt with logistics problems. For example, the first one was concerned with testing a series of new formulas for calculating inventory levels and procurement quantities. Further, such inventory levels were to be maintained centrally, and automatic resupply was to be used. To simulate the effect of such policies in the Air Force, the simulation model contained a number of aircraft, air bases, storage depots, repair depots, different transportation systems, and factories. In the second experiment, the problem studied was the organization of the logistics system for maintaining large numbers of highly dispersed intercontinental ballistic missiles on alert. This study was undertaken before such missiles were actually available. Questions such as the number of missiles to be supplied from one logistics center, the number of maintenance people, amounts of supplies and equipment, and the effectiveness of the logistics system in keeping missiles on alert, were all part of the study. In the third experiment, the study was concerned with the organization of Air Force depots, and particularly of the repair activities in these depots. The issue was one of more or less centralization and the degree of responsiveness to be required of the repair and resupply systems. From these problems, we can see that sophisticated questions were being asked of these studies, and that simulation is clearly a relevant technique. Also, because these problems each involved continuing complex decisions for which rules could not always be specified, we have used people as integral parts of the simulation model, and assigned to them specific jobs and functions in the simulated organizations. Thus, we have chosen to call this technique 'man-machine simulation'"(Geisler 1962, 243).

MURRAY = OR(LOGISTICS)

Logistics, in general, grew in importance and sophistication, and became a field of study and career path within the expanding field of OR (Geisler 1960). No one was more influential in the synergistic aspects of logistics and OR than Murray. He built the Logistics Department into RAND's largest group and, by his professional service, he played a key role in the early years of OR and management science (MS). He served as president of TIMS in 1961, chairman of the Military Operations Research Symposia (1963), and, from 1964 to 1967, he was a member of the Council of the Operations Research Society of America. For most of the 1960s, he was a member of the Editorial Board of the *Naval Research Logistics Quarterly*. From 1963 to 1975, he was editor for Logistics of *Management Science*. In 1972, he served as director and editor of the Proceedings of the NATO Conference on Application of Operations Research to Transport Problems. He edited *Logistics*, the first volume of a special MS series (Geisler 1975). During the 1955–1975 period, Murray authored or co-authored 24 articles in the *Naval Research Logistics Quarterly*, *Management Science*, *Journal of Forensic Sciences*, and *Operations Research*, as well as several independent compilations of analyses.

SHARING THE WISDOM

Growing recognition of the importance of logistics in modern warfare led President Nixon and Secretary of Defense Laird, in 1969, to call for a broad review of logistics as it was functioning in the Vietnam War. The effort was known as the Joint Logistics Review Board (JLRB). Army General Frank Besson was the chairman. The four other board members were of three-star military rank. They were supported by a staff of about 100 senior officers from all the services.

Murray was recruited temporarily from RAND as the research director of the JLRB. His job was to use the staff to do an in-depth review of all Vietnam logistics functions from 1965 on—planning, management, supply, maintenance, transportation, procurement, communications, manpower, ammunition, petroleum, construction. He was given supervision of a support contract that had been let to the management-consulting firm of Booz, Allen & Hamilton.

The review, much of it in the Far East, was extensive. Regular briefings were given to the JLRB and to the assistant secretary of Defense (Installations and Logistics). The final product was 20 volumes of findings—in the form of lessons learned—and recommendations. The main thrust, the need for integrated management of logistics functions, was not warmly received by the military departments. They took the position that logistics functions must be integrated with operations and, since operations were separated among the military departments, so must logistics functions be separated. Murray was once again confronting resistance to change. Nonetheless, the report was accepted and approved by the Secretary of Defense—the report became a valuable reference on the realities of logistics in war. After an absence of 15 months, Murray returned to RAND.

Upon his return, Murray was named assistant to the president. Despite his new title, Murray's primary interest at RAND was still logistics. He wanted to move beyond evaluating logistics performance through such criteria as cost, fill rates, and stockouts. He wanted to show the impact on readiness or combat effectiveness of having, or not having, the right support item at the right place at the right time. He followed with interest the development of the aircraft availability model by the Logistics Management Institute (LMI), as it judged inventory actions by their effect on system availability, not just on inventory measures. But the Air Force rejected RAND's proposals as too subjective. [LMI is a nonprofit consulting organization dedicated to improving management of the nation's public sector through research, analysis, and counsel. Originally, it was established to conduct logistics studies for the Department of Defense (DoD).]

In 1974, Murray accepted an invitation to spend a year as visiting professor of Management at MIT's Sloan School of Management and teach operations management to graduate students. In doing so, he

> PROFESSOR GEISLER
>
> "Shortly before the fall semester of 1974, I heard that Murray Geisler was to spend 1974–1975 at MIT as a visiting Professor of Management Science. As a second-year doctoral student at the Sloan School, I was woefully ignorant of his background, but realized that the operations management faculty mentioned him with much respect. In Spring 1975, he taught the graduate course in operations management and I was assigned as his teaching assistant for the course. This made me one of the very few people who were taught by Dr. Geisler in a classroom setting, something I understood he had not done before. He was a confident teacher who was well-prepared, but he stuck to the material and did not pepper it with any stories of his own. As I reflect on the class, it surprises me that he never spoke about his extensive experience in the field of logistics. Even when he included the 'flyaway kit' problem as a part of home work assignment, he did not mention his own role in its development. His course had a week or two devoted to simulation, a subject he knew in great depth and written his doctoral dissertation on. Given his usual reserve, it was all the more memorable for me when he mentioned, with a twinkle in his eye, that in defending his dissertation, he was able to teach the faculty on his dissertation committee a thing or two about simulations of inventory systems!" (Assad 2009).

mixed theory and application to show the complexities of inventory management, production scheduling, life cycle costing, and information management. He stimulated interest in those topics and advised in thesis efforts of several degree candidates.

While serving as a professor, Murray became a part-time consultant to LMI. By the time he had fulfilled his Sloan School appointment, Congress had terminated the Air Force program for use of computers in logistics. RAND was suffering spending cuts because of its support of that program. Murray extended his leave from RAND to continue his relationship with LMI and simultaneously to help the Army War College produce a history of the Vietnam War. He did so until February of 1976, when he formally concluded his 22-year affiliation with RAND.

Murray then joined LMI as senior logistician. His experience and stature made him a valuable asset to the Institute. He was known for his contributions, practical as well as theoretical, to OR analysis and logistics problem-solving. Increasingly, as OR burgeoned in academia, commerce, and the military, the significance of his early pioneering work was being recognized.

The Assistant Secretary of Defense (Installations and Logistics) tasked LMI to create a chairman of the board report on logistics. He observed that the logistics measures that he received were probably useful to managers of the various logistics functions, but they did not meet the needs of the DoD's senior logistics executive.

Murray was given the job of satisfying the Assistant Secretary's information needs. He understood how policy management differs from operational management in both role and function; he set out on a three-part plan to serve the policy manager. First, he developed a descriptive model of logistics, emphasizing levels, processes, inputs, and outputs. Next, he produced gross activity indicators of supply, maintenance, transportation, installations, and housing in the Air Force, and related them to weapon system status and activity. Finally, he moved to the level of the Office of the Secretary of Defense and produced a hierarchical framework that revealed the contributions of operational logistics to fulfill overall logistics objectives.

The well-documented 2-year study not only served the sponsor's purpose; it became an outstanding example of open-minded OR at the level of the executive suite. Despite Murray's vast logistics experience, there were no preconceived notions. He surveyed all promising quantitative techniques and concluded that the Analytic Hierarchy Process (AHP), developed by Thomas Saaty, was most appropriate for resolution of the decision problems encountered by the DoD's logistics activities (Saaty 1980). A discussion of the AHP was the basis for the third part of the study and for the product delivered.

Murray moved on to other projects for LMI's top-level sponsors. He produced an assessment of aircraft logistics planning models then in use, followed by an assessment of existing methods of measuring readiness. For LMI's own Board of Trustees, he evaluated the Institute's service to its sponsors, including new sponsors in such areas as energy, manpower, and foreign military assistance.

Once he had briefed the Board, Murray was asked to rescue a DoD project that was floundering. The director of Defense Research and Engineering had asked the Defense Systems Management College (DSMC) to produce a Guide to Management of Multinational Programs. Chapters assigned to various professors and consultants were arriving slowly or not at all. Murray, and John Fargher of DSMC, ended up writing almost all of the chapters. The sponsor decided to have the guide issued by the Joint Logistics Commanders. Several thousand copies were published and distributed (Fargher and Geisler1981). The document became a text for DSMC's Program Management Course. The effort led to additional LMI work on management of multinational programs, including codification of policies and procedures for U.S. representatives to follow in formulating memoranda of agreement with other countries.

Murray concluded his 7-year service to LMI with two smaller projects. He especially enjoyed one on installations management, an area he had barely touched in all his years of logistics analysis. He evaluated several proposals for change, investigated how some corporations managed their property, and produced recommendations capable of implementation in this politically sensitive area. In his final project, he reviewed DoD plans for a "Modeling information and support activity" and concluded that their adoption was not possible for the money that was available.

Throughout his LMI years, Murray made a significant contribution as mentor. He enjoyed the role and the recipients of his guidance appreciated it. His 12 LMI reports were all co-authored, mostly by analysts who were much younger and much less experienced than he.

Too Little Time

Murray retired from LMI in 1984, but agreed to write a book, *A Personal History of Logistics* (Geisler 1986). He did not see it in print—he succumbed to leukemia on August 6, 1985 at his home in Los Angeles. His *History* was published by LMI. Marge lives an active life in Los Angeles, enjoying travel and her grandchildren.

Murray did not live to see his two grandchildren, both of whom bear his middle name, Aaron; Alexandra Aaron Fite was born in 1993 and Aaron Aidan Fite was born in 1996. They live with their parents, Dr. Austin Fite and Lauren Sonia Geisler Fite, in Pacific Palisades, California. Lauren met her husband through her work as a physician's assistant after earning her B.A. in sociology from the University of California (Berkeley) and her B.S. in the physician's assistant program at Duke University.

Gary Evan Geisler and his wife, the former Linda Lawrence, live in retirement in Vero Beach, Florida. Gary earned his B.A. and M.A. in history at the University of California, Los Angeles. He was a Fulbright scholar at the University of Madrid and also studied at the University of Mexico City. After a few years teaching at the University of Cincinnati, he joined the Internal Revenue Service in Washington, D.C., earned a C.P.A., and enjoyed a full career as a tax law specialist.

The story of Murray's education, professional life, and his contributions and achievements—its twists and turns and fortunate decisions—is rather typical of the pioneers of the OR profession. A recurring theme is being there at the right time, but that has to be correlated with ability, drive, and vision, all of which Murray had in abundance. He accomplished much and we are fortunate to have his *History* and journal publications that record how he brought the field of logistics to any early fruition, the seeds of which have sprouted into major branches of OR.

References

Assad AA (2009) Personal communication

Beebe GW, Geisler MA (1942) Control of conception in a selected rural sample. Hum Biol 14(1):1–20

Dantzig GB (1963) Linear programming and extensions. Princeton University Press, Princeton, NJ

Dantzig GB (1985) Murray Geisler obituary. OR/MS Today 12(5):5–6

Fargher J, Geisler M (1981) Joint logistics commanders guide for the management of multinational program. Logistics Management Institute, Washington, DC

Geisler MA (1959) The simulation of a large-scale military activity. Manage Sci 5(4):359–368

Geisler MA (1960) Logistics research and management science. Manage Sci 6(4):444–454

Geisler MA (1962) Appraisal of laboratory simulation experiences. Manage Sci 8(3):239–245

Geisler MA (ed) (1975) Logistics. TIMS Studies in the Management Sciences, North-Holland, New York, NY

Geisler MA (1986) A personal history of logistics. Logistics Management Institute, McLean, VI

Geisler MA, Karr HW (1956) The design of military supply tables for spare parts. Oper Res 4(4):431–442

Jacobs W (1957) Air force progress in logistics. Manage Sci 3(2):213–224

Koopmans TC (ed) (1951) Activity analysis of production and allocation: proceedings of a conference. Wiley, New York, NY

Mertz RR (1971) Interview with Dr. Murray A. Geisler, February 22, 1971. Computer Oral History Collection, 1969–1973, 1977. Smithsonian National Museum of History, Washington, DC (Transcript 88 pages)

Saaty T (1980) The analytic hierarchy process. McGraw-Hill, New York, NY

Wood MK, Geisler MA (1951) Development of dynamic models for program planning. In: Koopmans TC (ed) Activity analysis of production and allocation: proceedings of a conference. Wiley, New York, NY, pp 189–215

Yater W, Traum A, Brown W, Fitzgerald R, Geisler M, Wilcox B (1948) Coronary artery disease in men eighteen to thirty-nine years of age: report of eight hundred sixty-six cases, four hundred fifty with necropsy examinations. Am Heart J 36(3):334–372

17

Hugh Jordan Miser

Michael H. Rothkopf

Hugh Miser's career path began as a university professor, but his World War II (WWII) service as a civilian military analyst brought him into the then embryonic field of operations research (OR). Once he was committed to OR, he helped shape this new profession in many important ways. He was a founding member of the Operations Research Society of America (ORSA), served as its secretary and president (1962), as well as the editor of its journal. By his lectures and writings, he was recognized internationally as the conscience of the field—a prime mover whose main concern was the establishment of OR as a science and a profession. He received the Military Applications Society's Jacinto Steinhardt Memorial Award for outstanding contributions to military OR, ORSA's Kimball Medal for distinguished service to the society and the profession, and was inducted into the International Federation of Operational Research Societies' (IFORS) Hall of Fame. He was awarded the U.S. Air Force medal for outstanding and exceptional civil service.

A Mathematical House and Beyond

Hugh Jordan Miser was born in Fayetteville, Arkansas, on May 23, 1917. He was the first of the four children of Nellie (Pyle) Miser, originally from Huron, South Dakota, and Wilson Lee Miser, born in 1886 in Pea Ridge, Arkansas. He had three younger siblings: brothers, John and Jim, and sister, Mary (Hamilton). Hugh's family was a mathematical one. Both parents taught mathematics at the college level. Wilson Miser earned a Ph.D. in mathematics from the University of Chicago in 1913; his dissertation title was "On linear homogeneous

A.A. Assad, S.I. Gass (eds.), *Profiles in Operations Research*, International Series in Operations Research & Management Science 147, DOI 10.1007/978-1-4419-6281-2_17,

differential equations with elliptic function coefficients." He was member of the faculty at Vanderbilt University. He was the co-author, with Claude Palmer, of *College Algebra*, a well-known text first published in 1928. Hugh was the only one of the Miser children to pursue a career in the family business of mathematics.

Hugh attended Vanderbilt and received a B.A. in mathematics, magna cum laude, in 1938. He then went to the Illinois Institute of Technology (IIT) where he earned his M.S. in mathematics in 1940. Hugh then held various teaching positions at IIT, Lawrence College in Wisconsin, where he served as acting head of the mathematics department, and Ohio State University. During this time, he joined with two members of the IIT faculty, Paul Andres and Haim Reingold, and co-authored the text *Basic Mathematics for Engineers* (Andres et al. 1945). It, along with its successor by the same authors, *Basic Mathematics for Science and Engineering*, sold over 200,000 copies through multiple revisions (Andres et al. 1955).

On New Year's Eve 1943, Hugh met Josephine (Jody) Spence Lehmann. Jody had graduated in 1942 from the University of Illinois, Phi Beta Kappa, and received a master's degree from the University of Minnesota in child development, with a minor in social work. When they met, she was working for the Red Cross in Chicago. They were married in Urbana, Illinois, on June 24, 1944. Hugh and Jody had four children. Their oldest son, James Spence Miser, graduated from Dartmouth College, and then received his M.D. from the University of Washington and became a pediatric oncologist. They had twin boys, Andrew and Wendell. Andrew received a Ph.D. in psychology from the University of Connecticut and is a psychologist. Wendell has a master's degree from the University of Illinois in environmental engineering and has worked for the Environmental Protection Agency since 1977. Their youngest child, Emily, graduated from Wesleyan College, has a master's degree in child development, and is an elementary school teacher.

INTO THE FRAY: MILITARY OR AND BEYOND

In 1945, Hugh followed through on a suggestion from an IIT friend that he should come to Washington, D.C. to help in the war effort by applying his mathematical skills as a civilian operations analyst for Headquarters (Pentagon), 20th Air Force. The work here, and later in Guam, gave Hugh his first exposure to applied mathematics and the emerging field of OR. It was a career-changing happening. (The 20th Air Force moved to Harmon Field, Guam, Mariana Islands, on July 16, 1945, assigned to the Pacific Air Command, U.S. Army. The unit's bombers, the Enola Gay and Bock's Car, dropped the first atomic bombs on Hiroshima and Nagasaki, respectively.)

One of Hugh's early military OR assignment was the investigation of the accuracy of the Eighth Air Force's WWII bombing missions over Germany. He was teamed at the Pentagon with the statistician William J. (Jack) Youden who had worked on this problem in both the European and Far Eastern theaters of war. Youden was an original member of the pioneering OR team that went to the U.K. in 1942 assigned to the Eighth Bomber Command. Hugh and Jack set out to reconcile the limited theory of bombing with the actual recorded bombing results so as to improve the accuracy of bombing missions (Miser 1992). It was Jack who had proposed earlier to the Eighth Air Force that instead of each bombardier releasing his aircraft's bombs independently of the others, that all aircraft release their bombs in a salvo based on the command of the lead bombardier. The salvo strategy was against regulations and forbidden—no one knew why! This change resulted in at least a 1,000% increase of bombs on target (Brothers 1954, 9; McArthur 1990). The no-salvo rule was quietly dropped. Based on theoretical considerations, a key analysis tool was the Youden (bombing) Chart that showed the expected percentage of bombs that fell within x feet of the target center as a function of accuracy and bombing pattern. The chart readily demonstrated the advantages of salvo bombing. Further, Hugh and Jack showed how the chart and its ramifications could be used to estimate force requirements for future missions. Armed with their resulting analysis, Hugh then developed a new scoring system that measured bombing accuracy for a bomber training school at Muroc Air Force Base, California. A key concern here was how to encompass outliers in each bombing pattern (Miser 1993a). Hugh's report found its way to Guam a few months later during his assignment there.

OR LESSONS LEARNED

In his paper "Craft in operations research" (Miser 1992), Hugh drew the following lessons learned from his experiences and those of other Air Force OR analysts during WWII:

"There are times when regulations can usefully be ignored in order to gain effective results. Genuine belief in one's results is a powerful aid to persuading others.

"When timely results are important, simple forms of analysis can be useful.

"The best way to get practical results may not be the standard one of writing a report and presenting its results to the high command.

"The author of the analysis results is usually the best person to present and discuss them,

"It is worth a great deal to shape one's explanatory theory into a form in which it speaks clearly, easily, and forcefully to operational people.

"Hands dirtied with the intimate and detailed sorting, inspection, and analysis of the evidence are a sign of a good analyst.

"Do not flourish your analysis mystique; make the presentation of the results as simple as they can be made honestly and effectively.

"All of science, and especially operations research, is an intensely human enterprise, as well as an intellectual one, in which a variety of human qualities and craft skills are engaged to enable analysts to survive as both well-balanced individuals and effective professionals" (Miser 1992, 634–639).

When WWII ended, Hugh returned to Ohio State where he completed his Ph.D. in mathematics in 1946 under the direction of Tibor Radó; his dissertation was on "Generalized conformal representatives of Fréchet surfaces." He then joined the mathematics department of Williams College as an assistant professor.

In 1949, Hugh decided to return to Washington where he accepted a position as an operations analyst in the recently formed U.S. Department of the Air Force (USAF), with headquarters in the Pentagon. By 1951, he was the USAF's Deputy Assistant for Operations Analysis, and from 1958 to 1959, he served as Acting Assistant for Operations Analysis. While much of the work that Hugh did during his 11 years at USAF headquarters was highly classified, he was undertaking and leading efforts to apply OR to Air Force operations and strategy. Hugh's effectiveness and leadership abilities were well recognized. He was awarded the USAF medal for outstanding and exceptional civil service.

Hugh's extensive OR Air Force experiences enabled him to broaden his scope of interest to related military areas, as well as being able to bring his talents to nongovernmental problems. He left the USAF in 1959 to become director of the Operational Science Laboratory, Research Triangle

Institute (RTI), Durham, North Carolina. (RTI is an independent, nonprofit research and development organization. It conducts research and provides technical services to government agencies, universities, foundations, private businesses, and other organizations.) In 1960, he joined the Massachusetts Institute of Technology's (MIT) U.S. Navy sponsored Operations Evaluation Group as director of its Applied Sciences Division. From 1962 to 1965, he was assistant to the director of systems planning and research at the Mitre Corporation, a nonprofit affiliate of MIT, in Bedford, Massachusetts. Then, from 1965 to 1968, he was a vice president for the Travelers Research Center in Hartford, Connecticut. While there, Hugh led a successful effort to reorient the way the automobile insurance industry addressed traffic safety.

INTO THE TRENCHES: OR SOCIETAL MATTERS

> WHAT IS OR?
>
> "Operations Research is the science that is devoted to describing, understanding, and predicting the behavior of ... man-machine systems operating in natural environments: [the] stabilities ... permit the science to be a fruitful one. Thus, operations-research workers are engaged in the three classical aspects of science:
>
> - Describing the behavior of these systems,
> - Analyzing this behavior by constructing theories (frequently called models) that account for the observed phenomena, and
> - Using these theories to predict future behavior, that is, the effects that will be produced by changes in the systems or in the method of operation" (Miser 1974a, 906).

Hugh was the lead civilian member of the group of ten analysts from the Operations Analysis Group, USAF, who attended the May 26, 1952, founding meeting of ORSA—all became founding members. So began his lifelong interest and participation in the professional, societal, and scientific aspects of OR. An early ORSA task was the writing of a brochure describing ORSA and OR. From 1958 to 1961, he was ORSA secretary, and, in 1961, he was elected vice president. He then served as ORSA's president from 1962 to 1963. Hugh also had a major impact on the international spread of OR. He attended the first international OR conference held at Oxford in 1957, and helped to organize the International Federation of Operational Research Societies (IFORS)—ORSA was one of the three founding members, along with the British and French OR societies. He attended every subsequent IFORS triennial meeting until his death in 1999. Hugh was an early advocate of the creation of the IFORS sponsored *International Abstracts in Operations Research* (IAOR). Started in 1961, the IAOR continues on the Web as IAOR Online (http://www.palgrave-journals.com/iaor/index.html). Later, Hugh played a key role in the cross fertilization of the different approaches to OR that developed in different parts of the world.

From 1963 to 1968, he edited "The Analyst's Bookshelf," the book review section of ORSA's journal, *Operations Research*. Hugh became editor-in-chief of *Operations Research* in 1968, a post he held until 1974. This period saw a great flowering and success of OR. Hugh was a superb editor, and it is no accident that many seminal papers appeared in *Operations Research* during his tenure. This is borne out by his article Miser (1974b), written when Hugh was retiring from being the editor-in-chief of *Operations Research*. Hugh recounts the operational aspects of the journal in terms of the response to the demand for publication space and the handling of time delays in the editorial process—all illustrated by insightful tabular data analyses that is the hallmark of an experienced operations researcher.

Hugh realized that a journal, especially one dedicated to the development and growth of a new field, is judged by the influential papers it publishes. He worked hard not to let the peer-review process turn away important, innovative papers. In one instance, he received a paper applying OR in a new area. Not knowing anyone working in that area, he decided to be the paper's referee. Although the paper was not perfect, he believed it was important, and accepted it. It was only after the paper was awarded ORSA's prestigious Lanchester Prize as the best paper of the year that he learned that it had previously been rejected by a rival journal. Hugh also believed that, in addition to publishing research results, the journal should be a forum for the discussion of issues of importance to the profession. He published a controversial letter to the editor by R. E. D. (Gene) Woolsey entitled "Operations research and management science today, or, does an education in checkers really prepare one for a life of chess?" (Woolsey 1972). Woolsey was critical of OR's apparent retreat from a problem orientation to a mathematical one. Hugh received several letters in response to Woolsey's commentary, but felt he had to reject some from leaders of the profession since they were ad hominem attacks on Woolsey; Hugh believed that only debate on the substantive issues raised by Woolsey's letter was appropriate.

> ON EDITORS AND AUTHORS
>
> "Scientific communities ask their leaders to take on the time-consuming and often onerous tasks of editorship, expecting that their broad and distinguished experience will inform the editorial processes and the material that they select for publication. The clear corollary is that the persons so selected should, in fact, use their best judgments throughout, thus helping to steer their fields in useful and venturesome directions. Among other things, this means deciding when to stick with the standard editorial processes and when to deviate from them, when to exercise unusual initiatives, and how to shape the contents of the journals they edit. To do less would be to fail to exercise their full knowledge, abilities, and perspectives
>
> "It follows that authors submitting papers should recognize that they are entering a partnership process that the professional community expects to have managed in this way. They should respect it and expect to occupy appropriate places in its outcomes" (Miser 1998, 122).

In his later years, he edited the Forum section of *Operations Research* that published discussions of ideas related to the nature of the profession. For the journal *Interfaces*, he wrote general articles and a column (The Easy Chair) about the profession.

Hugh was not only a successful editor, but a thoughtful one. He mentored other editors, including this author. He reflected on, wrote about, and understood the things that made a journal successful. His easy chair column (Miser 1998), "Journal editing as I see it," was reprinted and circulated to 1200 editors of journals in biology by the Council of Biology Letters (CBE).

University Life and New Ventures

In 1969, Hugh joined the University of Massachusetts (Amherst) as professor and chair of the Department of Industrial Engineering and Operations Research. He decided not to move to Amherst, but commuted from his well established home in Farmington, Connecticut. Because Hugh believed that students untrained in advanced OR methods could benefit greatly from analyzing real problems, he developed and taught a project-oriented course for lower division undergraduates.

Hugh left the University of Massachusetts in 1980 being honored with the title of professor emeritus. He did not, however, have in mind the idea of retiring. Over the years, based on his involvement in and broad view of problem solving, Hugh recognized that there were many important problems of modern society for which the application of OR methods was limited. This brought him to the study of the emerging field of systems analysis for which OR may be viewed as a precursor. As noted in Miser and Quade (1988, 5), Hugh's attention turned to "analysis to assist people responsible for policy or action to develop, understand, select, and implement what should be done in an uncertain environment to advance human welfare—and to evaluate the consequences of what was done." His view of the importance of systems analysis was strengthened during a 1979 visit to the International Institute for Applied Systems Analysis (IIASA) in Laxenburg, Austria. IIASA is supported by the academies of science of all the major western and eastern countries, and staffed by an international cadre of established researchers from many disciplines who are, in turn, supported by post-doctoral and doctoral students. IIASA's mission is to analyze, from a systemic view-point, major world problems such as energy, natural resources, agriculture, and world population. Hugh joined IIASA's staff in 1980 and remained there for 4 years. At IIASA, Hugh directed a systems analysis group and also served as IIASA's executive editor.

Hugh retired from IIASA with a mission to codify and explain what was then known about systems analysis. He joined with Edward S. Quade, a pioneer in the development and application of systems analysis for the RAND Corporation, to co-edit a three-volume interlocking exposition of systems analysis. The first two volumes are by Miser and Quade (1985, 1988)—*Handbook of Systems Analysis: Procedures, Applications, and Practice* and *Handbook of Systems Analysis: Craft Issues and Procedural Choices*. Due to Quade's death in 1988, the third volume, *Handbook of Systems Analysis: Cases*, was edited by Hugh (Miser 1995). These volumes not only combine into a major contribution to the field of systems analysis, but they also are of great value to the OR profession. Hugh wrote or co-authored a substantial fraction of the chapters in each of the volumes. From Miser and Quade (1988, 513), we cite the trenchant passage from Hugh's Chapter 12, "Underlying Concepts for Systems and Policy Analysis":

> Computer modeling in operations research and management science offers an example of overemphasis on rigor at the expense of relevance. Early models of modest size have been used successfully as tools in solving many operating problems, a fact that has led academic courses in these fields to give such models a very great emphasis—so much so that many regard the fields as devoted almost entirely to modeling. These successes have led to the creation of much larger computer models, which, however, have been much less satisfactory in yielding results about large-scale operating and policy problems, where the surrounding situation is more complex. However, the strong impulse—particularly in academia—behind the positivist ideal of rigor has given the construction of such large models a life of its own, increasingly divergent from real-world problems of practice, that has found its way into research institutes as well.
>
> This discussion is not intended to suggest that the working professional should reject the benefits of rigorous and analysis and careful thought about the problems he faces. Rather, the intent is to suggest that analysts should move toward the goal of relevance, thus moving toward the "swamp" wherein lie the problems of major social importance.

Hugh had a long-term scholarly interest in the history of OR, with especial interest in ensuring that the new generations of OR students, analysts, and academics had well-presented and detailed publications as sources. His OR history papers extend from Miser (1977)—"The history,

HISTORICAL ADVICE

In the late 1980s, I became interested in reading about the history of OR more systematically. As I looked for sources, I read what Hugh had written on the subject. I used to run into Hugh regularly at ORSA conferences. His thoughtful remarks at various fora, his measured tone, and his overall comportment (complete with the bow tie) reinforced my perception of Hugh as very much the éminence grise of our profession. At one meeting, I finally summoned up the courage to approach him to mention my interest in the history of OR. He listened carefully and remarked the amount of time and effort a good account of the subject would require. As an example, he cited one of his early articles on the subject (Miser 1977) and remarked that he had put more effort into that paper (on a per page basis) than some of his technical papers. He further remarked how he had to wait until his retirement to pursue this subject more fully. Finally, he said, "Rest assured, we'll put you to work, young man!" For me, these words proved to be prophetic. Over the years that ensued, Hugh was kind enough to send me an encouraging note when I published something on issues of practice and implementation in OR (Assad, 2009).

nature, and use of operations research"—to his last published article (Miser 2000)—"What OR/MS workers should know about the early formative years of their profession." Also, with OR being a rather new field, Hugh recognized that it would be appropriate and possible to gather and archive original material that describes the origins and development of OR. In the 1970s, working with a committee of members of ORSA and The Institute of Management Sciences (TIMS), and the librarians at the University of California, he helped to develop, at the request of the ORSA and TIMS councils, a plan to archive and preserve the papers of OR pioneers and related documents of OR's early days. A not-for-profit ORSA-TIMS Foundation was chartered. He was extremely disappointed when the councils decided to drop the archive initiative because of its cost. At one point, Hugh himself had extensive files of his own correspondence. Fortunately, the MIT Library agreed to preserve over 40 cubic feet of his files, but it has not yet catalogued them or made them generally accessible. (Ironically, the only reason MIT Libraries agreed to do this is Hugh's brief employment by MIT.)

The Sage of OR

From Hugh's extensive service to the OR profession, as evidenced by the societal and other positions he held, and ORSA's awarding him in 1975 the George Kimball Medal for distinguished service to the society and the profession, it would be easy to think of Hugh's contribution as primarily service. But, such a view would miss Hugh's most important contribution: his being a key philosopher and interpreter of the profession. Although he was a mathematician and understood the contributions that mathematics could and did make, he was part of the initial generation of operations researchers created by the necessities of WWII—a generation who understood that lives, and, perhaps, the survival of civilization were at stake. In that environment, decisions had to be made in a totally pragmatic way. Thus, in his practice of OR, in his teaching, and in his writings, Hugh saw OR as grounded in pragmatism. Although he valued theory, he was not impressed by theory for its own sake. For him, the primary value of and justification for theory was its ability to contribute to the solution of practical problems. He was unusual among leaders of the OR profession in writing about and encouraging thought about the craft issues of OR practice. Indeed, volume two of the three-volume *Handbook of Systems Analysis* is devoted solely to craft issues (Miser and Quade 1988).

When Hugh and Ed Quade were planning the second volume, they wanted to include "a foundational chapter on what the philosophy of science has to say about operational research" (Miser 1996a, 1322). Because they were unable to find a suitable author, Hugh took it upon himself to study the topic and write the chapter, thus becoming a leading scholar of the philosophy underlying OR. In Chapter 1, Hugh wrote (Miser and Quade, 1988, 6–7):

> The analyst's path is strewn with issues and difficulties that force him to make choices, some of which will prove to be crucial. He must settle on the nature of the problem that is to be considered; he must choose among alternative assumptions, data sources, methods of analysis, and forms of presentation; he must make decisions that reflect the resources of time and money available, the nature of the decisionmaker, and the properties of the decisionmaking environment; he must choose how to limit the extent of the investigation, how to rank the alternatives realistically, and how to estimate the likelihood that they can be implemented successfully. Such choices—and many others—occur throughout any systems analysis, and how they are made will have important bearings on the quality and success of the work.

When there was a need for someone to define or explain OR to a larger audience, Hugh was often the person to be called upon (Miser 1977, 1980, 1996b, c)—chapters in handbooks and encyclopedias, and the prestigious 100th anniversary issue of *Science*. For the latter, Hugh faced the challenge of describing to a general scientific audience the essence of the new science of OR. He does this with a short history of how OR evolved from the study of WWII operational problems, and then sets the concept by describing real-world OR applications: the analysis of a blood bank's operations, and the assignment of students to schools to produce a desegregated plan that would gain court approval. In a look to the future, he comments on how such OR succeses have caused the field to "study larger and larger systems or, in other words, to what is now called 'systems analysis'" (Miser 1980, 209).

By the 1980s, Hugh had developed broad and encompassing philosophical and professional views of OR (Miser 1991), much of which is detailed in the *Handbook(s) of Systems Analysis*. This is also evident by his *Science* article (Miser 1980) that he titled "Operations research and systems analysis." When the OR Forum section of *Operations Research* (of which Hugh was then editor) published the report, "Operations Research: The Next Decade," written by Committee on the Next Decade in Operation Research (CONDOR 1988), a National Science Foundation sponsored group of 24 OR professionals, Hugh and other readers felt that the report presented a restricted and narrow view of the field. Hugh recruited co-authors for a concerted response (Wagner et al. 1989). In his part of the response, Hugh addressed three issues to which he felt the CONDOR report had not given sufficient emphasis: the practice of OR, modeling, and the application of OR to strategic issues. He wrote about these issues, not only for a U.S. audience, but also for an international one served by the *European Journal of Operations Research* (EJOR), a publication sponsored by a consortium of European OR societies (Miser 1993b); and the *Journal of the Operational Research Society*, the flagship journal of the U.K. Operational Research Society (Miser 1991).

The list of professional societies to which Hugh belonged reflects on the breadth and depth of his interests. He was a member of ORSA, TIMS, the Canadian Operational Research Society (CORS), the American Mathematical Society, the Mathematical Association of America, the Society for Industrial and Applied Mathematics, the Institute of Mathematical Statistics, and the American Statistical Association. In addition, he was a member and fellow of the American Association for the Advancement of Science and the Connecticut Academy of Science and Engineering.

A Kind and Gentle Soul

Hugh was a kind and gentle soul who always wore a bow tie; he was a friend to those who would have him as one. He was a family man; he and Jody were blessed with their extended family of their four children and their spouses, 15 grandchildren, and one great grandchild. They were true partners; they each undertook leadership roles in the Congregational Church in Farmington, Connecticut, where they lived for 34 years, and Jody often accompanied him to professional meetings. Hugh served as the church Moderator, the primary lay leadership position in the church. He was one of the founders of the Connecticut Academy of Science and

AN APPRECIATION

"To receive his criticism of a draft paper was like being mauled by a friendly lion. You felt yourself being torn to pieces and then tenderly reconstructed in such a way that your pearls of wisdom could be seen and understood. In negotiation he was wise, charming, skilled at defusing explosive situations, but—if he felt strongly—immovable. . . .

"At first meeting, Hugh could appear to be just a folksy raconteur, with a story to match all occasions. You soon became aware of his wide ranging intellectual curiosity, his wisdom and his extraordinary thoroughness and hard work. He was an editor supreme, meticulous but always looking for new ideas. Beneath all this he was a man with wide cultural interests—particularly in music and in his church. And at the heart of it all there was an enormous pride in his wife, Jody, his lovely rambling home and his family. There may not be any theorems with his name attached, but his influence in the subject has been enormous" (Tomlinson 1999, 183).

HUGH AND JODY AT HOME

Engineering. Although a transplant to New England, he became an avid Boston Red Sox fan, but, unfortunately, did not live to see them win the 2004 World Series.

Hugh loved classical music, especially opera. During his 4 years at IIASA in Austria, he and Jody went to the opera once or twice a week. He loved to play the piano, especially the ragtime music of Scott Joplin. Hugh combined his love of music and his commitment to the church by being part of a number of church choirs. He had a strong bass voice that was always in demand by the directors. He also combined music, church commitment, and his engineering background to help several churches procure pipe organs for their sanctuaries. He also loved reading, and was seldom without a book in his hand. Another of his passions was trains—both real and models. He built from scratch several operational model railroad layouts, bringing great joy to his children.

Hugh died on June 22, 1999, alert and active until the end. Jody died on December 11, 2007.

Honors and Awards

Hugh's influence and importance to the OR profession were duly recognized. He served as ORSA'S president in 1962 and, over the years, he was awarded: ORSA's George Kimball Medal for distinguished service to the society and the OR profession (1975); The CORS Harold Larnder Prize that recognizes an individual who has achieved international distinction in OR (1990); and the INFORMS Military Applications Society's Jacinto Steinhardt Memorial Award (1999). In 2005, he was inducted into the IFORS' Operational Research Hall of Fame (Rothkopf 2005). He was awarded the USAF medal for outstanding and exceptional civil service, and received the 1952 Arthur S. Flemming Award of the U.S. Junior Chamber of Commerce for outstanding civil service for young civil servants.

His professional peers, as embodied by the membership of INFORMS, understood that Hugh had something special to say about OR. He was a bridge connecting 50 years of OR—from the embryonic OR of WWII to the world-wide OR profession at the end of the 20th century—and, most importantly, that he had helped define its course. In recognition of his pioneering contributions, INFORMS awarded him its 1999–2000 Philip McCord Lectureship that gave him the opportunity to travel world-wide to address OR audiences. Hugh was excited by this honor and was working actively preparing his lecture when he died.

Editors' Note

Mike Rothkopf died on February 18, 2008 while swimming his daily mile at the McCoy Natatorium at Pennsylvania State University. He was 68 years old. Mike was Smeal Chaired Professor of Supply Chain and Information Systems at Penn State's Smeal College of Business. Before joining Penn State, he was a professor at Rutgers University Business School and Center for Operations Research for 19 years. He had submitted a draft of his Hugh Miser profile in January. We were fortunate to have Mike as a friend and colleague, and very thankful to have his profile of his dear friend Hugh.

ACKNOWLEDGMENTS

This profile has benefited from discussions with Graham Rand; with Hugh's widow, Jody Miser; with Hugh's daughter, Emily Miser Welch, and his son James Miser; and, most of all, with Hugh himself during the 1980s and 1990s. I am grateful to Steven Gass, Associate Director for Research and Instructional Services, MIT Libraries, for extracting Hugh's curriculum vitae from his uncatalogued papers in the MIT library.

REFERENCES

Andres P, Miser HJ, Reingold H (1945) Basic mathematics for engineers. Wiley, New York, NY

Andres P, Miser HJ, Reingold H (1955) Basic mathematics for science and engineering. Wiley, New York, NY

Assad AA (2009) Personal communication

Brothers L (1954) Operations analysis in the U.S. Air Force. Oper Res 2(1):1–16

CONDOR (Committee on the Next Decade in Operation Research) (1988) Operations research: the next decade. Oper Res 36(4):619–637

McArthur C (1990) Operations analysis in the U.S. Eighth Air Force in World War II. History of mathematics, vol 4. American Mathematical Society, Providence, RI

Miser HJ (1974a) The editor's easy chair: what is operations research? Oper Res 22(4):903–909

Miser HJ (1974b) The editor's easy chair: the editorial operations. Oper Res 22(5):1122–1129

Miser HJ (1977) The history, nature, and use of operations research. In: Moder J, Elmaghraby S (eds) Handbook of operations research: models and applications, vol. 1. Van Nostrand Reinhold, New York, NY, pp 3–24

Miser HJ (1980) Operations research and systems analysis. Science 209(4452):139–146

Miser HJ (1991) Comments on OR as technology. J Oper Res Soc 42(5):429–431

Miser HJ (1992) Craft in operations research. Oper Res 40(4):633–639

Miser HJ (1993a) Coping with outliers in scoring bombing training results. Interfaces 23(5):70–76

Miser HJ (1993b) A foundational concept of science appropriate for validation in operations research. Eur J Oper Res 66(2):204–216

Miser HJ (ed) (1995) Handbook of systems analysis: volume three—cases. Wiley, New York, NY

Miser HJ (1996a) Comments prompted by 'On the nature of OR entering the fray'. J Oper Res Soc 47(10):1322–1323

Miser HJ (1996b) Practice of operations research and management science. In: Gass SI, Harris CM (eds) Encyclopedia of operations research and management science. Kluwer, Dordrecht, pp 504–508

Miser HJ (1996c) Operations research. In: Malcolm W (ed) International encyclopedia of business and management. Thompson Learning, London, pp 507–522

Miser HJ (1998) The easy chair: journal editing as I see it. Interfaces 28(5):115–123; Reprinted in CBE Views 22(3):71–75, 1999

Miser HJ (2000) The easy chair: what OR/MS workers should know about the early formative years of their profession. Interfaces 30(2):99–111

Miser HJ, Quade ES (eds) (1985) Handbook of systems analysis: volume one—overview of uses, procedures, applications, and practice. North Holland, New York, NY

Miser HJ, Quade ES (eds) (1988) Handbook of systems analysis: volume two—craft issues and procedural choices. North Holland, New York, NY

Rothkopf MH (2005) IFOR's operational research hall of fame: Hugh J. Miser. Int Trans Oper Res 12(2):259–262

Tomlinson R (1999) An appreciation of Hugh J. Miser. J Oper Res Soc 50(11):183

Wagner HM, Rothkopf MH, Thomas CJ, Miser HJ (1989) The next decade in operations research: comments on the CONDOR report. Oper Res 37(4):664–667

Woolsey R (1972) Operations research and management science today, or, does an education in checkers really prepare one for a life of chess? Oper Res 20(3):729–737

18

Abraham Charnes

Fred Y. Phillips and
Lawrence M. Seiford

Abraham (Abe) Charnes was an internationally acclaimed mathematician, a pioneer of operations research (OR), and one of the great figures in management science (MS). Abe's fundamental contributions in optimization, statistics, finance, marketing, and human resource management spanned a 50-year career. An outstanding characteristic of his research was his talent to create new mathematical formulations that were motivated by real-world applications, but entailed important theoretical and methodological advances.

He was co-recipient of the 1982 John von Neumann Theory Prize of the Operations Research Society of America (ORSA) and The Institute of Management Sciences (TIMS) with William W. Cooper (his long-time friend and academic collaborator) and Richard J. Duffin. He was awarded the Harold Larnder Prize (1989) from the Canadian Operations Research Society for achieving international distinction in OR, and the Notable Contributions to the Accounting Literature Award from the American Accounting Association. He received the Distinguished Public Service medal from the U.S. Department of the Navy (its highest civilian award) for his contributions as a research physicist and operations analyst during World War II (WWII). He was a fellow of the American Association for the Advancement of Science and the Econometric Society. He was a founding member and president (1960) of TIMS. His publications include eight books and over 400 journal articles.

A.A. Assad, S.I. Gass (eds.), *Profiles in Operations Research*, International Series in Operations Research & Management Science 147, DOI 10.1007/978-1-4419-6281-2_18,

HOPEWELL AND BEYOND

Abraham Charnes was born in Hopewell, Virginia, on September 4, 1917, the son of Harry and Rebecca Charnes. The family moved to Chicago when Abe was 6 years old and he attended Crane Technical High School, graduating in 1934. He entered the University of Illinois (Urbana-Champaign) that fall as an engineering student having received a Ceramic Engineering Scholarship of the Illinois Clay Manufacturers Association. He began his mathematical undergraduate studies the following spring under David G. Bourgin—partial differential equations, calculus of variations, fixed point theory, and functional analysis. He also studied algebra with Reinhold Baer.

Abe transferred from engineering to mathematics in the middle of his junior year. He received his B.A. in mathematics, with highest honors (Phi Beta Kappa), in 1938. He continued at Illinois and obtained his M.A. in mathematics in 1939. During the summer of 1941, he was a Fellow in the First National Program in Applied Mathematics at Brown University; he then moved to Harvard in the fall on a Gordon McKay scholarship in aeronautical engineering.

Abe's studies at Harvard were interrupted by the U.S. entering WWII. In the spring of 1942, Abe, as a civilian, joined the U.S. Navy Bureau of Ordnance. Later that year, he became an ensign in the U.S. Naval Reserve. During the war, he worked on electromagnetics, torpedo performance and control, supersonic flight, fire control, and weapon damage assessment (Phillips and Seiford 2006).

At the end of WWII, Abe returned to the University of Illinois and, under the direction of Bourgin, wrote his Ph.D. dissertation, "Wing-body interaction in linear supersonic flow." He was awarded his doctorate degree in 1947. Upon graduation, Abe received an Office of Naval Research

Postdoctoral Fellowship for research in linear topological algebra (1947–1948). In the fall of 1948, he began his teaching career as assistant professor of mathematics at Carnegie Institute of Technology (Carnegie Tech), Pittsburgh, Pennsylvania. It was there that he and William (Bill) W. Cooper "had a chance luncheon encounter" from which began their friendship and extremely productive collaboration (Cooper 2002, 36). It was also about this time that Abe met and later married Kathryn (Kay) Helen Francis. Kay was an assistant professor of education at Carnegie Tech teaching home economics. She and Abe were part of the faculty club lunch bunch. One day, Kay noticed that Abe was not at lunch and was told that he was sick in his room upstairs. Kay, being concerned that Abe was not going to eat, brought him lunch. They were married 2 years later and led a happy family life with their three children, Deborah, Daniel, and William (K. Charnes 2009).

Abe served on the faculty of four universities: Carnegie Institute of Technology (1948–1955); Purdue University (1955–1957); Northwestern University (1957–1968), where he was the Walter P. Murphy Professor of Applied Mathematics (1957–1968); and the University of Texas at Austin (1968–1992). At Texas, he was the Jesse H. Jones Professor and University System Professor (1968–1984), held the John P. Harbin Chair in the College of Business Administration (1984–1988), and was Professor Emeritus of Management Science and Information Systems (1988–1992).

Abe died on December 19, 1992, in Austin, Texas. He was survived by Kay, his three children, and four grandchildren (Phillips and Seiford 2006).

Research Contributions: The Many Faces of Abe

Abe's research activities had a major influence on a wide range of theory and industrial applications: mathematical and linear programming (LP), optimization, statistics, transportation, finance, marketing, and human resource management. Of particular import was his joint development with his long-time colleague Bill Cooper of goal programming (GP), data envelopment analysis (DEA), fractional programming, and multi-criteria programming. His early work in LP was instrumental in spreading the ideas of optimization and OR to a world-wide community of scientists and practitioners. In what follows, we describe the wide scope of Abe's research across many fields (Phillips 1993; Phillips and Seiford 2006).

Abe joined with Bill Cooper and Alexander Henderson to write the first book that described the mathematical, economic, and computational ideas of linear programming, *An Introduction to Linear Programming* (Charnes et al. 1953)—a book that brought these advances out from their U.S. Air Force and academic settings and made them available to the general, but embryonic OR community. Further, in 1960, with the publication of their two-volume *Management Models and Industrial Applications of Linear Programming*, Abe and Bill helped set the course of OR and MS for years to come (Charnes and Cooper 1961a).

Abe's initial industrial research began at Carnegie Institute and involved the blending of aviation gasolines for the Gulf Oil Company. Working with Bill Cooper and Bob Mellon (of the Gulf refinery staff), the three of them developed what can be considered the first killer application of both LP and OR, as described in their paper "Blending aviation gasolines—a study in programming interdependent activities in an integrated oil company" (Charnes et al. 1952). One can safely state that, as soon as this paper hit the street, just about all oil companies ordered computers and started to write LP computer programs for solving their related problems (Phillips and Seiford 2006; Symonds 1955). Today, most oil refineries plan their production with mathematical models that are descendants of this initial study. [A version of their paper was first presented at the Symposium on Linear Inequalities and Programming, June 14–16, 1951, Washington, DC, sponsored by the U.S. Air Force Directorate of Management Analysis Service, and appeared in the Symposium proceedings (Directorate of Management Analysis 1952)].

Abe and Kay

When solving the first small (10 equations and 22 variables) LP model of the blending formulation, it was noticed by Abe and associates that some of the simplex algorithm basic feasible solutions exhibited mathematical degeneracy, that is, a few of the basic solution variables were zero. As proven by Dantzig (1951), this could cause the algorithm to cycle between sub-optimal solutions and not find an optimal solution. Thus, Abe was motivated to resolve the degeneracy problem which he did in his paper "Optimality and degeneracy in linear programming" (Charnes 1952). Abe's solution introduced non-Archimedean field extensions into the mathematical programming literature, obtaining a complete regularization of LP problems.

The concept of constrained regression with preemptive or impossible goals was a ready extension of the non-Archimedean constructs. This first appeared in the constrained-regression salary formula problem for executive compensation and led to the important ideas and techniques of GP, with extensions to multi-objective programming problems (Charnes et al. 1955). The basic and important idea of GP stems from the desire to satisfy a set of $(m \times n)$ linear equations $AX = b,\ X \geq 0$, exactly, where the individual right-hand sides of the $(m \times 1)$ vector $b = (b_i)$ are considered goals to be met (personnel levels, desired salary, recruitment goals). But, the conditions of the problem form an infeasible problem. The desire then is to meet the conditions in an approximate fashion—"as close as possible" (Charnes and Cooper 1961a, 217)—by restating the problem as $AX + Y = b$, with $X \geq 0$, and $Y = (y_i)$ are a set of unrestricted deviation variables. The problem is then to minimize $\Sigma_i |y_i|$ subject to $AX + Y = b,\ X \geq 0$. The problem is restated in standard LP problem form by rewriting $Y = Y^+ - Y^-$, where $y_i = y_i^+ - y_i^-$ with (y_i^+, y_i^-) nonnegative deviation variables. The problem is now: minimize $\Sigma_i(y_i^+ + y_i^-)$ subject to $AX + Y^+ - Y^- = b,\ (X,\ Y^+,\ Y^-) \geq 0$. Besides having an untold number of business, industrial, and military applications, this important model leads to the solution of statistical-estimation problems such as minimizing the sum of absolute deviations and the Chebyshev problem of minimizing the maximum of absolute deviation (Charnes et al. 1968, 1988; Glover and Sueyoshi 2009; Wagner 1959).

Two-person, zero-sum games, particularly those against nature, require additional constraints to model real problems effectively, thereby posing major difficulties in analysis and computation. These concerns were resolved by reducing constrained matrix games, which are bi-extremal problems, to explicit pairs of dual linear-programming problems (Charnes and Cooper 1961a, 776–784). Other important computational developments include a modified simplex method, joint with Carlton Lemke, Abe's first Ph.D. student at Carnegie (Charnes and Lemke 1952). The general idea of developing effective traversals of extreme points for problems with local but non-global optima led to "Nonlinear power of adjacent extreme point methods" (Charnes and Cooper 1957), Lemke's solution of the bimatrix game problem, and Lemke's method for the solution of complementarity problems (Lemke 1965; Lemke and Howson 1964).

At the December 1953 meeting of the Econometric Society, Abe presented the formulation and ideas of chance constraints and stochastic decision rules that stemmed from a study of scheduling of heating-oil manufacturing and distribution. Here, the mathematical programming problem parameters are random variables and a solution must satisfy the constraints of the problem in a probabilistic sense, that is, the usual LP constraints are given as probability statements (Charnes and Cooper 1959; Charnes et al. 1958).

Over the years, Abe developed a strong interest in many operational problems associated with transportation. As part of the sequence of papers that were presented at the Transportation Research Symposium held at Purdue University on February 27–28, 1957, Abe joined with Tom Bartlett to discuss the cyclic scheduling, assignment, and routing of equipment to meet scheduling and maintenance requirements of the St. Louis and San Francisco (Frisco) Railroad (Bartlett and Charnes 1957). This work required the development of a practical solution method for the underlying mixed-integer programming problem. In another symposium paper, Abe and the Carnegie economist Merton Miller discussed a significantly different problem they had investigated for the Union Railroad. This problem was one of the earliest real-world example of covering or matching integer-programming problems (Charnes and Miller 1956). [All the Symposium papers appeared in the 1957 issue of *Naval Research Logistics Quarterly* 4(3). Abe was Chairman of the Symposium and, at that time, was professor of mathematics and industrial engineering at Purdue University.]

After transferring to Northwestern University in 1957, Abe continued his research on transportation problems, but extended his interests into broader engineering areas. He consulted with the Chicago Area Transportation Study and developed the first known extremal-principle model for the simulation of traffic flow on arterial networks without pre-specification of which links are to carry the major traffic flows and which are to carry the small cross-flow traffic (Charnes and Cooper 1958a,

1961b). He then investigated new applications that stemmed from sanitary and environmental engineering, electrical engineering, and mechanical engineering (he had joint appointments in the engineering sciences, mathematics, and economics departments). An unpublished memorandum, prepared with engineers J. Logan and W. Pipes in 1959 for the Office of Naval Research (ONR), discussed a model for the planning and evaluation of sewage treatment systems.

In a finance-oriented paper, "Intertemporal analysis and optimization of bank portfolios," Abe and David Chambers developed a basic model and related analyses for dynamic balance-sheet management of banks, an approach that is currently in use in many countries (Chambers and Charnes 1961). This work sparked collaborations with many of his Ph.D. students: a sequential estimation procedure for accounting tests of internal control (Charnes et al. 1964); on chance-constrained financial models (Charnes and Thore 1966); convex approximation and decentralization (Charnes et al. 1966); decentralization with preemptive goals (Charnes et al. 1967). Abe also worked with Merton Miller on financial flow models and opportunity costing of funds (Charnes et al. 1959) for which they were second place winners of the 1959 McKinsey Foundation Award. (Miller received the 1990 Nobel Prize in economic sciences for his pioneering work in the theory of financial economics.)

HOW TO GET A JOB WITH ABE

"In the summer of 1967, I [Fred Phillips] attended an NSF Summer Secondary Science Program in Iowa City with other high school students from around the country. I was put under the wing of Burton Conrad 'Connie' Gearhart, a grad student in industrial engineering. He and I worked on optimizing Iowa's legislative districts for geographic compactness, anti-gerrymandering and so on. Connie mentioned that there was a fellow named Charnes at Northwestern doing some great things with optimization. 'Oh,' I replied, 'I go to high school with his daughter Debie.' (Deborah spelled her nickname with one 'b.') Debie had made a presentation on LP in our freshman algebra class the year before, so I knew a little bit about her father. This was at Evanston Township High School, very close to Northwestern.

"As summer '68 approached, I knocked on the door of the Charnes home and asked Abe whether he might help me find a job at the Northwestern computer center. There were always a few consultants, called 'debuggers,' sitting in the center, helping users from all departments with their programming problems. I wanted to be one! Abe said he'd look into it, but he called back the following week to say that there would be room for me on some of his projects in the engineering science department. I wrote FORTRAN code for the dissertations of Steve Littlechild and Tony Walters, under their careful supervision, and first learned from Darwin Klingman not to give Abe my only copy of anything" (Phillips 2009).

For certain OR decision problems, the related mathematical model deals with ratios of the problem's variables. Such situations arise in economic equilibrium, sensitivity analyses, optimal maintenance and repair scheduling, DEA, as well as in other areas. With Abe at Northwestern and Bill Cooper at Carnegie, they solved the general case of the linear fractional-programming problem (also termed a ratio program) by reducing it to the solution of at most two non-Archimedean LP problems (Charnes and Cooper 1962). The basic fractional model can be stated as Maximize $(c^T x+\alpha)/(d^T x+\beta)$ subject to $Ax \leq b,\ x \geq 0$, (c^T and d^T transposed n-dimensional vectors, α and β are scalar constants, A an $(m \times n)$ matrix, x an $(n \times 1)$ variable vector, and the denominator positive over the feasible region. In another natural extension of linear (mathematical) programming, Abe pursued the study of semi-infinite programming, a structure that allows for either infinitely many constraints or infinitely many variables, but not both (Charnes et al. 1963).

The ideas of chance-constrained conditions that appeared earlier in the study of scheduling of heating-oil manufacturing and distribution were extended by Abe and his students in a number of areas. Chance-constrained games were developed with two of his Northwestern Ph.D. students, Michael Kirby and William Raike. In one paper, they analyzed two-person games with a payoff matrix whose elements are random variables (Charnes et al. 1968), while in a second paper, the zero-sum payoff matrix was fixed, but random perturbations with known distributions modify the strategy of each player before actual implementation of the strategies (Charnes et al. 1966). Research with Andrew Stedry set forth the first chance-constrained model for the management of conformance between plans and operations of an enterprise (Charnes and Stedry 1964a, b).

Abe, who worked on standard engineering control theory in WWII, considered its constructs inadequate for management control, and he studied it from his now broad perspective of management models (Charnes and Stedry 1966). He extended his basic views into the area of personnel management utilizing GP and generalized-network techniques. Starting with the ground-breaking study, "A model and a program for manpower management and planning," (Charnes et al. 1968), these developments formed the basis for an extensive collaboration in manpower planning and recruiting, equal employment opportunity, affirmative action, and management (Charnes et al. 1971, 1972, 1978b). (Most of Abe and Bill's manpower research was done for ONR; Richard Niehaus was an analyst with the Office of Civilian Management, U.S. Navy Department.)

Upon accepting a position at the University of Texas at Austin in 1968, Abe, joined by colleagues and Ph.D. students, extended his research activities in game theory and nonlinear programming. Game theory research included constrained n-person games and the calculation of the characteristic function (Charnes and Sorensen 1974); the disruption propensity of coalitions in an n-person game (Charnes et al. 1978e); and new solution notions for games with intertwined strategy sets and multiple payoffs to players (Charnes et al. 1990).

Abe's other research interests demonstrated his ability to contribute to a wide range of areas: a duality theory for LP over closed convex cones and a complete classification of duality (Ben-Israel et al. 1969); extremal principles and optimization dualities for embedding the Khinchin-Kullback-Liebler information-theoretic statistical estimation (Charnes et al. 1978d); a rigorous information-theoretic statistical justification for heuristic gravity potential estimates in interregional transfers in transportation (Charnes et al. 1972); and consumer purchase behavior and brand-shifting estimation techniques in marketing (Charnes et al. 1978a, 1984). Parallel developments occurred in GP and goal-interval programming in manpower planning, coastal land-use development, oil-spill management, and goal-arc approximations for large-scale models. The latter thrust led into goal-focusing replacement of utility functions (Charnes et al. 1979).

DATA ENVELOPMENT ANALYSIS

Over 25 years after their seminal blending of aviation gasoline application, Abe and Bill, along with Eduardo Rhodes, a Carnegie Ph.D. student, did it again—another killer mathematical programming model! Together, they developed the very important field of DEA, a new OR application area that expanded an organization's ability to evaluate quantitatively the performance (efficiency) of its operating (decision making) units (Charnes et al. 1978c, 1981; Cook and Seiford 2009; Cooper 2005; Cooper et al. 2000, 2004). DEA has been used to compare the efficiency of bank branches, public schools, hospitals, army recruiting districts, day-care centers, hospitals, courts, drug treatment centers, baseball batters, and many more (Seiford 1997).

The origins of DEA efficiency can be traced to the rationale for measuring of efficiency by M. J. Farrell (1957, 253):

> The problem of measuring the productive efficiency of an industry is important to both the economic theorist and the economic policy maker. If the theoretical arguments as to the relative efficiency of different economic systems are to be subjected to empirical testing, it is essential to be able to make some actual measurements of efficiency. Equally, if economic planning is to concern itself with particular industries, it is important to know how far a given industry can be expected to increase its output by simply increasing its efficiency, without absorbing further resources.

Rhodes, working on his Ph.D. dissertation, attempted to compare the performance of a sample of U.S. schools under a Federal government educational program (Program Follow Through) to a sample not under the program. As related by Cooper (2005, 5–6):

> The attempts by Rhodes to use statistical regression and correlation techniques on these data did not produce satisfactory results so he began to explore other alternatives. In the course of these explorations Rhodes uncovered a little known article by Farrell (1957) that described an approach which would now be identified with DEA. Calling this article to my attention led to the subsequent work that resulted in the formulations of DEA that are reported in Charnes et al. (1978c, 1981). This in turn led to new classes of models that could be used to implement the ideas underlying DEA in many additional contexts.

Comprehensive surveys of DEA are given in Emrouznejad et al. (2008) and Cook and Seiford (2009). The latter survey lists more than 4,000 DEA publications (articles, books, monographs) involving more than 2,500 authors in 42 countries. The first DEA paper, "Measuring the efficiency of decision making units" (Charnes et al. 1978c), is the most highly cited paper that has appeared in the *European Journal of Operational Research* with over 1,700 citations. The INFORMS Impact Prize, given in recognition of contributions that have had a broad impact on OR, was awarded in 2006 to Abe and Bill Cooper for their development of DEA.

On October 14–16, 1987, the University of Texas at Austin held a conference to honor and celebrate Abe's 70th Birthday. The conference program was organized around a variety of research topics which originated with Abe and in which he continued to be an important contributor. The conference was designed especially to recognize Abe as a valued teacher and educator. Thus, an author of every paper in the conference was either Abe's student or a student of a student, and so on. The conference had 300 invited guests, all of whom were connected to Abe through a similar lineal link. The conference papers are given in Phillips and Rousseau (1990).

2006 INFORMS IMPACT PRIZE

Data Envelopment Analysis (DEA) was first described in the article "Measuring the efficiency of decision-making units," *European Journal of Operational Research*, by A. Charnes et al. (1978c). It was selected as one of the 30 most influential papers published in the first 30 years of that journal.

Researchers and practitioners in many fields have recognized DEA's power and ease of use as a method for evaluating the performance of operational processes. DEA has been successfully applied in many organizations world-wide, including hospitals, HMO's, military units, universities, cities, courts, investment portfolio managers, and logistics and manufacturing firms. DEA helps identify peak performers in these organizations and suggests ways for the others to improve.

DEA's novel mathematical-programming-based, data-oriented approach has also been helpful in comparing the economies of nations and regions, due to its particularly effective means of accounting for the conversion of multiple inputs to multiple outputs.

Charnes and Cooper required very few assumptions in the development of DEA, so it has opened up possibilities for use in cases that were resistant to other OR approaches. For example, DEA helped improve pupil transportation in North Carolina, saving over 50 million dollars. This application was a finalist for the 1993 Edelman Prize.

The bibliography in Cooper, Seiford and Tone's *Data Envelopment Analysis: A Comprehensive Text with Models, Applications and References* (2000), lists over 2800 articles on applications and extensions of DEA. Bill Cooper (now in his 90s) continues to advance the DEA method and extend its monumental impact.

Abraham Charnes and William Cooper's decades-long friendship and collaboration ended with Dr. Charnes' death in 1992. Both have been inducted into the IFORS' Operational Research Hall of Fame. For their seminal work on DEA, INFORMS is delighted to award the 2006 Impact Prize to Abraham Charnes and William W. Cooper (INFORMS 2006).

THE GIFT (70TH BIRTHDAY CONFERENCE)

(Left to right): Bill Cooper, Kay, Abe, Fred Phillips

Painting by Amado Pena, Abe's favorite artist.

TEXAS ABE

ABE: A MULTIFACETED VIEW

FAMILY: DEBORAH CHARNES PEEL

"My daughters will learn about Grandpa through pictures, through stories from me and, I hope, from you, too.

"It's really kind of difficult to separate out, even grossly, his contributions to my life and my character from my mother's contributions because their influences over-lapped so much, since so many of their values, aspirations, and goals were so similar. These are the things that I want to say to my daughters about their grandfather, the things that I think of as most distinctly him.

"He was dedicated to the Truth, to Knowledge, and to Science—*all with capital letters*. He never spoke about these things; he lived them. His life was devoted to the pursuit of greater knowledge, and I think you all know he was intolerant of ignorance in all of its many forms; stupidity, mediocrity, bigotry, racism, but especially indifference to science.

"He revered great teachers, great thinkers, and great researchers, which many of you are, and he had a deep love of complex problems. (I think I got that from him also.) He felt that any efforts to understand the laws of nature and man were fully justified. Laws and rules had to be completely familiar to anyone attempting to solve problems before solutions could be either imagined or created.

"He loved his work (that's an understatement). His passion and need to pursue his beloved research were, of course, legendary. As my mother often said, his work was everything to him, and it was also his hobby. He truly was a lucky person to have found and pursued ideas in work that had such powerful meaning to him.

"He personified excellence and high achievement. In relationships at work and in his private life, he had mostly intense and long-lasting loyalties. There were, however, a few famous hatreds. But his students and colleagues knew, and you've heard today, that they really were, for him, an extended family, and he helped so many for all of his life. Some of his last wishes were, in fact, to help some Ph.D. students to finish their work.

"I suspect anyone who interfered with his teaching, his research, or his students, didn't much like being in that position for long. He championed not only his own work, but also that of his many talented students and colleagues. He championed merit and genius—he was wonderful to have backing you up. He certainly was always there to back me up."

(Deborah Charnes Peel—mother, physician, psychoanalyst. Abraham Charnes remembered: Eulogy. *Annals of Operations Research*, 1997, **73**, 373–374)

COLLEAGUE REMINISCENCES

"My memories of Abe Charnes are always accompanied by a vivid picture of him in explanatory mode e.g., standing with cocked head while answering a mathematical question I had put to him. The mathematical space we were discussing, no matter what its dimension, stood sharp and clear before Abe's eyes, and his slim fingered hands wove their rapid way through it, pointing out its key properties as they went. His marvelous geometric imagination was central to his mathematical thinking."

(Herbert A. Simon, Carnegie Mellon University. Abraham Charnes remembered. *Annals of Operations Research*, 1997, **73**, 376)

"I shall never forget Abe's ability to see through the mist of muddlesome discussions and reach the kernel of reality. There is one very special incident that I shall never forget. Abe was at Northwestern University and I was with Teledyne. It was early Sunday morning. Our Teledyne research team had worked with Abe straight through from Friday to Saturday, after midnight. It was 6:30 Sunday morning when I tried to explain to Abe what I thought the problem was that we were muddling over. I drew a typical George diagram on top of the cake carton; Abe's whole countenance lighted up and he exclaimed, 'I see it, we can solve it! When he asked me to explain the diagram to the team four hours later, I failed. He then said, 'George, this is what you were trying to say.' Abe's scholarship called the cake-box solution as 'chance constrained programming.' What a great morning Abe and I had that Sunday."

(George Kozmetsky, University of Texas at Austin. Abraham Charnes remembered. *Annals of Operations Research*, 1997, **73**, 378–379)

"At Texas, he [Abe] always kept a series of black and white photos on his wall near his desk. Students took one of his classes—Charnes 1 or Charnes 2. These consisted of a review on interesting applied math and programming problems in phase 1, and problems never solved and a review of promising solution strategies in phase 2. About six or seven of us took the course, three faculty and four graduate students, with the graduate students running tutorials to keep the faculty up to speed. You got used to his sudden stops and his pacing, with his hinged glasses being flipped up and then back down before he went to the board and made a dot with a piece of chalk and then, before anything was written, stopped to put his hinged glasses back up—he repeated this two or three times before a dam was released and a torrent of equations poured forth. At any rate, the reason for the black and white polaroids was that if we went to the board and started to develop something interesting, he would suddenly jump up and finish developing the model for you and say either 'interesting' or 'let me show you why that doesn't work.' However, often he would get so excited he would dash to his room and return with the camera to take a picture of the results. If 'interesting.' these pictures were posted in the office. From year to year, you could tell it was the same room and the blackboard because the blackboard had the same cracks."

(Kingsley E. Haynes, George Mason University. Abraham Charnes remembered. *Annals of Operations Research*, 1997, **73**, 383–384)

STUDENT REMINISCENCES

"Many of us remember the mountain of paper on Abe's desk back at Northwestern—it nearly achieved equilibrium when its height and slope caused new items to slide off into oblivion, but continued to slowly overflow the desk like the Antarctic ice shelf. You could never be sure, walking past the office door, whether Abe was seated behind the mound or not. I can recall the time that train tickets (Abe steadfastly refused to fly in those days) that had gone missing before a trip to some conference eventually turned up at the bottom of the pile—about a year and a half later!"

(Bill Raike, New Zealand. Abraham Charnes remembered. *Annals of Operations Research*, 1997, **73**, 383)

"Abe wasn't a religious Jew, but he was rather proud of his heritage and often used biblical and Talmudic verses to make his point. One of the verses he repeated quite often, especially when we were getting ready to launch a new application, was 'Na'ase Venishmaa' (in Hebrew) which translates into 'we shall do and we shall hear' (said by the Israelites when they finally decided to accept the ten commandments). The order in this short sentence indicates Abe's agenda. He was always concerned not to keep his ideas and models only at the theoretical level ('we shall hear'), rather, he wanted to pull up his sleeves and get involved in applications that will prove the viability of these ideas ('we shall do')."

(Boaz Golany, Technion, Israel. Abraham Charnes remembered. *Annals of Operations Research*, 1997, **73**, 380)

ABE ON THE SCIENCE OF MANAGEMENT

"To summarize and conclude, new mathematics has developed with, spurred, and consummated every important development in management and control of man's resources from the agricultural requirements of the Babylonians and Egyptians through the industrial revolution of Newton's time and the 'electro-mechanical revolution' of yesterday. These past achievements have been primarily for management and control of man's natural resources. Today we see emerging an 'electronic revolution' or technology applicable to the management and control of 'man's own technological and organizational creations. I see, therefore, not a *fad*, but an emerging new mathematics and thereby an emerging new science, the science of management, tomorrow" (Charnes 1955, 182).

ACADEMIC DESCENDANTS (ABE'S 70TH BIRTHDAY CONFERENCE)

References

Bartlett T, Charnes A (1957) Cyclic scheduling and combinatorial topology: generalization and analysis. Nav Res Log Q 4(3):207–220

Ben-Israel A, Charnes A, Kortanek K (1969) Duality and asymptotic solvability over cones. Bull Am Math Soc 75:318–324

Chambers D, Charnes A (1961) Inter-temporal analysis and optimization of bank portfolios. Manage Sci 7(4):393–410

Charnes A (1952) Optimality and degeneracy in linear programming. Econometrica 20(2):160–170

Charnes A (1955) Future of mathematics in management science. Manage Sci 1(2):180–182

Charnes A, Clower R, Kortanek K (1967) Effective control through coherent decentralization with preemptive goals. Econometrica 35(2):294–320

Charnes A, Cooper WW (1957) Nonlinear power of adjacent extreme point methods in linear programming. Econometrica 25(1):132–153

Charnes A, Cooper WW (1958a) Extremal principles for simulating traffic flow in a network. Proc Natl Acad Sci 44(2):201–204

Charnes A, Cooper WW (1958b) The use of model types in business planning. Proceedings, seminars on operations research. University of Michigan (1957–1958), vol 2

Charnes A, Cooper WW (1959) Chance-constrained programming. Manage Sci 6(1):73–79

Charnes A, Cooper WW (1961a) Management models and industrial application of linear programming, vols I and II. Wiley, New York, NY

Charnes A, Cooper WW (1961b) Multicopy traffic network models. In: Herman R (ed) Theory of traffic flow. Elsevier, Amsterdam, pp 85–96

Charnes A, Cooper WW (1962) Programming with linear fractional functionals. Nav Res Log Q 9(3 and 4):181–186

Charnes A, Cooper WW, Ferguson R (1955) Optimal estimation of executive compensation by linear programming. Manage Sci 1(2):138–151

Charnes A, Cooper WW, Henderson A (1953) An introduction to linear programming. Wiley, New York, NY

Charnes A, Cooper WW, Kortanek KO (1963) Duality in semi-infinite programs and some works of Haar and Caratheodory. Manage Sci 9(2):209–228

Charnes A, Cooper WW, Learner DB (1978) Constrained information theoretic characterizations in consumer purchase behavior. J Oper Res Soc 29(9):833–842

Charnes A, Cooper WW, Learner D, Phillips F (1984) An MDI model and an algorithm for composite hypotheses testing in marketing. Market Sci 3(1):55–72

Charnes A, Cooper WW, Mellon B (1952) Blending aviation gasolines—a study in programming interdependent activities in an integrated oil company. Econometrica 20(2):135–159

Charnes A, Cooper WW, Miller MH (1959) Application of linear programming to financial budgeting and the costing of funds. J Bus Univ Chic 32(1):20–46

Charnes A, Cooper WW, Niehaus RJ (1971) A generalized model for training and recruiting decisions in manpower planning. In: Batholomew DJ, Smith AR (eds) Manpower and management science. Lexington Books, Lexington, MA, pp 115–130

Charnes A, Cooper WW, Niehaus RJ (1972) Studies in manpower planning. Office of Civilian Manpower Planning, Department of the Navy, Washington, DC, July 1972, 205 pages

Charnes A, Cooper WW, Niehaus RJ (eds) (1978) Management science approaches to manpower planning and organization design, vol 8. TIMS Studies in the Management Sciences, North Holland, Amsterdam,

Charnes A, Cooper WW, Niehaus RJ, Sholtz D (1968) A model and a program for manpower management and planning. Management Sciences Research Report No. 132, Office of Civilian Manpower Management, Department of the Navy, Washington, DC, 1968, System Research Group Memorandum No. 206.

Charnes A, Cooper WW, Rhodes E (1978) Measuring the efficiency of decision making units. Eur J Oper Res 2(6):429–444

Charnes A, Cooper WW, Rhodes E (1981) Evaluating program and managerial efficiency: an application of DEA to program follow through. Manage Sci 27(6):668–697

Charnes A, Cooper WW, Schinnar AP, Terleckyj NE (1979) A goal focusing approach to analysis of trade-offs among household production outputs. Proceedings Annual Meeting of the American Statistical Association, Washington, DC, pp 194–199

Charnes A, Cooper WW, Seiford L (1978) Extremal principles and optimization dualities for Kinchin-Kullback-Leibler estimation. Mathematische Operationsforchung und Statistik 9(1):21–29

Charnes A, Cooper WW, Sueyoshi T (1986) Least squares/ridge regressions and goal programming/constrained regression alternatives. Eur J Oper Res 27 (2):146–157

Charnes A Cooper WW, Sueyoshi T (1988) A goal programming/constrained regression review of the Bell system breakup. Manage Sci 34(1):1–26

Charnes A, Cooper WW, Symonds G (1958) Cost horizons and certainty equivalents: an approach to stochastic programming of heating oil. Manage Sci 4(3):235–263

Charnes A, Davidson J, Kortanek K (1964) On a mixed-sequential estimating procedure with application to audit tests in accounting. Accoun Rev 39(2):241–250

Charnes A, Fiacco A, Littlechild S (1966) Convex approximations and decentralization: a SUMT approach. Systems research memorandum no. 165. Technological Institute, Northwestern University, Evanston, IL

Charnes, Huang AZ, Rousseau J, Wei Q (1990) Cone extremal solutions of multi-payoff games with cross-constrained strategy sets. Optimization 21(1):51–69

Charnes A, Kirby M, Raike W (1966) Zero-zero chance-constrained games. In: Hertz D, Melese J (eds), Proceedings of the fourth international conference on operational research. International Federation of Operational Research Societies, Wiley, New York, NY, pp 150–170

Charnes A, Kirby M, Raike W (1968) Chance-constrained games with partially controllable strategies. Oper Res 16(1):142–149

Charnes A, Lemke C (1952) Computational problems of linear programming: a modified simplex method for control of roundoff error in linear programming. Proceedings of the Association for Computing Machinery May 2–3, 97–98

Charnes A, Miller M (1956) A model for optimal programming of railway freight train movements. Manage Sci 3(1):74–92

Charnes A, Raike W, Bettinger C (1972) An extremal and information-theoretic characterization of some interzonal transfer models. Socioecon Plann Sci 6(6):531–537

Charnes A, Rousseau J, Seiford L (1978e) Complements, mollifiers and the propensity to disrupt. Int J Game Theor 7(1):37–50

Charnes A, Sorenson S (1974) Constrained n-person games. Int J Game Theor 3(3):141–158

Charnes A, Stedry A (1964a) Investigations in the theory of multiple-budgeted goals. In: Bonini CP, Jaedicke R, Wagner H (eds) Management controls: new directions in basic research. McGraw-Hill, New York, NY, pp 186–204

Charnes A, Stedry A (1964b). Exploratory models in the theory of budgetary control. In: Cooper WW, Leavitt H, Shelly M (eds) New perspectives in organization research. Wiley, New York, NY, pp 212–249

Charnes A, Stedry A (1966) Search-theoretic models of organization control by budgeted multiple goals. Manage Sci 12(5):457–482

Charnes A, Thore S (1966) Planning for liquidity in financial institutions: the chance-constrained method. J Finance 21(4):649–674

Charnes K (2009) Personal communication

Cook W, Seiford L (2009) Data envelopment analysis (DEA)—thirty years on. Eur J Oper Res 192(2):1–17

Cooper WW (2002) Abraham Charnes and W.W. Cooper (et al.): a brief history of a long collaboration in developing industrial uses of linear programming. Oper Res 50(1):35–41

Cooper WW (2005) Origins, uses of, and relations between goal programming and data envelopment analysis. J Multi Criteria Decis Anal 13(1):3–11

Cooper WW, Seiford L, Tone K (2000) Data envelopment analysis. Kluwer, Norwell, MA

Cooper WW, Seiford L, Zhu J (2004) Data envelopment analysis: models and interpretations. In: Cooper WW, Seiford L, Zhu J (eds) Handbook on data envelopment analysis. Kluwer, Norwell, MA, pp 1–39

Dantzig GB (1951) Maximization of a linear function of variables subject to linear inequalities. In: Koopmans TC (ed) Activity analysis of production and allocation. Wiley, New York, NY, pp 339–347

Directorate of Management Analysis (1952) Symposium on linear inequalities and programming, June 14–16, 1951. In: Orden A, Goldstein L (eds) Headquarters U.S. Air Force Washington, DC, April

Emrouznejad A, Parker BR, Tavares G (2008) Evaluation of research in efficiency and productivity: a survey and analysis of the first 30 years of scholarly literature in DEA. Socioecon Plann Sci 42(3):151–157

Farrell M (1957) The measurement of productive efficiency. J R Stat Soc–Series A 120(Part III):253–283

Glover F, Sueyoshi R (2009) Contributions of Professor William C. Cooper in operations research and management science. Eur J Oper Res 197(1):1–16

INFORMS (2006) www.informs.org/Prizes. Accessed 5 July 2009

Lemke C (1965) Bimatrix equilibrium points and mathematical programming. Manage Sci 11(6):681–689

Lemke C, Howson J Jr (1964) Equilibrium points in bimatrix games. J Soc Ind Appl Math 12(2):413–423

Phillips F (1993) Abraham Charnes, 1917–1992. OR/MS Today (April):44–45

Phillips F (2009) Personal communication

Phillips F, Seiford S (2006) IFORS' operational research hall of fame: Abraham Charnes. Int Trans Oper Res 13(3):273–277

Phillips F, Rousseau J (eds) (1990) Systems and management science by extremal methods: research honoring Abraham Charnes at age 70. Kluwer, Norwell, MA

Seiford L (1997) A bibliography of data envelopment analysis. Ann Oper Res 73:393–438

Symonds G (1955) Linear programming: the solution of refinery problems. Esso Standard Oil Company, New York, NY

Wagner H (1959) Linear programming techniques for regression analysis. Am Stat Assoc J 54(285):206–212

19

George Kozmetsky

Lawrence Secrest, David V. Gibson, and John Sibley Butler

George Kozmetsky played an essential role in the establishment and development of operations research (OR) and management science (MS) in the U.S. He was one of a small of number of visionaries who, during the early 1950s, recognized the importance of quantitative wartime research to business and organizations. He wrote an early treatise that spelled out the impact of computers on management and organizations. George played a key role in applying OR and MS to business and society by his uncanny talent for translating OR/MS academic research findings and theory into practical applications, as well as identifying promising new applications and fertile research areas for OR/MS researchers. He was instrumental in the founding of The Institute of Management Sciences (TIMS) and served as its president in 1958.

George is known for his multifaceted contributions to society: successful entrepreneur (cofounder of Teledyne, Inc.); theorist and writer in the fields of management, entrepreneurship and innovation, organizational behavior, and business education; Dean of the College of Business Administration, University of Texas at Austin, who transformed the College from a regional business school into a world-renowned educational institution; and the founder of the IC^2 Institute, a leading interdisciplinary organizations focusing on economic and societal growth and progress. In 1993, George was awarded the National Medal of Technology by President Clinton for being the driving force behind the commercialization of various technologies. Throughout his life, George has been at the right place at the right time with the needed intelligence, energy, talent, and vision to make a difference.

A.A. Assad, S.I. Gass (eds.), *Profiles in Operations Research*, International Series in Operations Research & Management Science 147, DOI 10.1007/978-1-4419-6281-2_19,

THE EARLY YEARS

George Kozmetsky was born in Seattle, Washington, on October 5, 1917, to George and Nadya (Omelanitz) Kozmetsky, both recent immigrants from Grodna, Belarus. He was 5 years old when his father died. To support the family, his mother worked cleaning Pullman cars and, when he grew old enough, George unloaded fishing boats at Seattle's docks. After graduating from Franklin High School in 1934, he enrolled at the University of Washington. There, George was exposed to a number of truly outstanding teachers, some of whom became role models: Clark Kerr, who later developed the University of California system, and Grant Butterbaugh, who asked George to run the statistics laboratory at the University. As a Reserve Officers' Training Corps (ROTC) student, he spent the summer of his junior year in 1937 training in Vancouver, Washington. As an early but minor example of being at the right place at the right time, George, who had learned Russian from his parents, was asked by the Barracks Commander, General George C. Marshall, to act as translator for two Russian pilots who had landed in Vancouver for refueling.

George was commended for his good work. In 1938, he graduated from the University of Washington at the age of 20 receiving a B.A. degree in political science. A year later, he and a fellow classmate opened the first certified public accounting firm in Olympia, Washington, where he worked until he joined the Army soon after Pearl Harbor was bombed.

George graduated from the University of Washington as an ROTC officer candidate, but he failed the physical exam and did not receive a commission. When the U.S. entered World War II (WWII), he felt strongly that he should support his nation. George enlisted in the Army as a private, passing the physical without mentioning that he had failed an earlier physical. Just before he was shipped to Europe in late 1943, he married his sweetheart, Ronya Keosiff, also a University of Washington graduate. George and Ronya had met 3 years earlier on a blind date. As children of Russian immigrant families, they shared the same culture, had similar values, and both had a deep appreciation for hard work and for education. According to George, on the day that they married—November 5, 1943—Ronya told him that she gave the marriage 5 years. Fortunately, Ronya's prediction proved incorrect, and they enjoyed a strong marriage and family life, while developing successful careers and making substantial and continued contributions to academia, business, and public service.

After his initial enlisted military training, George soon moved to the officer ranks. He served as an officer in the medical corps in Europe where he earned a Silver Star, Bronze Star with Oak Leaf Cluster, and Purple Heart. He was frequently on the front lines providing emergency medical care for injured soldiers. George was wounded while performing spot surgery. He rarely talked about his war experiences, but he did admit that he took the opportunity, while training for the medical corps and in the field of battle, to read every medical book and article that he could get his hands on. He retained an avid interest in medical research for the rest of his life. This was a pattern that George would follow throughout his professional life—he would identify an area that had great importance to him, become immersed in the literature, set forth a research agenda, and experiment with applications.

HARVARD, CARNEGIE, AND THE REVOLUTION IN MS

By the time George returned from Europe, he and Ronya had saved enough funds to support his graduate education. He was offered fellowships and assistantships to attend graduate school at several prominent universities, including Columbia and Michigan. He and Ronya concluded, however, that Harvard Business School was the place to pursue his graduate education. During his MBA studies, George prepared a number of Harvard Business School (HBS) case studies that focused on managerial accounting and control. He received his MBA as a Baker Scholar in 1947 and, thus, joined the ranks of the WWII military veterans who attended the HBS and who subsequently transformed American business in the postwar period (Callahan and Forbes, Inc. 2002). George decided to remain at Harvard to pursue a doctoral degree in a new teach and study program. In the summers, he taught and helped members and leaders of the United Steel Workers of America and other unions to understand contracts and financial reports, and how to improve their negotiating skills. George's dissertation focused on the financial aspects of labor unions and pension funds, the first time such an examination of this aspect of union administration had been undertaken. His dissertation was published by Harvard under the title *Financial Reports of Labor Unions* (Kozmetsky 1950). During this period, he taught a wide range of courses at Harvard's Graduate School of Business and at Radcliffe College: accounting, control, finance, labor economics, plus courses that addressed social, economic, and political problems. George was invited to join an informal, distinguished group of Boston area professors who met weekly to discuss the nascent field of what they had labeled the behavioral sciences. As a result of his research into administrative and behavioral sciences, he became quite familiar with the work of Carnegie's Professor Herbert Simon, a future Nobel Laureate in Economics, who laid the foundations for what is now called behavioral economics.

In 1950, Dean George Leland Bach and Professor Simon, from the new Graduate School of Industrial Administration (GSIA) at Carnegie Institute of Technology (now Carnegie Mellon University), visited Harvard to hire a talented case method professor to supplement the quantitative courses that Carnegie was offering. They hired George as an assistant professor of

THE GOOD OLD DAYS: THE CONTROLLER'S OFFICE IN THE 1960S

"After receiving my MBA in 1965, I (Larry Secrest) worked in the Ford Motor Company corporate controller's office and subsequently in the corporate controller's office of Lear Siegler, Inc, a rapidly growing conglomerate. At Ford, it typically took more than a month to close the monthly corporate books in those days—we were simply overwhelmed with accounting numbers and different accounting treatments. Management did not want to wait a month to find out how the company had performed financially in the prior month ... so we developed some pretty clever approximations to give management estimated profits and sales in a week or two after the end of the month. For example, we weighed invoices on a factory scale in order to estimate accounts payable (as I recall, 100 pounds of invoices ≈ \$60,000 of accounts payable—a rather interesting use of statistical inference).

"In both companies, the book by Simon et al. (1954) was a much used desk reference. Both of these large, rapidly growing companies were dealing with the many issues of consolidated financial reporting, and the need for decentralized management. Simon et al. was the best source of wisdom on how to maintain financial integrity, consolidate information in a timely manner, and allow decentralized decision making and, thus, improved performance. It was required reading for anyone who wanted to succeed in the modern large corporation" (Secrest 2008).

finance to start in the fall of 1950 (Cooper 2007). For his part, George recognized that a major change in the study of management was being led by the faculty at Carnegie. He quickly became intrigued with Carnegie's applications-research approach to management that focused on generating general theories of management and that drew on a broad base of knowledge including accounting, behavioral sciences, and organization theory, as well as emphasizing data-based field studies. At Carnegie, Herb Simon, George Kozmetsky, William (Bill) W. Cooper and their colleagues led in the application of behavioral science research to management. George became a strong advocate for the new discipline of MS; he was also a bridge to Harvard's more traditional case-study approach to teaching business. George obtained funding from the Controllership Foundation to support GSIA's research into the evolving nature of the controllership function which culminated in the book *Centralization vs. Decentralization in Organizing the Controller's Department* (Simon et al. 1954).

Business Leader and Entrepreneur

In 1952, George was offered an assistant controllership position at Hughes Aircraft by his Harvard classmate, Roy Ash. Initially, he indicated that he really did not have much interest and was very happy at Carnegie, but Ash persisted. George visited Hughes where several of his Harvard classmates and other talented managers were transforming Hughes from a failing aircraft company into a state of the art electronics company. He recognized that Hughes would be a place from which to refine his practical understanding of business and organizations, and also provide his family with a sounder economic base; George accepted the position. Hughes Aircraft had a truly extraordinary management team that included Roy Ash, Tex Thornton, Simon Ramo, and Dean Woolridge. George quickly realized, however, that although the company had successfully moved into defense contracting in record time, they had failed to set up an integrated accounting, control, and project management system that would allow the company to meet, on schedule, its commitments for delivering a major new airborne radar system. The urgency of the situation caused him to conclude that the establishment of traditional accounting controls would take too long. With the assistance of MS academic colleagues, especially Bill Cooper and Abe Charnes, a new management and control system was designed and installed; it was based on statistical measurement and provided the tools for managing and delivering the radar project on time. George subsequently transferred from the controller's office to Hughes Advanced Electronics Laboratory where he learned to program and build computers (Cooper 2007).

At this point in his career, George was faced with a number of attractive career opportunities, including job offers from Robert McNamara at Ford Motor Company and Robert Trueblood at Carborundum. He decided he wanted a position that would allow him to focus on using computers in management. Roy Ash, who had left Hughes for Litton Industries, offered George the chance to develop a computer system for Litton. He took the challenge and, in a relatively short period, proved his vision by transforming Litton's computer department.

THE DOUBLE-K CRYSTAL BALL

"Mathematics and electronics will have far-reaching effects on business operations through their impact on information-processing systems, on automation of the office and on management planning and control. But electronics and mathematics also have important applications in other major areas—automation of production facilities, simulation of problems, and scientific computation for engineering and experimental purposes" (Kozmetsky and Kircher 1956, 190).

Recognizing George's entrepreneurial abilities, Tex Thornton, now the head of Litton, offered him the management of his own division. Shortly thereafter, the division received a major contract from the Navy for the development of a large, airborne computer system which evolved into the Department of Defense Airborne Warning and Control System. George always identified Thornton's offer to head a stand-alone division at Litton as an important beginning in his development as an entrepreneur. In his new position, he worked directly with Henry Singleton who was in charge of inertial guidance systems for Litton. Together, they were a tremendously talented management team: Henry was extremely capable in science and engineering; George was equally capable in business systems and finance. In less than 6 years they and their colleagues built a division with an order backlog of more than $1 billion. When it became clear, however, that they would not be able to maintain the Electronic Equipments Division as an integrated group within Litton, George and Henry, now corporate vice presidents, left the company. Within 6 months, they founded Teledyne, Inc. in 1960.

The founding of Teledyne was in many respects a classic application of MS. The company's strategic plan built on the strengths of each founder: Henry's prowess in inertia guidance, robotics, and information theory, and George's expertise in management systems, computers, government contracting, and finance. They studied future markets, growth rates, and profit margins for a number of products and components. In the end, they carefully selected the product-market areas with the best potential for growth and profit and in which they were likely to have a competitive advantage. The goal was to grow their new company to $1 billion in 10 years.

Ronya and George

As Teledyne was being founded, Ronya, who was a social worker, decided to change her career. She took classes in education at the University of California, Los Angeles and became a schoolteacher. Since George and Henry had agreed to initially forgo a salary from their new company, the schoolteacher temporarily became the primary breadwinner for the Kozmetsky family.

At first, Teledyne's business model did not progress as the founders had envisioned. In less than a year, the company was close to running out of working capital. At this moment, as fate would have it, a fire swept through the town of Bellaire destroying several hundred homes, including the Kozmetsky's. When the insurance check arrived, Ronya, George and family decided to continue living in their motel room and to invest the insurance money in Teledyne. These funds helped keep the company solvent until George and Henry were able to raise additional capital. After a shaky first year, the company began to thrive, becoming very successful in government contracting and in the acquisition of promising technology companies. Teledyne developed into an amalgam of over 130 companies that made products that included electronics, engine tools, acoustic research, stereo speakers, and seismic systems to monitor earthquakes. The founders exceeded their original goal of building a $1 billion company in 10 years by accomplishing it in eight.

When Teledyne was founded, George told Henry that one day he hoped to return to academia to promote advanced teaching methods and curricula in management. Teledyne was prospering and George was eager for another challenge. As his son Greg noted in Walters (2003, 14), "He felt it was time for him to bring academia and the corporate world closer together."

MS at the Longhorn Corral: UT-Austin

George received very attractive offers from Carnegie Mellon and the University of California, Berkeley (which offered a joint appointment in engineering and business), but ultimately, The University of Texas (UT) at Austin presented him with the most attractive offer: Dean of one of the nation's largest collegiate schools of business. Judson Neff, a former teacher of George at Harvard, and Foster Parker, an executive with Brown and Root (an engineering and construction company), both influential members of the UT Business Advisory Council, were strong advocates for George's appointment. George was particularly impressed with the quality and commitment of the University's Board of Regents and Chancellor Harry Ransom and Provost Norman Hackerman. In the summer of 1966, the Kozmetsky family drove from Los Angeles to Texas. In September, George began his new job as Dean of the College of Business Administration, The University of Texas at Austin.

In typical Kozmetsky fashion, George set a number of important goals prior to becoming Dean. One of his first and highest priorities was to create a first class MS/OR faculty. To achieve this goal, he turned to Carnegie Mellon and to his many MS/OR colleagues. His major hires included three highly respected professors: Abe Charnes, Tim Ruefli, and some years later, Bill Cooper. In conjunction with his hiring goals, George brought leading academicians from Harvard, Stanford, SRI International, NASA, and other leading institutions to conduct research into technology management (including management of research organizations), technology forecasting, and to generate practical applications from these research efforts. Leaders in the technology sector who were recruited included Albert Shapero, James Bright, and Eugene Konecci. Also, under George, particular emphasis was placed on technology entrepreneurship and innovation. Within a very short period of time, UT-Austin became internationally recognized as a leading center for OR/MS, management of technology, entrepreneurship, and practical applications.

George used his business contacts to revitalize the College of Business Administration Advisory Board, and to enhance close communication between the Texas and global business communities and the College's professors. Shortly after his arrival, the College received its first endowed chair which was soon followed by almost two dozen additional endowed chairs. Business leaders in Texas and elsewhere increasingly began to turn to UT-Austin for help with problems that they encountered in their businesses.

George encouraged the College of Business faculty to work more closely with colleagues in other academic fields to solve important problems. Joint appointments and interdepartmental research and teaching by the College's staff included cooperative activities with faculty and students in engineering, architecture, history, philosophy, economics, political science, psychology, social work, and other academic disciplines. Particularly, strong relationships were forged with the LBJ School of Public Affairs and the Law School where joint MBA degrees were offered for the first time. George also served as an advisor to The University of Texas System Offices, a position which enabled him to promote interdisciplinary research throughout the entire university system; he was instrumental in the establishment of the Advanced Robotics Research Center at UT-Arlington, The Center for Energy and Economic Diversification at UT-Permian Basin, and a variety of cross-border programs with Mexico.

During his fifth year at UT-Austin, George undertook a major revision of the graduate business curriculum that emphasized MS/OR, management of research and development, management of technology, technology forecasting, entrepreneurship, new product development, technology venturing, program management, and seminars for husband–wife teams focusing on career and life planning—all concepts that students could use in their careers. Simultaneously, he introduced Classroom 2000 that incorporated state of the art information technologies into the classroom. In conjunction with the College of Engineering, he established an experiential laboratory course focusing on the dynamics of fast-growth new technologies, successful commercialization of promising new technologies, and methods for accelerating regional economic development.

The new curriculum and subsequent modifications focused on producing creative and innovative managers, a theme that George followed throughout the remainder of his career. As George noted: "[The] University of tomorrow must get prepared to research and teach creative and innovative management as a new discipline requiring understanding and implementation of solutions to generalized as well as specific problems of society" (Kozmetsky 1984, 3; [also Kozmetsky 1985; Smilor et al. 1988, xiii–xiv]).

IC^2

In 1982, George resigned as Dean of the College of Business Administration to devote full time to the Institute for Constructive Capitalism (IC^2) that he had founded as an arm of The University of Texas at Austin in 1977. George was deeply concerned with the seeming inability of many parts of the world to participate in the economic and technical growth that was occurring in the U.S. His vision was to have a Think and Do center that would seek ways to improve business, government, and academic relations at home and abroad, and to accelerate economic development throughout the world through technology commercialization.

The Institute gave George the freedom and opportunity to work on large-scale unstructured problems that required transdisciplinary and collaborative research and methodologies. He again drew on his national and international colleagues in academia, business, and government to found a virtual network of IC^2 Institute Fellows, ultimately totaling over 200 leading scholars and practitioners. The underlying guidelines of IC^2 were to foster academic, business, and government collaboration; to be global and multidisciplinary; to focus on unstructured problems; to appreciate both technology and ideology; to extend academic and professional boundaries; and to engage in theory application as well as theory development—Think and Do was the motto.

George catalyzed, networked, and supported Institute Fellows research in areas within and outside of OR: Data Envelopment Analysis (DEA); evolutionary economics; innovation economics; medicine; telecommunications; diffusion of innovation; network methods (both the classical OR type and the newer MySpace type); strategic alliances and consortia; Post-Cold War politics and nuclear disarmament; demographics; lean manufacturing; statistical information theory and optimization; and Japanese management methods, among others (Cooper et al. 1997; Kozmetsky et al. 1994; Ronstadt and Kozmetsky 1999; Thore et al. 1995).

The Institute, subsequently renamed the Institute for Innovation, Creativity and Capital (still IC^2), was a key catalyst at UT-Austin and with greater Austin's business and government sectors. It helped implement the grand plan that transformed Austin from a university and government town into a world recognized technopolis.

In 1989, IC^2 launched the Austin Technology Incubator with city, county, and university support. Simultaneously, IC^2 also established The Texas Capital Network to provide a virtual meeting place for venture capital providers, angels, and entrepreneurs. In the early 1990s, IC^2 was instrumental in the establishment of additional community entrepreneurial and networking organizations such as the Austin Software Council. In 1996, IC^2 launched UT-Austin's first Internet-based degree program, the M.S. in Science and Technology Commercialization. Based on these and other creative and innovative programs, IC^2 engaged in a range of state, national, and international research, education, and technopolis building programs. Over the years, IC^2 has become recognized as a key contributing institution in the areas of entrepreneurship, technology commercialization, and services to accelerate wealth and job creation through technology-based entrepreneurship and economic development.

> TECHNOPOLIS
>
> "'Techno' reflects an emphasis on technology and 'polis' is Greek for city-state and reflects the balance between the public and private sectors. Three factors are especially important in the development of a Technopolis and provide a means to measure the dynamics of a modern technology city-state: the achievement of scientific preeminence; the development and maintenance of new technologies for emerging industries; and the attraction of major technology companies and the creation of homegrown technology companies" (Smilor et al. 1988, xiii–xiv).

IMPACT ON TIMS AND MS

The Operations Research Society of America (ORSA) was founded in 1953 mainly by OR analysts who were introduced to the field due to their work for the various branches of the U.S. military during WWII. George, Bill Cooper, and Mel Salveson, as well as a few others, decided that a professional society, one that emphasized a more inclusive approach to management, was needed; it would encompass all disciplines, especially behavioral sciences, that could contribute to increased understanding of the practice of management. They felt very strongly that the focus of ORSA was too narrow (tending to military applications) and that requirements

TIMS SECOND ANNUAL BANQUET, NYC, 1955

Head table (left to right): Roger Crane (president 1957), Ezra Glazer (past secretary), George Kozmetsky (past secretary; president 1958), Herbert Simon (vice president), William Cooper (president 1954), Merrill Flood (president 1955), Melvin Salveson (vice president; president 1959), C. West Churchman (chairman, editorial board; president 1962), Gifford Symonds (vice president; president 1956), Alex Orden (secretary-treasurer), Andrew Vazsonyi (past president 1954), David Hertz (president 1964).

for membership (at that time) too constraining. Thus, from its inception, The Institute of Management Sciences (TIMS) was global in membership and open to all scholars in disciplines that impacted the practice of management. Particular emphasis was placed on interdisciplinary research and applications. Founding TIMS President Bill Cooper recalled that George, as a Hughes Aircraft executive, provided space and financial support for the first meetings of the group that would later become TIMS. In 1953, TIMS was formed in New York City with around 200 initial members (Cooper 2007). George stayed active in TIMS serving as its first secretary-treasurer in 1954 and its president in 1958.

As a trained accountant, George ran his enterprises by the numbers and was interested in anything that could make the numbers tell a more complete story. OR/MS fit the bill perfectly, as did the advent of the computer. He and many of his university colleagues believed that a revolution in management education was underway, and that this revolution consisted of transforming management and administration from an art to a science. These themes remained major forces throughout George's academic and corporate activities. He demonstrated that OR/MS provided business organizations with invaluable tools for improving efficiency and effectiveness. He was a prime force in moving OR/MS from academia to business practice.

VISION OF THE FUTURE

Both George and Ronya were children of immigrants and recognized how fortunate they were to have been born in the U.S. They also believed that immigrants and minority populations provided a disproportionate role in new business development and entrepreneurship which resulted in America becoming the world's principal entrepreneurial power. To the end of his life, George undertook to explain the critical role that immigrants and minorities had in the economic development of the nation. The book *Immigrant and Minority Entrepreneurship: The Continuous Rebirth of American Communities* (Butler and Kozmetsky 2004) was published after George died. In many ways, it is a reflection of George's personal background and the importance he and John Butler placed on immigrants as a source of entrepreneurship and innovation in America.

Although George's most visible accomplishments rested in the fields of economic development, entrepreneurship, business leadership, technology investments, and mentoring, he also had a profound impact on the teaching and study of management through his extensive writings and published works. Throughout his life, George could be found on the boundaries and intersections of knowledge, identifying critical issues facing society, trying to understand the key factors driving large, ill-defined problems and seeking solutions to those issues.

Shortly after the birth of the modern electronic computer, George became convinced that the computer would have a huge transformational impact on the practice of management and management controls. Thus, one of his earliest publications is a seminal discourse on the pervasive role computers would play in the future development of management, management control systems, and the management of organizations (Kozmetsky and Kircher 1956).

In partnership with Herb Simon and others, George also made a major contribution to the practical application of OR/MS in a pioneering study of the controllership function in large organizations. This study discussed the behavioral and quantitative issues impacting large organizations, generally, and the impact of both behavioral sciences and computers on the controllership function in large, complex organizations (Simon et al. 1954). It thus provided a viable framework which allowed decentralized management of increasingly large and complex modern organizations, while retaining the integrity of financial controls and reporting at the corporate level.

By the time he became Dean of the College of Business at The University of Texas at Austin in 1966, George was convinced that unparalleled changes were taking place which would rapidly transform and drive the 21st-century global economy and society:

> First, we are on the threshold of a great technology era in the U.S. and throughout the world. Technology is dramatically altering the shape and direction of society and the way people think and act. The rapid increase in and diversity of new technologies are changing the nature of economic competition. How communities, regions, and nations anticipate and respond to this new competitive environment will largely determine the health and viability of their economies.
>
> Second, the nature of economic development has fundamentally and permanently changed. New institutional alliances are altering the strategy and tactics of economic development and diversification. New relationships

> between the public and private sectors—especially among business, government, and academia—are having far-reaching consequences on the way we think about and take action on economic development (Smilor et al. 1988, xiii–xiv).

George would often emphasize the impact of these changes stating, "You can't have a government that doesn't trust business, you can't have business that doesn't trust government, and academia can't proceed on a tangent that has no reference to business or government" (Walters 2003, 15).

At UT-Austin, George reorganized the College of Business based largely on the role of technology and economic development. Upon joining IC^2, a major focus of his became the impact and management of technology commercialization on economic development and wealth creation. Much of the research and publications during the early years of IC^2 reflected this emphasis. As George noted, "Today's environment for managing change is fundamentally different from even a decade ago. . . . International competition is taking the form of a worldwide scientific, technological and economic race for preeminence" (Kozmetsky 1985, xv). Notable publications inspired by George that focused on technology commercialization and economic development include *Creating the Technopolis* (Smilor et al. 1988), *Technology Companies & Global Markets* (Gibson 1991), and *The Technopolis Phenomenon* (Gibson and Smilor 1992).

Simultaneously, George recognized that existing management philosophies and concepts were inadequate to deal with dramatically increasing rates of change in society, multidisciplinary issues and solutions, and the need to blend the best elements of professional and entrepreneurial management into a new discipline: creative and innovative management. This recognition gave rise to an additional research focus: What management characteristics and training are required by the changes that will be occurring in the 21 century? In George's opinion, a new type of manager was required with the ability to identify and solve complex, unstructured problems that typically cross functional and disciplinary boundaries:

> In my definition, creative management involves abilities to take a problem or crisis and develop its issues, generate alternative solutions, and select feasible initiatives from among the alternatives. Furthermore, creative and innovative

> management includes the ability to use initiatives as a first step to solutions. These initiatives need to be monitored to determine that the actions are indeed solving the problem and not creating new ones (Kozmetsky 1984, 4).

Charnes and Cooper sum up the impact for academia: "The end result, we believe, will be a new and important academic discipline that presently seems to be missing or at least underemphasized in our schools of management" (Charnes and Cooper 1984, xvi).

During his career, George edited, co-authored or authored over 25 books and hundreds of articles. More relevant, his vision and ideas continue to inspire a growing number of publications from other authors. Abe Charnes and Bill Cooper co-edited *Creative and Innovative Management: Essays in Honor of George Kozmetsky* (1984), a collection of writings in the many areas of interest to George. This book, perhaps better than George's own writings, illustrates the breadth and depth of his influence on society.

GIVING BACK

Ronya, George, and the Kozmetsky family strongly believed in the importance of giving back to society and in striving to improve the world. Just after moving to Texas, in 1966, Ronya and George acted on their longtime belief in and support of philanthropy by creating the RGK Foundation to help fund research and solutions to important societal problems in which they and society had interest. Over the years, the RGK Foundation has distributed tens of millions of dollars to worthy causes, contributing to such diverse social needs as mathematics and science education programs, medical research, women's rights, and housing for hurricane victims in Honduras. Ronya and George were especially committed to children-oriented causes and Ronya was a founding board member of the Austin Children's Advocacy Center. Ronya was inducted into the McCombs School of Business (the renamed UT-Austin School of Business) Hall of Fame in 1995. She sponsored the annual Women in Business Leadership Conference through the RGK Foundation. Ronya's car has long displayed a bumper sticker that states: "If you think education is expensive, try ignorance." The RGK Foundation also gave generously to Austin art and civic organizations, as well as to the University of Texas and St. Edwards University. The month before he died, George gave $6 Million to Stanford University to found the Kozmetsky Global Collaboratory to work with IC^2 at UT-Austin on large-scale unstructured global challenges.

HONORS, AWARDS, AND REMEMBRANCES

George Kozmetsky is remembered for his splendid disregard of disciplinary boundaries, for his steadfast vision and mental agility, for his successes in business, and for his unstinting mentorship and inspiration of generations of students and faculty. On September 30, 1993, George was awarded the National Medal of Technology by President Clinton "for his commercialization of various technologies through the establishment and development of over one hundred technology-based companies that employ tens of thousands of people and export over one billion dollars worldwide" (U. S. Department of Commerce, 1993). George's comments regarding his selection for the Medal probably best captures what he valued most:

THE MEDAL OF TECHNOLOGY

Left to right: George Kozmetsy, President Clinton, Vice-president Gore

"As a first-generation American, I find it very humbling—indeed beyond my imagination—to be given this award. Whatever I have accomplished is the result of being an American plus great good fortune. I had wonderful parents. I have had a wife and children who were always with me, an excellent education, wise mentors, friends who inspired and supported me, exciting entrepreneurial ventures, and the opportunity to participate in cutting-edge programs and projects that link the academic, business, and government sectors for the general welfare. Mrs. Kozmetsky and I feel very strongly that individuals who have been so privileged ought to contribute to society. To be honored for simply being a responsible citizen is overwhelming, and I am deeply grateful." (Kozmetsky in U. S. Department of Commerce, 1993).

George's other awards and honors include: the 1987–1988 Dow Jones Award from the American Assembly of Collegiate Schools of Business for his outstanding contributions to management education, the 1988 Thomas Jefferson Award from the Technology Transfer Society for his work in advancing technology transfer; and induction into the Texas Business Hall of Fame for his business contributions to the State of Texas.

In 1989, he received the University of Washington Alumnus Summa Dignatus Award. He was Austin's 1992 Entrepreneur of the Year. He is the first recipient of the Entrepreneurial Leadership Award from the MIT enterprise Forum of Cambridge, Inc. The YWCA of Austin honored George among its first Mentors and Allies for his sensitivity and gender blindness that goes beyond the definition of mentor to promoter of women in leadership positions.

George always took his greatest pride in the success of the many students and colleagues that he worked with and mentored. Even after failing health required him to use a wheelchair (he suffered from amyotrophic lateral sclerosis, often referred to as Lou Gehrig's Disease), he continued to work with students and colleagues on his and their favorite projects like the Cross-Border Institute for Regional Development initiative with South Texas and Northern Mexico education, business, and government communities. Young scholars and entrepreneurs had a special appeal for George, and he was always available to advise the interested as long as they would come by his office anytime between 4:30 and 6:30 a.m.

George died on April 30, 2003. Reflecting on his father's impact, George's son Greg stated:

> I think my father's legacy, besides his family which always came first, is all the students he touched over the almost 50 years of teaching. He loved teaching; he loved the university; he loved meeting and talking to students. I really think that first and foremost, he would say he was an educator (Knight Rider 2003).

During George's memorial service, Larry Faulkner, President of UT-Austin, captured George in the following word portrait:

> George was a creative force of very rare power and quality, not only in this university but also in the business community worldwide. His institutional legacy here is extraordinary, and his influence will be felt for generations. And at a personal level, all who knew him will miss his generous spirit and remarkable imagination and vision (Knight Rider, 2003).

In a late-in-life conversation with the authors, George lamented the fact that "so many exciting things are happening. . . there are so many urgent needs that I won't be around to watch and help." In short, to George, life was a quest to better the world by learning everything he could about an area that he saw as having a great impact on mankind, generating hypotheses regarding how best to achieve useful goals, and then proceeding to test and implement these hypotheses. Life was an opportunity to conduct experiments, to learn more about the key elements of the world, and to accelerate the pace of change.

Such was the life of George Kozmetsky, always on the cutting edge of knowledge, always a fast ride with goals well beyond what anyone would normally dream, coupled with a pervasive excitement and energy of actually making a difference.

ACKNOWLEDGMENTS

The authors especially thank our following colleagues and friends of George Kozmetsky for their insights, revelations, and comments that provided important content for this chapter: Kenneth Walters, professor emeritus, University of Washington; Bill Cooper, professor emeritus, UT-Austin, one of the original six IC^2 fellows; Fred Phillips, associate provost for research, Alliant International University, and formerly research director at IC^2; and Tim Ruefli, professor, UT-Austin, the first associate director of IC^2.

REFERENCES

Butler J, Kozmetsky G (eds) (2004) Immigrant and minority entrepreneurship: the continuous rebirth of American communities. Praeger, Westport, CT

Callahan D, Forbes, Inc. (2002) Kindred spirits: Harvard Business School's Extraordinary Class of 1949 and How They Transformed American Business. Wiley, New York, NY

Charnes A, Cooper W (eds) (1984) Creative and innovative management: essays in honor of George Kozmetsky. Ballinger, Cambridge, MA

Cooper W (2007) George Kozmetsky's impact on management science and operations research. Tape recorded interview conducted by L. Secrest. The University of Texas at Austin (September 21)

Cooper W, Thore S, Gibson D, Phillips F (eds) (1997) Impact: how IC^2 research institute affects public policy and business practices. Quorum, Westport, CT

Gibson D (ed) (1991) Technology companies and global markets: programs, policies and strategies to accelerate innovation and entrepreneurship. Rowman and Littlefield, Savage, MD

Gibson D, Kozmetsky G, Smilor R (eds) (1992) Technopolis phenomenon: smart cities, fast systems, global networks. Rowman & Littlefield, Lanham, MD

Gibson D, Smilor R (eds) (1992) Technology transfer in consortia and strategic alliances. Rowman & Littlefield, Lanham, MD

Knight Rider/Tribune Business News (2003) Entrepreneur who gave Austin its tech wings dies at 85 (May 1)

Kozmetsky G (1950) Financial reports of labor unions. Division of Research, Graduate School of Business Administration, Harvard University, Boston, MA

Kozmetsky G (1984) Creative and innovative management: a new academic frontier. In: Charnes A, Cooper W (eds) Creative and innovative management: essays in honor of George Kozmetsky. Ballinger, Cambridge, MA, pp 3–25

Kozmetsky G (1985) Transformational management. Ballinger, Cambridge, MA

Kozmetsky G, Kircher P (1956) Electronic computers and management control. McGraw-Hill, New York, NY

Kozmetsky G, Thore S, Phillips F (1994) DEA of financial statements data: The U. S. computer industry. J Product Anal 5(3):229–248

Ronstadt R, Kozmetsky G (eds) (1999) Newer perspectives on the generation of knowledge. IC^2 Institute, The University of Texas at Austin. Austin, TX

Secrest L (2008) Personal communication

Simon H, Kozmetsky G, Guetzkow H, Kyndall G (1954) Centralization vs. decentralization in organizing the controller's department. Controllership Foundation, New York, NY

Smilor R, Kozmetsky G, Gibson D (eds) (1988) Creating the Technopolis: linking technology commercialization and economic development. Ballinger, Cambridge, MA

Thore S, Kozmetsky G, Ruefli T, Phillips F, Yue P (1995) DEA and the management of the product cycle: the U.S. computer industry. Comput Oper Res 23(4):341–356

U.S. Department of Commerce (1993) The National Medal of Technology Awarded by The President of the U.S. to George Kozmetsky. The U.S. Department of Commerce, Technology Administration. Washington, DC

Walters K (2003) George Kozmetsky: American innovator. A life at the intersection of technology and ideology. Univ Wash Bus (Fall):14–15

20

JAY WRIGHT FORRESTER

DAVID C. LANE AND
JOHN D. STERMAN

JAY WRIGHT FORRESTER'S distinguished career at the Massachusetts Institute of Technology (MIT) began with pioneering work in servomechanisms and digital computation. Drawing on that work, he then developed and founded the field of system dynamics. By integrating concepts of feedback control theory and digital computation, systems dynamics offered a new modeling approach for simulating the behavior of social systems. Such models provide insights into the behavior of complex systems and help the user to design effective policies to improve system performance. Jay established system dynamics as an academic discipline at MIT's Sloan School of Management. He led path-breaking research applying system dynamics to critical business and public policy problems, from high-tech start-ups to urban policy and global development. System dynamics is now one of the most widely used systems approaches in the world, with academics and practitioners on every continent pursuing work in diverse fields. His writings continue to inspire the field and he remains actively involved in its development.

Jay is a member of National Academy of Engineering and a fellow of the Institute of Electrical and Electronic Engineers (IEEE). His honors include: U.S. National Medal of Technology; Medal of Honor (IEEE); and the Pioneer Award, IEEE Aerospace and Electronic Systems Society. He was inducted into the International Federation of Operational Research Societies' (IFORS) Operational Research Hall of Fame, and has honorary degrees from nine universities.

A.A. Assad, S.I. Gass (eds.), *Profiles in Operations Research*, International Series in Operations Research & Management Science 147, DOI 10.1007/978-1-4419-6281-2_20,

FRONTIER YEARS AND BEYOND

Jay Wright Forrester was born on July 14, 1918 on a cattle ranch near Climax, Nebraska, to Ethel Pearl Wright Forrester (1886–1958) and Marmaduke (Duke) Montrose Forrester (1883–1975). Their daughter, Barbara Francis, was born in 1921. Both parents attended Hastings College, Nebraska. They were the original homesteaders of this land close to the American frontier. When they arrived in Nebraska around 1910, both worked as country schoolteachers. Jay was taught at home by his mother for his first 2 years of schooling. After that, he rode his horse one and a half miles to a one-room schoolhouse. There, for the first 2 years, he was taught by his father.

Jay developed an early interest in electricity, tinkering with doorbells, batteries, and telegraphs. He recalls that being raised on a Nebraska cattle ranch offered plenty of opportunities to get his hands dirty finding practical solutions to real problems such as building a wind-powered generator to provide the first electricity to the ranch (Forrester 1992). He was offered a scholarship to an agricultural college, but decided that the life bucolic was not for him and, instead, enrolled in the University of Nebraska to study electrical engineering.

After earning a bachelor's degree in electrical engineering in 1939, Jay moved to MIT. He worked as a research assistant with Gordon Brown, a pioneer in servomechanism theory and applications (Brown and Campbell 1948). During World War II, Jay worked on feedback-and servo-control systems for radar. For his master's thesis, he designed and built a servo to stabilize radar antennae on naval ships. In 1943, the prototype was installed on the aircraft carrier Lexington and Jay subsequently traveled to Pearl Harbor to ensure its continued functioning. Though a civilian, he volunteered to stay on board when the fleet was ordered to sea to make sure the servo (and thus the ship's radar) worked. During the mission, the Lexington participated in the retaking of the Marshall Islands and survived a torpedo strike. He received an S.M. degree in Electrical Engineering from MIT in 1945; his thesis was titled "Hydraulic Servomechanism Developments."

From 1940 to 1951, Jay was Associate Director of MIT's Servomechanism Laboratory and then Head of the Digital Computer Division in MIT's Lincoln Laboratory (1951–1956). He directed the ASCA project (Airplane Stability and Control Analyzer), aimed at developing flight simulators to test new aircraft designs. Originally envisioned as an analog computer, Jay realized that the real-time requirements of the planned simulator could not be met with analog components. Jay learned of digital computation through MIT alumnus Perry Crawford. He then visited the computing centers at Harvard and the University of Pennsylvania's Moore School of Electrical Engineering where the ENIAC was being built. There he met the Princeton mathematician John von Neumann, who was helping to design the Moore School's next generation computer, the EDVAC, and J. Presper Eckert, one of the developers of the ENIAC. These visits convinced Jay that the ASCA project would be based on digital computation, a bold decision given that all existing digital computers were far too slow and limited to meet the requirements of ASCA. As director (from 1951) of the MIT digital computer laboratory, Jay led the development of the Whirlwind computer, which was, for years, the only machine fast enough for real-time simulation of complex dynamical systems such as an aircraft.

ON THE LEXINGTON

The hydraulic servomechanism to stabilize radar antennae built by Forrester and Brown. The prototype was installed on the Lexington. This one is on display in the Forrester Conference Room at the MIT Sloan School of Management. Photo: John Sterman.

Whirlwind became the central element of the SAGE (Semi-Automatic Ground Environment) system and became the first computer produced in volume. SAGE was built to defend North America from Soviet bomber attack and consisted of a network of digital computers and long-distance communication systems that sent target-tracking information from radar stations to computers. The Whirlwind-based computers in each center processed the data and computed flight plans for interceptor aircraft and missiles, a demanding real-time application requiring high reliability (Jacobs 1986). With roughly 80,000 vacuum tubes in each of the approximately three-dozen SAGE centers, reliability was an immense technical challenge. Jay's legendary drive for quality and reliability led to design improvements and manufacturing standards yielding unprecedented results—when the last of the SAGE centers was decommissioned in 1983, the system-wide uptime over their roughly 25 years of service was 99.8%, making it probably the most reliable military command and control system ever implemented. Jay's colleagues and students during this period went on to major accomplishments. Robert Everett, Jay's second in command in the Whirlwind project, went on to lead the MITRE Corporation (originally, MIT Research and Engineering), created by the spinoff of the digital computer division of MIT's Lincoln Laboratory. Kenneth Olson, one of Jay's students, went on to found and lead the Digital Equipment Corporation.

Computer memory became a major bottleneck early in the development of Whirlwind. Memory cost $1 per digital bit per month. Electrostatic tubes provided some storage, but were expensive and unreliable. Mercury delay lines were tried, with the bits stored as sound waves traveling down a tube of mercury (Slater 1987). Jay even considered leasing a microwave relay line between Boston and Buffalo, New York, to store bits in a pulse train of electromagnetic radiation. The need for fast, high-capacity data storage spurred Jay to create coincident-current magnetic core memory in 1949 (Forrester 1951, 1953, U.S. Patent Office 1956). Core memory was cheap, stable, and reliable. It was the industry standard for decades (Evans 1983) and was the memory device that flew to the moon on the Apollo missions. For its invention, the IEEE awarded Jay its Medal of Honor in 1972. In 1979, he was made a member of the U.S. National Inventors' Hall of Fame, and, in 1989, along with Robert Everett, was awarded the Presidential Medal of Technology.

WHIRLWIND

Whirlwind: Jay (center, standing) and on his left Robert Everett at Whirlwind I test control in 1950. Used with the permission of The MITRE Corporation.

NATIONAL MEDAL OF TECHNOLOGY

In 1989, along with Robert Everett, Jay received the National Medal of Technology, the nation's highest award for technical achievement. Photo: George Bush Presidential Library and Museum.

CORE MEMORY

Coincident Coil Magnetic Core Memory: Jay holding a 64 × 64 core memory plane, 1954. Picture used with the permission of The MITRE Corporation.

As described in Slater (1987), Jay's achievements during the period of 1940–1955 were extraordinary (for Jay's perspective, see Forrester 2007a). Jay's experiences proved of great value in the next chapter of his scientific life—the development of his major work, the field of system dynamics and its application to critical problems in business and public policy. Naturally, his life also had personal dimensions of no less significance. It was early in this period that Gordon Brown introduced Jay to Susan Swett. They married on July 27, 1946 and went on to have three children. In 1952, they moved into a brown-shingled house in Concord, Massachusetts, their home until 2007.

The Emergence of System Dynamics

By the mid-1950s, Jay felt that "the pioneering days in computers were over" and, ever seeking new frontiers, was looking for new challenges (Forrester 1992, 343). His work with servomechanisms, digital computation, and SAGE had provided extensive experience in the management of complex organizations and large-scale high-technology projects. He relates a conversation with the then-president of MIT James Killian, who

> ... brought a group of visiting dignitaries to see us at the Lincoln Laboratory. While walking down the hall with Killian, he told me of the new management school that MIT was starting, and suggested that I might be interested. The Sloan School of Management had been founded in 1952 with a grant of 10 million dollars from [MIT alumnus] Alfred Sloan, the man who built the General Motors Corporation. The money was given on the expectation that a management school in a technical environment like MIT would probably develop differently from one in a liberal arts environment like Harvard, Columbia, or Chicago. Maybe better, but in any case different, and it was worth 10 million dollars to run the experiment (Forrester 2007a, 347).

Jay joined the Sloan faculty in 1956. He spent the first year considering what contribution digital computation and control theory might make to management. A 1956 memo to the faculty research seminar titled "Dynamic models of economic systems and industrial organizations" laid out his initial thinking, and became the first in a series of D-Memos (dynamic modeling memos). Jay organized an industrial dynamics group and he and its members began to log their models, reports, class assignments, papers, and musings in the D-Memo series, which continued through the 1990s, when computers and the

Internet made it unnecessary to keep such paper files. The D-Memos, now in the MIT archives, are a remarkable record of the evolution of a new field (most are available on a DVD distributed by the System Dynamics Society).

Jay's first dynamic model arose through chance conversations with executives at the General Electric Corporation (GE) (Forrester 1992, 2007a). GE managers were puzzled by large fluctuations in production, inventories, employment, and profit. These oscillations endured despite the managers' best efforts, and were attributed to outside forces, specifically business cycle fluctuations in incoming orders. By talking to the managers and observing how the different departments were run, Jay elicited an account of how individual managers, from the retail level, through distribution channels, to the factories, responded to the information locally available to them as they tried to control their piece of the organization. Rather than attributing the fluctuations to exogenous events, he saw the production and distribution of appliances as a system of interacting units. The managers in each link in what today is called a supply chain were responding in a locally rational fashion to the incentives and information they faced; for example, the need to provide good customer service while avoiding excessive inventories. The resulting changes in orders, production, hiring, and other decisions then fed back to alter inventories, backlogs, prices, and advertising, creating a system consisting of multiple feedback loops, just as a servomechanism consisted of a closed-loop control system. Managers at each link of the supply chain altered the orders they placed with suppliers to compensate for variations in orders and inventories, just as his antenna stabilization servo adjusted the position of the antenna to compensate for the pitch, roll, and yaw of the Lexington. Where, however, the servo damped out the variations in the environment, the feedback structure of the supply chain amplified them into persistent cyclical swings.

In building this first model, Jay retained several vital features of the situation, including an explicit stock and flow network for resources such as inventories and labor, the long-time delays between actions and outcomes such as shipping and production delays, and nonlinearities such as the impact of inventory on shipments and nonnegativity constraints on production. Retaining these features meant that the system was not analytically tractable. Simulation was required. Jay carried out the first simulation of this system by hand, calculating production, shipments, hiring, and other flows from inventory, work in progress, workforce, and other system states, then updating these stocks, week by simulated week. The results, recorded in a lab notebook, showed how the management policies of the firm generated robust oscillations even when demand was constant. Although inventories and backlogs are intended to absorb temporary fluctuations in orders so that costly production changes can be minimized, Jay found that the firm's own policies, sensible and rational from the perspective of the managers at each decision point, led to substantial amplification of perturbations in orders, and instability for the system as a whole, a phenomenon now known as the bullwhip effect.

Jay soon moved to computer simulations of this problem. Further work showed how feedback control theory could be adapted to understand puzzling, counterintuitive behavior in a range of management and human systems (Forrester 1956). He called the approach industrial dynamics (Forrester 1958).

INDUSTRIAL DYNAMICS

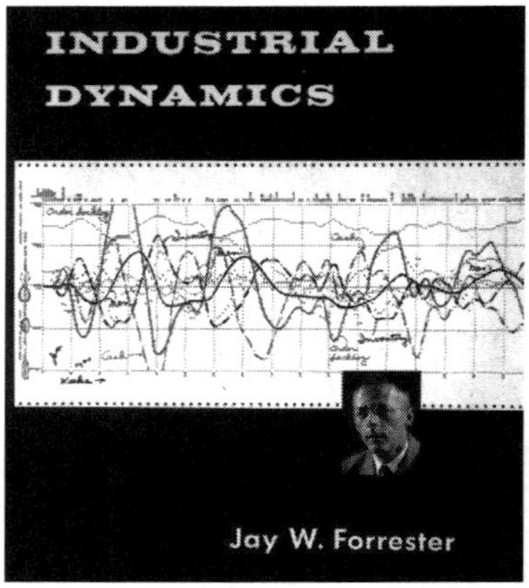

Jay recruited talented MIT undergraduates as research assistants—Willard Fey, Jack Pugh, Edward Roberts, and others—and spent the next years developing applications and laying out a vision of the contribution that they could make to management (Forrester 1959, 1960). These ideas led to his classic book, *Industrial Dynamics* (Forrester 1961).

As described in Richardson (1991), systems concepts including feedback control, mutual causality, deviation-correcting, and deviation-amplifying processes were in the air during the middle of the last century. Jay's unique contribution, detailed in *Industrial Dynamics*, was to develop ideas about systems, feedback, control, and dynamics that were previously restricted to engineering and physical contexts into a rigorous yet practical method for enterprise design, a method designed to "find management policies and organizational structures that lead to greater success" (Forrester 1961, 449). Richardson (1991) documents the connections and parallels between Jay and other pioneers of cybernetics, systems theory, and operations research (OR). Many of the principles Jay articulated to guide effective modeling and policy design for complex systems had no precedent in the work of others in systems theory and OR, and were revolutionary in their implications—on principles for modeling (Forrester 1960, 1968b, c), on the design of corporations (Forrester 1965), and on the counterintuitive behavior of social systems (Forrester 1971b).

Industrial Dynamics contributed a set of four principles for effective modeling of complex systems: counterintuitive system behavior is driven by system structure, structure involves nonlinear relationships, computer simulation is necessary to explore behavior, and that applying the previous three ideas provides a rigorous yet pragmatic way for managers to improve the design of organizations.

Jay's first principle was that the puzzling (counterintuitive) behavior of companies, economies, indeed, all systems, whether physical, physiological, economic, or social, emerged endogenously from their structure. That structure includes physical elements such as stocks of inventory, labor, capital, order backlogs; information systems that determined what information was available to each decision maker and the extent to which that information is delayed, smoothed, aggregated, biased, or otherwise corrupted by processes of measurement, reporting, and subjective adjustment; and, most importantly, by the policies and decision processes of the actors at each decision point in the system. He stressed the importance of discovering and representing the mental model of the decision maker. Similar to, but largely independent of, the view of Herbert Simon and his colleagues (Cyert and March 1963, Simon 1957), Jay emphasized that effective models of human systems must capture the bounded rationality of the agents' decision processes. Decision

> POLICIES AND DECISIONS
>
> "... understanding of decision making has been greatly handicapped by the presumption that it is a more subtle and more sophisticated process than it actually is.... It is my feeling that in a dynamic information-feedback system the human decision maker is usually using a great deal less than the total amount of information available to him. Furthermore, the information available to him is a great deal less than that commonly presumed. In general, his actions with respect to any given decision stream will be almost entirely conditioned by less than ten information inputs" (Forrester 1961, 100).

making should be represented in models as it is, warts-and-all, and not presumed to be the fully rational optimizing behavior of the mythical homo economicus.

The interaction of the physical structure, information flows, and decision processes creates a network of feedback loops that generates the dynamics of the system. People use information about system states such as inventory, labor, order backlogs, and the company's reputation for service quality to make decisions; those decisions then condition production, shipments, hiring, orders, and other rates of flow that alter the system states. These processes form closed loops, some of which constitute control processes (negative feedbacks) such as the loop whereby excessive inventories led firms to cut production below shipments, thus lowering inventory levels. Some form self-reinforcing processes (positive feedbacks), such as the loop whereby customers react to an increase in supplier lead times by increasing their safety stock targets and ordering farther ahead, actions that further deplete supplier inventories and swell order backlogs, causing still longer delivery delays, a process known today as phantom ordering and one that played a major role in the tech bubble of the late 1990s (Sterman 2000).

The second principle is that nonlinearity plays a central role in the dynamics of complex systems. Jay knew from his experience with electromechanical systems that nonlinearities decisively conditioned their structure and behavior. Jay saw that economic and social systems were also intrinsically nonlinear and could not be adequately approximated with linear methods. Structurally, nonlinearities abound in the real world. Product shipments are generally determined by orders, until inventory is depleted, at which point shipments necessarily fall to zero. Production increases with work hours up to a point, then peaks and falls as fatigue cuts productivity, boosts errors, and triggers accidents. Behaviorally, linear systems cannot exhibit locally unstable behavior and global stability, cannot exhibit bifurcations, endogenous shifts in their modes of behavior, and cannot evolve. Yet, with few exceptions, such as the Lotka-Volterra predator-prey model (Murray 1989), OR, economics, and dynamical theory were dominated at that time by linear models. Linear theory dominated because it was analytically tractable. Even after the computer became widely available, nonlinearity was slow to penetrate these disciplines. Jay designed his modeling method from the start to incorporate nonlinearities easily and intuitively. Subsequent developments have shown the prescience of Jay's focus on nonlinearity (Mosekilde 1996; Strogatz 1994). Physical and social scientists now recognize the intrinsically nonlinear character of physical, biological, and socio-economic systems. Though the terms chaos and self-organization were unknown at the time, Jay's early models are among the first models of human behavior ever developed to exhibit phenomena such as deterministic chaos, self-organization, and increasing returns.

Jay's third principle, that simulation was needed to explore system behavior, led to the development of a practical computer simulation methodology for business, economic, and social systems. He and his first students created tools for simulating complex human systems that included diagramming conventions and general purpose computer simulation languages. The first compiler for such simulations, developed by Richard Bennett, was dubbed SIMPLE (Simulation of Industrial Management Problems with Lots of Equations). SIMPLE was followed by DYNAMO (DYNAMic MOdeling), which remained the standard for system dynamics modeling for several decades. Subsequently, the personal computers and graphical user interfaces triggered rapid growth in the number of software packages for dynamic modeling, including iThink, Powersim, Vensim, and many others.

The fourth and perhaps most radical of Jay's innovations was his focus on system dynamics as both a rigorous tool to develop scientific knowledge and a practical tool to improve the performance of organizations. He consistently argued that senior managers should build models to understand and improve their organizations. Jay believed that a manager's role is not merely captain of the ship, but designer of the ship (Keough and Doman 1992). This corporate designer role was an innovative approach to both modeling and management, one he has advocated throughout his career. To carry out rigorous scientific modeling and to develop models that made a difference required engagement with the mental models of managers and other stakeholders; while understanding might develop without the active participation of key decision makers, implementation of new policies could not. There are practical and theoretical links here with recent developments in OR (Lane 1994, 1999). Jay's focus on implementation and the need to engage the decision makers in the modeling process underlies subsequent work on organizational learning (Forrester 1971c, Senge 1990) and the development of protocols for group model building (Lane 1992, Richmond 1997, Vennix 1996, Vennix et al. 1997).

MANAGEMENT LABORATORIES FOR ENTERPRISE DESIGN

"Industrial dynamics is the investigation of the information-feedback character of industrial systems and the use of models for the design of improved organizational form and guiding policy.

"It is only through costly experience and errors that managers have been able to develop effective intuitive judgment. We need to expedite this learning process. Other professions in similar circumstances have turned to laboratory experiments. ... Controlled laboratory experiments on industrial and economic situations are now possible with computers to do the work. ... The manager, like the engineer, can now have a laboratory in which he can learn quickly and at low cost the answers that would seldom be obtainable from trials in real organizations.

"Industrial dynamics is an approach that should help in important top-management problems. ... The attitude must be one of enterprise design. ... The goal should be to find management policies and organizational structures that lead to greater success" (Forrester 1961, 13, 43, 449).

THE MODEL VERSUS A MODELING PROCESS

"In any real-life applications of modeling to the generation of policy ... the models are always in a continuous state of evolution. Each question, each reaction, each new input of information, and each difficulty in explaining the model leads to modification, clarification, and extension.

"I believe we are proposing the 'Process' of modeling rather than particular frozen and final models. The difference in viewpoint becomes especially important as we move into the implementation phase. It seems to me that the average person will be greatly concerned if he feels that the future and alternatives are being frozen once and for all into a particular model. Instead, we are suggesting that models will help to clarify our processes of thought: they will help to make explicit the assumptions we are already making and they will show the consequences of the assumptions. But as our understanding, our assumptions, and our goals change, so can the models.

"Rather than stressing the single-model concept, it appears that we should stress the process of modeling as a continuing companion to, and tool, for, the improvement of judgment and human decision making" (Forrester 1971c).

DEVELOPING A DISCIPLINE

Throughout the 1960s, Jay and his students applied system dynamics to a growing range of problems through teaching, research, consulting, and practical management applications (Roberts 1978a, Richardson 1996, Sterman 2007). Jay, as a member of the original DEC board of directors, built a series of models examining the growth of high-technology start-ups and used them to inform his position on key issues facing the company (Forrester 1964, 1968a, 1975). DEC became the second largest computer firm in the world; much of its early success can be attributed to the policies that Jay—informed by his models—advocated as a member of the board. The corporate growth model included production capacity, inventory, shipments, financial management, product development, and the other key tangible assets and processes. A key aspect of the model, however, was its portrayal of intangible elements such as the knowledge, skills, and attitudes of managers, engineers, and salespeople; standards for product quality and the pressures altering them; organizational routines for pricing and resource allocation; and the ability of top management to project its goals throughout the organization (Forrester 2007a). Such intangibles are now known as dynamic capabilities and constitute an active focus of research in strategy (Teece et al. 1997). The story of DEC's later demise (long after Jay left the board) is a fascinating tale consistent with many of the dynamics Jay described in his early corporate growth models (Schein 2003).

Jay's co-workers went on to contribute to the spread of these ideas. Will Fey taught industrial dynamics as a professor at Georgia Institute of Technology. Ed Roberts became the David Sarnoff Professor at the MIT Sloan School of Management, where he contributed important work applying system dynamics to the management of technology, health care, and public policy (Roberts 1978a, b). He and Jack Pugh founded Pugh-Roberts Associates, the first of many consulting firms to apply industrial dynamics.

URBAN DYNAMICS

Toward the end of the 1960s, Jay and his students began to address public policy issues and the more general term system dynamics replaced industrial dynamics. *Urban Dynamics* (Forrester 1969) was a study of the processes underlying the development, stagnation, decline, and recovery of cities. The project began when the mayor of Boston, John F. Collins, chose not to run for a third term and became a visiting professor at MIT with an office next to Jay's. In the former mayor's struggles with urban problems, Jay recognized the same policy resistance and unintended consequences he had so often observed in corporate contexts. He suggested to Collins that they develop a systems dynamics model of the problem situation. Characteristically, the model was developed not merely by reference to theory, but in conjunction with Collins and others with first-hand experience managing large cities. The model endogenously generated the dynamics of urban growth and stagnation over several hundred years. As the simulated city evolved, population growth, and the crowding and aging of the housing stock and industrial base, gradually shifted the city from an engine of upward mobility to a poverty trap. The model explained why so many policies implemented during the 1960s and 1970s to alleviate urban poverty failed, and in some cases made the problems of the cities and their citizens, particularly the poor, worse. Subsequent events have shown this analysis to be largely correct, but it was enormously controversial at the time. The hostile reactions to the work, as well as examples in which support arose from unexpected quarters, are discussed in Forrester (1992). The key seemed to be spending sufficient time with the model to understand its assumptions and the source of its dynamics and policy insights. The account of Jay's testimony to a U.S. House of Representatives sub-committee on urban growth gives an idea of how he went about explaining his ideas; an edited version was published as "Counterintuitive behavior of social systems" (Forrester 1971b).

COUNTERINTUITIVE TESTIMONY TO CONGRESS

"The nation exhibits a growing sense of futility as it repeatedly attacks deficiencies in our social system while the symptoms continue to worsen. Legislation is debated and passed with great promise and hope. But many programs prove to be ineffective. Results often seem unrelated to those expected when the programs were planned. At times programs cause exactly the reverse of desired results.

"It is now possible to explain how such contrary results can happen. There are fundamental reasons why people misjudge the behavior of social systems. There are orderly processes at work in the creation of human judgment and intuition that frequently lead people to wrong decisions when faced with complex and highly interacting systems.

"People would never attempt to send a space ship to the moon without first testing the equipment by constructing prototype models and by computer simulation of the anticipated space trajectories. No company would put a new kind of household appliance or electronic computer into production without first making laboratory tests. Such models and laboratory tests do not guarantee against failure, but they do identify many weaknesses which can then be corrected before they cause full-scale disasters.

"Our social systems are far more complex and harder to understand than our technological systems. Why, then, do we not use the same approach of making models of social systems and conducting laboratory experiments on those models before we try new laws and government programs in real life? The answer is often stated that our knowledge of social systems is insufficient for constructing useful models. But what justification can there be for the apparent assumption that we do not know enough to construct models but believe we do know enough to directly design new social systems by passing laws and starting new social programs?" (Forrester 1971b, 52–53).

World Dynamics

In 1970, Jay began work with the Club of Rome to apply system dynamics to perhaps the most important issues of social policy: the dynamics of global development. ["The Club of Rome is independent of any political, ideological and religious interests. Its essential mission is 'to act as a global catalyst for change through the identification and analysis of the crucial problems facing humanity and the communication of such problems to the most important public and private decision makers as well as to the general public'" (Club of Rome 2009).] Jay developed a model capturing feedbacks among population, natural resources, pollution, agricultural and industrial production, capital investment, and quality of life.

The resulting book, *World Dynamics* (Forrester 1971a), posed sharp questions about the relationship between growth and quality of life, generating heated discussion in popular and scholarly forums worldwide. *World Dynamics* led to a more detailed modeling study directed by Dennis Meadows, who had just received his Ph.D. under Jay and had joined the MIT faculty. Described in *The Limits to Growth* (Meadows et al. 1972), that study triggered worldwide controversy and debate (Forrester et al. 1974). More important, because the world models were fully documented and easily replicated, they led to a wide range of critiques and extensions (Meadows et al. 1982, 1992, 2004).

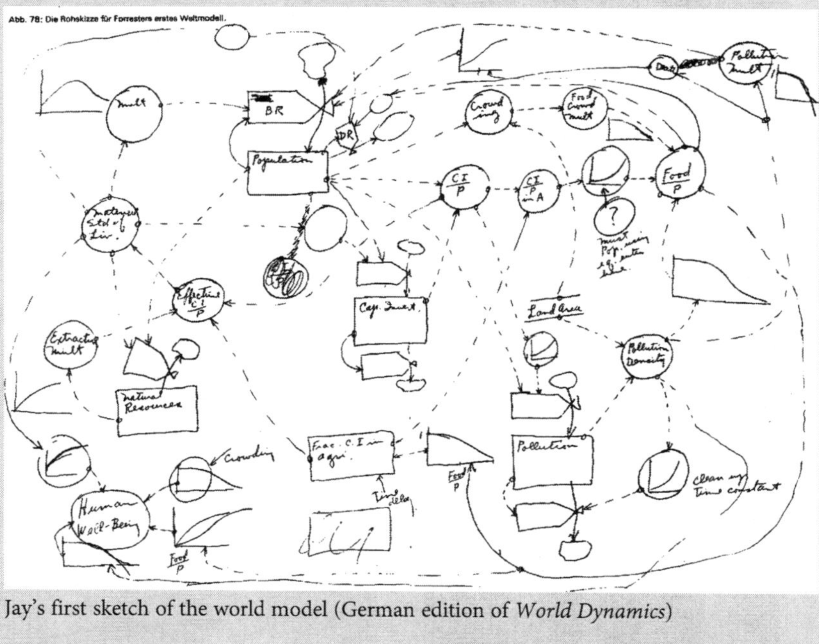

Jay's first sketch of the world model (German edition of *World Dynamics*)

With this stream of work, Jay launched the field of global modeling (de Steiguer 1997), an activity that also provoked storms of criticism. His willingness to endure stems from the intellectual courage and integrity to face the consequences of his analysis, no matter how unpalatable or unpopular they may be. Jay did not shrink from questioning conventional wisdom, from the role of churches in influencing population growth (Forrester 1973) to the belief that unending economic growth is both desirable and necessary (Forrester 1971a, b).

Consequently, Jay drew the wrath of left and right, conservatives, and progressives, often at the same time. His work has been instrumental in shaping public thinking on the interactions among the environment, development, pollution, and natural resources. The economist, Paul Ormerod, commenting on the contribution of *World Dynamics* and *Limits to Growth*, observed that its "true and lasting significance ... was the development of a fundamentally different approach to understanding the workings of the economy to that of orthodox economics" (Ormerod 1994, 36).

Developments in more than 35 years since the publication of *World Dynamics* have only underscored the importance of Jay's insights. Perpetual growth of population and material production on a finite planet is impossible. Long-time delays in the response of the economy and technology to resource scarcity and environmental degradation cause human activity to overshoot the carrying capacity of the planet. Research, unavailable when Jay formulated his world model, now shows clearly that humanity has already overshot the global carrying capacity and is rapidly consuming and degrading the natural capital stocks upon which our civilization depends, from groundwater to soils to fish stocks to the climate (Meadows et al. 2004, Wackernagel et al. 2002). An article (*Wall Street Journal* 2008), headlined, "New limits to growth revive Malthusian fears," observed, "the resource constraints foreseen by the Club of Rome are more evident today than at any time since the 1972 publication of the think tank's famous book, *The Limits of* [sic] *Growth*." Yet, Jay's most important insight in *World Dynamics* is not about how much oil remains in the ground, how much CO_2 we can dump into the atmosphere, or the potential for technology to find alternative energy sources or reduce pollution. It is that there is no purely technical solution to the challenge of creating a sustainable society. Technological innovation, market forces, and government policies are all aimed at ameliorating the symptoms of stress—pushing back the limits to growth by finding more energy, reducing greenhouse gas emissions, irrigating marginal lands, and designing new cultivars to boost food production, thus allowing growth to continue until another limit is reached. In a series of e-mail posts to the system dynamics community discussion list (Forrester 2008), Jay stated "obvious and self-evident courses of attacking symptoms rather than underlying causes will be futile. ... [T]reating one symptom can unleash a different overwhelming reaction." "[G]rowing population and industrialization will overwhelm the short-term efforts if we do not restrain these forces that are exceeding the carrying capacity of the earth."

To the Present

System dynamics began to coalesce into an academic field in the 1970s. Programs were started at universities in the U.S. and around the world. Conferences were organized and textbooks written. The System Dynamics Society was created in 1983, with Jay its first president. A dedicated journal was created, the modern form of which, *The System Dynamics Review*, appeared in 1985. In the following year, IBM's Thomas Watson, Jr. endowed the Jay W. Forrester Chair in Management at MIT. In his book, *The Fifth Discipline*, Peter Senge (1990) explored the relationship between system dynamics and organizational learning, attracting a new generation of managerial interest. The MIT System Dynamics Group continues research into a wide range of complex systems issues, from organizational change to climate change. System dynamics is one of the most popular electives at the MIT Sloan School of Management, attracting over 400 students per year (compared to an MBA program of about 375 per year). Jay's students and those he inspired, including the authors of this profile, went on to found or lead academic programs in system dynamics around the world (Sterman 2007).

Over the years, Jay remained active and began a large modeling study of economic dynamics, which integrated endogenous accounts of business cycles, inflation and stagflation, the growth of government, and the great waves of economic expansion and depression (Forrester, 1997, 1979, 1980, Forrester et al. 1976, Sterman 1985).

Jay formally retired from the MIT Sloan School in 1989, an event which he said "has had no effect whatsoever on my work" (Forrester 1997). Ever focused on the high leverage points to foster enduring change, he has for some years devoted most of his time to catalyzing the education of young people in the principles of systems (Forrester 1990, 1993). Interest in the education of young people and how they could learn systems thinking is a long-established area of application for system dynamics (Roberts 1978b). A new wave of experiments to develop the systems thinking and modeling capabilities of young people began in the late 1980s when Gordon Brown (then retired) introduced Frank Draper, a middle school teacher in Tucson, Arizona, to system dynamics. The enthusiastic response of Draper and his students soon spread; Jay himself was involved in the resulting K-12 project, which developed new teaching materials and introduced dynamic modeling into schools across the U.S. (Creative Learning Exchange 2009).

Consistent with his early focus on engaging managers in the modeling process, Jay not only believes young people should and can learn system dynamics and modeling, but also calls for a revolution in pedagogy as well. He believes that effective education requires learner-directed learning in which teachers are not the source of answers, but guides and coaches who help learners develop the inquiry skills they will need to become systems citizens (Creative Learning Exchange 2009). The theory of political and social change these beliefs represent is fundamentally optimistic, hopeful, and empowering. It is a view that, if begun early enough, everyone can gain an appreciation for the complex dynamics of natural and human systems, and then use that insight to design policies to create a better world.

Legacy and the Next Frontier

In person, Jay is quiet, imposingly tall, and faultlessly courteous. He speaks slowly and confidently, producing analyses of a complexity seldom found in conversation. He is direct and unambiguous with both praise and criticism. He is also often hospitable and convivial, happy to enjoy a joke and quick to share humorous stories himself.

He speaks of his parents, Gordon Brown, and his wife Susan as those to whom he feels most indebted. His discharging of this debt has produced work the legacy of which is immense.

Along with other pioneers of computer science, Jay's innovations in hardware, software, and computer simulation ushered in the digital age. Simulation is now used routinely throughout the natural and social sciences, hailed as a third branch of science, standing alongside theory and experiment as a unique and vital method to advance human knowledge (Pool 1992).

The field of system dynamics is healthy and growing. System dynamics is increasingly used in corporations, government, and other organizations. It is taught in a growing number of universities and schools. It is applied to issues from organizational change to climate change, from physiology to fiscal policy. On a lighter note, *Urban Dynamics* inspired video-game designer Will Wright to create SimCity (Seabrook 2006). Only a few years ago, simulation was difficult, expensive, and scarce. Today children in elementary school routinely create and manage simulated worlds of stunning complexity through interactive computer games. Of course, system dynamics is much more than a method for computer simulation, more than mathematical models grounded in control theory and nonlinear dynamics. It is also a practical tool for policymakers by which they can help to solve important problems. It is qualitative and quantitative, hard and soft, a theoretical discipline and a pragmatic approach for group modeling and policy design. Key concepts of system dynamics, including feedback, counterintuitive behavior, limits to growth, nonlinearity, tipping points, and many others are now integrated into the discourse of management, social theory, and everyday life. Discussions of critical public policy issues routinely refer to unintended consequences and policy resistance. Scientists, policymakers, and the media discuss the many positive feedbacks that can cause runaway climate change and debate whether we have passed the tipping point leading to irreversible melting of polar ice sheets.

Yet, Jay ever questing, ever focused on the important problems, is not satisfied. Speaking at the 2007 International System Dynamics Conference celebrating the 50th anniversary of the founding of the field, Jay, rather than reviewing the achievements of a half-century, challenged the field to move boldly into the next frontier, to tackle the most important problems no matter their difficulty:

> The first 50 years of system dynamics have established an introduction to the field. We have shown the importance of achieving a better understanding of complex systems in nature and human affairs. Now, the field is on a plateau ready to launch the next great thrust forward. . . . We are now at about the same state of advancement that engineering was when MIT first opened its doors in 1865. . . . System dynamics started 50 years ago with academic programs that focused on the outside world with emphasis on major issues outside of academia. However, the pressures inherent in academic institutions are driving our field back into academic journals and away from the public that we should be serving. . . . System dynamicists must go behind the symptoms of

> trouble and identify the basic causes. At first, such arguments will be met with disbelief, scorn, and ridicule. To prevail, the battle must be sustained until public understanding begins to change" (Forrester 2007b, 359–360, 370).

No one should be surprised. From the Sand Hills of Nebraska to the MIT servomechanism laboratory, from the Marshall Islands to the dawn of the computer age, from *Industrial Dynamics* to *World Dynamics*, from corporate boardrooms to elementary school classrooms, Jay Wright Forrester has lived his entire life on the frontier.

ON THE FRONTIER

Jay Wright Forrester, 1940 © JWF

ON HIS WAY

Jay at a celebration of the 100th anniversary of the MIT Electrical Engineering Department in 2003. He stands on a Segway, a nonlinear dynamic system of sensors, servos, and real-time digital feedback controls, directly descended from innovations Jay helped pioneer. Photo: Dan Bricklin.

HONORS AND AWARDS

Jay is a member of National Academy of Engineering, a fellow of the IEEE, the American Academy of Arts and Sciences, Academy of Management, and the Royal Society of Arts (London). His honors include: Medal of Honor, IEEE; Systems, Man, and Cybernetics Award for Outstanding Accomplishment, IEEE; Howard N. Potts Medal, The Franklin Institute; Inventor of the Year, George Washington University; U.S. National Medal of Technology (1989); Pioneer Award, IEEE Aerospace and Electronic Systems Society; Valdemar Poulsen Gold Medal, Danish Academy of Technical Sciences. He is a member of the IFORS' Operational Research Hall of Fame (Lane 2006). Jay has received honorary Doctorate of Engineering degrees from the University of Nebraska, Newark College of Engineering, and the University of Notre Dame; and honorary Doctorate of Science degrees from Boston University and Union College. He also received honorary doctorate degrees from the University of Mannheim (Political Science), State University of New York (Humane Letters), University of Bergen (Dr. Philosophy), and the Universidad de Sevilla (Honoris Causa).

REFERENCES

Brown GS, Campbell DP (1948) Principles of servomechanisms. Wiley, New York, NY

Club of Rome (2009) www.clubofrome.org. Accessed 19 Feb 2009

Creative Learning Exchange (2009) http://clexchange.org. Accessed 19 Feb 2009

Cyert R, March J (1963) A behavioral theory of the firm. Prentice Hall, Englewood Cliffs, NJ. (2nd edn, 1992, Blackwell, Cambridge, MA)

de Steiguer JE (1997) The age of environmentalism. McGraw-Hill, London

Evans C (1983) Conversation: Jay W. Forrester. Ann Hist Comp 5(3):297–301

Forrester JW (1951) Digital information storage in three dimensions using magnetic cores. J Appl Phys 22(1):44–48

Forrester JW (1953) Coincident-current magnetic computer memory developments at M.I.T. In: Chu JC (ed) Argonne national laboratory computer symposium, August 3–5, 1953. Argonne National Laboratory, Lemont, IL, pp 150–158

Forrester JW (1956) Dynamic models of economic systems and industrial organizations. Note to the Faculty Research Seminar, MIT System Dynamics Group D-Memo 0 (5 November). (Reprinted as Forrester, J. W. 2003). Syst Dyn Rev 19(4):331–345

Forrester JW (1958) Industrial dynamics: a major breakthrough for decision makers. Harv Bus Rev 36(4):37–66

Forrester JW (1959) Advertising: a problem in industrial dynamics. Harv Bus Rev 37(2):100–110

Forrester JW (1960) The impact of feedback control concepts on the management sciences (in Forrester 1975, 45–60)

Forrester JW (1961) Industrial dynamics. MIT Press, Cambridge, MA

Forrester JW (1964) Modeling the dynamic processes of corporate growth. In: Proceedings of the IBM Scientific Computing Symposium on Simulation Models and Gaming, York Town Heights, NY, pp 23–42

Forrester JW (1965) A new corporate design. Ind Manage Rev (now the Sloan Manage Rev) 7(1):5–17

Forrester JW (1968a) Market growth as influenced by capital investment. Ind Manage Rev 9(2):83–105

Forrester JW (1968b) Industrial dynamics – after the first decade. Manage Sci 14(7):398–415

Forrester JW (1968c) Principles of systems. MIT Press, Cambridge, MA

Forrester JW (1969) Urban dynamics. MIT Press, Cambridge, MA

Forrester JW (1971a) World dynamics. Pegasus Communications, Waltham, MA

Forrester JW (1971b) Counterintuitive behavior of social systems. Technol Revi 73(3):52–68

Forrester JW (1971c) "The" model versus a modeling "process." Dept. memo D1621-1, system dynamics group, MIT. (Reprinted 1985). Syst Dyn Rev 1(1):133–134

Forrester JW (1973) Churches at the transition between growth and world equilibrium. In: Meadows DL, Meadows DH (eds) Towards global equilibrium. Wright-Allen Press, Cambridge, MA, pp 337–353

Forrester JW (1975) Collected papers of Jay W. Forrester. Wright-Allen Press, Cambridge, MA

Forrester JW (1977) Growth cycles. Dè Economist (Q Rev R Netherlands Econ Assoc) 125(4):525–543

Forrester JW (1979) An alternative approach to economic policy: macrobehavior from microstructure. In: Kamrany NM, Day RH (eds) Economic issues of the eighties. Johns Hopkins University Press, Baltimore, MD, pp 80–108

Forrester JW (1980) Information sources for modeling the national economy. J Am Stat Assoc 75(371):555–566

Forrester JW (1990) System dynamics – adding structure and relevance to pre-college education. In: Manning KR (ed) Shaping the future. MIT Press, Cambridge, MA, pp 118–131

Forrester JW (1992) From the ranch to system dynamics. In: Bedeian AG (ed) Management Laureates: a collection of autobiographical essays, vol 1. JAI Press, Greenwich, CT, pp 335–370

Forrester JW (1993) System dynamics as an organizing framework for pre-college education. Syst Dyn Rev 9(2):183–194

Forrester JW (1997) Personal communication

Forrester JW (2007a) A recollection of the history and reflections for the future at the golden anniversary of the field. Syst Dyn Rev 23(2–3):345–358

Forrester JW (2007b) System dynamics—the next fifty years. Syst Dyn Rev 23(2–3):359–370

Forrester JW (2008)Posting to online System Dynamics Mailing List: www.vensim.com/sdmail/sdmail.html, 7 April, SD 6994, and 14 April, SD 7034. Accessed 2 Oct 2010

Forrester JW, Low GW, Mass NJ (1974) The debate on world dynamics: a response to Nordhaus. Policy Sci 5(2):169–190

Forrester JW, Mass NJ, Ryan CJ (1976) The system dynamics national model: understanding socio-economic behavior and policy alternatives. Technol Forecast Soc Change 9(1–2):51–68

Jacobs JF (1986) The SAGE Air defense system: a personal history. MITRE Corporation, Cambridge, MA

Keough M, Doman A (1992) The CEO as organization designer: an interview with Professor Jay W. Forrester, the founder of system dynamics. McKinsey Q (2)3–30

Lane DC (1992) Modeling as learning: a consultancy methodology for enhancing learning in management teams. Eur J Oper Res 59(1):64–84

Lane DC (1994) With a little help from our friends: how system dynamics and "soft" OR can learn from each other. Syst Dyn Rev 10(2–3):101–134

Lane DC (1999) Social theory and system dynamics practice. Eur J Oper Res 113(3):501–527

Lane DC (2006) IFOR's operational research hall of fame: Jay Forrester. Int Trans Oper Res 13(5):483–492

Meadows DH, Meadows DL, Randers J (1992) Beyond the limits. Chelsea Green, Post Mills, VT

Meadows DH, Meadows DL, Randers J (2004) Limits to growth: the thirty-year update. Chelsea-Green, White-River Junction, VT

Meadows DH, Meadows DL, Randers J, Behrens W (1972) The limits to growth. Universe Books, New York, NY

Meadows DH, Richardson J, Bruckmann G (1982) Groping in the dark. Wiley, Chichester

Mosekilde E (1996) Topics in nonlinear dynamics: applications to physics, biology and economic systems. World Scientific, Singapore

Murray JD (1989) Mathematical biology. Springer, Berlin

Ormerod P (1994) The death of economics. Faber & Faber, London

Pool R (1992) The third branch of science debuts. Science 256 (April 3):44–47

Richardson G (1991) Feedback thought in social science and systems theory. University of Pennsylvania Press, Philadelphia, PA

Richardson G (1996) Modelling for management, 2 vols. Dartmouth, Aldershot

Richmond B (1997) The strategic forum: aligning objectives, strategy and process. Syst Dyn Rev 13(3):131–148

Roberts E (ed) (1978a) Managerial applications of system dynamics. Pegasus Communications, Waltham, MA

Roberts N (1978b) Teaching dynamics feedback systems thinking: an elementary view. Manage Sci 24(8):836–843

Seabrook J (2006) The game master. The New Yorker. November 6. www.newyorker.com/archive/2006/11/06/061106fa_fact. Accessed 14 Mar 2008

Schein E (2003) DEC is dead, long live DEC: the lasting legacy of the digital equipment corporation. Berrett-Koehler, San Francisco, CA

Senge PM (1990) The fifth discipline: the art and practice of the learning organization. Doubleday/Currency, New York, NY

Simon H (1957) Administrative behavior; a study of decision-making processes in administrative organizations, 2nd edn. Macmillan, New York, NY

Slater R (1987) Portraits in silicon. MIT Press, Cambridge, MA

Sterman JD (1985) A behavioral model of the economic long wave. J Econ Behav Organ 6(1):17–53

Sterman JD (2000) Business dynamics: systems thinking and modeling for a complex world. Irwin/McGraw-Hill, New York, NY

Sterman JD (ed)(2007) Exploring the next frontier: system dynamics at 50. Special Iss Syst Dyn Rev 23(2–3)

Strogatz S (1994) Nonlinear dynamics and chaos. Addison-Wesley, Reading, MA

Teece D, Pisano G, Shuen A (1997) Dynamic capabilities and strategic management. Strateg Manage J 18(7):509–533

U.S. Patent Office (1956) Multicoordinate digital information storage device. U.S. Patent 2,736,880. U.S. Patent Office, Washington, DC

Vennix JAM (1996) Group model-building: facilitating team learning using system dynamics. Wiley, Chichester

Vennix J, Richardson G, Andersen D (eds) (1997) Group model building. Syst Dyn Rev (special issue) 13(2)

Wackernagel M, Schulz NB, Deumling D, Linares AC, Jenkins M, Kapos V, Monfreda C, Loh J, Myers N, Norgaard R, Randers J (2002) Tracking the ecological overshoot of the human economy. Proc Natl Acad Sci 99(14):9266–9271

Wall Street Journal (2008) New limits to growth revive Malthusian fears (24 March 2008)

21

Russell Lincoln Ackoff

Maurice W. Kirby and
Jonathan Rosenhead

Russell Lincoln Ackoff had a distinguished and influential career in two disciplines and in (at least) two continents. With his mentor, colleague, and friend, West Churchman, he established the path-breaking Operations Research Group at Case Institute of Technology. In 1957, they were joint authors (with Leonard Arnoff) of the famous textbook *Introduction to Operations Research*, the first integrated operations research (OR) text. A founding member of the Operations Research Society of America (ORSA), Russ served as its president (1956) and was the recipient of ORSA's Kimball Medal. He was awarded the Silver Medal of the U.K. Operational Research Society (ORS) and was inducted into the International Federation of Operational Research Societies' (IFORS) Hall of Fame.

Recognized as one of the pioneering developers of OR, Russ gradually became disenchanted with the discipline he had helped to form. His outspoken criticisms of what he perceived to be OR's narrow focus, and his radical position on what was needed for a more socially relevant practice, received scant support from the OR profession in the U.S. Elsewhere, especially in the U.K., and within the worldwide systems movement, his perspective has remained hugely influential and led to changes in the practice of OR.

A.A. Assad, S.I. Gass (eds.), *Profiles in Operations Research*, International Series in Operations Research & Management Science 147, DOI 10.1007/978-1-4419-6281-2_21,

EARLY YEARS: PHILADELPHIA AND THE DEPRESSION

Russ grew up in Philadelphia, Pennsylvania, the city where he was born on February 12, 1919. His father, Jack, emigrated from Russia to the U.S. with his parents when he was 2 years old. His mother, Fannie Weitz, was born in Philadelphia. Jack was a manufacturer's sales representative and, as soon as Russ's younger sister Elaine was old enough, Fannie managed his office. The family had severe financial difficulties during the Great Depression, and Russ helped out by taking whatever work was available—selling razor blades door-to-door, delivering pamphlets, and working in an iron mill.

Russ was particularly influenced by his father; his maternal uncle, David Wait; and by his maternal grandfather, Saul Wait (who had emigrated from Russia with his wife around 1880), all of whom he adored. His grandfather was a butcher who delivered meat in a Model T Ford and in a horse-drawn sleigh when it snowed. His grandfather was highly intelligent, self-educated, and spoke a number of languages, but he could not read or write in English. Russ had fond memories of learning to read and write along with his grandfather and grandmother. In junior high school, he received the American Legion award as the best all-around student, and he was president of his high school class.

ROUND ROBIN: U. PENN, U.S. ARMY, WAYNE U., CASE INSTITUTE, U. PENN

The extended Ackoff family placed a high value on learning—Fannie's older brother, David, was a successful Philadelphia architect. There was no question about Russ going to college; even though his parents could not afford the tuition, they borrowed money to get him started. Russ enrolled in the 1937 freshman class at the University of Pennsylvania (U. Penn) and, following his uncle, specialized in architecture. Russ did receive a scholarship soon after enrolling but worked nights and weekends throughout his undergraduate studies. He graduated in 1941 with a B.A. in architecture and received five awards, the largest number for any graduate from the university that year.

Russ decided to go on to graduate school at U. Penn and, strangely, accepted a teaching fellowship in philosophy. This was a considered decision, but its outcome was not what he anticipated. As Russ explained (Novak 2000):

> Doing philosophy was an opportunity to learn something new. I expected this to be an adjunct to my practice of architecture. But it turned out the other way. It turned out that the philosophy of science gave me the opportunity to design social systems, and I was more interested in people-oriented systems than in buildings. They were both design, but different kinds of design. I like creating things.

Russ entered the Ph.D. program in September 1941, but his studies were soon interrupted. In early January 1942, he was drafted into the U.S. Army, and it was not until 1946 that he returned to U. Penn.

Russ started his war service with basic training, after which he was assigned first to the headquarters of an armoured medical battalion and then to the engineering section of the headquarters of the Fourth Armoured Division. In 1943, he went overseas with that unit to the Pacific theatre. Later that year, he was sent to Officer Candidate School (OCS) in Brisbane, Australia, where one of his legs was broken in a self-defense exercise. He finished OCS with his leg in a walking cast.

FIRST ARCHITECTURAL ASSIGNMENT

"I was in the invasion of the Philippines, on the island of Leyte. After the occupation of that island was completed, General Seibert asked me (then a sergeant) to design and build a recreation center for the troops. I did that using only materials at hand, bamboo, grass (nipa, a long grass), trunks of palm trees, and abandoned signal corps wire. Labor was provided by natives whom I hired, both men and women. The center involved a long dock that extended out into the Barugo River to be used for swimming. The dock was a large cage for large abandoned empty oil drums, and therefore it could rise and fall with the tide. While installing the bridge, I was in the river in the nude when the general suddenly appeared in his motor launch. I jumped out of the water onto the dock and saluted him. He broke out into laughter and explained that he had never been saluted by a nude soldier before" (Ackoff 2009).

On his return to U. Penn in 1946, he resumed his research under the supervision of C. West Churchman; Russ was West's first doctoral student. He completed his doctorate after only four semesters in the program. His dissertation title was "An experimental definition of personality" (Kirby and Rosenhead 2005; Mason 2004; Ulrich 2004).

The journey that both Russ and West then made, which took them from philosophy to OR, was due to the influence of the ideas of one of their U. Penn teachers, Edgar A. Singer, Jr. (Russ had two semesters in Singer's seminar.) In fact, when Singer became ill, Russ and assistant professor West Churchman were asked to take over the seminar. The result was that Russ was a student in a class in which he was also an instructor. Singer had studied at Harvard with William James, but he had developed a somewhat different version of James' pragmatist philosophy. In general, pragmatism takes the stance that the meaning and value of all human endeavors, including philosophy and science, must be judged by the way they serve the practice of human life.

HOW TO DEFEND A DISSERTATION

"Because the content of my [doctoral] thesis crossed a number of disciplines, the examining committee that was formed was larger than any previously use by the department. ... The examination began at 7:00 PM one evening; I was dismissed at about 9:00 PM, after a very hostile grilling by a committee whose members had only a slight idea of what I had done. From then until 1:00 AM they debated on whether it was a thesis in philosophy or some other subject. West fought the committee relentlessly and wore it down. It finally accepted my thesis because of exhaustion, not approval of its content" (Ackoff 1988, 353).

"... we discovered that in the archives of the graduate school of Penn there was a rule which said that a candidate and his thesis advisor could coauthor the thesis, which we did. (This rule upset some of the faculty, who said they could not decide how much Ackoff had done and how much I had done. Couldn't we identify one or more chapters which were primarily his? Of course, we could: no problem. Using random numbers, we selected one which was Ackoff's alone.) ... [The thesis] was 'telephone book' size, so it is questionable whether other faculty members even tried to read it and certain that none of them studied it, with the exception of the Singerians. ... Russ, of course, did not help me at all in the task of getting him passed. He wasn't shy then any more than he is now. When one professor ... asked Russ a question, Russ responded that he, the well-established professor, was not qualified to ask the question. I suspect this was a unique event in the history of Ph.D. oral" (Churchman 1990, 128).

In 1945, West, though only 32 years old, was elected chairman of the Philosophy Department at U. Penn. Together, West and Russ tried to establish within the department an Institute of Experimental Method that would take Singer's philosophical ideas forward and apply them practically to current social issues. The department, however, was not persuaded; Russ's teaching appointment was not renewed. In 1947, Russ accepted an appointment in philosophy and mathematics at Wayne University (now Wayne State University) in Detroit, Michigan, and West followed a year later as an associate professor in philosophy. Once again, however, the idea of a practically-oriented institute as part of the department of philosophy could not be brought off. In 1951, they both moved to the department of engineering administration at Case Institute of Technology, with Russ as an associate professor and West a professor (Ulrich 2004). The basis for this move was that they saw the congruence of their thinking with the incipient field of OR.

At Case, they were able to put their ideas to the test. There they established the first M.S. and Ph.D. programs in OR, held a series of major annual OR conferences, and began to teach short courses for industrial practitioners. They formed the Operations Research Group (ORG), directed by Russ, in which they undertook corporate consulting assignments (Arnoff 1957; Dean 2004). They joined with E. Leonard Arnoff, a Case professor hired by West, to write the textbook *Introduction to Operations Research* (Churchman et al. 1957). This book provided a categorization and description of the developing tools and techniques of OR, but not only that. The techniques were situated within an articulated framework of Singer's philosophy and of the systems approach; chapter two is titled "An Operations Research Study of a System as a Whole." The book's insights, examples, and overall view of OR were influenced by the ORG's consulting activities in such areas as production and inventory control (Ackoff 1955) and integrated process control (Arnoff et al. 1958). It was the first internationally recognized OR textbook.

Shortly afterwards, West accepted a professorship at the Graduate School of Business Administration, University of California, Berkeley. In 1964, Russ moved back to U. Penn, which was to be, in one form or other, his home for the rest of his career, especially at the Wharton School. Before taking up his new post, Russ held a visiting chair at the University of Birmingham where his well-attended short courses established a rapport between him and a generation of OR practitioners in the U.K. While there, he acted as marriage broker between the U.K. Operational Research Society and the prestigious Tavistock Institute of Human Relations in the setting up of the Institute for Operational Research (IOR). IOR's mission was to extend the applications of OR in the public sector, but it also provided an interface between OR and the social sciences (Kirby 2003). (A history of IOR and Russ's involvement is given in Friend et al. [1988].)

THE TASK OF THE OR WORKER

"It can be said with justification that his task is to advise executives and management but not to make the decisions himself. The ultimate responsibility for the decisions still rests with the executive or the manager. With this we agree, provided that certain conditions are met by the OR worker. First of all, he must feel fully committed to the decision to which he has come. He must be prepared, if invited to do so by the executive concerned, to say what decision he would undertake if he were sitting in the manager's chair. . . . He must be willing to be judged by the manager in exactly the same way that the manager has to be judged, namely, on the basis of the performance of his recommendations when they are put into operation" (Rivett and Ackoff 1963, 98).

This was not the first, or indeed the last, significant encounter by which Russ influenced the direction of British OR. In the 1950s, Pat Rivett, head of the National Coal Board's OR group (at the time the U.K.'s premier practitioner unit) spent six months at Case. When, in 1964, he became the first professor of OR in Britain, he modeled his department at Lancaster University on the arrangements at Case. Russ and Pat Rivett set up a long-lasting student exchange program between Lancaster and U. Penn. Their joint book, *A Manager's Guide to Operational Research*, was a key text for explaining OR accessibly to potential users (Rivett and Ackoff 1963). Russ also became very close to the leading British operational researcher Stafford Beer.

Central to Russ's educational philosophy was that M.A. students undertake engagements with practical problems while studying. For this purpose, Russ was extraordinarily successful in generating a wide range of contract work with major organizations—with the added advantages that both students and additional faculty positions could be supported out of the income. At U. Penn, Russ set up not only an academic department of statistics and OR but also a research center to house the projects. The last statement is a simplification, as over the years there was to be a series of departments and research centers—the Management Science Center became the Management and Behavioral Science Center; but Russ left this behind to set up the Busch Center (named after the founding dynasty of the Anheuser Busch brewing corporation with which he had a long-term relationship). A similar evolution took place in his establishment of and attachment to a series of teaching programs. From 1964 on, Russ continually redefined his view of OR, and, in the process, he found it necessary to shed old forms and to relieve some of the people who had been recruited to them. (He himself did not believe in academic tenure, and, early on, deposited a signed but undated resignation letter with the dean of the Wharton School to be used if and when the dean chose.)

ACKOFF'S FABLES: VERY SHORT STORIES WITH VERY SHARP POINTS

Russ noted that that older he became, the more he used stories to help make points in his lectures, discussions, and writing. All the stories (tales) but a few were essentially true (some hearsay) and tell of many experiences. Of course, Russ dramatized them a bit. He called these tales fables (Ackoff 1989, 375). His most famous fable follows, with others collected in Ackoff (1978, 1991).

THE UPS-AND-DOWNS STORY

"The manager of a large office building had been receiving an increasing number of complaints about the building's elevator service, particularly during rush hours. When several of his larger tenants threatened to move out unless this service was improved, the manager decided to look into the problem.

"He called on a group of consulting engineers who specialized in the design of elevator services. After examining the situation, they identified three possible courses of action: (1) add elevators, (2) replace some or all of the existing elevators with faster ones, or (3) add a central computerized system so that the elevators could be 'routed' to yield faster service.

"The engineers then conducted cost-benefit analyses of these alternatives. They found that only adding or replacing elevators could yield a large enough improvement of service, but the cost of doing either was not justified by the earnings of the building. In effect, none of the alternatives was acceptable. They left the manager with a dilemma.

"He called a meeting of his staff and presented the problem to them in the format of what he called a 'brain-storming' session. Many suggestions were made, but each was demolished. The discussion slowed down. During a lull the new young assistant in the personnel department, who had been quiet up to this point, timidly made a suggestion. It was immediately embraced by everyone present. A few weeks later, after a relatively small expenditure, the problem had disappeared.

"Full-length mirrors had been installed upon all the walls of the lobbies on each floor.

"Moral: With reflection it becomes apparent that there is more than one way to look at a problem" (Ackoff 1978, 53–54).

In 1968, Russ coauthored a textbook of conventional OR techniques that in many ways paralleled those in his seminal book with Churchman and Arnoff (Ackoff and Sasieni 1968). This was followed 2 years later by his influential book, *A Concept of Corporate Planning*, an attempt to persuade OR to be more ambitious in tackling management's more strategic issues (Ackoff 1970a). But, by the mid-1970s, Russ's disquiet with what OR had become constituted a central focus of his work. Russ was by then operating through the Busch Center and the Social Systems Science (S^3) program.

At the Busch Center, he generated and conducted studies which were, by this time, beyond the established boundaries of OR. One celebrated example is the project carried on over a long period working with community leaders in the West Philadelphia district of Mantua. In 1968, Mantua was 80 deprived city blocks just to the north of the university, whose population of around 22,000 was 98% black. Instead of researching the problems of the ghetto, he brought ghetto members into the university to research how the resources of the university could help achieve development. The rationale was that only the ghetto itself could solve its problems, and it needed to do so in its own way. Achievements included the establishment of a community credit union, an industrial complex housing manufacturing firms, and an architectural and planning center together with a joint workshop (with university staff and students) to produce neighborhood development plans (Ackoff 1970b; Jackson 2000, 241–243). This work is widely recognized as the first example of what is now known as community OR (Midgley and Ochoa-Arias 2004).

The S^3 program was also highly unconventional. It was strikingly different in the breadth of student intake, in staff, and in curriculum than the, by now, standard OR program, heavy with mathematics and algorithms. He attracted star faculty members with strikingly diverse orientations—Hasan Osbekhan from planning; Thomas Saaty, who would shortly develop his Analytic Hierarchy Process; and Eric Trist from the U.K., originator of the socio-technical-systems approach. There was, however, very little team orientation; each of the stars tended to operate independently and recruited his or her own constellations of supporters.

It was during the 1970s that Russ's disillusionment with the course and conduct of OR was registered publicly in ever more ringing tones. There was widespread discussion of what many people perceived as a lack of practical relevance of OR's research directions and a lack of penetration of OR's practice into strategic areas of public and private decision making. These views were prevalent worldwide, but found their strongest expression in Britain where it was commonly referred to as a crisis (Tomlinson 1974). Similar pessimistic views were expressed by Beer (1970), Simpson (1978), and others (Kirby (2003, 1135). None of the remedies offered matched Russ's in its root-and-branch radicalism.

Ackoff's Apostasy and Beyond

To explain what has been called Ackoff's apostasy, we need to identify the key philosophical and practical elements that by now were underpinning Russ's approach to OR. It is reasonable to say that these ideas, rather than his role in launching OR as an academic subject, constitute his most lasting contribution to the discipline.

The seeds of Russ's disquiet can already be seen in two aspects of *Introduction to Operations Research*. There the authors state that "OR tries to find the best decisions relative to as large a portion of a total organization as is possible" (Churchman et al. 1957, 6). This statement leads to the first principal characteristic of the discipline, namely, its consistency with a systems approach. The second key element was the need for a team approach based upon an eclectic choice of disciplines: ". . . the team approach lies in the fact that most man-machine systems have physical, sociological, economic, and engineering aspects. These phases of the system can best be understood and analyzed by those trained in the appropriate fields" (Churchman et al. 1957, 10).

The emphasis on the interdisciplinary approach to decision making was to be one of Ackoff's continuing concerns. By 1961, he was already registering his unease at the undue narrowness of OR models—they too often failed to incorporate "psychological and social variables" (Ackoff 1961, 26). He also expressed strong reservations about the trend towards suboptimization and, more critically, the fact that OR models were being applied to problems of limited scope. In his speech as retiring president of ORSA, he specifically highlighted the potential for OR to contribute to raising living standards in the Third World (Ackoff 1957). He foresaw that a continuing emphasis on specialized modeling techniques could fuel a search for problems to match those techniques, rather than the strong problem orientation that was needed.

A series of publications through the 1970s laid the groundwork for the more overtly critical publications that emerged toward the end of the decade. His coauthored book, *On Purposeful Systems* (Ackoff and Emery 1972), provided an erudite and thought-provoking account of how systems thinking can be applied to human behavior. To the extent that a system is purposeful, understanding of its aims can only be gained by taking into account the mechanisms of social, cultural, and psychological systems (Ackoff and Emery 1972). The following year, he distinguished the problems of the machine age from those of the systems age that was already emerging. The introverted OR discipline, he said, was not broad enough to research effectively the key operating characteristics of our society (Ackoff 1973). In a paper based on an address to a joint meeting of ORSA and The Institute of Management Sciences, he critiqued OR's focus on optimization and objectivity, saying that this constituted a guaranteed opting out from reality (Ackoff 1977a).

Other heavyweight papers emerged in this period. One, on the social responsibility of OR (Ackoff 1974a), stimulated a vigorous debate (Ackoff 1975; Rosenhead 1976; Chesterton et al. 1975). Russ advanced the position that operational researchers have a broader responsibility than to a single employer. Fortunately, however, in taking up this responsibility, they would in fact be advancing the employer's interests at a meta-level. His critics disputed the idea that conflicts of interest between a powerful organizational actor and those who would experience the consequences of action were always resolvable to their mutual benefit. In another of these papers, Russ revisited an earlier paper on Third World development (Ackoff 1957, 1977b). The latter paper took a very different line than his paper 20 years earlier: He now felt that only if people carried out the planning of their own future themselves (with some help from professional planners) would that planning be effective. This paper reflected the ideas which he had already put into action in the Mantua project, while the exposition of them reflected a recent sabbatical period in Mexico.

The definitive statement of Russ's separation from the main body of OR came in two papers he gave in the U.K. at the ORS annual conference in 1978. One paper was given at the beginning of the conference and the other at the end. Of course, the several hundred attendees had not been following Russ's intellectual journey, so the opening paper, "The future of OR is past" (Ackoff 1979a), hit the conference like a bombshell. He had tremendous influence and respect in Britain, and the uncompromising message of the paper was deeply uncomfortable. The themes already mentioned were marshalled into a powerful polemic on the collapse of inter-disciplinarity, the limited purchase of context-free mathematical methods, and the irrelevance guaranteed by optimization and objectivity. The result was that OR had and would continue to move down the organizational hierarchy and be admitted only to relatively insignificant problems. Russ was proclaiming, in fact, the complete irrelevance of OR (in its production and distribution output-oriented guise) to the resolution of society's key problems. The effect of this barrage of connected argumentation was reinforced by Russ's reputation, his elegant mastery of language, and his skill as an orator.

The conference was buzzing with discussion about what Russ had said. And the British OR community continued buzzing after the conference was over. This ferment was by no means calmed by Russ's closing speech. In it, he sketched out his philosophy for a systems-based, transdisciplinary, and participatory approach which he called Interactive Planning (as embodied in his S^3 program at U. Penn) to the redesign of businesses, organizations, and management. The critique evidently resonated with many of the OR conference attendees' concern (perhaps partially formed) about the limitations of the approach and methodology of their chosen profession. Russ's prescription, however, was less immediately attractive. Something more homegrown was required.

There is no question that the impact of Russ's apostasy on OR has been most significantly transmitted through the U.K. OR community rather than that of the U.S. Neither his 1979 papers (Ackoff 1979a, b) nor a subsequent trenchant broadside in *Operations Research* itself, provocatively titled "OR, a post-mortem" (Ackoff 1987), produced a U.S. response. It was as if when confronted with Russ's withdrawal from OR in favor of an alignment with the systems movement (a move paralleled by both Churchman and Beer), U.S. OR preferred to ignore rather than confront the challenge.

Perhaps, the reason for that response was, partly, that U.K. operational researchers have never been as enthusiastic about mathematical optimizing techniques as their transatlantic cousins. The preceding years had seen a number of perceptive, if less sharply focussed, criticisms by senior U.K. operational researchers of the methods and standing of the discipline (Beer 1970; Simpson 1978; Tomlinson 1974). And some of the ideas in Russ's publications had already been picked up for discussion at meetings (Kirby 2003, 1135–1136). Quite independently, in response to the perceived crisis in conventional OR, a number of participative methods had already been developed by U.K. academics and practitioners, collected together in Rosenhead (1989).

The Ackoff papers, as they became known, produced a surge of activity. A series of meetings were organized to discuss them, and some of the papers were subsequently published, notably, the paper "The changing role of O.R.?" by Tobin et al. (1980). There were some who took exception to particular aspects of Russ's critique—George Mitchell's 1980 president's address to the OR Society, and Sam Eilon's writing in the journal *Omega* (Eilon 1980; Mitchell 1980). The emerging consensus in the U.K., however, was that the practice and, especially, the academic study of OR had indeed fumbled its way into a cul-de-sac whose contours Russ had dramatically illuminated and that the mode of escape was not precisely through the door marked Interactive Planning (Ackoff 1974b).

Over the following three decades, the British school of model-based participatory planning, known alternatively as Problem Structuring Methods (PSMs) and as soft OR, has developed to the point where it is a standard part of the curriculum in U.K. masters courses in OR and is among the most used OR methods in the U.K. civil service (Turner 2008). The Ackoff papers debate was a major milestone in this significant development. PSMs have moved beyond the goal-seeking focus of the established paradigm to embrace the possibility of intervention in ill-structured problem situations (Checkland 1981)—which is entirely in accordance with Russ's critique of conventional OR.

Here we have been concerned mainly with Russ's life in and influence on OR. Since his terminal break with the mainstream of OR, there has been three decades of highly productive Russ in which he has lived predominantly in the systems discipline and community. His contribution to systems ideas and practice—his writings both early and late—have been enormously influential to the worldwide systems community. One measure of this is the recent four-volume set of readings on systems thinking (Midgley 2003). When leading international experts in the systems field were asked to nominate articles for inclusion, more papers by Russ were proposed than by any other author.

Strictly within the OR context, how should one summarize Russ's lasting heritage? His great gift of establishing rapport with powerful clients enabled him to direct a great number of path-breaking practical studies. He has certainly left us a range of challenging written work which continues to resonate—he continued to write into his ninth decade. In total he has authored 24 books, the latest in 2008 (Ackoff and Greenberg 2008). His writing style was both muscular and readable (paralleling his outstanding skills as a public speaker). Commonly, the approach is polemical—directed to establishing a conclusion from clear premises and definitions, rather than to a discursive examination of possible viewpoints. He liked an argument, but he expected to win. A friendly commentator notes "the withering critical judgements for which he has become well known" (Friend 1990, 199).

Russ resigned (not retired) from U. Penn in 1986 as the Anheuser Busch Professor Emeritus of Management Science. For the next 20 years, he maintained an independent research center, INTERACT: the Institute for Interactive Management. In 2003, he did return to U. Penn as Distinguished Affiliated Faculty in the Organizational Dynamics

MANAGEMENT F-LAWS: HOW ORGANIZATIONS REALLY WORK

Russ's book includes over 80 *f*-Laws (flawed laws) that "are truths about organizations that we might wish to deny or ignore--simple and more reliable guides to managers' everyday behaviour than the complex truths proposed by scientists, economists, sociologists, politicians and philosophers." For example:

"Managers who don't know how to measure what they want settle for what they can measure."

"The higher their rank, the less managers perceive a need for continuing education, but the greater their need for it."

"You rarely improve an organization as a whole by improving the performance of one or more of its parts."

"Good teachers produce skeptics who ask their own questions and find out their own answers; management gurus produce only unquestioning disciples" (Ackoff et al. 2007).

degree program in the School of Arts and Sciences. There he taught a graduate course in "Systems thinking applied to management" and advised graduate students. Despite his sustained efforts, when it comes to the institutionalization of his precepts in the U.S., the record is patchy: No effective academic homes have emerged for people working in his spirit. Perhaps one reason for this is that ever since his collaboration with Churchman, Russ did not establish continuing partnerships with colleagues of comparable stature. As a result, among the younger generations of U.S. operations researchers, his name is little known.

Elsewhere, though, the picture is different. As we have seen, his impact on the OR scene in the U.K. has been and continues to be profound. Russ's work is highly regarded, for example, among operations researchers in Sweden, and in Denmark, where the early relationship he established with the leading hard-OR researcher Arne Jensen opened up a space for people to work in a more socially conscious framework. Through other key relationships, he maintained a continuing OR presence in such countries as Mexico, Peru, India, and New Zealand—despite the fact that Russ formally severed his ties with OR 30 years ago.

That said, his chief honors must be as a pioneer developer of the academic study of OR, and as a fearless promoter of stringent debate about the subsequent direction of the discipline he helped to set in motion.

Russ died on October 29, 2009 from complications following hip replacement surgery. He was survived by his wife of 22 years, Helen Wald Ackoff; three children from his first marriage, Alan Ackoff, Karen Ackoff, and Karla Ackoff Kachbalian; and his stepson, Richard Wald.

Honors and Awards

Russ was president of ORSA (1956) and, in 1975, received ORSA's George E. Kimball Medal for service to the profession and the society. In 1971, he was awarded the U.K. OR Society's Silver Medal (its highest honor). He was made a Member of the Russian Academy of Natural Sciences in 1993. He was a fellow of the Institute for Operations Research and the Management Sciences, the American Statistical Society, and the International Academy of Management. He received honorary doctorates from the University of Lancaster (U.K.), Washington University, University of New Haven, the Pontifical University of Peru at Lima, the University of Lincolnshire and Humberside (U.K.), and Florida International University.

In September 2000, he was honored by U. Penn by the establishment of the Ackoff Center for Advancement of Systems Approaches in the School of Engineering and Applied Science, the result of generous contributions of Ray Stata (Chairman of the Board, Analog Devices), the Anheuser-Busch Foundation, and the General Motors Foundation. In 2002, the Russell Ackoff Doctoral Student Fellowship for Research was established in the Wharton School. In 2008, the Russell L. Ackoff Systems Thinking Library and Archive was created in the U. Penn Organizational Dynamics program in the School of Arts and Sciences.

References

Ackoff RL (1955) Production and inventory control in a chemical process. Oper Res 3(3):319–338

Ackoff RL (1957) Operations research and national planning. Oper Res 5(4):457–468

Ackoff RL (1961) The meaning, scope and methods of operations research. In: Ackoff RL (ed) Progress in operations research, vol 1. Wiley, New York, NY, 1–34
Ackoff RL (1970a) A concept of corporate planning. Wiley, New York, NY
Ackoff RL (1970b) A black ghetto's research on a university. Oper Res 18(5):761–771
Ackoff RL (1973) Science in the systems age: beyond IE, OR and MS. Oper Res 21(3):661–671
Ackoff RL (1974a) The social responsibility of OR. Oper Res Q 25(3):361–371
Ackoff RL (1974b) Redesigning the future: a systems approach to societal problems. Wiley, New York, NY
Ackoff RL (1975) A reply to the comments of Keith Chesteron, Robert Goodsman, Jonathan Rosenhead and Colin Thunhurst. Oper Res Q 26(1):96–98
Ackoff RL (1977a) Optimization + objectivity + opt out. Eur J Oper Res 1(1):1–7
Ackoff RL (1977b) National development planning revisited. Oper Res 25(2):207–218
Ackoff RL (1978) The art of problem solving: accompanied by Ackoff's Fables. Wiley, New York, NY
Ackoff RL (1979a) The future of OR is past. J Oper Res Soc 30(2):93–104
Ackoff RL (1979b) Resurrecting the future of operational research. J Oper Res Soc 30(3):189–199
Ackoff RL (1987) President's symposium: OR, a post-mortem. Oper Res 35(3):471–474
Ackoff RL (1988) C. West Churchman. Syst Pract 1(4):351–355
Ackoff RL (1989) Ackoff's fables: very short stories with very sharp points: all Mantua is divided into seven parts. Syst Pract 2(4):375–377
Ackoff RL (1991) Ackoff's Fables. Wiley, New York, NY
Ackoff RL (2009) Personal communication
Ackoff RL (2010) Memories (posthumous publication). Triarchy Press, Axminster, Devon
Ackoff RL, Addison H, Bibb S (2007) Management f-laws. Triarchy Press, Axminster, Devon
Ackoff RL, Emery F (1972) On purposeful systems. Tavistock, London
Ackoff RL, Greenberg D (2008) Turning learning right side up: putting education back on track. Wharton School Publishing, Upper Saddle River, NJ
Ackoff RL, Sasieni M (1968) Fundamentals of operations research. Wiley, New York, NY
Arnoff E (1957) Operations research at case institute of technology. Oper Res 5(2):289–192
Arnoff E, Kania E, Day E (1958) An integrated process control system at the Cummins Engine Company. Oper Res 6(4):467–497
Beer S (1970) Operational research as revelation. J Oper Res Soc 21(1):9–21
Checkland P (1981) Systems thinking, systems practice. Wiley, Chichester
Chesterton K, Goodsman R, Rosenhead J, Thunhurst C (1975) A comment on Ackoff's "The social responsibility of operational research." Oper Res Q 26(1):91–95

Churchman CW (1990) Ackoff comes of age. Syst Pract 3(2):125–130

Churchman C, Ackoff RL, Arnoff E (1957) Introduction to operations research. Wiley, New York, NY

Dean B (2004) West Churchman and operations research: case institute of technology. Interfaces 24(4):5–15

Eilon S (1980) The role of management science. J Oper Res Soc 31(1):17–28

Friend J (1990) Planning concepts, planning contexts. Syst Pract 3(2):195–206

Friend J, Norris M, Stringer J (1988) The institute for operational research: an initiative to extend the scope of OR. J Oper Res Soc 39(8):705–713

Jackson M (2000) Systems approaches to management. Kluwer, New York, NY

Kirby M (2003) The intellectual journey of Russell Ackoff: from OR apostle to OR apostate. J Oper Res Soc 54(11):1127–1140

Kirby M, Rosenhead J (2005) IFORS' operational research hall of fame. Russell. L. Ackoff. Int Trans Oper Res 11(1):129–134

Mason R (2004) IFORS' operational research hall of fame: C. West Churchman, Int Trans Oper Res 11(5):1585–588

Midgley G (2003) Systems thinking. Sage, London

Midgley G, Ochoa-Arias A (2004) Community operational research: OR and systems thinking for community development. Kluwer, New York, NY

Mitchell G (1980) Images of operational research. J Oper Res Soc 31(6):459–466

Novak C (2000) Interview with Russell L. Ackoff. Technos Q 9(3). http://www.ait.net/technos/tq_09/3ackoff.php. Accessed 28 Dec 2009

Rivett B, Ackoff RL (1963) A manger's guide to operational research. Wiley, New York, NY

Rosenhead J (1976) Some further comments on the social responsibility of operational research. Oper Res Q 27(1:ii):266–272

Rosenhead J (ed) (1989) Rational analysis for a problematic world: problem structuring methods for complexity, uncertainty and conflict. Wiley, Chichester

Simpson M (1978) Those who can't. Oper Res Q 29(6):517–522B

Tobin N, Rapley K, Teather W (1980) The changing role of O.R.? J Oper Res Soc 31(4):279–288

Tomlinson R (1974) OR is. Oper Res Q 25(3):347–360

Turner H (2008) Government operational research service: civil OR in U.K. central government. J Oper Res Soc 59(2):148–162

Ulrich W (2004) Obituary: C West Churchman, 1913–2004. J Oper Res Soc 55(11):123–1129

22

David Bendel Hertz

Edward K. Baker and
Robert T. Plant

David Bendel Hertz, through his writings and his work at the management consulting firms of Arthur Andersen and Company and McKinsey and Company, helped bring the ideas and methods of operations research (OR) to the corporate world. In particular, he defined the field of modern risk analysis with his 1964 seminal paper, "Risk Analysis in Capital Investment," that appeared in the *Harvard Business Review*. This paper achieved the status of a *Harvard Business Review Classic* in 1979. He is a founding member of both the Operations Research Society of America (ORSA) and The Institute of Management Sciences (TIMS); he is one of only five individuals who served as president of both organizations (TIMS in 1964; ORSA in 1974). David received the George E. Kimball Medal for outstanding service to the profession from ORSA and is a fellow of the Institute for Operations Research and the Management Sciences (INFORMS). He served as president of the International Federation of Operational Research Societies (IFORS) from 1977 to 1979.

Texas to New York

David Bendel Hertz was born in Yoakum, Texas, on March 25, 1919 to Emanuel and Wilhelmina (Schmulen) Hertz. Emanuel was a businessman who was born and raised in New York City. He had moved to Texas to start a mercantile business thinking that Yoakum, situated at the junction of two railroad lines 110 miles southwest of Houston, would become an important business center. Wilhelmina was born in Lafayette,

A.A. Assad, S.I. Gass (eds.), *Profiles in Operations Research*, International Series in Operations Research & Management Science 147, DOI 10.1007/978-1-4419-6281-2_22,

Louisiana. She was a niece of Henri Bendel, who established his famous women's specialty fashion store in 1896 at 10 West 57th Street, New York City, which was relocated in 1990 to 712 Fifth Avenue. David's mother had strong Louisiana family connections and was renowned as a cook with a definite Cajun influence. (When a *New York Times* food critic discussed the merits of a certain New York restaurant's Cajun Maque Choux [a sautéed vegetable dish], David responded with a letter containing his mother's recipe for the dish that was a family favorite.)

The Yoakum schools proved no challenge for young David. He finished the curriculum by age 14. He was then sent to live with his uncle in Great Neck, New York, to get a more rigorous high school education. David graduated in 1935 at the top of his class in mathematics.

David then enrolled at Columbia University where he received a B.A. in Science in 1939 and a B.S. in Engineering in 1940. While at Columbia, David met Barbara Valentine, then a high school senior, who would go on be to a political science major at Barnard College. The two were married on December 20, 1941, while David awaited assignment in the U.S. Navy.

OBSERVATIONS OF A YOUNG MANAGEMENT SCIENTIST

"As a young student of management and a newly commissioned officer in the Navy at the start of World War II, I found myself constantly surprised at the effectiveness of the management processes of the vast enterprise of ships, men, and shore stations. In fact, considering all the difficulties I knew must exist—difficulties of coordination, direction, scheduling, assignment, allocation—it was surprising that it operated at all. But operate it did—and the accepted witticism explaining it all was that the Navy was a system designed by geniuses to be operated by morons. And what this meant, of course, was that the situational logic had been thought through beforehand and that the necessary instructions and training for each of the parts had been provided, so that under specific circumstances the apparatus worked as smoothly and effectively as any well-designed machine—in fact better, since to a certain extent it was self-repairing" (Hertz 1965, B-93).

During World War II (WWII), David served

in the U.S. Navy as a Lieutenant Commander in the Caribbean and the Pacific, and took part in the landings and the surrender in Japan. While on active duty, but between assignments, David attended the U.S. Naval Postgraduate School (then located in Annapolis, Maryland), where he received an M.S. in 1944. After WWII, David remained active in the U.S. Naval Reserve and retired with the rank of a commander.

Upon returning to civilian life, David enrolled in Columbia University's engineering Ph.D. program, receiving his degree in 1949. His dissertation title was "The Theory and Practice of Industrial Research." Professor Walter Rautenstrauch was chair of his dissertation committee, which also included professors R.T. Livingston and S. B. Littauer of the Industrial Engineering Department, as well as R.P. Piperoux, Director of Engineering Research at the Celanese Corporation of America.

Academia, Professional Societies, and Management Consulting

From 1949 through 1953, David was an assistant professor and then a tenured associate professor of industrial engineering at Columbia University. His first book was based on his dissertation and had the same title (Hertz 1950). In it, David developed and synthesized the newly emerging scientific theories for modern industrial management and their application in practice that included many of the methods that are now part of operations research (OR) and management science (MS). It was only natural that David became very interested and involved in the formation of two new related professional societies, ORSA and TIMS.

ORSA was founded in 1952 with David, a founding member. He would serve as editor of the ORSA series, Publications in Operations Research, from 1959 to 1973; he was elected president of the society in 1974. Although an enthusiastic member of ORSA, David, as well as many other early OR practitioners, felt that the society, in its publications and meetings, overemphasized its military origins. Thus, in early 1953, he and Barbara hosted a meeting in their East 74th Street apartment in New York to see what might be done about it:

> "There were about ten of us," David recounted. "Bill Cooper was there. Merrill Flood was there. We spent a long night talking about the creation of (what would become) The Institute of Management Sciences. The big question, with ORSA already up and running, was, Why are we doing this? We were doing it because a lot of people, myself included, were not satisfied

> with the idea that ORSA didn't seem to be concerning itself with the management world. We believed that we could make important contributions in this area by applying the techniques of operations research, which we called management science"(Horner 1993, 41–42).

On December 1, 1953, TIMS was formed, with William (Bill) Cooper of Carnegie Institute of Technology, its first president.

Soon thereafter, David left Columbia University to take management positions at the Radio Corporation of America and then at the Celanese Corporation. In 1957, he became principal in charge of OR at Arthur Andersen and Company. It was in management consulting that David would find his stride. In 1963, David accepted a position in the New York office of McKinsey and Company where he would spend the next 20 years, rising to the position of Senior Director. While at McKinsey, David's best-known work, "Risk analysis in capital investment," would appear in the Harvard Business Review (HBR) (Hertz 1964).

THE STYLISH NEWSBOY

Early OR practitioners, when they encountered a new application area, had to either extend the few available proven basic models or they had to invent new ones, in each case adapting them to fit real-world requirements. A case in point was the study of the management of seasonal style-goods inventories, reported in Hertz and Schaffir (1960)—an application that typifies the early OR situations that David and his consulting associates were asked to resolve. As the paper notes: "The primary difference, in an economic sense, between management of seasonal style-goods inventories and other inventories, lies in the 'sudden-death' character of the highly styled and highly seasonal merchandise involved." The authors recognized that the general situation is described by the classical newsboy problem. Their extension for a textile manufacturer showed how to determine "the duration and timing of the selling seasons, and for forecasting total sales for the season for each individual item in the line, at different probability levels," and "criteria for weekly re-evaluation of inventories" (Hertz and Schaffir 1960, 45).

In the HBR paper, David calls for a new concept in the valuation of the capital investment. Rejecting the heretofore generally accepted method of most likely net present value of future cash flow, David set forth arguments and examples for the incorporation of the probability distributions of the salient factors in the investment decision. Market share, price, and cost now all became random variables with estimates of their probablity distributions. David then proposed the use of computer simulation to produce a distribution of possible discounted cash flows.

Early in the paper, David delineated what he termed "the Achilles Heel" of the accepted method then in use:

> The fatal weakness of past approaches thus has nothing to do with the mathematics of rate-of-return calculation. We have pushed along this path so far that the precision of our calculation is, if anything, somewhat illusory. The fact is that, no matter what mathematics is used, each of the variables entering into the calculation of rate of return is subject to a high level of uncertainty. . . .
>
> The "expected rate" of return represents only a few points on a continuous curve of possible combinations of future happenings. It is a bit like trying to predict the outcome of a dice game by saying that the most likely outcome is a "7" (Hertz 1964, 97).

In closing the paper, David pragmatically notes about the nature of decision making and risk:

> Lastly, the courage to act boldly in the face of apparent uncertainty can be greatly bolstered by the clarity of the portrayal of the risks and possible rewards. To achieve these lasting results requires only a slight effort beyond what most companies already exert in studying capital investments (Hertz 1964, 106).

In 1979, the HBR article became an HBR Classic, having sold over 153,000 reprints (Hertz 1979).

The article generated a wealth of interest. Two recurrent themes emerged from the responses: (1) How are the uncetainites in each of the key elements of the investment to be determined, and (2) once the uncertainites were quantified and displayed, how was the best decision to be made? David addressed these concerns in a sequel and companion HBR article, "Investment policies that pay off" (Hertz 1968). These issues were addressed further in two books, *Risk Analysis and its Applications* (Hertz and Thomas 1983) and *Practical Risk Analysis, An Approach Through Case Histories* (Hertz and Thomas 1984).

David's collaboration with Howard Thomas started in the early 1970s when David was a visiting professor at the London Business School (LBS) where he met Thomas, a faculty member and director of its Decision Analysis Group. David's LBS assignment was to teach a special graduate-level elective course on risk analysis and planning under uncertainty. Their first book evolved as the two became good friends and pursued their joint research.

Hertz and Thomas (1983) was written to satisfy three basic purposes: The book allowed the ideas of David's HBR articles to be more fully developed and expanded; the authors sought to develop a broader understanding of the concepts of uncertainty and risk analysis such that managers might apply similar decision methodology to all their areas of decision responsibility; and the book's applied perspective could be further developed through the use of realistic detailed case analyses and examples. The publisher agreed to a second volume that would enable managers and students to master the methods through a variety of case histories (Hertz and Thomas 1984).

In the second volume, the authors, with assistance from their colleagues at the LBS, present a number of interesting cases concerning the use of risk analysis and decision making under uncertainty. Most of the cases involve firms based in the U.K.; they are typically rich in detail and cover a wide range of situations. Such well-known companies as Sainsbury, National Feeds, and Austrans Limited are featured. Ever concerned with the implementation of the methods, the authors include a final postscript providing guidelines for getting started in using risk analysis.

In addition to his seminal work in risk analysis, David was also at the forefront of the use of computers in MS. In his 1969 book, *New Power for Management, Computer Systems and Management Science*, he writes the following:

> In an increasingly complex business environment, management science techniques are providing a growing number of firms with decisive competitive advantages. These techniques are achieving even greater effectiveness as they are increasingly coupled to the expanding power of the computer. In effect, they supply answers to the three most basic management questions: (1) Where are we heading? (2) Where should we be heading? And (3) How do we get there? Management science techniques can furnish corporate decision makers with realistic, dollars-and-cents evaluations of alternative courses of action aimed toward specific goals, and help shape their decisions accordingly (Hertz 1969, 44).

Written at a time when mainframes ruled the world of computing—personal computers were below the planning horizon—the book devotes a significant amount of time to information systems and planning, as well as to the field of MS. Overall, the information systems chapters of the book presage the modern computer era and present a useful contemporary picture of the modern chief information officer. The advantage of hindsight enables readers to understand why the migration of the computer to the central business core has propelled information technology to the highest levels of the management hierarchy, while MS has been relegated to mostly technical staff positions within other departments.

As David's responsibilities at McKinsey became more diverse, he felt the need to have a better understanding of the legal issues that are often encountered by clients when making decisions. In 1983, at the age of 64, he received a Juris Doctorate from the New York University Law School. He became a member of the New York Bar and later would serve as of counsel to the law firm of Leva, Hawes, Symington, and Martin, Washington, DC.

Back to Academia and New Adventures

David relocated to Miami in 1984 where he accepted an appointment as Distinguished Professor of Management Science and Computer Information Systems (CIS) in the School of Business Administration, University of Miami. Within the school, David helped influence the direction of the newly formed CIS Department and created the Intelligent Computer Systems Research Institute as a vehicle through which he could pursue one of his lifelong interests, artificial intelligence (AI). David continued to publish papers in collaboration with his doctoral students and the department's junior faculty members, whom he mentored with his usual Socratic method of inquiry leveraged by his keen wit.

David's enthusiastic but cautious views on AI, expert systems, and how the business manager can benefit from their promise are contained in his article "Artificial intelligence and the business manager" (Hertz 1983) and his book *The Expert Executive* (Hertz 1988). He was a proponent of using AI to help make planning, production, marketing, and distribution decisions. He also recognized that as software projects became increasingly complex, controlling software development cost overruns and schedule delays became increasingly important. He and colleagues proposed that software development could be modeled as an economic production process and set forth a theoretical approach to software-cost modeling derived from economic production theory and systems optimization (Hu,

Plant, and Hertz 1998). The proposed Minimum Software Cost Model (MSCM) was compared with other widely used, and fancifully named software cost models, such as COCOMO (COnstructive COst MOdel) and SLIM (Software LIfecyle Management). David and colleagues showed that the MSCM model was comparable to, if not superior to, the SLIM model and significantly better than the rest of the models. Additionally, the MSCM model was shown to provide insights about the behavior of software development processes and environments that could be used to formulate guidelines for better software project management policies and practices.

David retired from the University of Miami in 1991 as Professor Emeritus. But, never one to turn down a challenge, David, at the age of 74, became an entrepreneur and founded the company Identification Technologies International (ITI). ITI's mission was to investigate and expand current technologies to allow facial recognition to become an easily integrated part of a secure identification system. This technology had important applications in the banking industry where, for example, facial recognition technology could be used in conjunction with a personal identification number at an automated teller machine to reduce fraudulent use. David provided intellectual leadership, capitalized the company, raised venture capital, and, with his team, developed an advanced prototype system, ready for production. The intellectual property, the technology, and the company were subsequently acquired by a technology security company and relocated to Texas.

SERVICE, AWARDS, AND HONORS

Early in his OR career, David became involved with the establishment of international OR societies. He was the TIMS representative on the U.S. committee that helped to organize the first international conference on OR held in Oxford, England, September 2–6, 1957. At that meeting, he presented the results of a survey of OR activities in the U.S. (Hertz 1957). David was also a member of the initial committee that led to the founding of IFORS. David attended the first IFORS meeting in Aix-en-Provence, September 5–9, 1960; he gave a paper with the unusual title "On elegance in Operations research" (Hertz 1961). His theme: "When O.R. work is done professionally by a good craftsman in the field who uses imagination and avoids fetishism of method and techniques, there will occasionally emerge that beauty which his fellow scientists calls elegance" (Hertz 1961, 33). He was president of IFORS from 1977 to 1979.

For the OR/MS profession in the U.S., David served as president of TIMS (1964) and president of ORSA (1974). He served for 14 years as the first editor of ORSA's Publications in OR Series (PORS). For PORS, David sought more specialized and scholarly monographs that conventional publishers might be hesitant to publish because of the possible financial risk involved. In 1984, he received ORSA's Kimball Medal for distinguished service to the profession. He is a fellow of INFORMS.

Throughout his professional career, David was committed to public service. He was a trustee of Columbia University, Columbia-Presbyterian Hospital, and Columbia University Press. During the years that John Lindsay was mayor of New York City, David was named to the 17-person Operations Research Council for the City of New York to advise and reorganize the municipal government of the city. Among those also named to the OR Council were Russell Ackoff, West Churchman, Peter Drucker, Merrill Flood, and Ralph Gomory. A Mayor's Chart Room was designed by David and Deputy Mayor Timothy W. Costello in which decision-making data were to be gathered, organized, and presented (*New York Times*, 1967). Mayor Lindsay, after his first use of the Chart Room, was quoted as saying, "This is the most informative hour I've spent on this job."

In an op-ed piece that appeared in the *New York Times* on January 14, 1971, David called for the technological and regulatory creation of an economic and thermodynamic equilibrium to save the planet from its impending environmental crisis. In December of 1985, while a Visiting Scholar at the Harvard Community Health Plan, David presented his forward-thinking ideas on the use of quantitative processes and procedures to assist in medical decision making. David's algorithms for clinical decisions, clinical data banks, mathematical models, and Bayesian decision models, and the use of artificial intelligence within diagnostic expert systems, are tools and ideas that are still coming to fruition.

For David Bendel Hertz, the world was his stage, and in his career, he played many important parts. This prescient polymath was at the leading edge of the development and creation of the fields of OR, AI, and management information systems. His vision for the future vitality and safety of the planet and the health of its inhabitants involves the use of the technologies which he helped pioneer.

BARBARA AND FAMILY

David and Barbara

It is often said that extraordinary men marry extraordinary women. This was certainly the case with David. Barbara Valentine Hertz was a gifted author, editor, and publisher. During the 1960s, Barbara rose to become Managing Editor of *Parents* Magazine. In 1980 and 1981, Barbara was the publisher of the national magazine *Primetime*. Barbara also loved campus life. In 1968, she returned to Barnard College as the Director of Development. Later, after she and David moved to Miami, Barbara would serve as the Director of Foundation Relations at the University of Miami for 20 years.

Barbara and David have two daughters, Barbara Bendel Hertz Burr, born in 1944, and Valentine Hertz Kass, born in 1947. Dr. Burr is a graduate of the Harvard Medical School and a pediatric/adolescent psychiatrist, practicing in Boston, Massachusetts. Ms. Kass is a Program Director in the Informal Science Education program at the National Science Foundation. David and Barbara have four grandchildren and three great-grandchildren. Now both retired, they live in Coral Gables, Florida.

THE UNITY OF SCIENCE AND MANAGEMENT

"There is an old French saying, 'With two lovers there is one who loves and one who lets himself be loved. In analyzing the affair—dare I call it a love affair?—between science and management it would hardly occur to anyone to put management in the driver's seat. Yet this is just what I believe has been the case. We scientists think we have been the aggressive innovator and management the all-too-passive recipient. On the contrary, I suggest that management has seduced science to its side by leading it away from truly scientific approaches to the kind of wish-fulfillment attacks on trivial problems we are all apt to deplore when we see them in cold print. And the kind of writing and exposition we do that management complains it cannot understand is simply the protective coloration of respectable clothing that any sensible seduced professional would want to wear when appearing before his own family.

"However, the time has come instead to assert the equality of these two partners and to capitalize on the unity of the pair. It is time we began to speculate about management and use our understanding of 'operations' (largely technical matters, in management's view) to develop appropriate and useful descriptions and explanations of the management world. Whatever the essence of management is, it has led science to revolutionize the physical environment of the manager. Thus, we can point to the revolution in information handling and transmission that results in order-of-magnitude changes in our ability to store, manipulate and retrieve data. However, new forms of energy, new technologies, and new political frontiers and applications of management science have obviously not brought success to management of the problems of feeding the hungry, administering to the ill, reducing pollution, congestion, and ugliness in the cities, or conserving what is left of land, forests, and wildlife in the face of encroaching industry" (Hertz 1965, B-95).

David's worldview of OR and MS, and science in general, and his hoped-for positive impact they would have on management, was the theme of his talk given at the end of his term of office as president of TIMS (Hertz 1965). In it, he laments the apparent subservient role of OR and MS vis-à-vis management, but recognizes that the proper unity of these professions can contribute much to the betterment of mankind. Words delivered in 1965 are still meaningful today.

REFERENCES

Hertz DB (1950) The theory and practice of industrial research. McGraw-Hill, New York, NY

Hertz DB (1957) Progress of industrial operations research in the U.S. In: Davies M, Eddison R, Page T (eds) Proceedings of the first international conference on operational research (Oxford 1957). The English Universities Press, London, pp 455–467

Hertz DB (1961) On elegance in operations research. In: Banbury J, Maitland J (eds) Proceedings of the second international conference on operational research (Aix-en-Provence 1960). English Universities Press, London, pp 33–43

Hertz DB (1964) Risk analysis in capital investment. Harv Bus Rev 42(1):95–106

Hertz DB (1965) The unity of science and management. Manage Sci 11(6):B-89–97

Hertz DB (1968) Investment policies that pay off. Harv Bus Rev 46(1) 96–108

Hertz DB (1969) New power for management, computer systems and management science. McGraw-Hill, New York, NY

Hertz DB (1979) HBR classic: risk analysis in capital investment. Harv Bus Rev 57(5):169–181

Hertz DB (1983) Artificial intelligence and the business manager. Computerworld 17(43):20–21

Hertz DB (1988) The expert executive; using AI and expert systems for financial management, marketing production, and strategy. Wiley, New York, NY

Hertz DB, Schaffir KH (1960) A forecasting method for management of seasonal style-goods inventories. Oper Res 8(1):45–52

Hertz DB, Thomas H (1983) Risk analysis and its applications. Wiley, New York, NY

Hertz DB, Thomas H (1984) Practical risk analysis, an approach through case histories. Wiley, New York, NY

Horner P (1993) TIMS turns 40. ORMS Today 20(2):40–43

Hu Q, Plant R, Hertz DB (1998) Software cost estimation using economic production models. J Manage Inf Syst 15(1):143–163

New York Times (1967) Chart room to aid Lindsay by listing data for decisions. September 23.

23

Richard E. Bellman

Arjang A. Assad

The post-World War II (WWII) development of operations research (OR) owes much to two organizations: Princeton University and the RAND Corporation. Richard (Dick) Bellman's formative years were spent at both—associations that shaped his research interests and led him to areas of applications that included well-known OR problems. His interest in mathematics developed early; he trained as a mathematician and always regarded himself as one. The Princeton mathematics department offered him an excellent graduate program, his first teaching opportunity, and, especially, a research environment that introduced him to key areas of his future research program. The exceptional group of scientists at RAND exposed him to game theory, optimization theory, and many classical applied problems in OR. Dick's invention and development of dynamic programming (DP) in the early 1950s marked the beginning of a new era in the analysis and optimization of large-scale systems. His research established him as a towering figure among the contributors to modern optimization, control theory, and systems analysis. Dick was among the first to realize the transformative impact of computers, not just in terms of the computational power required for obtaining numerical solutions, but in the very nature of how decision problems would be approached and solved. He was noted for his prolific output—he published over 600 research papers and over 40 books.

Dick was a recipient of the John von Neumann Theory Prize of the Operations Research Society of America (ORSA) and The Institute of Management Sciences (TIMS), the Norbert Wiener Prize in Applied Mathematics from the American Mathematical Society and the Society

A.A. Assad, S.I. Gass (eds.), *Profiles in Operations Research*, International Series in Operations Research & Management Science 147, DOI 10.1007/978-1-4419-6281-2_23,

for Industrial and Applied Mathematics, and the Medal of Honor from the Institute of Electrical and Electronics Engineers (IEEE). He was elected a member of the National Academy of Sciences.

EARLY LIFE AND EARLY COLLEGE DAYS

Richard Ernest Bellman was born on August 26, 1920 in Brooklyn, New York. At the time of his birth, his parents, John James Bellman and Pearl Saffian Bellman, were 20 and 18 years old, respectively. Dick did not know his paternal grandparents. His grandfather had emigrated to the U.S. from Kiev to escape the threat of political persecution in Czarist Russia and died before Dick was born. He ran a small grocery store on Bergen Street near Prospect Park in Brooklyn where John was born. His grandmother died when Dick was a year old.

Dick's maternal grandfather, Samuel Saffian, was born in Poland and emigrated to the U.S. at the age of 14, arrived in New York City, and found his way to Philadelphia to the home of his brother. Samuel married Mildred Apple who had emigrated from Germany at the age of 4 and was brought up in Philadelphia. They later moved to New York City.

When Dick was 10 years old, John moved the family from the Bronx to Brighton Beach in Brooklyn. The family was well off financially, and Dick enjoyed childhood comforts until John decided to go into business for himself. He became a victim of the Great Depression, compounded by a series of unwise job choices; the family eventually went on welfare. John died of a heart attack in 1943. Mildred died a few years later when Dick was in his twenties (Bellman et al. 1986).

Dick's interest in mathematics developed early. At the age of 11, he had his first serious encounter with mathematics by working through Schultze's *Elementary and Intermediate Algebra* (1925). He attended Brooklyn's Abraham Lincoln High School where his mastery of algebra made him a star pupil, with, of course, his favorite subject being mathematics. He was a member of the mathematics team and ranked the city's top mathematics student during his senior year. Dick's superb memory enabled him to excel in Latin, English, and history. He retained a lifelong interest in languages and history.

Dick graduated from high school in January 1937 and enrolled in the tuition-free City College of New York (CCNY). At that time, CCNY was noted for its very strong faculty and its excellent academic student body. All entering freshman were required to take a college aptitude test. Dick attracted some attention by scoring 360, as compared to the CCNY average of 220 and the national average of 180. A year later, he was asked to take the test again and scored 598 out of 600!

The one-way daily commute to CCNY involved a 1-hour train trip. Dick recalled: "My chief memory of this period was one of hunger. My grandfather had given me one dollar a week for carfare and lunch. One dollar a week covered carfare, leaving nothing for lunch" (Bellman 1984, 40–41).

In 1938, Dick transferred to Brooklyn College which was located only 20 minutes away from his home by subway. An additional advantage, compared to CCNY, was that Brooklyn College had girls; CCNY was effectively an all-boys school—women were allowed to enroll only in its graduate program. Dick had entered college intending to become a theoretical physicist. He pursued this goal at Brooklyn College while majoring in mathematics. Because the mathematics courses took little effort on his part, he was able to concentrate on his physics courses. He found the physics classes to be dull, plus they required experimental work that turned him off. He decided to choose a career in mathematics.

Dick's interest in languages remained strong and he took four semesters of Greek, reading the classics. It was in these classes where he recognized the power of his memory—he never took notes. He also took 2 years of German, knowing that it would be useful to read important mathematical works in the original German. He also enjoyed writing and was the editor of the *Brooklyn College Mathematics Mirror* for 2 years. In his junior and senior years, he was a member of the college's three-person mathematics team, which on March 4, 1939, won the William Lowell Putnam Mathematical Competition for colleges in the U.S. and Canada (sponsored by the Mathematical Association of America). Each member of the winning team received an award of $50. The second prize went to MIT whose team included the future renowned physicist Richard Feynman (Cairns 1939).

Dick graduated in 1941 with a B.A. in mathematics from Brooklyn College. He was to receive an honorary medal upon graduation, but, instead, he asked to be given a mathematics text; his gift was a copy of the famous book by Whittaker and Watson (1927), *A Course of Modern Analysis.* He followed this up with his first mathematics book purchase, *Theory of Functions* by Titchmarsh (1932).

Dick had many friends while in college, but the first girl he was drawn to was his high-school sweetheart, Betty Jo Kates, who studied history and literature at Brooklyn College. They dated and made plans for the future. Dick once said that Betty Jo had the quickest mind of anyone he had met and meant this as a purely factual statement (K. Bellman 2009). They decided to get married after completing their doctoral degrees, but circumstances were to change this decision.

Dick chose Johns Hopkins University for his graduate studies and started in September, 1941. Johns Hopkins had a strong faculty in mathematics (Zariski, Van Kampen, Wintner, and Kershner) and excellent graduate students. He roomed with two students and lived very frugally at a cost of 26 cents a day (Bellman 1984, 57). Due to the "uncertainties in the world situation," Dick and Betty Jo got married on November 22, 1941—he was 21 and she was 18. "We were both too young," he later commented (Bellman 1984, 60).

WWII interrupted Dick's plan of becoming a mathematician. To avoid being drafted, he searched for a war-related job. He found one in 1942, located in Belleville, Wisconsin, as an instructor of electronics. Because he did not

> MOCK TURTLE
>
> While he was a graduate student at Johns Hopkins University, Dick lived very cheaply in an apartment with two roommates. Their landlord had very strict rules that disallowed pets. This caused Dick to launch an involved practical joke. He bought a small turtle at a pet shop and begged the landlord that he be allowed to keep it. When this was agreed to, Dick went back to the pet shop each week and exchanged the current turtle for a larger one. After a while, the landlord was flabbergasted by the size of the turtle and insisted that it be removed. Dick pleaded again, promising that he could shrink the turtle back to a more acceptable size. The process of exchanging turtles was now reversed. The landlord could never understand how Dick accomplished the shrinking (K. Bellman 2009).

know the subject, he was sent to nearby Scott Air Force Base, where he trained as instructor in radio and electronics. After completing the course, he was transferred to Truax Air Field, Madison, Wisconsin, where he taught electronics to soldiers—to his surprise, Dick enjoyed the subject. Betty completed her semester at Brooklyn College and decided to become a nurse at the onset of WWII. She joined Dick in Madison in June.

Dick has termed his Madison, Wisconsin, period as his Horatio Alger (rags to riches) phase. It started with both Dick and Betty Jo enrolling in the University of Wisconsin. Dick's teaching schedule was the night shift from 11:00 p.m. to 7:00 a.m., which allowed him to attend graduate school during the day. This grueling schedule eventually proved detrimental to his health and he repeatedly fell ill. During a three-week period, when he was quarantined for one of these illnesses, he read mathematics incessantly and even wrote a paper, "On almost orthogonal series" (Bellman 1944). In Madison, Dick completed what he called his "first piece of good mathematics" (Bellman 1984, 74). While investigating a result in stability theory (Bellman 1943a), he established the Bellman-Gronwall inequality (Bellman 1943b; Beckenbach and Bellman 1965, 134–135), a result that proved highly useful in his later research. Dick obtained his M.A. degree in mathematics from the University of Wisconsin at the end of 1943.

PRINCETON TO LOS ALAMOS TO PRINCETON

At Madison, Dick met Stan Ulam whom he described as "a first-rate mathematician" (Bellman 1984, 74). The two had many conversations about scientific philosophy, mathematics, and modeling. When the war broke out, Dick wanted to move back East and appealed to Ulam for help. Ulam used a "Machiavellian way" to bring Dick to the attention of Solomon Lefschetz (chairman of Princeton University's Mathematics Department). He wrote to Lefschetz saying that he "had a very able student who was so able that he deserved considerable financial support," adding that he "doubted that Princeton could afford it" (Ulam 1976, 132). Thus challenged, Lefschetz called Dick to offer him a position in July 1943. Dick and Ulam were to reunite 2 years later at Los Alamos.

Dick was concerned that he could not accept Lefschetz's offer to pursue graduate studies at Princeton for fear of being drafted immediately, but Lefschetz suggested that he join Princeton as an instructor in the Army Specialized Training Program (ASTP). Dick had always dreamt of going to Princeton but had considered the dream unattainable. He accepted immediately and left Madison in September 1943. Most valuable to Dick was the friendship he developed with Lefschetz: "Lefschetz often expressed strong views . . . he was occasionally wrong but never in doubt. Probably, it was because our temperaments were so similar that we became such close friends" (Bellman 1984, 119).

Dick's duties at Princeton were to teach five sections of calculus, each meeting three times a week. This amounted to 15 hours of teaching at Princeton compared to 42 hours teaching electronics at Madison. His salary was $2600 a year. Along with his teaching, he took graduate mathematics courses, including a course in point-set topology from A. W. Tucker. Life at Princeton as a graduate student suited Dick, but was not to last. The Army disbanded the ASTP in March 1944. The men enlisted in the program were sent to basic training in preparation for joining the European theater of war.

In April 1944, Glen Camp offered Dick a position at the Naval Radio and Sound Laboratory located in Point Loma, San Diego. Camp was a mathematical physicist and a former student of Robert Oppenheimer at Berkeley. He ran the theoretical section of the laboratory. Dick accepted knowing that he could still be drafted. He did some mathematical work on sonar and also some experimental work. He was not to excel in this latter area and "ruined several thousands of dollars of equipment" (Bellman 1984, 85). Dick was drafted in December 1944 and assigned to the Los Alamos National Laboratory. He and Betty Jo were united on Christmas Day 1944 in Albuquerque and then headed for Los Alamos.

Dick spent 2 years working in Los Alamos's Theoretical Physics Division, where he reported to the numerical section. Initially, his supervisor thought of using him as a computer, performing calculations required for solving differential equations on desk calculators. But Dick was looking for a more challenging problem. He was referred to Hans Bethe who sent him to his collaborator, Robert Marshak.

Marshak asked Dick to solve a supposedly difficult equation. This equation was of considerable interest to the physicists as it estimated the yield of the first nuclear-bomb test at the Alamogordo Bombing Range conducted in August 1945. The fact that Dick was able to solve this equation analytically in fifteen minutes caught the attention of the physicists who had considered the equation difficult.

At Los Alamos, Dick's close circle of friends included Norman Greenspan (whom he knew as a teenager) and the mathematicians Peter Lax and Milt Wing, whom he met at Los Alamos. Dick was discharged from the Army on March 1, 1946 and returned to Princeton immediately. He studied for his qualifying examinations with the intention of quickly completing his doctoral studies. The qualifying examiners were Lefschetz, John Tukey, and Max Shiffman. Initially, Dick had intended to work with Salomon Bochner, a pure mathematician he respected and taken a course from in analysis, and pursue his early research interest in analytic number theory. Lefschetz, however, urged him to write up his results on stability theory. This provided Dick with the quickest path to earning his Ph.D., which he received in June 1946. His dissertation was "On the boundedness of solutions of non-linear differential and difference equations," with his results appearing in Bellman (1947a, b, 1949).

Lefschetz had secured funds from Mina Rees, head of the mathematics branch of the Office of Naval Research (ONR), to support work in differential equations. These funds supported Dick as a postdoctoral fellow. The research program at Princeton in nonlinear differential equations also included Joseph LaSalle and Peter Lax. Dick's summer was spent gathering material for an ONR monograph on solutions of nonlinear differential and difference equations. He also gathered notes for his first book, *Stability Theory of Differential Equations* (Bellman 1953b), now considered an early classic in the field. This marked the start of his seminal work in control theory.

Number theory remained a strong interest of Dick's while he was at Princeton. His frequent collaborators were Hal Shapiro and Ernst Straus. His interest in analytic number theory was to persist: He published an introductory volume on the subject late in his career (Bellman 1980). He also taught himself enough Russian to read papers in mathematics and became a frequent reviewer of Russian papers for *Mathematical Reviews*.

A major event was the birth of his son, Eric, on October 13, 1947. This led Dick to develop a technique for stopping infants from crying. He would "hold him up with one hand and let him examine the ceiling. For some reason, this always worked" (Bellman 1984, 125).

STANFORD AND RAND 1948–1951

Because Princeton had a tradition of not hiring its recent graduate students, Lefschetz advised Dick to look for a permanent position in November 1947. He interviewed at a few places and was offered a position at Stanford. This prompted Lefschetz and Tukey to break with tradition and offer him a permanent position at Princeton as an assistant professor. Although he had always thought of this as his dream position, and knew that Lefschetz would be hurt if he refused, he chose Stanford over Princeton when Gabor Szego, chairman of the mathematics department, offered him the position of associate professor at Stanford. His salary was $5000 per year, and he was 28 years old.

Having lost his summer support at Princeton, Dick was offered a position with General Electric Corporation near Boston. About the same time, Ted Harris, a recent Princeton Ph.D. in mathematics and then at the RAND Corporation (a U.S. Air Force–sponsored research organization in Santa Monica, California), arranged a summer job for Dick at RAND at a salary of $209 a week, about twice of what General Electric had offered. Dick chose California over Boston, a decision which led to his long association with RAND. Intellectually, Dick found RAND to be a highly exciting place.

The summer of 1948 at RAND turned out to be a stimulating and formative period for Dick. As he noted, "I was exposed to a number of significant mathematical ideas. Large systems, effective numerical solution, the application of mathematics to the social sciences, mathematical model making, theory of games and branching processes" (Bellman 1984, 142). RAND also provided a perfect environment for his introduction to OR. Dick specifically mentions George Dantzig's algorithm for linear programming (LP) as his "first exposure to effective numerical solution, which subsequently became a central theme of my research" (Bellman 1984, 135). At the time, RAND's staff included Abe Girschick, George Brown, Ted Harris, Alec Mood, and Lloyd Shapley. Its summer program attracted the Princeton economist Oskar Morgenstern, as well as future luminaries George Dantzig, Samuel Karlin, John Tukey, Dave Blackwell, Kenneth Arrow, and Martin Shubik.

Game theory was a key research area of RAND's mathematics division, where John von Neumann made frequent visits. Dick also collaborated with the mathematician David Blackwell on game theory. They published a paper on bluffing (Bellman and Blackwell 1949), which was communicated to the National Academy of Sciences by von Neumann and was the subject of an article the New York Times ran on December 25, 1949. They also wrote an article on two-person poker that was published as the cover story for *Scientific American* (Bellman and Blackwell 1951). And, completing a busy summer, he also worked on branching processes and probabilistic models with Ted Harris (Bellman and Harris 1951, 1952).

PRINCETON MOVERS, INC.

Harold Kuhn, a Princeton mathematics student, whose financial situation was rather perilous, found himself in the need of money to buy a pair of shoes. As Harold relates:

"... I accepted a job moving the furniture of Dick Bellman, the father of 'dynamic programming' (Bellman 1957f). It happened this way. One morning at breakfast in the boarding house where I took my meals, a moving man appeared. He explained that his assistant was drunk and unable to carry furniture from a house to a moving van. He offered $10 to anyone who was willing to replace his drunk helper. The house moving seemed to be an easy one; it involved the contents of a small house in the project known as the 'barracks,' *temporary* housing built in 1946 for military personnel returning to the university after the war. ...

"From the moving man, I learned it was Dick Bellman's furniture that I would be moving. He was leaving Princeton for an appointment as an associate professor at Stanford. It seemed like an easy way to earn $10 because the house was so small. But I was wrong! Dick Bellman's wife had been a contestant on a television quiz show in which the prizes were household appliances. Thus, the 'small house' contained two refrigerators, two stoves, two dishwashers, two television sets, etc. In short, I worked very hard for my pair of shoes" (Kuhn 2002, 132).

Upon arrival at Stanford, Dick bought a house in the Searsville Lake district, a few miles West of the university. At Stanford, Dick and Betty Jo cultivated a social life, especially with the economist/mathematician Kenneth Arrow and his wife Selma, the statistician Abraham Girschick and his wife Mary, and Albert and Betty Bowker (he was chairman and key architect of Stanford's department of mathematical statistics). They also visited the mathematicians Rafael and Julia Robinson in Berkeley (Julia was at RAND during the summer of 1948). His daughter, Kirstie, was born on July 20, 1949.

At Stanford, Dick taught courses on applied mathematics and algebra. He also set out to strengthen the graduate offerings, teaching a graduate course on stability theory. Dick's first Ph.D. student and the only one from Stanford was Kenneth Cooke. Dick suggested Cooke's 1952 dissertation topic, "The asymptotic behavior of the solutions of linear and non-linear differential-difference equations." They later coauthored a number of papers and books. Dick also continued his research on number theory, working with his Princeton collaborators Ernst Straus and Hal Shapiro (Bellman 1950, 1951; Bellman and Shapiro 1952; Bellman and Straus 1949). In addition to Shapiro who was working at RAND, Dick found a number of collaborators among other RAND mathematicians.

The important shift in Dick's research program occurred in summer of 1949, when he went back to RAND. Ed Paxson, a RAND research engineer, who was interested in weapons system analysis, suggested that he work on multistage decision processes. Earlier, Dick had studied multistage games with Joseph LaSalle, but he now made this area the key focus of his future research program. This decision was to lead to the creation of dynamic programming (DP) as the central strand in a three-pronged research agenda that occupied most of his attention in the early 1950s. The other two strands were control theory and time-lag processes. It is important to remember the weak status of these areas when Dick defined his research program:

> At the time, dynamic programming was barely known to a few people; control theory was not the fashionable field it is now; differential difference-equations did not occupy the important position they do now ... (Bellman 1984, 154).

Apart from the time spent at RAND, Dick's stay at Stanford was interrupted by an invitation to Princeton from the theoretical physicist John Wheeler. Wheeler was working on the hydrogen bomb project, considering it as a necessary national security measure and wanted Dick on his team. Taking a leave of absence from Stanford for 1951–1952, Dick joined with Wheeler but, after several months, did not find much to do for the project. He left the project to arrive early at RAND, where he had planned to spend the summer of 1952.

RAND 1952–1965

In 1952, Dick faced a major decision: to continue teaching at Stanford (where he was tenured) or take a full-time position at RAND. A key attraction of Stanford was his freedom to pursue his own research agenda such as analytic number theory, a subject he had wanted to investigate ever since he was sixteen. While at RAND, he would continue to develop the theory and applications of DP. He chose RAND, moved to Los Angeles in 1952, and bought a house in West Los Angeles for $32,000.

Dick at RAND

At RAND, Dick worked closely with several collaborators. These included Irving Glicksberg and Oliver Gross on control theory, and Kenneth Cooke on differential-difference equations. Dick also identified Stuart Dreyfus as an able collaborator. Stuart, a Ph.D. mathematician from Harvard, was initially employed as a computer programmer, working the night shift on the JOHNNIAC computer [named after John von Neumann, (Gruenberger 1968)]. He soon became a frequent coauthor and close associate. During this period, Dick worked on the shortest path problem (Bellman 1958a), invariant embedding (Bellman and Kalaba 1959), and scheduling (Bellman 1956c). He also gave a series of lectures at the Hughes Aircraft Company covering his research on control processes. These lectures were edited and expanded into the book *Adaptive Control Processes: A Guided Tour* (Bellman 1961).

Early in his career (1948), Dick thought of using mathematics to model physiological processes related to cancer. The impact of his having worked on the atomic bomb always weighed heavily upon him—he viewed his research in the medical sciences as a way of serving society (K. Bellman 2009). As computers proved to be increasingly powerful, he felt that computational testing of realistic models was within reach. In 1958, he visited the Sloan-Kettering Institute looking for problems and collaborators in the area of cancer therapy. There, he met John Jacquez, a medical doctor, who was also a researcher in cancer chemotherapy (Roth 1986, 411). Together, they developed models of how a drug injected into the bloodstream distributes itself in the body over space and time (Bellman et al.

1960a, b) . Dick's early work in this area resulted in RAND's first grant from the National Institutes of Health (Jacquez 1985, 3).

It was not all work at RAND. Dick played kriegspiel (blind chess—neither player can see the other's board) against Ray Fulkerson, Stuart Dreyfus, Lester Ford, and Lloyd Shapley. Dick was an excellent tennis player and often played with (or against) Merrill Flood and Ray Fulkerson. With Ray as his partner, they won the 1959 West Los Angeles Industrial Tennis League Doubles Championship without dropping a single set.

On a more personal front, Dick and Betty Jo separated in 1962. Dick married Nina Day—a native Californian who was 17 years his junior. In the 1960s, when Dick made a number of international trips to present his research, Nina was often his traveling companion. One of Dick's major trips took him around the world in 1963, the first time he left the U.S.. This trip was so tiring and taxing that he vowed never to leave Santa Monica again.

TRIPLE APPOINTMENT AT USC: 1965–1984

In 1965, Dick became Professor of Mathematics, Electrical Engineering, and Medicine at the University of Southern California (USC), Los Angeles, an appointment in which he excelled by applying his talents for teaching, research, writing, and the mentoring of the next generation of students. He started a program in applied mathematics that included invariant imbedding techniques in physics and the numerical solution of functional equations. He also taught a 2-year sequence of courses that covered DP, control theory, invariant imbedding, and mathematical biosciences. His research in these areas, based on lectures given at the University of Kentucky in 1966, is described in the book *Some*

Dick and Nina

Vistas of Modern Mathematics (Bellman 1968). In its preface, the optimistic Dick writes: "To the chairman of a department, or to a dean, I hope it will be useful in planning the new curriculum for the department of mathematics or in explaining the many different kinds of interaction that can exist between a computer installation and the Department of Mathematics" (Bellman 1968, viii).

At USC, Dick's research focused increasingly on the application of mathematics to medicine and the biological sciences, including the scanning processes for the location of a tumor. He continued his work in artificial intelligence and simulation in psychotherapy (Bellman and Smith 1973), and also developed an interest in neurophysiology. He founded the journal *Mathematical Biosciences*, one of the premier journals in mathematical biology, and authored the book *Mathematics in Medicine* (Bellman 1983).

Dick continued to accept international lecture engagements. He visited Moscow and Leningrad in 1966, and France in 1968 where he gave lectures at the Sorbonne University, and the French Institute for Computer Science and Automatic Control (INRIA) in Versailles. In 1971, he visited several universities in New Zealand, and in 1972, he lectured at the Tatung Institute in Taiwan. In Taiwan, the highlight of the trip was the meeting Dick and Nina had with Madam Chiang Kai-shek as Nina's mother had been born in China (Bellman 1984, 316–318).

In the early 1970s, Dick started to have problems with his balance and found his tennis game deteriorating. His difficulties were diagnosed as being due to a benign tumor of the 8th cranial nerve (neuroma). He was operated on in 1973, but was crippled from postsurgical complications. This resulted in serious disabilities in speech and in movement and required him to use a wheelchair—a most difficult situation for anyone, but even more so for someone who had led an active athletic life. Dick could not talk for almost a year. Remarkably, during this period he planned the entire content of a book in his head so that it could be put to paper later (K. Bellman 2009). Despite his handicaps, he continued to work and collaborate with others and retained his sense of humor. Remarkably, he published close to 100 papers after the operation (Dreyfus 2003). Even in his last month, he wrote a paper on fuzzy DP for the *TIMS Studies of Management Sciences* series (Esogbue and Bellman 1984). He died in Los Angles of a heart attack on March 19, 1984. He was survived by his wife Nina and his children, Eric and Kirstie.

Together, they wrote a touching tribute to Dick (N. Bellman et al. 1984). Nina Bellman passed away in August 2007. Eric Bellman is a psychotherapist working on mental health and juvenile justice issues in California. He originated the "Bellman Syndrome"—a way of treating borderline personality disorders. Kirstie, who has a Ph.D. in neuropsychology, is a principal scientist at Aerospace Corporation, where she works on the interface of cognitive sciences and artificial intelligence, developing new models and architectures for systems integration (K. Bellman and Landauer 2000).

THE BELLMAN CONTINUUM

Given the scope of this profile, it is impossible to reflect the extent of the total contributions Dick made to various fields of mathematics and related application areas. A sample that spans the main strands of his research has been collected in *The Bellman Continuum* (Roth 1986), a compilation of 65 papers and several speeches. Here, we review those of his contributions that are most relevant to OR, optimization theory, and decision making.

DYNAMIC PROGRAMMING

Dick's name will be forever associated with DP. How did DP get its name? It is hard to avoid the impression that the choice of programming was a nod to LP, as Dick seems to have indicated to Harold Kushner (2004, 5). The word dynamic appealed to him for two reasons: It had a precise meaning in the physical sciences, and, as an adjective, it was impossible to use it in a pejorative sense. Even if the words dynamic and programming appealed to Dick, the specific conjunction—dynamic programming—was suggested by Betty Jo (K. Bellman 2009). More broadly, Dick's work helped to establish much of the standard terminology of multistage decision processes, giving it such terms as imbedding and policy approximation (Bellman 1984, 159; Dreyfus 2002).

For many, Dick is best known for his principle of optimality:

> An optimal policy has the property that whatever the initial state and the initial decision are, the remaining decisions must constitute an optimal policy with regard to the state resulting from the first decision (Bellman 1957f, 83).

The simplest and most intuitive example of optimal policy is DP applied to the shortest path problem. Here, the shortest path from node p to node t also provides the shortest path from node q to node t, where q is the node chosen as the first node to visit after node p. For Dick, this principle served an important function:

> My first task in dynamic programming was to put it on a rigorous basis. I found that I was using the same technique over and over to derive a functional equation. I decide to call this technique, "The Principle of Optimality." Oliver Gross said one day, "The principle is not rigorous." I replied, "Of course not, it is not even precise." A good principle should guide the intuition (Bellman 1984, 174).

Another DP example is an allocation problem. An amount p of a single resource is to be deployed successively in pursuit of some gain. Let $f(p)$ denote the optimal gain if the best allocation policy is followed. Further, let q be the amount of the resource deployed in the current move which is known to yield a gain of $g(q)$. And, as a result of this move, the amount of the resource that remains is $T(p,q)$. (We assume that both g and T are known functions.) Then the principle of optimality allows us to state the functional relation

$$f(p) = \mathrm{Max}_q\{g(q) + f[T(p,\ q)]\}. \tag{1}$$

Dick believed that the subject area of DP "can be partitioned in several ways, either with respect to the precise form of (1), or with respect to the process giving rise to (1), or with respect to deterministic, stochastic, or adaptive features, and so on" (Bellman and Dreyfus 1962). Viewed in this way, the DP-based approach could be applied to a wide spectrum of problems in subjects ranging from optimization to control theory, as well as stochastic processes.

A central contribution of Dick's was his use of functional equations for the analysis and solution of multistage problems. Current familiarity and wide use of functional equations should not obscure the novelty of the technique when Dick first introduced it. He was particularly adept in using this technique in unexpected settings (Dreyfus 2008). He took great pleasure in showing how one could derive the maximum height of a projectile shot directly upward from a simple functional (Bellman and Kalaba 1965, 17).

WHERE TO MINE

An example of a DP problem and its associated functional equation is Dick's gold-mining problem. He discussed it in several publications—in an early paper (Bellman 1954b) and in his text (Bellman 1957f); it is reviewed by Danskin (1959). The resulting functional equation is not trivial, but the optimal policy can be derived with some effort.

Two gold mines, Anaconda and Bonanza are mined with a single mining machine. Suppose that Anaconda contains x units of gold, while Bonanza has y units. The mining machine is rather delicate and subject to breakdowns. If it is used to mine Anaconda, there is a probability α that it will end up in good working order. If so, it will mine a fraction r of the gold. With probability $(1- \alpha)$, however, it will break down and come up with no gold. The corresponding parameters for Bonanza are probability β and fraction s of the gold. The decision problem is as follows: Choose an initial mine to send the machine to. If it comes up unbroken, decide the next mine to send it to. Continue until a breakdown occurs. We seek the optimal policy (choices at each stage) that yields the maximum expected gold.

To derive the functional equation, let $f(x, y)$ be the expected amount of gold resulting from an optimal policy when the two mines contain x and y units of gold, respectively. If the machine is sent to Anaconda, then with probability α it will produce rx units of gold in this stage and the amount of gold left in Anaconda will be reduced to $(1 - r)x$. So future decisions in the optimal policy will yield $f((1 - r)x, y)$ The sum of these two terms gives a yield of $\alpha[rx + f((1-r)x, y)]$ if the machine is assigned to Anaconda and, by similar reasoning, $\beta[sy + f(x, (1 - s)y)]$ if assigned to Bonanza. The optimal decision selects the maximum of these two expressions, with the functional equation being

$$f(x, y) = \text{Max } \{\alpha[rx + f((1-r)x, y)]; \beta[sy + f(x, (1-s)y)]\}.$$

Bellman (1957f) develops the solution to this equation in a few pages and arrives at a result that can be stated elegantly. The end machine should be assigned to Anaconda if $[\alpha/(1-\alpha)]rx > [\beta/(1-\beta)]sy$, to Bonanza if the reverse inequality holds, and to either if equality holds (Danskin 1959).

Dick was the best expositor and disseminator of DP, especially with papers that provided the foundations for the subject. His first DP paper, "On the theory of dynamic programming" appeared in the *Proceedings of the National Academy of Sciences* (Bellman 1952). To introduce the new area more widely to researchers, Dick published DP papers in *Econometrica* (Bellman 1954a, 1955a), *Journal of the Operational Research Society of America* (Bellman 1954b, 1957b), *Management Science* (Bellman 1956b, 1957c), and *Science* (Bellman 1966). The task of providing a rigorous mathematical foundation for the theory was carried out in a series of papers culminating in Bellman (1957a). In parallel, Dick wrote several papers illustrating the use of DP in a wide range of OR applications: optimal allocation (Bellman 1954b), smoothing problems (Bellman et al. 1954), equipment replacement theory (Bellman 1955b), warehousing

(Bellman 1956b), scheduling theory (Bellman 1956c; Bellman and Gross 1954), tactical air-warfare (Bellman and Dreyfus 1958a), reliability theory (Bellman and Dreyfus 1958b), routing (Bellman 1958a), and location–allocation problems (Bellman 1965).

Dick also introduced the concept of Markovian decision processes (Bellman 1957e). His well-known label-correcting algorithm for shortest path problems appeared in Bellman (1958a) and Bellman and Kalaba (1960). The interface between DP and mathematical programming was explored in Bellman (1956d, 1957d) and Bellman and Kalaba (1962) and reviewed by Kagiwada and Kalaba (1986).

Dick's first text on DP, *Dynamic Programming*, was published by Princeton University Press (Bellman 1957f); every chapter captures aspects of Dick's DP research. The import of this book was captured 46 years later by Yale University professor Eric Denardo:

> Typically, a good mathematics book—and this is a very good mathematics book—surveys a field whose fundamentals are deeply understood. By contrast, in this book Bellman described a research program that he was avidly pursuing. By writing it, he generously invited other researchers to jump in. Many did. It is no overstatement to say *Dynamic Programming* launched a thousand professorships [(Dover Publications reprint 2003, v; Bellman (1957f)].

The publication of *Dynamic Programming* illustrates another facet of Dick's personality. Being confident that the book would do well, he was anxious to have it published by Princeton University Press—it was his way of saying thank you for his Ph.D. (Bellman 1984, 178). The book turned out to be a best seller, with sales of more than 13,000 copies by 1983—the Press made a profit. Dick's second book on DP, *Applied Dynamic Programming*, was coauthored with Stuart Dreyfus (Bellman and Dreyfus 1962). Extensive discussions of DP also appeared in his books on control theory (Bellman 1961; Bellman and Kalaba 1964, 1965).

TIME-LAG PROCESSES AND DIFFERENTIAL-DIFFERENCE EQUATIONS

Given Dick's earlier interest in stability theory for differential equations, it was natural for him to conduct similar analyses of differential-difference equations, work he started in the early 1950s. These are differential equations in which the original function or its derivatives appear with time dependence, not just on the usual time parameter t, but also on a discrete set of time lags $(t - t_1)$ for t_i $(i = 1, \ldots, k)$ with $0 < t_1 < t_2 < \ldots < t_k$. In addition to applications in physics, engineering, and biomedical processes, these equations also arise in control processes due to the lags in the system dynamics. This resulted in one of his earliest publications in this area (Bellman 1947b). At RAND, he continued this interest by writing a survey of the field with the mathematician John Danskin (Bellman and Danskin 1954). Dick continued this line of research by investigating existence theorems, asymptotic behavior, stability questions, and series expansions for differential-difference equations (Bellman and Cooke 1959). Later, his extensive collaboration with Kenneth Cooke expanded into the book, *Differential-Difference Equations,* the first comprehensive exposition of this class of equations (Bellman and Cooke 1963).

INVARIANT IMBEDDING

A major strand of Dick's research is his seminal work on invariant imbedding (Bellman 1962; Bellman et al. 1960), initially inspired by the work of the Nobel laureate Chandrasekhar (1960). The idea of imbedding a particular problem within a larger class of problems is employed within mathematics and mathematical physics and is a key factor in the functional equation approach employed by DP (Bellman and Kalaba 1959). The idea is to formulate functional equations, defined for a broader set of values of the problem variables, that yield the solution to the original problem when the functions are

> EXPAND AND CONQUER BY IMBEDDING
>
> "Consider a discrete random walk for a particle with jumps of size one (in either direction). The position of the particle is restricted to the integer points $\{0, 1, \ldots, N\}$, and we treat 0 and N as absorbing barriers. What is the probability $\pi(k)$ of the particle starting in position k reaching 0 before it reaches N? (This question is familiar as one posed in the gambler's ruin problem.) The invariant embedding approach involves extending the search to evaluate a more general function $\varphi(h, k)$ defined as the probability of a particle starting in position k reaching h before reaching N. Note that $\pi(k) = \varphi(0, k)$; we originally were interested in only one value of the function φ (for each k). Instead, the imbedding approach requires that we evaluate the function φ at a larger set of points. It turns out that computing the values of $\varphi(h, h+1)$ for various values of h allows us to compute all other values of interest" (Bellman and Wing 1975, 67–72).

evaluated for specific values of the variables. Dick called this DP without optimization.

CONTROL THEORY

Dick's research interests in differential equations and analysis naturally led him to variational problems in optimization theory. He first studied such problems using standard calculus of variations procedures. His joint work, with RAND associates Irving Glicksberg and Oliver Gross, investigated nonclassical problems and aspects of control theory (Bellman et al. 1956a, b). Some of their control theory results were published in a RAND monograph (Bellman, Glicksberg, and Gross 1958); it was translated into Russian and had a great influence on Russian researchers. Dick, however, soon convinced himself that calculus of variations did not provide an effective tool for solving variational optimization problems. Instead, he relied on the techniques of DP in his study of such problems (Bellman 1953a, 1954b, c, 1956a, 1957b).

His application of DP to control problems also highlighted a key insight—the fundamental importance of policies:

> How did one obtain the numerical solution of optimization problems? . . . I did not wish to grapple with this thorny question, and I had certainly not contemplated the application of dynamic programming to control processes of deterministic types. Originally, I had developed the theory as a tool for stochastic decision processes. However, the thought finally dawned on me that the desired solution in a control process was a policy: "Do thus-and-thus if you find yourself in this portion of state space with this much time left." Conversely, once it was realized that the concept of policy was fundamental in control theory, the mathematization of the basic engineering concept of "feedback control," then the emphasis upon a state variable formulation became natural. We see then a very interesting interaction between dynamic programming and control theory (Bellman 1984, 181).

> Basic to this procedure [DP] is the concept of a policy, a rule for telling what decision to make in terms of the current position of the system. The major advantage of this new control concept over the classical ideas of control lies in its flexibility. We are prepared for all eventualities. No matter what the current position, a policy informs us what control to exert. No longer are we bound by preconceived notions of the nature of the most desirable path. . . . implicit in the idea of a policy is the basic notion of learning from experience.

> Multistage decision making is regarded as the repeated application of a policy. A policy which is most efficient in the sense of minimizing time, or fuel, or cost or of maximizing profit is called an optimal policy (Bellman 1966, 35).

DP had a major influence on control theory research and application (Larson 1979). First, in the study of dynamic control problems, DP was the impetus for shifting the field from frequency-domain-based techniques to time-domain techniques; second, DP proved to be equally useful when the principle of optimality was extended to deal with uncertainty and applied to uncertain stochastic control problems. More specifically, his strategy of replacing two-point boundary problems with initial value problems offered significant computational advantages (Bellman et al. 1960). Dick's major exposition of control theory is the two-volume set, *Introduction to the Theory of Control Processes* (Bellman 1967, 1971).

Dick's research in control theory was groundbreaking as well as extensive—his inquisitive mind led him to contribute to diverse application areas (Leondes 1980; Panossian 1986). In his last technical book, *Mathematical Models in Medicine*, he described how medical treatments in drug therapy or radiation therapy can be formulated as optimal control problems that can be solved by DP or related techniques (Bellman 1983).

COMPUTERS

As a developer of solution techniques for DP, and variational and control problems, Dick was quite concerned about computational effectiveness. He was particularly sensitive to a major obstacle that he called the "curse of dimensionality," (Bellman 1957f , ix). The curse arises in DP, control theory, integer programming, combinatorial problems, and, in general, time-dependent problems in which the number of states and/or data storage requirements increases exponentially with small increases in a problem's parameters or dimensions. Although computer speed and storage, as well as analytical and computational advances, have eased the curse somewhat, it is appropriate to recall Dick's admonishment: "In view of all that we have said in the forgoing sections, the many obstacles we appear to have surmounted, what casts the pall over our victory celebration? It is the curse of dimensionality, a malediction that has plagued the scientist from the earliest days" (Bellman 1961, 94).

> SPADEWORK
>
> "When facing new classes of mathematical problems which cannot be treated by means of existing analytical devices, it is extremely important to be able to examine classes of numerical solutions in the hope of discerning patterns of behavior. . . . If we do not wish to suffer the usual atrophy of armchair philosophers, we must occasionally roll up our sleeves and do some spadework. With the aid of dynamic programming and digital computers we can methodically engage in mathematical experimentation" (Bellman and Dreyfus 1962, v).

Dick viewed the computer as the "single biggest single advance" in scientific methodology (Bellman 1984, 87). He was a proponent for using computers as a tool for exploration from which new mathematical conjectures could be formulated and as means of conducting mathematical experimentation. He believed that the discernment of patterns provides important clues to the mathematician for formulating propositions, constituting the crucial step in mathematical discovery (Bellman and Brock 1960).

Dick was especially drawn to new problem areas, whose resolution required the analysis of data and related information that were less precise or given in nonquantifiable terms. This motivated his early interest in fuzzy systems, a generic problem area, first proposed by the electrical engineer Lotfi Zadeh. Together, they addressed multistage decision processes in a fuzzy environment in which the system under control is either deterministic or stochastic (Bellman and Zadeh 1970).

Similarly, Dick pursued artificial intelligence from the broad perspective of whether computers can think—the subtitle of his book on the subject (Bellman 1978). In addition to decision making, simulation, and uncertainty, the book covers learning models, mathematical models of the mind, humor, consciousness, and learning. His contributions to computer science are reviewed by Lew (1986). Earlier, Bellman wrote:

> Much to his distress, the mathematician is confronted with the problem of analyzing "thinking." . . . As might be expected, there is no simple, or even unique, explanation of this phenomenon of the human mind. Let us consider ways in which a mathematician can approach this thorny subject. Do machines think? It is not surprising that a great deal of controversy, much of it emotional and visceral, has arisen over this issue. What is surprising is that many people who should know better are not aware of the fact that the question is devoid of meaning. Until we have defined what we mean by a

> "machine," what is meant by "think," and, particularly, what is meant by "can," all we can agree on is that a question—some question—is simplified (Bellman 1966, 36–37).

Simulation

One of Dick's early projects in simulation was the construction of a business game. The project came about when a senior partner at the management consulting firm of Booz, Allen, and Hamilton asked Dick to construct a business game for executive training at the American Management Association (AMA). At that time, such games were relatively new—the paper by Bellman et al. (1957) was one of the first to describe the deterministic business simulation game in detail. The game did not have an explicit objective function, but instead, the players were told to play the game as if this were an actual business operation (the formulas governing the evolution of state variables were purposely hidden from the players).

Dick's other major simulation effort was of a different nature. He wanted to investigate "how simulation as a technique for rational decision-making could be applied to areas in which qualitative features are important ones" (Bellman and Smith 1973, vii). With characteristic boldness, the area chosen was psychotherapy! The project, which started in the early 1960s, involved designing a simulated interview process between a therapist and an uncooperative subject, with the subject's behavior measured parametrically. The results of the study are reported in the book *Simulation in Human Systems* (Bellman and Smith 1973).

Author and Editor

During his career, Dick wrote nearly 50 books and monographs. The importance of writing texts was impressed upon him by the mathematician Aurel Wintner at Johns Hopkins, who required the graduate students to read the entire Borel series (the series of books on analysis written by Émile Borel). Dick wrote: "I was surprised that a mathematician of Borel's stature would busy himself with editorial duties. After reading these volumes, I understood how important it was to the progress of the mathematics that research results be widely available" (Bellman 1984, 57–58). At Princeton, Lefschetz reinforced the notion by encouraging Dick to write books. "He said that one could write a large number of papers and no one would see the themes. If I wanted to make things clear, I had to write books" (Bellman 1984, 118). Later, this inspired Dick to start the well-known series of books *Mathematics in Science and Engineering.*

INTERNATIONAL AUTHOR

In his National Academy of Engineering Memorial Tribute to Dick, Solomon Golomb, a USC professor in electrical engineering and a colleague of Dick's, commented on how popular and influential Dick's books were in Russia: "Dick's name is probably cited more at Soviet scientific meetings than at American ones. As an indirect proof of this, let me just mention that Dick was invited to be the main speaker at the first, and only, International Congress of Mathematicians held in Russia, in 1966. More than ten of his books have been translated and published in the Soviet Union. No other American scientist has been given such honors in the USSR" (Golomb 1989, 27).

Most of Dick's books were expositions of areas of research he was personally involved in and often included results from recent research or new research problems. He also wrote a number of textbooks. Apart from the books devoted to DP, the textbooks most relevant to OR were *Algorithms, Graphs, and Computers* (Bellman et al. 1976) and *Mathematical Aspects of Scheduling and Applications* (Bellman et al. 1982). Both cover a variety of well-known OR problems. To many students, Dick was known through his texts on differential equations (Bellman and Cooke 1968), matrix analysis (Bellman 1970), and his books on inequalities (Beckenbach and Bellman 1961, 1965).

Dick's personality shined through his books in many ways. Dick dedicated some books to famous people, usually mathematicians: Bellman and Dreyfus (1962) to John von Neumann; Beckenbach and Bellman (1965) to Hardy, Littlewood, and Polya; and Bellman (1971) to Stan Ulam. *Adaptive Control Processes* is dedicated to Lyapunov, Poincaré, and Lefschetz—readers are likely to be perplexed by the strange pictograms on the book's spine (Bellman 1961). Here, Dick wanted to "go Wiener one better." Norbert Wiener's term "cybernetics" was derived from the Greek word for steersman. Dick placed the Egyptian hieroglyphic for oarsman on the spine (Bellman 1984, 197).

Dick also founded and launched the *Journal of Mathematical Analysis and Applications* and the new journal *Mathematical Biosciences.* He served as the editor of the latter journal for the first 22 volumes. Dick's edited volumes related to OR include Bellman (1963, 1964).

BELLMAN AND SOCIETY

At various points of his life, Dick became engaged in societal issues at the local and extended levels:

> What is perhaps least known is Bellman's concern about problems facing society. In 1966, barely a year after the Watts riots in Los Angeles, when the country was in major social turmoil, Dick invited the Bootstrap Director to send students to computer classes given at night at USC Medical School and the staff classes given in the day. . . . That effort led in the next summer, to the first computer literacy course ever given at the high school level (Adomian and Esogbue 1984, 7).

From 1969 to 1975, Dick became involved with the Center for the Study of Democratic Institutions, Santa Barbara, California. Headed by the educator Robert M. Hutchins, the Center was established in 1959 with the objective of confronting and attempting to resolve the vital issues facing American society. In 1969, Hutchins invited Dick to propose new directions for the center. True to form, Dick proposed mathematics as an approach to understanding societal systems. He remained involved as an associate of the Center until it was reorganized in 1975.

According to Adomian and Lee (1986, 634), Dick, during his handicapped years, diverted his attention to a project designed to bring microcomputers into the institutions and homes of the handicapped where the handicapped were intelligently defined to include the gifted, the elderly, and the institutionalized.

ON THE PERSONAL SIDE

At age 18, Dick was 6 feet 1 inch tall, and weighed 174 pounds (a weight he maintained for most of his life). He was left-handed and a gifted athlete, especially at tennis, and could beat most people at table tennis.

Dick was a person of strong opinions and was not afraid of expressing himself, even in the face of what most would consider formidable opposition. He faulted the four bold predictions by Simon and Newell (1958) on what computers can achieve in 10 years as unscientific sensationalizing (Bellman 1958b). He attacked the conclusions stated by the futurist and political scientist Herman Kahn in his book *On Thermonuclear War* (Kahn 1962).

Dick felt that Kahn (with whom he was friendly on a personal level) had done "a very poor job of operations research." Dick stated that "he was willing to challenge any number in the book except the page numbers" (Bellman 1985, 210).

Dick was an indefatigable letter writer. He wrote letters to newspapers, friends, and businesses that he felt had done him wrong—he was not one to hold back his feelings. He was also an inveterate prankster and enjoyed playing practical jokes.

Dick had many collaborators drawn from his students and friends. Dick's USC students specialized in a range of areas: Austin Esogbue—OR in hospitals; Arthur Lew—computer science; Edward Angel—DP and invariant embedding; Christine Shoemaker—agricultural ecology; and John Casti—author of popular books on mathematics and systems. Kirstie Bellman, an accomplished scientist in her own right, gives us a glimpse of her father's influence on her integration science research in K. Bellman and Landauer (2000).

HONORS AND AWARDS

In 1970, Dick received the first Dickson Prize in Science awarded annually to a U.S. scientist who had been judged by Carnegie Mellon University to have made the most progress in the scientific field. Also, in 1970, he was the first recipient of the Norbert Wiener Prize in Applied Mathematics, awarded jointly by the American Mathematical Society and the Society for Industrial and Applied Mathematics, for his pioneering work in the area of dynamic programming and for his related work on control, stability, and differential-delay equations. In 1976, he was the second recipient of the ORSA and TIMS's John von Neumann Theory Prize for sustained contributions to theory in operations research and management science. He was elected a member of the National Academy of Engineering in 1977 and member of the National Academy of Sciences in 1983.

He was awarded the Institute of Electrical and Electronics Engineers (IEEE) 1979 Medal of Honor for contributions to decision processes and control system theory, particularly the creation and application of dynamic programming. He was elected a fellow of the American Academy of Arts and Sciences in 1975 and of the Society for

Mathematical Biology in 1980. He was awarded the 1977 Gold Medal of IEEE for DP and the Heritage Medal of the American Council for Control in 1983. He received honorary doctorates from the University of Aberdeen (1973), University of Southern California (1974), and the University of Waterloo (1975).

A REMINISCENCE

Hired as a programmer, I arrived at RAND in 1955, a math major 2 years out of college with no illusions of doing anything resembling mathematical research. Dick Bellman had just completed his book on various mathematical aspects of dynamic programming and had decided that the time was ripe for establishing its computational usefulness. By luck, I was assigned to work, I assumed, *for* him. He seems to have had other ideas. Rather than merely handing me a formula to be programmed, he led me into a conference room and lectured me on dynamic programming, of which I had never heard, and the reasoning behind the formula. He made it clear that I was to work *with* him. This led to a series of papers that Dick could easily have authored solely with perhaps an acknowledgement of my programming assistance, but which he insisted we co-author. With his encouragement I gradually began to think about other applications of dynamic programming, and he taught me how to write up and submit for publication a few papers on my own.

Then, after a couple years of this collaboration, if you can call it that, I was walking down a hallway at RAND one day and encountered Dick walking with a stranger. He introduced me to his companion, A. W. Tucker, and said "Stuart wants to go to graduate school at Harvard, but I think he should go to Princeton, don't you?" Tucker, I learned, was Chairman of the Princeton Mathematics Department and he immediately took the bait, encouraging me to come, instead, to Princeton. This whole conversation was news to me since I had given no thought to going to any graduate school at all, let alone Harvard or Princeton.

Off I went to Princeton, supported by RAND. (For the sake of full disclosure, I had neither interest nor ability for the kind of pure mathematics required by the Princeton mathematics department and soon transferred to applied mathematics at Harvard. Dick good naturedly once remarked to me that by leaving his alma mater, Princeton mathematics, to go to Harvard applied mathematics, I had raised the level of both departments.)

I still don't know what it was Dick saw in me that I had never seen in myself, but I am forever grateful. Often seen as distant and arrogant, I am living proof that deep inside Dick was capable of being a warm, caring and helpful individual. I am certain that I am far from being the only such recipient of his kind support (Dreyfus, 2008).

ACKNOWLEDGMENTS

I am indebted to Dr. Kirstie Bellman for her valuable input, her responses to our queries, and the pictures of Dick Bellman used in this profile. I am grateful to Professor Stuart Dreyfus for reviewing an early version of this profile and for supplying his reminiscences.

REFERENCES

Adomian, G, Esogbue A (1984) The contributions of Richard Ernest Bellman (1920–1984) to operations research and management science. OR/MS Today 11(4):6–7

Adomian G, Lee E (1986) The research contributions of Richard Bellman. Comput Math Appl 12A(6):633–651

Beckenbach E, Bellman R (1961) Introduction to inequalities. Random House, New York, NY

Beckenbach E, Bellman R (1965) Inequalities. Springer, New York, NY

Bellman K (2009) Personal communication

Bellman K, Landauer C (2000) Towards an integration science: the influence of Richard Bellman on our research. J Math Anal Appl 249(1):3–31

Bellman N, Bellman E, Bellman K (1986) Richard Ernest Bellman. In: Roth R (ed) The Bellman continuum. World Scientific, Singapore, pp 3–11

Bellman R (1943a) The stability of solutions of linear differential equations. Duke Math J 10(4):643–647

Bellman R (1943b) An integral inequality. Duke Math J 10(3):547–550

Bellman R (1944) Almost-orthogonal series. Bull Am Math Soc 50 (1):517–519

Bellman R (1947a) The boundedness of solutions of linear differential equations. Duke Math J 14 (1):83–97

Bellman R (1947b) On the boundedness of solutions of nonlinear differential and difference solutions. Trans Am Math Soc 62(3):357–386

Bellman R (1949) On the existence and boundedness of solutions of nonlinear differential-difference solutions. Ann Math Series 2 50(2):347–355

Bellman R (1950) Generalized Eisenstein series and non-analytic automorphic functions. Proc Natl Acad Sci 36(6):356–359

Bellman R (1951) On the functional equations of the Dirichlet series derived from Seigel modular forms. Proc Natl Acad Sci 37(2):84–87

Bellman R (1952) On the theory of dynamic programming. Proc Natl Acad Sci 38(8):716–719

Bellman R (1953a) Some functional equations in the theory of dynamic programming. Proc Natl Acad Sci 39(10):1077–1082

Bellman R (1953b) Stability theory of differential equations. McGraw-Hill, New York, NY

Bellman R (1954a) Some problems in the theory of dynamic programming. Econometrica 22 (1):37–48

Bellman R (1954b) Some applications of the theory of dynamic programming: a review. J Oper Res Soc Am 2(3):275–288

Bellman R (1954c) Dynamic programming and a new formalism in the calculus of variations. Proc Natl Acad Sci 40(4):231–235

Bellman R (1955a) Bottleneck problems, functional equations and dynamic programming. Econometrica 23(1):73–87
Bellman R (1955b) Equipment replacement policy. J SIAM 3(3):133–136
Bellman R (1956a) A variational problem with constraints in dynamic programming. J SIAM 4(1):48–61
Bellman R (1956b) On the theory of dynamic programming—a warehousing problem. Manage Sci 2(3):272–275
Bellman R (1956c) Mathematical aspects of scheduling theory. J SIAM 4(3):168–205
Bellman R (1956d) Dynamic programming and Lagrange multipliers. Proc Natl Acad Sci 42(10):767–769
Bellman R (1957a) Functional equations in the theory of dynamic programming-VI: a direct convergence proof. Ann Math 65(2):215–223
Bellman R (1957b) Dynamic programming and the numerical solution of variational problems. Oper Res 5(2):277–288
Bellman R (1957c) On a dynamic programming approach to the caterer problem. Manage Sci 3(3):270–278
Bellman R (1957d) On the computational solution of linear programming problems involving almost-block-diagonal matrices. Manage Sci 3(4):403–406
Bellman R (1957e) A Markovian decision process. J Math Mechan 6(4):679–684
Bellman R (1957f) Dynamic programming. Princeton University Press, Princeton, NJ. Dover Publications reprint 2003
Bellman R (1958a) On a routing problem. Q Appl Math 16(1):87–90
Bellman R (1958b) On 'heuristic problem solving,' by Simon and Newell. Oper Res 6(3):448–449
Bellman R (1961) Adaptive control processes: a guided tour. Princeton University Press, Princeton, NJ
Bellman R (1962) Invariant imbedding and random walk. Proc Am Math Soc 13(2):251–254
Bellman R (ed) (1963) Mathematical optimization techniques. University of California Press, Berkeley, CA
Bellman R (ed) (1964) Stochastic processes in mathematical physics and engineering. American Mathematical Society, Providence, RI
Bellman R (1965) An application of dynamic programming to location-allocation problems. SIAM Rev 7(1):126–128
Bellman R (1966) Dynamic programming. Science 153(3731):34–37
Bellman R (1967) Introduction to the mathematical theory of control processes. Volume 1: linear equations and quadratic criteria. Academic, New York, NY
Bellman R (1968) Some vistas of modern mathematics: dynamic programming, invariant imbedding, and the mathematical biosciences. University of Kentucky Press, Lexington, KY

Bellman R (1970) Introduction to matrix analysis, 2nd edn. McGraw-Hill, New York, NY

Bellman R (1971) Introduction to the mathematical theory of control processes. Volume 2: nonlinear processes. Academic, New York, NY

Bellman R (1978) An introduction to artificial intelligence—can computers think? Boyd and Fraser, San Francisco, CA

Bellman R (1980) Analytic number theory—an introduction. Addison-Wesley, Reading, MA

Bellman R (1983) Mathematics in medicine. World Scientific, Singapore

Bellman R (1984) Eye of the Hurricane: an autobiography. World Scientific, Singapore

Bellman R, Blackwell D (1949) Some two-person games involving bluffing. Proc Natl Acad Sci 35(10):600–605

Bellman R, Blackwell D (1951) Red dog, blackjack, and poker. Sci Am (January):44–47

Bellman R, Brock P (1960) On the concept of a problem and problem-solving. Am Math Mon 67(2):119–134

Bellman R, Clark C, Malcolm D, Craft C, Ricciardi F (1957) On the construction of a multi-person, multi-stage business game. Oper Res 5(4):469–503

Bellman R, Cooke K (1959) Stability theory and adjoint operators for linear differential-difference equations. Trans Am Math Soc 92(3):470–500

Bellman R, Cooke K (1963) Differential-difference equations. Academic Press, New York

Bellman R, Cooke K (1968) Modern elementary differential equations. Addison-Wesley, Reading, MA

Bellman R, Cooke K, Lockett J (1970) Algorithms, graphs, and computer. Academic, New York, NY

Bellman R, Danskin J (1954) A survey of the mathematical theory of time-lag, retarded control, and hereditary processes. Report R-256, RAND Corporation, Santa Monica, CA

Bellman R, Dreyfus S (1958a) On a tactical air-warfare model of Mengel. Oper Res 6(1):65–78

Bellman R, Dreyfus S (1958b) Dynamic programming and the reliability of multicomponent devices. Oper Res 6(2):200–206

Bellman R, Dreyfus S (1962) Applied dynamic programming. Princeton University Press, Princeton, NJ

Bellman R, Esogbue A, Nabeshima I (1982) Mathematical aspects of scheduling and applications. Pergamon Press, Oxford, England

Bellman R, Glicksberg I, Gross O (1954) Some problems in the theory of dynamic programming—a smoothing problem. J SIAM 2(2):82–89

Bellman R, Glicksberg I, Gross O (1956a) Some nonclassical problems in the calculus of variations. Proc Am Math Soc 7(1):87–94

Bellman R, Glicksberg I, Gross O (1956b) On the bang-bang control problem. Q Appl Math 14(1):11–18

Bellman R, Glicksberg I, Gross O (1958) Some aspects of the mathematical theory of control processes. Report R-313, RAND Corporation, Santa Monica, CA

Bellman R, Gross O (1954) Some combinatorial problems arising in the theory of multistage processes. J SIAM 2(3):175–183

Bellman R, Harris T (1951) Recurrence times for the Ehrenfest model. Pacific J Math 1(2):179–193

Bellman R, Harris T (1952) On age-dependent binary branching processes. Ann Math 55(2):280–295

Bellman R, Jacquez J, Kalaba R (1960a) Some mathematical aspects of chemotherapy, 1: one-organ models. Bull Math Biophy 22(2)181–198

Bellman R, Jacquez J, Kalaba R (1960b) Some mathematical aspects of chemotherapy, 2: the distribution of a drug in the body. Bull Math Biophy 22(3):309–322

Bellman R, Kalaba R (1959) Functional equations, wave propagation, and invariant imbedding. J Math Mechan 8(5):683–704

Bellman R, Kalaba R (1960) On k-th best policies. J SIAM 8(4):582–588

Bellman R, Kalaba R (1962) Mathematical programming and the maximum transform. J SIAM 10(3):550–567

Bellman R, Kalaba R (eds) (1964) Selected papers on mathematical trends in control theory. Dover, New York, NY

Bellman R, Kalaba R (1965) Dynamic programming and modern control theory. Academic, New York, NY

Bellman R, Kalaba R, Wing G (1960) Invariant imbedding and the reduction of two-point boundary value problems to initial value problems. Proc Natl Acad Sci 46(12):1646–1649

Bellman R, Shapiro H (1952) On the normal order of arithmetic functions. Proc Natl Acad Sci 38(10):884–886

Bellman R, Smith C (1973) Simulation in human systems: decision-making in psychotherapy. Wiley-Interscience, New York, NY

Bellman R, Strauss E (1949) Continued fractions, algebraic functions, and the Pade table. Proc Natl Acad Sci 35(8):472–476

Bellman R, Wing G (1975) An introduction to invariant imbedding. Wiley, New York, NY

Bellman R, Zadeh L (1970) Decision-making in a fuzzy environment. Manage Sci 17(4):141–164

Cairns W (1939) The William Lowell Putnam Competition. Am Math Mon 46(5):307–308

Chandrasekhar S (1960) Radiative transfer. Dover, New York, NY

Danskin J (1959) Review of dynamic programming by Richard Bellman. Oper Res 7(4):536–540

Dreyfus S (2002) Richard Bellman on the birth of dynamic programming. Oper Res 50(1):48–51

Dreyfus S (2003) IFORS' operational research hall of fame: Richard Ernest Bellman. Int Trans Oper Res 10(5):543–545

Dreyfus S (2008) Personal communication

Esogbue A, Bellman R (1984) Fuzzy dynamic programming and its extensions. In: Zimmermann H, Zadeh L, Gaines B (eds) Fuzzy sets and decision analysis (TIMS Studies in the Management Sciences: Volume 20). North-Holland, Amsterdam, pp 147–167

Golomb S (1989) Richard E. Bellman. National academy of engineering memorial tributes, vol 3. The National Academies Press, Washington, DC, pp 22–29

Gruenberger F (1968) The history of the JOHNNIAC. RM-5654-PR. The RAND Corporation, Santa Monica, CA

Jacquez J (1985) Richard Bellman. Math Biosci 77(1):1–4

Kagiwada H, Kalaba R (1986) The work of Richard Bellman-I: Optimization. Comput Math Appl 12A(6):785–790

Kahn H (1962) On thermonuclear war. Princeton University Press, Princeton, NJ

Kuhn H (2002) Being at the right place at the right time. Oper Res 50(1):132–134

Kushner H (2004) Remarks of Harold J. Kushner upon receiving the 2004 Bellman Heritage award. http:///www.a2c2.org/awards/bellman/index.php. Accessed 30 Jul 2009

Larson R (1979) Editorial: Richard Bellman wins the IEEE Medal of Honor. IEEE Trans Automat Control AC-24(3):385–386

Lew A (1986) Richard Bellman's contributions to computer science. J Math Anal Appl 119(1–2):90–96

Leondes CT (1980) An appreciation of Professor Richard Bellman. J Optim Theor Appl 32(4):399–406

Panossian H (1986) Richard Bellman and stochastic control systems. Comput Math Appl 12A(6):825–829

Roth R (ed) (1986) The Bellman continuum. World Scientific, Singapore

Simon H, Newell A (1958) Heuristic problem solving: the next advance in operations research. Oper Res 6(1):1–10

Schultze A (1925) Elementary and intermediate algebra. Macmillan, New York, NY

Titchmarsh E (1932) Theory of functions. Oxford University Press, Oxford, England

Ulam S (1976) Adventures of a mathematician. Charles Scribner's Sons, New York, NY

Whittaker E, Watson G (1927) A course of modern analysis. Cambridge University Press, Cambridge, England

24

Egon Balas

Graham K. Rand

For the past 40 years, Egon Balas has been one of the most distinguished contributors to the theory of integer programming. He was recognized for his work by being awarded the John von Neumann Theory Prize by the Institute of Operations Research and the Management Science (INFORMS) and the European Operational Research Societies' EURO Gold Medal. He was further honored as a fellow of INFORMS and inducted into the International Federation of Operational Research Societies' (IFORS) Hall of Fame. The impact of his contributions has been enhanced by his extensive research collaborations: His speech on receipt of the EURO Gold Medal acknowledged the contributions of his 50 or so coauthors, listing them by name, with the number of joint papers.

Egon is Professor of Industrial Administration and Applied Mathematics and Thomas Lord Professor of Operations Research at Carnegie Mellon University. The story of his life, as will be told, is split into two almost equal parts. Marked by political turmoil of historical significance, the first part of his life makes for a remarkably eventful story. It is this first part that made the man, while the account of his mathematical achievements in the second part constitutes an equally impressive and important chapter in the history of operations research (OR).

A.A. Assad, S.I. Gass (eds.), *Profiles in Operations Research*, International Series in Operations Research & Management Science 147, DOI 10.1007/978-1-4419-6281-2_24,

THE FIRST PART

Egon Balas was born on June 7, 1922, to Ignác and Boriska Blatt in Cluj, Romania, a city of 110,000 people, which had been part of Hungary until 1918. He was the first child of a middle-class family of Hungarian Jews. When he was six, however, his father went bankrupt, and Egon had a relatively difficult childhood and adolescence. During a successful school career, he took up table tennis, ping-pong as he calls it, and achieved a high standard, competing in tournaments. As this was forbidden by his school, he used the name Balázs, a common Hungarian family name, to avoid his own name appearing in the papers. (After World War II [WWII], as a Communist Party activist, he was advised to change his name because it sounded Jewish—he reverted to the name under which he had played in the ping-pong tournaments.) He finished high school in June 1941 and received his baccalaureate. He attended a short series of lectures on calculus, given by an assistant professor of mathematics at the University of Kolozsvár, that was offered to Jewish would-be students who were excluded from the university (Cluj had become Kolozsvár when taken over by Hungary in September 1940).

Egon first started to work in the foundry of the local Iron Works in February 1942 and joined the Iron and Steelworkers' Union. A month later, he became a member of the undercover Hungarian Communist Party, having been introduced to Marxism in 1940, while still at school. By this time, Hungary had entered WWII on the German side. Because the job at the Iron Works "was not attractive, the pay was miserable, and the political opportunities were limited" (Balas 2000, 54), Egon arranged a move to a metal-working factory as a lathe operator. There he helped organize work stoppages and the distribution of antiwar leaflets throughout Kolozsvár. Of necessity, he was leading a clandestine life, and, in the spring of 1943, he was asked by the party to go underground, a risky undertaking which involved procuring false identity papers. He assumed the identity of a first-year law student and managed, with some difficulty, to find a safe hiding place. It did not remain safe for long, and he was soon on the run again, taking up a new identity, this time of a medical student.

His luck ran out on August 26, 1944, when he was arrested at gunpoint and taken by train to Budapest the following day. In his memoir, *Will to Freedom: A Perilous Journey through Fascism and Communism* (Balas 2000), he offers a compelling account of his imprisonment and ordeal under torture, followed by his escape, at the end of the year. During a long period of hiding, he took the name András Hegedűs, a fellow prisoner who had escaped and whose birth certificate Egon had. (Hegedűs became Hungary's prime minister from April 1955 to October 1956.) Egon's objective was to reach the Russian lines, which he did in April 1945, only to find himself a prisoner of the dreaded SMERSH, the military branch of the Soviet Secret Police. At first, they did not believe his story and interrogated him for a week. His release allowed him to return home to discover that his parents and his brother Bobby had not survived the deportation of the Jews from Kolozsvár. Only 7 out of the 30 of Egon's extended family who were alive in May 1944 survived the war.

Egon rose to high positions in postwar Communist Romania. His wartime experiences, particularly the fact that he had not betrayed anyone under torture, and his organizing ability gave him great credibility in the Communist Party, into which he poured all his energies. In October 1945, he was sent to Bucharest as one of two dozen delegates from the Cluj region (Kolozsvár had reverted to its Romanian name) to the national conference of the Romanian Communist Party. The whole delegation was locked in a guest house for the duration of the conference, without permission to go out! For a brief period in 1947, he was in the Agrarian Section of the party's Central Committee, whose deputy director was Nicolae Ceauşescu (the notorious leader of Communist Romania from 1965 until his execution in 1989), with whom he had few contacts, and those he had were not friendly.

In 1948, he was moved to the Ministry of Foreign Affairs and was appointed secretary of the Romanian Legation in London. The spelling of his name was changed from the Hungarian Balázs to the more Romanian-sounding Balaş. He was required to familiarize himself with the British political situation by reading the press carefully, getting acquainted with journalists and politicians, and attending public meetings and reporting what he discovered. Later he was required to handle contacts with the

British Communist Party. He attended sessions of Parliament and the national conference of the Labour Party in Scarborough. His time in London came to an end when he was expelled by the British government in retaliation for the Romanian government's expulsion of two British diplomats.

Egon had taken advantage of a law, enacted to rectify the injustice done to those whose education had been interrupted during the war years for political or racial reasons, by enrolling as a student in the Economics Department of Bólyai University. During 1946 and 1947, he took all the first- and second-year exams. In 1948, he returned to Cluj to take exams and received his degree, a license in economics, in the spring of 1949.

On December 21, 1948, while on a visit to Bucharest from London, he married Edith Lővi, whom he had met in 1946. She had attended the same school as Egon, though after him. She and her family had been deported to Auschwitz in May 1944 where she had briefly encountered Egon's brother Bobby. She later was transferred to Bergen-Belsen and was liberated by the British army. Of the several thousand Jewish families deported from Cluj, the Lővi family of three was the only one to survive in its entirety. After a honeymoon in Paris, on the way back to London, they had only a few weeks before his expulsion. Egon describes Edith as "my wife and companion of a lifetime" (Balas 2002, 1).

Edith and Egon (1949)

On return from London in March 1949, he was appointed head of the Directorate of Economic Affairs in the Foreign Ministry and lecturer at the Institute of Economic Studies and Planning (ISEP), teaching world economics. In 1952, an increasingly hostile political atmosphere developed as a result of a campaign against right-wing deviation. Egon was dismissed from his job at the Ministry of Foreign Affairs in June and arrested in August, in the middle of the night, by the Securitate. He spent 2 years in solitary confinement, being interrogated for weeks at a time. His account of this experience is harrowing (Balas 2000). During the interrogations, he was told that Edith had started divorce proceedings, but his trust in Edith was such that he knew it could not be true. He describes how once or twice a week he "went to the opera or to a concert" by recreating in his mind the music he had heard in better circumstances (Balas 2000, 258). Unbroken, he was suddenly released in November 1954, taken blindfolded to his old neighborhood, where he was left on the pavement. He found his way to friends and learned that his second daughter, Vera, had been born early in 1953. Edith had become aware of her pregnancy a couple of weeks after his arrest. Anna, his elder daughter, was now four-and-a-half years old.

At this time, Egon's political perspective was that, although he had become aware of the anomalies of the communist system, he believed that the system could be reformed and a better society built. He was allowed to keep his teaching position at ISEP on a part-time basis and started working at the Institute for Economic Research in March 1956, setting up a section on the economics of capitalist countries. During this period, he published several papers and articles on economic matters. Several caused controversy and one, in particular, caused real trouble after the editorial board had to apologize publicly for its ideological blindness in having accepted a revisionist piece of writing. Egon had been working on a book, *Contributions to a Marxist critique of Keynesianism*, since early 1956, which was published in April 1958. At first it was well received, but, towards the end of the year, several hostile reviews branded the book as alien to Marxist thinking; it was subsequently withdrawn from circulation. In September, he had lost his teaching position. He was amused to find that his employment record stated that he had been associate professor without leave while he had been in jail! He was finally expelled from the party and lost his job at the Institute in the spring of 1959. These events in his personal life and the prevailing attitude in Romania caused him to cease to believe that the system could be redeemed by reform.

THE SECOND PART

Egon started a new life as a self-educated applied mathematician. He was first attracted to OR when he came across a recently published book, *Linear Programming and Economic Analysis*, by Dorfman et al. (1958). The contents caught his imagination. He managed to obtain a job at the Design Institute for Forestry and the Timber Industry in Bucharest. His first work in OR, jointly with Ladislau Ivănescu (the late Peter Hammer), concerned transportation planning for a type of lumber that had to be shipped regularly throughout the year from 36 production sites to seven consumption centers. They used linear programming (LP) and network flow theory and, because they had no access to a computer, employed a team of students to do hand calculations.

The results indicated a potential 8% saving and were deemed of such value that Egon and Ladislau were given access to the one and only available (Romanian-built) computer. They soon became involved with a second application, this time in the textile industry. Between spring 1960 and winter 1961, they wrote a series of five papers that were published in English in two parts (Balas and Ivănescu 1962a, b). These publications provided efficient methods for solving parametric transportation problems of several types. Their results brought them to the attention of the European and American OR communities. In addition, a paper on the generalized transportation problem appeared in *Management Science* (Balas and Ivănescu 1964).

IN MEMORY OF PETER

"In the late 60s, a colleague at Carnegie Mellon University asked me what happened to my former collaborator, Ivănescu. I told him that Ivănescu was the same as Peter L. Hammer, and explained the double name change. My colleague's reaction was, 'What a pity!' 'Why?' I asked. 'Well, Ivănescu was so exotic, so romantic. Hammer? There are thousands of them.' There may have been thousands, but few were like this one, as it soon became clear to everybody" (Balas 2000, 388).

At first, Egon was not interested in problems with integer-constrained variables. This changed, when, in 1962, he confronted the problem of optimizing the harvesting plan from an area of forests. The forests consisted of sets of plots containing trees with different age and quality characteristics, with the harvest limited by the total quantity of trees that could be cut. A road network had to be built to get access to the plots, and this translated into a major, nonlinear cost item. This caused Egon to formulate the associated nonlinear programming problem in terms of variables that could be either 0 or 1, thus making him aware how such variables could be used to formulate many logical (yes or no) conditions present in real-world problems.

Because there was no computational method available to him to solve this type of problem, Egon devised his own approach. He named it the additive algorithm, as it involved only additions and comparisons; it became more widely known as implicit enumeration (Balas 1965). It is based on a series of logical tests that explore the implications of setting certain variables equal to either 0 or 1. It was easy to implement and did not require an LP-based code; Egon was able to solve several versions of the forest management problem that contained about 40 variables.

He presented this work at a statistical conference in Bucharest at the end of 1963. To get it into the Western literature as quickly as possible, he arranged to have it published as a short communication in the April 1964 Bulletin of the French Academy of Sciences, the *Comptes Rendus de l'Académie des Sciences, Paris*. He also submitted the full English version to the journal *Operations Research* in February 1964, and to the Mathematical Programming Symposium held in London in July 1964. Of course, there was no possibility of his attending that meeting, but his paper (Balas 1965) was distributed to the participants—it became the most frequently cited of all articles published in that journal between 1954 and 1982.

In 1960, Egon decided to emigrate; a loosening of restrictions on Jewish emigration had made it possible. Permission was exceedingly difficult to obtain and just applying could have severe consequences. He first attempted to arrange a ransom, a process that enabled emigration permits for Jews to be obtained for money. He raised about $7000, but this attempt was blocked, no doubt because of his past. Despite the considerable risks involved, he then decided to act openly and started the emigration process. In the spring of 1964, recognizing that the time to successful conclusion, if ever, could not be predicted, he moved to a new job with the Institute of Mathematics' Center of Mathematical Statistics. In early 1965, on hearing that it was possible to defect through Yugoslavia, he applied for passports for the family, ostensibly for a 3-week vacation there. This resulted in an interview with the Securitate in which he was offered the possibility of obtaining passports with conditions—he would go to Yugoslavia, defect to the West, and then serve as a spy. He declined the opportunity to train for a new profession, explaining that he liked his present one and he was good at it.

In October 1965, he and Edith intensified their attempts to emigrate and resubmitted their formal application. Their friends thought they would never be allowed to leave the country. At the end of December, he received a letter advising him that he had been fired from the Center of Mathematical Statistics; he was without a job. To make something happen, he arranged to see various key players, including several members of the Governmental Commission on Passports and Visas, the Minister of Foreign Affairs, the Minister of Commerce, the head of the Securitate, and the Chairman of the Commission of the National Assembly for the Application of the Constitution. Finally, in July 1966, he and his family were granted permission to leave the country. Thus, after 6 years of frustration, the Balas family (Egon, Edith, Anna, and Vera) landed in Naples in the early morning of July 27.

The Rest of the Story

While waiting in Italy for an entry visa to the U.S., he obtained a job at the International Computing Center in Rome, whose director was Claude Berge, later to be a fellow Euro Gold Medalist. He also enrolled in doctoral programs at the Universities of Brussels and Paris, which subsequently led to Ph.D.s in economics and mathematics (Doctor of Science in Economics, University of Brussels, 1967, and Docteur de l'Université, Mathematics, University of Paris, 1968). When their visas were granted in April 1967, while in Toronto, Egon and family moved to Stanford, where he spent 5 months, from May to September, as visiting professor with George Dantzig's group. In the fall of 1967, Egon accepted a professorial appointment with the Graduate School of Industrial Administration (GSIA) of the Carnegie Institute of Technology, which had just merged with the Mellon Institute to become Carnegie Mellon University (CMU). Since then, at CMU, he has been Ford Distinguished Research Professor (1967–1968), Professor of Industrial Administration and Applied Mathematics (since 1968), the GSIA Alumni Chair (1980–1996), University Professor (since 1990), and Thomas Lord Professor of Operations Research (since 1996).

In the late 1960s, theoretical work in integer programming (IP) was focused mainly on cutting-plane theory. When it came to solving practical problems, however, the only useful computer algorithms were of a branch-and-bound type, the ideas of which had led Egon to IP. Despite the considerable popularity of his additive algorithm in the 1960s and 1970s, he was aware of its limitations. He thus turned to the study of convex analysis techniques: polarity, maximal convex extensions, and projection. He was able to derive a rich family of cutting planes from any convex set that contained the feasible set, but whose interior contained no feasible integer point, by solving the LP relaxation and working with the polyhedral cone defined by the optimum. Intersecting each extreme ray of this cone with the boundary of the convex set yields the required number of affinely independent points to define a hyperplane, the intersection cut, which cuts off the LP optimum but no feasible integer point. It soon became clear that any intersection cut could be viewed as a disjunctive cut, that is, a cutting plane derived from a disjunction. The disjunctive condition can be recast as including the LP constraint set on both sides of the disjunction. Clearly, each side of the above disjunction defines a polyhedron, and optimizing subject to this disjunction is equivalent to optimizing over the union of two polyhedra. Thus was born disjunctive programming, the study of optimization over unions of polyhedra.

> MINI BILL OF RIGHTS
>
> Soon after arriving at CMU, a colleague gave Egon a card listing his constitutional rights if he were ever arrested: the right to remain silent, to be represented by a lawyer, to have an open, public trial. He imagined himself telling his interrogators in Romania that he wanted to see a lawyer! (Balas 2000, 445).

Two basic results on disjunctive programming have immediate relevance to 0–1 programming, pure or mixed. First, there is a compact representation of the convex hull of a union of polyhedra in a higher-dimensional space. Second, an important family of disjunctive sets, called facial, can be convexified sequentially by imposing the disjunctions one by one, each time deriving the convex hull of the points satisfying the current set of constraints. Zero–one programs are facial disjunctive programs, which means that they are sequentially convexifiable; the integer hull of a (pure or mixed) 0–1 program with p 0–1 variables can be obtained in p steps, by imposing the 0–1 conditions one at a time, deriving at each step the inequalities defining the convex hull of the current disjunctive set. General integer programs when viewed as disjunctive programs are not facial, and they are not sequentially convexifiable. Thus one of the first insights yielded by the disjunctive programming approach was to identify the main characteristic that distinguishes 0–1 programming (pure or mixed) from general IP.

ON MATHEMATICAL DISCOVERY

"The above sketch of my research career may have conveyed the wrong impression that the life of a scientist is a sequence of pleasant ventures with happy endings. Nothing could be further from the truth. My life as a researcher was and is characterized by frequent periods of tension: whenever a new discovery is on the horizon, a tremendous amount of effort and concentration is needed in order to snatch the secret from nature by digging deeper and deeper. One day the new result seems at hand; you just have to write it down. Next day it all seems to have been an illusion; nothing seems to work when you want to put it together. When you finally manage to crystallize your findings into a theorem that you think you have proved, upon going the third or fourth time through the logical steps of your proof you suddenly discover a minor, insignificant-looking crack. But when you set out to repair it, the crack widens into a gaping hole. It turns out that what you thought was true is only true under certain circumstances. Maybe there is something of the same flavor that is true under more general conditions, but it is not what you had originally thought it was. So now you nail down your result for the more limited situation for which it is true, and you go on groping for a more general result that holds beyond those specific circumstances. And so on . . . . This is the flavor of mathematical discovery. It is an uneven process that often becomes hectic, with periods of sleepless or half-sleepless nights. It requires the kind of passionate concentration in the grip of which you forget about everything else for a while. To be successful at it, you must have 'fire in your belly.' And it certainly helps if your basic inclination is to persist and not give up in the face of difficulties, not to become dejected in case of setbacks, but to try again and again until you manage to find the right way" (Balas 2000, 451–452).

These results and others were described in a CMU technical report (Balas 1974). As they were not backed up by computational results, the work was not well received. The report, which Egon was unwilling to rewrite to meet the requirements of a referee, was not published in a journal at the time. Over the next 25 years, he received several requests for the report, but it did not appear in print until 1998, when, as an invited paper (with a foreword by Gerard Cornuéjols and William Pulleyblank) it was published as Balas (1998).

Although disjunctive programming had initially been largely ignored, when later the same results were produced in a new framework, called lift and project, the reaction was quite different (Balas et al. 1993). The return to the ideas of disjunctive programming was prompted by new developments on matrix cones. It was soon discovered that some of the matrix cone procedures were in fact isomorphic to the disjunctive programming procedure for generating the integer hull of a 0–1 program. The work of Balas et al. (1993, 1996) concentrated on algorithmic aspects and was accompanied by the development of an efficient computer algorithm, Mixed Integer Program Optimizer (MIPO). Ceria (2007) gives a vivid rendering of the atmosphere that surrounded the birth of lift-and-project; Cornuéjols (2007) gives his perspective of this work. MIPO was used to demonstrate that a particular version of disjunctive cuts, called lift-and-project cuts, could solve most of the mixed integer programs given in the Mixed Integer Program Library that had been impervious to solution by direct application of just branch-and-bound algorithms (Balas et al. 1996). This development in IP theory is now well integrated into computational practice. In particular, after the discovery of a radically improved method for generating lift-and-project cuts (Balas and Perregaard 2003), these ideas have become a standard feature of discrete optimization systems—the Xpress-Optimizer software uses lift and project as its default cut-generating procedure. An open source implementation is available (Balas and Bonami 2007).

Although the ideas of intersection cuts, disjunctive programming, and lift-and-project have their origin in the geometry of mixed 0–1 programs and the mathematics of convex analysis, much of Egon's research was inspired by real-world problems and situations such as the forest management problem that originally started him on a life in OR. For example, in the late 1960s, job shop scheduling and machine-sequencing problems triggered his interest in implicit enumeration on disjunctive graphs (Balas 1969, 1970) and later led to the development of the shifting bottleneck procedure, which is widely used as an efficient scheduling heuristic (Adams et al. 1988; Balas and Vazacopoulos 1998).

In the late 1980s, cutting planes were finally found to be useful in practice when combined with enumerative methods under the names of branch and cut, or cut and branch. By the mid-1990s, they were shown to be able to solve many, if not most, of the problems on which branch and bound alone failed. This progress, a decisive change in the state of the art, was made possible by the advent of faster computers and more efficient LP codes. But, it was primarily brought about by improved use of cutting planes—improvements in the cutting-plane techniques themselves and the combination of convexification with enumeration. In this context, it was with considerable satisfaction that, during the early 1990s, Egon experienced the success of the approach that became known as lift and project, whose roots stem from his work on disjunctive programming in the early 1970s. An overview of the significance and main uses of projection, lifting and extended formulation in integer, and combinatorial optimization is given in Balas (2005).

The concept of lift and project also originated from a practical situation. In 1980–1981, while on sabbatical at the University of Köln, Egon was approached by a Dutch OR group that was trying to solve a municipal bus-driver-scheduling problem. The group planned to use a set-covering model that had more than 100,000 variables. This was due to the need to explicitly generate all compatible pairs of combined morning and afternoon duties. Egon realized that there were morning duties and afternoon duties that could be scheduled separately, with each of them giving rise to a set-covering problem with a few hundred variables. But, this required the imposition of conditions that morning and afternoon duties had to be compatible in order to be assignable to the same driver. If the morning and afternoon duties were represented as the two node sets of a bipartite graph, then the conditions that were needed were the linear inequalities defining the convex hull of incidence vectors of perfectly matchable subgraphs of this bipartite graph. This characterization was developed with Pulleyblank, who happened to be on sabbatical in Bonn from the University of Waterloo. The combined ideas are known as extended formulation and projection (Balas and Pulleyblank 1983), and their significance in integer and combinatorial optimization is described in Balas (2005).

A final example of how real-world problems often lead to interesting theoretical developments concerns Egon's research on the traveling salesman problem (TSP). In the mid-1980s, he was approached by the OR group at LTV Steel about the scheduling of their steel rolling mill. A rolling mill processes steel slabs into sheets by passing the hot slabs between two layers of rolls which flatten them. The order in which the slabs are processed is crucial—it affects both the quality of the product and the efficiency of the process. Scheduling the rolling mill consists of selecting the slabs and putting them into an appropriate sequence. This was formulated as a prize-collecting TSP (Balas 1989, 1995). Several heuristic algorithms were developed and incorporated into a software system named ROLL-A-ROUND. It was in continual use at the Cleveland Works of LTV Steel for at least a decade. This practical problem led to the study of the polyhedral structure of the prize-collecting TSP, to the identification of facet-defining inequalities, and, later, to a partial characterization of the cycle polytope of a directed graph. It also stirred Egon's interest in the facial structure of the asymmetric traveling salesman polytope and started the chain of events that led to his joint TSP research with Matteo Fischetti, University of Padova (Balas and Fischetti 1993, 1999; Balas et al. 1995).

Egon and Edith.

Euro Conference, Prague, July 2007

Graham Rand and Egon.

THE BALAS FAMILY (2002)

Left to right: John, Vera, Robert, Edith, Egon, Sherwood, Anna, Alex.

Since arriving in the U.S. some 40 years ago, Egon Balas has had a remarkable career and has made valuable contributions to the theory and practice of OR. But what of his family? Edith accompanies him to OR meetings, particularly those in Europe such as the EURO Conference in Prague, July, 2007. She also has had a remarkable professional career as an art historian, having obtained a Ph.D. from the University of Pittsburgh and becoming professor of art history at Carnegie Mellon. She is the author of seven books and numerous articles. Anna is a practicing psychoanalyst. She is married to Sherwood Waldron, also a psychoanalyst, and they have a son, Alex. Vera teaches high school physics and mathematics and has two sons, John and Robert. Thus, Egon and Edith are doting grandparents, their most satisfying achievement.

Honors and Awards

At the age of 45, Egon's OR and applied mathematical research career that began in Romania had the opportunity to flourish in a new environment. His contributions to OR have earned him numerous honors and awards: John von Neumann Theory Prize (1995); EURO Gold Medal (2001); INFORMS Fellow (2002); IFORS' Hall of Fame (2006); Senior U.S. Scientist Award, von Humboldt Foundation; Honorary Doctorate, Miguel Fernandez University, Spain; Honorary Doctorate in Mathematics, University of Waterloo; Honorary Doctorate University of Liege (Honoris Causa); external member of the Hungarian Academy of Sciences; member, U.S. National Academy of Engineering. His story of courage and tenacity is an inspiring one; indeed, a life from adversity to acclamation.

REFERENCES

Adams J, Balas E, Zawack D (1988) The shifting bottleneck procedure for job shop scheduling. Manage Sci 34(3):391–401

Balas E (1965)_ An additive algorithm for solving linear programs with zero-one variables. Oper Res 13(4):517–546

Balas E (1969) Machine sequencing via disjunctive graphs: an implicit enumeration algorithm. Oper Res 17(6):941–957

Balas E (1970) Machine sequencing: disjunctive graphs and degree-constrained subgraphs. Nav Res Logistics Q 17(1):1–10

Balas E (1974) Disjunctive programming: properties of the convex hull of feasible points. MSRR No. 348, Carnegie Mellon University, July

Balas E (1989) The prize collecting traveling salesman problem. Networks 19(6):621–636

Balas E (1995) The prize collecting traveling salesman problem: II polyhedral results. Networks 25(4):199–216

Balas E (1998) Disjunctive programming: properties of the convex hull of feasible points. Disc Appl Math 89(1–3):3–44

Balas E (2000) Will to freedom: a perilous journey through fascism and communism. Syracuse University Press, Syracuse, NY. (Also translated into Hungarian, Romanian, French and Italian)

Balas E (2002) Some thoughts on the development of integer programming during my research career – lecture delivered upon receiving the EURO Gold Medal, July 9, 2001, Rotterdam. Eur J Oper Res 141(1):1–7. (Also published in 2007 in Ann Oper Res 149:19–26)

Balas E (2005) Projection, lifting and extended formulation in integer and combinatorial optimization. Ann Oper Res 140:125–161

Balas E, Bonami P (2007) New variants of lift-and-project cut generation from the LP tableau: open source implementation and testing. In: Fischetti M, Williamson DP (eds) Integer programming and combinatorial optimization: Proceedings of the 12th IPCO conference. Springer, Berlin, pp 89–104

Balas E, Ceria S, Cornuéjols G (1993) A lift-and-project cutting plane algorithm for mixed 0-1 programs. Math Program A 58(3):295–324

Balas E, Ceria S, Cornuéjols G (1996) Mixed 0-1 programming by lift-and-project in a branch-and-cut framework. Manage Sci 42(9):1229–1246

Balas E, Fischetti M (1993) A lifting procedure for the asymmetric traveling salesman polytope and a large new class of facets. Math Program A 58(3):325–352

Balas E, Fischetti M (1999) Lifted cycle inequalities for the asymmetric traveling salesman problem. Math Oper Res 24(2):273–292

Balas E, Fischetti M, Pulleyblank WR (1995) The precedence-constrained asymmetric traveling salesman polytope. Math Program A 68(3):241–265

Balas E, Ivănescu PL (Hammer) (1962a) On the transportation problem - part I. Cahiers du Centre d'Études de Recherche Opérationelle 4(2):98–116

Balas E, Ivănescu PL (Hammer) (1962b) On the transportation problem - part II. Cahiers du Centre d'Études de Recherche Opérationelle 4(3):131–160

Balas E, Ivănescu PL (Hammer) (1964) On the generalized transportation problem. Manage Sci 11(1):188–202

Balas E, Perregaard M (2003) A precise correspondence between lift-and-project cuts, simple disjunctive cuts, and mixed integer Gomory cuts for 0-1 programming. Math Program B 94(2–3):221–245

Balas E, Pulleyblank WR (1983) The perfectly matchable subgraph polytope of a bipartite graph. Networks 13(4):495–516

Balas E, Vazacopoulos A (1998) Guided local search with shifting bottleneck for job shop scheduling. Manage Sci 44(2):262–275

Ceria S (2007) A brief history of lift-and-project. Ann Oper Res 149:57–61

Cornuéjols G (2007) Revival of the Gomory cuts in the 1990's. Ann Oper Res 149:63–66

Dorfman R, Samuelson P, Solow R (1958) Linear programming and economic analysis. McGraw-Hill, New York, NY

25

E. Leonard Arnoff

David F. Rogers

After completing his World War II (WWII) military service, E. Leonard (Len) Arnoff returned to college and obtained a Ph.D. in mathematics in 1952. He then took a position as a mathematician with the precursor of the U.S. space agency. Soon afterwards, an out-of-the-blue telephone call set him on a distinguished career in operations research (OR); he joined the OR Group at the Case Institute of Technology, the first university to offer M.A. and Ph.D. degrees in OR. At Case, Len organized and directed seminal OR programs, led applied research projects, and co-authored the first comprehensive OR text, *Introduction to Operations Research* (1957), with C. West Churchman and Russell Ackoff. He then went on to create and direct one of the earliest management science groups within a management consulting organization. Len then became Dean of the College of Business, University of Cincinnati, and transformed it into a modern teaching and research institution. He served as president (1968–1969) of The Institute of Management Sciences (TIMS), president of the Omega Rho International Honor Society for Operations Research and Management Science, and was a fellow of the American Association for the Advancement of Sciences.

The Early Days: Student and the Soldier

Len was born in Cleveland, Ohio on October 15, 1922, to David and Lena (Mentz) Arnoff. David, an excellent carpenter, emigrated from Russia as a young teenager, joining his brother in the residential building contractor business. He married Lena from Windsor, Ontario, Canada, and they became lifelong residents of Cleveland. Len was the

A.A. Assad, S.I. Gass (eds.), *Profiles in Operations Research*, International Series in Operations Research & Management Science 147, DOI 10.1007/978-1-4419-6281-2_25,

middle of seven children. Preceded by three older sisters, Len was followed by three younger brothers, making for interesting family dynamics. The Arnoffs lived in Glenville, on the east side of Cleveland, where the majority of the city's Jewish population resided for 50 years (1890–1940). Glenville of the 1920s and 1930s was described by the historian John J. Grabowski as a crucible in which a generation of giants in intellectual and entrepreneurial activities were forged (Hanft 2007): among many examples are former U.S. Senator Howard Metzenbaum and Julius (Julie) Kravitz who turned Julie's Master Mart into the Pik-n-Pay grocery chain. Len attended neighborhood public schools and graduated from Glenville High School at the age of 15. Notable alumni include Jerry Siegel and Joe Shuster, who created Superman, the comic book hero, while attending Glenville High.

Len considered both law and mathematics and attended Western Reserve University (a precursor to Case Western Reserve University). His propensity for mathematics won out and he received a B.S. in Mathematics in 1943. After graduation, with the U.S. in the thick of WWII, he tried to enlist in the U.S. Navy but, because he was color blind, Navy regulations ruled it out. Shortly afterwards, in 1943, Len was drafted into the U.S. Army and sent to Lehigh University in Bethlehem, Pennsylvania under the Army Specialized Training Program. Upon graduation with an M.S. in Mechanical Engineering, he was assigned to the 84th Infantry Division, Company C, 335th Regiment; the division was sent to England in September, 1944. In November, Len's division entered combat in Germany, fought in the Battle of the Bulge, crossed the Rhine in April, and ended the war on the west bank of the Elbe River. Len received the Bronze Star medal for bravery, and the Purple Heart with two oak leaf clusters for three shrapnel injuries.

Personal and Intellectual Pursuits

After being discharged from the Army, Len returned to Cleveland to continue his education and was awarded a teaching scholarship at Case Institute of Technology. He received his M.S. in mathematics at the end of the 1948 spring semester. About this time, Len was set up on a blind date by a distant cousin who was also a friend of Ann Edith Kolisch, an Ohio State University undergraduate and native of Cleveland. Ann was on an internship assignment at the Cleveland Rehabilitation Center. She chose not to waste a weekend night with an unknown, but the blind date with Len on Monday, March 9, 1948 led to their engagement in less than a month on April Fool's Day. They were married August 21, 1948. Len decided to pursue doctoral studies and applied to MIT in Cambridge, Massachusetts and the California Institute of Technology (Caltech) in Pasadena. He was accepted at both schools, but he chose a friendlier climate and enrolled at Caltech in September, 1948. Len majored in mathematics with a minor in physics and received his Ph.D. in June, 1952. His dissertation, "Partial Differential and Difference Equations," was done under the direction of H. Frederic (Henri) Bohnenblust. During his Ph.D. studies, Len worked as a hydrodynamicist for the Naval Ordnance Test Station in Pasadena.

After receiving his doctorate, Len and Ann returned to Cleveland where he took a position as mathematician and aeronautical research scientist at the Lewis Flight-Propulsion Laboratory, National Advisory Committee for Aeronautics, now part of the National Aeronautics and Space Administration. His Ph.D. research on differential and difference equations was put to good use in the study of the dynamics of aircraft (Ribner and Arnoff 1952). Soon afterwards, Len received a call from C. West Churchman, Director of the Operations Research Group at Case Institute of Technology. Churchman learned of Len by a call to Caltech in search of applied mathematicians, and enticed him to return to Case. Len joined the faculty at Case as assistant professor in the fall of 1953 and his hire, with Russell Ackoff already on board, completed the basic cadre of the Case Operations Research Group (ORG). The ORG was one of the first U.S. OR academic units, having started in 1951 as part of the Department of Engineering Administration. Case was the first university to offer both masters and doctoral degrees in OR. Len greatly enjoyed the academic life of teaching, serving on dissertation committees, and the administration of graduate programs. He was at home teaching linear programming (LP) to students, and lecturing in OR short courses for industrial and business executives. In Arnoff (1957), Len described the then unique Case ORG's approach to OR education, one that integrated education, research (both basic and sponsored), and dissemination. His interests in applied research were quite broad—a typical early analysis developed a general approach to product search with the aim of obtaining a group of acceptable alternate possibilities, coupled with a product-evaluation procedure to determine an optimum choice (Stillson and Arnoff 1957). It was also at Case where Len was able to demonstrate his innate understanding of the importance of social capital, his empathy for students, colleagues, and community, which continued through his academic life. Besides, he and Ann loved to host parties for colleagues and their families!

CASE ORG—THE EARLY DAYS

Sid Hess received his Ph.D. from Case in 1960 and noted in his reminiscences about first discovering OR:

"In 1956, Fortune magazine published a sexy article about industrial applications of OR, including counting nodes to help Seabrook Farms grow peas more efficiently and a linear programming application Heinz used to distribute catsup

"I also learned of Case Institute of Technology's pioneer OR graduate program. ORSA [Operations Research Society of America] was meeting in Washington, D.C., that year so several of us [soldiers] persuaded the Army [Chemical Corps] to send us. At the meeting Len Arnoff made a sales pitch for the Case Group and virtually offered us research assistantships on the spot. Jim Fisher and I accepted

"In those early days Case was a heady place. Research was primarily industrially sponsored, pioneering work in inventory control, distribution, competitive bidding for oil leases, R & D management, and marketing science. Course work was largely based on these research applications plus lots of statistics, sampling, philosophy of science and, of course, OR methodology as being developed by Wes[t] Churchman, Len Arnoff and Russ Ackoff. Other faculty included Mike Leyzorek, Fred Hanssmann and Bill Abendroth, later joined by John Little, Pat Rivett [visiting from the U.K.] and Glen Camp among others. Eli Naddor was their first Ph.D. graduate; Jim Fisher, Larry Friedman, George Summers and I comprised the second wave" (Hess 2002, 38).

The Seminal Book

Beginning in 1952, an intensive 2-week summer class, "Short Course in Operations Research," was offered annually at Case for leaders in industry, business, and government. Based on the course lecture material, the trio of Case researchers, C. West Churchman, Russell Ackoff, and Leonard Arnoff, wrote the first integrated text in OR, *An Introduction to Operations Research* (1957). The book covered the then main models considered to be essential in the practice of OR: inventory, allocation, waiting-time, replacement, and competitive bidding. Key contributions were the authors' discussions of the essential characteristics of OR, the use of the team approach, and the viewing of an OR study within the systemic aspects of the problem environment. They offered their view of how to conduct an OR study as a process in six phases. The modeling phase emphasized developing a measure of effectiveness, E, as a function of controllable variables x_i and uncontrollable variables y_j within the general form of an OR model, $E = F(x_i, y_j)$. It is clear that their consulting work at Case greatly influenced the contents by the inclusion of chapters on "Model Testing" and "Controlling and Implementing the Solution," with a final chapter on "Selection, Training, and Organization for Operations Research."

THE MAJOR PHASES OF AN OR PROJECT

1. Formulating the problem.
2. Constructing a mathematical model to represent the system under study.
3. Deriving a solution from the model.
4. Testing the model and the solution derived from it.
5. Establishing controls over the solution.
6. Putting the solution to work: implementation.

(Churchman et al. 1957, 13).

Although Len arrived at Case after the other authors, "He wrote a good deal of it, particularly the mathematical sections," stated co-author and good friend Russ Ackoff (2007). "Ackoff and Churchman were the salesmen and had figured out the approach. Len was the nuts and bolts and the problem solver," noted Ann Arnoff (2007). The book was translated into French, German, Japanese, Russian, Slovak, and Spanish and had 27 printings. It served the OR academic community as the basic classroom text for many years. The authors' royalties from the book's first 2 years of sales were a faculty gift to establish the T. Keith Glennan (President of Case) Laboratory of Industrial Electronics, and all subsequent royalties were donated for Case OR graduate scholarships. (The three co-authors were selected as delegates to the First International Conference in Operational Research that was held in Oxford, England, September 2–6, 1957.)

When Churchman left for a teaching position at the University of California, Berkeley in 1958, Russ Ackoff became Director of the ORG and Len was appointed Assistant Director in charge of all administrative duties for the OR program. Len, however, continued to teach all of the LP courses. His comprehensive review article of LP included an extensive bibliography and reflects his thorough style (Arnoff and Sengupta 1961). "Len had more knowledge of mathematics; he understood it. He taught understanding and thus made the subject become alive" (Ackoff 2007). Len headed up or assisted in over 100 applied research projects for corporations while at

A CLASSIC STUDY: THE CUMMINS ENGINE COMPANY

This study described the results of an operations-research team working in the areas of order processing, production, and inventory control for the Cummins engine manufacturing company. Cummins manufactured lightweight high-speed diesel engines for automotive, industrial, and marine use. It also manufactured the components used to assemble or to supply replacement parts. The company's management had become interested in OR based on reported case studies and wanted to consider the "application of OR to their specific problem of reducing costs associated with parts shortage and, in turn to the more general problem of economic material control" (Arnoff et al. 1958, 468). An OR team was established with two representatives from the Case ORG and two from Cummins—an industrial engineer from the Manufacturing Division and a budget analyst from the Control Division. The Cummins study was a successful one that followed the how to modeling advice and basic inventory ideas given in the classic text of Churchman et al. (1957).

A FAMILIAR GRAPH

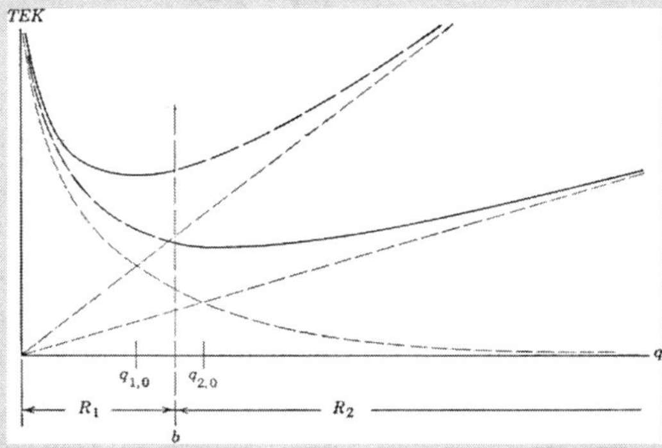

R_1, R_2 = price ranges; b = price break; TEK= total relevant monthly cost; $q_{1,0}$, $q_{2,0}$ = economic order quantities.

Economic Lot-size Curves (Churchman et al. 1957, 239).

Case. Although such studies were usually cloaked in proprietary secrecy, two such examples were published in the open literature. One was for an electric utilities company: the determination of when additional generating capacity should be added to the electrical utility system and the optimum size of the units to be added (Arnoff and Chambers 1956). The other application was for the Cummins Engine Company and dealt with its overall production activities (Arnoff et al. 1958). If all of the Case projects were performed in a like manner, there must have been a great many very pleased clients.

In the summer of 1959, Len took sabbatical leave from Case and served as an OR consultant for the U.S. State Department in Japan where he taught at Waseda University in Tokyo, the largest privately endowed university in Japan, and traveled the country with Waseda faculty. With Ann and their daughters, 6-year-old Janice and 4-year old Susan, Len took that opportunity to travel the world. He lectured and consulted in India, Switzerland, and Israel, and visited Hawaii, Hong Kong, Bangkok, England, and France.

CONSULTING AT ERNST AND ERNST

Working and directing applied OR projects at Case whetted Len's appetite for the practice of OR and prepared him for his next position. In 1960, he joined the management consulting firm of Ernst and Ernst in Cleveland, officially resigning from Case in 1961. His mission at Ernst—create and direct a management science consulting group. His official titles were Principal, Partner, and National Director of Planning and Operations Services. Also, included under his direction were the application of management sciences, decision-oriented management information systems, and financial modeling. In 1975, Len was placed in charge of the Professional Development Program for all consulting personnel of what had then become the firm of Ernst and Whinney. For this program, he developed a detailed set of curricula for the courses taught at their Education Center in Cleveland. Len maintained contact with the academic community by his participation on committees regarding OR at Case, and by being a member of the Advisory Council for the Weatherhead School of Management, the business school for Case Western Reserve University. (Case Institute of Technology, a school of science and engineering, and Western Reserve University, known for its strong liberal arts and professional programs, joined together in 1967 to form Case Western Reserve University.)

While at Ernst and Whinney, Len maintained his OR leadership activities by participating in conferences and publications. He continued on the editorial staff of the TIMS journal, *Management Science*, a task started in 1955 while at Case and which lasted 15 years. He was an invited featured speaker at the Second Annual Meeting of the American Institute of Decision Sciences (now the Decision Sciences Institute) in 1970, and at the Winter Simulation Conference in Miami Beach, Florida in 1978. As Len became removed from direct project work, his publications tended to be of a more general nature and focused on aspects of OR modeling: Arnoff (1970) where he stressed applying the basic OR approach to the developing field of management information systems; Arnoff (1971a) where he described the essential characteristics that determine the success or failure of models in real applications; and his advice in the use of mathematical and simulation models to, respectively, management advisers (Arnoff 1971b), internal auditors Arnoff (1973), and managers (Arnoff 1976). Len took early retirement from Ernst and Whinney in 1983, with a look toward future opportunities.

> A PAPER FOR ALL SEASONS
>
> In his presentation at the 1978 Winter Simulation Conference, Len describes how the then rather new corporate financial models could prove to be a panacea for executives "searching for effective ways to cope with mounting financial problems." He introduced his ideas with the following statement that seems familiar as we write this in 2009:
>
> "As we know, the economy has experienced some rather severe shocks during the past year or so. We are in the midst of rampant inflation ... accompanied by a deep and lingering recession. Unemployment has risen rapidly and in the main, corporate profits have plunged. The energy crisis certainly hasn't helped. The economic outlook is indeed bleak."
>
> Len then goes on to describe what constitutes a corporate financial model and enumerates a full range of corporate financial problem situations that he seems to have encountered and resolved in his consulting activities. He concludes by noting that he will now concentrate on "Applications of Financial Modeling in Health Care" (Arnoff 1978, 33–36).

Back to School: University of Cincinnati

Samuel J. Mantel, Joseph S. Stern Professor of Operations Management in the College of Business at the University of Cincinnati, had previously been in the Economics department at Case and served with Len on several OR projects. Len contacted Sam and inquired about the open College of Business dean's position, and, after deciding to apply, they spent many hours on the phone discussing things such as the college's successes, challenges, politics, departments, and even individual faculty members. When interviewing for the position, upon meeting individual faculty, Len was readily able to discuss their research efforts. "Len was prepared like nobody I'd seen before. He was up on everyone in the college before he came for the interview. We were amazed on the homework he had done," noted Mantel (2008). Len was appointed the fourth dean of the College of Business, University of Cincinnati on September 1, 1983. He consistently espoused clear initiatives of making the academic programs in the College among the best in the nation by positioning the curriculum to emphasize that all graduates of the College were computer literate, a fairly new concept in the 1980s (Daly 1983). Len also challenged the faculty to develop creative use of information technology, respond to technological changes with curriculum modifications, consider the impact of automation upon the work force, and stay in touch with the needs and opportunities in the real world through applied business research (Arnoff 1985). With enthusiasm and drive, he helped transform the College of Business into a well-respected, full range, modern teaching and research institution. Len implemented a much-improved reappointment, promotion, and tenure process, and instituted faculty summer release time and support, special monetary support for newly hired faculty, and competitive college grants for all faculty. His strategic plan emphasized graduate education and research, the implementation of which enabled the College to greatly enhance its relative national standing.

The College of Business had been sprawled over six campus locations when Len arrived. He spearheaded a successful building drive to meet the $4 million in challenge grants of business magnate Carl H. Lindner, Chairman of the Board, founder, and principal shareholder of American Financial Group, Inc. and former President of the Board and Chief Executive Officer of Chiquita Banana. Lindner contributed about $5 million and the faculty moved into Lindner Hall in 1986. And, by some coincidence, the Quantitative Analysis and Information Systems (QAIS) Department ended up on the top floor in a prime location! Len well-supported the current OR faculty in the College and appointed OR Professor Dennis J. Sweeney as Associate Dean, who, along with colleague David R. Anderson, were co-authors of the still very popular text on management science (Anderson, et al. 2008). Although Len might have favored the quantitative side of the College, he appreciated the importance of all of the functional areas by supporting new ventures such as the Center for Family Owned Business.

On another level, Len showed that he understood the importance of the unifying aspects of socialization within a university community. Soon after his arrival, he invited the college faculty, staff, and their families to his home for a "Have Dinner with the Dean," resulting in over 600 guests in a 3-month period. Evelyn Schott, Assistant Dean of Business Affairs, remembers interviewing with Dean Arnoff for a position and being somewhat taken aback when he asked her when she was born. "I later learned he made a point of sending every employee in the college a personal birthday card. To

Len Arnoff, the workplace was an extension of your family and he wanted the employees to feel that," she explained (Schott 2008).

RETIREMENT AND . . .

Len retired from the University of Cincinnati in 1988 and moved to Sanibel Island, Florida. There he enjoyed boating to nearby islands, strolling the beaches for shells, fishing, playing tennis several times a week, and the general island lifestyle. Len and Ann were charter members and on the advisory committee of Bat Yam, Temple of the Islands in Sanibel, which held its first service on January 11, 1991 with 20 families; it has grown to over 150 families. For Len, compared to his earlier driven days of accomplishment, life in Sanibel was just going with the flow. The flow went well until June, 1990, when he was diagnosed with final stage renal cell carcinoma. Len underwent experimental treatments, but succumbed at his home in Blue Ash, Ohio on November 3, 1991.

Ann still resides in Sanibel, Florida and continues to pursue her and Len's common interest of volunteering, travel, and shells. Their daughter, Janice, had died in 1965 of sudden liver complications at age 13. Daughter Susan received an undergraduate degree in nursing, worked for 7 years as an intensive care nurse, received a Master of Business Administration degree from the University of Cincinnati bestowed to her by Len, and is now a Certified Public Accountant and the mother of Alex and Rachel.

HONORS AND AWARDS, AND SERVICE TO THE PROFESSION AND SOCIETY

Len was ecumenical in his giving of himself to both professional and community service organizations. He belonged to ORSA, TIMS, the British Operational Research Society, the American Mathematical Society, the Mathematical Association of America, the North American Society of Corporate Planning, the Institute of Management Consulting, the Greater Cleveland Growth Association, Sigma Xi Scientific Research Society, and Omicron Delta Kappa. Len served as a consultant member of several committees and task forces for the American Institute of Certified Public Accountants. His professional stature and service were recognized by his being elected as the President of TIMS (1968–1969) and the president of the Omega Rho International Honor Society for Operations Research and Management Science (1982–1984). He was a fellow of the American

Association for the Advancement of Sciences, and honorary member of Beta Gamma Sigma, Delta Sigma Pi, and Beta Alpha Psi.

His community service included being a Past Master of the Forest City Masonic Lodge in Cleveland, and Co-President (1964–1965) of the Cleveland Heights Parents and Teachers Association. He was on the National Board of Directors for Camp Fire U.S. (formerly Camp Fire Girls) for 6 years, and the National Board of Directors for Panorama Services and Products, Inc., the Camp Fire for-profit subsidiary. Len was an Official for the U.S. Figure Skating Association and a member of the Cleveland Skating Club and Plaza Figure Skating Club, Shaker Heights, Ohio, serving the latter as its president (Susan was an accomplished figure skater). He was a Member of the Hamilton County (Ohio) Blue-Ribbon Task Force on Child Abuse and Neglect, on the cabinet for the 1990 Cincinnati United Way campaign, and was active with the Carpenter's Ridge Homeowners Association (Blue Ash, Ohio) and Sanibel Community Association.

MOLSON TROPHY (CANADA)

Season's (1979) Largest;
21 Pound Northern Pike.

Len was an ardent fisherman and often fished the northern waters of Canada. He enjoyed boating and cruises—he and Ann took over 20 cruises. Len often played softball at both Caltech and Case, enjoyed photography, was fluent in French, and savored an occasional excellent cigar.

A Most Suitable Memorial

Professor Gene Woolsey, a long-time Colorado School of Mines faculty member and devoted supporter of OR, is the world's most prominent proponent of getting your hands dirty for effective practical problem solving. Gene was first exposed to OR in a course at the University of New Mexico—Churchman et al. (1957) was the text, affectionately referred to as the Big Red Book (Woolsey 2002). Gene became hooked on OR. Len had more than once indicated his desire to have Gene visit the College of Business and give a lecture. But, a visit by Gene never occurred during Len's tenure as Dean. Several years later, I was in charge of arranging speakers for the QAIS Department's Seminar Series through our University of Cincinnati Institute for Operations Research and the Management Sciences Student Chapter and Omega Rho Honor Society Chapter. We wanted to upgrade our typical speaker offerings and Gene Woolsey came to mind.

I called Gene and he said he would love to come visit gratis, except for one thing—he wanted us to name the lecture. After pondering it for a couple of days, the first E. Leonard Arnoff Memorial Lecture on the Practice of Management Science was born in 1992. Jerome Schnee, Dean of the College of Business at that time, supported the idea and queried if we were planning to hold the lecture annually. I hesitated to commit to a yearly lecture—doing it once was fine, but committing to it annually was daunting. But now, in its 19th year, the Arnoff memorial lectures have developed into a most distinguished series of lectures on the practice of management science. Len would be quite pleased with the lecture's emphasis upon contemporary applications in OR such as for crime and criminal justice, guns, airline safety, HIV prevention, enterprise risk, and health care. Besides Gene, the lecture has featured a prominent list of other senior OR speakers—Thomas Cook, Saul I. Gass, Russell L. Ackoff, Alfred Blumstein, and John D. C. Little—and a host of other OR path-breakers. It has been very satisfying to see the impact the lecture series has had on local practitioners and the College's students, faculty, and staff.

REFERENCES

Ackoff R (2007) Personal communication

Anderson D, Sweeney D, Williams T, Martin K (2008) An introduction to management science: quantitative approaches to decision making. Thomson/South-Western, Mason, OH

Arnoff AEK (2007) Personal communication

Arnoff EL (1957) Operations research at Case Institute of Technology. Oper Res 5(2):289–292

Arnoff EL (1970) Operations research and decision-oriented management information systems. Manage Account 51(12):11–16
Arnoff EL (1971a) Successful models I have known. Dec Sci 2(2):141–148
Arnoff EL (1971b) Cost cutting that works. Manage Adv 8(5):35
Arnoff EL (1973) Cost-cutting that works. Int Aud 30(2):34
Arnoff EL (1976) Managers should marry models. Strategy Leadersh 4(4):11–15
Arnoff EL (1978) Applications of corporate financial models. In: Proceedings of the 10th conference on winter simulation, vol 1. IEEE Press, Piscataway, NJ, pp 33–36
Arnoff EL (1985) Challenge grant goal. . .academic excellence. CBA Bus J 1(1):2
Arnoff EL, Chambers J (1956) On the determination of optimum reserve generating capacity in an electric utility system. Oper Res 4(4):468–480
Arnoff EL, Kania E, Day E (1958) An integrated process control system at the Cummins Engine Company. Oper Res 6(4):467–497
Arnoff EL, Sengupta S (1961) Mathematical programming. In: Ackoff R (ed) Progress in operations research, vol I. Wiley, New York, NY, pp 106–210
Churchman C, Ackoff R, Arnoff EL (1957) Introduction to operations research. Wiley, New York, NY
Daly A (1983) E. Leonard Arnoff—business administration dean. University of Cincinnati, The News Record September 28
Hanft L (2007) The connection that a place brings: talk revisits. Cleveland Jewish News, August 3. http://www.clevelandjewishnews.com/articles/2007/08/05/news/local/talk0803.txt. Accessed 4 Dec 2008
Hess S (2002) I had never heard of OR. ORMS Today 29(5):38
Mantel SJ (2008) Personal communication
Ribner H, Arnoff EL (1952) Interaction between a supersonic stream and a parallel subsonic stream bounded by fluid at rest. Technical Note 2860, National Advisory Committee for Aeronautics, Lewis Flight Propulsion Laboratory, Cleveland, OH, December
Schott E (2008) Personal communication
Stillson P, Arnoff EL (1957) Product search and evaluation. J Market 22(1):33–39
Woolsey G (2002) Frank Parker Fowler Jr. made me do it. ORMS Today 29(5):34

26

Berwyn Hugh Patrick Rivett

Graham K. Rand

Although there were formal operations research (OR) groups in the U.K. conducting pioneering work on military problems during World War II (WWII), there were other researchers who, individually or collectively, contributed to the war effort in other areas that would now be considered OR. A case in point is Berwyn Hugh Patrick (Pat) Rivett. From his military and subsequent government employment, Pat unknowingly segued into one of the U.K.'s earliest and most important civilian OR establishments. The move into the new job turned out to be a stroke of good luck—Pat became the head of the largest U.K. OR department, the National Coal Board's Field Investigation Group (FIG). FIG pioneered the application of new OR techniques within the U.K. and expanded the scope of OR by investigating new problem areas. He was appointed to the U.K.'s first university chair in OR (Lancaster University). Pat was recognized as an early and leading expositor of OR. He brought his extensive practical experience into the classroom with his texts, while, in more general books and articles, he proselytized executives and managers by expounding on the nature of OR and how the application of OR methods could be of service to their organizations.

Pat was president of the Operational Research Society (ORS) and received the ORS Silver Medal, the Society's most prestigious award. He was awarded the Canadian Operational Research Society's Harold Larnder Prize, presented annually to an individual who has achieved international distinction in OR.

A.A. Assad, S.I. Gass (eds.), *Profiles in Operations Research*, International Series in Operations Research & Management Science 147, DOI 10.1007/978-1-4419-6281-2_26,

BEFORE OR

Berwyn Hugh Patrick "Pat" Rivett was born in Oswestry, Shropshire, on April 2, 1923, the son of a former Baptist minister, who had become an inspector for the National Society for Prevention of Cruelty to Children. His family moved to London when he was only 3 months old, because an older brother had obtained a place at Kings College. Pat later followed in his brother's footsteps, with the intention of becoming a schoolteacher, but, after he graduated in Mathematics (First class) at Kings College, London, in 1943, he was drafted into a statistics research group within the Ministry of Supply. He worked on quality control and inspection schemes of ammunition and bomb fragmentation patterns. When war ended, he stayed on in the Ministry of Supply as a statistician, first, as a scientific officer and then a senior scientific officer. Still on armaments research, he did a lot of work on the tracking of aircraft and ground control of fighters and antiaircraft systems. During this time, he obtained an M.Sc. at Birkbeck College (1947) and also lectured for two nights a week at Battersea Polytechnic. He married Ann, a Welsh girl, in 1949. His in-laws were Welsh miners, and he went round coal-mines in Wales and became very much a part of the mining culture—an Englishman who was tolerated in the South Wales' anthracite region.

THE NATIONAL COAL BOARD

Created in 1947, the National Coal Board (NCB) was the nationalized British coal mining company. In 1951, Pat saw two newspaper advertisements for jobs at NCB, one for a statistician at NCB headquarters and one for someone to head up what was called a Field Investigation Group (FIG) that was carrying out OR in the mining industry. He asked his boss about the difference between OR and statistics and was told OR was just the same as statistics but it paid more. He decided to apply for the OR job and, greatly to his surprise, he got it; "a decision that changed the face of operational research in the U.K. as well as in the NCB" (Tomlinson 1971, 5). When he started at FIG, the group was about seven strong. It included at least two others who were to become well known in OR, Steve Cook, the deputy head of the group, and Maurice Sasieni.

The projects undertaken at that time were all carried out for NCB headquarters and were very large scale; typically, a project took three or four people up to 4 years. There was no information, so all data had to be collected and all calculations had to be done by hand. Pat recalled that "if you wanted to know the number of men working on a haulage system you went round and counted them because the books were not in the state that you could find out the number" (Rivett 1994b).

An early study arose from a mine disaster at Derbyshire's Creswell Colliery in September 1950, where a conveyor-belt had caught fire, and the fumes had gone round the pit and killed 80 men (Rivett 1953). The fumes came because it was a rubber belt and the stone loading onto the belt had jammed, and, with the friction, the belt had caught fire. There was no means of warning the men of an emergency. The FIG team investigated the layout of emergency communication systems in mines, and, in so doing, created what was probably the first ever major practical study to use simulation. They did the original studies with paper and pencil and, when a random number was required to decide whether or not a call was answered, they looked up tables of random numbers. Interestingly, when this work was reported some years later at the first international OR conference, it was Pat who presented the paper that was written by two of his colleagues (Clapham and Dunn 1957).

The FIG staff rapidly expanded to 60, and, indeed, it eventually grew to become the largest OR group in the U.K. Kirby (2003, 259) believes this was due to Pat's infectious enthusiasm for OR, complemented by his considerable skills in exposition, as well as what Pat called "some startling results." A 3-week underground training course was started by NCB at the adult training centre in Scotland, where new

OR recruits learned how to operate rope haulage, how to test a roof to determine if it was safe, and other basic mining skills. Pat was very keen on this, because he wanted the FIG staff to know what people were doing and what they were observing. The staff did not enjoy it, and life was fairly tough as they had to go into the pit. Pat recalls going down over a hundred mines: "You would go down with the 6 o'clock shift and you would have a cup of tea and bread and dripping, I remember at one colliery in Yorkshire, to see you through the shift and you would come up at two o'clock when the shift ended, and you operated the same hours as the mine workers and were very much part of the context of the industry. That was in those days very much the sort of thing. To be a miner was a great thing and we were very proud to be associated with coal miners" (Rivett 1994b).

In 1953, Pat went to the U.S. and spent 2 months touring OR groups in the military, industrial sectors, and universities, including attending the two-week training course at Case Institute of Technology with Russ Ackoff. He discovered "inventory control and linear programming and the transportation algorithm." He describes how he "came back to the Coal Board Group, bounded around like a retriever dog telling the lads all about this" (Rivett 1994b). For once, however, his enthusiasm met little response. The first international OR conference (held in Oxford, 1957) revealed how far the U.K. was behind the U.S. technically (Davies et al. 1957), so Pat revisited the States in 1958, spending 3–4 months at Case Institute—this time on his return everyone was eager to be trained. Study groups were organized within FIG, where individuals learned a subject and then taught it to everyone, once a month on Saturday mornings. The excellence of the work became widely known, and Pat was delighted when his staff moved to other jobs, so spreading the gospel of OR; he was particularly pleased that so many subsequently obtained chairs in British universities.

Pat later sought to convey the many lessons that he had learned during his time at FIG, especially in his writings. One series of ten articles, published in the *Operational Research Society Newsletter* between December 1998 and September 1999, was entitled "Defining Moments in OR." Collectively, they illustrate Pat's practical approach to OR and his engaging writing style.

DEFINING MOMENTS IN OR: THE BEAUTY FACTOR

"There was an investigation of the incidence of accidents in coal mines and the factors which might cause them. Part of this work was a study of minor 'non reportable' injuries which could be dealt with in the first aid centre at the mine. The study was carried out at a selection of mines, each with a different accident rate for minor injuries. In all the mines except one, the accident rate had been fairly constant over a period of time. But at one mine the researcher was interested to note that the level of injuries had increased markedly from one calendar year to the next.

"He visited the mine and spent some time in the first aid centre with the (very attractive) nurse who outlined the different injuries suffered by the miners who presented and their possible causes. They had certainly increased significantly since January last. Discussions with the mine staff revealed no change at all in mining conditions or processes before or during January and the increase was inexplicable.

"After work that day the researcher talked with the mine manager in the miners' club bar. He outlined his total frustration and told the manager that the only bright spots in the whole day had been the gorgeous nurse.

"'Yes she really is lovely,' said the manager. 'But you should have seen the one we had before her, a real dragon,' he continued.

"The researcher's nostrils began to twitch and a shiver went down his spine. 'How long has the new nurse been here?'

"'Started just before Christmas. The best Christmas present the lads ever had,' replied the manager.

"The researcher had a big smile. 'Have a pint of the best,' he replied.

"*Moral of the story*. No desk bound analysis, taking data through the letter box, would have solved this problem. No software would have helped. No analyst visiting the mine would have solved it from the data in the books and records—no matter how good a mathematician or statistician he or she was.

"It had to be a friendly extrovert researcher who liked meeting people and *listening* to them" (Rivett 1994a, 12–13).

AFTER NCB

Pat recalls that an "American friend of mine, David Hertz, who was .... in charge of OR within Arthur Andersen... suggested I might go to Arthur Andersen and open up OR in their London office. The deal was very nice, the salary was good and I started with 8/9 months . . . in New York with all my expenses paid for, for Ann and for my daughter and a flat provided with my salary paid in full in London. . . ." He went to Andersen in 1960, spent some time in New York and then came back and opened the U.K. office. He discovered that he was learning a lot. Coal had been a one-product industry and now he was exposed to a wide variety of industries: Mars and Petfoods, British Petroleum, textiles in Lancashire, and so on. Though it was invigorating, he was not keen on being a consultant, because he did not like being judged on his chargeable time rather than the quality of his work. He complained, "No one ever looked at the reports I'd written" (Rivett 1994b).

THE OPERATIONAL RESEARCH SOCIETY

The OR Club was formed in April 1948 with Sir Charles Goodeve as president. The genesis of the Club was the need of a mutual support group for introducing OR into British industry. The Club's informal nature allowed only one representative from an organization, and Donald Hicks, NCB's first Director of Scientific Control, was the one from NCB (Hicks was responsible for inaugurating OR in the mining industry). Thus, Pat was not able to join the Club. The actual number of members was not clear, but appeared to be less than 100. Pat commented: "Amazingly, there were 76 organizations which had somebody who was doing OR, or thought they were, thus membership was limited to about 76" (Rivett 1994b). A decision was taken that the Club should stop acting in a club-like way (it used to hold its meetings once a month at the Royal Society) and that it should be reconstituted as a society, the Operational Research Society (ORS). This occurred in November 1953. Hicks volunteered Pat as the first honorary secretary of the ORS; Pat claimed to be the only member of the ORS who never actually applied to join. He was secretary for about 8 years.

It was decided that there would be some form of loose criterion for membership which was that "people should be doing OR and if they thought they were doing it, they virtually were" (Rivett 1994b). In the beginning, membership spread through the heavy industries and the military. In 1958, an annual conference of the ORS was started and went round all the watering places—Harrogate, Leamington, and Cheltenham—but not for the water! Pat estimated that they were attended by maybe 50 or 80 people, a very high proportion of the membership. A constitution was formed with a 2-year presidential term of office. After several notable presidents, such as the Earl of Halsbury, the committee thought it would be a good thing if the distinguished national figures alternated with what Pat refers to as "horny-handed sons of toil" as president, resulting in Pat becoming president (1962–1963) (Rivett 1994b). The wish to alternate OR people with national figures did not materialize—all presidents since Pat have been drawn from within the profession. During his 2-year term as president, Pat's insatiable quest for new initiatives led to the formation of both regional and study groups.

The International Federation of Operational Research Societies (IFORS)

The first international OR conference, held in Oxford in 1957, was the result of a proposal sent from Russ Ackoff to Pat in January 1955. The ORS, the Operations Research Society of America (ORSA), and The Institute of Management Sciences (TIMS) became the sponsoring societies. Committees were set up early in 1956, led by Sir Charles Goodeve and Pat (U.K.) and Thornton Page and David Hertz (U.S.).

THE FRONT ROW AT OXFORD

(Left to right) Charles Goodeve, Thornton Page, Philip Morse, George Dantzig, Pat Rivett, David Hertz, Len Arnoff

Twenty-six of the senior delegates met at *The Bear*, in Woodstock, on the evening of September 5. The countries represented were the U.K., the U.S., France, Australia, Canada, India, Japan, the Netherlands, Norway, and Sweden. Thornton Page proposed that another conference should be held in 3 years' time in Europe of the same size and nature as the Oxford conference, and, in the meantime, an international federation of national operational research societies should be formed. It was agreed that Charles Goodeve and his committee would continue to act as a secretariat until a federation was established and that they should prepare a draft charter. It was understood that the federation would consist of member societies, not individuals; that the secretariat (committee) should be elected on a rotating basis; that votes would be on some proportional basis; that finances would be raised by subscription; that new member societies would be elected on evidence of qualification; and that the charter could be amended by a majority vote. As a result of continued deliberations, preparing the statutes, and creating the working arrangements for the federation, IFORS came into existence on January 1, 1959. Pat often remarked that "it is the only constitution I think still which has a square root mentioned in it," referring to the voting formula: It is proportional to the square root of the number of qualified members of the individual society (Rivett 1994b).

There were initially three member societies: ORSA, ORS, and the Société Française de Recherche Opérationelle (SOFRO). The statutes gave control of the Federation to a Board of Representatives of one member per society: The initial members were John Lathrop (ORSA), Germain Kreweras (SOFRO), and Pat (ORS). The Statutes provided for a Secretariat consisting of a Secretary (an executive officer) and Treasurer, to be nominated by a designated Foster Society. The first such society was ORS, with Charles Goodeve (British Iron and Steel Research Association) as Secretary and Donald Hicks (NCB) as Treasurer. It had been agreed at the Woodstock meeting that another conference should be held in spring 1960. In fact, Germain Kreweras' suggestion that it should be held in Aix-en-Provence in France was later approved, though the dates turned out to be September 5–9, 1960. Pat became the chairman of the scientific program committee for that conference (Banbury and Maitland 1961), a role he also carried out for the 5th conference, held in Venice in 1969. Rand (2000) gives a more detailed history of the foundation and beginnings of IFORS.

Pat was a frequent visitor to South Africa, as his sister lived in Natal. He actively supported the Operations Research Society of South Africa

(ORSSA), which was launched in 1969. Pat was a major driving force in its establishment and was present at a meeting whose purpose was to inaugurate ORSSA. He was instrumental in persuading the attendees to establish ORSSA as an independent society—a strong contingent of statisticians was adamant that it should be under the wing of the Statistics Society. But Pat convinced them that OR interests were broader.

He was very instrumental in the international OR conference held in Pretoria in 1984. Not only was he one of the main international speakers, but he also encouraged Roger Collcutt, Brian Haley, Heiner Müller-Merbach, Hans-Jürgen Zimmermann, and others to attend the conference. This was of great importance for the South African OR community—the conference was held during the apartheid era. Pat's involvement in South Africa at that time was not universally popular in the U.K. Pat continued his visits to South Africa after the demise of apartheid, usually at the time of conferences organized by ORSSA.

LANCASTER UNIVERSITY

The 1960s were exciting times for universities in Great Britain. Seven new universities were established: Sussex, Essex, East Anglia, Kent, Warwick, York, and Lancaster. Although Lancaster was the last of these new universities to be announced, it preceded both Kent and Warwick in its date of opening; the University's first students arrived in October 1964. While President of ORS, Pat visited a number of universities, including Reading, Exeter, and Manchester, to arouse their interest in OR. During his visit to Manchester, he met the economist, Charles Carter, soon to be the first Vice-Chancellor of the University of Lancaster. Carter, who had previously met Russ Ackoff, made the creation of a Department of Operational Research, the first in the U.K., an early academic priority. Carter wrote to Ackoff to say that he would like to have a chair of OR as one of his first appointments. On Ackoff's recommendation, Carter got in touch with Pat. Carter, along with Noel Hall, the chairman of the Lancaster University planning board, met Pat at the Athenaeum Club in London. Noel Hall, Principal of Brasenose College, Oxford, had previously been Principal of the Administrative Staff College at Henley, which was, at that time, the premier management college in Britain. Pat recalls, "there was Sir Noel Hall and Charles Carter and myself having a cup of tea, and I talked about what I wanted to do, which was very much in the image of

Case Institute because I had been to Case Institute and I knew Russ well, and I always thought this was the way we should do it—project-based, earning the money, demonstrating that you can do it in anger [in practice] and earn money and creating posts and having studentships, and all this sort of thing which appealed to them both so they offered me the job and I took it" (Rivett 1994b). Those who knew Charles Carter would be very surprised to hear that this business was concluded in quite such a casual way.

A few weeks later, Carter wrote to Pat, concerned that a salary had not been agreed upon. The die was cast—Pat's appointment had been announced. His name was reported to the University Grants Committee in March 1963. At that time, Pat was working for Arthur Andersen and getting £5500. He took £3200 to go to Lancaster, so it was a very big drop in salary. But, he said, "I never regretted it, it was something I wanted to do and so that's how it started, and then I immediately got in touch [with Mike Simpson]. I used to meet Mike Simpson when I was in Andersen's and he was at Touche in a pub and we would exchange horror stories of how ghastly it was being consultants in the city and so Mike leapt like a shot at the thought of coming aboard." He also invited Alan Mercer and Harry Dunn to join them and was pleased that "the four people who started OR at Lancaster were all Coal Board, we'd all worked together in the Coal Board so it was an offshoot of the Coal Board OR Group" (Rivett 1994b).

Thus, Pat took up the first established chair in OR at a British university. Once again, he was in at the beginning of something new. He set about the work with enormous enthusiasm and was heavily involved in the establishment of the University. He laid the foundations for the highly regarded department that the Lancaster OR group soon became. The department was the first at Lancaster to press for higher degrees and, in the autumn of 1964, the Masters Degree program in OR was inaugurated with 10 students. The OR department was also the first to mount a short course at the university on simulation (April 13–17, 1964, conducted by guest lecturer K. D. Tocher) for people from industry, commerce, and the public services. Alan Mercer recalls:

> From 1964 to 1967, Lancaster was simply an incredible place to be, with both Pat and Mike on different occasions describing the period as the most stimulating in their lives. Lancaster was unashamedly modelled on Ackoff's ideas, with industrial practice by postgraduate students being of paramount importance. Clients were charged significant fees for these services, so that in

> the first academic year of 1964–1965, sufficient monies were raised to pay the salaries of newly recruited staff and to support entirely all the Masters' students, no Research Council studentships being available. As well as generating this large consultancy income, Rivett, Simpson and Mercer gave virtually all the Masters' lectures, some of which were thrown open to outside participants to increase the Department's revenue. None of the three had any lecture material to draw upon, so that course planning, preparation and delivery was one mad rush. In between, there were one week courses given by eminent lecturers from the U.S. and the U.K. Pat seemed to know everybody and nobody ever declined his invitation! It was wonderful to listen to Russ Ackoff, Abe Charnes, West Churchman, Bill Cooper, George Feeney, Dave Hertz from the States, Keith Tocher and Edward de Bono from Britain, and many others (Mercer 2005, 1120).

Research and Publications

Pat was a leading expositor of OR. His string of texts extol the OR method with emphasis on applications (Rivett 1968, 1972a, 1980, 1994a; Rivett and Ackoff 1963). He was also an evangelist of OR, always keen to explain the usefulness of OR to those in other fields. Pat's mission as ORS President was "to get universities interested in OR and to engage the interest of the wider community" (Rivett 1994b). During his period of office, he addressed many chambers of commerce and even appeared on the BBC, presenting four programs on OR in prime-time slots. He published in non-OR/MS journals in the fields of library management (Rivett 1957), management studies (Rivett 1967), organizational theory (Rivett 1972b), and physics (Rivett 1976).

> PAT'S NINE PRINCIPLES
>
> "1. It is not enough to think you understand what you read. Ask also what other people will understand.
> 2. Always question the data.
> 3. Think before you analyse.
> 4. Do not expect all distributions to be normal.
> 5. Objectives are not absolutes—they change even during a study.
> 6. Be ultra cautious in handling estimates of probabilities where there is no possibility of validation.
> 7. Examine the problem boundaries.
> 8. On s'engage et puis on voit. [You engage and then you see (Napoleon).]
> 9. Go for the jugular"
>
> (Rivett 1994a, 246–247).

He was certainly not interested in publication for publication's sake. He would have found the present culture in British universities, dominated as it is by the Research Assessment Exercise (RAE), to be irksome. (RAE is a periodic national exercise which assesses the quality of U.K. research and informs the selective distribution of public funds for research by the four U.K. higher-education funding bodies.) But publish he did, as testified by an interesting series of papers on multivariate statistics, focusing on structural mapping and multidimensional scaling (Clarke and Rivett 1978; Rivett 1977a, b, 1978).

One of Pat's loves was football (soccer), to which he applied his analytical skills to determining, inter alia, where new league teams should be located (Rivett 1975). His paper includes the suggestion that "if the eager football fan could dematerialize himself in science fiction fashion and arrive painlessly on the terraces of any Football League club in the country, then it is likely that we would all finish up at Liverpool, Everton, Manchester United, Leeds and Millwall" (Rivett 1975, 810). The surprising inclusion of Millwall in the list, the team Pat supported, caused the editor to add "some bias in the last named?" (Rivett 1994b). Pat's mischievous sense of fun is also seen in an article (Rivett 1982) in which he discusses the possibilities arising from the value of pi being allowed to float. Müller-Merbach (1982) entered the spirit of the debate by discouraging Pat to allow pi to increase because, he argued, public projects cost pi times the planned budget.

After Lancaster

Pat left Lancaster for the University of Sussex in 1967 to become Professor of Operational Research. This move did not meet his expectations. The University did not like his contacts with industry; indeed, there were student demonstrations against what he was trying to do, and his filing cabinets were broken into. His wife, Ann, died in 1981, and he was left with his young daughter Caroline. He worked part-time, before retiring from Sussex in 1988.

Shortly after retirement, in 1989, he found great happiness in his second marriage, to Mary, and returned to the North of England to live in the Lake District. He renewed his contacts at Lancaster University and became an honorary professor. For many years, he was involved in consultancy work with the National Health Service in Lancashire on delivery of health care for the frail elderly and the preventive management of coronary heart disease (Rivett and Roberts 1995; Rivett et al. 1998).

A MAN OF THE WORD.

During the 1990s, Pat made several visits to the U.S., including visiting Keith Womer at the University of Mississippi twice—first for a full year, shortly after his marriage to Mary, returning a couple of years later for the fall semester. Keith recalls what delightful company they were, and the enjoyment of Pat's tales of early OR work in England, often related at *The Gin*, an Oxford (Mississippi) watering hole that substituted for an English pub. Pat was writing his book of applications, *The Craft of Decision Modelling* (Rivett 1994a), during his first visit and used parts of it in his courses. He told Keith that one of the cases, "Competitive Tendering for Conner Mining," was named for Conner Hall, where he had taught at Ole Miss (Rivett 1994a, 193–212).

He and Mary returned south to Arundel, in Sussex, where he became very involved in the local Anglican church, as a lay reader. During the summer of 2006, over a year after his death on July 3, 2005, I visited the church and talked briefly to a lady working in the office. I was pleased, but not surprised, that she not only recalled Pat with affection but remembered details of his funeral service. It was a marvellous occasion, with much music and eulogies from several clergymen. His warmth and enthusiasm made an impact on all who were privileged to come into contact with him. I, like so many others, remember him bringing joy and fun at all times, whether in a formal lecture or in a snatched conversation.

Honors and Awards

Pat was the first honorary secretary of the Operational Research Society (ORS) and later served as its president (1962–1963). In 1968, Pat received the ORS Silver Medal, the Society's most prestigious award as well as being named an ORS Companion of OR in 1983 for sustained support and encouragement for the development of operational research. He was awarded the Canadian Operational Research Society's 1987 Harold Larnder Prize, presented annually to an individual who has achieved international distinction in OR [Harold Larnder Memorial Lecture (Rivett 1988)].

Acknowledgements

It was my privilege to interview Pat Rivett on June 8, 1994, in Lancaster for the Operational Research Society's archives (Rivett 1994b). Much of the material mentioned earlier is based on that interview. Those who knew Pat well may detect his distinctive style—one aspect of his considerable talent for communication. Of necessity, there has been editing; it was a very long interview, and much of the material is not relevant to this profile. Furthermore, where possible, facts have been checked as Pat was recalling events from many years previously. I acknowledge the help I have received in this regard by Pat's friend and former colleague, Alan Mercer, who contributed Pat's obituary to the *Journal of the Operational Research Society* (Mercer 2005).

References

Banbury J, Maitland J (eds) (1961) Proceedings of the second international conference on operational research (Aix-en-Provence 1960), English Universities Press, London

Clapham J, Dunn H (1957) Communications in colleries. In: Davies M, Eddison R, Page T (eds) Proceedings of the first international conference on operational research (Oxford 1957). The English Universities Press, London, 291–305

Clarke D, Rivett BHP (1978) Structural mapping approach to complex decision-making. J Oper Res Soc 29(2):113–128

Davies M, Eddison R, Page T (eds) (1957) Proceedings of the first international conference on operational research (Oxford 1957). The English Universities Press, London

Kirby M (2003) Operational research in war and peace: the British experience from the 1930s to 1970. Imperial College Press, London

Mercer A (2005) Patrick Rivett. J Oper Res Soc 56(10):1119–1121

Müller-Merbach H (1982) On Rivett's free market geometry, or: Pi as a financial constant. J Oper Res Soc 33(11):1047

Rand G (2000) IFORS: the formative years. Int Trans Oper Res 7(2):101–107

Rivett BHP (1953) Underground communications. Oper Res Q 4(4):61–65

Rivett BHP (1957) The use of information in improving operations. ASLIB Proc 9(10):315–322

Rivett BHP (1967) What is operational research? J Manage Stud 4(2):176–188

Rivett BHP (1968) Concepts of operational research. C.A. Watts, London

Rivett BHP (1972a) Principles of model building: the construction of models for decision analysis. Wiley, London

Rivett BHP (1972b) Art of operations research - where it has been - where it is – where it might go. Organ Dyn 1(1):32–42

Rivett BHP (1975) The structure of league football. Oper Res Q 26(4, ii):801–812

Rivett BHP (1976) Operational research: (2) perspective and prologue. Contemp Phys 17(1):95–102

Rivett BHP (1977a) Policy selection by structural mapping. Proc R Soc Lond Ser A 354:407–423

Rivett BHP (1977b) Multidimension scaling for multiobjective policies. Omega 5(4):367–379

Rivett BHP (1978) Structural mapping applied to single value policies. Omega 6(5):407–417

Rivett BHP (1980) Model building for decision analysis. Wiley, Chichester

Rivett BHP (1982) Free market geometry. J Oper Res Soc 33(6):515–516

Rivett BHP (1988) The sixth honest serving man. INFOR 26(2):79–90

Rivett BHP (1994a) The craft of decision modelling. Wiley, Chichester

Rivett BHP (1994b) Interview June 8, 1994 (unpublished)

Rivet BHP (1998) OR Newsletter, December

Rivett BHP, Ackoff L (1963) A manager's guide to operational research. Wiley, London

Rivett BHP, Anglin T, Meyer A (1998) The transformation of the probability of a heart fatality into a queue position. Health Care Manage Sci 1(2):95–100

Rivett BHP, Roberts P (1995) Community health care in Rochdale Family Health Services Authority. J Oper Res Soc 46(9):1079–1089

Tomlinson RC (1971) OR comes of age: a review of the work of the Operational Research Branch of the National Coal Board 1948–1969. Tavistock, London

27

HOWARD RAIFFA

RALPH L. KEENEY

HOWARD RAIFFA has been a pioneer in all aspects of the decision sciences. His path breaking research has advanced the forefront of statistical decision theory, game theory, decision analysis, behavioral decision theory, risk analysis, and negotiation analysis. While serving as Harvard University's Frank P. Ramsey Professor of Managerial Economics, Howard authored and co-authored a series of seminal books that formed the basis of the decision sciences. Howard helped establish the Kennedy School of Government at Harvard University, had a key role in the negotiations that created the International Institute for Applied Systems Analysis (IIASA), and served as its first director. Among other honors, he received the Operations Research Society of America (ORSA) Lanchester Prize for best publication in Operations Research (OR), and the Decision Analysis Society Frank P. Ramsey Medal for outstanding contributions to decision analysis. He was elected to the American Academy of Arts and Sciences and the National Academy of Engineering.

GROWING UP IN NEW YORK CITY

Howard Raiffa was born on January 24, 1924 in New York City. He was raised in the Bronx during the Great Depression in a strong loving family of five that included an older sister and younger brother. His father emigrated to America from Russia (from the region of Minsk) about 1912 and had no schooling in the U.S. He worked as a jobber, or as a middleman, buying and selling woolens without ever holding any inventory of his own. His mother was born in the U.S. and lived her earlier years in the lower East

A.A. Assad, S.I. Gass (eds.), *Profiles in Operations Research*, International Series in Operations Research & Management Science 147, DOI 10.1007/978-1-4419-6281-2_27,

Side of Manhattan where most of the Jewish refugees from Europe lived when they first came to the U.S. She had to quit school in the seventh grade to take care of her ailing mother.

Howard's family did not stress intellectual pursuits. Their home had very few books and he was neither encouraged, nor discouraged, to read good books. As Howard states in his unpublished memoir (Raiffa 2006), "I never in my pre-college career read anything I was not required to read for school." Nevertheless, it was critically important in the Raiffa family that the children do well in school. In this regard, Howard excelled in mathematics, a fortuitous talent, as it did not require much reading, and, thus, it took little time from his real passion of sports.

At age seven, the Raiffa family moved to a four room apartment in the Bronx that overlooked a school yard where he would spend all daylight hours playing city sports. From about the age of 12, he played three-on-a-side basketball incessantly. His skill was recognized.

When it came time to attend high school, Howard had to make a decision. He could apply to a prestigious academic school (Townsend Harris or the Bronx High School of Science), a new school (Columbus) near his home, or Evander Childs High School, three subway stops away, that had a good basketball team and essentially had recruited him. He chose Evander, a behemoth of a school with 14,000 students. He made the varsity team in his freshman year and actually played as a substitute, at age 14, in a high school playoff game in Madison Square Garden. In his senior year, Howard was captain of the team.

During high school, Howard found a part-time job working 10 hours on Saturday and 5 hours on Sunday in a wholesale men's clothing establishment. His job was to run the elevator and keep the stock in order (that is, hanging in the right places). His pay, which started at 20 cents an hour and increased to 35 cents an hour after 4 years, was used for clothes and various small items. Between playing basketball and working, Howard had little time for any intellectual or cultural activities, even if he had wanted such stimulation.

In 1940, Howard met Estelle Sophie Schwartz. As Howard states, "[he] immediately fell in love. She was beautiful with a great figure and an infectious laugh" (Raiffa 2006). They were married 5 years later and have celebrated 65 years of marriage.

College and Military Service

Howard knew in high school that he would surely attend college to get a bachelor's degree, but he was not sure in what area. When the U.S. entered World War II, Howard realized that it was only a matter of time before he would be drafted. Hence, he decided to go to the free City College of New York (CCNY), not to play basketball, but to take a lot of mathematics in his freshman year to better position himself for the Army and to remain close to Estelle. He then arranged a so-called voluntary induction in January of 1943 just before his 19th birthday.

During his sophomore year at CCNY, Howard enlisted in the Army Air Force's special program for meteorologists. The idea was that upon completion of the meteorology program in about 15 months, he would get a commission as a second lieutenant. After several false starts, and what must be the Guinness Book of Records for the number of times in basic training, Howard spent 6 months in an intensive training program in pre-meteorology at the University of Iowa. However, at the graduation ceremony, graduates were informed that the Air Force already had enough meteorologists. They were given the option of becoming officers with special training in radar. Howard chose the program in Ground Controlled Approach (GCA), a blind-landing system that tracked planes by radar to help them land in terrible weather conditions. The last instruction to the pilot was: "You are now fifty feet over the edge of the runway. Take over visually and land" (Raiffa 2006).

After the Japanese surrender, Howard and his team of 16 enlisted men, operators as well as mechanics, were transferred to Atsugi Air Base (later at Tachikawa) in the Tokyo-Yokahama area. About this assignment, Howard wrote:

> Now that was an exciting period of my life! Lots of action. Many a time we would instruct a pilot to take over visually and land and he would respond: "Can't see a thing. Are you sure we are where you say we are?" and up he would go. A sort of blind-landing touch and go. It was frequently so murky that the pilots could not see the huge floodlights at each side of a narrow 90 foot wide runway. Weather reports, especially from the Russians over Siberia, were almost non-existent and planes constantly kept arriving in Japan from Hawaii and Alaska, low on fuel, with no alternate place to land, only to encounter low visibility conditions due to weather and smog arising from the burning of low-grade fuels.
>
> I worked long hours (90-hour weeks) and had no time to see Japan. Never got off the base. Gradually my team of 16 enlisted men started to disappear because of automatic discharges from the Army. I, in desperation, set up a school to train fighter pilots—war heroes, recently arriving in Japan after duty in Europe—as GCA operators to talk down pilots in emergencies. My students all outranked me but they were a terrific bunch. My first role as teacher (Raiffa 2006).

While in the military, Howard decided, upon his discharge, to enroll at the University of Michigan and study actuarial science. Because of the college credits received for the courses taken in the Army, he needed only a year to receive his bachelor's degree in actuarial mathematics in 1946. During this time, Howard came to realize that he wanted to study something more cerebral and theoretical. He had particularly enjoyed probability in his actuarial program, so he enrolled in Michigan's master's degree program in statistics and received his M.S. degree in 1947. This program also turned out to be disappointing, as it was short on theory and depth and long on computational manipulations. One course, however, taken during Howard's statistics program, turned out to have a significant influence. As he elaborated:

> Along with the courses I took in statistics, I also took, for cultural curiosity, a course in the foundations of mathematics by Professor Arthur Copeland who later became one of my mentors. Copeland taught the course using the R. L. Moore pedagogical style. Moore was a topologist (a branch of mathematics) at the University of Texas and he had a proselytizing set of disciples. Copeland's first assignment was weird: "Here are some seemingly unrelated mathematical curiosities. Think about them. Try to make some conjectures about them. Try to prove your conjectures. Try to discover something of interest to talk about."

> I, along with twenty other students, drew a blank and so I came to the next class with nothing to contribute. So did the other twenty students.
>
> At the beginning of class on that next day, Professor Copeland asked, "Does anybody have any contributions to make?" We sat and sat and sat and ten minutes went by and he said, "Class dismissed." He added, "The same assignment tomorrow." The following day he started the class with, "Anybody have anything to say?" Finally, someone raised a hand and asked a question. The course was pure R. L. Moore. No books were used, absolutely no books. It was taboo to look at the literature because you might find hints. You should act as if you were a mathematician in the 17th century trying to prove something new. No matter that we were "discovering" well-known results; it was new to us. We students did not study mathematics; we did mathematics. The R. L. Moore method of teaching turned me on. I knew then that I wanted to become a mathematician because it was so much fun and, to my surprise, I found out that I was pretty good at it. I think of the R. L. Moore pedagogical style as somewhat akin to the case method: both require an active involvement in the learning process (Raiffa 2006).

While studying for his Ph.D., Howard received the prestigious Rackham Fellowship awarded to the outstanding University of Michigan graduate student for two consecutive years. In addition, he worked 10 hours per week as a research assistant on an Office of Naval Research (ONR) sponsored research program, jointly administered by Michigan's Mathematics Department and the School of Engineering. His responsibility included the mathematical formulation of important real problems. As submarine warfare was one such important topic, he read the theory of games book by John von Neumann and Oskar Morgenstern (1947) and became quite interested in the field. Subsequently, in 1949 and 1950, he wrote a two-part report addressing some mathematically structured issues of two-person, non-zero-sum games.

Part A of the report-concerned strategies for repeated plays of what later became known as the two-person prisoner's dilemma game. Howard included what is now well-known as a Bayesian analysis of the prescriptive choices that one player should make, given a judgmental probabilistic description of how the other player would choose under all the possible circumstances.

Part B of the report concerned complex non-zero-sum games with no solution. Howard sought an arbitration rule that would provide a fair compromise solution for any non-zero-sum game. Howard wrote: "I remember distinctly how I started my research on arbitration rules. I attended a lecture by an economics professor and labor arbitrator, William Haber. During his talk about arbitration, I experienced an 'aha' inspiration; I jumped out of my chair while the lecture was going on, went back to my study, and wrote vigorously for hours without a break on how I would arbitrate non-zero-sum games. That constituted Part B of my ONR report on non-zero-sum games. I used the Kakutani Fixed Point Theorem to show the existence of equilibria strategies" (Raiffa 2006).

Howard's report was available in 1950 and printed as a technical report in the Engineering Department (Raiffa 1951). It was not peer reviewed; it was simply an informal report. At this time, April 1951, Howard was preparing to take his oral qualifying exam and searching for a thesis topic in mathematics, when "... a surprising thing happened" (Raiffa 2006):

> My wife, Estelle, received a telephone call from the famous algebraist, Richard Brauer, who was the chairman of my oral examining committee. He informed my wife that, on the basis of my written report, the committee decided to excuse me from my oral exam. And then he said: "By the way, the committee would like to talk to Howard about his thesis." I came in the next day all excited about the fact that I didn't need to take my oral exam and was told that the committee thought it appropriate that I slightly revise my recently completed Engineering Report as my doctoral dissertation. I was stunned. So I ended up not having to take an oral exam, not having to write a thesis, and I was through before I thought I started (Raiffa 2006).

Thus, in 1951, Howard received his Ph.D. in mathematics. His dissertation, "Arbitration schemes for generalized two-person games," included seminal research that provided useful insights for cooperative games. Arthur Copeland was Howard's dissertation advisor.

In April 1951, it was too late to enter the academic job market for the next academic year. Thus, Howard accepted a post-doctorate position with a new interdisciplinary seminar on mathematics in the social sciences initiated by the Departments of Mathematics and Psychology. He was hired to be the rapporteur of the seminar. In addition, he learned psychological measurement theory working with Clyde Coombs from psychology and Larry Klein from economics, organized the classroom notes of two mathematics professors to be distributed to students, gave a series of seminar talks to the statistics faculty and doctoral students on Wald's newly published book, *Statistical Decision Theory* (Wald 1950), and continued research related to his dissertation on game theory focused mainly on the prisoner's dilemma game. Howard's intellectual activity related to game theory in this period is reviewed in Raiffa (1992a).

ACADEMIC CONTRIBUTIONS

Howard's main academic interest dealt with how analysis can help an individual or a group of individuals make better decisions. Irving LaValle, in an exceptionally thoughtful summary of Howard's academic contributions, described Howard's message as follows:

> Unless one is an exceptionally good holistic thinker, some systematic analysis can be of considerable help in making better decisions and in striking better bargains. Such analysis does not require superhuman rationality. It rests on certain principles of consistent behavior that most people find perfectly reasonable in the simple contexts in which they are asked to evaluate them. It calls upon them to constructively supply their own preferences and their own judgments, so that the resulting implications for action are ones that are good according their own lights. And Raiffa warns of the common cognitive traps, such as overconfidence in quantifying judgments (Alpert and Raiffa 1982) and too-hasty overweighting of anticipated feelings such as regret (Raiffa 1985). The analytical framework is broad enough to deliver on the claim that, if something matters to one, it can and it should be systematically taken into account. In the realm of negotiations, it implies that one should make a real effort to see things from the other parties' points and interests, one should seek ways to enlarge the pie at least as much as one strives to attain a goodly share of it, one should never neglect the linkages of the present situation with other current and future situations (the reputation you save may be your own), and one should explore mutually beneficial ways of utilizing outside parties as mediators or arbitrators, all the while keeping in mind the observed realities of how people actually behave in negotiation settings (LaValle 1996, 419–420).

Howard has made substantial contributions to the areas that are generally referred to as the decision sciences. His path breaking research in all aspects of the decision sciences has pushed the forefront of statistical decision theory, game theory, decision analysis, behavioral decision theory, risk analysis, and negotiation analysis.

In 1952, Howard accepted a faculty position with the Department of Mathematical Statistics at Columbia University. While there, with Duncan Luce, he wrote the book, *Games and Decisions: Introduction and Critical Survey* (Luce and Raiffa 1957), which organized and communicated the concepts, results, and relevance of game theory to a wide audience including social scientists and political analysts. It is still the classic source for the basic concepts and results of game theory, as well as the original foundations for decisions under uncertainty. This book is deep in substance and yet accessible for many readers, a common trait of all of Howard's authored and co-authored books.

ANATOMY OF A BOOK

In a 1988, Duncan Luce described how he and Howard joined together to write *Games and Decisions: Introduction and Critical Survey* (Luce and Raiffa 1957):

"In 1953 I became managing director of Columbia University's Behavioral Models Project . . . led by the late Paul L. Lazarsfeld and including, among others, Howard Raiffa. Its charge was expository pieces about mathematical modeling of behavioral and social problems. Raiffa opted to cover 2-person games and statistical decision theory, and I focused on both n-person games, which had attracted my interest while studying small groups at the Massachusetts Institute of Technology, and information theory. Later that year, Raiffa and I recognized that we should combine our game-theory essays and that a book, not an essay was in the making. . . .

"By the end of the year (1956) we had a fairly complete manuscript with the working title Conflict, Collusion, and Conciliation (our ultimate title was suggested by Albert Tucker) on which we received comments, the most extensive being from Harold W. Kuhn, and early in 1957 we considered it finished" (Luce 1988, 16).

In 1957, Howard moved to Harvard University. This directly led to Howard's next major area of work, statistical decision theory, done at the Business School jointly with Robert Schlaifer and John Pratt. The books, *Applied Statistical Decision Theory* (Raiffa and Schlaifer 1961) and *Introduction to Statistical Decision Theory* (Pratt et al. 1965), provide the foundations for Bayesian analysis of the standard statistical problems. This approach combines prior knowledge about problems with subsequent specific data gathered to lend operational insight for real decisions. *Applied Statistical Decision Theory* was republished by Wiley in 2000 as a classic in mathematics and statistics. An interesting interview of Howard discussing his academic career through 1967 is found in Feinberg (2008).

With Pratt and Schlaifer, Howard authored a fundamental paper that jointly axiomatized utility and subjective probability as a prescriptive theory to guide decision making (Pratt et al. 1964). Howard illustrates the operational use of these ideas in his book, *Decision Analysis* (Raiffa 1968), which covered the foundations of decision analysis and introduced many advanced topics such as risk sharing, group decisions, and multiple objectives. This book was the first to widely provide the foundation of decision analysis, and was the source for many students, academics, and lay persons becoming acquainted with the field.

Beginning in the mid-1960s, Howard's research interests expanded from problems focused on management or economics to problems associated with the public domain. This included problems of governmental policy, science policy, public health, and clinical medicine. Such problems usually involve multiple objectives, and require, for their resolution, the application of value tradeoffs to the objectives. Howard spent the Summer of 1965 at the RAND Corporation where he greatly furthered his thinking about the analysis of problems with multiple objectives. His RAND report, "Preferences for multi-attributed alternatives" (Raiffa 1969), outlined how to address these problems using decision analysis.

In the fall of 1966, I began graduate school at the MIT Operations Research Center. Soon after, I had the good fortune to meet Howard. In December, I saw an announcement at the OR Center stating that Professor Howard Raiffa of Harvard University was giving an evening presentation at MIT on decision analysis sponsored by the local chapter of a professional organization. I attended and was engrossed in the presentation. Afterwards, I asked Professor Raiffa about potential master's theses topics. The topic that he mentioned that most appealed to me concerned multiple objectives. He also told me that he was teaching a course on decision analysis in the coming spring term in the Economics Department at Harvard. As an MIT student, I could cross enroll in the course, which I did. During that spring, I also worked on my master's thesis on two-objective utility functions, which was completed in the summer of 1967. Soon after, even though Howard was a Harvard professor and I was an MIT student, he agreed to be my doctoral advisor, and MIT approved this arrangement. I completed my dissertation on multiple-objective utility functions in 1969.

Right after my graduation in June 1969, I proposed to Howard that we write a book together on decision making involving multiple objectives. [The discussion process that led to his agreement is described in Keeney (1992,

Howard Raiffa Ralph Keeney

385–386).] Subsequently, in the book, *Decisions with Multiple Objectives* (Keeney and Raiffa 1976), the concepts and procedures of decision analysis were extended to decisions involving multiple objectives. Specifically, the book focuses on the theory and practice of developing and assessing multi-attribute utility functions and applying them to resolve important problems. As LaValle (1996, 441) stated, this book "made a convincing case for the wide, practical applicability of structured multi-attribute utility in the real world."

Howard's next book founded the discipline of negotiation analysis. *The Art and Science of Negotiation* (Raiffa 1982) developed the concepts and procedures that rendered negotiation analysis as a field with substance relevant for serious study. It was awarded the 1985 Melamed Prize of the University of Chicago Business School for the most significant published work by a faculty member in a school of business in the preceding 2 years. Before that time, much of the knowledge about negotiations was basically art and folklore, with essentially no science. This book brilliantly communicates and illustrates a simple message, namely, that the analysis of negotiations can provide very useful insights for negotiators. It also stresses the value of cooperative negotiations by indicating how all sides can better achieve their objectives by various cooperative strategies. It also includes Howard's well-known asymmetric prescriptive/descriptive approach that probabilistically describes how your negotiating partner (or adversary) may behave, and then prescribes what decisions you should make to achieve your objectives for these potential partner behaviors. The book, *Negotiation Analysis* (Raiffa 2002), integrates much of Howard's earlier work and includes developments in negotiation analysis from 1982.

In 1998, Howard joined with the management consultant John Hammond and myself to write *Smart Choices* (Hammond et al. 1999).

This book synthesizes the major ideas and procedures from decision analysis research and applications and communicates them in a manner that anyone facing important decisions can understand and use. *Smart Choices* has sold over 100,000 copies and has been translated into 16 different languages. It was awarded the annual book prize of the Institute for Conflict Prevention and Resolution in 1999 and the Annual Publication Prize of the Decision Analysis Society in 2001.

> THE ELEMENTS OF GOOD DECISION-MAKING
>
> "Address the right decision *problem*
>
> Identify your real *objectives*
>
> Develop a range of creative *alternatives*
>
> Understand the *consequences* of the alternatives
>
> Make appropriate *tradeoffs* among conflicting objectives
>
> Deal sensibly with *uncertainties*
>
> Account for your *risk-taking attitude*
>
> Plan ahead for *linked decisions* over time"
>
> (Hammond et al. 1999, Chapter 1).

LEADERSHIP CONTRIBUTIONS

Howard's academic achievements have been extensive, but it is important to note his other contributions. He has always believed that international cooperation on problems common to different societies is critical. From 1967–1972, he had a key role in the negotiations that created the first non-governmental international research institute, the International Institute for Applied Systems Analysis (IIASA), located near Laxenburg, Austria. It was initially sponsored by the National Academies of Science of 12 countries, including the U.S. and the former Soviet Union. IIASA came into existence on October 4, 1972, with Howard serving as its first director from 1972 to 1975. The original motivation for its establishment was to enable scientists from East and West to work together on problems of common

> INITIAL INVOLVEMENT WITH IIASA
>
> In a talk he delivered at IIASA on September 23, 1992, Howard described how his involvement with IIASA began:
>
> "On a Saturday afternoon early in 1967, I got a call from Bundy [McGeorge Bundy, an advisor to U.S. President Johnson] at home, saying that he was in Cambridge and could he meet me the next day; he would like me to do some consulting. I said, 'What kind of consulting?' He said, 'It's pro bono but it won't take long.' I have figured out that since then, I have worked on IIASA affairs for about 15,000 hours" (Raiffa 1992b).

concern. Howard represented the American Academy of Arts and Sciences as its IIASA counsel member from 1980 to 1987. IIASA is still a vibrant intellectual research institution. It is now sponsored by 18 national organizations in Africa, Asia, Europe, and North America, and currently conducts interdisciplinary scientific studies on environmental, economic, technological, and social issues in the context of human dimensions of global change (IIASA 2008).

In the late 1960s, Howard played a key role in creating Harvard's Kennedy School for Government. His original suggestion included "a school with its own faculty, primarily stressing a professional masters program, with say 50% to 75% of its instruction by the case method, and with its own mini-campus" (Raiffa 2006). He then became part of a committee of four who designed the Kennedy School and who are recognized as the School's four founding members. At Harvard, Raiffa held academic appointments in the Departments of Economics and Statistics, the Kennedy School of Government, and the Graduate School of Business. In addition, he developed and taught courses on decision making in the Schools of Law, Medicine, and Public Health, and for undergraduates in Harvard College.

In addressing all of the numerous issues in creating and then guiding IIASA as its first director, Howard's interest in negotiations increased. When he returned to Harvard in 1975, he began pursuing teaching, research, and negotiations in earnest. His elective course on negotiation analysis at the Harvard Business School was partially an experimental research laboratory. He included insights from this course and analytical results developed over the next few years in his book *The Art and Science of Negotiation* (1982) that initially provided a systematic and logical foundation for negotiations. During this time, Howard, along with Roger Fisher of the Harvard Law School, established Harvard's Program on Negotiation (PON). Howard was a principal sponsor of a PON project called the Negotiation Roundtable that focused on the study of negotiations in managerial settings.

THE EDUCATOR

Howard's initial academic position was an Assistant Professor of Mathematical Statistics at Columbia University from 1952 to 1957. He accepted an unsolicited offer from Harvard University in 1957 as an Associate Professor of Statistics with a joint appointment in the Statistics Department and the Harvard Business School. He became a Professor of

Statistical Decisions in 1960 and was appointed the Frank P. Ramsey Professor of Managerial Economics in 1963. He retired from Harvard as Professor Emeritus in 1994.

Howard has been the main dissertation advisor or committee member of approximately 90 doctoral students. One of Howard's doctoral students, Irving LaValle, has written a revealing view of how Howard interacted with his students:

> Raiffa has been exceptionally generous with his ideas and his time. Seeing him on short notice was never a problem, and he gave his complete attention and participation to his students' intellectual struggles. In due course, he assisted with good placements and, unlike most mentors, refused co-authorship of publications based on the dissertation. This generous and modest refusal, together with his insistence on alphabetical listing of his coauthors, makes it impossible ever to take a complete inventory of Raiffa's original ideas. Only glimpses appear, typically in footnotes, of his role in the work of others (LaValle 1996, 418).

It should be clear, however, that the value of Howard's contributions and intellectual influence is due to both his research publications and the research and leadership of his doctoral students such as Daniel Ellsberg, Harvey Fineberg, Victor Fung, C. Jackson Grayson, John Hammond, Gordon Kaufman, Joseph Pliskin, James Sebenius, James Vaupel, Milton Weinstein, Robert Wilson, and Richard Zeckhauser.

Several of Howard's books, described earlier, have defined and organized the foundational substance that is taught today in the basic courses of their respective fields: game theory, Bayesian statistical decision theory, decision analysis, multiple-objective evaluation, and negotiation analysis. Prior to these books, the fields had a few articles that very knowledgeable and dedicated teachers could use in their courses. In addition, Howard was at least a decade ahead of others in producing another type of educational product.

In the early 1970s, when Howard was traveling a great amount for meetings that eventually led to the creation of IIASA, he developed an audiographic course on decision analysis titled "Analysis for Decision Making." This material was used in his absence for distance learning by numerous students at Harvard in public health, education, economics, and the Kennedy School; it was also distributed by the *Encyclopædia Britannica Educational Corporation.* Subsequently, Howard developed additional audiographic lectures on topics such as auctions, game theory, and the foundations of statistical reasoning that were widely used as parts of courses at Harvard.

Honors and Awards

Howard was elected a member of the American Academy of Arts and Sciences in 1972, and in 2005, he was elected to the National Academy of Engineering. He has received recognition for exceptional contributions by several professional societies including the Distinguished Contribution Award of the Society of Risk Analysis in 1984, the Frank P. Ramsey medal for distinguished contributions in decision analysis of the Institute for Operations Research and Management Sciences (INFORMS) Decision Analysis Society in 1984, the Gold Medal of the International Society for Multiple Criteria Decision Making in 1998, and the INFORMS Expository Writing Award in 2002. Howard and I shared the ORSA Lanchester Prize for the best OR publication in 1976 for our book, *Decisions with Multiple Objectives: Preferences and Value Tradeoffs.*

In 2000, Howard received the Carnegie Mellon University Dickson Prize for Science, awarded annually to individuals who have made outstanding contributions to science in the U.S. He has received honorary doctorate degrees from Carnegie Mellon University, the University of Michigan, Northwestern University, Ben Gurion University of the Negev, and Harvard University.

Summing Up

Succinctly, I can sum up my understanding of Howard's life as follows: 1924–1932, growing; 1932–1942, growing and playing basketball; 1942-present, life with Estelle, his wife and companion since 1945. Estelle and Howard have two children, Mark born in 1952 and Judy born in 1955, and three grandchildren. Estelle is a retired social worker, active in local politics, and retired recently from the Board of Directors of Travelers Aid Family Services. She is a loving friend to me and my family.

Howard is a renowned researcher and author, a superb teacher, an ideal dissertation advisor, a wonderful colleague, and a loyal friend. I am very fortunate to know him in all of these roles.

REFERENCES

Alpert M, Raiffa H (1982) A progress report on the training of probability assessors. In: Kahneman D, Slovic P, Tversky A (eds) Judgment under uncertainty: heuristics and biases. Cambridge University Press, New York, NY, pp 294–305

Feinberg SE (2008) The early statistical years: 1947–1967. A conversation with Howard Raiffa. Stat Sci 23(1):136–149

Hammond JS, Keeney RL, Raiffa H (1999) Smart Choices. Harvard Business School Press, Boston, MA

IIASA (2008) About IIASA. http://www.iiasa.ac.at. Accessed 15 Apr 2008

Keeney RL (1992) Value-focused thinking. Harvard University Press, Cambridge, MA

Keeney RL, Raiffa H (1976) Decisions with multiple objectives: preferences and value tradeoffs. Wiley, New York, NY

LaValle I (1996) The art and science of Howard Raiffa. In: Zeckhauser RJ, Keeney RL, Sebenius JK (eds) Wise choices. Harvard Business School Press, Boston, MA, pp 417–456

Luce RD (1988) Citation classic: games and decisions: introduction and critical survey. Citation Classic (15) April 11:16

Luce RD, Raiffa H (1957) Games and decisions: introduction and critical survey. Wiley, New York, NY

Pratt JW, Raiffa H, Schlaifer R (1964) The foundations of decision under certainty: an elementary exposition. J Am Stat Assoc 59(306):353–375

Pratt JW, Raiffa H, Schlaifer R (1965) Introduction to statistical decision theory (preliminary edition). McGraw-Hill, New York, NY. Final edition published in 1995 by MIT Press, Cambridge, MA

Raiffa H (1951) Arbitration schemes for generalized two-person games. Engineering Research Institute, University of Michigan, Ann Arbor, MI. Report No. M720-1, R30

Raiffa H (1968) Decision analysis: introductory lectures on choices under uncertainty. Addison-Wesley, Reading, MA

Raiffa H (1969) Preferences for multi-attributed alternatives. RM-5868-DOT/RC, RAND Corporation, Santa Monica, CA. Reprinted in 2006 in Multi Criteria Decis Anal 14(4–6):115–157

Raiffa H (1982) The art of science and negotiation. Harvard University Press, Cambridge, MA

Raiffa H (1985) Back from prospect theory to utility theory. In: Grauer M (ed) Decision processes. Springer, Heidelberg, pp 100–113

Raiffa H (1992a) Game theory at the University of Michigan, 1948–1952. In: Roy Weintraub E (ed) Toward a history of game theory. Duke University Press, Durham, NC, pp 165–175

Raiffa H (1992b) How IIASA began.http://www.iisa.ac.at/docs/history.html. Accessed 14 Apr 2008)

Raiffa H (2002) Negotiation analysis. Harvard University Press, Cambridge, MA

Raiffa H (2006) A memoir: analytical roots of a decision scientist (unpublished)

Raiffa H, Schlaifer R (1961) Applied statistical decision theory. Division of Research, Graduate School of Business Administration, Harvard University, Boston, MA. Republished in 2000 by Wiley-Interscience

Von Neumann J, Morgenstern O (1947) Theory of games and economic behavior, 2nd edn. Princeton University Press, Princeton, NJ

Wald A (1950) Statistical decision functions. Wiley, New York, NY

28

D. Ray Fulkerson

Robert G. Bland and
James B. Orlin

Ray Fulkerson's seminal work in network flows, large-scale linear programming (LP), combinatorial optimization, and combinatorics has had an enormous influence on the practice of operations research (OR). His seminal book, *Flows in Networks*, co-written with Lester R. Ford, Jr., was instrumental in bringing network flow theory and algorithms to the domain of OR practice—in communications, transportation, supply systems—and in hastening the development of academic courses in networks, graph theory, and combinatorics. Much of his most influential work began with an application, a puzzle, or a specific computational obstacle. Pursuit of the underlying mathematical structures led Ray and his collaborators to broad and profound methodological innovations, such as cutting planes and column generation, and to the foundations of network flow theory and polyhedral combinatorics.

Ray's career spanned 20 years with the RAND Corporation, followed by 6 years at Cornell University. In 1967, he received the Lester R. Ford Award of the Mathematical Association of America for his expository paper on network flows and combinatorial OR. He was the recipient of the Southern Illinois University Award for Outstanding Professional Achievement in 1972. He was made a member of the International Federation of Operational Research Societies' Hall of Fame in 2005.

A.A. Assad, S.I. Gass (eds.), *Profiles in Operations Research*, International Series in Operations Research & Management Science 147, DOI 10.1007/978-1-4419-6281-2_28,

THE FULKERSONS

Ray Fulkerson was born on August 14, 1924, in Tamms, Illinois. Tamms is in the southernmost part of Illinois, near the confluence of the Mississippi and Ohio Rivers, almost 400 miles south of Chicago. It is just over the borders with both Missouri and Kentucky, and not far from Tennessee and Arkansas. Ray's entire childhood was spent in southern Illinois, although his family moved several times.

Ray's parents, Elbert Fulkerson and Emma Grace Mossberger, were also born and raised in southern Illinois. Elbert was the high school principal in Tamms, and later in Carterville, where Ray attended grade school and then high school. Emma was valedictorian in the eighth grade graduating class of her primary school and was later a teacher in that school. She was a deeply religious woman, who was very active in church and community services throughout her adult life. She was also a volunteer nurse's aid. Ray was the third of Elbert and Emma's six children. His two brothers, Glen and Dick, and three sisters, Merle, June and Grace, all became educators. All three Fulkerson sons obtained doctorates.

Because Ray was so much like his father, Elbert's story, interesting in its own right, is especially relevant. Elbert was born in a one-room log cabin on a farm and was the sixth of seven children, one of whom died in infancy. Elbert's father barely eked out a living from farming the clay hills. He was not a very ambitious man, but Elbert's mother was eager for her children to become educated and have a better life than the farm provided. She worked hard, and although she could barely read, she taught Elbert to start reading. She had several very serious illnesses and young Elbert spent many nights crying himself to sleep at the foot of his mother's bed, fearing that she would not survive the night. The attitude of the family was always to expect the worst, and Elbert received little comfort from his father or siblings.

At age seven, Elbert suffered a severe injury to one eye, caused by measles. This resulted in considerable pain and impaired vision for his entire life. Because of his eye problems and the family's financial circumstances, Elbert was not able to begin grade school until he was 13 years old. Even then, he was not able to attend in cold weather because he lacked proper shoes for the long walk to school. He was embarrassed at being so much older and bigger than his classmates, but he was eager to learn, and he completed the 8 years of grade school in 3 years. There was no opportunity for Elbert to attend high school; there were none in the rural area where he lived. Instead, he took the written exam to qualify as a grade school teacher and passed it, but he was too young to be eligible to teach. At age 18, he received a teaching certificate. Prior to beginning his remarkable career as an educator, Elbert spent the summer doing manual labor at 50 cents per day, the proceeds from which enabled him to buy his first suit and wristwatch—basic necessities for a teacher. In his first teaching jobs, Elbert was, essentially, a one-man school; he taught all of the classes and even did the janitorial work. He also began taking summer courses at what is now Southern Illinois University (SIU), then called Southern Illinois Normal University. His reputation as a teacher spread, and he was offered an opportunity to start a 2-year high school in an area that previously had not had any high schools. He accepted, taught all of the courses, and, in his first year, taught himself the material just in time to teach it to his students. Elbert continued with his summer studies and received a bachelor's degree at SIU and then a master's degree at the University of Illinois. In 1944, he became a faculty member and Secretary of the Faculty at SIU. Today there is a Fulkerson Hall at SIU named in his honor.

Ray and his four eldest siblings grew up during the Great Depression. The Fulkerson family squeezed by on Elbert's modest salary as a teacher. Occasionally, there was not enough money for the school to pay him in cash. Instead, he received scrip that was accepted at some local stores. To supplement the family's food supply and finances, Ray helped Elbert with a vegetable garden and clerked in a small grocery store. All of the children adopted their parents' love of learning. Elbert taught them at home, as well as in high school, where he taught everything from Latin to math. His reputation for being fair but firm was clearly recognized at school and at home. Ray played the baritone horn in the school band. He had lead roles in school plays, including the role of one of the attorneys in a play based on the transcripts of the Dred Scott trial, which took place in pre-Civil War St. Louis, only about 100 miles from Carterville. Ray played varsity basketball, and, after every basketball game, he and Elbert would carefully dissect each play. Ray was also an avid and skilled tennis player.

Ray regarded Elbert to be one of three truly influential teachers from his school years. It was certainly from Elbert that Ray learned to appreciate clarity of thought and expression. Ray developed tendencies toward perfectionism, austerity, and loneliness that may have also come from Elbert. Ray and Lester Ford dedicated their book, *Flows in Networks* (1962), to their fathers, who were both mathematicians. At the memorial service for Elbert on April 5, 1972, Ray said that his part of the dedication "was a heart-felt tribute to an inspirational teacher of mathematics, and to a father I loved, honored and respected" (Fulkerson 1972).

THE FULKERSON FAMILY (1944)

Front row (Left to right) Elbert, Dick, Emma

Back row (Left to right) June, Grace, Glen, Ray, Merle

Glen had been missing in action in the Pacific and surprised everyone by showing up at home unannounced. Elbert called the Red Cross to get Ray home for a couple of days of leave from the Army Air Corps.

FROM SIU TO WISCONSIN, TWICE

In 1942, Ray began his freshman year at SIU—he had graduated at the top of his high school class at the age of 16. He played for the SIU varsity tennis team and was a member of the student theatre group. Ray left SIU in January 1943 to serve in the U.S. Army Air Corps. He attended the Army Specialized Training Program at the University of Wisconsin, and then studied meteorology at Chanute Field, near Urbana, Illinois. In 1946, Ray was honorably discharged as a first lieutenant and returned to SIU. He graduated in 1947 with a B.A. in mathematics, again at the top of his class. In 1947, Ray returned to the University of Wisconsin, this time as a graduate student in mathematics. He received an M.S. in 1948 and a Ph.D. in January 1951 (major in algebra and minor in analysis). Ray's Ph.D. dissertation, "Quasi-Hermite forms of row-finite matrices," was written under the supervision of Cyrus C. MacDuffee. Shortly after beginning his Ph.D. studies, Ray married Eleanor White, who had also graduated from SIU in 1947. Their first son, Guy, was born in Madison in 1950.

> FORECASTS WITH NO UNCERTAINTY
>
> Lt. Fulkerson was a meteorologist, responsible for briefing pilots before they went out on training missions. In spite of the vagaries of weather forecasting, Ray's supervisor had ordered the meteorologists to not allow any uncertainty to creep into their forecasts. Ray was extremely uncomfortable about sending pilots off without any expression of doubt when he regarded his forecast to be speculative. He solved the problem by walking to the front of the briefing room, ceremoniously flipping a coin, and then, upon carefully examining the outcome of the flip, announcing a certain forecast to the pilots.

Ray's dissertation topic was not a good indicator of where his mathematical pursuits would soon lead. It was in algebra, it was non-algorithmic, and dealt with infinite dimensional spaces. Almost all of Ray's later work would be in optimization and combinatorics (broadly interpreted), it was often algorithmic, and only once (Folkman and Fulkerson 1970) dealt with the infinite. There were exciting developments in combinatorics at Wisconsin during Ray's time there, and one of Ray's fellow graduate students, Herb Ryser, played a central role. Ryser, along with Professor R. H. Bruck, wrote one of the fundamental papers on finite projective planes (Bruck and Ryser 1949). Herb and Ray remained lifelong friends, and wrote four papers together in the early 1960s, while Ray was at RAND and Herb was on the faculty at Ohio State and then Syracuse. Herb moved to the California Institute of Technology in 1967.

THE RAND YEARS: 1951–1971

After completing his Ph.D., Ray joined the Mathematics Department of the RAND Corporation, Santa Monica, California, at the invitation of the department head, John Williams. RAND was set up in 1948 as a think tank "to promote scientific, educational, and charitable purposes," with its initial support in logistics and systems analysis from the U.S. Air Force (RAND 1948). Williams had assembled a stellar group of talented mathematicians and, basically, turned them loose. Many of Ray's colleagues in the department also became celebrated figures in the development of the mathematical foundations of OR, including: Richard Bellman, George Dantzig, Merrill Flood, Lester Ford, Selmer Johnson, Lloyd Shapley, and Philip Wolfe. The RAND Mathematics Department also included emerging stars in other areas of the mathematical sciences, including: Wendell Fleming, Ted Harris, and Herb Scarf. Ray also worked extensively with Herb Ryser, his friend and fellow graduate student at Wisconsin, and a consultant at RAND. In addition, he collaborated with several young scholars who worked at or visited RAND, most notably Jon Folkman and Jack Edmonds. Other consultants and visitors included: Ken Arrow, David Gale, Ralph Gomory, Sam Karlin, John Nash, Al Tucker, and Bill Tutte.

Fleming (2009) recalls that during the glory years of RAND there was no shortage of appreciation among the Mathematics Department members for the excellence of the group and its range of interest. Legend has it that one of the department seminars began with the hypothesis: "Consider two worlds, W_1 and W_2."

Prior to coming to RAND, Ray had studied neither LP nor graphs and networks. Merrill Flood, head of logistics research at RAND, quickly steered him in those directions. Among his first assignments at RAND were to translate Theodore S. Motzkin's 1936 Ph.D. thesis, "Contributions on the theory of linear inequalities," from German to English, and to write notes based on graph theory lectures given at RAND in 1951 by Albert W. Tucker. Tucker (1976) reported that Ray's notes were, " much more thorough than the talks" they were based on. Within a few years of these initial assignments at RAND, Ray had made major contributions to both areas.

One of Ray's first publications and most lasting contributions arose from a computational challenge concerning the traveling salesman problem (TSP): find a shortest traveling salesman tour through 49 cities, a major city from each of the (then) 48 states and Washington, D.C. His collaborators in this endeavor were George Dantzig and Selmer Johnson. Ray had worked previously with both, but in very different circumstances. Ray's first publication was a collaboration with Dantzig on the use of LP to solve a tanker-scheduling problem (Dantzig and Fulkerson 1954a). Ray had learned meteorology from Selmer Johnson during the war at Chanute Field. Len

Berkovitz, another RAND colleague and friend, had also studied meteorology at Chanute Field at that time, but they never met until later at RAND.

With respect to the origins of the TSP, Dantzig et al. (1954b, 393) note as follows: "The origin of the problem is somewhat obscure .... It appears to have been discussed informally among mathematicians ... for many years. Surprisingly little in the way of results has appeared in the mathematical literature." They go on to say that Albert Tucker and Merrill Flood recalled Hassler Whitney mentioning the TSP in a seminar at Princeton in the early 1930s (Whitney did not recall that), and that Flood "should be credited with stimulating interest .... As early as 1937, he tried to obtain near-optimal solutions in reference to the routing of school buses." (A detailed history of the TSP is given in Applegate et al. [2006].) It seems that the conversations about the TSP at Princeton in the 1930s included the puzzle of finding an optimal tour through 48 cities, one from each state. Dantzig, Fulkerson, and Johnson aimed to solve that puzzle (with a 49th city, Washington, D.C., added). RAND's Julia Robinson (1949) had observed that the assignment problem is a relaxation of the TSP. Isidor Heller (1953) and Harold Kuhn (1955, 1991) had already begun to examine the facets of the convex hull of all tours, and observed that the number of facets grows very rapidly with the number of cities. Even for a TSP on five or six cities, the number of linear inequalities needed to pose the TSP as an LP was too big to solve with the computing resources available at that time.

Dantzig et al. (1954b) had a brilliant insight that enabled them to overcome this difficulty—perhaps Dantzig's simplex method could start the computation with a workably small subset of the inequalities, and others could be generated as needed. Their overall approach, which engendered the area of polyhedral combinatorics, was highly innovative. They considered the convex hull of the integer points in the feasible region of what we now call an LP relaxation of the TSP. They invented what are now known as cutting planes, generating them only as needed, to try to determine the convex hull in the vicinity of an optimal integer solution. The specific cutting planes they used for the 49-city TSP are what are now called subtour elimination constraints. These constraints were not quite strong enough to determine an optimal integer solution for this 49-city example, so they used what they called "a combinatorial approach" to prove optimality of their solution (Dantzig et al. 1954b, 397, 405, 408; 1959, 62–65). Based on the values of dual variables from the LP solution, they deduced that certain edges must be included, and others excluded, from every optimal tour. In addition, with a limited enumeration, they could speculatively assess the effects of forcing other edges in or out and prove their tour optimal. This combinatorial approach has the essence of what is now termed branch-and-bound. They even managed, after the fact, to generate two additional constraints, which, together with the subtour elimination constraints, allowed for a proof of optimality that was entirely polyhedral—they gave optimal dual variables for the LP relaxation with the added constraints (cutting planes). As pointed out by Applegate et al. (2006), the two additional ad hoc constraints are closely related to what are now known by TSP researchers as comb inequalities.

TOUR DE FORCE

Early in 1954, Dantzig, Fulkerson, and Johnson invented cutting planes and employed them to find an optimal traveling salesman tour through Washington D.C. and a representative city from each of the (then) 48 states (total optimal distance: 12,345 miles). This work laid the foundations of polyhedral combinatorics and integer programming.

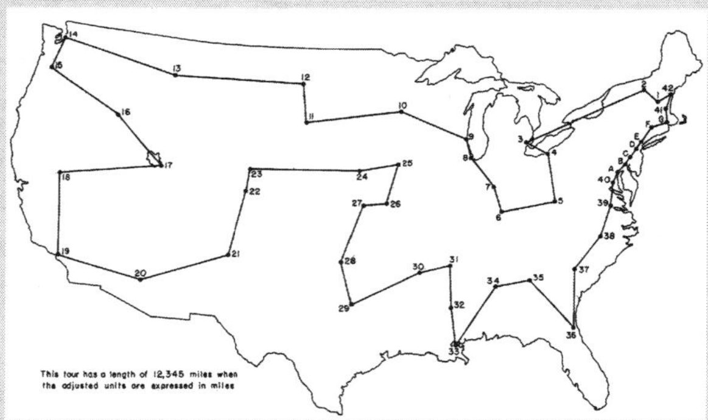

(Dantzig et al. 1954b)

Prior to solving the 49-city TSP, the three researchers discussed how many added constraints might be needed. Hoffman and Wolfe (1985, 8) recall:

> Dantzig, Fulkerson, and Johnson had nevertheless to face the possibility that billions of cuts might be needed. An optimistic Dantzig wagered one dollar with a pessimistic Fulkerson: Dantzig contended that the number of cuts needed would be at most 25, Fulkerson that it would be at least 26. The result was very close: Dantzig remembers that it turned out to be 26, but the published paper says only 25 were needed.

Newsweek Magazine ran a story on this "ingenious application of linear programming" in the July 26, 1954 issue. It is highly unlikely, however, that even Dantzig, Fulkerson, and Johnson could have anticipated then the practical impact that this work would ultimately have. Gomory's work on general cutting planes, motivated by the Dantzig, Fulkerson, and Johnson paper, suggested broad mathematical implications, but computational progress was, initially, very slow.

Much later, computational work began to focus on cutting planes for the TSP and other combinatorial problems with special structure. These techniques have led to astonishing improvements in our ability to find optimal solutions, or to get provably near-optimal solutions to problems like the TSP and general integer-programming problems.

Ray is probably best known for his work on network flows. His collaboration with Lester R. Ford, Jr. began in 1954 and ultimately laid the entire foundation of the subject. It started with a specific application to military operations first brought to Ray's attention over lunch at RAND by Frank S. Ross, a retired army general, and Ted Harris. They were working on a project to evaluate interdiction strategies on the capacity of the Eastern European rail network to support a conventional war (Harris and Ross 1955). Ross had been in charge of the U.S. Army's transportation corps in Europe. Les and Ray concentrated on what to them was the important related mathematical and practical problem of interest: how to determine the maximum flow through the network. This led to their first paper on network flows in which they proved the now famous max-flow min-cut theorem: the maximum-flow value from the source (origin) to the sink (destination) in a capacitated network is equal to the minimum capacity of a cut separating the source from the sink. This work first appeared as a RAND research memoRANDum (Ford and Fulkerson 1954) and then as Ford and Fulkerson (1956a). There followed a stream of papers on maximum flows, minimum cost flows, dynamic flows, and their classic book, *Flows in Networks* (Ford and Fulkerson 1962), which was translated into French, Japanese, Polish and Russian. It proved to be the standard textbook and reference in network flows for many years. It is still an excellent resource, full of elegant insights about network flows, and extremely well written. The 1993 text, *Network Flows: Theory, Algorithms, and Applications* (Ahuja et al. 1993, iii), is dedicated in part "as a tribute to Lester Ford and Ray Fulkerson, whose pioneering research and seminal text in network flows have been an enduring inspiration to us and to a generation of researchers and practitioners."

The original proof of the max-flow min-cut theorem in Ford and Fulkerson (1954, 1956a) is not constructive. It is based on packing (0,1)-incidence vectors of source-sink paths into the capacity vector. The only algorithm in the 1956a paper is for planar graphs embedded with the source and sink on the outside face. The famous augmenting path algorithm of Ford and Fulkerson first appeared in Ford and Fulkerson (1957). From the preface to the Ford and Fulkerson (1962, viii):

> Throughout the book the emphasis is on constructive procedures, even more, on computationally effective ones. Other things being nearly equal, we prefer a constructive proof to a non-constructive one, and a constructive proof that leads to an efficient computational scheme is, to our way of thinking, just that much better.

Thus, the book presents the constructive proof based on the augmenting path algorithm, rather than the earlier proof. Nonetheless, the original proof is quite interesting, and foreshadows Ray's later work on blocking polyhedra and integrality properties. Hoffman (1978) gives an interesting discussion of the original non-constructive proof.

The paper on multicommodity flows is another gem (Ford and Fulkerson 1958). In the conventional maximum-flow problem, there is a single commodity with specified source and sink. Here there are *k* different commodities, each with a specified source and sink, which share fixed capacities on the arcs. They modeled the multicommodity flow problem as an LP problem, with a decision variable for every source-sink path for a commodity. To solve this linear program with an exponentially large number of variables, they developed the first column generation algorithm. They avoided explicit consideration of all but a small number of columns by generating columns as needed, using duality and shortest path computations. This approach, which is dual to the constraint-generation approach from Ray's TSP papers with Dantzig and Johnson, inspired the important papers of Dantzig and Wolfe (1961) on decomposition and Gilmore and Gomory (1961, 1963) on the cutting-stock problem.

A problem from the natural gas industry and the computational challenges it posed motivated Ray's development of the out-of-kilter algorithm (Fulkerson 1961a) [independently developed by George Minty (1966)]. Dantzig (1959) had formulated a gas-balance problem for Southern California Gas Co. as a dynamic network-flow problem. The flow problem was time-staged and, potentially, too large to be solved by available methods of the day. Ray developed the out-of-kilter method as a practical method for solving it. His approach was to first subdivide the problem into manageable pieces, combine the solutions from those pieces, and then put the combined flows into kilter. The algorithm alternates between primal steps of sending flow and dual steps of adjusting prices, all the time bringing the solution closer to primal and dual feasibility by reducing kilter numbers. The analysis in terms of kilter numbers and kilter steps motivated work that has led to very fast minimum-cost flow algorithms using successive approximation techniques. Similarly, the earlier Dantzig et al. (1956) paper on the primal-dual algorithm for LP (a generalization of Kuhn's Hungarian algorithm for the assignment problem) underlies a substantial stream of work initiated 40 years later by Goemans and Williamson (Williamson 2002) on primal-dual approximation algorithms for network design and other combinatorial optimization problems.

Ray always enjoyed finding interesting applications of network-flow theory. Sometimes they were applications in a conventional, practical sense such as using LP duality and network flows for critical path method/program evaluation review technique (CPM/PERT) analyses of project cost curves [Fulkerson (1961b) and Kelley (1961) independently discovered the same approach]. Often they involved using network flows to obtain constructive proofs of known results in combinatorics (Fulkerson 1966), such as Dilworth's Theorem that the minimum number of chains to cover the elements of a partially ordered set is equal to the maximum size of an antichain (subset of mutually unrelated elements) (Dilworth 1950). Ray was intrigued by the observation that many important results in extremal combinatorics had counterparts with which they formed symmetric pairs. For example, Menger's Theorem tells us that the maximum number of edge-disjoint source-sink paths in a two-terminal network is equal to the minimum size of a source-sink cut (Menger 1927). (The max-flow min-cut theorem provides a more general, weighted version.) This result on packing paths has a symmetric result, due to Robacker (1955), which interchanges the roles of paths and cuts. Similarly, Dilworth's Theorem on covering with chains has a symmetric result that interchanges the roles of chains and antichains. Much of the work that Ray did in the last 10 years of his life focused on developing unifying theories for these phenomena. His paper with Jack Edmonds on bottleneck extrema (Edmonds and Fulkerson 1970), his introduction of the theories of blocking and antiblocking pairs of polyhedra (Fulkerson 1971), and his less noticed, but lovely paper,

"Networks, frames, blocking systems" (Fulkerson 1968), are all in this vein. The blocking theory, motivated by results on weighted packing of two-terminal paths and cuts, and the antiblocking theory, motivated by results on weighted covering with chains and antichains, explore these phenomena from a geometric, polyhedral viewpoint, and examine when the theorems hold in a stronger form in which integer-optimal solutions always exist.

Given the importance of his research, it is remarkable that those who worked or studied with Ray tend to think of him first as a gentleman and only second as a scholar. He was a man of high principles, who held himself to the highest standards in both scholarship and personal conduct. During his years at RAND, Ray enjoyed the company of his many friends and collaborators. There were friendly rivalries in tennis, poker, and kriegspiel (blind chess—neither player can see the other's board), and Ray was very skilled at all three. The regular kriegspiel players included Richard Bellman, Stuart Dreyfus, Les Ford, and Lloyd Shapley. They were all good, but Ray was the best. George Dantzig often refereed the matches and found Ray's ability to re-construct his opponent's board uncanny. Richard Bellman was Ray's regular tennis partner, and, in 1959, they won the West Los Angeles Industrial Tennis League Doubles Championship without dropping a single set.

There were occasional contests to see whose algorithm was fastest by sitting down and hand-calculating (1950s time frame) solutions to a small problem presented to the contestants by a neutral referee. In one such contest, which Les Ford describes as a face-off, Ray and Les were matched up against George Dantzig and a partner to see whether the their primal-dual algorithm (Ford and Fulkerson 1956b) could beat Dantzig's transportation simplex algorithm (Dantzig 1951) on a small transportation problem. Ray and Les won, but Ray conceded later that they were bound to win, independent of the merits of the algorithm, because Les was a whiz at hand computation—there was some evidence that their victory also had something to do with the merits of their algorithm. In a letter dated April 4, 1959, Ray wrote to Les, who had left RAND, mostly to discuss progress on their book. The last paragraph includes information on a re-match of the earlier duel (Fulkerson 1959):

> ... our Hitchcock method [the primal-dual method for the transportation problem] has been programmed .... Programmer was Kurt Spielberg of IBM. He quotes the following times for our method vs. stepping stone (simplex):

Matrix	Simplex (min/sec)	Our (min/sec)
130 × 30	2'13"	1'30"
160 × 30	4'56"	3'34"
190 × 30	7'5"	4'
220 × 30	11'58"	4'58"

RAY AT WORK AND PLAY

When Elbert Fulkerson was young, his family could not afford proper clothes for church, so they never went. He was embarrassed by his ragged clothes and shoes, even in his rural grade school. As an adult, Elbert never felt properly dressed until he had on a tie and jacket. In the breezy atmosphere of the Math Department at RAND, everyone except Ray dressed casually. Others wore lumberjack shirts or dressed Hawaiian-style, but Ray, in spite of his utterly unpretentious manner, wore a tie and jacket.

Ray was always for the underdog. Mostly, this manifested itself in private, in the form of a generous donation, or a quiet, but heartfelt expression of support. At least once, in 1966, it took him to a very public forum, armed with his mathematical toolbox. Along with Lloyd Shapley and two other RAND scientists, Ray made a public presentation to the Los Angeles School Board showing how the school system could simultaneously remedy de facto segregation and under-utilization of facilities by efficient routing of buses.

Ray was a huge fan of basketball. One of the highlights of his time at RAND came when he looked out of an office window and saw the legendary basketball star, Wilt Chamberlain, playing volleyball on the beach. Ray raced home and back, returning with a basketball, and breathlessly approached Wilt for his autograph.

CORNELL: 1971–1976

Ray's 20 years at RAND were exciting, rewarding, and professionally defining. During the late 1960s, RAND budgets tightened and support for basic research declined. The negative consequences for the Mathematics Department were difficult for Ray, because he took great pride in the accomplishments of his colleagues, and he felt that some of them were not treated fairly—he fought vigorously for them and for their mission. Ultimately, he left RAND for Cornell University and the position of Maxwell M. Upson Professor of Engineering and Professor of Operations Research and Applied Mathematics.

The academic environment at Cornell was very much to Ray's liking. His sensibilities were ideally suited to being a professor at a university. This was, however, a very difficult time for him. He had recently been divorced after 22 years of a mostly unhappy marriage.

Ray at the blackboard (Cornell).

The abrupt separation from his two beloved sons, Guy and Lee (born in Santa Monica in 1955), his colleagues, and friends of 20 years, and the life he had in southern California was painful. Ray had suffered from serious medical problems for years, including two abdominal surgeries, one for a perforated bowel. Not long after his arrival at Cornell, Ray was diagnosed with Crohn's disease, which caused a great deal of discomfort for the remainder of his life. His love for the elegance of good mathematical research, and the opportunity to share it with his students, advisees, and new colleagues must have been something of a tonic for his woes. Ray taught Ph.D. courses in network flows and extremal combinatorics. His lectures were meticulously prepared, and his enthusiasm for the subject was infectious. He exuded a quiet dignity, and became the intellectual leader of the OR department at Cornell. His office door was always open for anyone who wished to talk with him.

Just about every day, Ray would go to lunch with some of his students. Because of his Crohn's disease, he always ordered the same meal: a cheese sandwich, a milkshake with a raw egg in it, and rice pudding. He was happy to talk about politics, sports, math, wherever the conversation wandered. When the topic turned to research, his enthusiasm for good work and the people who did it was palpable. He took exceptional pleasure in the accomplishments of young scholars.

Students

Ray was a visiting professor at Stanford in 1966. Among the Ph.D. students in mathematics at Stanford was Alan C. Tucker. Ray had learned graph theory in 1951 from writing the notes of lectures given at RAND by Alan's father, Al Tucker. Ray, in turn, taught graph theory to Alan, and served (unofficially) as Alan's thesis advisor; George Dantzig was the official advisor. Alan finished his thesis on proper circular arc graphs in 1969. (The thesis topic came from Fred Roberts, who was also in the RAND Mathematics Department.)

At Cornell, two Ph.D. students completed their theses under Ray's direction, David B. Weinberger (1973) and Robert G. Bland (1974). Weinberger's thesis resolved several interesting questions raised by Ray's theory of blocking pairs of polyhedra. It determined the pairs of blocking polyhedra associated with various network-flow problems and answered questions raised by Ray concerning blockers related to tours in complete graphs and the intersection of matroids (Fulkerson and Weinberger 1975). Bland's thesis used oriented matroids to isolate the combinatorial structure that underlies LP duality. This work was strongly motivated by Ray's paper (Fulkerson 1968), and by two closely related papers of Paul Camion (1968) and R.T. Rockafellar (1969), which hinted strongly (explicitly in Rockafellar's case) that there ought to be a generalization of LP duality in matroidal structures. [All three of the motivating papers were, in turn, rooted in Al Tucker's approach to LP (Nering and Tucker 1993).] The anticycling rules for the simplex method given in Bland (1977) were a natural by-product of the oriented matroid approach to LP duality. A third student, Gary C. Harding, was working under Ray's supervision at the time of Ray's death. Professor Les Trotter oversaw the completion of the thesis (Harding 1977). Two joint publications emerged from that work: one on edge-disjoint branchings (Fulkerson and Harding 1976), and another on parametrically maximizing shortest paths (Fulkerson and Harding 1977).

Ray was an excellent advisor. Everyone knew his standards were high, but some students were initially unaware of how broadly those standards would be applied. He was just as concerned with the proper use of dependent and independent clauses as dependent and independent variables. The that-versus-which debate was a particularly bloody affair.

A Dual Tribute

At a memorial service 3 days after Elbert Fulkerson's death in April 1972, Ray read a tribute to his late father (Fulkerson 1972). He began by apologizing for reading from written notes, explaining that under those highly emotional circumstances it would be difficult to be precise without reading. "And my father was a man who appreciated precision in expression, mathematics, and all things," he said. He went on to describe how Elbert had instilled in each of his six children "a love of excellence, an appreciation of scholarship, and a keen sense of justice." Elbert Fulkerson loved "clarity of thought" and "was a man of quiet moral courage, impeccable integrity, and enormous intellectual curiosity," Ray continued. Elbert was deeply loved and profoundly admired by Ray as both his father and his teacher, and Elbert's imprint on Ray was unmistakable. Ray had, in a sense, written his own eulogy. He died on January 10, 1976.

Ray published 52 papers, one book, and edited a collection of graph theory papers during his 20 years at the RAND Corporation and 6 years at Cornell. In 1967, his paper, "Flow networks and combinatorial operations research" (Fulkerson 1966) was recognized for expository excellence by the Mathematical Association of America (MAA) which presented Ray its Lester R. Ford Award. (The award is named for Lester R. Ford, Sr., a former president of the MAA, and the father of Ray's frequent collaborator.) Ray received the Southern Illinois University Award for Outstanding Professional Achievement in 1972. In 2005, he was made a member of the International Federation of Operational Research Societies' Hall of Fame (Bland and Orlin 2005).

Ray was a mathematical scientist first. He loved the elegance and grace of mathematics, and he loved good mathematics for its own sake. The remarkable breadth of his influence on OR methodology is very much a reflection of the broad and rigorous mathematical perspective that he brought to all of his work.

Ray's death in 1976 was mourned by all who knew him. It was noted by the establishment of the D. R. Fulkerson Prize in Discrete Mathematics by the American Mathematical Society and the Mathematical Programming Society, memorial statements in several journals (Billera and Lucas 1978; Chvátal 1976; Hoffman 1978; Ryser 1977), and a special issue of *Mathematics of Operations Research* that began with a dedication to Delbert Ray Fulkerson by the editor (Veinott 1976, 1) "It can be said about a few men that they profoundly influenced the fields in which they labored. And some of these also enriched the lives of others with their grace and warmth. Ray Fulkerson was such a man." More than 30 years have passed since Ray Fulkerson's death, but his work still remains fresh, and it continues to inspire scholars today.

ACKNOWLEDGMENT

We are extremely grateful to several of Ray's family members, friends, and admirers who assisted in providing information for this chapter, including: Michel Balinski, Len Berkovitz, Louis Billera, Wendell

Fleming, Les Ford, Dick Fulkerson, Lloyd Shapley, David Shmoys, Alan Tucker, David Weinberger, Allen Ziebur, and, especially, Merle Fulkerson Guthrie. The nice presentation of Ray's personal story in Billera and Lucas (1978) and the source material they gathered provided background material for this chapter. Some of the content here is from Bland and Orlin (2005).

REFERENCES

Ahuja R, Magnanti T, Orlin J (1993) Network flows: theory and applications. Prentice Hall, Upper Saddle River, NJ

Applegate D, Bixby R, Chvátal V, Cook W (2006) The Traveling salesman problem: a computational study. Princeton University Press, Princeton, NJ

Billera L, Lucas W (1978) Delbert Ray Fulkerson. Math Oper Res 1(4):298–310

Bland R (1974) Complementary orthogonal subspaces of R^n and orientability of Matroids. Unpublished doctoral dissertation, Cornell University, Ithaca, NY

Bland R (1977) New finite pivoting rules for the simplex method. Math Oper Res 12(3):103–107

Bland R, Orlin J (2005) IFORS' operational research hall of fame: Delbert Ray Fulkerson. Int Trans Oper Res 12(3):367–372

Bruck R, Ryser H (1949) The nonexistence of certain finite projective planes. Can J Math 1:88–93

Camion P (1968) Modules unimodulaires. J Comb Theory 4:301–362

Chvátal V (1976) D. Ray Fulkerson's contributions to operations research. Math Oper Res 1(4):311–320

Dantzig G (1951) Application of the simplex method to a transportation problem. In: Koopmans TC (ed) Activity analysis of production and allocation: proceedings of a conference. Wiley, New York, NY, pp 359–373

Dantzig G (1959) Optimum gas balance (unpublished report)

Dantzig G, Ford L Jr, Fulkerson DR (1956) A primal-dual algorithm for linear programs. In: Kuhn H, Tucker AW (eds) Annals of mathematics study, vol 38. Princeton University Press, Princeton, NJ, pp 171–181

Dantzig G, Fulkerson DR (1954) Minimizing the number of tankers to meet a fixed schedule. Nav Res Log Q 1(3):217–222

Dantzig G, Fulkerson DR, Johnson S (1954) Solution of a large-scale traveling salesman problem. Oper Res 2(4):393–410

Dantzig G, Fulkerson DR, Johnson S (1959) On a linear-programming, combinatorial approach to the travelling salesman problem. Oper Res 7(1):58–66

Dantzig G, Wolfe P (1961) The decomposition algorithm for linear programming. Econometrica 29(3):767–778

Dilworth R (1950) A decomposition theorem for partially ordered sets. Ann Math 51:161–166

Edmonds J, Fulkerson DR (1970) Bottleneck extrema. J Comb Theory 8:299–306
Fleming W (2009) Personal communication
Folkman J, Fulkerson DR (1970) Flows in infinite graphs. J Comb Theory 8:30–44
Ford L Jr, Fulkerson DR (1954) Maximal flow through a network. Research Memorandum RM-1400. The RAND Corporation, Santa Monica, CA
Ford L Jr, Fulkerson DR (1956a) Maximal flow through a network. Can J Math 8:399–404
Ford L Jr, Fulkerson DR (1956b) Solving the transportation problem. Manage Sci 3(1):24–32
Ford L Jr, Fulkerson DR (1957) A simple algorithm for finding maximal network flows and an application to the Hitchcock problem. Can J Math 9:210–218
Ford L Jr, Fulkerson DR (1958) A suggested computation for maximal multicommodity network flows. Manage Sci 5(1):97–101
Ford L Jr, Fulkerson DR (1962) Flows in networks. Princeton University Press, Princeton, NJ. Reissued 2010, Princeton University Press
Fulkerson DR (1959) Letter to Lester Ford
Fulkerson DR (1961a) An out-of-kilter method for minimum cost flow problems. J Soc Ind Appl Math 9:18–27
Fulkerson DR (1961b) A network flow computation for project cost curves. Manage Sci 7(2):167–178
Fulkerson DR (1966) Flow networks and combinatorial operations research. Am Math Mon 73(2):115–138
Fulkerson DR (1968) Networks, frames, blocking systems. In: Dantzig GB, Veinott AF Jr (eds) Mathematics of the decision sciences. American Mathematical Society, Providence, RI, pp 303–335
Fulkerson DR (1971) Blocking and anti-blocking pairs of polyhedra. Math Progr 1(2):168–194
Fulkerson DR (1972) In memoriam: Elbert Fulkerson, April 5, 1972. Unpublished
Fulkerson DR, Harding G (1976) On edge-disjoint branchings. Networks 6(2):97–104
Fulkerson DR, Harding G (1977) Maximizing the minimum source-sink path subject to a budget constraint. Math Program 13(1):116–118
Fulkerson DR, Weinberger D (1975) Blocking pairs of polyhedra arising from network flows. J Comb Theory Ser B 18:265–283
Gilmore P, Gomory R (1961) A linear programming approach to the cutting stock problem: part I. Oper Res 9(6):849–859
Gilmore P, Gomory R (1963) A linear programming approach to the cutting stock problem: part I. Oper Res 11(6):863–887
Harding G (1977) Some budgeted optimization problems and the edge-disjoint branchings problem. Unpublished doctoral dissertation, Cornell University, Ithaca, NY
Harris T, Ross F (1955) Fundamentals of a method for evaluating rail net capacities. Research Memorandum RM-1573. The RAND Corporation, Santa Monica, CA

Heller I (1953) On the problem of the shortest path between points. I. Abstract 664t, Bull Am Math Soc 59:551–551

Hoffman A (1978) D. R. Fulkerson's contributions to polyhedral combinatorics. Math Program Stud 8(1):17–23

Hoffman A, Wolfe P (1985) History. In: Lawler EL, Lenstra JK, Rinnooy Kan AHG, Shmoys DB (eds) The traveling salesman problem: a guided tour of combinatorial optimization. Wiley, New York, NY, pp 1–15

Kelley J Jr (1961) The critical path planning and scheduling: mathematical basis. Oper Res 9(3):296–320

Kuhn H (1955) On certain convex polyhedra. Abstract 779t, Bull Am Math Soc 61:557–558

Kuhn H (1991) On the origin of the Hungarian method. In: Lenstra JK, Rinnooy Kan AHG, Schrijver A (eds) History of mathematical programming: a collection of personal reminiscences. CWI and North Holland, Amsterdam, pp 77–81

Menger K (1927) Zur allgemeinen Kurventhoerie. Fundam Math 10:96–115

Minty G (1966) On the axiomatic foundations of the theories of directed linear graphs, electrical networks, and network programming. J Math Mechan 15(3):485–520

Nering E, Tucker A (1993) Linear programming and related problems. Academic, Boston, MA

RAND Corporation (1948) http://www.rand.org/about/history/. Accessed 10 Apr 2008

Robacker J (1955) On network theory. Research Memorandum RM-1498. The RAND Corporation, Santa Monica, CA

Robinson J (1949) On the Hamiltonian game (a traveling salesman problem). Research Memorandum RM-303. The RAND Corporation, Santa Monica, CA

Rockafellar R (1969) The elementary vectors of a subspace of R^n. In: Bose RC, Dowling TA (eds) Combinatorial mathematics and its applications. University of North Carolina Press, New York, NY, pp 104–127

Ryser H (1977) In memoriam. D. Ray Fulkerson, 1924–1976. J Comb Theory Ser B 23:1–2

Tucker A (1976) Personal letter to L.J. Billera and W.H. Lucas

Veinott A (ed) (1976) Math Oper Res 1(4)

Weinberger D (1973) Investigations in the theory of blocking pairs of polyhedra. Unpublished doctoral dissertation, Cornell University, Ithaca, NY

Williamson D (2002) The primal-dual method for approximation algorithms. Math Program B 91(3):447–478

29

Harold W. Kuhn

Saul I. Gass and
Guillermo Owen

Shortly after World War II, many college mathematics students pursuing their educational programs had no way of knowing that a new, mathematically-based applied field, operations research (OR), had originated from the exigencies of military planning and operations. OR did not enter college curricula until the mid-1950s, and then, only rarely in mathematics departments. Occasionally, through mainly fortuitous circumstances, mathematical problems that were to have a significant impact on OR were brought to the attention of a select group of such students. Remarkable mathematical, computational, and applied advances resulted from this early exposure. One of the most influential of this new generation of mathematical researchers was Harold W. Kuhn.

In the late 1940s, the Princeton mathematics department was the only such department whose faculty and graduate students had the unique and seminal opportunity to do research in two new and important fields, the theory of games and linear programming (LP). The results would change not only mathematics, but also economics, OR, and the way organizations and individuals make decisions. The theory of games had just been launched by professors John von Neumann and Oskar Morgenstern, and a research project on LP was being initiated by professor Albert Tucker. Harold and David Gale, both graduate students, were asked by Tucker to be members of the project. By joining this research effort, Harold changed his mathematical interests from the subjects of his doctoral thesis (topology and algebra) to optimization and decision theory. Thus, he began a career in which he made seminal contributions to mathematical programming, the theory of games, and combinatorial optimization. In recognition of his contributions to OR, Harold was a

A.A. Assad, S.I. Gass (eds.), *Profiles in Operations Research*, International Series in Operations Research & Management Science 147, DOI 10.1007/978-1-4419-6281-2_29,

co-recipient with Tucker and Gale of the 1980 Operations Research Society of America and The Institute of Management Sciences John von Neumann Theory Prize.

THE EARLY YEARS

Harold was born in Santa Monica, California, on July 29, 1925. His father, William Sebastian Kuhn, was born in San Francisco in 1880. William's father, Charles Kuhn, had immigrated from Germany to marry in New York in 1867; he is listed in the 1870 city directory of Denver, Colorado as an upholsterer, but moved on to settle in San Francisco's German community. William spoke only German until he entered school at the age of six. Harold's mother, Mignon Lorenz, was born in Benicia (a small town 35 miles north-east of San Francisco) in 1890. Both her parents were immigrants, her father from Germany and her mother from England. Neither of Harold's parents went to school beyond the fifth grade. Harold's father worked for Pacific Mutual Life for 45 years, starting when he was 14 years old. He made up for lost schooling by taking evening classes and eventually became a claims adjuster. After the 1906 San Francisco earthquake, Pacific Mutual moved to Los Angeles and William moved with them. Mignon's family, after her farther died, also moved to Southern California to be with her mother's family. William met Mignon when she worked for him at Pacific Mutual Life. Harold had one brother, Clifford Lorenz Kuhn, who was born in 1919 and died in 1940.

The Kuhn family moved from Santa Monica to south central Los Angeles when Harold was 2 years old. There he attended public schools, graduating in 1942 from Manual Arts High School at age 16 (he skipped a grade in elementary school). For college, he applied to the

California Institute of Technology (Caltech) and was accepted. A science day visit to Caltech had left Harold with a positive impression of the school: "I knew it was the place for me" (Kuhn 2008). Whether he could afford to attend was questionable; the family was extremely poor. Harold's father had suffered a heart attack in 1939 and his disability insurance payments were barely enough to live on. Anxious for Harold to start college, and to save money by having Harold live at home, the family moved to Pasadena where Caltech was located. He enrolled in 1942 and elected to major in mathematics and physics. He was the only undergraduate who did not live on campus. Of the 160 students in the entering freshman class, Harold maintained the highest grade point average into his junior year.

In July 1944, he was drafted into the U.S. Army and was sent for infantry basic training to Camp Roberts, California, about 225 miles north of Los Angeles. While there, he took and passed an Army qualifying examination in linguistic proficiency and, after finishing basic training, was transferred to the Army Specialized Training Program in Japanese at Yale University. The graduates of this program were to be sent to Japan to supervise interpreters in future war crimes trials. Due to a knee operation in the spring of 1946, Harold was unable to go to Japan. He was discharged in early summer of 1946 and finished his junior year at Caltech in the summer session. In 1947, Harold received a B.S. degree (major in mathematics with honors)—he decided to pursue a Ph.D.

BEING IN THE RIGHT PLACE AT THE RIGHT TIME (KUHN 2002)

For his doctorate, Harold was admitted to both Yale and Princeton, two schools he had been exposed to already. While studying Japanese at Yale, Harold had audited a graduate course in algebra with Oystein Ore (who, in the 1960s, wrote the books *Theory of Graphs* and *Graphs and their Uses*). Also, while at Yale, Harold had visited a student friend at Princeton and sat in on a few mathematics classes. Because he felt its mathematics faculty was stronger, Harold chose Princeton (although it offered him less financial support than Yale). His admission to Princeton was based on the recommendation of Frederic Bohnenblust who had recently joined Caltech from Princeton.

Although the G.I. Bill (Servicemen's Readjustment Act of 1944) paid Harold's full tuition and a small stipend while he was at Caltech, and would do the same for 2 years of graduate school, he continued to find himself "with practically no financial resources" (Kuhn 2002, 132). For the academic year of 1948–1949, the Princeton mathematics department awarded him its $700 J.S.K. Fellowship that was paid in 10 monthly installments. [J.S.K. stands for John S. Kennedy (1830–1909), a commissioner for the Union Pacific Railroad who gave railroad stocks to the Princeton mathematics department to sponsor a fellowship.] Even so, as Harold noted, "Thus, after paying room and board, there was not enough left to buy a pair of shoes!" (Kuhn 2002, 132). In the spring of 1948, the "impecunious graduate student" visited professor Albert Tucker to ask for summer employment, a career-changing event (Kuhn 2002, 132). Harold did not know then that 1948 was the right time and Princeton, with Al Tucker in the mathematics department, was the right place.

George Dantzig had recently (October 3, 1947) visited the Institute for Advanced Study in Princeton to meet with John von Neumann to discuss computational aspects of the new field of LP. His encounter with the "great Johnny" is a fascinating story unto itself (Dantzig 1991, 24). At this meeting, von Neumann made the important and seminal observations that LP must have duality relationships similar to what he had discovered for zero-sum two-person games and that the LP and game models were equivalent.

George, in a subsequent visit to von Neumann to discuss a university-based research project to study LP, met Al Tucker for the first time (May 1948). Al, who was not familiar with George's research, offered to drive him to the train station at Princeton Junction for his return to Washington. During the drive, George described his U.S. Air Force sponsored research that led to his LP model formulation and the development of the simplex method (Albers and Alexanderson 1985). The importance of related open research questions intrigued Al and, with encouragement from George, he was able to obtain funding from the Office of Naval Research (ONR) to set up a research project that summer, a project that lasted over two decades. Al chose two graduate students, Harold and David Gale to join his research team. Thus, Harold started on his distinguished research career.

Not knowing anything about the theory of games, Al, Harold, and David took turns lecturing to each other on the chapters from von Neumann and Morgenstern's seminal book, *Theory of Games and Economic Behavior* (1944). Their studies resulted in the statement and proof of the duality theorem for LP problems and LP's equivalence to zero-sum two-person games (Gale et al. 1951). As Harold noted: "A major result of this work was the duality theorem for linear programming, which says that to each minimization problem there is associated a maximization problem constructed on the same data with a number of properties that relate them," especially that the minimum value is equal to the maximum value, if solutions exist (Kuhn 2002, 132). Dantzig, based on his earlier discussion with von Neumann, had proved duality relationships between the LP problem and the zero-sum two-person games that were first published in Dantzig (1951a). Of historical interest is that the papers (Gale et al. 1951) and Dantizg (1951a) were both published in Koopmans (1951), the proceedings of what turned out to be a most important conference on "Activity Analysis of Production and Allocation" sponsored by the Cowles Commission for Research in Economics, located at the University of Chicago (June 20–24, 1949). The proceedings included, among other important papers, the first published statement of Dantzig's simplex algorithm, Dantzig (1951b). This conference is referred to as the 0th International Symposium on Mathematical Programming (Balinski 1991).

In the fall of 1949, when Al was on a year's sabbatical leave at Stanford, he decided to pursue some of his initial thoughts about the relationship between LP and the theory of electrical networks. This led to a problem that dealt with minimizing heat loss in a network as a quadratic program (a new term at that time). He then proposed that the ONR team of three investigate duality and related issues for quadratic programming; David declined, but Harold accepted. The work developed via letters exchanged between Al in Stanford and Harold in Princeton. ["Remember: These were the times in which there was no e-mail and one used carbon copies, not Xeroxes" (Kuhn 2002, 133).] Based on Harold's suggestion, they decided, instead, to study the more general problem of what they termed nonlinear programming (NLP, also a new term). The form of the problem they considered can be stated as follows (Kuhn and Tucker 1951, 483):

> Let the mapping Fx be an m-vector whose components $f_1(x), \ldots, f_m(x)$ are differentiable functions of x defined for $x \geq 0$. Let $g(x)$ be a differentiable function of x defined for $x \geq 0$. The *maximum problem* is then: To find an x^0 that maximizes $g(x)$ constrained by $Fx \geq 0, x \geq 0$.

Their research on this problem was completed in the spring of 1950. Al presented the results at a RAND conference in May 1950. It was there that an example, due to Charles Tompkins, raised the need for a "constraint qualification" that ruled out "singularities on the constraint set, such as an outward pointing 'cusp' " (Kuhn and Tucker 1951, 483).

Al was invited by Jerzy Neyman to give a paper at the Second Berkeley Symposium on Mathematical Statistics and Probability that was held from July 31 to August 12, 1950, and he chose to speak on his NLP research with Harold. It was the proceedings of this symposium that brought the topic of NLP to the theoretical and applied OR and mathematical worlds (Kuhn 1976). Of importance to their results were conditions for the solution of the original maximizing nonlinear programming problem and its dual (saddle-point problem). These conditions were later named (by others) the Kuhn-Tucker (KT) conditions: for a problem with "a nonlinear objective function and nonlinear constraints, . . ., there appear as necessary conditions for a local optimum the existence of generalized Lagrange multipliers that satisfy conditions that are 'dual' to the original constraints" (Kuhn 2002, 133). The KT conditions have been renamed the Karush-Kuhn-Tucker (KKT) conditions in recognition of William Karush's unpublished University of Chicago master's thesis in which results similar to the KT conditions were obtained (Karush 1939; Kuhn 1976, 1991b).

> KKT+NLP
>
> The joint research of Harold and Al that led to the KKT conditions and NLP had an "extraordinary rapid and widespread application" to OR applications, algorithm development, and related computations. In explaining their influence, Harold offered the following comments:
>
> "First, the model of nonlinear programming was flexible enough to encompass a large class of real-life problems that had not been adequately treated by the techniques then available. In societal terms, after the successes of operations research in the Second World War, a number of major industries were willing to try out this new model.
>
> "Second, the necessary conditions established by Karush, Kuhn, and Tucker formed the starting point for a large number of algorithms to solve nonlinear programs.
>
> "Third, and perhaps the most necessary factor, the first half of the 1950s saw the development and rapid expansion of computers that could be programmed to solve this sort of problem" (Kuhn 2002, 133–134).

In parallel with his ONR-based research, Harold obtained an M.S. in 1948 and a Ph.D. in 1950, both in mathematics. Under the supervision of Ralph Fox, Harold wrote his dissertation on "Subgroup theorems for groups presented by generators and relations." He was an instructor in mathematics from 1949 to 1950. In addition, Harold married Estelle Henkin on Christmas Eve, 1949. A busy and productive period!

After receiving his Ph.D., Harold was a Fulbright Research Scholar at the Institut Henri Poincaré in Paris (1950–1951). He returned to Princeton as a lecturer in mathematics for 1 year. Then, from 1952 to 1958, he was an assistant professor of mathematics at Bryn Mawr College, Bryn Mawr, Pennsylvania, where he taught a full range of courses as a member of a three-person mathematics department. He spent a sabbatical year as a Senior Postdoctoral Fellow at the London School of Economics, after which he returned (1959) to Princeton as an associate professor of mathematics and economics. Harold was promoted to full professor in 1963, and spent the rest of his academic career there. He taught both undergraduate and graduate courses, and served as chair of the Ph.D. committee for a number of mathematics students, as well as many economics students. He also served (1961–1983) as Scientific Director and Board Member for Mathematica, Inc., a Princeton-based corporation dedicated to mathematical, economic, and policy research.

For Mathematica, Harold directed ground-breaking studies in "Safety and reliability for nuclear weapons" for the U.S. Atomic Energy Commission and "Inspection models" for the U.S. Arms Control and Disarmament Agency (ACDA). Additional Mathematica projects with ACDA (1965–1968) dealt with formal studies of conflict and negotiation with respect to arms control agreements. For this research, Harold assembled a stellar cadre of mathematicians and economists: Robert

OPTIMAL PUBLISHING STRATEGY?

Class notes of Harold's (spring 1952) game theory course, the first such course taught at Princeton, were recorded by a student, Courtney S. Coleman. Harold compiled them into a mimeographed report, *Lectures on the Theory of Games*, that was published in this form by ONR in 1952; it was widely distributed by ONR and Harold, and was often cited by other authors. Plans were to publish *Lectures* in the Princeton University Press series *Annals of Mathematics Studies* as volume 37. For many reasons, volume 37 was not published until 2003. In the preface to Kuhn (2003, viii), Harold explains: "The reader is deserved an explanation as to why these lectures are published nearly 50 years after they were taught as a course in the mathematics department at Princeton University. The text was submitted, and you could read in the Annals Studies of that time: 'Annal Study 37, in press, $3.00.' At that time, I withdrew the manuscript for alterations, primarily hoping to add something on the rapidly developing theory of n-person games. The revisions were never made, and the lectures were never published." (The Dresher, Tucker and Wolf, volume 38, was published in 1956!). As noted in the Princeton University Press's advertisement: "The book opens by addressing 'matrix games,' a name first introduced in these lectures as an abbreviation for two-person, zero-sum games in normal form with a finite number of pure strategies. It continues with a treatment of games in extensive form, using a model introduced by the author in 1950 that quickly supplanted von Neumann and Morgenstern's cumbersome approach. A final section deals with games that have an infinite number of pure strategies for the two players" (Princeton Press 2009).

Aumann, Gerard Debreu, John Harsanyi, Michael Maschler, John Mayberry, Herbert Scarf, Reinhard Selten, Martin Shubik, and Richard Stearns. It is unclear whether the impact of their research was of practical value to the real-world of arms control policy and negotiations. But, the results led to important theoretical advances in game theory, especially in bargaining under incomplete information, infinitely repeated games, cooperative solutions to n-person games, and game theoretical aspects of gradual disarmament (Harsanyi et al. 1992). This work did bring together four future Nobel Prize laureates in economics: Aumann, Debreu, Harsanyi, and Selten!

ADDITIONAL RESEARCH CONTRIBUTIONS: GAME THEORY, HUNGARY, AND BEYOND

Harold's first major contribution to game theory was a serious revision of the theory of von Neumann and Morgenstern. Further, in his writings, he proved to be a sharp expositor who was able to clarify, explain, and augment the detailed mathematical material and arguments as given in von Neumann and Morgenstern (1944).

It is of interest to follow Harold's approach (Kuhn 1950). He first notes that von Neumann and Morgenstern considered two descriptions of a general n-person game: (1) the extensive form, an all-inclusive formal characterization, and (2) the normal form, in which the concept of a pure strategy enables a more tractable and simplified form to represent the game. He then states: "Since all games are found in extensive form while it is practical to normalize but a few, it seems reasonable to attack the completion of a general theory of games in extensive form. . . . This note presents two new results in this theory which appear to have far-reaching consequences in the computational problems of normalized games" (Kuhn 1950, 570–571). He then offers a new formulation of games in extensive form and then carefully defines "certain concepts associated with a game which are confused and ambiguous in common use," in particular, behavior strategies and perfect recall (Kuhn 1950, 571). His discussion ends with statements of two theorems dealing with the solution of general n-person games: (1) A sufficient condition that an n-person game have an equilibrium point among the pure strategies for all possible assignments of the pay-off function is that the game have perfect information, and (2) a game with perfect recall can be solved using behavioral strategies in place of mixed strategies. The proofs are given in Kuhn (1953). The Kuhn (1950) paper was published in the *Proceedings of the National Academy of Sciences* and was communicated by von Neumann. Harold's geometric reformulation of the extensive form of a game has replaced von Neumann's original formulation in all of the subsequent literature.

In Kuhn (1961), Harold developed a method for finding equilibrium strategies for bimatrix games that was constructive, as opposed to the non-constructive existence proof given by Nash using Kakutani's fixed-point theorem. He also studied games of fair division, generalizing the classical two-person result to the n-person case (Kuhn 1967a). His research activities in game theory were augmented by his editorial interests in ensuring that the important early contributions to the field were duly collected and made readily available. He joined with Al Tucker to compile and edit, as well as to contribute to, the latest research developments in game theory (Kuhn and Tucker 1950, 1953). [Al later joined with other co-editors to edit three additional volumes in the *Annals of Mathematics Studies* (Dresher et al. 1957, 1964; Luce and Tucker 1958).] Al and Harold also wrote the theory of games entry for the 1957 edition of the *Encyclopædia Britannica* (Kuhn and Tucker 1957). With George P. Szego, Harold co-edited a book on mathematical applications to economics (Kuhn and Szego 1969), the proceedings of a NATO Advanced Workshop, which they co-directed in Varenna, Italy in 1967. And, in Kuhn (1997), he compiled important early papers in game theory, many of which, even in the age of websites and digitalization, are difficult to find. As a further sign of his stature in the field,

A BEAUTIFUL FRIENDSHIP

Harold and John Nash were graduate students together in the late 1940s. Nash received his Ph.D. in 1950 under Al Tucker with his thesis on "Non-cooperative games," the topic that was recognized in 1994 by the Nobel Prize committee to be of central importance in modern economic theory. Although Nash went to MIT as an instructor in mathematics from 1951 to 1959, Harold, over the years, maintained informal contact with him and considered John a friend and colleague. It was in 1959 that Nash began his long bout with schizophrenia (Nasar 1998, 1995). In 1994, John and his wife Alicia were living a rather quiet and secluded life in Princeton. For the preceding 10 years, during which John's illness was in a state of remission, Harold had been nominating John for the Nobel Prize in Economics. In the spring of 1993, Harold was invited to Stockholm to press the case for John; it was hinted then that 1994 would be the year. Harold was informed in August 1994 that the selection committee would nominate John and co-recipients John Harsanyi and Reinhard Selten. In Kuhn and Nasar (2002, vii), Harold recounts how he broke the news to John "on a bench in front of a minimalist Japanese fountain at the Institute for Advanced Studies [sic] in Princeton on October 1994." For Harold, this encounter was "the defining moment that divided the period when he (John) was in the depths of his despair from his reentry into a world that he always deserved" (Kuhn and Nasar 2002, vii).

Harold was chosen to write an introduction to the Commemorative Edition (60th anniversary) of von Neumann and Morgenstern's *Theory of Games and Economic Behavior*. Harold, a friend and colleague of John Nash, a co-recipient of the 1994 Nobel Prize in economics, joined with Sylvia Nasar, to edit Nash's papers (Kuhn and Nasar 2002).

In Kuhn and Tucker (1951), Al and Harold showed how the problem Al wanted to originally study, the quadratic-programming problem, is equivalent to a saddle-point problem, and, under very special conditions, to the solution of a linear-programming problem. Further, they discuss the extension of their results to the vector maximum problem, which calls for the simultaneous maximization of multiple objectives subject to nonlinear or linear constraints, and prove the theorem that states the conditions for which such a problem has a solution. Here, the solution concept has to deal with efficient (Pareto optimal) solutions that are often encountered in activity analysis and related economic competitive equilibrium problems (Koopmans 1957). The influence of their research on vector optimization problems is described in Cohon (1978) and Chankong and Haimes (1983). The results of Kuhn and Tucker (1951) helped to motivate research and applications in multi-criteria decision making, and, especially, in the sub-field of multi-objective LP in which all of the objective functions and constraints are linear.

A SERMON

"This sermon will be short. We have seen that the same results, which is central to the subject of nonlinear programming, was found independently by mathematicians who found their inspiration in the calculus of variations, geometrical inequalities, the theory of games, duality in topology, network theory, and linear programming. This result which has proved to be useful, at least in the sense of suggesting computational algorithms, was sought and found first with no thought given to its application to practical situations. It was rediscovered and recognized as important only in the midst of the development of the applied field of mathematical programming. This, in turn, had a beneficial effect. With the impetus of evident applicability, the mathematical structure of the subjects neighboring mathematical programming has deepened in the last quarter century. A scattering of isolated results on linear inequalities has been replaced by a respectable area of pure mathematics to which this symposium bears witness. Notable achievements have been recorded in the subjects of convex analysis, the analysis of nonlinear systems, and algorithms to solve optimization problems. This has been possible only because communication has been opened between mathematicians and the potential areas of application, to the benefit of both. The historical record is clear: the lines of communication between applied fields such as mathematical programming and the practitioners of classical branches of mathematics should be broadened and not narrowed by specialization. This symposium is a constructive step in this direction" (Kuhn 1976, 19–20).

An important OR and economic problem is that of finding the location of, say, a factory such that the sum of the straight-line distances of the factory to a fixed set of warehouses is minimized. This problem has a long history that starts with the French mathematician Fermat in 1643; much later it was introduced to a more general audience by the German economist Weber in 1909. In a sequence of papers, Harold investigated the general Fermat problem (in which positive weights are associated with each factory–warehouse combination) and stated a maximizing dual problem for which the maximum is equal to the Fermat minimum (Kuenne and Kuhn 1962; Kuhn 1965, 1967b). For the Fermat primal-dual problems, Harold notes the "remarkable manner" in which they parallel the duality of LP, that is, the solution of either problem provides a solution to the other problem (Kuhn 1967b, 51).

In 1953, Harold attended a summer workshop held at the Institute for Numerical Analysis (INA) located at the University of California, Los Angeles (UCLA) that was organized to study a range of combinatorial-optimization problems such as integer programming and the traveling salesman problem. While there, he met Charles B. Tompkins, an INA mathematician. Tompkins was trying to solve an assignment problem: find an optimal assignment of n persons to n jobs given numerical ratings for each person's performance on each job. He was using the INA's computer to solve a 10×10 assignment problem. It was well-known that this assignment problem could be cast as a linear program in 20 equations and 100 non-negative, 0–1 variables (one equation is redundant, and due to the structure of the equations, the problem solved as a linear program results in an optimal solution with 0–1 variables). But, INA's Standards Western Automatic Computer (SWAC) could not handle problems of this size. Tompkins was using the SWAC to enumerate the $10! = 3{,}628{,}800$ permutations of ten objects. He failed to do so.

Harold, at that time, was reading Dénes König's (1950) book on graph theory and realized that the matching problem on a bipartite graph on two sets of n vertices was equivalent to an $(n \times n)$ assignment problem in which the ratings are 0 and 1. König gave a combinatorial algorithm for solving such problems. Harold decided to see whether it could be adapted to general assignment problems. He noticed that König referred to a paper by E. (Jenö) Egerváry (1955). When Harold returned to Bryn Mawr in the fall, he obtained a copy of Egerváry's paper, and, with a "large Hungarian dictionary and grammar," translated the paper. "As I had suspected, the paper contained a method by which a general assignment problem could be reduced to a finite number of 0–1 assignment problems. Using Egerváry's reduction and König's maximum matching algorithm, in the fall of 1953 I solved several 12 by 12 assignment problems (with 3-digit integers as data) by hand. Each of these examples took under two hours to solve and I was convinced that the combined algorithm was 'good'" (Kuhn 1991a, 78). Harold named the new algorithm the Hungarian Method (HM) to make it clear that his research was inspired by König and Egerváry (Kuhn 1955, 1956, 1991a). The influence of Harold's (1955, 1956) papers has been quite diverse. Andras Frank (2004) provides a short tour of how far and wide the HM has traveled. A feature of the HM that was soon recognized was that it was a good algorithm in the sense of having a polynomial time bound on the number of its operations. While Harold contented himself with the observation that HM has a finite number of steps, James Munkres (1957) was able to provide an explicit polynomial expression on the maximum number of operations. Further details are given in Schrijver (2003, 298–300, 436–437).

Estelle, Harold, Andras Frank
(At the 50th anniversary celebration of the Hungarian Method, Budapest, October 31, 2005; courtesy Jack Edmonds.)

HM = THE BEST PAPER

The *Naval Research Logistics* citation for the best paper published in its first 50 years read as follows:

"This pioneering paper set a style for both content and exposition of many other algorithms in combinatorial optimization, and also launched and inspired the primal–dual algorithm for more general linear optimization problems. The journal is also pleased to recall that the research was sponsored by the Office of Naval Research Logistics Project at Princeton University. Professor Kuhn's enduring contributions to optimization, discrete and continuous, linear and nonlinear, from its earliest days in the 1950s, are legendary. He is a man who was in the right place, at the right time, with the right stuff" (Simchi-Levi 2005,1).

In 2005, the *Naval Research Logistics* (NRL) journal awarded a prize to the best paper it had published in the 50 years since its founding. The selection committee chose Harold's paper (Kuhn 1955) on the HM as the best! Further, the journal established a yearly prize for the best article written over the previous 3 years; Harold was the eponym of this prize. Harold was also honored in 1992 by the Hungarian Operational Research Society by being elected an honorary member. On October 31, 2005, the Egerváry Research Group of the Hungarian Academy of Sciences held a 50th anniversary celebration day in Budapest to honor the HM and the publication of Kuhn (1955).

Although Harold's research contributed greatly to economics, OR, and related fields, he has always considered himself to be a mathematician. As his research contributions here are quite extensive, we call attention to a just few topics that have interested him: linear inequalities and related systems (Kuhn and Tucker 1956); the approximation of fixed points (Kuhn 1968, 1969); and a new constructive proof of the fundamental theorem of algebra (Kuhn 1974).

Harold's undergraduate and graduate teaching cut across both the departments of mathematics and economics and included courses in price theory, managerial economics, micro-economics, mathematical economics, trade theory, game theory, and linear and nonlinear programming. His undergraduate course in mathematical programming was a top-rated course that attracted as many as 90 students and influenced a number of them to take graduate degrees in OR. His doctoral students include several contributors to game theory—Dilip Abreu and Charles Frank (Princeton), Guillermo Owen (Naval Post Graduate School), and

HOW TO CHOOSE A CLASS

"I first heard of Harold Kuhn when, as a beginning undergraduate, I read an article in the student newspaper, *The Daily Princetonian*, on the results of student course evaluations (which had been introduced a short time earlier). To my (and everyone else's) surprise, the class with the highest student ratings was not in one of the popular disciplines, such as English, history, or politics, but was in a field that most students avoided: mathematics. That most highly rated course at the University was 'Mathematical Programming,' taught by Professor H. W. Kuhn. Despite having no concept of what the course title meant, I felt compelled to enroll in the course based on the strong endorsement of my fellow students and was eager to discover what 'mathematical programming' was. In taking the class, thanks to Professor Kuhn's inspiring lectures, full of visual images of algorithmic steps and the meaning of optimality, I discovered not just what mathematical programming was but, more importantly, I gained a glimpse of what it could be and an ongoing thirst to discover more. In addition to stirring my interest in the field, Professor Kuhn's classes are the benchmark I set for my own teaching. His clarity in describing the path of that which is known while also imparting the potential of what is yet to be learned is indeed an example for all to follow" (John Birge 2009).

Richard Stearns (University at Albany, State University of New York)—and economic theorists Oliver Hart (Harvard) and Alvin Klevorick (Yale).

Harold, as a concerned member of Princeton's academic community, participated in several key faculty committees that had a major impact on the school's governance. He served as secretary of the Faculty Advisory Committee to the President and wrote a key document on "Students and the University" that was adopted by the Princeton faculty in 1968. This led to broad changes in the participation by students in the governance of the University. He was also a member of the Committee on the Governance of the University that was established to examine Princeton's governance and enhance its inclusiveness. This committee introduced some of the most sweeping administrative changes in the University's history among which the establishment of the Council of the Princeton University Community.

Harold and his wife Estelle have three sons: Clifford (born 1952), Associate Professor of History at Georgia State University; Nicholas (born 1955), Professor of Mathematics at the University of Virginia; and Jonathan (born 1958), Director of Arts and Antiquities of the New York City Parks Department. They have six grandsons and one granddaughter. Estelle is the sister of the late mathematical logician Leon Henkin, who at the time she and Harold met, was Fine Instructor at Princeton, having received his Ph.D. in 1947. Estelle received a B.A. in 1947 from New York University and worked for the New York City Welfare Department. She managed the American Civil Liberties Union's Trenton office from 1972 to her retirement in 1990.

Harold retired voluntarily from Princeton on July 1, 1995 and was appointed Professor of Mathematical Economics Emeritus. Since 2005, he has resided in New York City. His major activity is giving invited lectures at national and international meetings that have intriguing titles, such as, "57 Years of Close Encounters with George" Kuhn (2009).

Honors and Awards

In recognition of his contributions to the advancement of the theory of OR, Harold was awarded, jointly with Al Tucker and David Gale, the 1980 Operations Research Society of America and The Institute of Management Sciences John von Neumann Theory Prize. He was elected a fellow of the Econometric Society (1961), the Institute for Operations

Research and the Management Sciences (2002), and the Society for Industrial and Applied Mathematics (SIAM) (2009). He was elected a member of the American Academy of Arts and Sciences (1992). Harold's service to professional organizations includes: the first council of SIAM and its third President (1954–1955); Executive Secretary, Division of Mathematics, National Research Council (1957–1960) and a member of the Council (1961–1964); and a council member of the American Association of University Professors (1959–1962). He was elected an Honorary Member of the Hungarian Operational Research Society in 1992, and received an Honorary Doctorate in Economics from the University of Bergamo, Italy in 2001.

REFERENCES

Albers D, Alexanderson G (1985) Mathematical people. S. B. Maurer, interview with Albert Tucker. Birkhäuser, Boston, MA, pp 339–348

Balinski M 1991. Mathematical programming: journal, society, recollections. In: Lenstra J, Rinnooy Kan A, Schrijver A (eds) History of mathematical programming. North-Holland, Amsterdam, pp 5–18

Birge J (2009) Personal communication

Chankong V, Haimes Y (1983) Multiobjective decision making. North-Holland, New York, NY

Cohon J (1978) Multiobjective programming and planning. Academic, New York, NY

Dantzig G (1951a) A proof of the equivalence of the programming problem and the game problem. In: Koopmans TC (ed) Activity analysis of production and allocation. Cowles Commission Monograph 13, Wiley, New York, NY, pp 330–335

Dantzig G (1951b) Maximization of a linear function of variables subject to linear inequalities. In: Koopmans TC (ed) Activity analysis of production and allocation. Cowles Commission Monograph 13, Wiley, New York, NY, pp 339–347

Dantzig G (1991) Linear programming. In: Lenstra J, Rinnooy Kan A, Schrijver A (eds) History of mathematical programming. North-Holland, Amsterdam, pp 19–31

Dresher M, Shapley L, Tucker A (eds) (1964) Advances in game theory. Annals of Mathematics Studies 52. Princeton University Press, Princeton, NJ

Dresher M, Tucker A, Wolfe P (eds) (1957) Contributions to the theory of games, vol III. Annals of Mathematics Studies 39. Princeton University Press, Princeton, NJ

Egerváry E (1955) On combinatorial properties of matrices (trans: Kuhn H). Logistics Papers (Issue 11), Paper 4, George Washington University, Washington, DC, pp 1–11

Frank A (2004) On Kuhn's Hungarian method—a tribute from Hungary. Nav Res Logistics 52(1):2–5

Gale D, Kuhn HW, Tucker A (1951) Linear programming and the theory of games. In: Koopmans TC (ed) Activity analysis of production and allocation. Cowles Commission Monograph 13, Wiley, New York, NY, pp 317–329

Harsanyi J, Mayberry J, Scarf H, Selten R (1992) Game-theoretic models of cooperation and conflict. Westview Press, Boulder, CO

Karush W (1939) Minima of functions of several variables with inequalities as side conditions. Master's Thesis, Department of Mathematics, University of Chicago, Chicago, IL

König D (1950) Theorie der endlichen und unendlichen Graphen: Kombinatorische Topologie der Streckenkomplexe. Chelsea Publishing Company, New York, NY, 1950; (originally published in 1936, Mathematik in Monographien 16, Akademische Verlagsgesellschaft. Leipzig)

Koopmans T (ed) (1951) Activity analysis of production and allocation: proceedings of a conference. Cowles Commission Monograph 13. Wiley, New York, NY

Koopmans T (1957) Three essays on the state of economic science. McGraw-Hill, New York, NY

Kuenne R, Kuhn HW (1962) An efficient algorithm for the numerical solution of the generalized Weber problem in spatial economics. J Reg Sci 4(2):21–33

Kuhn HW (1950) Extensive games. Proc Natl Acad Sci. 36(10):570–576

Kuhn HW (1953) Extensive games and the problem of information. In: Kuhn HW, Tucker A (eds) Contributions to the theory of games, II. Annals of Mathematics Studies 28. Princeton University Press, Princeton, NJ, pp 193–216

Kuhn HW (1955) The Hungarian method for the assignment problem. Nav Res Logistics Q 2(1–2):83–97

Kuhn HW (1956) Variants of the Hungarian method for assignment problems. Nav Res Logistics Q 3(4):253–258

Kuhn HW (1961) An algorithm for equilibrium points in bimatrix games. Proc Natl Acad Sci 47(10):1657–1662

Kuhn HW (1965) Locational economics and mathematical programming. In: Proceedings of the colloquium on the application of mathematics to economics, Budapest, 1963, Publishing House of the Hungarian Academy of Sciences, pp 235–242

Kuhn HW (1967a) On games of fair division. In: Shubik M (ed) Essays in mathematical economics. Princeton University Press, Princeton, NJ, 29–37

Kuhn HW (1967b) On a pair of dual nonlinear programs. In: Abadie J (ed) Nonlinear programming. North-Holland, Amsterdam, pp 38–54

Kuhn HW (1968) Simplicial approximation of fixed points. Proc Natl Acad Sci 61(4):1238–1242

Kuhn HW (1969) Approximate search for fixed points. In: Zadeh L, Neustadt L, Balakrishnan A (eds) Computing methods in optimization problems. Academic, New York, NY, pp 199–211

Kuhn HW (1974) A new proof of the fundamental theorem of algebra. Math Program Stud 1:148–158

Kuhn HW (1976) Nonlinear programming: a historical view. In: Cottle R, Lemke C (eds) Nonlinear programming: proceedings of the SIAM-AMS Symposia, New York, March, 1975, vol 9. American Mathematical Society, Providence, RI, pp 1–26

Kuhn HW (1991a) On the origin of the Hungarian method. In: Lenstra J, Rinnooy Kan A, Schrijver A (eds) History of mathematical programming. North-Holland, Amsterdam, pp 77–81

Kuhn HW (1991b) Nonlinear programming: a historical note. In: Lenstra J, Rinnooy Kan A, Schrijver A (eds) History of mathematical programming. North-Holland, Amsterdam, pp 82–96

Kuhn HW (ed) (1997) Classics in game theory. Princeton University Press, Princeton, NJ

Kuhn HW (2002) On being in the right place at the right time. Oper Res 50(1):132–134

Kuhn HW (2003) Lectures on the theory of games. Annals of Mathematics Studies 37. Princeton University Press, Princeton, NJ

Kuhn HW (2008) 57 years of close encounters with George. http://www2.informs.org/History/dantzig/articles_kuhn.html. Accessed 12 Aug 2009

Kuhn HW, Nasar S (eds) (2002) The essential John Nash. Princeton University Press, Princeton, NJ

Kuhn HW, Szego G (eds) (1969) Mathematical systems theory and economics. Lecture notes in operations research and mathematical economics, Springer, New York, NY

Kuhn HW, Tucker A (eds) (1950) Contributions to the theory of games, vol I. Annals of Mathematics Studies 24. Princeton University Press, Princeton, NJ

Kuhn HW, Tucker A (1951) Nonlinear programming. In: Neyman J (ed) Proceedings of the second Berkeley symposium on mathematical statistics and probability, University of California Press, Berkeley, CA, pp 481–492

Kuhn HW, Tucker A (eds) (1953) Contributions to the theory of games, vol II. Annals of Mathematics Studies 28. Princeton University Press, Princeton, NJ

Kuhn HW, Tucker A (eds) (1956) Linear inequalities and related systems. Annals of Mathematics Studies 38. Princeton University Press, Princeton, NJ

Kuhn HW, Tucker A (1956) Games, theory of. Encyclopædia Britannica 10:6–11

Luce R, Tucker A (eds) (1959) Contributions to the theory of games, vol IV. Annals of Mathematics Studies 40. Princeton University Press, Princeton, NJ

Munkres J (1957) Algorithms for the assignment and transportation problem. J Soc Ind Appl Math 5(1):32–38

Nasar S (1998) A beautiful mind. Simon and Schuster, New York

Nash J (1995) Autobiography. In: Frängsmyr T (ed) The Nobel prizes 1994. The Nobel Foundation, Stockholm; also see Kuhn and Nasar (2002) 5–11

Princeton Press (2009) http://press.princeton.edu/titles/7560.html. Accessed 24 Feb 2009

Schrijver A (2003) Combinatorial optimization: polyhedra and efficiency, vol A. Springer, New York, NY

Simchi-Levi D (2005) Editorial: The Harold W. Kuhn Award. Nav Res Logistics 52(1):1

von Neumann J, Morgenstern O (1944) The theory of games and economic behavior. Princeton University Press, Princeton, NJ

30

Saul I. Gass

Arjang A. Assad

The founders of operations research (OR) all trained and practiced in other fields before coming to OR. As OR expanded to a wider range of application areas after World War II (WWII), they turned to address OR's new seminal challenges. If we view these founders as the first OR generation, then the second generation are those individuals who learned from and worked directly with the first generation during the late 1940s and early 1950s. By this reckoning, Saul Gass commands a senior ranking within the second generation. In 1952, as a civilian government mathematician, he joined Project SCOOP (Scientific Computation of Optimal Programs), the U.S. Air Force's Pentagon-based organization that, under George Dantzig, was the first to develop and apply linear programming (LP). This position shaped much of his subsequent OR research and practice, and led to his 1958 book, *Linear Programming: Methods and Applications*, the first textbook on LP.

Saul was among the first OR professionals having become a member of the Operations Research Society of America (ORSA) in 1954. His career bridges the academic-practitioner divide and spans what he calls the three P's of OR: practice, process, and professionalism. First, as a practitioner of OR, and then an academic at the University of Maryland, he has contributed to the field as a teacher, scholar, author, consultant, involved citizen, ambassador, expositor, and chronicler of OR. He served as president of ORSA (1976) and was awarded ORSA's George E. Kimball distinguished service medal. He is a fellow of the Institute for Operations Research and the Management Sciences (INFORMS), and received the INFORMS Expository Writing Award for publications in OR and MS that have set an exemplary standard of exposition.

A.A. Assad, S.I. Gass (eds.), *Profiles in Operations Research*, International Series in Operations Research & Management Science 147, DOI 10.1007/978-1-4419-6281-2_30,

EARLY YEARS: BOSTON, ARMY, COLLEGE, FIRST JOB

Saul Irving Gass was born on February 28, 1926 in Chelsea, Massachusetts, to Louis and Bertha (Kotker) Gass, both of whom, as youngsters, emigrated from Russia (Louis in 1910 and Bertha in 1914). They each were from a family of seven children. Saul's parents met in Boston and were married in 1920. Louis was an insurance agent for Metropolitan Life and sold what was called weekly insurance—low cost insurance whose small premiums had to be collected on a weekly basis. He also wrote a column in Yiddish for the *Jewish Daily Forward* newspaper, hosted a Sunday Yiddish radio show, and wrote poetry in Yiddish and English. During the Great Depression, family finances were usually tight. Bertha and two of her sisters worked as part-time waitresses (unionized) at their brother-in-law's catering hall. She was a lifetime member of the Pioneer Women, the Labor Zionist women's organization in the U.S. Most members of the extended Gass and Kotker families lived in the Boston area—all four grandparents were alive through the late 1930s. Saul was the second of two children; his brother Gerald (Jerry) A. Gass was born October 21, 1922.

Saul's family lived in Chelsea occupying the first floor of grandfather Kotker's three story wood-framed house until it was lost by foreclosure in 1931. They then moved to the Roxbury section of Boston and lived there through the mid-1950s. Saul attended the local schools and graduated in 1943 from Roxbury Memorial High School for Boys.

His favorite high school subjects were mathematics and physics. He also took navigation and aeronautics (WWII special courses) in his senior year, and did well in military drill, a then Boston tradition that culminated each year in a high school marching competition during a downtown Boston parade. He was a member of both the track and baseball teams.

Saul tried to enlist in the Navy and Air Force college training programs, but his color blindness kept him out. He then enlisted in the Army Specialized Training Program (ASTP), a program setup to identify, train, and educate academically talented enlisted men as a specialized corps of Army officers during WWII. He was part of the enlisted reserve corps and expected to enter the program when he turned 18. Meanwhile, he managed to complete a freshman year of engineering at Northeastern University. But, in early 1944, the Army canceled the ASTP, and soon after he turned 18, Saul was inducted into the Army (March 17, 1944) at Fort Devens, Massachusetts.

He underwent basic infantry training at Camp Blanding, Florida, and then, at Camp Shelby, Mississippi, he trained as a machine gunner with the newly formed 65th Infantry Division. On January 10, 1945, the Division shipped out from New York City, landed in Le Havre, and stayed in France (Camp Lucky Strike) and trained for a month. The Division, part of General Patton's Third Army, moved into action on March 17, 1945, continued through southern Germany, crossed the Danube River at Regensburg, and then into Austria. Saul's unit, Company M, 261st Regiment, stopped at the west bank of the Enns River (the Russians occupied the opposite side). This was their location on V-E Day, May 8, 1945, making the 65th Division the Army unit that had gone the furthest east. During occupation, Saul wrote for and became editor of his battalion and regimental newspapers (he was promoted to Staff Sergeant/Information and Education); he was stationed in St. Florian, Austria and Pfaffenhoffen, Germany. He shipped home in April 1946.

After his military discharge on May 23, 1946, Saul hoped to resume his university studies. He also planned to marry Trudy Candler, a local girl whom he had met when he was 15 and courted while on leave before going to Camp Shelby. Trudy's family had moved to Los Angeles in January 1945; she finished her senior high school year at Hamilton High. They married on June 30, 1946 in Los Angeles and came back to Boston and lived with Saul's parents. With the support of the G.I. Bill that paid college tuition and $75 a month stipend, and Trudy working as a secretary/bookkeeper, Saul re-enrolled at Northeastern. He transferred to Boston University in January 1947 to major in education and mathematics, intending to be a high school teacher. He graduated with a B.S. in education (major in mathematics) in June 1949. Because Saul had taken extra mathematics courses as an undergraduate, he was able to earn his M.A. in mathematics in August 1949 by taking courses that summer. During his senior year, Saul had spent a semester as a student teacher in one of Boston's high schools, an assignment he did not find particularly exciting. He decided against a career as a high school teacher and searched for a job as a mathematician by writing letters to many organizations and registering with the U.S. Civil Service.

In November 1949, Saul was offered a position (sight unseen) as a mathematician with the U.S. Air Force as a GS-7 ($3725 a year) and joined the Aberdeen Bombing Mission (ABM) in Los Angeles. This civilian Air Force group's main task was to analyze "photographic plates and high-speed camera film of high-altitude aircraft and bomb drops that took place at Edwards Air Force Base, north of Los Angeles in the desert country" (Gass 2002a, 64):

> At ABM, we read the plates and film by eye on a Mann Comparator, recorded the results by hand, and processed the readouts on the Marchant and Monroe desk calculators. The old-fashioned way! I did become deeply involved in bomb ballistic work and was given the task of investigating new machine readers that could automatically record the positions of both the aircraft and bomb images and punch the results on IBM cards.

The Los Angeles group sent its results to the Aberdeen Proving Ground in Aberdeen, Maryland, where bombing tables were being developed. His work at ABM exposed him to the need for computational accuracy, constant checking, and data validation.

From Project SCOOP to Project Mercury

Saul was not too keen on bomb ballistic work or enamored with Los Angeles, and applied for a job transfer with the U.S. Civil Service. He accepted a position (again, sight unseen) as a mathematician (GS-9 at $5,060 a year) with the Air Force's Pentagon-based Directorate of Management Analysis, Washington, D.C.

Saul was already a father—Ronald S. Gass was born in Los Angeles on June 3, 1951. The family of three drove to the Washington D.C. area in the first week of January 1952. The next addition to the family, Joyce A. Gass, was born in Arlington, Virginia on June 22, 1955.

At the Pentagon, Saul joined the staff of Project SCOOP. This U.S. Air Force research office had been formed in June 1947 and officially designated as Project SCOOP in October 1948. Headed by the economist Marshall Wood, with George Dantzig as chief mathematician, the main objective of Project SCOOP was to plan the requirements for Air Force programs. In the Air Force, the words programs and programming were used in the specific military sense (computer programs were barely known and called codes at that time). For the Air Force, a typical programming exercise involved constructing a time-phased plan of requirements of materials for supporting a specific war scenario. According to Dantzig:

> The military refer to their various plans or proposed schedules of training, logistical supply, and deployment of combat units as a *program*. When I first analyzed the Air Force planning problem and saw that it could be formulated as a system of linear inequalities, I called my paper *Programming in a Linear Structure* (Dantzig 2002, 46).

At the core of Project SCOOP was the interpretation of an economy or organization based on Dantzig's mathematical statement of the LP model. The structure used a triangular or rectangular technology matrix to specify the requirements and their interrelationships. This extended the Leontief input–output model of an economy: the triangular case involved no optimization and was readily solved, while the rectangular case involved the new concept of optimizing a linear objective function that would guide the computation, if one had a method and computational power to solve it. With keen foresight, Wood and Dantzig identified the promise of Project SCOOP:

> To compute programs rapidly with such a mathematical model, it is proposed that all necessary information and instructions be systematically classified and stored on magnetized tapes in the "memory" of a large scale digital electronic computer. It will then be possible, we believe, through the use of mathematical techniques now being developed to determine the program which will maximize the accomplishment of our objectives within those stated resource limitations (Wood and Dantzig 1949, 17).

Historically, Project SCOOP marks two momentous developments: it is considered the birthplace of LP and the locus of its computer-based application. Calling the Pentagon activity "the first linear-programming shoppe" (Gass 2002, 61), Saul wrote:

> All of us in OR are indebted to Project SCOOP. The linear-programming model, the simplex method, the first computer-based solution of LP problems, much of the theory of linear and mathematical programming, the basic computational theory of linear programming, and the extension of LP to industry and business all stemmed, wholly or in part, from the research and developments of Project SCOOP (Gass 1997a, 246).

While George Dantzig, Alex Orden, and other Project SCOOP analysts were developing the key algorithmic procedures for the simplex method, the computational challenges came into sharper focus. In the Mathematical Formulation Branch, Saul worked on the formulation of Air Force problems and helped develop and test new procedures for solving the resulting LPs. He recalls his entry into this dynamic and heady research environment:

> I was assigned to the Mathematical Formulation Branch. Walter Jacobs, a mathematician, was branch chief. He introduced me to linear programming by suggesting that I read reprints of Dantzig's three seminal papers [Dantzig 1951a, 1951b, 1951c]. Even though I was a fairly recent mathematics graduate, the concepts and ideas described in these papers were new to me and rather complex. What does the uninitiated make of such things as zero-sum games and the solving of hundreds of equations in hundreds of variables, especially in pre-computer days? Fortunately, I had a course in numerical calculus and knew something about Gaussian elimination and how to solve (3x3) systems of equations! (Gass 2002a, 64).

Project SCOOP was intimately tied to the development of computers. The National Bureau of Standards (NBS) built, with SCOOP funds, the Standards Eastern Automatic Computer (SEAC). Saul would often drive to the NBS Washington, D.C. campus from the Pentagon to deliver problems to be solved on the SEAC. Early computational tests on the SEAC compared the simplex method with other approaches (relaxation method and fictitious play) and demonstrated the effectiveness of the simplex method (Gass 2002; Hoffman et al. 1953; Orden 1952a, b). Project SCOOP also installed the second production unit of the UNIVAC–I computer in April 1952—it was formally turned over to the U.S. Air Force in June 1952. The UNIVAC's simplex code could solve problems of dimensions 250×500. Saul helped to check out the code—could it detect an unbounded problem?—as well as solve Air Force LP planning problems.

George Dantzig left the Pentagon in June 1952 for the RAND Corporation. By 1955, Project SCOOP was starting to wind down, but it had already made its mark on OR by assembling and supporting a remarkable network of researchers that extended from the Pentagon, NBS, Princeton University, Carnegie-Mellon University, and the RAND Corporation. In addition, Project SCOOP ran two LP symposia of great historical importance in 1951 and 1955. Saul attended the second symposium (held in the Pentagon) and presented a paper on finding first feasible solutions to LP problems (Gass 1955).

Saul left Project SCOOP in May 1955 to join IBM as an Applied Science Representative. His job was to help the salesmen, most of whom had little knowledge of digital computers, sell and install the new IBM computers designed for either commercial or scientific computation (702, 704, 705 vacuum-tube computers). Saul attended the standard 3-week sales training class in Endicott, New York where he learned commercial applications and sang songs from the IBM songbook! He also attended a 3-week scientific computing course in New York City at the T. J. Watson Laboratory, Columbia University. He was assigned to the Washington commercial sales office located in downtown Washington on Connecticut Avenue, and, later, also worked with IBM's federal government marketing office.

Saul left IBM in 1959 to join the Washington, D.C. consulting services company, Corporation for Economic and Industrial Research (CEIR), as Director of Operations Research. CEIR was one of the first companies to provide computer-based OR consultation services; they had purchased an IBM 650 drum-memory computer and an IBM 709 computer both of which Saul helped to install. There he worked briefly with William Orchard-Hays, who had come from RAND where he designed some of the first simplex-method codes for George Dantzig.

Saul rejoined IBM the following year when he was offered a position to work on the expanding U.S. space program for which IBM was a major subcontractor. Saul became manager of the simulation group of IBM's portion of the National Aeronautics and Space Administration's (NASA) Project Mercury Man-in-Space Program. He was responsible for the development of real-time simulation procedures used to validate the computational and data flow equipment system that IBM developed for Project Mercury. Initially, IBM's key task was to calculate the orbit of the space capsule based on radar telemetry data collected from tracking stations across the globe, with the data to be processed simultaneously by twin back-to-back IBM 7090 (transistor and core memory) computers located at the Goddard Space Center, Greenbelt, Maryland. IBM's tasks were soon

AT THE MERCURY COMPUTER CONSOLE

(Left to right) Chelsea Crater, Saul, Richard Hanrahan

expanded to include the processing of the lift-off data to determine if the spacecraft would enter into a viable orbit, determining the time to fire the re-entry rockets, and calculating the resulting path to ocean splashdown. In essence, IBM had to develop the analysis and required computer programs, run computer centers at Goddard, Cape Canaveral, and a Bermuda back-up, and operate an engineering, and communications subsystem that enabled the NASA flight controllers to monitor all phases of a Project Mercury mission. Saul noted that it was "the first real-time decision-making system with a man-in-the-loop" (Gass 1961b, 1999; Visco and Sheldon 2005, 55).

On May 1, 1961, Saul was appointed project manager of IBM's total Project Mercury program. On May 5, the first U.S. manned-capsule sub-orbital flight was scheduled to occur with Astronaut Alan Shepard on board. That day, the scene before lift-off at the Goddard Space Flight Center was tense:

> The crowds of NASA and IBM VIPs that gathered were kept back [from the computers] by a set of stanchions and ropes. We felt like gladiators who would be fed to the lions if something went wrong. All went well . . . (Gass 1999).

After this flight, Saul went to Cape Canaveral to observe all the manned orbital missions. From the VIP grandstand at Cape Canaveral, Saul watched John Glenn's lift-off on February 20, 1962, the first U.S. manned orbital flight. The success of Glenn's historic flight brought top management's attention to IBM's efforts—the next week Saul briefed IBM's board of directors in New York City, and, a few weeks later, he and Trudy traveled to Vienna, Austria, where Saul described IBM's support of Glenn's mission to a meeting of IBM's top European salesmen.

BACK TO SCHOOL AND OR PRACTICE

While at Project SCOOP, Saul had taken courses in emerging OR topics. In 1952–1953, he took a two-semester game-theory course that Albert Tucker and Harold Kuhn taught at American University on Thursday nights. One or the other would make a day-trip from Princeton University to Washington, D.C. to consult for the George Washington University Naval Logistics Research Project (supported by the Office of Naval Research), teach the night course, and then rush to Union Station to catch the last train back to Princeton. Saul formally enrolled in the doctoral program in mathematics at American University and took several other classes including: "Linear programming and game theory" from Alex Orden, "Methods of operations research" from Joseph McCloskey, and "Linear programming" from Alan Hoffman.

In 1963, NASA centralized its space activities at the Johnson Space Center, Houston, Texas. Saul did not intend to move to Houston, where IBM's space activities were to be relocated. He applied for a 2-year leave of absence under IBM's resident graduate fellowship program. His application was approved; it came with full pay and tuition.

While American University would have been a convenient choice, Saul decided on the University of California (UC), Berkeley for his doctoral studies. Having maintained contact with George Dantzig, who had joined UC in 1960, Saul visited the UC campus in March 1963 and discussed the situation with Dantzig, who was encouraging. The Gass family drove to their rented house in Berkeley in August 1963.

Saul started his Berkeley studies in September 1963, as a student in the Industrial Engineering and Operations Research Department. In addition to George Dantzig, the professorial OR staff included Robert Oliver and William Jewell, both MIT OR graduates. Because he had a substantial background in LP, Saul was not allowed to take Dantzig's LP course for credit; but he did audit the course. (Saul's text, *Linear Programming: Methods and Applications*, had been published in 1958.) There is an amusing story about how uneasy some of his classmates felt when they found out that Saul had already published an LP text. Dantzig taught the course from notes from his in-press text, *Linear Programming and Extensions*—the print version arrived on campus in October 1963. The Dantzig and Gass families socialized and often went to dinner with their children.

When Saul started looking for a dissertation topic, Dantzig suggested that he contact Roy Harvey, an OR analyst at Esso in San Francisco (Dantzig consulted with Esso). Harvey had developed a large-scale LP model for a production-inventory application whose structure was multi-stage but weakly coupled due to a set of tie-in variables. Saul devised a decomposition approach working with the dual variables to exploit the block structure of the individual stages. He named the resulting algorithm the dualplex method. His dissertation, "The dualplex method of large-scale linear programs," was completed in mid-summer 1965 (Gass 1966, 1972a); Saul received his Ph.D. in Engineering Science in September 1965.

Saul was one of George Dantzig's earliest doctoral students. Before him, Dantzig's Ph.D. students were Samirendra Sinha (1963) and Richard Cottle (1964). Dantzig's students, who also earned their degrees in the same year as Saul, were Earl Bell, Jr., Mostafa el-Agizy, Ellis Johnson, Paul Rech, Richard van Slyke, Roger Wets, and Richard Wollmer.

When Saul returned to IBM in the summer of 1965, he joined its recently organized Federal Systems Division, located in Gaithersburg, Maryland, as manager of Federal Civil Programs, responsible for urban problem contracts and consulting. Typically, this work did not involve much OR modeling or LP, but Saul did get a chance to apply OR to urban problems as a full-time member of the Science and Technology Task Force of the President's Commission on Law Enforcement.

The Commission was created by President Lyndon Johnson in 1965, partly in reaction to the issue of crime in the streets, an issue that Barry Goldwater had raised in the 1964 election campaign. It was mainly comprised of lawyers and sociologists. To augment the work of the Commission, a Science and Technology Task Force, directed by Alfred Blumstein (Ph.D. in OR, Cornell University), was formed at the Institute for Defense Analyses. Its charge was to bring scientific thinking to bear on crime (Blumstein 2002). Al recruited Saul to join his staff in 1966. Another recruit was Richard Larson who was completing his undergraduate degree in electrical engineering from MIT. Saul was responsible for developing the task force's approach to how science and technology can best serve police operations.

Saul left IBM in 1969 to help form a consulting firm, World Systems Laboratories, Inc., as Senior Vice-President. Its initial clients were a steel manufacturing company and governmental organizations in the Washington, D.C. area. But, economic conditions in 1970 forced the financial backers of World Systems to retrench and close the company. Saul then joined Mathematica, Inc., the well-known OR and economics-consulting firm led by Tibor Fabian, headquartered in Princeton, New Jersey (Harold Kuhn and Oskar Morgenstern were on the board of directors). Saul was director of its Washington, D.C. area office located in Bethesda, Maryland.

At Mathematica, Washington, Saul became involved and managed projects that were mainly public sector studies: a student aid model for the Department of Education; development and analysis of an educational data bank for the Appalachian Regional Commission; measuring telecast effectiveness for the Corporation for Public Broadcasting; consulting for the Chief of Naval Operations; development of a simulation model for the D.C. police department; and a National Science Foundation project to evaluate policy-related research in police protection.

Mathematica also had a contract with the Environmental Protection Agency (EPA) to conduct a survey of modeling in the non-military governmental area. Saul and Roger L. Sisson compiled and edited a collection of such applications (with chapters written by Saul and Roger and other subject matter experts) and published it in book form for the EPA as *A Guide to Models in Governmental Planning and Operations.* Feeling that the book's material was of value to the academic and consulting communities, Saul and Roger published it privately and distributed it from Saul's basement, with Trudy as bookkeeper (Gass and Sisson 1975). A total of 2000 copies were distributed (Visco and Sheldon 2005). Much later, Saul was to review public sector models in Gass (1994f).

AN ACADEMIC HOME

Saul began teaching OR early in his professional career. He first taught "An introduction to linear programming" in 1954 at the U.S. Department of Agriculture (a course George Dantzig initiated in 1950), and then for the American University mathematics department and the George Washington University school of engineering. In 1973 and 1974, he taught an evening course in OR to business students at the University of Maryland, College Park. Rudy Lamone, an OR Ph.D., was named dean of Maryland's College of

(Left to right) Rudy Lamone, Arjang Assad, Saul, Alan Goldman at Saul's 80th birthday symposium, February 25, 2006, University of Maryland, College Park.

Business and Management in 1973. Rudy recruited Saul for a full-time position and he joined the College in September 1975 as professor and chair of the management science and statistics department with the task of building a high-quality OR department. Saul was to spend the next 26 years at the university. He had "found a home" (Visco and Sheldon 2005, 57).

During his tenure as chairman (1975–1979), Saul recruited several new faculty, including Frank Alt, Michael Ball, Larry Bodin, Bruce Golden, and myself. Saul remained on the faculty for the next 22 years. During these years, he supervised doctoral and masters students, as well as teaching graduate courses in LP and related topics, and introductory OR courses to undergraduate business and MBA students.

Research in OR Methods and Applications

Saul's contributions to OR methodology may be divided into two broad areas: LP and its extensions, and decision-aiding methods. A large part of Saul's interests in the first area can be traced back to his days at Project SCOOP.

Saul's first major contribution to the theory of LP was his algorithmic procedure for solving a parametric linear-programming problem. The concept arose in 1952 at SCOOP when Walter Jacobs introduced Saul to the problem in the context of production smoothing: a decision maker faces the conflicting objectives of simultaneously minimizing monthly fluctuations in production and inventory carrying costs. By attaching weights to the two objectives, one can express this as a single objective LP and determine the trade-offs between the two objectives. This can be viewed as a single-parameter problem, where the parameter measures the ratio of the cost of a unit change in production to the cost of carrying one unit of inventory (Gass 1958, 158–164).

Using a modified simplex tableau, Saul first solved some test problems by hand. When Thomas Saaty (Ph.D. student from Yale) joined SCOOP for the summer of 1952, he and Saul worked out the details with help from the mathematicians Leon Goldstein (SCOOP) and Alan Hoffman (NBS). This resulted in three papers by Gass and Saaty that addressed the parametric linear-programming problem: Min $(c + \lambda d)x$ subject to $Ax = b,\ x \geq 0$, where c and d are the cost vectors of the two conflicting objectives and λ is the parameter to be varied. In their first paper, Saaty and Gass (1954), they showed, using the simplex method, how the optimal solution behaves as a function of λ. Related papers dealt with computational aspects and generalizations to more objective functions (Gass and Saaty 1955a, 1955b). Further development and extensions of parametric programming for LPs are discussed (Gal 1980, 1983) and in Osei-Bryson (2006).

When Walter Jacobs gave Saul the task of solving the parametric problem, a conceptual link with multi-objective linear programming (MOLP) was already present. One way to balance two competing objectives was to use a weighted objective function. This meant using a single parameter that reflected the relative weights. As Saul noted: "We recognized that the scheme could be generalized to more objectives and described a multi-parameter approach where two objectives meant one parameter, three objectives, two parameters and so on." (Gass 1997c, 13). Saul later returned to MOLP in several papers that dealt with the problem of how to select an efficient (Pareto) optimal solution (Dror and Gass 1987; Dror et al. 1988; Gass and Dror 1983, 2003).

Another area of interest that can be traced to Saul's Project SCOOP experiences is what Saul later called encounters with degeneracy (Gass 1993b) and the related problem of cycling using the simplex algorithm (Gass 1979; Gass and Vinjamuri 2004). Magnanti and Orlin (1988) show how parametric programming can be used to avoid cycling, thus making a connection between two of Saul's research strands.

Early on, Saul was intrigued by Dantzig's paper that showed the equivalence of a zero-sum two-person game and the linear-programming problem (Dantzig 1951a). One method proposed for the former problem was the method of fictitious play (Brown 1951). Fictitious play was known to converge, but ever so slowly. Could you speed it up? Much later, he posed the problem to his Ph.D. student, Pablo Zafra. They developed an easily implemented iterative restarting procedure for the original fictitious play method that causes it to converge rapidly. They also showed how their method could be used to find an initial solution to an LP problem by solving the equivalent skew-symmetric matrix game (Gass and Zafra 1995; Gass et al. 1996).

In a problem that stemmed from quality control concerns for manufactured parts, Saul investigated the simple question: How does one fit a circle to a given set of points in the plane? This question arises where a coordinate measuring machine is used to measure a manufactured part (drilled circle or shaped disc, spherical-shaped body) and a set of measured data points is analyzed to determine if the part meets stated quality control standards. This measurement problem evolved from a project supported by the U.S. National Institute of Standards and Technology (NIST). Following an idea proposed by Christoph Witzgall (NIST), Saul, Chris, and Howard Harary (NIST) developed a modified Chebyshev minimax objective as the criterion of fit (Gass et al. 1998). Given n points P_i with coordinates $(x_i,\ y_i), i = 1, \ldots, n$ on the plane and a circle of radius r_0 with its center located at (x_0, y_0), let r_i denote the radial distance from the center to P_i. If $r_i = r_0$ then P_i lies on the constructed circle and a perfect fit obtains if this is true for all i. Otherwise, the absolute difference of the squared radial distances $\left|r_i^2 - r_0^2\right|$ can be viewed as an error term. The objective considered was to

$$\text{Min Max}_{i=1,\ldots,n}\left|r_i^2 - r_0^2\right| \text{ or Min}\left\{\text{Max}_{i=1,\ldots,n}\left|(x_i - x_0)^2 + (y_i - y_0)^2 - r_0^2\right|\right\}.$$

With the (x_i, y_i) fixed, the problem is to determine the values of (x_0, y_0) and r_0. This can be done by transforming the problem into a linear program (Gass and Witzgall 2004). Extensive computational experience is reported in Gass et al. (1998) that shows that the squared formulation provides a very good approximation to the Chebyshev minimax criterion without squares (minimizing the maximum absolute difference), that is, Min Max $|r_i - r_0|$. Saul and his Ph.D. student, Pallabi Guha Roy, applied this concept in a novel way to select an efficient solution (Pareto optimum) from a subset of such solutions of a MOLP (Gass and Roy 2003).

In addition to optimization models, Saul has maintained a general interest in decision-aiding methods, chief among these being the Analytic Hierarchy Process (AHP) (Saaty 1980). Saul started to teach the subject at the University of Maryland soon after it was developed and has since written several expository articles on the subject (Bodin and Gass 2003; Forman and Gass 2001; Gass 2005a). Saul also used the AHP in several new applications:

- Assigning weights to thousands of deviation variables in large-scale goal programming models for personnel planning (Gass 1986, 1987c),
- Rating the complexity of conferences planned by NIST (Gass and Torrence 1991),
- Numerical rating for model accreditation (Gass 1993a).

Saul's theoretical contribution to AHP methodology appears in his papers with the Hungarian OR mathematician Tamás Rapcsák (Gass and Rapcsák 1998, 2004). They proposed a different method for deriving the vector of weights (or priorities) of the AHP alternatives based on the singular value decomposition (SVD) of the pairwise comparison matrix A.

Starting in 1984, military personnel planning models were analyzed by Saul as part of a consultant team to extend and implement the Army Manpower Long Range Planning System that was developed and made operational in 1982–1983. This led to a goal-programming optimization model "to determine by grade, skill, and/or years-of-service, the number of soldiers (or officers) the Army can put in the field ... so as to meet manpower goals over 7- to 20-year planning horizons" (Gass et al. 1988, 5). Because the goal-programming model was large-scale (9060 equations and 28,730 variables), Saul proposed procedures to help the user derive the weights in the objective function for skill-grade combinations goals by time period (Gass 1986). Large-scale military and other manpower planning models are reviewed in Gass (1991b).

THE EXPOSITOR OF OR

Many in the OR community likely first came across Saul's name through his expository works—his LP text or one of his Model World columns published in the journal *Interfaces*. Throughout his career, a substantial part of Saul's efforts have gone into expositions of OR.

Saul wrote his first expository work on LP at the request of Walter Jacobs at Project SCOOP. Jacobs asked Saul to prepare a non-technical booklet on LP aimed at an Air Force audience. This led to a 27-page internal Air Force report, "The application of linear programming techniques to Air Force problems" (Gass 1954). The report highlighted Air Force applications such as the transportation problem in the context of shipping units of equipment required for the B-47 aircraft from three Air Force depots (sources) to five Air Force bases (destinations) (Gass 1970, 18–22). Another application was the contract-awards problem faced by the U.S. Armed Forces that required them, by law, to award contracts for supplies at the lowest cost (Gass 1970, 104–108).

Saul's book, *Linear Programming: Methods and Applications* (Gass 1958), was the first text on LP. Prior to 1958, the only publicly available accounts of LP were the monographs by Charnes et al. (1953) and Vajda (1956). The origins of Saul's text go back to notes he prepared for the introductory LP course he taught at the Department of Agriculture Graduate School. The notes were transformed into the text during his first year at IBM (1955–1956). Each Saturday, Saul went to the Library of Congress for quiet writing time. Accordingly, the book's dedication to Trudy mentions "lost weekends" (Gass 1958, v). Saul had difficulty marketing a text on the little known subject of LP; based on a couple of hand-written chapters, it was rejected by a number of publishers, until McGraw-Hill decided to take a chance. Saul then had the full manuscript typed, and McGraw-Hill sent it out for review. All reviewers were reasonable (if not enthusiastic). One wrote: "I don't know how many people are now fascinated by 'min cX subject to $X \geq 0$, AX=b.' If there are enough of them, this seems a decent place for them to learn how to do it" (Gass 1990c, 132).

The first edition of *Linear Programming* was translated into Russian in 1961, the first book on the subject in the Russian language—25,150 copies were printed and sold for 80 kopeks each (U.S. price was $6.50); the Russians did not pay any royalties. Translations into several other languages followed, again being the first LP text in those languages. Subsequent editions appeared in 1964, 1969, 1975, and 1985. The fifth edition (Gass 1985) was almost 2.5 times the length of the first. It was reprinted by Dover Publications in 2003. Over 57,000 copies of the English language editions have sold.

Saul was co-author of an historically important LP reference, *Linear Programming and Associated Techniques: A Comprehensive Bibliography on Linear, Nonlinear, and Dynamic Programming*, published by the Johns Hopkins University Operations Research Office (ORO) (Riley and Gass 1958). This bibliography cited, with annotations, over 1000 items that included articles, books, monographs, conference proceedings, and theses completed through June 1957. Saul's co-author, Vera Riley, was a staff bibliographer at ORO. Saul also reviewed the advances in LP and LP codes in Gass (1961a).

Next to Saul's LP text, his most widely known publication is the primer, *An Illustrated Guide to Linear Programming* (Gass 1970). Similar to what was done in the introductory game-theory book, *The Compleat Strategyst* (Williams 1954), the *Guide* employed illustrations involving graphic characters to bring the concepts to life. The caterer problem is described in terms of a management science consultant's report on how to replace dirty napkins at the lowest cost for the Mad Hatter's tea parties. [The caterer problem was a disguised (secret?) version of the U.S. Air force problem of determining the trade-off cost between overhauling and procuring new aircraft engines (Jacobs 1954)].

PASS THE SUGAR

(Gass 1970, 37) © Saul I. Gass

In 1985, the *Illustrated Guide* was partially reincarnated in Saul's introductory text on OR modeling, *Decision Making, Models and Algorithms: A First Course*, complete with its original illustrations (Gass 1985). Retaining 16 chapters from the *Guide*, *Decision Making* adds new material on modeling, decision trees, and the AHP.

In 1994, Saul embarked on a major project with Carl Harris. They edited the *Encyclopedia of Operations Research and Management Science*, the first of its kind (Gass and Harris 1996). The second edition was published in 2001, a year after Carl's sudden death (Gass and Harris 2001)—it is dedicated, in memoriam, to both Carl and their dear friend and OR pioneer, Hugh Miser.

MANAGING THE MODELING PROCESS

For over 30 years, Saul has studied not just the technical contents of models, but also the total environment of decision-aiding with models. He refers to this as "managing the modeling process" (Gass 1987b, 1; 1991a). His interest started in the early 1970s when he was the principal investigator of a National Science Foundation contract to Mathematica. His task was to evaluate 50 police protection models based on the documentation provided and the accompanying research papers. Reflecting on the challenge of developing a framework for evaluation, Saul wrote:

> Our first problem was to determine what was meant by evaluation and how you do it. The literature gave little guidance, so we developed our own evaluative process. I quickly learned that analysts do not document, cannot, or will not write well, do not state their modeling assumptions, are unclear as to their data sources, maybe perform sensitivity analyses but do not tell you, and so on. . . . For me, the outcome was to start thinking about the problem of what we really mean by good modeling practices and implementation, and how do you evaluate model-based projects (Gass 1987b, 2).

Saul defines model evaluation or assessment as "the process by which interested parties, who were not involved in the model's origins, development and implementation, can assess the model's results in terms of its structure and data inputs so as to determine, with some level of confidence, whether or not the results can be used in decision making" (Gass 1977, 28).

> **THE MODELING POLICE FORCE**
>
> "From our perspective, good modeling practice assumes that someone, someday, will knock at your door and shout: 'Open up, it's the modeling police force! We're here to take your model down to Headquarters, question it to see what makes it tick, and plug it into the lie detector machine to determine if it tells the truth. You have three minutes to call your analyst, programmer, and lawyer.' . . . If you were the poor soul behind the door, what would you wish you had done so that the evaluation of your model comes out excellent?" (Gass 1997a, 252).

The need for research on how to perform model evaluation and assessment arose in the 1970s when the Department of Energy (DOE) developed an integrated set of LP and other models (Project Independence Evaluation System) to aid the Nixon administration in developing its future energy plans. Congress, the U.S. General Accounting Office (GAO), and other interested parties, questioned the assumptions and details of the modeling system and the analysis (U.S. GAO 1976a, 1976b). How does one respond to such questions when the results are generated by a very complex computer-based system?

Saul's interest in model evaluation and assessment continued through a series of studies sponsored by GAO, DOE, and the NIST (Gass 1981a, 1983a). He assisted the GAO in developing an evaluation procedure for complex models (U.S. GAO 1979, Gass and Thompson 1980). Saul summed up his own synthesis of model validation and assessment in his feature article in *Operations Research* (Gass 1983). Drawing on his examination of many models, Saul wrote about the documentation required for a complex model to be understood, used, and maintained successfully (Gass 1984).

THE OR PROFESSIONAL, STATESMAN, AND AMBASSADOR

Throughout his career in OR, Saul has been involved in the profession as an active and vocal citizen. His care and concern for the well-being of OR has found many outlets. Chief among these is his collection of Model World articles. These articles address such issues as models and the modeling process (Gass 1990a, 1990b); the profession and its ethics (Gass 1994d, 1994e; Gass et al. 1997); the history of OR (Gass 1990c, 2002b; Gass and Assad 2005b); the publication practices of OR/MS journals (Gass 1996); and how can you tell if you are an OR person (2007). Collectively, these articles, along with his invited articles and plenary addresses, reveal several recurrent themes that constitute Saul's professional credo:

> THE MAKING OF AN OR PERSON
>
> "It would be nice if we had proven and tested psychological and/or aptitude tests for determining what makes an OR person. However, I do not know of any. In the meantime, we can only ponder the following:
>
> "On what basis would you advise a youngster to consider OR as a profession?
>
> (a) Good in mathematics,
>
> (b) Solves Sudoku puzzles,
>
> (c) Mows the lawn over a shortest route,
>
> (d) All the above" (Gass 2007, 581).

OR: its past, present and future. In his article for ORSA's President's Symposium, Saul reflected on "what OR has done and should do." He noted that while OR should take pride in its success in solving a wide variety of operational problems that the OR pioneers took as their challenge, the operational outlook is too "restrictive a view of what OR can and should be" (Gass 1987a, 320). In the future, OR will have to tackle complex organizational problems, meet the challenge of decision-making in real-time, and contribute to policy analysis (Gass 1997b);

Professional identity and the OR process: Saul has voiced his concerns about how, as a profession, OR runs the risk of being diluted as diverse professions adopt and use OR techniques (Gass 1987a, 1990b). For Saul, this risk is heightened by the identification of OR by its methods alone and calls for a shift in perspective;

The science of modeling: Saul has challenged the OR profession to develop a science of modeling (Gass 1987b, 1988). He feels that OR should emulate computer science for treating its practice an object of inquiry: "Unlike those working in the field of computer science, we have not attracted the behavioral and psychological researchers to the study of the practice and implementation of our professional endeavors" (Gass 1991a, 10);

Ethics: Saul has stated his position on ethics in OR clearly: It is essential for the future well-being of the operations research profession that its ethical concerns and problems be investigated and discussed in a more demanding fashion by its practitioners, academics, and related professional societies. The problems will not go away. They will become more prevalent with the wider use of operations research methodology. Operations research needs a code of ethics and professional practice (Gass 1994a, 223). Saul continues to voice this concern (Gass 2009a).

AMBASSADOR TO THE OR WORLD

Saul's extensive international travels have served to build personal and professional relations in various parts of the world. In 1977, under the U.S./USSR Academy of Sciences Exchange Program, Saul (and Trudy) spent 1 month in the Soviet Union visiting scientific institutions in Moscow, Kiev, Tbilisi, and Novosibirsk. This led to his participation in other U.S.–USSR exchanges. One such exchange was a workshop on large-scale optimization, organized by Saul in 1990 at the University of Maryland, with the participation of researchers from the Central Economic Mathematical Institute of the USSR Academy of Sciences (Moscow). In

May 1993, he led a delegation of 15 OR professionals to Russia and Hungary under the auspices of the Citizen Ambassador Program. As a Fulbright Scholar, Saul spent 2 months at the Computer and Automation Research Institute, Hungarian Academy of Sciences in 1995–1996. In 2001, Saul was selected for a 5-year term as a Fulbright Senior Specialist and visited and lectured at universities in Chile, Japan, New Zealand, Spain, and Taiwan. In his writings, he described his views on the technology transfer of OR from developed countries to developing countries and concluded: "each country must develop its unique approach to the practice of OR, a practice that fits within and is part of a country's cultural and managerial decision-making framework" (Gass 1988, 7, 1991a).

BEYOND THE TECHNICAL STUFF

Saul has a strong sense of humor, which pervades his writings and enlivens his speeches. He is the perfect master of ceremonies and a capable organizer of professional gatherings and events. His flair for showmanship shined through the three OR/MS East vs. West Knowledge Bowls he conducted for ORSA/TIMS and INFORMS national meetings. In these contests, two teams of prominent OR/MS citizens competed to answer questions that tested their OR/MS cultural literacy. I was a judge at two of these events and witnessed the pleasure Saul took in preparing and pitching the questions. Sample questions appear in Gass (2002d). Saul organized the first ORSA/TIMS 10 K race when he was chair of the 1988 joint TIMS/ORSA national meeting, Washington D.C. He was honored by having subsequent races at OR meetings, both in the U.S. and in Europe, named the Saul Gass 10 K Race. (Arjang Assad)

Saul and Fred Glover, IFORS conference, Lisbon 1993

THE CHRONICLER OF OR

According to Saul, the first generation of OR professionals were the 71 persons who attended the founding meeting of ORSA on May 26 and 27, 1952 at the Arden House, Harriman, New York, together with a dozen pioneers who were not founding members (Gass 2002b). Thirteen members of this group served as ORSA Presidents through 1974.

> One might say that starting in 1975, the administrative and organizational aspects of ORSA were managed by a new, second generation of OR professionals. . . . My emphasis on the generational aspects of OR lets me segue into my main theme: the training and experiences of my generation of OR professionals were shaped by the early decades of remarkable scientific and management advances by the now somewhat forgotten first generation (Gass 2002b, 46).

In his historical writings, Saul has chronicled the achievements of these pioneers of OR. His papers in this area fall into two groups. The first group are historical pieces that cover the early history of the field: on the founding of ORSA (Gass 2002b); the definition of OR (Gass and Assad 2005); the early development of LP and the diet problem (Gass 1989; Garille and Gass 2001). The second group has to do with key periods of historical interest in which Saul was involved personally: Project SCOOP and the origins of LP (Gass 1989, 1990c, 1997c, 2002a); Project Mercury experiences (1961b, 1999); the Washington OR scene (Gass 1997a); OR pioneers and colleagues (Gass 2003, 2005b; Gass and Gross 2000).

Saul published a number of articles to commemorate the 50th anniversary of the founding of ORSA (2002a, b, c). One is a timeline of key events in the history of OR—it lists 241 events that captured the development of the field (Gass 2002c). Together, Saul and I expanded this list into a book with 417 annotated entries: *An Annotated Timeline of Operations Research: An Informal History* (Gass and Assad 2004).

In 1996, Saul summarized his career in OR as follows:

> My career spans both sides of the OR equation. For 25 years I worked as a mathematician and an OR analyst, and I directed OR for a couple of consulting firms; I had the good fortune to work in some important areas: Linear programming, the first man-in-space program, criminal justice, energy modeling and manpower systems ... For the last 20 years, I have been an academic (Gass 1996, 105).

Saul's career in OR now approaches six decades. His publications include six books, over 80 journal articles, nine book chapters, four edited volumes and reports, and 2 dozen occasional pieces. As a committed and vocal citizen of the OR community, Saul has never shied away from

THE GASS FAMILY

(Left to right) Ellie, Joyce, Arianna, Ron, Trudy, Joey, Saul
(Granddaughter Arianna's high-school graduation, June 22, 2009)

Ron Gass, his wife Johanna (Joey) Pfund, and their daughter Arianna reside in Connecticut. Ron is an attorney with a national insurance and reinsurance law practice. He is a reinsurance dispute resolution expert who frequently serves as an umpire or arbitrator in industry arbitrations. Joey received a B.A. in music and a Master of Musicology from the University of Maryland, College Park. She is a website project manager and developer with a significant regional corporate clientele. Arianna is currently attending Vassar College and is a member of the Class of 2013.

With degrees in special education and mental health, Joyce Gass teaches and practices massage in the Los Angeles area. She is the Executive Director of the Institute for Psycho-Structural Balancing, a holistic school for massage therapy. Joyce is a certified instructor for *"Prepare for Surgery, Heal Faster"*™ healing workshops which provide useful mind/body techniques for those facing surgery, cancer treatment, or other illnesses. She and her partner Ellie Josephs live in Topanga, California. Ellie found her passion in music playing the French horn and piano. She is a graduate of The Manhattan School of Music. After years of performances and a career as the Production Coordinator of the Los Angeles Philharmonic, Ellie was drawn to the healing arts. She is a massage therapist and has a private practice in Los Angeles.

expressing his opinion on key issues of the OR profession. At times, Saul has gone against the grain of majority opinion (Gass 1972b, 1994b, 1994c). In such cases, his conduct sets an example of how professional debate can be carried out with equanimity.

Not one to rest on his laurels, Saul continues to remain actively engaged in the practice and profession OR. On a personal note, I have had the privilege of being his colleague for over three decades and one of his collaborators for the last seven years. Like many of his friends and colleagues, I continue to marvel at the way he has embraced and enriched his chosen profession.

HONORS AND AWARDS

Saul served as president of ORSA (1976) and was awarded the George E. Kimball Medal for distinguished service to ORSA and the profession (1991). He was made a fellow of the Institute for Operations Research and the Management Sciences (INFORMS) (2002), and was the recipient of the INFORMS Expository Writing Award (1997). In 1996, he received the Jacinto Steinhardt Memorial Award of the INFORMS Military Applications Society for outstanding contributions to military operations research. He served as President of Omega Rho, the international operations research honor society (1985–1986); vice-president for the North American Operations Research Region of the International Federation of Operational Research Societies (1995–1997); and vice-president for international activities of INFORMS (1996–1997). His University of Maryland honors include: Westinghouse Professor (1983–1992); Distinguished Scholar-Teacher (1998); Dean's Lifetime Achievement Professor (2000); Professor Emeritus (2001).

ACKNOWLEDGMENTS

This profile makes liberal use of the extended interview with Saul published by the Military Operations Research Society (Visco and Sheldon 2005), and the invited chapter (Assad 2006) I wrote for Saul's Festschrift (Alt, Fu, and Golden 2006). I am indebted to Saul for providing additional information for this paper by making some of his earlier papers available and for answering queries put to him.

REFERENCES

Alt F, Fu M, Golden B (eds) (2006) Perspectives in operations research: papers in honor of Saul Gass' 80th birthday. Springer, New York, NY

Assad A (2006) Four score years of Saul I. Gass: Portrait of an OR professional. In: Alt F, Fu M, Golden B (eds) Perspectives in operations research: papers in honor of Saul Gass' 80th birthday. Springer, New York, NY, pp 24–72

Blumstein A (2002) Crime modeling. Oper Res 50(1):16–24

Bodin L, Gass S (2003) On teaching the analytic hierarchy process. Comput Oper Res 30(10):1487–1498

Brown G (1951) Iterative solution of games by fictitious play. In: Koopmans TC (ed) Activity analysis of production and allocation. Wiley, New York, NY, pp 374–376

Charnes A, Cooper W (1953) An introduction to linear programming. Wiley, New York, NY

Dantzig G (1951a) A proof of the equivalence of the programming problem and the game problem. In: Koopmans TC (ed) Activity analysis of production and allocation. Wiley, New York, NY, pp 330–335

Dantzig G (1951b) Maximization of a linear function of variables subject to linear inequalities. In: Koopmans TC (ed) Activity analysis of production and allocation. Wiley, New York, NY, pp 339–347

Dantzig G (1951c) Application of the simplex method to a transportation problem. In: Koopmans TC (ed) Activity analysis of production and allocation. Wiley, New York, NY, pp 359–373

Dantzig G (1963) Linear programming and extensions. Princeton University Press, Princeton, NJ

Dantzig G (2002) Linear programming. Oper Res 50(1):42–47

Dror M, Gass SI, Yellin A (1988) Experiments with an interactive procedure for MOLP given weak orders on variables and objectives. Eur J Oper Res 34(1):78–85

Forman E, Gass SI (2001) The analytic hierarchy process: an exposition. Oper Res 49(4):469–486

Gal T (1980) A "historiogramme" of parametric programming. J Oper Res Soc 31(5):449–451

Gal T (1983) A note on the history of parametric programming. J Oper Res Soc 34(2):162–163

Garille S, Gass SI (2001) Stigler's diet problem revisited. Oper Res 49(1):1–13

Gass SI (1954) The application of linear programming techniques to air force problems. Paper AFAMA 3-4-52, Air Force, Directorate of Management Analysis Service, Headquarters USAF (December)

Gass SI (1955) A first feasible solution to the linear programming problem. In: Antosiewicz H (ed) Proceedings of the second symposium in linear programming.

Directorate of management analysis, Headquarters, USAF, Washington, DC, pp 495–508
Gass SI (1958) Linear programming: methods and applications. McGraw-Hill, New York, NY
Gass SI (1961a) Recent developments in linear programming. In: Alt F (ed) Advances in computers, vol 2. Academic, New York, NY, pp 295–377
Gass SI (1961b) The role of digital computers in Project Mercury. Computers: key to total systems control, proceedings of the eastern joint computer conference, vol. 20. Macmillan, New York, NY, pp 33–46
Gass SI (1966) The dualplex method of large-scale linear programs. ORC Report 66-15. Operations Research Center, University of California, Berkeley, CA
Gass SI (1970) An illustrated guide to linear programming. McGraw-Hill, New York, NY. (Reprinted by Dover Press, 1990)
Gass SI (1972a) The dualplex method applied to special linear programs. Proceedings of the international federation of information processing societies, 1971. North-Holland, New York, NY, pp 1317–1323
Gass SI (1972b) Reactions to the guidelines for the practice of operations research. Oper Res 20(1):224–225
Gass SI (1977) Evaluation of complex models. Comput Oper Res 4(1):27–35
Gass SI (1979) Comments on the possibility of cycling with the simplex algorithm. Oper Res 27(4):848–852
Gass SI (1983) Decision-aiding models: validation, assessment, and related issues for policy analysis. Oper Res 31(4):603–631
Gass SI (1984) Documenting a computer-based model. Interfaces 14(3):84–93
Gass SI (1985a) Decision making, models and algorithms: a first course. Wiley-Interscience, New York, NY
Gass SI (1985b) Linear programming: methods and applications, 5th edn. McGraw-Hill, New York, NY (Dover Publications reprint 2003.)
Gass SI (1986) A process for determining priorities and weights for large-scale linear goal programs. J Oper Res Soc 37(8):779–785
Gass SI (1987a) A perspective on the future of operations research. Oper Res 35(2):320–321
Gass SI (1987b) Managing the modeling process: a personal reflection. Eur J Oper Res 31(1):1–8
Gass SI (1987c) The setting of weights in linear goal-programming problems. Comput Oper Res 14(3):227–229
Gass SI (1988) Operations research – supporting decisions around the world. In: Rand G (ed) Operational research '87, proceedings of the 11th international conference on operational research (IFORS). North-Holland, New York, NY, pp 3–18
Gass SI (1989) Comments on the history of linear programming. IEEE Ann Hist Comput 11(2):147–151
Gass SI (1990a) Model world: have model, will travel. Interfaces 20(2):67–71

Gass SI (1990b) Model world: danger, beware the user as modeler. Interfaces 20 (3):60–64
Gass SI (1990c) Model world: in the beginning there was linear programming. Interfaces 20(4):128–132
Gass SI (1991a) The many faces of OR. J Oper Res Soc 42(1):3–15
Gass SI (1991b) Military manpower planning models. Comput Oper Res 18(1):65–73
Gass SI (1993a) Model accreditation: a rationale and process for determining a numerical rating. Eur J Oper Res 66(2):250–258
Gass SI (1993b) Encounters with degeneracy: a personal view. Ann Oper Res 46/47(1–4):335–342
Gass SI (1994a) Ethical concerns and ethical answers. In: Wallace W (ed), Ethics in modeling. Elsevier, Tarrytown, NY, pp 207–225
Gass SI (1994b) Not this merger proposal. OR/MS Today 21(1):44–46
Gass SI (1994c) Don't merge, restructure. OR/MS Today 21(2):69–70
Gass SI (1994d) On ethics in operational research. J Oper Res Soc 45(8):965–966
Gass SI (1994e) Model world: ethics in the not so real MS/OR world. Interfaces 24 (6):74–78
Gass SI (1994f) Public sector analysis and operations research/management science. In: Barnett A, Pollock SM, Rothkopf M (eds) Handbook of OR: operations research and the public sector. North-Holland, New York, NY, pp 23–46
Gass SI (1996) Model world : On academics, applications, and publications. Interfaces 26(6):105–111
Gass SI (1997a) The Washington operations research connection: the rest of the story. Soc-Econ Plan Sci 31(4):245–255
Gass SI (1997b) Model world: OR is the bridge to the 21st century. Interfaces 27(6):65–68
Gass SI (1997c) On the education of a multi-criteria researcher: a personal view (Abraham Charnes Distinguished Lecture). In: Caballero R, Ruiz F, Steuer R (eds) Lecture notes in economics and mathematical systems, #332: advances in multiple objective and goal programming. Proceedings of the second international conference on multi-objective programming and goal programming, Torremolinos, Spain, May 16–18, 1996, Springer, Berlin, pp 5–26
Gass SI (1999) Project Mercury Man-in-Space real-time computer system: "You have a go, at least seven orbits!" IEEE Ann Hist Comput 21(4):37–48
Gass SI (2002a) The first linear programming shoppe. Oper Res 50(1):61–68
Gass SI (2002b) Model world: reflections on the 50th anniversary of the founding of ORSA. Interfaces 32(4):46–51
Gass SI (2002c) Great moments in history. ORMS TODAY 29(5):31–37.
Gass SI (2002d) Not a trivial matter. ORMS Today 29(5):46–48
Gass SI (2003) IFORS' operational research hall of fame: George B. Dantzig. Int Transact Oper Res 10(2):191–193

Gass SI (2005a) Model world: the great debate: MAUT versus AHP. Interfaces 35(4):308–312

Gass SI (2005b) In memoriam: the life and times of the father of Linear Programming. ORMS Today 32(4):40–48

Gass SI (2007) Model world: the hidden ingredient and nature vs. nurture. Interfaces 37(6):577–581

Gass SI (2009a) Ethical guidelines and codes in operations research. OMEGA 37(6):1044–1050

Gass SI, Assad A (2004) An annotated timeline of operations research: an informal history. Springer, New York, NY

Gass SI, Assad A (2005) Model world: Tales from the timeline: the definition of operations research and the origins of Monte Carlo simulation. Interfaces 35(5):429–435

Gass SI, Collins R, Meinhardt C, Lemon D, Gillette M (1988) Army manpower long range planning system. Oper Res 36(1):5–17

Gass SI, Dror M (1983) An interactive approach to multiple-objective linear programming involving key decision variables. Large Scale Sys 5(20):95–103

Gass SI, Gross D (2000) In memoriam: Carl M. Harris 1940–2000. INFORMS J Comput 12(4):257–260

Gass SI, Harris C (eds) (1996) Encyclopedia of operations research and management science. Kluwer, Boston, MA

Gass SI, Harris C (eds) (2001) Encyclopedia of operations research and management science, 2nd edn. Kluwer, Boston, MA

Gass SI, Nahmias S, Harris C (1997) Model world: the academic midlife crisis. Interfaces 27(5):54–57

Gass SI, Rapcsák T (1998) A note on synthesizing group decisions. Decis Support Sys 22(1):59–63

Gass SI, Rapcsák T (2004) Singular value decomposition in AHP. Eur J Oper Res 154(3):573–584

Gass SI, Roy P (2003) The compromise hypersphere for multiobjective linear programming. Eur J Oper Res 144(3):459–479

Gass SI, Saaty T (1955a) The computational algorithm for the parametric objective function. Naval Res Log Quart 2(1 and 2):39–45

Gass SI, Saaty T (1955b) Parametric objective function (part 2)—generalization. J Oper Res Soc Am 3(4):395–401

Gass SI, Sisson R (1975) A guide to models in governmental planning and operations. Sauger Books, Potomac, MD

Gass SI, Thompson B (1980) Guidelines for model evaluation: an abridged version of the U.S. General Accounting Office exposure draft. Oper Res 28(2):431–439

Gass SI, Torrence S (1991) On the development and validation of multicriteria ratings: a case study. Socioecon Plann Sci 25 (2):133–142

Gass SI, Vinjamuri S (2004) Cycling in linear programming problems. Comput Oper Res 31(2):303–311

Gass SI, Witzgall C (2004) On an approximate minimax circle closest to a set of points. Comput Oper Res 31(4):637–643

Gass SI, Witzgall C, Harary H (1998) Fitting circles and spheres to coordinate measuring machine data. Int J Flex Manufact 10(1):5–25

Gass SI, Zafra P (1995) Modified fictitious play for solving matrix games and linear-programming problems. Comput Oper Res 22(9):893–903

Gass SI, Zafra P, Qiu Z (1996) Modified fictitious play. Nav Res Logistics 43 (7):955–970

Hoffman A, Mannos M, Sokolowsky D, Wiegmann N (1953) Computational experience in solving linear programs. SIAM J 1(1):1–33

Jacobs W (1954) The caterer problem. Nav Res Logistics Q 1(2):154–165

Magnanti T, Orlin J (1988) Parametric linear programming and anti-cycling pivoting rules. Math Program 41(3):317–325

Orden A (1952a) Solution of systems of linear inequalities on a digital computer. In: Proceedings of the association of computing machinery. Pittsburgh, PA, (May 2)

Orden A (1952b) Applications of the simplex method to a variety of matrix problems. In: Symposium on linear inequalities and programming. Headquarters, U.S. Air Force, Washington, DC, pp 28–55

Osei-Bryson K-W (2006) Towards exposing the applicability of Gass & Saaty's parametric programming procedure. In: Alt F, Fu M, Golden B (eds) Perspectives in operations research: papers in honor of Saul Gass' 80th birthday. Springer, New York, NY, pp 235–246

Riley V, Gass SI (1958) Linear programming and related techniques: a comprehensive bibliography. Operations Research Office, Johns Hopkins Press, Chevy Chase, MD

Saaty T (1980) The analytic hierarchy process. McGraw-Hill, New York, NY

Saaty T, Gass SI (1954) The parametric objective function – part I. J Oper Res Soc Am 2(3):316–319

U.S. GAO (1976a) Review of the 1974 Project Independence Evaluation System. OPA-76-20, Washington, DC (April 21)

U.S. GAO (1976b) Ways to improve management of federally funded computerized models. LCD-75-111. Washington, DC (August 13)

U.S. GAO (1979) Guidelines for model evaluation. PAD-79-17, Washington, DC (January)

Vajda S (1956) The theory of games and linear programming. Methuen, London

Visco G, Sheldon B (2005) Military Operations Research Society (MORS) oral history project interview of Saul I. Gass. Mil Oper Res 10(4):39–62

Williams J (1954) The Compleat Strategyst. McGraw-Hill, New York, NY

Wood M, Dantzig G (1949) Programming of interdependent activities, I, general discussion. Econometrica 17(3–4):193–199. (Also, Koopmans TC (ed) (1950) Activity analysis of production and allocation. Wiley, New York, NY, pp 15–18

31

Thomas L. Saaty

Luis G. Vargas

Most OR researchers tend to focus their research programs on one central area and contribute to its further development by deepening the theoretical understanding or extending the range of applications. Tom Saaty is different. Since encountering OR in the early 1950s, about the time he completed his formal education as a mathematician, he has involved himself in diverse subfields of OR, making significant contributions to the mathematics of OR, queueing, networks, linear programming (LP), arms control, and decision analysis (DA). Tom has also applied his talent as an expositor by writing seminal books on these topics.

Tom is widely known for his development of the Analytic Hierarchy Process (AHP)—an adaptive and appropriate approach for the resolution of a wide range of complex, multicriteria, real-world decision problems. The AHP addresses the central problem of DA: to select from among alternative choices one that is best with respect to stated criteria. This requires the decision maker to make comparisons between the choices, incorporating both quantifiable and intangible factors in the process. Tom pioneered the theory of relative measurement necessary in building meaningful numerical scales associated with intangibles, which led to the framework he introduced as AHP.

Tom is a fellow of the American Association for the Advancement of Science, a member of the International Academy of Management, and the National Academy of Engineering. He was awarded the 2008 Institute for Operations Research and the Management Sciences (INFORMS) Impact Prize for the development of the AHP and its wide impact on OR.

A.A. Assad, S.I. Gass (eds.), *Profiles in Operations Research*, International Series in Operations Research & Management Science 147, DOI 10.1007/978-1-4419-6281-2_31,

EARLY UPBRINGING

Tom Saaty is descended from Assyrian Christians from northern Iraq. His father, David Saaty, lived in Rhode Island for a number of years, but returned to his hometown of Mosul in the early 1900s. David was a well-known entrepreneur whose ambition was to modernize Iraq; he opened the country's first ice-making factory in 1923. David married Tom's mother, Dola, in 1925, and Tom, their first child, was born in Mosul on July 18, 1926. They had three other children, John, Ben, and Grace.

When Tom was born, Iraq was administered by the British. Mosul remained in Ottoman hands until World War I ended in the Middle East with the signing of the Armistice of Mudros (October 30, 1918). David's life was saved by British troops who liberated Mosul just before he, along with other prominent western-oriented citizens, were about to be hanged by the Turks (Joseph 1919). Many of his friends were British and he was openly in favor of the British staying forever, as he expressed in a letter to his relatives in Rhode Island (Joseph 1919).

Many Christian families sent their children to the Quaker Brummana High School in Lebanon, as was Tom when he was 15; his brothers and sister followed in subsequent years. He rarely returned home except for brief vacations. After Tom graduated from Brummana, he attended the American University of Beirut for 2 years.

When he was 19, Tom came to the U.S. to attend Columbia Union College (a Seventh-day Adventist college) in Takoma Park, Maryland, just north of Washington, DC. He received his B.A. from Columbia Union (1948), and then attended the Catholic University of America, receiving an M.S. degree in Physics (1949). A mathematics teacher and friend from Columbia Union, George G. O'Brien, suggested to Tom that he should continue his studies, not in physics, but in mathematics, a subject in which

Tom displayed a natural talent. He applied and was accepted into the mathematics Ph.D. program at Yale University. There he received an M.A degree in mathematics (1951), and his Ph.D. in mathematics (1953), with Professor Einer Hille supervising his dissertation "On the Bessel Tricomi [Differential] Equation." Tom then spent a year doing postgraduate work at La Sorbonne under the direction of Henri Cartan.

DISCOVERING OPERATIONS RESEARCH

Tom's interest in OR was initiated by George O'Brien who was a consultant to the U.S. Air Force's Project SCOOP (Scientific Computation of Optimal Programs). Project SCOOP, with headquarters in the Pentagon, was a major Air Force scientific task force established in 1948 for formulating and solving a wide range of Air Force planning and programming problems. George Dantzig was SCOOP's chief mathematician.

Tom worked at SCOOP as a summer intern in 1951 and 1952. There he joined with other SCOOP mathematicians, especially Saul Gass, to analyze a range of Air Force linear-programming problems and their solution by Dantzig's recently developed simplex method. Their research on parametric programming yielded a sequence of papers that demonstrated how such problems could be readily solved by the simplex method for the full range of values of a parameter in the objective function (Gass and Saaty 1955a, b; Saaty and Gass 1954).

After graduating from Yale, and, despite efforts by Hille who wanted Tom to stay at Yale to do research, he joined Melpar, Inc. (1953–1954) as a scientific analyst studying submarine defense. At that time, Melpar was a division of Westinghouse Air Brake Company, specializing in classified military and civilian government contracts in electronics, chemistry, bio-medicine, and related areas. (The name came from combining the first syllables of the names of its founders, Thomas Meloy and Joseph Parks. Since 1995 it has been a subsidiary of Raytheon Corporation, Bedford, Massachusetts.)

In 1954, Tom joined the Operations Evaluation Group (OEG) in Washington DC. OEG was formed in 1945 from the World War II Operations Research Group that evolved from the Antisubmarine Warfare Operations Research Group organized by Philip Morse in 1942.

While at OEG, he became interested in Navy messages and how to process different categories; he was intrigued by the related but diverse mathematical and statistical research, especially the work of Agner K. Erlang. Tom did extensive research in the subject and amassed material that appeared as a paper in *Operations Research* titled "Résumé of useful formulas in queu[e]ing theory" (Saaty 1957). His objective was to provide "a convenient summary of some results in queu[e]ing, which in the author's opinion would be of value to investigators applying the theory to operational problems" (Saaty 1957, 162). Evidently, the editor of the journal agreed, as the paper spread over 39 pages! This paper helped to bring waiting-line concepts to OR researchers and motivated Tom to write *Elements of Queueing Theory with Applications* (Saaty 1961a). To write this book, he collected copies of 450 papers (in the days when it was difficult to make copies) and spent a year in London (1958–1959) where he gradually absorbed and arranged the material, thus making them accessible to the OR community.

At OEG, Tom worked on classified submarine detection problems and the mathematics of radar reconnaissance, a topic of interest due to U.S. aircraft flights (RB48 and U-2) over the Soviet Union. A major contribution was his analysis on the spread of biological agents for the Army Chemical Corps and a stochastic model with absorbing barriers that was actually used to develop bacteria for military use (Saaty 1961b). As Tom stated, "I am not proud of that and did it because of the mathematical challenge" (Saaty 2008). As he never felt at home at OEG, he left in 1957 and joined the Navy Management Office.

In 1958, Tom was appointed scientific liaison officer to the U.S. Embassy in London. In that role, he traveled the world and visited many countries to determine the current status of their mathematical, statistical, and OR developments. During these travels, he wrote the seminal, graduate-level text *Mathematical Methods of Operations Research* (Saaty 1959). This was the first OR book that described the field's mathematical foundations and techniques covering optimization, game theory, and statistical and probabilistic methods. In returning to the U.S. in 1959, he was named Director of Advanced Planning in the Office of Naval Research (ONR), a post he held until 1961. His successful bacteriological warfare model led to an appointment as Scientific Analyst in the Arms Control and Disarmament Agency (ACDA), Department of State, where he stayed from

1963 till 1969. He also served as the Executive Director of the Conference Board of the Mathematical Sciences (1965–1967) in Washington, DC.

ARMS CONTROL, MATHEMATICS, AND SOCIAL ISSUES

At ACDA, Tom was responsible for technical projects involving mathematicians and economists who were working on arms control problems by applying models and mathematics such as game theory and conflict resolution. In his book, *Mathematical Models of Arms Control and Disarmament*, arms control is defined as "the attempt of nations to impose arbitrary limits on the instruments and consequences of conflict" (Saaty 1968, 4). The book coalesced related research in areas such as problems of equilibrium and stability, the role of game theory and its applications, negotiations with incomplete information, models of treaty enforcement, and conflict resolution. Tom's experiences at ACDA and, as an observer at negotiations held in Geneva between the U.S. and Russia, started him thinking about how to solve such complex decision problems. This was the genesis of the AHP. As Tom later recalled (Saaty 2001a, xii):

> Two things stand out in my mind from that experience. The first is that the theories and models of the scientists were often too general and abstract to be adaptable to particular weapon tradeoffs. It was difficult for those who prepared the U.S. position to include their diverse concerns within this framework and to come up with practical sharp answers. The second is that the U.S. position was prepared by lawyers who had a great understanding of legal matters, but were no better than the scientists in assessing the value of the weapon systems being traded off. It was nobody's fault. What was needed was a realistic and a genuine way to derive priorities that reflect the relative benefit, opportunities, cost, and risks in giving up one system or part of a system in return for what the other side was willing to give.

He also proposed creating an International Center for Conflict Resolution whose members would be retired leaders and negotiators with the aim to resolve conflicts creatively without the use of force. No action was taken at that time, but today such conflict resolution programs reside in many places, in particular, at the Carter Center, founded by President Jimmy Carter in 1982.

While at ACDA, Tom also investigated basic mathematical problems, and, over time, wrote a number of books: *Finite Graphs and Networks* (Busacker and Saaty 1965), *Optimization in Integers and Related Extremal Problems* (Saaty 1970), and *The Four Color Problem: Assaults and Conquest* (Saaty and Kainen 1977), and two volumes on nonlinear mathematics (Saaty and Bram 1964) and (Saaty 1967). Tom was awarded the 1972 L. R. Ford Prize in mathematics for his paper on the four-color problem (Saaty 1972).

Tom worked with leading scientists and mathematicians, writing some works and editing others on topics in modern, social, biological, and behavioral mathematics (Saaty 1973; Saaty and Alexander 1981; Saaty and Boone 1990; Saaty and Weyl 1969). Of especial importance was the three-volume set of *Lectures on Modern Mathematics* (Saaty 1963, 1964, 1965) that he edited; the volumes were based on lectures he organized, sponsored by George Washington University and the ONR. The lecturers included such eminent mathematicians as Paul Halmos ("A glimpse of Hilbert spaces," volume I), John Milnor ("Differential topology," volume II), and Paul Erdös ("Some recent advances and current problems in number theory," volume III). But the work that departs most from Tom's mathematical research and demonstrates his creativity and imagination is the book *Compact City*, co-authored with George B. Dantzig (Dantzig and Saaty 1973). This book describes a novel approach to urban planning that can be viewed as an OR and systemic way to designing a city.

Compact City is an attempt to answer the question: "Can the U.S. preserve and extend a high and spacious standard of living in a world where populations are ever increasing, and where all seek to match our standards by taking more and more of their fair share of the world's nonrenewable resources?" (Dantzig and Saaty 1973, v). The book describes how to create a better environment by designing a city for one million people that would fit into a one-mile square eight-story structure, and how such a city's functions and facilities would be organized. It has many innovative suggestions: the design and use of elevators, location of transportation networks so that no point in the city would be more than a 5-minute commute from any other point, and continuous 24/7 work and leisure periods.

In 1969, Tom left ACDA and started his academic career as a professor at the University of Pennsylvania. There he held appointments in the Wharton School Graduate Groups of Operations Research and

Social Systems Sciences. In 1979, he moved to the University of Pittsburgh where he is a University Professor and a member of the Department of Decision, Operations, and Information Technology, Joseph M. Katz Graduate School of Business.

THE ANALYTIC HIERARCHY PROCESS (AHP)

Throughout his many activities, especially his work at ACDA, Tom found himself concerned with developing procedures and methods for analyzing and resolving a central and ubiquitous decision problem in OR: the ranking or prioritizing alternative choices that are compared and evaluated against stated criteria. His efforts here led to the development of the AHP (Saaty 1977a, 1980, 1986, 2001a). Although the AHP has a unique and valid mathematical basis, it is simple and elegant. Its application departs from the traditional decision analysis school where preferences are expressed using utilities whose values depend on one's aversion or proneness to risk. The AHP is based on three principles of human behavior: (1) decomposition of a complex problem to understand it, (2) comparison of its parts to determine the degree or intensity of their interaction and influence on the whole, and (3) synthesis to assemble the understanding and knowledge gathered when studying the parts and their interactions.

These three principles are implemented by:

Structuring decision problems in a hierarchy from a goal at the top to criteria (and often sub-criteria of several levels) to alternatives of choice at the bottom;

Eliciting judgments from decision makers about their preference, likelihood, or importance on pairs of elements in the structure using the fundamental scale of the AHP, which is an absolute scale;

Deriving a scale of measurement (or priorities) from the judgments. These scales can be derived for both tangibles and intangibles by using human judgment and values for the particular decision being made. The mathematics to derive the scales involves reciprocal matrices and eigenvector theory that is mathematically interesting and original in its own right;

Synthesizing the derived scales using the hierarchic structure to give an overall best choice. It is usually simpler to make a judgment on one pair at a time than to cope with the entire complexity all at once. The overall result is then a natural consequence of combining the results of many simple basic comparisons. While not requiring it, the method provides a way of checking and improving the consistency of the decision maker's comparisons (total consistency is rare in human affairs).

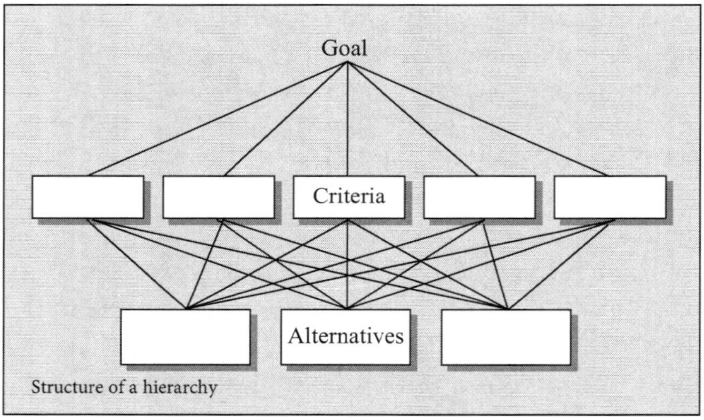

Structure of a hierarchy

An early and successful AHP project was the Sudan Transport Study (Saaty 1977b). This work was the result of over a 2-year collaboration of 40 scientists, engineers, and economists and produced a comprehensive transport plan for the Sudan. The study was sponsored by the Kuwait Fund for Arab economic development. Here, rival parties agreed on the development of the Sudan to transform it into the breadbasket of the Middle East (but due to a coup d'état in 1985, the entire plan collapsed). This study was deemed one of the best of the year (1977) by the annual prize competition committee of the College on the Practice of Management Science, The Institute of Management Sciences (Bleuel 1977).

During the mid-1970s, the implementation of the AHP was difficult as the computation of the required principal right eigenvector of a matrix is complex and best done on a computer, and in those days, only mainframe computers were available. With the introduction of the PC, the use of the AHP spread worldwide. In 1983, Tom co-founded the company Expert Choice to market the AHP software.

GRANDMA AND HOW IT BEGAN

"In the early 1970s the Egyptian government asked Tom Saaty, a pioneering mathematician with a fistful of awards, to help clarify the Middle East conflict. The Egyptians needed a coherent, analytical way to assess the pros and cons of their less than cozy strategic relationship with the Soviet Union. Saaty, a Wharton professor with a background in arms-control research, tackled the question with game theory, a mathematical process used to analyze complex problems. 'The Egyptians drew the conclusion that they should ask the Russians to leave Egypt,' says Saaty, now a professor at the University of Pittsburgh's Katz School. The Egyptian government was pleased with his work (and eventually did ask the Russians to leave), but Saaty himself wasn't satisfied with the process. He felt his conclusion was incomplete—that important but intangible information was left out of the final equation because game theory was too rigid. 'I couldn't use it to solve a real-life problem,' he says.

"So the academic started brainstorming in his sweltering Cairo hotel room. He flashed on youthful memories of growing up in Mosul, Iraq, where he lived until he was 14. 'My grandmother was a very handsome, very pesky woman,' the 72-year-old professor recalls. And she had her own method of figuring out thorny problems, says Saaty. 'You'd say, "Grandma, you don't have a Ph.D. in mathematics." But she was well informed and would do better than any scientist.' She identified sets of options in pairs, then weighed options against each other. Nothing revolutionary here. But Grandma went a step further. 'She said, "I like that better than that—a lot better," 'squeaks Saaty in a granny-squeaky falsetto, his thinning white hair flopping about. It dawned on Saaty that what he needed—and what Grandma had come up with in a rudimentary way—was a process that captured intensities of human feeling and assigned them numerical values" (Palmer 1999, 153–154).

The creation of the AHP has been surrounded by controversy and criticism by members of the more traditional decision analysis school who grew up under the umbrella of a set of axioms that were mathematically convenient but unrealistic when dealing with human behavior (Luce and Raiffa 1957). The AHP is a departure from traditional schemes, as paired comparisons are central to the method's determination of priorities. Tom and others have fought many academic battles over the foundations and use of the AHP vis-à-vis traditional decision analysis methods. Forman and Gass (2001) provide a background of the controversy and offer a wide range of real-world decision situations in which the AHP has been applied.

There have been literally thousands of successful applications of the AHP, many of which have been reported in both academic and general publications. Of particular importance is the study conducted by IBM Rochester's AS/400 Division that received the prestigious Malcolm Baldrige National Quality (1990) award from the U.S. Department of

Commerce for using the AHP to "articulate what it means to be the best" in its pursuit of being the best-of-breed in computer-integrated manufacturing (Eyrich 1991, 40).

Starting in 1988, researchers and practitioners of the AHP have presented and shared their ideas in a series of symposia held around the world. To date, there have been ten such symposia with the first in Tianjing, China, and the tenth in Pittsburgh.

THE ANALYTIC NETWORK PROCESS

Although the theoretical developments of the Analytic Network Process (ANP) were already set by the time the first AHP book was in print, it has taken almost 20 years before the necessary computational capabilities were in place for this theory to be usable to the average decision maker. The ANP is a natural extension of the AHP for a wide class of problems in which the hierarchical levels feedback and influence other levels and themselves. That is, not only does the importance of the criteria determine the importance of the alternatives as in a hierarchy, but also the importance of the alternatives determines the importance of the criteria (Saaty 2001b, 2001c). The ANP yields the same result as the AHP when the lower levels of a hierarchy are independent of the upper levels. An example of an ANP structure is Leontief's input–output model of an economy that shows how industries utilize and feedback their resources to one another.

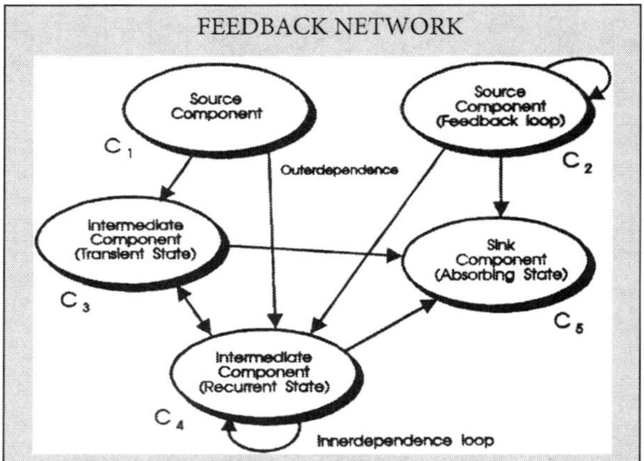

(Saaty 2001a, 85)

HUMOR AND CREATIVITY

Tom loves music and humor. He possesses CDs and records of all of Beethoven's works and, at one point in his life, could identify any piece of Beethoven's music. In a different vein, when he was younger, it was clear to him he was not good at telling jokes. As a way of his studying how to be a raconteur of jokes, he has, over the years, compiled hundreds of jokes into many booklets under a wide variety of pseudonyms. He shares and recounts this collection with colleagues and friends, and with his classes.

> FROM THE *THINKING MAN'S JOKE BOOK*
>
> "Three sons bade farewell to their mother and went out west to go into ranching. They found a nice spread and formed a partnership with each son specializing in a particular animal. One son raised cattle, the second son raised sheep, while the third raised pigs. They became very prosperous and had a steady and growing market for their products.
>
> "Their ranch also grew in size and it dawned on them that they had not given their ranch a name. After discarding a number of possibilities, they wrote to their mother asking her to suggest an appropriate one. She immediately wrote back that they should name the ranch FOCUS. What a strange name, thought the sons. As they could not figure out why this name, they wrote again asking their mother, 'Why FOCUS?'
>
> "She replied, 'Why not FOCUS? That's where the sons raise meat'. (Act and Steria 1967, 28).

Creativity in just about all fields, especially mathematics, has been a continuing area of Tom's research. The first instance of this appeared in his OR text as its last chapter: "Some Thoughts on Creativity" (Saaty 1959, 281–399). At the Katz Graduate School, he teaches a business course that deals with all aspects of creativity, not just decision making. Here, as with all his endeavors, he assembles, studies, and integrates previous work by others with his own contributions and presents the material in a masterful fashion. His book, *Creative Thinking, Problem Solving and Decision Making* (Saaty 2001d), ties together his main pursuits and interests in this area.

THE SAATY FAMILY

Tom has five children: Linda, Michael and Emily with his first wife, Bernadine, and John and Daniel with his present wife, Rozann.

Linda Mike Emily and Pat Harker Tom Rozann John Daniel

Courtesy of the Saaty family (2004)

Linda is an advertising consultant and lives in Virginia. Michael has a degree in electrical engineering from the University of Pittsburgh and lives in New Mexico. Emily and Pat Harker: Emily is a graduate of the University of Pennsylvania and teaches high school mathematics; Pat has a Ph.D. in industrial engineering from the University of Pennsylvania—he was the former dean of the Wharton School and is president of the University of Delaware. John is a graduate of Northwestern University and the Wharton School; he is CEO of Decision Lens, Inc., a company that supports group decision making. Daniel is a graduate of the University of Pittsburgh and is vice-president of Decision Lens, Inc. John and Daniel both live in Virginia. Tom and Rozann have 10 grandchildren.

Honors and Awards

Tom was elected a fellow of the American Association for the Advancement of Science (AAAS) in 1959 and, in 1970, to the Real Academia de Ciencias Exactas, Fisicas y Naturales (Royal Academy of Mathematical, Physical and Natural Sciences) of Spain. In 1982, he was named Alumnus of the Year by Columbia Union College. The Thomas L. Saaty Prize was established in 1981 by the *American Journal of Mathematical and Management Science* to recognize the best research paper in applied advances in the mathematical and management sciences. He was inducted into the Beta Gamma Sigma Honor Society in 1984.

Tom was elected a member of the International Academy of Management in 1998, and a member of the National Academy of Engineering in 2005. He received the Gold Medal of The International Society on Multiple Criteria Decision Making in 2000. In 2007, he was awarded the International Quality Function Deployment Akao Prize from Japan, given to individuals who have demonstrated excellence in their practice and dissemination of an integrated approach to quality systems thinking over many years. And, in 2008, Tom was awarded the INFORMS Impact Prize for his development and application of the AHP and for its broad impact on the field.

SUMMING UP

Tom Saaty Luis Vargas

Based on his broad mathematical talents, Tom has devoted his professional life to the advancement of knowledge across many fronts with great productivity. He has contributed a few bookshelves of books and papers that collectively address a wide array of OR and mathematical topics, many of which were seminal and helped to advance the maturing OR field. His major developments, the AHP and ANP, can be used by decision makers to resolve the universal and vexing decision problem, the setting of priorities and the ranking of alternative choices.

I first met Tom in the fall of 1975 when I went to Wharton to do research under his supervision. I was recommended to him by Professor Sixto Rios, whom Tom had met in Madrid in 1959. After more than 30 years of knowing Tom Saaty as my mentor, colleague, and friend, I can say that there never has been a dull moment.

REFERENCES

Act M, Steria H (1967) The thinking man's joke book. G & S Publishers, Washington, DC
Bleuel W (1997) The practice of management science. Interfaces 8(1):2–5
Busacker R, Saaty TL (1965) Finite graphs and networks. McGraw-Hill, New York, NY
Dantzig G, Saaty TL (1973) Compact city. W. H. Freeman, New York, NY
Eyrich H (1991) Benchmarking to become the best of breed. Manufact Sys 9(4):40–47
Forman E, Gass S (2001) The analytic hierarchy process: an exposition. Oper Res 49(4):469–486
Gass S, Saaty TL (1955a) The computational algorithm for the parametric objective function. Nav Res Logistics Q 2(1–2):39–45
Gass S, Saaty TL (1955b) Parametric objective function (part II)—generalization. Oper Res 3(4):395–401
Joseph I (1919) News from Mesopotamia. The New York Times, February 9
Luce R, Raiffa H (1957) Games and decisions: introduction and critical survey. Wiley, New York, NY
Palmer B (1999) Staying smart: managing companies and careers in the new economy. FORTUNE May 10, 153–156
Saaty TL (1957) Résumé of useful formulas in queuing theory. Oper Res 5(2):162–200
Saaty TL (1959) The mathematical methods of operations research. McGraw-Hill, New York, NY
Saaty TL (1961a) Elements of queueing theory with applications. McGraw-Hill, New York, NY
Saaty TL (1961b) Some stochastic processes with absorbing barriers. J R Stat Soc 23(2):319–334
Saaty TL (ed) (1963) Lectures on modern mathematics, vol I. Wiley, New York, NY
Saaty TL (ed) (1964) Lectures on modern mathematics, vol II. Wiley, New York, NY
Saaty TL (ed) (1965) Lectures on modern mathematics, vol III. Wiley, New York, NY
Saaty TL (1967) Modern nonlinear equations. McGraw-Hill, New York, NY. Reprinted by Dover Publishers, 1982
Saaty TL (1968) Mathematical models of arms control and disarmament. Wiley, New York, NY
Saaty TL (1970) Optimization in integers and related extremal problems. McGraw-Hill, New York, NY
Saaty TL (1972) Thirteen colorful variations on Guthrie's four color conjecture. Am Math Mon 79(1):2–43
Saaty TL (1973) Topics in behavioral mathematics. Math Assoc Am, Washington, DC
Saaty TL (1977a) A scaling method for priorities in hierarchical structures. J Math Psychol 15(3):234–281
Saaty TL (1977b) The Sudan transport study. Interfaces 8(1):37–57
Saaty TL (1980) The analytic hierarchy process. McGraw-Hill, New York, NY

Saaty TL (1986) Axiomatic foundation of the analytic hierarchy process. Manage Sci 32(7):841–855

Saaty TL (2001a) The analytic hierarchy process. In: Gass SI, Harris CM (eds) Encyclopedia of operations research and management science. Kluwer, Boston, MA, pp 19–28

Saaty TL (2001b) The analytic network process. In: Gass SI, Harris CM (eds) Encyclopedia of operations research and management science. Kluwer, Boston, MA, pp 28–35

Saaty TL (2001c) The analytic network process. RWS Publications, Pittsburgh, PA

Saaty TL (2001d) Creative thinking, problem solving and decision making. RWS Publications, Pittsburgh, PA

Saaty TL (2008) Personal communication

Saaty TL, Alexander J (1981) Thinking with models. Pergammon Press, Oxford

Saaty TL, Boone LW (1990) Embracing the future. Praeger, New York

Saaty TL, Bram J (1964) Nonlinear mathematics, McGraw-Hill, New York, NY. Reprinted by Dover, 1981

Saaty TL, Gass SI (1954) Parametric objective function (part I). Oper Res 2(3):316–319

Saaty TL, Kainen P (1977) The four-color problem; assaults and conquest. McGraw-Hill, New York, NY

Saaty TL, Weyl F (eds) (1969) The spirit and uses of the mathematical sciences. McGraw-Hill, New York, NY

32

Anthony Stafford Beer

Jonathan Rosenhead

Stafford Beer was one of the most remarkable figures that operational research (OR) in any country has produced. A charismatic, even flamboyant character, he founded two major and pioneering OR groups in the U.K., wrote some of the best books about the subject, and was a world leader in the development of systems thinking. He is widely acknowledged as the founder of the field of management cybernetics. This international stature led to his engagement as advisor to several national presidents. His charm, conviction, and force of character opened up opportunities in industry and business, and later in government, which would have been denied to a more conventional figure.

Stafford was a leading practitioner of OR who also made significant intellectual contributions to the discipline. As a writer of elegance and power, he produced some of the most compelling texts on the nature of OR practice. His 1966 book, *Decision and Control*, won the Operations Research Society of America's (ORSA) Lanchester Prize for the best contribution to OR published in that year. He served as president (1970–1971) of the Operational Research Society (ORS), the Society for General Systems Research, and the World Organization for Systems and Cybernetics. As a systems guru, Stafford was concerned with designing appropriate feedback loops into social systems. His later work was on participative methods designed to enable large groups to arrive at solutions to their own problems. What united these aspects of his work was an early and consistent commitment to an holistic approach to complexity.

A.A. Assad, S.I. Gass (eds.), *Profiles in Operations Research*, International Series in Operations Research & Management Science 147, DOI 10.1007/978-1-4419-6281-2_32,

FROM LONDON TO INDIA AND INTO OR

Anthony Stafford Beer was born on September 25, 1926 at Putney, South West London, the son of William John Beer, chief statistician at Lloyd's Register of Shipping, and Doris Ethel Beer. His younger brother, Ian, went on to become headmaster of one of Britain's most celebrated public (private) schools, Harrow.

Stafford went to Whitgift, a public school in Croydon, South London. He was 15 years old when he decided he wanted to become a philosopher, but this would have spanned across the various educational streams from which he was supposed to choose just one. At 16, frustrated by this straitjacket, he got himself expelled—in effect by rejecting each of the options for study which the school offered. He went to University College, London (which, during World War II, had been evacuated to the Welsh coastal town of Aberystwyth) to study philosophy and psychology for a year before he was old enough to enlist in the Royal Artillery in 1944. He never went back to full-time education.

Stafford was commissioned in the Royal Fusiliers, then transferred to the 9th Gurkha Rifles in India. (The Gurkhas were British Army regiments which recruited regular soldiers from Nepal and North India, but with British officers.) After the war he stayed in the Army, serving, at first, in the Punjab as Staff Captain, Intelligence. It was a complex situation in the approach to India's independence and the partition of the sub-continent into India and Pakistan. There were Indian Army troops, British troops, state troops, with many maharajahs having their own armies. To keep track of their relationship, force structure, and location, Stafford reverted to his university studies of philosophy and developed a symbolic logic model of the situation. Later, when he returned to England and described his model to various people, he was told, "that's interesting, what you've been doing, that's operational research" (Capey 1996, 1). It was the first time Stafford had heard of OR.

FAMILY HISTORY—A GENERATIONAL GAP

Interviewed in 1996 (Capey 1996, 32), Stafford said that there was a consensus among his friends and colleagues that he should write some kind of autobiography. In fact he did write a tantalizing autobiographical fragment produced in just a few copies (Beer 2001). Tantalizing, because it ends, more or less, with his birth, and, because it is constructed quite deliberately as a patchwork quilt. We learn quite a lot about the stages of Stafford's inferred development as a fetus.

From this fragment we can, however, learn a great deal about Stafford's family background. His father and mother, known as Will (or Bill) and Doll, were born on the same day in the same year. As children they came to live on opposite sides of the same street in the Fulham district of London. It was a respectable neighborhood and social life centered on the church.

Will's father, Henry, was a character, a good deal older than his wife who was a seamstress. Henry was a painter, whose uncertain living was based on restoring old paintings rather than painting new ones. He had bohemian habits, tended to dress extravagantly (frock coat and top hat) and chewed cheroots the whole time. He was also a fitness fanatic. Perhaps that explains how on one occasion it took 13 policemen to escort him home. He also had a disconcerting tendency to accidents in church. Carrying a candle on Palm Sunday, he managed to set fire to a ceremonial palm. At midnight mass on Christmas Eve, his carriage of the ceremonial cross was so wayward that it tipped up the ornamental votive lamp.

Will enlisted in the Navy during the First World War (by lying about his age) and served as a wireless operator. After the war, he joined Lloyd's Register of Shipping (the recorder of ships and their seaworthiness) starting in the drawing office. Six years passed before Will and Doll married, and, even so, started married life in her parent's house. Stafford was born 2 years later, in the downstairs room. When Stafford was two, the new family moved to more commodious accommodation on what was then the southern fringe of London's built-up area.

Will stayed for the rest of his working life at Lloyd's Register, rising to a senior position. He seems to have reacted to his extravert and extravagant father by adopting a rather severe attitude to life, so that this rather formal institution with its strong sense of probity and respectability suited him well.

It is hard not to think that some of the buccaneering spirit of Grandfather Beer skipped a generation and resurfaced in Stafford.

Stafford returned to Britain from India in early 1947. He didn't fancy "marching troops up and down" in peacetime and managed to get appointed as an Army psychologist (Capey 1996, 1). He was helped in this by his former psychology professor, who must have drawn a veil over his lack of a degree. Stafford persuaded the Army to set up an education center where, as commanding officer, he undertook research into psychopathology and illiteracy. He instituted crash courses, which markedly raised the educational age of the disturbed soldiers passing through. Again he was told, "I was doing operational research [human factors]" (Capey 1996, 2). It should be said that the limits on the scope of what was understood by OR then were a good deal wider than they are today.

In 1947, and while still in the army, he married Cynthia Hannaway with whom he would have five children. It was in discussion with Cynthia that he decided that he should leave the Army and find a civilian job. In 1949, he was decommissioned from the Army with the rank of captain and joined the largest U.K. steel firm of its day, United Steel, as a management apprentice.

STEEL AND CYBERNETICS

United Steel had resulted from an amalgamation in the inter-war years of a number of large companies across the north of England, each of which retained its original name and a degree of separate identity. Stafford, now living in Sheffield, soon made an impression on one of the directors of Samuel Fox and Company, a plant situated 10 miles outside Sheffield making alloys and stainless steels. This was the first of a number of times that senior managers in United Steel were impressed by Stafford's creativity, confidence, and outsize personality, and who took him under their wings. Stafford was allowed to invent his own job. He proposed that he work on production control, which, as a concept, did not exist at Fox, and set about developing a suitable system. He clearly saw this as another operational research enterprise; he joined the OR Club (later the OR Society) that had been founded in 1948, and attended virtually all its meetings in London.

Eventually, Stafford was able to set up a small OR group at Fox with himself as head, and was then given the responsibility of implementing the production control system he had designed. Central to the system was "a simple apparatus and a method of standardising control charts, with the aid of which a large number of charts can readily be kept under surveillance" (Beer 1953, 169). His group's expanded role was achieved over the strong objection of Fox's General Works Manager. The resulting row brought him to the favorable attention of Fox's chief executive, and, before too long, word of his group's work reached the Managing Director of the United Steel group. This resulted in Stafford being given the job of organizing an OR group for the whole of United Steel. It was 1955, and he was 29.

Stafford was always a lateral thinker and a wide reader. It was in 1950 that he read and was impressed by Norbert Wiener's *Cybernetics* that dealt with "the entire field of control and communication, whether in the machine or in the animal" (Weiner 1948, 11). He wrote to Wiener to say "I think I am a cybernetician" and enclosed some of his recent work: Wiener responded enthusiastically and invited him at once to MIT (Capey 1996, 7). Stafford's 1959 book, *Cybernetics and Management*, would be written in his spare time over the next several years despite the heavy responsibilities of his day job. The book was eventually translated into 13 languages. As its title suggests, it took the still nascent subject of cybernetics and explored its relevance for management processes and structures. [It was Wiener, the originator of cybernetics, who would later identify Stafford as the father of management cybernetics (Rose 1993).]

More immediately, this interest was relevant to the naming of the new department and its home. Stafford was given permission to purchase a large house near the United Steel head office in Sheffield, which he called Cybor House—representing the name he gave the group: Department of Operational Research and Cybernetics. Cybor House became a name to conjure with in British OR.

To give the new group a presence around the company (whose component firms were as much as 200 miles apart), Stafford bought half a dozen caravans which could be left on site. They had Department of Operational Research and Cybernetics emblazoned on the side. The caravans were equipped as offices to give the group their own base office while working on location (staff stayed in hotels if away from home). But they also raised the profile, and became highly visible talking points.

Stafford was not a mathematician, but he compensated by attracting numerically gifted people through a series of innovative and intriguing advertisements. One of those who joined in this way was Keith D. Tocher, already an established mathematician at Imperial College London. Stafford was in luck. In 1956, a new head of the Mathematics Department there closed down Tocher's project of building the Imperial College Computing Engine, so Tocher was in the mood to move (Lehman 1993). Indeed the advertisements attracted a wide range of new staff, from philosophers and a doctor of divinity through anthropologists and biologists to experienced industrialists. Both in his writings and in his practice, Stafford was one of the earliest advocates and post-war enactors of inter-disciplinary teams as the way to tackle complex organizational problems.

Keith D. Tocher

Although United Steel was a progressive company by the standards of the British steel industry in the 1950s, this is a very relative accolade. The OR group and its expansive head were widely viewed in the company as verging on the eccentric. Stafford attracted a highly diverse group of individualists many of whom stood out in the rather conventional world of a northern steel town and company for their attire, background, and perspectives. In these circumstances, the success of the group can be seen as remarkable and a tribute to Stafford's vision and drive.

Stafford himself reinforced the unconventional image of the group by some cybernetically inspired projects. In collaboration with the noted cybernetician Gordon Pask, he experimented with colonies of various small insects and with pond-life, regarding them as self-organizing systems that would not require their circuitry to be designed in detail to form a fabric of communication. Light-sensitive microorganisms were trained to solve equations. [This strand of his work is described in Beer (1962).] There was a tank of green algae in his office, which he was reported to feed with iron filings. More directly on company business, he conducted experiments at one of the company's rolling mills to move toward what he called a cybernetic factory—installing automatic control mechanisms and largely or completely dispensing with the need for workers (Beer 1962). This attempt to develop a different kind of model of how a factory might operate was not greeted with uniform enthusiasm even inside Cybor House.

A quite different approach to the modeling of production started with a map of the hot end flow of the steel works and the game of tiddlywinks. In a unique application of simulation, Stafford stayed up all night playing the game using the randomness of the destination points of the winks to explore the possible flows of hot ingots.

Stafford's next development was the invention of the stochastic analogue machine (SAM), an elaborate 6-foot high construction. In this, tiddlywinks were replaced as the source of variance by randomly released ball bearings. Empirically-observed delay or process statistical distributions could be specified for as many as 10 networked activities, and the in-process stocks, which resulted were visible as accumulations of ball bearings. The machine is described in Stafford's paper given at the First International Conference on Operational Research (Beer 1957, 166–175). In effect, this was a special purpose analogue computer. A particular advantage of SAM over the digital computers just becoming available was that it allowed hands-on interactive use. Managers could watch the way in which the machine replicated steel works processes, enabling them and the operational researchers to "get a feeling" for the task of controlling this complexity (Andrew 1993, 62).

From SAM, the next step was from analogue to digital. In 1956, Cybor House took delivery of a Ferranti Pegasus computer, perhaps the first computer in Britain installed for the sole use of management science (MS). It occupied most of the ground floor of one of the two buildings that now made up Cybor House. Many applications ran on it, including interactive gaming with steel industry managers as players. But most significantly, it became the test-bed for the development of digital simulation. Here, Tocher took over and made the subject his own. Rather than programming the Pegasus in its machine code, he developed a special generalized simulation language (GSL) from which the widely used control and simulation language (CSL) was derived. And he wrote the classic text on the subject, *The Art of Simulation* (Tocher 1963). Although Tocher is commonly credited as a (if not the) father of the simulation field, it was Stafford who opened the way.

Over time, the managerial regime at United Steel became less sympathetic to OR, and, in the end, the ingrained conservatism of the firm set limits on the problems Stafford and his group would be allowed to tackle. Stafford resigned from United Steel in 1961, leaving a Cybor House OR staff of 70 professionals. With the publicly owned National Coal Board's Field Investigation Group, it was one of the two largest U.K. OR groups of the period, and it helped to train many of the upcoming generation of British operational researchers. [Kirby (2000) describes United Steel's OR experiences in the context of that of other U.K. steel industry OR groups of the same period.]

STAFFORD AND ME

Here is a personal story of Stafford's high-energy impact as a recruiter for OR. The year was 1958, and he had been invited to speak to the Cambridge University student mathematical society, known as the Archimedians. His talk was a stunning tour de force, the first time that I or (I think) any of the student audience had heard of OR. But that was only the beginning. We had booked an overnight visitor's room for him in St. Johns, which was also my college. After his talk, he and several of the audience came back to the rooms I shared with David, a fellow student. Stafford talked and talked about the work his group was doing with boundless enthusiasm and good humor. One by one the visitors left until only David and I remained. And still Stafford held forth about the future scope for OR, chain smoking cigars, and with frequent recourse to a hip flask. When he left to go to his room he still seemed fresh—but we were exhausted by his sheer exuberance. Speaking to David (now a successful property developer) 50 years later, he still remembered the encounter vividly, using such phrases as larger-than-life, real salesman, bon viveur, articulate, convivial.

When I finished my studies, I opted for OR. And that was why I joined Cybor House in 1961, only to find that Stafford had left 2 days previously. Two years later, I went to work for SIGMA, his new consultancy firm.

(Jonathan Rosenhead)

OR Consultantcy I: U.K.

On leaving United Steel, Stafford set up Science in General Management, or SIGMA, a specialized OR consultancy. It was in effect, the first such consultancy in the U.K. (predecessors having been in effect just sole practitioners). Stafford was Managing Director, but ownership lay with a larger group of companies under French control. A notable appointment as a Director was Roger Eddison, who, like Stafford, became President of the ORS. They formed a strong, complementary, and very friendly partnership—Stafford's inclinations did not extend to some of the detailed management issues necessary for survival. [At least one member of the Cybor House team has suggested that some of the obstructions to his proposed initiatives at United Steel could be put down to the senior management's determination "to keep him away from the money" (Ludley 2008).]

Once again, an OR group with highly varied backgrounds was recruited, and, by 1966, SIGMA's professional staff had grown to 50. Stafford was brilliant at marketing projects to clients, with staff recruitment based on work for public utilities, public agencies, and private firms, with a smattering of work for government. A strong feature was repeat business, indicating that somehow the project teams were managing to deliver on the imaginative promises that Stafford (and other senior salesmen) had made. One engagement, which was to have major consequences was with Chile. A permanent, ten-member office was set up there, first working with the publicly owned CAP steel company and then for the Chilean railways.

The portfolio of SIGMA's consultancy engagements was necessarily eclectic, and perhaps offered less opportunities for the deployment of what one participant has described as "Stafford's special insights" (Andrew 1993, 63). Nevertheless, Stafford always managed to convey a sense of excitement. Attention was always paid to the possibility of innovative, off-beat solutions. SIGMA itself was the canvas for a range of Stafford's organizational innovations. When the firm expanded into new premises, he took the opportunity to design the internal layout. A central feature was the library, which Stafford preferred to call the Interaction Space. There was no question of consultants working through the coffee break. Whenever he was there, Stafford would open every office door and shoo the workers to the area regardless of how tight their deadlines.

> LIFE AT SIGMA
>
> Stafford's immediate SIGMA staff of secretary, personnel assistant, and project assistant was good-humoredly called his harem. All still remember their time working for Stafford with affection. The only drawback was the smell of tobacco that clung to them as a result. One day, his project assistant, upon opening his office door, confronted cigar smoke so thick, she wondered if Stafford was there!
>
> Another SIGMA institution, Sigmoot, was held every month or two. This was an informal meeting after work with drinks and nibbles for all staff from cleaners to the managing director. Anybody with a point to raise could bang their shoe on the table top and make it. Sigmoot interaction ranged from light chatter through heavy-weight discussion, but it was the latter that would predominate as the evening lengthened. In keeping with the general ethos was the provision of sabbatical entitlements, which staff could accumulate; sabbaticals could be taken for any purpose—so long as it had nothing to do with serious work (Capey 1993, 20). It can be gathered that an extremely strong sense of camaraderie developed—and staff reunions are still taking place more than 40 years after the group, in effect, ceased to exist.

Stafford's progression through life was always something of a high-wire act. This was true at SIGMA as at United Steel. It emerged almost accidentally that the market value of the firm was far higher than anyone had realized. Having grown the firm for 5 years, Stafford thought he should have some small stake in the equity. The French owners thought otherwise, though they did offer to double his salary, which Stafford thought insulting. He discussed this situation with the Managing Director of the International Publishing Corporation (IPC), then the largest publisher in the world, for whom SIGMA was engaged in a major project. The upshot was that Stafford moved to IPC in 1966 as Development Director.

Though SIGMA continued in various amalgamated versions for many years afterwards, in effect its significant period ended with Stafford's departure (Andrew 1993, 64). Stafford moved on to new challenges and opportunities.

As IPC's Development Director, Stafford tried to prepare the firm, comprising several book publishers, as well as mass circulation newspapers and a major magazine portfolio, for the technological challenges and opportunities that were just coming over the horizon. He sponsored research into innovative technologies, many of them information technology-based, and bought companies which could help to provide them. He set up the New Enterprises Division whose component units included a company called International Data Highways, a phrase he coined 30 years before the term Information Highway came into vogue. With this work, he positioned IPC to be one of the leaders on the path toward electronic publishing (Andrew 1993, 65).

Retrospectively, it seems clear that Stafford and a mammoth corporation such as IPC would not be able to form a proper working relationship. There was a boardroom disagreement about development policy, and, during 1970, he gradually phased himself out of IPC and set up as an independent consultant. This became not a retreat, but an advance—Stafford entered into perhaps the most productive period of his life, both intellectually and in practical terms.

AUTHOR: OR, CYBERNETICS, AND BEYOND

In 1966, Stafford's book, *Decision and Control* (Beer 1966), was published—it is his inspirational tour of the terrain of OR and cybernetics written in clear, stylish prose, and accompanied by his own elegant illustrations. *Decision and Control* had been dictated onto tapes in the small hours of the morning, and typed by his SIGMA secretary. It won ORSA's Lanchester Prize for the best book of its year. The following year, his book, *Management Science* (Beer 1967), came out, a slim but lucid and seductive introduction to the area for non-specialists. Once Stafford was his own master, the flow of major books became continuous. These included (but were not limited to) *Brain of the Firm* (1972), *Platform for Change* (1975), and *The Heart of Enterprise* (1979). His books were translated into 13 languages, giving his ideas a world-wide reach. In all, he wrote ten books, all but one single-authored.

From well before his writing *Cybernetics and Management* (Beer 1959), Stafford was wrestling with the issues arising in the management and control of complex organizations, indeed of complexity itself. Further, his progressive orientation led him to concern himself with the autonomy of those whom others might see simply as subjected to control. The culmination of this work, most fully captured in *Brain of the Firm* and in *The Heart of Enterprise*, and summarized in a journal article Beer (1984), is the Viable System Model (VSM).

The VSM is an intellectual structuring of what it takes for a system to be viable, that is, capable of independent existence. The model, which he developed could have been described in various ways, but in *Brain of the Firm* he chose a neurophysiological terminology (rather than the set-theoretic version he had used previously). Stafford insisted, however, that this choice of terminology should not to be taken as an analogy.

OR Consultancy II: The World

In parallel with his post-1970 theoretical strand was Stafford's consultancy practice. In all, after going independent, he carried out work in 23 countries. These included commissions for the presidential offices of Mexico, Uruguay, Venezuela, and Chile, answering directly to the President in the latter three cases.

Most notable of his projects was his assistance to the government of President Salvador Allende in Chile, a country where SIGMA had done extensive work a few years earlier. More significant, however, in the genesis of this now legendary work, was that his writings on a systems view of organizations had been read avidly by a group of young industrial engineers and mathematicians in Chile who saw in his writings the hope for fairness and justice in their society. In July 1971, Stafford received a letter inviting him to help them in their work.

VIABLE SYSTEM MODEL (VSM)

The VSM approach is based on findings of earlier cyberneticians that for complexity to be managed, two key strategies are the attenuation of environmental variety (for example, by some type of filtering) and the amplification of the variety of managerial control to match it. What is needed is requisite variety (Ashby 1956, 202–218), implying an appropriate balance between central control and peripheral autonomy.

The VSM describes five different functions that need to occur if a system is to be viable, termed Systems One through Five. System One consists of an organization's primary activities; indeed it will normally have a number of such System Ones, each interacting with its own environment. System Two performs the coordination function, avoiding clashes or instabilities that might arise. System Three performs the control function, conveying organizational policy to Systems One and Two and ensuring that implementation takes place. System Four is concerned with gathering relevant information on outside developments, and, to identify what is relevant, it needs to have an awareness of the organization as a whole. System Five establishes policy, the organization's ethos and its identity: it will need to arbitrate in the debates and conversations between Systems Three and Four, it will represent the organization to the outside world, and it will need to recognize and take action in extreme situations. For System Five to perform its functions, it needs to receive suitably filtered signals from the rest of the organization. Beer (1984) provides an account of the development of VSM. A summary of VSM is given in Mingers and Rosenhead (2001, 267–276).

These Systems need to exist and to work well in any viable organization. Furthermore, any viable system will contain other viable systems, and itself be contained within such a system, making the model indefinitely recursive.

The VSM model can be used in either of two ways—to diagnose weaknesses and problems in an existing organization, and to design more effective structures. Examples of these uses are discussed in Espejo and Harnden (1989). VSM has developed an enthusiastic following, both applying the approach and developing it, but this influence is to be found more in the systems community than within OR.

Allende had become President of Chile in 1970, the first democratically elected Marxist head of state anywhere. He instituted a massive nationalization of key sectors of the means of production, distribution, and exchange. These were vested in CORFO (Corporación de Fomento de la Producción), and it was from CORFO that Stafford's invitation had come. On his first visit to Chile, he took with him the draft manuscripts of *Brain of the Firm* (Beer 1972), and all members of the CORFO team read it during his 8-day stay. By the end of that time, they had developed an outline plan for the cybernetic regulation of the social economy of Chile.

For over 2 years, until Allende was overthrown in 1973 in the bloody Pinochet coup, Stafford and a substantial project team worked on developing these ideas into an implementable control system. As workers occupied privately owned factories, in many cases abandoned by their proprietors, the sector itself was growing by leaps and bounds. Project Cybersyn was to be a real-time computerized information system—a breathtakingly ambitious project given the technology of the time. Its scope went from the local workshop to the national economy in 11 recursions of Stafford's Viable System Model. The elements that had to be developed at a breathless pace included the Cyberstride Program Suite to monitor and filter information flows at each level of recursion, the Checo programs to model the Chilean economy, and the Operations Room with multiple displays of information to be used for decision making at the highest level or by worker committees. (The Cyberstride Program Suite was in a direct line of descent from the ideas developed by Stafford for steelworks control; Project Cybersyn's control room inspired much of the subsequent development of the boardroom of the future.)

STAFFORD AND ALLENDE

"When I first expounded the cybernetic model of any viable system . . . to President Allende, I did so on a piece of paper lying between us on the table. I drew for him the entire apparatus of interlocking homeostats, in terms of the neurophysiological version of the model—since he is by profession a medical man. It consists of a five-tier hierarchy of systems. I worked through the first, second, third and fourth levels. When I got to the fifth, I drew an histrionic breath—all ready to say 'And this, compañero presidente, is you.' He forestalled me. 'Ah,' he said, with a broad smile, as I drew the topmost box, 'at last—the people'" (Beer 1975, 447).

In Chile during the 1970s, the still incomplete version of the system, making use of a network of about 500 telex machines, was effective in defeating the CIA-funded truckers strike of October to November 1972 (Treverton 1988, 141–143). On September 8, 1973, Allende ordered the control room to be moved into the La Moneda, the presidential palace. The military took over 3 days later, and Allende died that afternoon—shooting himself rather than be captured. It is fitting that in March, 2008, a replica of the control room was installed as an exhibit in a museum beneath La Moneda (Barrionuevo 2008). A full account of the Chilean experience is given in the additional five chapters of the second edition of *Brain of the Firm* (Beer 1981).

Stafford had always regarded himself as aligned with the progressive cause in politics. From at least the 1970s, he certainly called himself a socialist. At United Steel, he worked collaboratively with the trade unions, and cited the anti-union stance of the senior managers at United Steel as one of his major sources of friction while working there (Capey 1996, 11–12). The same perspective is evident in the Chilean work in the explicit statement that science and technology were to be at the service of the people rather than the other way around. This was to be accomplished by ensuring that the information reported to higher levels was limited to issues that required a more comprehensive view, while allowing people on the ground to include their own algedonic indices (measures of their satisfaction or dissatisfaction).

World View

In the 1990s, this perspective surfaces again in a different form in his book *Beyond Dispute: the invention of Team Syntegrity* (Beer 1994). Team Syntegrity is a participatory method for enlisting the creativity of substantial groups in order to develop innovative solutions to shared issues. It can thus be seen as complementary to the control perspective of VSM. Non-hierarchical and democratic, it is in tune with the growing pressure to involve all stakeholders in the decision-making process (Bunker and Alban 1997). It has been widely adopted, with a growing international network.

In the later decades of his life, Stafford Beer's increasing concentration on a systems perspective tended to take him away from the OR community—a re-orientation he shared with his old friend Russ Ackoff in the U.S. (There was, however, no acrimonious breach with the OR community, as there was in the case of Ackoff.) Another close friend for many years was Pat Rivett, one of the central figures of the U.K. OR community. When, without prior notice, Rivett published an article, which in effect was highly critical of Beer's approach, their relationship was undoubtedly affected (Rivett 1977).

Before the gradual distancing of relations occurred between Stafford and the OR community, he became president of the ORS (1970–1971). As one might have expected, his term as president of ORS was turbulent. He initiated a root-and-branch rethink of the Society's structure. One constituent proposal, for a Register of Practitioners to guarantee professional competence, provoked the most major controversy of any in the Society's history—though it fell to Tocher, his successor as president, to handle the resulting dissent. In the end, the proposal was overwhelmingly rejected by the membership.

The breadth of Stafford's interests is indicated by his many society presidencies, visiting chairs and honorary degrees—remarkable achievements for someone with no conventionally earned degree at all! This most singular distinction was removed by the awarding of a higher doctorate, a D.Sc. from the University of Sunderland in recognition of his published work—he was then 73 years old.

His impact on the way we think about management and systems was due both to his magnetic personality and to the power and eloquence of his writing. In his books and other writing, he always takes an expansive view of his subject matter. His approach was always challenging, subversive even, to conventional modes of decision making. Radically then, unfashionably now, he believed in the benefits of a scientific approach, though he railed against reductionism. Unlike other management writers, he saw science as freeing thought and action, not trapping it in narrow procedures and techniques. Indeed, it was a constant theme of his that the greatest possible autonomy of action should be maintained at all levels of the organization, not just at the top.

Stafford Beer was a larger-than-life character. He was tall, broad, and brimful of energy. In earlier life he dressed and lived the part of an entrepreneurial businessman, in well-cut suits and chain-smoking cigars. After the move to London, he drove a Rolls Royce, and for many years lived in Firkins, a rather grand house in a well-heeled suburb of London. In the sitting room, known by his family as The Posh Room, Stafford had designed a waterfall down one slate wall, controlled by an ultrasonic switch in his pocket. The light switch, set in a cork wall, was a piece of etched copper, which you stroked.

TRANSISTION

Much changed thereafter. In fact the process of internal change had started earlier. A convert to Catholicism in 1943 at the age of 17, he left the church in 1968. He and Cynthia divorced, and in 1969 he married Sallie Steadman, a widow with a young daughter. There were two children from this marriage, and Stafford always counted all eight as his—five sons and three daughters.

Following the Chilean coup, Stafford made a conscious decision to change his life. He sold Firkins and renounced almost all his material possessions in favor of his family. He was now separated from Sallie, and moved in 1974 to a remote and very basic two-room stone cottage in the Welsh hills, for many years without running water or a telephone. There, between consulting assignments and fulfilling the duties of his visiting professorships that helped to meet his family obligations, he lived in very simple fashion. Clean-shaven in his early career, and suavely bearded in his SIGMA days, he now grew a full beard, giving him the presence of an Old Testament prophet.

From 1984, he divided his time between the cottage and an alternative base in Toronto that became a center of interest in his work. He lived there with his new partner, Allenna Leonard, also a noted figure in the systems movement. She is a past-president of the American Society for Cybernetics and specializes in Stafford's work, especially his Viable System Model and Team Syntegrity Process. Stafford died in Toronto on August 23, 2002.

Stafford's sheer enthusiasm for life in its many aspects could be overpowering and quite non-Anglo-Saxon. There were those who armored themselves against the combined demands of his larger-than-life personality and the off-center intellectual territory that he explored. This became clear to me when I joined United Steel in 1961 just days after he had left. The internal debate about Stafford was still going on when I followed him to SIGMA 18 months later. There were critics (why do I think they mostly had an engineering background?) who doubted what of solidity had been achieved, while others felt that they had been taking part in an outstanding intellectual and practical voyage. What you could not find was anyone who was neutral.

This polarization tended to be repeated elsewhere, between those who were distrustful of what they saw as his showmanship, and those who were converted into lifelong admirers and supporters. This was of course the 1960s when hard mathematical OR was in the ascendancy. With the growth of soft OR and the systems movement, a more natural constituency for Stafford and his work developed both within and outside OR. He himself was deeply loyal and affectionate to his friends. It was typical that Stafford spent the year following the Pinochet coup concentrating on helping to find safe places outside Chile for members of the project team.

His range was extraordinary. His management writings spanned the fields of human cognition and communication. He painted and wrote poetry throughout his career. In 1977, a book of his poems, *Transit*, was published (Beer 1977); in 1992 and 1993, nine large paintings, a meditation on Requiem, were exhibited in an apse of the Roman Catholic Cathedral in Liverpool; and he illustrated many of his own books with informative diagrams of great elegance. Beyond that, he taught yoga; spoke French, Gurkhali, classical Greek, Hindi, and enough Spanish to translate Chilean poetry; and played several instruments including the mandolin. He is responsible for the design of the powerful and impressive medals that ORS awards for excellence. When, as President, I presented the ORS Silver Medal to George Dantzig in 1986, his reaction was that it was the finest medal he had ever been given. The ORS has set up the annual Stafford Beer Medal for the best publication in its journals on information systems and on knowledge management.

ORS SILVER MEDAL—DESIGNED BY STAFFORD BEER

AWARDS AND HONORS

Stafford Beer held honorary doctorates from: Concordia University, Montreal, Canada; University of St. Gallen, Switzerland; University of Valladolid, Spain; University of Glamorgan, Wales. He was an honorary fellow of the University of St. David's, Wales, and of Liverpool John Moores University, where he was named Honorary Professor of Organizational Transformation, and where his collection of books and papers resides; and an honorary fellow of the International Institute for Social Invention. He was a fellow of the World Academy of Art and Science and the Royal Society for Arts and Science. He served as president of the ORS, the Society for General Systems Research (now the International Society for the Systems Sciences), and the World Organization for Systems and Cybernetics, a post he held at the time of his death. He was a quondam trustee of the American Society for Cybernetics (and was awarded its McCulloch Plaque), and was for 20 years a Governor of the International Council for Computer Communication. A commemorative issue of *Kybernetes* was published following his death (Espejo 2004).

Stafford Beer's place in the OR firmament is assured, but he has left a legacy which is far wider than OR alone. In particular, the systems community is also a member of his intellectual family, and in his later years the leading member. In post-war Britain, he was an innovator, arguing for and implementing inter-disciplinarity in the influential practitioner OR groups which he founded. His powerful and eloquent books developed a systems perspective on the management of complex organizations, and his VSM provides a generic blueprint for designing the structure of organizations so that they can survive in a changing environment. He explored in his work the implications of a holistic approach to organizational and social problems, and exemplified this approach in his own life. His ideas and the methods based on them continue to have a world-wide reach, and indeed to inspire new generations with the relevance and indeed the indispensability of a systems perspective.

ACKNOWLEDGMENTS

I am writing this profile as someone who was influenced by Stafford while still a student, not least in my choice of career, and who went on to work in two of the OR units, which he established. So, this is an engaged account. My personal experience has been doubly useful in that I know or knew many of the other people who worked with Stafford in those earlier days. I am grateful to a remarkably large number of them for the generous contribution of their recollections.

REFERENCES

Andrew A (1993) Stafford Beer—personal reminiscences and reflections. Kybernetes 22(6):60–73

Ashby W (1956) Introduction to cybernetics. Chapman and Hall, London

Barrionuevo A (2008) Before '73 Coup, Chile tried to find the right software for socialism. New York Times, March 28.

Beer S (1953) A technique for standardising massed batteries of control charts. Appl Stat 2(3):160–165

Beer S (1957) The mechanical simulation of stochastic flow. In: Davies M, Eddison R, Page T (eds) Proceedings of the first international conference on operational research. English Universities Press, London, pp 166–175

Beer S (1959) Cybernetics and management. English Universities Press, London

Beer S (1962) Towards the cybernetic factory. In: Foerster H, Zopf G (eds) Principles of self-organization. Pergammon Press, Oxford, pp 25–89

Beer S (1966) Decision and control: the meaning of operational research and management cybernetics. Wiley, Chichester

Beer S (1967) Management science: the business use of operational research. Aldus Books, London

Beer S (1972) Brain of the firm. Allen Lane, London

Beer S (1975) Platform for change. Wiley, Chichester

Beer S (1977) Transit. CWRW Press, Wales. Limited Edition, Private Circulation. Second expanded edition 1983

Beer S (1979) The heart of enterprise. Wiley, Chichester

Beer S (1981) Brain of the firm, 2nd edn. Wiley, Chichester

Beer S (1984) The viable system model: its provenance, development, methodology and pathology. J Oper Res Soc 35(2):7–26

Beer S (1994) Beyond dispute: the invention of team syntegrity. Wiley, Chichester

Beer S (2001) Think before you think: learning an outlook, mimeograph (unpublished)

Bunker B, Alban B (1997) Large group interventions: engaging the whole system for rapid change. Jossey-Bass, San Francisco, CA

Capey R (1996) Interview with Stafford Beer. Operational Research Society Archive (MSS.335), Modern Records Centre, University of Warwick, Coventry, December 10

Espejo R (2004) Special issue: tribute to Safford Beer. Kybernetes 33(3)

Espejo R, Harnden R (1989) The viable system model: interpretations and applications of Stafford Beer's VSM. Wiley, Chichester

Kirby M (2000) Spreading the gospel of management science: operational research in Iron and Steel, 1950–1970. J Oper Res Soc 51(8):1020–1028

Lehman MM (1993) Interview 178, IEEE History Centre, Institute of Electrical and Electronic Engineers. http://www.ieee.org/portal/cms_docs_iportals/iportals/aboutus/history_center/oral_history/pdfs/Lehman178.pdf. Accessed 30 Mar 2009

Ludley J (2008) Personal communication

Mingers J, Rosenhead J (2001) An overview of related methods: VSM, system dynamics, and decision analysis. In: Rosenhead J, Mingers J (eds) Rational analysis for a problematic world revisited: problem structuring methods for complexity, uncertainty and conflict. Wiley, Chichester, pp 267–288

Rivett BHP (1977) The case for cybernetics—a critical appreciation. Eur J Oper Res 1(1):33–7

Rose J (1993) A tribute to Stafford Beer: reflections and reminiscences. Kybernetes 22(6):6–7

Tocher K (1963) The art of simulation. English Universities Press, London

Treverton G (1988) Covert action: the limits of intervention in the postwar world. Basic Books, New York, NY

Wiener N (1948) Cybernetics: control and communication in the animal and the machine. Wiley, New York, NY

33

John F. Magee

Saul I. Gass

The total number of operations analysts who served as such during World War II (WWII) probably numbered in the low hundreds. Of these, a lesser number decided not to return to their prewar professions, but chose to start a new career that utilized and extended their wartime operations research (OR) experiences. They were the first generation of OR professionals. John F. Magee is a notable example of the second generation. These individuals helped define OR as a new science of decision making, and significantly expanded the application of OR to non-military settings.

John's indirect route to OR led him to a pioneering, nongovernmental OR consultancy organization, the Operations Research Group at Arthur D. Little, Inc. (ADL); he was its first full-time member. His consultant assignments in logistics and related research and practice produced seminal work in production planning and inventory control, and logistics in general. He remained with ADL for 48 years, a company he led as president, chairman, and CEO. He was president of the Operations Research Society of America (ORSA) in 1966 and The Institute of Management Sciences (TIMS) in 1971–1972, one of only five individuals who served as president of both organizations.

Early Life

John was born in Bangor, Maine, on December 3, 1926, the son of John H. Magee and Marie (Frawley) Magee. John's father, seriously injured in World War I, was a successful insurance broker, regional director of the Federal Housing Administration, member of the economics faculty of the

A.A. Assad, S.I. Gass (eds.), *Profiles in Operations Research*, International Series in Operations Research & Management Science 147, DOI 10.1007/978-1-4419-6281-2_33,

University of Maine, a nationally recognized insurance authority, and author of several widely used insurance texts. John's early years were spent in Bangor where he attended parochial elementary and high schools. He was an excellent student, played on the varsity high school basketball team, and picked up an enjoyment of mathematics due to a pair of "great mathematics teachers" (Magee 2007). Because of his father's handicap due to a war injury, John worked closely with his father on such tasks as editing galley proofs for his father's books. When John was in his junior year in high school, the draft demands of WWII caused his father to suggest that he apply to the University of Maine's summer school, take courses in mathematics and English, and then possibly transfer there in the spring term of his senior high school year. Thus, he would be able complete a full college year before he turned 18 and was drafted, an accomplishment that his father thought would motivate him to continue college when the war was over. The University turned him down as unqualified.

To demonstrate his potential to handle college level work at summer school, John took the College Board examination in the spring of 1943 in his junior year of high school. The registration form required the name of the primary college (University of Maine) to which the results would be sent, plus the name of an alternate school. John had no alternate in mind, but he entered the name of Bowdoin College, as it was the University of Maine's principal sports rival. Maine still turned him down for summer school, but the Bowdoin Dean of Admissions called and offered John immediate admission as a freshman with a scholarship! In June of 1943, after completing his high school junior year, John enrolled at Bowdoin. Thus, one's life and career are shaped.

At Bowdoin, John majored in mathematics. As was the case for most colleges, the student body was depleted by the war and small classes encouraged close contact with excellent faculty members, some who became lifelong friends. John played on the varsity basketball and baseball teams, was involved in student government and campus politics, and "enjoyed the house parties" (Magee 2007). Though majoring in mathematics, he was not interested in a scientific or engineering career. Having no firm idea of what he wanted to do, John took introductory courses in as many fields as his schedule allowed. He managed a heavy course load throughout the 12-month wartime college year. He completed four semesters and course work equivalent to mid-junior year before his 18th birthday and the coming military service.

Military Service, Back to Bowdoin and Beyond

Being drafted into the Army in late 1944 usually meant being sent to an infantry basic training camp (for about 3 months) and then being shipped to an infantry unit in the European war zone, not a pleasant prospect. Learning that a Navy language program at the University of Colorado had openings, John applied and was accepted for a 13-month immersion course in Japanese, the youngest of his class of six. He would enter as a civilian, become an apprentice seaman when he turned 18, and if he survived the course, be commissioned as an ensign. As things turned out, he was made an ensign midway through the course. The war ended while he was still in Colorado. He then spent 6 months assigned to the Office of Naval Intelligence in Washington, D.C. where one of his tasks was to study captured Japanese military logs in an attempt to discern clues concerning U.S. and allied aircraft that were missing in action.

John returned to Bowdoin for the 1946 summer semester. By convincing the Dean of the College to give him academic credit for his Japanese language studies and correspondence courses in mathematics and literature taken while in the Navy, he graduated *summa cum laude* in September 1946, with honors in mathematics and a membership in Phi Beta Kappa. He was 3 months shy of his 20th birthday.

Upon graduation, John was not clear as to a choice of a profession. He decided to continue his college education by applying to the Harvard Business School (HBS). He first had to persuade the School that his Navy career should be considered adequate work experience; he was accepted for the March 1947 class. Because he had the fall and winter of 1946–1947 to spend in Bangor, he took a job as an assistant in the University of Maine mathematics department doing related chores and grading papers. Taking advantage of free tuition, he enrolled as a master's degree candidate in mathematics and economics, and, with much going on in between (course work at Maine, Harvard degree, first jobs), he received an M.A. in mathematics and economics in 1953. His thesis was on the economic applications of linear programming (LP), certainly one of the earliest of such theses.

At Harvard, he again found himself the youngest of his class of 300. He took a number of courses in finance and economics, and received his MBA, with high distinction, in June, 1948. It was here that he met his wife Dorothy, a Wellesley student, who had grown up in Santiago, Chile, the daughter of a Texas metallurgical engineer. They were married in New York City (NYC) in 1949.

John's career choices now focused on those jobs that had an analytical or quantitative content. He did not want to teach and felt that he was too young to be hired as a consultant; an actuary for an insurance company seemed a possibility. Since he still had some GI Bill tuition support available, he enrolled in Columbia University's mathematical statistics Ph.D. program. The program was then under the leadership of Abraham Wald and Jacob Wolfowitz. John "loved" the mathematics courses, but found mathematical statistics "dull." In a conversation with Wald, John asked about career opportunities in mathematical statistics. Wald replied, "You can always teach statistics" (Magee 2007). John decided to move on.

The year 1949 was one in which jobs were, in general, difficult to find. John circulated his résumé (with one going to the graduate placement office at Harvard) and did receive a few finance-related job offers from NYC-based companies. But, an offer from Johns-Manville Corporation to join their financial analyst program that meant working at a plant with factory (floor-level) personnel seemed to be a valuable opportunity. He accepted, and after a short training period, he was assigned to the Building Products Division in Manville, New Jersey. It was here that he first gained knowledge of warehousing and inventory requirements to support seasonal shipments of building products and the concept of efficient cycling of product runs across manufacturing lines. He also helped the clerks in the accounting office close the books at month end and got to know how people on the front line, whether factory or office workers, think. As John recalled: he and Dottie "were rather poor but very happy, and I enjoyed the work" (Magee 2007).

OR AT ARTHUR D. LITTLE, INC.

Founded in 1886, Arthur D. Little (ADL), Inc. was the world's first management consulting firm, pioneering in contracted technology research. During WWII, several members of ADL had worked in or had become familiar with military OR and, upon their return, felt that OR activity should be a part of ADL's growing management consulting practice. In 1949, anxious to bring new areas of service to ADL, Raymond Stevens, an ADL Senior Vice President, bought into their idea. He asked staff member Harry Wissman, an expert in industrial economics, to build an OR group. Wissman was the first MBA to work at ADL, having joined the company in the early 1930s, soon after receiving his MBA from Harvard. In looking for a junior helper, he visited the HBS graduate placement office (ADL was located on Memorial Drive in Cambridge, Massachusetts, next to the Massachusetts Institute of Technology (MIT). There he found John's résumé—an MBA with a strong mathematics background and a good academic record—but with a new job at Johns-Manville. Wissman was not deterred. He called John to explore his interest.

John did interview at ADL headquarters and, after discussions with Wissman, Stevens, and other ADL staff, he accepted their job offer. Although he was not sure what ADL and OR were all about, the position appeared to give him an opportunity to use his mathematical and statistical skills in business problem solving. Equally important, it meant a 30% increase in salary. He reported to work at ADL on January 15, 1950, the first member of the first consulting industrial OR group. Dottie managed to find an apartment in the tight postwar housing market in nearby Arlington.

In 1954, John and Dottie moved to a newly built house in Concord where they raised three children: Catherine M. Milligan, a graduate from Middlebury College, housewife, mother, and a rider in competition horse shows; Andrew Magee, a graduate of Stanford University and the University of California Berkeley Law School, who practiced law, but now teaches science at a private secondary school in San Francisco; and John H. Magee, a graduate of Bowdoin College who went on to receive a Ph.D. in organic chemistry, University of Minnesota, and is vice president of a private biotechnology company. The Magee clan includes five grandchildren and two great grandchildren.

Dottie died at home on May 31, 2009. John still lives in the Concord house.

THE ADL OR GROUP

Over the next few years, under the guidance of Harry Wissman, the ADL OR group assembled a stellar staff of experienced OR analysts, with many coming from the U.S. Navy's Operations Evaluation Group (formerly the wartime Operations Research Group)—David Boodman, Arthur Brown, Martin Ernst, Sherman Kingsbury, John Lathrop—and consultants that included Philip Morse and George Wadsworth from MIT, and George Kimball from Columbia University. In 1956, Kimball became a full-time employee of ADL as Scientific Advisor and was made a vice president in 1961. Over the next decade, the group grew to over 40 professionals, with the experienced members from the military groups complemented by young Ph.D.s in physics, chemistry, electrical engineering, and chemical engineering, plus a few MIT graduate students working full or part time while completing their dissertations.

THE ADL OR STAFF (1952)

Sherman Kingsbury, Harry Wissman, Jack Lathrop, John Magee, Charles McGee.

An opportunity to test the usefulness of OR in an industrial setting had arisen just before John was hired. It concerned the distribution of Sears, Roebuck and Company's mail order catalogs (Magee 2002, 2007). ADL had recently completed a very valuable technical assignment for Sears, Roebuck, so Raymond Stevens used the opportunity this success provided to suggest to Theodore Houser, Senior V.P. of merchandising, and himself an MIT graduate, that Sears support an experiment to see what the ADL OR activity could do at Sears. Mr. Houser agreed and posed the following. Sears had been producing and distributing its catalogs since

1893, with its eagerly awaited spring and fall catalogs starting in 1896. By 1950, at the beginning of each season, these two very expensive catalogs were sent to about five million names (from a list of ten million). The question posed was: "Which names on their customer list should get the catalog to maximize revenue? Catalog costs were high enough that a small loss in the efficiency of catalog distribution would seriously affect mail order profits" (Magee 2002, 149). Sears had been studying this issue using field tests for decades.

Over the years, Sears first used a simple file card/stencil system and progressed to an electro-mechanically controlled Addressograph plate system, with each record holding a customer's history of orders for the prior 18 months. The millions of records were maintained by clerks in ten regional mail order processing plants. To determine whether a customer was sent a catalog, a set of rules related to the number, size, and timing of the customer's orders was applied by a clerk (later by processing the records through the wired-board Addressograph machines). If the customer met the conditions of the rules, an address label was printed and a catalog sent. The hope was that ADL's OR staff could improve the catalog mailing rules; a challenging test case for the embryonic group.

Due to the proprietary nature of the Sears work, the details were not allowed to be published for many years. A discussion of ADL's analysis is given in Magee (2002). A key result was that the behavior of a customer group could be described by assuming that an individual customer had an unknown inherent characteristic called the mean ordering rate, and the probability that customer would order n times in a 6-month period would follow a Poisson distribution with a mean equal to the mean ordering rate. Further, by assuming that the distribution of the values of this mean ordering rate in the population followed an exponential distribution, one could then calculate the number of customers expected to order zero, once, twice, . . . in a 6-month period; the calculated distribution of

> LEARNING ON THE JOB
>
> "The observations concerning ordering patterns and distributions gave the basis for some relatively elegant mathematical modeling of the customer population, which I proudly wrote up in a memorandum for Mr. Stevens to forward to Mr. Houser, to show what great progress we were making. Mr. Houser's response was clear and firm, 'If your team cannot learn to communicate with us in Midwestern English, perhaps you should not try' " (Magee 2007).

customers by number of orders closely matched the observed distribution. These observations plus others, for example, that the receipt of one of the main catalogs effectively doubled the observed order rate for the season and that the value of an individual order had no value in predicting future purchases, led to improved decision rules that were implemented by Sears. A long-term relationship between ADL and Sears was established with further studies of catalog-related issues resulting in practical and theoretical results, such as the interaction between the retail store and catalog businesses and an analysis of customer behavior as a Markov process (Herniter and Magee 1961).

Logistics: Production Planning and Inventory Control

The ADL OR group made an early entry in applying OR methods to industrial logistics, with its first such project for the Baby Products Division of Johnson & Johnson in 1951. The division produced a variety of products—baby oil, shampoo, talcum powder—on highly automated mixing and packaging lines. The constraints of the production system defined a classical situation: fixed line capacity, overtime at premium rates, full-time crew augmented by temporary workers at high cost, seasonal demand, closing of the plant for summer vacation, and the need for high inventories. The problem: what products to produce, when, and how much?

Working with George Kimball, John was responsible for providing a description of the plant and its operations and to collect data on product demand, cost, plant flexibility, and constraints, while Kimball explored various modeling approaches. But first, a literature search on production planning and inventory control had to be made. Not much turned up. There was some work by Bell Laboratories on economic order quantities and a book, *Quantity and Economy in Manufacture* by Fairfield E. Raymond (1931). It was on this project that Kimball introduced John to LP, though not by that term which was generally unknown in 1951. Eventually, Kimball described an approach that fit the facts John had gathered: minimizing a linear cost function subject to a series of linear constraints, a classic example of a linear-programming problem. This turned out to be a fruitful approach; based on it, Kimball and John developed a process for the Johnson & Johnson problem that determined production and inventory levels to meet peak demand while minimizing inventory investment and carrying cost. This and subsequent projects in distribution inventory management for the General Electric Company and production scheduling for the American Thread Company enabled John to "get a full perspective on the role of inventories in the management of the supply-production-distribution system" which led to a modern theory of inventory management "as a useful economic asset, to be managed for optimal performance of the logistical system," instead of viewing inventory as a "necessary evil" (Magee 2002, 151).

John described this new approach to inventory management in a series of three papers that was published in the *Harvard Business Review* (HBR) under the titles "Guides to inventory policy, I, II, III" (Magee 1956a, 1956b, 1956c). This was the first time the HBR published such a series on a single topic. John expanded these articles into a book-length practical guide to the subject for managers and engineers, *Production Planning and Inventory Control* (Magee 1958). The book was translated and published in six languages and sold actively world wide. In 1967, David Boodman revised the book as co-author, resulting in several more years of active sales around the globe.

John followed the inventory papers with another HBR publication, "The logistics of distribution" (Magee 1960). This article was designed to introduce business managers to the concept of the logistics system; the use of the term logistics had hitherto been largely reserved to the military. A reviewer characterized this article as "the undisputed classic writing" in business logistics and "very influential in achieving the explosive growth and acceptance of logistics by the business community" (Thompson 1976, 403). The article later became the kernel of two books, *Physical Distribution Systems* (Magee 1967), and *Industrial Logistics* (Magee 1968), both of which were also translated and published abroad.

In 1964, the HBR published two related articles by John, "Decision trees for decision management" (Magee 1964a) and "How to use decision trees in capital investment" (Magee 1964b). The term decision tree was drawn from a quotation by Pierre Massé, a French economist: "The decision problem is not posed in terms of an isolated decision.... The problem is posed in terms of a tree of decisions" (Massé 1962, 250). The articles introduced the business community to the concepts of uncertainty, alternative outcomes, and consequential decisions, in such diverse contexts as plant investment, support of R&D programs, or marketing initiatives.

John continued to write and speak on management issues throughout his career at ADL. After 1970, he tended to focus on broader issues of interest to management, rather than on specific subjects related to OR, such as logistics. Topics covered a wide range of issues of interest to general management: business organization, especially as potentially affected by evolving information technology; risk and crisis management; new approaches to R&D management; impact of information technology on specific industries such as publishing or banking; and implications to American business of the evolving European Community.

EARLY ADVICE AND CONCERN

Prior to the John's HBR inventory articles, John and Cyril Herrmann, an ADL consultant from the MIT faculty, published an article in the HBR, "Operations research for management" (Herrmann and Magee 1953). This seminal article attracted interest in OR among a wide spectrum of managers; it was based on an internal piece John wrote to help explain to ADL clients the basic ideas of OR and how OR might be of use to them. In it, the authors noted their concern about the future of OR:

"If operations research is to have a future, the professional groups and research workers in the field must ultimately establish 'operations research' as something more than a catchword, by proving its continued usefulness to management in the solution of important business problems. . . .

"There is, unfortunately, already evidence of growing attempts to capture the term, subordinate it to other established fields in the general area of statistics or engineering, and to apply it to activities of operators on the fringes of the established and reputable management and engineering fields. The growth of such tendencies, if unchecked by education and publicity, may well threaten to send the term 'operations research' along the way of others, like 'efficiency engineering,' which sooner or later became victims of indiscriminating acceptance and careless usage" (Herrmann and Magee 1953, 112).

Up the Organizations and Honors

As the ADL OR staff and consultancy work grew, John continued to work on projects, but started to move into management positions. In 1958, ADL hired General James M. Gavin, recently retired chief of R&D of the U.S. Army, as executive vice president and potential future CEO. As his initial assignment, Gavin took on leadership of the Management Services Division and asked Wissman to be his principal staff assistant; John was promoted to lead the OR Group. In 1961, Gavin was appointed president of ADL and, in 1962, he made John head of the Management Services

Division, a position he held until 1968. (By then, Gavin was chairman and CEO of ADL.) John was then elected executive vice president and a member of the Board of Directors. He joined ADL's president's staff to do professional work and special projects, including the implications of establishing a public market for the company's stock. This last task was accomplished in 1969, when public trading of ADL stock began. In 1971, ADL was not doing well financially; its stock price had fallen and profit sharing was at an all time low. John accepted the request by Gavin and Royal Little, Chairman of the Memorial Drive Trust that held the majority of ADL shares, to take over operating responsibility of the company. He was elected president of ADL a few months later.

John reorganized and streamlined ADL's management structure and, within a year, the financial situation improved. With Gavin retiring in 1974 and his anticipated replacement as CEO, Board of Director's member Eli Goldston, dying before taking office, John became CEO and president. From 1986 until 1988, John was also Chairman of the Board of Directors. After retiring as CEO in 1988, he continued as non-executive Chairman until 1998.

In the face of growing management responsibilities, John continued active professional consulting. He led ADL's work in support of the President's Commission on Postal Organization, established in 1967, that resulted in the Postal Service being set up as a government corporation outside the federal bureaucracy. During the 1970s, John served as a member of a small, active board of directors of the Pennsylvania Company that held the non-rail assets of the Penn Central Railroad during the its bankruptcy; the board was responsible for building and managing these assets, including large real-estate holdings, a network of amusement parks, an oil and gas pipeline company, and a professional sports team, among others. The Pennsylvania Company became the core of the reorganized Penn Central Corp. He studied the information management issues of the National Bank of Greece and worked on logistics projects in France and Japan. At the time of his retirement in 1998, he had completed some years of consulting on organization and governance issues for a large Indian steel company.

During John's leadership as head of Management Services and later as president, ADL dramatically expanded its presence around the world, opening management consulting offices in several parts of Europe, Latin

America, and Asia. A technical contract R&D laboratory was established in Cambridge, England, and project offices in the Middle East and Africa. He also oversaw the establishment and growth of the ADL Management Education Institute, unique as a subsidiary of a profit-making firm authorized and accredited to grant M.S. degrees in management; its student participants were drawn from throughout the free world. John's management responsibilities, consulting, and leisure travel took him and Dottie to every continent including Antarctica.

> AT THE BEGINNING
>
> John attended the first national meeting of ORSA that was held in Washington D.C., November 17–28, 1952, where he presented a paper on "Linear programming in production scheduling." He also authored "The effect of promotional effort on sales" that appeared in the February, 1953 issue of *Operations Research*, **1**(2) 64–74 (this issue was the first to contain technical papers). And, along with ADL associates George Kimball, Philip Morse, and Martin Ernst, John attended the second meeting of the International Federation of Operational Research Societies that was held in Aix-en-Provence, France, where he was "introduced to white Châteauneuf-du-Pape in large quantity" (Magee 2007).

As John notes, the OR professional society, ORSA, "was possibly conceived, and certainly later incubated until it gained strength, at the offices of the Arthur D. Little Operations Research Group. . . . The company subsidized the work of the organization and later provided the early home of the Society's administrative activities" (Magee 2002, 153). The ADL staff, especially George Kimball and ADL's consultant Philip Morse, were prime movers in the founding of ORSA. ADL staff member John Lathrop served as secretary of the organizing committee and later as first secretary of ORSA. Founding members of ORSA from ADL were Sherman Kingsbury, John Lathrop, John Magee, Harry Wissman, as well as consultants Morse and Kimball, with Morse becoming ORSA's first president. Staff members of ADL who served as ORSA president were John Lathrop, Martin Ernst, George Kimball, and John Magee (1966), with Magee also serving as president of TIMS (1971–1972).

John began his OR career with ADL with the questions "What is OR?" and "What is Arthur D. Little?" (Magee 2002). His scientific and management training, innate ability, and the fortuitous association with such people as George Kimball and Harry Wissman enabled him to become one of the first OR practitioners to address industrial problems. He extracted from his practice new theoretical results in logistics and made them available through his writings. He not only found out about OR, but helped to form the subject, and was instrumental in establishing ORSA, the U.S. professional OR society. He also found out about ADL and, through his management skills, enabled the company to become a pioneer and industry leader in OR consulting services through the last four decades of the 20th century.

John received ORSA's Kimball Medal for distinguished service in 1978. In 1996, he was awarded the honorary degree of L.L.D. from Bowdoin College. He was elected a Life Fellow of Phi Beta Kappa. He has served on the Board of Directors of Houghton Mifflin Company, John Hancock Mutual Life Insurance Company, and the Bank of New England. He has been associated with many nonprofit organizations: Woods Hole Oceanographic Institution (Trustee), New England Aquarium (Trustee and Governor), Boston Museum of Science (Overseer), Bowdoin College (Overseer, Trustee and Chairman of the Board), Boston University Medical Center (Trustee Council), Thompson Island Education Center (Trustee and Chairman of the Board), and Emerson Hospital (Director and Secretary of the Board). He served as a member of the Visiting Committees of the Harvard Business School and the School of Management of the University of Southern California, on the Dean's Council of the Harvard Graduate School of Education, and, for 3 years, as President of the Commercial Club of Boston whose membership included business and institutional leaders of the region.

John and Dottie at the Taj Mahal (1986)

In a justly deserved retirement mode, John enjoys golf and downhill skiing, and, as he describes in Magee (2007), he is a watercolorist who has sold a few paintings, exhibited in local art association shows, with a one-man show held in of 2007.

References

Herniter J, Magee JF (1961) Customer behavior as a Markov process. Oper Res 9(1):105–122

Herrmann C, Magee JF (1953) Operations research for management. Harv Bus Rev 31(4):100–112

Magee JF (1953) The effect of promotional effort on sales. Oper Res 1(2):64–74

Magee JF (1956a) Guides to inventory policy I. Harv Bus Rev 34(1):49–60

Magee JF (1956b) Guides to inventory policy II. Harv Bus Rev 34(2):103–116

Magee JF (1956c) Guides to inventory policy III. Harv Bus Rev 34(3):57–70

Magee JF (1958) Production planning and inventory control. McGraw-Hill, New York, NY

Magee JF (1960) The logistics of distribution. Harv Bus Rev 38(4):89–101

Magee JF (1964a) Decision trees for decision making. Harv Bus Rev 42(4):126–138

Magee JF (1964b) How to use decision trees in capital investment. Harv Bus Rev 42(5):79–96

Magee JF (1967) Physical distribution systems. McGraw-Hill, New York, NY

Magee JF (1968) Industrial logistics. McGraw-Hill, New York, NY

Magee JF (2002) Operations research at Arthur D. Little, Inc.: the early years. Oper Res 50(1):149–153

Magee JF (2007) Unpublished autobiographical notes

Massé P (1962) Optimal investment decisions: rules for action and criteria for choice. Prentice-Hall, Englewood Cliffs, NJ

Raymond F (1931) Quantity and economy in manufacture. McGraw-Hill, New York, NY

Thompson H (1976) The great writings in marketing. The Commerce Press, Plymouth, MI

34

PHILIP STARR WOLFE

ALAN J. HOFFMAN

In the early 1950s, even though there were few university departments of operations research(OR), some students found the subject attractive due to the influence of a faculty advisor, specific work experience, or the intellectual challenges OR presented. In Phil Wolfe's case, all played a role.

Returning to college after military service, Phil became a leader in optimization theory. He is known principally for his research in mathematical programming (MP), particularly on extensions of linear and nonlinear programming. But his contributions extend beyond pure research. He worked hard to ensure that the MP community of scholars developed the professional bonds through a professional society, journals, regular symposia, informal meetings, and newsletters. His commitment to this cause earned him the respect and affection of the MP community.

Phil's research accomplishments and influence within the MP community were recognized with the John von Neumann Theory Prize of the Operations Research Society of America (ORSA) and The Institute of Management Sciences (TIMS), and the Distinguished Service Award given by the Mathematical Programming Society (MPS). His research was conducted at the U.S. Air Force (Pentagon), Princeton University's mathematics department, the RAND Corporation, and IBM's Thomas J. Watson Research Center.

A.A. Assad, S.I. Gass (eds.), *Profiles in Operations Research*, International Series in Operations Research & Management Science 147, DOI 10.1007/978-1-4419-6281-2_34,

Shaped by Science Fiction

Phil Wolfe, the second son of Sidney and Dorothy Anderson Wolfe, was born in San Francisco, California, on August 11, 1927. At that time, his mother wrote human-interest stories for the *San Francisco Chronicle*, and his father ran a trade publication for the gift and art business on the West Coast. Phil's only sibling was his older brother David, born in 1921. Sidney's parents came to San Francisco from Poland in the 1870s. Dorothy's maternal grandfather, Henry Nikolas Bolander, was California's first state botanist. Dorothy's mother was one of twin sisters born to his wife in Guatemala, where Henry had been sent on a mission for the German government.

Courtesy Michael J. Johnson.

Phil's family moved frequently around California in his early years—he remembers living in Bakersfield, Los Angeles, San Jose, Mill Valley, Sausalito, Oakland, and Alameda. His interest in science began with the gift of a microscope for his seventh birthday, followed by studying his brother's high-school physics textbook. Phil's early school years went well; he was selected valedictorian for his sixth grade graduation. But his first year at Alameda High School was a disaster, except for his German and general science classes. He received the lowest passing grades in his algebra courses and understood neither the manipulations nor their motivations.

The next year he studied plane geometry, also a troubling course. One day, however, he experienced the epiphany of many mathematicians; "I recall deciding that it must make sense somehow, and spent the whole day reading Euclid's axioms and early theorems and putting it all together. It was a profound experience. By the end of it, I knew what Euclid was doing ... we became colleagues" (Wolfe 2009). Phil was the star of that class, and the teacher, who was retiring, gave him her collection of mathematics books. He used them to learn differential and integral calculus on his own—he was a top student in science during the rest of his high-school years.

But these were unhappy years for Phil's family. His parents divorced and his brother David, who had joined the Army in 1939, was killed in the Japanese invasion of the Philippines. Phil lived with his mother in Alameda and later in Berkeley, where he entered the University of California in 1943. He did reasonably well as a student until he fell in love and more or less gave up on science and academic life. Eventually, he withdrew from the university and was drafted into the Army shortly after his 18th birthday and just as World War II was ending.

Love affair over, Phil liked the Army. He had various assignments, all in the U.S., the last one teaching German to intelligence agents for work in Germany. When he was discharged in 1947, he returned to Berkeley and received his A.B. in physics and mathematics in 1948—and a mathematics department prize for the best undergraduate record. Phil wanted to go on for a Ph.D. in physics, but (like many other mathematicians) was not fond of the way physicists reasoned about mathematical concepts. He concluded physics was not for him and continued on in mathematics and received his M.A. in 1950.

He was interested in fundamental mathematical topics, especially set theory and logic. He wanted to study under the famous Berkeley logician Alfred Tarski. But, in 1949, he was diverted by a story, "The Finan-seer," that had appeared in the October issue of *Astounding Science Fiction* (Locke 1949). In this story, professors, using something called the theory of games, have amazing success in the stock market. Phil quickly bought the book, *Theory of Games and Economic Behavior* (von Neumann and Morgenstern 1944). He was intrigued by the idea that the theory might be a branch of mathematics that could be used in the real world of competitive activities; he aimed at writing a doctoral dissertation on the subject.

Since dissertations required an interested faculty advisor, Phil became the pupil of the statistician Edward Barankin. Barankin, who taught courses in optimization, had Phil read reports from the University of California at Los Angeles OR project and from the RAND Corporation, a center of game theory research. For the summer of 1951, Barankin arranged for an internship for Phil with Barankin's friend, the mathematician George Dantzig, at the Air Force's Project SCOOP in the Pentagon. (SCOOP is an acronym for Scientific Computation of Optimal Programs, an Air Force project for the analysis and computation of Air Force plans and programs. Dantzig had devised and formulated the basic and general linear-programming (LP) mathematical model and invented

(Cover by Alejandro Canedo)

TRUE SCIENCE FICTION STORY

"Not far before the end of Al's [Tucker] life, I told him, for the first time, what attracted me to game theory, namely the story in *Astounding Science Fiction*. Al said, 'Oh, yes. The author got the idea from a public lecture that I gave in early 1949. As a matter of fact, I've got a copy of the magazine. Would you like to have it?' I still have that copy. It was a wonderful closed circle."

Wolfe interview in Lustig (2001).

THE FINAN-SEER

BY E. L. LOCKE

The speaker went on. And on. And on. ". . . I want to remind you that I am the only real business man among you. I am the only one here that ever met a payroll and so I know just how to handle this situation. Now when I was running my butcher shop in Carteret, New Jersey, I found ways and means to keep my business solvent. I tell you it was a tough job because it was a credit store and you should have seen how many of my customers were deadbeats. I—"

A quiet but sarcastic voice cut in, "I don't doubt that the Professor of Meat Cutting was able to devise ways and means. After all, I understand the high incidence of corpulence among butchers is not entirely accidental. Large stomachs have their uses in the weighing process."

A beet-red color suffused the face of the first speaker and strangling sounds came out of his throat. Before he could collect himself to retort, a silver-haired, dignified-looking man got up from the head of the

132 ASTOUNDING SCIENCE-FICTION

the simplex algorithm for solving it. Dantzig had received his Ph.D. from Berkeley in 1946.) George challenged Phil to find a way to resolve the problem of cycling (circling) that could cause the simplex algorithm not to converge to an optimal solution.

The simplex method finds the lowest point on a polyhedron in n-dimensional space, as measured by a minimizing linear objective function, by moving from a vertex to an adjacent lower valued vertex, and so on, until a lowest vertex (minimum) is reached. The mathematician Fourier proposed this method in the first decade of the 19th century—it is intuitively obvious that it should succeed (Fourier 1826, 1827; Grattan-Guinness 1970). But, the algebra needed to carry out these moves may have difficulties if the current vertex lies on more than n hyperplanes, a situation termed degenerate. It is conceivable that the required algebraic transformation could not cause a move to another vertex and not prove that a best vertex had been found, but only generate successive algebraic descriptions of the same vertex, and this cycle of transformations would continue. I believe the first person to recognize (but not resolve) the problem was Frank Hitchcock (1941).

J. Harvey Edmondson, who was taking Dantzig's 1951 LP class (the first of its kind) at the U.S. Department of Agriculture Graduate School, in response to a class exercise, wrote an unpublished paper that resolved the situation for the general LP problem with m equations and n variables. His approach was to slightly perturb the polyhedron so that each vertex was on exactly n hyperplanes (Dantzig 1963). At the time I was a mathematician at the National Bureau of Standards (NBS) in Washington, D.C., conducting research sponsored by Project SCOOP. Shortly before Phil came to Washington, I had constructed (in 1951) the first example of an LP problem for which cycling was shown to occur (Dantzig 1963; Hoffman 1953; Micchelli 2003). Phil found what Dantzig was looking for—an algebraic way of executing the Edmondson perturbation scheme. Phil's idea was to replace the real numbers used as coordinates in n-dimensional space by m-dimensional vectors, which are lexicographically ordered, that is, $x > y$ if, in the first coordinate where x and y differ, say the jth coordinate, $x_j > y_j$. This lexicographic ordering of vectors in m-space was something Phil learned from courses in logic, verifying the adage that nothing learned is ever wasted. Phil described his idea in an Air Force memorandum, but its first appearance in a mathematical journal was a paper by Dantzig et al. (1954). Dantzig, who seems to be the actual writer of the three-author paper, calls the resulting calculations a generalized simplex method. Phil's idea had legs. It was an essential part of the proof of Gomory's epochal papers establishing the scaffolding for integer linear programming (Gomory 1958, 1963). It inspired many generalizations of the simplex method, as well as the duality theorem in a variety of circumstances (Wolfe 1963a). Phil and I become lifelong friends.

> LIFE IN THE PENTAGON
>
> "When I showed up in the summer of 1951 at that job at the Pentagon, the first thing I did was briefly interview other people in Project SCOOP. I went into the office of one economist, introduced myself and asked if I could sit down and talk to him. He said, 'Look, I'm making a detailed model of the American economy. So you're going to have to excuse me because I'm terribly busy'" [Wolfe interview in Lustig (2001)].

Phil's summer at Project SCOOP was abbreviated by an impatient letter from his new girl friend in Berkeley. He returned early, and they married in 1952.

Berkeley and Princeton: 1952–1957

In 1952, Phil was a graduate student in mathematics, working toward a Ph.D. at the University of California, Berkeley. By 1965, he would become a leading authority in the field of optimization, the central theoretical and applied mathematical framework that was in the forefront of OR's major advances in industry, business, and government.

His dissertation, presented in 1954, consisted of two papers. The first was a version of his Air Force memorandum on a generalization of the simplex method. The second paper grew from Phil's desire to do something original in game theory; he proved a conjecture of Gale and Stewart (1953) about circumstances which would imply that an infinite win–lose game had a winning strategy (Lustig 2001). Phil was pleased with this result for a couple of reasons: it used material he had learned in a topology course from John Kelley, a professor he admired very much, not only for his pedagogy, but also for his vigorous protests of the loyalty oath that the California regents imposed on academics. Further, Phil had proved the conjecture before the Gale–Stewart paper appeared, showing he could raise research questions as well as solve them. This impressed Phil's adviser Barankin, although he was not warmly inclined to game theory. Barankin, however, defended Phil's work vigorously when the departmental chairman, Griffith Evans, also not warmly inclined to game theory, expressed some misgivings ["Where is the mathematics, Mr. Wolfe?" (Wolfe 2009)]. The title of Phil's dissertation reflected the two contributions: "I. Games of infinite length. II. A non-degenerate formulation and simplex solution of linear programming problems."

Ph.D. advisors sometimes help their students to find their first job, usually an academic one. In Phil's case, Barankin's efforts set the course of Phil's career. Barankin wrote to Princeton professor Albert Tucker about Phil's status, and Tucker offered Phil an instructorship.

Although the RAND Corporation had offered a job at twice the Princeton salary, Phil chose Princeton. He and his wife drove across the country in an old car whose maximum speed was below the legal minima of some toll roads, arriving in dense fog on the night of September 10, 1954. When he visited Fine Hall (the home of the mathematics department) the next morning, he was thrilled to see that Tucker had already posted P. Wolfe on the faculty directory.

The most important work Phil accomplished during his stay at Princeton was his research on quadratic programming, but many themes of his later career can be seen taking shape during this period. He took an interest in computing, and visited the Institute for Advanced Study where Julian Bigelow, its chief engineer, helped Phil write programs for the machine John von Neumann had designed. He helped Tucker administer the Office of Naval Research's sponsored Princeton Logistics Research Project, the pioneering research center in game theory and LP. Besides his own work, Phil wrote and distributed occasional reports on meetings and conferences to the Princeton mathematics community. He enjoyed the stream of visitors—George Dantzig, David Gale, Harold Kuhn, Theodore Motzkin, John von Neumann, and many others.

Marguerite Frank, who had written her Ph.D. thesis on Lie algebras with Adrian Albert at Harvard, was visiting Princeton and working in the ONR project. She and Phil joined forces and began studying nonlinear optimization under linear constraints. Using an observation of Barankin and Robert Dorfman, they developed a procedure for quadratic programming. Phil wrote their joint paper during a summer vacation at Big Sur, California, with his typewriter at the edge of the Pacific Ocean.

FINE HALL

I asked Phil to describe the architecture of Fine Hall of that period (the current Fine Hall is a modern large building in a different part of the Princeton campus, the name following the department, and the old building is now known as Jones Hall, honoring its donor).

"Fine Hall was neo-Gothic, brick with lots of white ornamentation like gargoyle rainspouts. Three stories: classrooms, office, big library, . . . bow windows, wood paneling everywhere. Each bathroom stall had an excellent reading light. At the end of the second floor was the common room, where tea was served around four almost every day, and almost everyone came. There was a big fireplace on which was inscribed Einstein's remark in German that 'God is subtle, but not malicious'" (Wolfe 2009).

[Henry B. Fine, Princeton class of 1880, started, in 1885, as an assistant professor of mathematics at Princeton (College of New Jersey until its name change in 1896), becoming the senior member of the department by 1900, and then chairman from 1904 to 1928.]

Their completed manuscript was submitted to *Naval Research Logistics Quarterly* during the time I was the journal's managing editor. I found that the paper had a conceptual and intellectual depth which caused me to worry about finding an appropriate referee. Fortunately, about the same time, Harry Markowitz submitted a paper with a similar theme, portfolio selection by parametric quadratic minimization. I sent each the other's paper for refereeing. Since neither found anything objectionable to warrant rejection, they were both accepted and published in the same issue (Frank and Wolfe 1956; Markowitz 1956).

Phil also taught undergraduate courses in calculus and a graduate course in game theory, and wrote papers in game theory (Sion and Wolfe 1957; Wolfe 1956). In 1957, inspired by Markowitz's paper, Phil developed a procedure based on the simplex method for solving quadratic-programming problems that only required making minor modifications in a simplex algorithm computer code. He sent a copy to Dantzig who replied, "This is a terrific result, if it's true" (Lustig 2001).

Princeton now offered Phil an assistant professorship, a 3-year, non-tenure track appointment. Instead, he accepted an offer from RAND at a salary twice as large as what was offered earlier. So, back to California; it was 1957.

At RAND: 1957–1966

The RAND corporation, located in Santa Monica, had been created in 1946 by the U.S. Air Force and, in 1948, was incorporated as a non-profit organization to improve policy and decision making through research and analysis. It has had many distinguished scholars in a wide array of fields and was a leader in OR research methods. Oddly, Phil was not assigned to the mathematics department, but to the computing group. Initially, Phil thought he was hired to replace William Orchard-Hays, the premier developer of LP computer codes, who had recently left RAND to join the Corporation for Economic and Industrial Research, a Washington, D.C. consulting firm (Mapstone 1972). George Dantzig had joined RAND in 1952, and

> CATTLE FEED
>
> "For a couple of years I worked with the Kern County Land Company. They had to feed a herd of 10,000 to 20,000 cows every day. They came to RAND, having heard about linear programming, and asked if we could help with the classic diet problem. The dieticians have a whole set of requirements about how much salt, protein, carbohydrates, and a bunch of other ingredients their cows should ingest. They wanted to go out in the market every week, look at the current foodstuff prices, and make the best deal they could. I took their weekly data for about a year, put it into our computer, and gave them answers about how they should feed the cows. It was kind of fun, my first practical computation" [Wolfe interview in Lustig (2001)].

Orchard-Hays, working with Dantzig, had developed simplex algorithm-based codes for the IBM card-programmed calculator and the IBM 701 and 704 digital computers. But Phil was not a master computer programmer—most of his time was spent on algorithmic research, finding ways to improve the simplex algorithm. He did, however, influence RAND computing in other ways. He persuaded members of the RAND computer programming staff to try FORTRAN, the new high-level computer language/compiler to see if it assembled computer code more deftly than the programmers could accomplish using machine/assembly code and their personal skills, which it did. (This was in contrast to Orchard-Hays writing elaborate LP codes in machine language.) He undertook with Leola Cutler a series of computational tests of various LP ideas, which was the beginning of the collection of test problems that have been useful to the MP community (Wolfe and Cutler 1963). And he continued the practice begun at Princeton of writing reports and giving lectures on the state of the art of computation in various optimization venues. His principal RAND associates were George Dantzig, Ray Fulkerson, and Lloyd Shapley. Phil also took advantage of the pleasant yearlong weather to run on Santa Monica beach and swim and bodysurf in the warm Pacific Ocean.

What is probably Phil's best known work, the Dantzig–Wolfe decomposition method, came from this environment (Dantzig and Wolfe 1960). Building on an idea of Ford and Fulkerson for multi-commodity network flow problems (Ford and Fulkerson 1958), Dantzig and Wolfe observed that the methods of the simplex method could apply even when the columns of the LP matrix were not explicitly available. What was needed was a way of generating them when they had to be tested for insertion into the basis. In particular, if you had a LP model where several smaller linear-programming problems were subjected to only a few equations that linked variables of the separate smaller problems, you could,

A DECOMPOSABLE LP MODEL

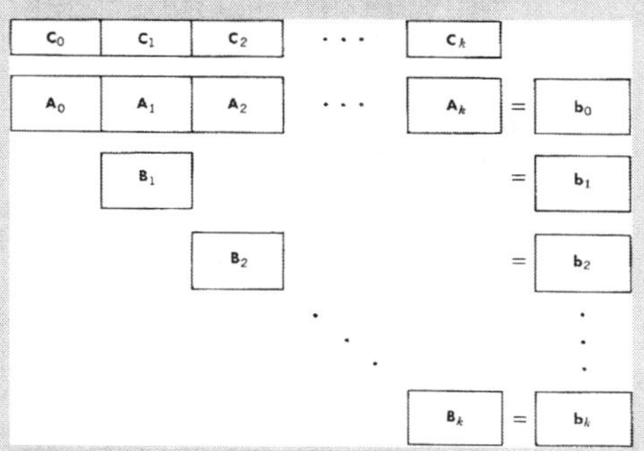

in principle, solve the whole problem by shifting the focus alternately between the smaller problems and the interconnected problem over vectors that were convex combinations of the vectors of the smaller problems. The idea had broad implications in both mathematical and economic contexts, and it inspired such applications as Gilmore and Gomory's analysis of the cutting-stock problem (Dantzig 1963; Gilmore and Gomory 1961, 1963). Phil continued his interest in nonlinear programming and published reviews of the field (Wolfe 1961, 1963b, 1967).

In early 1964, Phil and his wife divorced, at her request, which he eventually came to see was justified by his concentration on work and neglect of all else. His personal life was at a low point. His friend Ralph Gomory, who was Director of the Mathematical Sciences Department at IBM's T. J. Watson Research Center, Yorktown Heights, N.Y., arranged to have Phil spend a 6-month sabbatical at the IBM's Zurich research laboratory. He enjoyed the experience and wondered why he had absorbed from his RAND colleagues an anti-IBM prejudice. When Ralph later offered him a regular position at the Yorktown Heights research center, Phil agreed. There were also other reasons for leaving RAND: Dantzig had left to join the faculty at the University of California, Berkeley; Ray Fulkerson was contemplating leaving for academia (which he did by going to Cornell University); and RAND was beginning to urge researchers to find their own funding rather than rely on Project RAND. In April, 1966, Phil, with the Porsche he bought in Zurich, arrived in Southampton, England where he boarded the P&O cruise ship Chusan which stopped at various resorts, passed through the Panama Canal, and went on to San Francisco where Phil visited his mother. Later that year, he drove to New York to join the research staff of the Mathematical Sciences Department (MSD) at IBM Yorktown. He bought a tiny house not far from work and turned its basement into a compact carpentry shop.

IBM and Beyond

I had been a member of the MSD since 1961, and I was thrilled to learn that Phil was joining our group. He was, in the opinion of many of us, the second leading figure in the mathematical optimization community (after George Dantzig, of course). Apart from integer programming, Phil was an authority on every aspect of optimization. He also had the distinction of being the only person to have received paychecks from each of the three shrines of MP: Project SCOOP at the Air Force, the Logistics Research Project at Princeton, and the RAND Corporation. Phil spent the rest of his career in the Mathematical Sciences Department of IBM's Thomas J. Watson Research Center, Westchester County, just 45 miles north of Times Square.

It did not take long for Phil to learn that it was not easy for a single man living in northern Westchester County to meet women. So he joined the Chappaqua Drama Group soon after his arrival, and was chosen for a part in their next production, *Look Homeward, Angel*, based on the book by Thomas Wolfe (which Phil had read years earlier while in the Army, attracted first by the author's name, but more profoundly by identifying with the hero, Eugene Gant). Phil played Ben, Eugene's brother, and I thought he did rather well; his voice resonated, his posture was graceful, and he had reason to be pleased with his work. He played in several other productions, but then decided to try his hand at stagecraft. In February, 1968, Phil contacted the Beechwood Players, a local theatre company that was in need of set builders. The young woman manager, Hallie Flanagan, explained to Phil that she had not been able to find a place suitable for building sets. Phil offered the use of his shop. Hallie agreed—they soon established a professional and social relationship, and were married in June. (Hallie was named after her father's mother, who was a prominent theatrical producer, director, playwright, and author. From 1935 to

A BUCOLIC RESEARCH SETTING

The building that housed the Watson Research Center is one of the last structures designed by Eero Saarinen. It has a mostly stone exterior and an entrance dominated by a metal sculpture by Seymour Lipton, somewhat resembling a rocket, which we informally gave several ribald names. In our time, about 1500 people worked there, of whom about a third were Ph.D.s. The public rooms (auditorium, cafeteria, library) are very attractive; the individual offices are small (even the large ones would seem cramped in a typical university setting) and have no view of the outside. The corridors have beautiful fenestration, but not even the Director of Research has a window. A reminder of the date of construction (the building was completed in 1961) is an atomic bomb shelter in the basement. In the rear is an enormous parking lot for employees. During cold weather, after smoking inside the building was banned, the few remaining addicts would sometimes huddle outside the rear doors, provoking the sign: "Please do not smoke near the door. This building sucks." In warm weather, there is tennis, basketball, and soccer on the back lawn for the fit and hearty. And you can lunch on a patio outside the dining room, and enjoy the magnificent view of the front lawn.

1939, she was the director of the Federal Theater Project, part of Roosevelt's New Deal Works Project Administration.)

In 1970, Phil and Hallie moved to their present home some 20 minute's walk to work. Phil has made the round trip on foot every day the weather allowed it. Their daughter Sarah was born in 1974. Sarah showed early talent for mental arithmetic and independent thinking. She moved to Alaska in 1996 and attended the University at Fairbanks, earning a B.S. in mathematics, minor in geology, and an M.A.T. (Master of Arts in Teaching) in secondary education. She teaches high-school mathematics in Fairbanks and is a volunteer firefighter, an emergency medical technician, and a curling enthusiast.

By the time Phil joined IBM, Ralph Gomory had succeeded Herman Goldstine as head of MSD. Although IBM was deeply involved in the development of LP and related software in many parts of its organization, IBM Research had no group specifically identified with optimization (Spielberg 2007). For many reasons, especially anticipated growth, it became clear that MSD needed such a group, and a group leader had to be chosen. There were several MSD members who had made notable contributions to optimization (and to OR)—Paul Gilmore, T. C. Hu, Dick Karp, Alan Hoffman—but no one was as qualified as Phil, by interests or temperament or achievement to lead MSD's optimization research. Within a few years, he was given that responsibility and he organized a small, outstanding group that included Ellis Johnson, Earl Barnes, and Harlan Crowder.

Besides administration, Phil's work continued to deal with aspects of nonlinear programming: globally convergent methods for unconstrained optimization (Wolfe 1969, 1971); nondifferentiable optimization (Wolfe 1970, 1974); and constrained optimization (Wolfe 1966). A major achievement, in the spirit of other work in the department on limits of computation, is given in Brent et al. (1973) in which the authors prove the fundamental result: if d is the order of the highest derivative used in an algorithm to find a simple root of an analytic function of one variable, then the order of convergence cannot exceed $d + 2$.

Phil also applied his leadership abilities to the establishment of a professional community for furthering the developments of the rather new analytical and computational field of MP. There was probably no one more active in furthering research and applications of MP within the Association for Computing Machinery and its Special Interest Group in Mathematical Programming. In 1970, Phil and Michel Balinski started the journal *Mathematical Programming.* When the MPS was started in 1971—to formalize the selection of sites for the triennial mathematical programming symposia, to supervise the journal and other publications, and to generally promote the subject—Phil was one of the principal founders. He was the MPS chairman in 1978–1980. Friends of Optimization (FoOp) was an informal organization started by Phil as a way of getting the MP community in the New York metropolitan area together. It held meetings at various venues, with speakers who were in the forefront of developments. FoOp disbanded in the late 1980 when its functions were served by other forums.

On the occasion of his 65th birthday, Phil was honored by his friends and colleagues for his fundamental contributions to the mathematical programming field with a two-issue Festschrift of *Mathematical Programming* (Cottle et al. 1993). A text and video interview of Phil is among those conducted by Irv Lustig in his interview series with optimization trailblazers (Lustig 2001).

17TH INTERNATIONAL MATHEMATICAL PROGRAMMING SYMPOSIUM, GEORGIA INSTITUTE OF TECHNOLOGY, ATLANTA, GEORGIA, 2000—FOUNDERS AWARD

(Left to right) Philip Wolfe, Harold Kuhn, Harry Markowitz, Ralph Gomory, George Dantzig, Alan Hoffman, Guus Zoutendijk, William Davidon.

Because of his wide knowledge of optimization—theoretical, computational, and algorithmic—Phil was often invited to speak at prestigious scientific groups. In 1974, he addressed the International Congress of Mathematicians, a most distinct professional honor. When the Russian mathematician L.G. Khachiyan's polynomial time ellipsoid method algorithm for LP (Khachiyan 1979) was first announced in the U.S., it was hilariously misunderstood by a *New York Times* reporter—"Soviet mathematician is obscure no more" and "the mystery author of a new mathematical theorem that has rocked the world of computer analysis" (*New York Times* 1979)—Phil was invited all over the world to explain the mistakes and put in perspective the theoretical value of Khachiyan's work, as well as its weak computational aspects. Phil also served as an adjunct professor, principally for Columbia University's Industrial Engineering and Operations Research Department, and also for the mathematics departments of the City University of New York and the New York (Brooklyn) Polytechnic Institute.

Phil retired from IBM in 1996, and continued teaching for several years at Polytechnic and Columbia. Now he enjoys full retirement, helping Hallie run her landscaping business, and volunteering as a tutor at a local college

Honors and Awards

Phil is a fellow of the American Association for the Advancement of Science (1972), the Econometric Society (1983), and the Institute for Operations Research and the Management Sciences (2002). He was awarded the ORSA and TIMS John von Neumann theory prize in 1992. He received a Distinguished Service Award and a Founders Award from the MPS in 2000.

Acknowledgments

In preparing this profile, I have been aided by correspondence with Andrew Conn, Richard Cottle, and, especially, Phil Wolfe.

References

Brent R, Winograd S, Wolfe P (1973) Optimal iterative processes for rootfinding. Numerische Mathematik 20:327–341

Cottle R, Goldfarb D, Hoffman A (eds) (1993) Festschrift in honor of Philip Wolfe. Math Program 62(1), 62(3)

Dantzig G (1963) Linear programming and extensions. Princeton University Press, Princeton, NJ

Dantzig G, Orden A, Wolfe P (1954) The generalized simplex method for minimizing a linear form under linear inequality constraints. The RAND Corporation, Research memorandum RM-1264, April; also, 1955, Pac J Math 5(2):183–195

Dantzig G, Wolfe P (1960) Decomposition principle for linear programs. Oper Res 8(1):101–111

Ford L Jr, Fulkerson D (1958) A suggested computation for maximal multicommodity network flows. Manage Sci 5(1):97–101

Fourier J (1826) Solution d'une question particulière du calcul des inégalités. Bulletin des Sciences par la Société Philomatique, Paris 99-100, Oeuvres 2, pp 315–319

Fourier J (1827) (Extract from) Analyse des travaux de l'Académie Royale des Sciences, pendant l'année 1824. Partie mathématique. Histoire l'Académie Royale des Sciences 7:xlvii–lv

Frank M, Wolfe P (1956) An algorithm for quadratic programming. Nav Res Logistics Q 3(1 and 2):95–110

Gale D, Stewart F (1953) Infinite games with perfect information. In: Kuhn H, Tucker A (eds) Contributions to the theory of games, vol II. Annals of Mathematics Studies Number 28. Princeton University Press, Princeton, NJ, pp 245–266

Gilmore P, Gomory R (1961) A linear programming approach to the cutting stock problem. Oper Res 9(6):849–859

Gilmore P, Gomory R (1963) A linear programming approach to the cutting stock problem -- part II. Oper Res 11(6):863–888

Gomory R (1958) Outline of an algorithm for integer solutions to linear programs. Bull Am Math Soc 64(5):1958

Gomory R (1963) An algorithm for integer solutions to linear programs. In: Graves R, Wolfe P (eds) Recent advances in mathematical programming. McGraw Hill, New York, NY, pp 269–302

Grattan-Guinness I (1970) Joseph Fourier's anticipation of linear programming. Oper Res Q 21(3):361–364

Hitchcock F (1941) The distribution of a product from several sources to numerous localities. J Math Phys 20:224–230

Hoffman A (1953) Cycling in the simplex method. National Bureau of Standards Report, No. 2874, December 16, Washington, DC

Khachiyan LG (1979) A polynomial algorithm in linear programming (in Russian). Doklady Akedamii Nauk SSR 244:1093–1096. [English translation: 1979. Soviet Mathematics Doklady 20:191–194]

Locke E (1949) The Finan-seer. Astounding Science Fiction (October)

Lustig I (2001)Interview with Phil Wolfe (text and video). http://www.e-optimization.com/directory/trailblazers/wolfe. Accessed 2 Oct 2010

Mapstone R (1972) Interview with Philip Wolfe. Computer oral history collection, 1969–1973, 1977. Archives Center, Smithsonian National Museum of American History, Washington, DC. http://invention.smithsonian.org/downloads/fa_cohc_tr_wolf721128.pdf. Accessed 6 May 2009)

Markowitz H (1956) The optimization of a quadratic function subject to linear constraints. Nav Res Logistics Q 3(1 and 2):111–133

Micchelli CA (ed) (2003) Selected papers of Alan J. Hoffman with commentary. World Scientific, River Edge, NJ

New York Times (1979) A Soviet discovery rocks world of mathematics. November 7, 1

Sion M, Wolfe P (1957) On a game without a value. In: Dresher M, Tucker A, Wolfe P (eds) Contributions to the theory of games, vol III. Annals of Mathematics Studies Number 39. Princeton University Press, Princeton, NJ, pp 299–306

Spielberg K (2007) IP over 40+ years at IBM scientific centers and marketing. Ann Oper Res 149(1):195–208

Von Neumann J, Morgenstern O (1944) Theory of games and economic behavior. Princeton University Press, Princeton, NJ (2nd edn, 1947; 3rd edn, 1953)

Wolfe P (1956) Determinateness of polyhedral games. In: Kuhn H, Tucker A (eds) Linear inequalities and related systems. Annals of Mathematics Studies Number 38. Princeton University Press, Princeton, NJ, pp 195–198

Wolfe P (1961) Accelerating the cutting plane method for nonlinear programming. J Soc Ind Appl Math 9(3):481–488

Wolfe P (1963a) A technique for resolving degeneracy in linear programming. J Soc Ind Appl Math 11(2):205–211

Wolfe P (1963b) Methods of nonlinear programming. In: Graves R, Wolfe P (eds) Recent advances in mathematical programming. McGraw-Hill, New York, NY, pp 67–86

Wolfe P (1966) On the convergence of gradient methods under constraints. Report RZ-204. IBM T. J. Watson Research Center, Yorktown Heights, New York, NY

Wolfe P (1967) Methods of nonlinear programming. In: Abadie J (ed) Nonlinear programming. North-Holland, Amsterdam, pp 97–131

Wolfe P (1969) Convergence conditions for ascent methods. SIAM Rev 11(2):226–235

Wolfe P (1970) Convergence theory in non-linear programming. In: Abadie J (ed) Integer and non-linear programming. North Holland, Amsterdam, pp 1–36

Wolfe P (1971) Convergence conditions for ascent methods, II: some corrections. SIAM Rev 13(2):185–188

Wolfe P (1974) Note on a method of conjugate subgradients for minimizing nondifferentiable functions. Math Program 7(1):380–383

Wolfe P (2009) Personal communication

Wolfe P, Cutler L (1963) Experiments in linear programming. In: Graves R, Wolfe P (1963) (eds) Recent advances in mathematical programming. McGraw-Hill, New York, NY, pp 177–200

35

Harry Markowitz

John B. Guerard, Jr.

A few researchers, as a result of their training, experiences, and possibly, their inherent curiosity, work in a number of somewhat unrelated fields and discover seminal and far-reaching results in each of those fields. Harry Markowitz has proved to be such a researcher. He has made major contributions to diverse areas of operations research (OR), as well as to the fields of economics and finance. His research contributions in the latter fields include the theory and practice of financial investments and stock portfolios, and applying this research to the money-on-the-line real-world of investments. His OR contributions range across the theory and application of linear programming (LP), quadratic programming, computer simulation, and OR modeling.

Harry was awarded the 1989 Operations Research Society of America (ORSA) and the Institute of Management Sciences (TIMS) John von Neumann Theory Prize. In 1990, he was a co-recipient of the Nobel Prize in economics. He has excelled in both the academic and practitioner worlds, and his research has helped to transform Wall Street as much as any other individual in the past 50 years.

From Comic Books to Darwin

Harry Markowitz was born in Chicago on August 24, 1927, the only child of Morris and Mildred Markowitz who owned a small grocery store (Markowitz 1990). Although Harry grew up during the Great Depression, his family lived in a nice apartment and always had enough to eat—Harry even had his own room. He had what many people would consider a normal childhood, playing baseball and touch football in a neighborhood

A.A. Assad, S.I. Gass (eds.), *Profiles in Operations Research*, International Series in Operations Research & Management Science 147, DOI 10.1007/978-1-4419-6281-2_35,

park, and playing the violin in the high school orchestra. Harry also enjoyed reading, particularly comic books and adventure magazines, including *The Shadow*. In late grammar school and throughout high school, he enjoyed popular accounts of physics and astronomy. At Carl Schurz High School, Harry began to read classics of philosophy and science, especially those of David Hume and Charles Darwin's *On the Origin of Species*. Even at that early age, Harry enjoyed Hume ["my philosopher" (Markowitz 1993, 3)] because his views allowed for the development of theories or models; he recognized that empirical testing was possible by building upon Hume's empirical emphasis (Markowitz 1993). With respect to Darwin, Harry, in his Nobel Prize autobiography, stated that he "was moved by Darwin's marshalling of facts and careful consideration of possible objections" (Markowitz 1990).

Having grown up in Chicago, it was assumed that Harry would attend the University of Chicago, as did his astronomer uncle, William Markowitz, who, while working for the U.S. Naval Observatory, developed the ephemeris time scale that was adopted as the international time standard. In 1945, Harry, with no scholarship assistance, did enroll in the University of Chicago's special 2-year bachelor's degree program. (Later, he received scholarships and fellowships.) He found all the courses of interest due to their emphasis on reading original materials. He was especially interested in the philosophers he read in a course called OII: Observation, Interpretation, and Integration (Markowitz 1990).

Becoming an economist was not one of Harry's childhood dreams. After receiving his B.A. (philosophy) in 1947, he had to choose an upper division concentration and he decided on economics. As he noted in an interview (Yost 2002, 3), "I liked the applications that economics had; the theoretical structure to the discipline." Microeconomics and macroeconomics were possible choices, but eventually it was the economics of uncertainty which excited his interests. Harry spent much of his time studying the theory of games and expected utility theory of John von Neumann and Oskar Morgenstern, augmented by Jacob Marschak's arguments concerning expected utility, the Friedman-Savage utility function, and Savage's defense of personal probability (Markowitz 1990). He had the good fortune to have Friedman, Marschak, and Savage among his Chicago graduate teachers.

Professor Tjalling Koopmans taught a course on activity analysis that emphasized the economic notion of efficiency and the analysis of efficient and inefficient production activities, concepts that proved crucial to Harry's future research and accomplishments. Koopmans was also director of the Cowles Commission for Research in Economics and invited Harry to become one of its student members. (The Cowles Commission was founded in Colorado Springs in 1932 by Alfred Cowles, a businessman and economist. It is dedicated to the pursuit of linking economic theory to mathematics and statistics. In 1939, the Cowles Commission moved to the University of Chicago. Both Friedman and Koopmans received the Nobel Prize in economics.)

When it was time for choosing a topic for Harry's dissertation, a chance conversation changed the course of events. As Harry related (Yost 2002, 4):

> Now several years later, I am at the stage where I have to choose a dissertation. I am now at a Masters and I am working towards my Ph.D. I went to my advisor, Professor Jacob Marschak, to ask him if he had any suggestions about a dissertation topic. He was busy, so I sat out in his anteroom. There was another gentleman there and we got to talking. He was a broker and suggested that I apply mathematical statistical techniques to the stock market. So when I got in to see Professor Marschak I said, 'The guy out there suggested I do a dissertation on the stock market.' At the time I was a student member of the Cowles Commission and Marschak had been formerly the head of the Cowles Commission. Marschak explained that Alfred Cowles, who had endowed the Cowles Commission, was particularly interested in the application of econometric techniques to the stock market. Marschak did not know the financial literature, and he suggested I see Professor Marshall Ketchum in the Business School. [Marshall Ketchum, was Professor of Finance in the Graduate School of Business at the University of Chicago, was editor of *The Journal of Finance* from 1946 to 1955, and served as president of the American Finance Association.]

> TO DIVERSIFY OR NOT TO DIVERSIFY
>
> As Harry noted in his article on the early history of portfolio theory (Markowitz 1999, 5):
>
> "In the *Merchant of Venice*, Shakespeare has the merchant Antonio say:
>
> My ventures are not in one bottom trusted.
> Nor to one place; nor is my whole estate
> Upon the fortune of this present year;
> Therefore, my merchandise makes me not sad.
>
> Act I, Scene 1
>
> Clearly, Shakespeare not only knew about diversification but, at an intuitive level, understood covariance."

Harry often tells the story of his epiphany, how the basic concepts of portfolio theory came to him one afternoon in the library while reading the 1938 book, *Theory of Investment Value*, by John Burr Williams (Markowitz 2002, 154). Williams proposed that the value of a stock should equal the present value of its future dividends. Because future dividends are uncertain, Harry interpreted Williams's proposal as an investor should value a stock by its expected future dividends. But if an investor were only interested in expected values of securities, the investor would only be interested in the expected value of the portfolio. And, to maximize the expected value of a portfolio, an investor would need to invest only in a single security. This, Harry knew, was not the way investors did or should act—one does not put all their eggs in one basket. Investors diversify because they are concerned with risk as well as return. Thus, action based on expected return only (like action based on certainty of the future) must be rejected as descriptive of actual or rational investment behavior. Williams's seminal text was written shortly after Graham and Dodd's *Security Analysis* (1934), and drew heavily from their valuation approach. Moreover, the Graham and Dodd low price-to-earnings and net current asset value (buying stocks for their liquidation or break-up value) strategies were included in the Williams monograph.

To Harry, an obvious measure of risk came to mind, the variance of the portfolio. The fact that portfolio variance depended on covariances of the securities added to the plausibility of the approach. Since there were two criteria, risk and return, it was natural to assume that investors selected from the set of Pareto optimal (non-dominated) risk–return combinations. The fact that the variance of the portfolio, that is the variance of a weighted sum, involved all covariance terms reinforced reasonableness of the approach. The riskiness of the portfolio was composed not only of the riskiness of the individual securities, as measured by their standard deviations, but also by the relative movements of the securities to one another, as measured by the covariance or correlation coefficient of the securities.

To minimize risk, one seeks to identify securities with lower, if not negative covariance or correlation coefficient. Since there were two criteria—risk and expected return—the natural approach for an economics student was to imagine the investor selecting a point from the set of Pareto optimal combinations of expected return and variance of return. Harry labeled the combinations of risk and return that were nondominated by other combinations as efficient and named the set of these points the efficient frontier. Over the succeeding years, the Markowitz approach to portfolio analysis has undergone various modifications, but it has always been concerned with finding the maximum return for a given level of risk, or the minimum risk for a given level of return. The efficient frontier traces out the optimal points along the risk–return frontier. For an investor, the choice of a portfolio's expected return and standard deviation is determined by the investor's tolerance of risk (Markowitz 1952a, 1956, 1959, 1970, 1987).

The concept of the efficient frontier can now be found in most textbooks on financial management or investment management and analysis. The efficient frontier is used to discuss why many investors prefer stocks to cash or bond investments, why smaller-capitalized stocks may be preferred to larger-capitalized stocks, and why portfolios composed of global stocks are very often preferred to portfolios composed of domestic-only stocks. Before Harry's work, investors described in words what was meant by the risk–return trade-off; after Harry, investors were able to precisely quantify, in mathematical terms, the risk–return trade-off (Rubinstein 2002). The risk–aversion of the investor identifies the portfolio produced from the efficient frontier. If one seeks to maximize the long-run rate of return of the portfolio and maximize the expected utility of (final) terminal wealth, then one selects the portfolio that maximizes the excess return-to-risk measure, known as the Sharpe Ratio (Sharpe 1994). This portfolio maximizes the geometric mean of the strategy, as well (Bloch et al. 1993; Markowitz 1959).

HARRY AT RAND

Although Harry had finished his course work, but not his dissertation, he felt that it was time to find a job. While attending a meeting of the American Economic Association, he met members of the

RAND Corporation's economics department. (At the time, RAND was a research organization in Santa Monica, California, supported by the U.S. Air Force.) They expressed interest in Harry's research and offered him a job at 50% more than an offer he already had from a university—Harry left the University of Chicago for RAND in early 1952 (Yost 2002, 7–8). Harry's initial RAND research was applying LP to economic problems.

Harry's first exposure to LP occurred at RAND when he was asked to read George Dantzig's paper on the simplex method (Dantzig 1951) and to supervise the computer programming and the running of RAND's first simplex code on an IBM Card Programmed Calculator (CPC). (The CPC consisted of a tabulator [punched-card accounting machine], an electronic calculator, and a bank of 16 mechanical storage registers, all strung together with cables; its input and output was via punched cards.) The LP problem under investigation had 30–40 equations, but the CPC was capable of only doing up to four simplex iterations (steps) per day (Yost 2002, 8). George Dantzig, the developer of LP and the simplex method, joined the RAND staff in June 1952, and over time, with the programming talent of William Orchard-Hays, RAND's ability to solve LP problems made great advances as computer speeds increased and computer-based LP algorithms improved (Markowitz 2002; Orchard-Hays 1984). Dantzig's work contributed to the solution of the portfolio selection problem in the following ways:

> Markowitz (1956) defines the portfolio selection problem as that of finding mean-variance efficient portfolios subject to linear equality and inequality constraints. This is the same constraint set as that of linear programming, but with mean-variance efficiency rather than the optimization of a linear function as the objective. The portfolio with maximum expected return, when it exists, is the natural starting point in tracing out the set of efficient portfolios. Since expected return is a linear function of portfolio investments, finding the portfolio with maximum expected return is a linear-programming problem. Dantzig's simplex algorithm not only provides the solution to this problem, but also provides the critical line algorithm [for generating the efficient frontier] with various other services. In particular, it determines whether or not the constraint set specified by the analyst is feasible, whether or not feasible portfolio expected return is bounded and, if the model is rank deficient, it provides an equivalent model which is not rank deficient (Markowitz 2002, 155).

WOLFE'S GENERALIZATION

"My work at the RAND Corporation did not include 'portfolio analysis.' But no one objected to my taking the time to write my [Markowitz] 1952[a] and [Markowitz]1956 articles. I submitted the latter to the *Naval Research Logistics Quarterly* [*NRLQ*] edited by Alan J. Hoffman. Elsewhere, Phil Wolfe [joint with Marguerite Frank, (Frank and Wolfe 1956)] had been working on the quadratic-programming problem, to minimize a quadratic function (Q[x] - λL, Q[x] is positive semidefinite, L linear) subject to linear constraints. [Q[x] is composed of the product of the decision variables x, the portfolio weights, and the covariance terms. Thus, one seeks to minimize risk, as measured by the portfolio standard deviation, the square root of the portfolio variance.] Wolfe also submitted his work to *NRLQ*. Hoffman sent Wolfe's paper to me and my paper to Wolfe for refereeing. We both recommended that the other paper be published, and both were [published in the same joint issue of *NRLQ*, 1956, 3 (1 and 2)]. As a by-product of tracing out the efficient frontier, the critical line algorithm minimizes Q[x] - λL (for variance Q[x] and expected return L) for all $\lambda \geq 0$. Thus the critical line algorithm is, incidentally, a quadratic-programming algorithm. It struck Phil Wolfe that the critical line algorithm solves the quadratic-programming problem in a sequence of steps which are precisely the same as the steps by which the simplex algorithm solves the linear-programming problem, with one exception. The variables of the quadratic program come in pairs X_i, η_i. When one of these pairs is IN the linear programming 'basis,' the other is OUT.[In the King's English, the portfolio weights were either zero (the security was 'OUT', or not included in the efficient portfolio), positive (if 'IN' or included in the efficient portfolio), or negative (if one could short a security) and the weights summed to one.] Wolfe thus defined quadratic programming as an example of linear complimentarity programming. At first it seemed that the practical use of this observation was to easily convert a linear-programming code into a quadratic-programming (or portfolio selection) code. Subsequently, it was found that other problems satisfied the linear complimentarity format, e.g., non-zero-sum games (Lemke 1965)" (Markowitz 2002, 155).

THE BOOK: *PORTFOLIO SELECTION: EFFICIENT DIVERSIFICATION OF INVESTMENTS*

In 1954, Harry finished his dissertation on "Portfolio analysis" and received his Ph.D. in economics from the University of Chicago. During that year, he received a call from Yale Professor James Tobin inviting him to spend the 1955–1956 academic year at Yale—the Cowles Commission was moving there. Tobin, a Harvard-educated Ph.D. in economics, was also director of the newly named Cowles Foundation for Research in Economics at Yale. He invited Harry to write a Cowles Foundation monograph on portfolio theory.

A draft of the monograph was finished during Harry's visit; it was reviewed by Tobin and the economist Gerard Debreu (a Cowles research associate—both Tobin and Debreu would later receive the Nobel Prize in economics), revised by Harry, and published as Cowles Foundation Monograph 16 under the title *Portfolio Selection: Efficient Diversification of Investments* (Markowitz 1959). The book has justly gained its place of significance in the economic literature, particularly following the publication of the research of Harry's student, William (Bill) Sharpe, on the Capital Asset Pricing Model CAPM (Sharpe 1964). The importance of *Portfolio Selection* to financial analysts was readily recognized, but, possibly due to its title, its value did not become apparent to researchers in mathematical programming and OR.

The monograph built upon Harry's 1952a and 1956 papers and their extensions. Brealey (1991) describes four of Harry's extensions: "First, Markowitz analyzes the utility implications of alternative selection rules and in the case of semi-variance criterion, he discusses the solution procedure. Second, Markowitz includes a discussion of portfolio selection under logarithmic utility and points out that the portfolio that maximizes the geometric mean return is approximately mean-variance efficient. Third, he showed that as the number of securities in the portfolio is increased, portfolio variance approaches the average covariance. Finally, he suggested that the covariance matrix may be simplified if the correlation between security returns result simply from the common influence of the market" (Brealey 1991, 8–9). As Breasley further notes: "He [Harry] is at all times concerned with the problems of practical implementation" (Breasley 1991, 9). Capital market theory assumes that investors make decisions on the basis of means and standard deviations, as well as the semi-variance. Markowitz has never believed that investors ignore, or should ignore, non-normal distributions such as positively skewed stock returns. Portfolio managers often calculate Sortino ratios of downside risk and discuss non-normality of stock and portfolio distributions. (The Sortino ratio, a modification of the Sharpe ratio, measures the risk-adjusted return of an investment asset, portfolio or strategy.) Markowitz was very much aware of minimizing the underperformance of an index benchmark in 1959.

In the monograph's Appendix A: The Computation of Efficient Sets, and in Appendix B: A Simplex Method for Portfolio Selection, Harry clarifies the use of Frank and Wolfe's (1956) quadratic-programming algorithm and the role of his critical line algorithm (Markowitz 1956). Harry notes: "[Frank and] Wolfe was primarily concerned with minimizing a quadratic, and incidentally noted that the amended simplex algorithm would solve the portfolio selection problem; as compared with Markowitz, who was primarily concerned with the portfolio selection problem and incidentally noted that the critical line algorithm would minimize a quadratic" (Markowitz 1959, 222). He further asserts the equivalence of the two algorithms and concludes by stating: "The practical significance of the above result is that any of the linear programming codes for high speed, internally programmed computers can be conveniently converted to a quadratic programming code for solving, among other things, the portfolio selection problem" (Markowitz 1959, 339).

ON SHARING KNOWLEDGE

"One day in 1960, having said what I had to say about portfolio theory in my 1959 book, I was sitting in my office at the RAND Corporation in Santa Monica, California, working on something quite different, when a young man presented himself at my door, introduced himself as Bill Sharpe, and said that he also was employed at RAND and was working toward a Ph.D. degree at UCLA. He was looking for a thesis topic. His professor, Fred Weston, had reminded Sharpe of my 1952 article, which they had covered in class, and suggested that he ask me for suggestions on a thesis topic. We talked about the need for models of covariance. This conversation started Sharpe out on the first of his (ultimately many) lines of research, which resulted in Sharpe (1963).

"For all we know, the day Sharpe introduced himself to me at RAND could have been exactly 10 years after the day I read Williams. On that day in 1960, there was *no* talk about the possibility of using portfolio theory to revolutionize the theory of financial markets, as done in Sharpe (1964), nor was there any inkling of the flood of discoveries and applications, many by Sharpe himself, that were to occur in investment theory and financial economics during the next four decades" (Markowitz 1990, 14).

Return to RAND: Pivoting and Simulation

On his return to RAND in 1956, Harry continued his earlier work with the economist Alan Manne and extended their previous research to develop industry-wide and multi-industry activity analysis models of industrial capabilities (Markowitz 1990). Harry and the industrial engineer Alan Rowe, a RAND consultant and professor at the University of California's Anderson School of Management, co-authored five chapters on metalworking in the Manne and Markowitz (1963) edited volume *Studies in Process Analysis*. The resulting LP model they encountered was rather large—had many equations and variables—and the available LP computer codes were restricted to about 200 equations. They looked for ways to modify or reduce the structure to make it computationally tractable. As their model's equations were such that few variables had non-zero coefficients, they were able to solve small problems by carefully selecting the simplex method's pivot elements so that the application of the usual Gaussian elimination introduced as few non-zeros as possible. Using this idea, Harry developed an approach and a pivot selection rule, when applied to a large-scale problem, greatly reduced the generation of the transformed matrix's non-zero elements (Markowitz 1957). Harry described such systems as having sparse matrices, a term and method that has had application to similar computational problems. This idea was carried out at RAND, programmed by Orchard-Hays, within a product form of the inverse LP computer code to solve large models with relatively few non-zero coefficients (Dantzig et al. 1954; Markowitz 2002; Orchard-Hays 1984). ". . . that was the hardest thing he had to program," reported Orchard-Hays (Yost 2002, 10).

In the 1950s, RAND was using computer-based simulation procedures to evaluate military situations, with special emphasis on war games. To that end, RAND created a logistics laboratory within its Economics Department. Although Harry was familiar with simulation ideas and techniques as applied to industrial operations and warfare research, his first hands-on exposure to such simulations was when he was assigned to the laboratory to coordinate the development of its computer-based simulation models (Markowitz 2002; Yost 2002). The laboratory's first simulation model, called LP1, "was a man-machine simulation in which actual air force logistics officers played the role of air force logistics officers. The computer flew simulated missions, generated part failures and other maintenance requirements, and kept track of parts supplies and aircraft status" (Markowitz 2002, 157).

Some time after LP1 was finished, Harry received a job offer from the General Electric (GE) Company and accepted a position within the Manufacturing Services Department located in New York City. Alan Rowe was now with GE and was the supervisor of the programming of a large, detailed job-shop simulator (Markowitz 1993). Harry, based on Rowe's experiences and his own understanding of the problems in building a simulator, had developed ideas as to how to reduce programming time and increase a simulator's flexibility. He applied his ideas to the building of the GE Transformer Department's shop simulator, the General Electric Manufacturing Simulator (GEMS). Although GEMS was well received at GE, it was not as flexible as Harry hoped it would be.

Harry gave much thought to what attributes a simulation should have and decided to write a nonproprietary simulation language. He did not want to develop it at GE because it would then be proprietary. He went on the job market and returned to RAND to put his ideas to work. At RAND, Harry teamed up with Bernard Hausner and, with the help of Herbert Karr, developed SIMSCRIPT, a very powerful, influential, and long-lived computer-based simulation system (Markowitz et al. 1963). Hausner was a RAND computer programmer and was responsible for the development of the computer-based simulation language using Harry's novel simulation concepts of entities, sets, and events. The initial name for the language was Simulation Programming System One (SPS1). Herb Karr, a friend of Harry's, was hired to write the SPS1 manual. Herb wrote the manuscript for the evolving language, which became known as SIMSCRIPT. SIMSCRIPT introduced some novel concepts. It was "designed to facilitate the programming of 'discrete event' simulation models, especially 'asynchronous' discrete event simulators, as compared to continuous time or difference equation models" (Markowitz 2002, 157).

WHAT IS SIMSCRIPT?: ENTITIES, ATTRIBUTES, AND SETS

"The objective of SIMSCRIPT was to allow the simulation programmer to describe the world to be simulated, and relieve said programmer from implementation details insofar as we could. The SIMSCRIPT world view is as follows: As of an instant in time the system to be simulated has a *status* that changes at points in time called *events*. Status is described in terms of how many of various types of *entities* exist, what are the values of their *attributes*, and what entities belong to the *sets* which other entities own. Early 21st Century programming languages are likely to refer to Entities, Attributes, and Sets as Objects, Properties, and Collections (or Child–Parent relationships). Programming languages at the beginning of the 1960s spoke instead of variables and arrays.

"The SIMSCRIPT [I] programmer described the entities, attributes, and sets of the system to be simulated on a Definition Form. In those days, the computer input was typically the punched card. The data written on the Definition Form, to be keypunched and placed in the SIMSCRIPT source program deck, included names of entity types; names of attributes, their data types, and precision information; the names of sets plus information as to what type of entity owns the set, what type belongs to it, and how the set is organized.

"Changes in status were described in event routines written in the SIMSCRIPT programming language. The language included commands to CREATE and DESTROY entities, FILE entities into or REMOVE them from sets, FIND set members meeting specified tests, DO some action(s) FOR EACH member of sets, CAUSE or CANCEL subsequent event occurrences, etc., as well as perform arithmetic operations on attributes. We sought to make the commands English-like, 'self-documenting'" (Markowitz 2002, 157).

The RAND Corporation owned SIMSCRIPT and used it for its logistics simulations (Johnson 1986). SIMSCRIPT was made available without charge through the SHARE organization, a nonprofit, volunteer-run user group for IBM mainframe computers that was founded in 1955 for exchanging technical information about programming languages, operating systems, database systems, and user experiences.

Once again, Harry left RAND, this time to join with Herb Karr to form the California Analysis Centers, Inc. (CACI) to provide SIMSCRIPT consultant services and courses. They each put up $6,000 as the initial capitalization; CACI was incorporated on July 17, 1962. Harry was chairman of the board and technical director, and Herb was president (Yost 2002). Harry then joined with RAND simulation researchers, Philip Kiviat and Richard Villaneuva, to develop and design a new version of SIMSCRIPT—SIMSCRIPT II—a user-oriented, general-purpose simulation programming language (Markowitz et al. 1969). It was released by RAND in 1969. Under contract to IBM, Harry, based on SIMSCRIPT II, helped to develop an experimental programming language integrated with a database management system, the EAS-E system (Malhotra et al. 1983; Markowitz 2002). EAS-E (pronounced EASY) was built around the entity, attribute, and set (EAS) view of application development. IBM used EAS-E for an internal application, but it was never released as a product. CACI developed and marketed a proprietary version of SIMSCRIPT II, SIMSCRIPT II.5. In 1968, Harry left CACI and, for the first time, joined academia as a professor at the business school of the University of California-Los Angeles (UCLA).

GOING PUBLIC

"By the beginning of 1968 CACI had grown from Herb and me to a small but growing company planning to 'go public.' CACI's initial public offering did in fact take place during the second half of 1968. That was the good news. The bad news was that Herb and I had a major disagreement over the pricing of a new product, then a disagreement over how to settle disagreements. This was finally settled on March 15—the Ides of March—of 1968 when Herb Karr, with about 47% of CACI stock and Jim Berkson, vice president of finance, with about 5% of the stock, fired me with about 47% of the stock" (Markowitz 2002, 159).

EPILOGUE RE RAND

Although Harry was probably first hired at RAND based on his University of Chicago research on portfolio analysis, his RAND research activities did not involve such analysis. Harry published four of the seven papers of which he is most proud while at RAND (Markowitz 2009). These articles show the diversity of his mind. The first two are of his University of Chicago research. "Portfolio selection," the basis of his Nobel Prize (Markowitz 1952a), and "The utility of wealth," a standard reference in utility theory and the economics of uncertainty (Markowitz 1952b). The third paper, "The optimization of a quadratic function subject to linear constraints," was the initial application of quadratic programming to the portfolio construction problem (Markowitz 1956), while the fourth paper, (Markowitz 1957), "The elimination form of the inverse and its application to linear programming," represents his research into sparse-matrix methods, the only one of the four that was relevant to RAND. Harry's three other most favorite publications are the SIMSCRIPT monograph (Markowitz et al. 1963), the discussion of the EAS-E system (Malhotra et al. 1983), and "Financial market simulation" (Jacobs et al. 2004).

THE PERIPATETIC PROFESSOR CUM CONSULTANT

During his academic career, Harry taught MBA and Ph.D. level courses in investments and portfolio management. He was a professor at UCLA (1968–1969), University of Pennsylvania's Wharton School (1972–1974), and Rutgers University (1980–1982). From 1974 to 1983, he was a staff member at IBM's T. J. Watson Research Center, Yorktown Heights, New York. In 1982, Harry was appointed the Marvin Speiser Distinguished Professor of Finance and Economics at Baruch College, City University of New York. In 1993, he retired from Baruch College as Distinguished Professor Emeritus. He was a visiting Professor at Hebrew University, Jerusalem; University of Tokyo; and the London Business School. Harry moved to San Diego in 1993, where he lives with his wife Barbara. He is an adjunct professor at the Rady School of Management, University of California, San Diego. Harry and Barbara support the nonprofit Rational Decision Making Research Institute, where he developed a new nonproprietary version of EAS-E (Yost 2002, 33–34; EAS-E.ORG 2009).

Harry continued to consult during his academic tenure, serving, from 1984, as President of the Harry Markowitz Company, and, from 1990 to 2000, as Director of Research, Global Portfolio Research Department (GPRD), for the Daiwa Securities Trust Company, the U.S. affiliate of Japan's Daiwa Securities. GPRD develops research models and manages

Harry and Barbara

money for institutional clients using the Daiwa (Markowitz) portfolio mean-variance optimization system (Bloch et al. 1993). Harry created the GPRD as a start-up research department, and allowed researchers considerable flexibility in terms of suggesting models to be tested; Harry, however, insisted on using academically rigorous standards in a Wall Street atmosphere. The GRPD published several journal articles in addition to managing institutional assets. In the early 2000s, Harry worked in joint research with Bruce Jacobs and Ken Levy of Jacobs and Levy Equity Management, a provider of quantitative equity strategies for institutional clients, where he helped to construct the JLM Market Simulator (Jacobs et al. 2004). The JLM simulator is an asynchronous simulation that investors can use to create a model of the market using their own inputs. The investor's portfolio selection choice comes from the risk–aversion coefficient parameter that helps the client choose from a desired portfolio on the efficient frontier.

Harry has made great intellectual contributions to the worlds of finance, investment management, LP, sparse matrices, and computer simulation, many of which are discussed in his collected papers (Markowitz 2008). He has been a successful endowed professor, consultant, research staff member, and entrepreneur. Modern portfolio theory is now a standard topic in college courses and texts on investments, and widely used by institutional investors and by many quantitative money managers for stock selection for equity portfolios.

Honors and Awards

In 1982–1983, Harry served as president of the American Finance Association. He was awarded the ORSA and TIMS John von Neumann Theory Prize in 1989 for his ground-breaking work in portfolio selection, mathematical programming, and simulation. In 1990, he was the joint recipient (with Merton Miller and William Sharpe) of the Nobel Prize in economics for his for pioneering work in the theory of financial economics.

Harry has been awarded Honorary Doctorates from the Université D'Orléans (1992), University of Athens (1992), Aristotle University of Thessaloniki (2003), and the American International University in association with the Royal College of Physicians and Surgeons of the U.S. of America (2003).

REFERENCES

Bloch M. Guerard J Jr, Markowitz H, Todd P, Xu GL (1993) A comparison of some aspects of the U.S. and Japanese equity markets. Jpn World Econ 5(1):3–26

Brealey R (1991) Harry M. Markowitz's contribution to financial economics. Scand J Econ 93(1):7–17

Dantzig G (1951) Maximization of a linear function of variables subject to linear inequalities. In: Koopmans TC (ed) Activity analysis of production and allocation: proceedings of a conference. Wiley, New York, NY, pp 339–347

Dantzig G, Orden A, Wolfe P (1954) The generalized simplex method for minimizing a linear form under linear inequality constraints. RAND Corporation Report RM-1264, Santa Monica, CA

EAS-E.ORG (2009) www.eas-e.org. Accessed 3 Feb 2009

Frank M, Wolfe P (1956) An algorithm for quadratic programming. Nav Res Logistics Q 3(1 and 2):95–110

Graham B, Dodd D (1934) Security analysis: principles and technique. McGraw-Hill, New York, NY

Jacobs B, Levy K, Markowitz HM (2004) Financial market simulation in the 21st century. J Portfolio Manage 30 (30th anniversary issue):142–151

Johnson L (1986) Oral history of Harry M. Markowitz. Computer History Museum. CHM Reference number X4804.2008, Mountain View, CA. http://archive.computerhistory.org/resources/text/Oral_History/Markowitz_Harry/102658332.05.01.acc.pdf. Accessed 14 Mar 2009

Lemke C (1965) Bimatrix equilibrium points and mathematical programming. Manage Sci 11(7):681–689

Malhotra A, Markowitz HM, Pazel DP (1984) EAS-E: an integrated approach to application development. ACM Trans Database Syst 8(4):515–542

Manne A, Markowitz HM (1963) Studies in process analysis: economy-wide production capabilities. Wiley, New York, NY

Markowitz HM (1952a) Portfolio selection. J Finance 7(1):77–91

Markowitz HM (1952b) The utility of wealth. J Pol Economy 60(2):151–158

Markowitz HM (1956) The optimization of a quadratic function subject to linear constraints. Nav Res Logistics Q 3(1 and 2):111–133

Markowitz HM (1957) The elimination form of the inverse and its application to linear programming. Manage Sci 3(3):255–269

Markowitz HM (1959) Portfolio selection: efficient diversification of investments. Cowles Foundation for Research in Economics Monograph 16, Yale University, New Haven, CT

Markowitz HM (1970) Portfolio selection: efficient diversification of investments. Wiley, New York, NY

Markowitz HM (1987) Mean-variance analysis in portfolio choice and capital markets. Basil Blackwell, New York, NY

Markowitz HM (1990) Autobiography. http://nobelprize.org/nobel_prizes/economics/laureates/1990/markowitz-autobio.html. Accessed 29 Jan 2009

Markowitz HM (1991) Foundations of portfolio theory. J Finance 46(2):469–477

Markowitz HM (1993) Trains of thought. Am Econ 37(1):3–9

Markowitz HM (1999) The early history of portfolio theory: 1600–1960. Financ Anal J 55(4):5–16

Markowitz HM (2002) Efficient portfolios, sparse matrices, and entities: a retrospective. Oper Res 50(1):154–160

Markowitz HM (2008) Harry Markowitz: selected works. World Scientific, Hackensack, NJ

Markowitz HM (2009) Personal communication

Markowitz HM, Hausner B, Karr H (1963) SIMSCRIPT: a simulation programming language. Prentice-Hall, Englewood Cliffs, NJ

Markowitz HM, Kiviat P, Villanueva R (1969) The SIMSCRIPT II programming language. Prentice Hall, New York, NY

Orchard-Hays W (1984) History of mathematical programming systems. Ann Hist Comput 6(3):296–312

Rubinstein M (2002) Markowitz's "Portfolio Selection": a fifty-year retrospective. J Finance 57(3):1041–1045

Sharpe W (1963) A simplified model for portfolio analysis. Manage Sci 9(2):277–293

Sharpe W (1964) Capital asset prices: a theory of market equilibrium under conditions of risk. J Finance 19(3):425–442

Sharpe W (1994) The Sharpe Ratio. J Portfolio Manage 21(1):49–58

Williams J (1938) The theory of investment value. Harvard University Press, Cambridge, MA

Yost J (2002) An interview with Harry Markowitz. http://special.lib.umn.edu/cbi/oh/pdf.phtml?id=297. Accessed 29 Jan 2009

36

John D. C. Little

John R. Hauser and
Glen L. Urban

The development of operations research (OR) after World War II (WWII) was greatly influenced by OR pioneers who applied their knowledge and experiences to the problems of business and industry. In parallel, they brought OR methods and practices into academia, and created the environment for the development and training of a new cadre of OR practitioners and researchers. In the forefront of this new group, we find John D. C. Little. From being the first to receive a Ph.D. in OR, John went on to leave his own lasting imprint on the field as a progenitor of the field of marketing science and its applications, the person behind the eponymous Little's Law for queues, an influential academic leader, and a highly successful OR researcher, practitioner, and entrepreneur.

For his innovative and seminal research in marketing, John received the American Marketing Association (AMA) Charles Parlin Award for contributions to the practice of marketing research (1979), the AMA's Paul D. Converse Award for lifetime achievement (1992), and MIT's Buck Weaver Award for outstanding contributions to marketing (2003). He was president of the Operations Research Society of America (ORSA) in 1979, The Institute of Management Sciences (TIMS) in 1984–1985, and first president (1995) of the Institute for Operations Research and the Management Sciences (INFORMS), one of only two individuals who have served as president of the three organizations.

A.A. Assad, S.I. Gass (eds.), *Profiles in Operations Research*, International Series in Operations Research & Management Science 147, DOI 10.1007/978-1-4419-6281-2_36,

BOSTON BORN, MASSACHUSETTS BRED, AND ON TO MIT

John D. C. Little was born in Boston, Massachusetts, on February 1, 1928, the son of John D. and Margaret J. Little, and grew up in Andover, Massachusetts. Both his father and mother were natives of Massachusetts, he was born in Malden and she in Gloucester. Margaret graduated from Smith College and, after her three children had grown up, became an English teacher and then the principal of a private school in Andover. John D. attended Malden public schools and Dartmouth College, but left without graduating to become an ambulance driver in France during World War I. Later, he held various positions including reporter for the *Boston Herald*; editor for a financial journal in Washington, D.C.; bond salesman for a Boston brokerage firm; writer for the Office of War Information in Washington during WWII; and a credit manager.

John (D. C.) lived in the West Parish part of Andover, which was then quite rural, an exurbia from which his father commuted to Boston by train. John attended Andover's elementary and middle public schools and was a good student, especially in mathematics and science. For high school, he obtained a scholarship to the independent Phillips Academy in Andover and won most of the science-related prizes. He graduated Andover in 1945 and started college at the Massachusetts Institute of Technology (MIT). Due to the wartime acceleration of MIT's academic program, his freshman year began in the summer. He decided to major in physics, which appealed to him as a worthy intellectual challenge. John did more than study—he became editor-in-chief of *Voo Doo*, the MIT humor magazine. He also "took a minor in hitchhiking to Wellesley [a women's college]" (Little 2008). He graduated in 3 years, receiving an S.B. degree in 1948. Tired of school and not yet wanting to enter the working world, John hitchhiked around the country for 10 months. This brought him to the point where work seemed better than poverty, so he joined the General Electric Company as an engineer. In 1951, he went back to MIT, enrolling as a graduate student in physics. Although he passed his general exams in physics, his intellectual curiosity caused him to search out other areas—a research assistantship (RA) on an unclassified air defense project with the psychologist J.C.R. Licklider, and a course in the new field of OR taught by George Wadsworth of the mathematics department.

John, inspired by the challenges offered by the embryonic field of OR, obtained an RA working for the physicist Philip Morse. Morse, a WWII pioneer in OR, is recognized as the founder of OR in the U.S.—he had established the first U.S. OR group for the Navy in 1942. After WWII, Morse founded and directed MIT's inter-departmental Operations Research Center. John's RA was for a Navy-sponsored project, "Machine Methods of Computation and Numerical Methods."

Machine computation meant Whirlwind, one of the earliest digital computers—it had been built in the late 1940s by MIT personnel to support the Navy's research program. John's task was to learn about Whirlwind, how to program it, and to compute a book of tables for spheroidal wave functions—rather esoteric functions used for calculations in theoretical physics. John did not think the resulting book was going to be a big seller, but, in assuming the task, he thought "an RA is an RA"; one usually does not question the professor, especially when the job helps pay your tuition. More importantly, John became one of the few people in the world with access to a digital computer. The knowledge of what a computer could accomplish was central to his future research and consulting activities. He felt that "computers are cool" and he can often be found "playing" with the latest technological devices (Little 2008).

When John asked Morse about possible thesis topics, Morse mentioned a few from physics and then wondered if John would be interested in an OR topic. Physics or OR? John faced the decision and concluded, "Physics is fine, but look at it this way. Bohr solved the hydrogen atom and it's beautiful. Then somebody named Hylleraas solved the helium atom. It took him 7 years on a hand-crank calculator and it's ugly. Beyond helium there are another 100 or so elements with bad prognoses. I'm searching for a field with a lot of unsolved hydrogen atoms. OR looks good" (Little 2007, 3).

John's dissertation research dealt with the study of water-flow management in a hydroelectric reservoir and dam system (Little 1955). In particular, his analysis dealt with Washington State's section of the Columbia

River, its Grand Coulee hydroelectric plant, and the Franklin Delano Roosevelt Lake (the reservoir that formed behind Grand Coulee Dam).

The problem was how best to schedule the amount of water flow used to generate electricity. The system is naturally dynamic with the available water to generate power a function of seasonal rainfall and the runoff from snow melting in the mountains during the spring and summer. In the fall and winter, when precipitation falls as snow in the mountains, the natural river flow drops drastically. Water leaves the reservoir from spill-over (wasted energy) and when the water is drawn down to generate electricity. The problem is interesting because the power is proportional to the head, the height of the water behind the dam relative to the water below the dam, and to the rate of water flow. A greater flow generates more electricity, but also reduces the head at a faster rate.

The tradeoff was challenging. As John notes, decisions have to be made: ". . . in the spring and summer, the right decisions about water use are obvious—indeed they are hardly decisions—whereas in the fall and winter such decisions require a balancing of the benefits of future against immediate water use in the face of uncertain future flow" (Little 1955, 188).

John formulated the problem as a dynamic program, although, at the time, he did not know it was a dynamic program. John faced the dynamic-programming "curse of dimensionality," the rapid increase in computing time with the number of state variables (Bellman 1957, ix). The simplest credible formulation required two state variables: one for the amount of water in the reservoir (which determines the head) and the other for the current river flow. Thanks to Whirlwind, John was able to finish his thesis before his RA ran out (Little 2007, 4).

The thesis was very likely the first non-defense application of dynamic programming to a problem of practical importance. Real data were used for the historical stream flows. The models of Grand Coulee and its reservoir were simplified to save computing time, but were based on actual physical dimensions (Little 1955).

John received his Ph.D. in OR in 1955, the first person in the world to receive such a degree—his dissertation title was the "Use of Storage Water in a Hydroelectric System," and his advisor was Philip Morse. John, to be precise, describes his Ph.D. as being in physics and OR, since his general exams were in physics and his thesis in OR (Little 2008).

In 1953, John married Elizabeth (Betty) Alden, an MIT physics graduate student who worked on ferroelectrics under Professor Arthur von Hippel, a pioneer in dielectrics, especially ferromagnetic and ferroelectric materials. She received her Ph.D. in physics from MIT in 1954, one of very few women to do so at the time. After their marriage, Betty and John moved to Marlborough Street in Boston's Back Bay. The rent was high "but not too bad and we could split it" (Little 2007, 4). Beginning a lifetime of exercise (John jogs and bicycles daily at age 80), they had to walk up five flights to reach their tiny, top floor, two-room apartment. It had sloping ceilings to conform to the roof, but it suited their purposes admirably.

Army, Case Institute, and Laying Down the Law

In early 1955, 3 months after completing his thesis, John was drafted into the Army. He was stationed for 2 years at Ft. Monroe, Hampton, Virginia, where he served as an operations analyst working on military OR problems that included probabilistic models of land mine warfare. As an antidote to the Army, John and Betty bought a sailboat and had a wonderful time getting in and out of trouble on the Chesapeake Bay. Their first child, John N. Little, was born in the Ft. Monroe Army hospital. Upon his discharge from the Army in 1957, John began his academic career as an assistant professor at Case Institute of Technology (now Case Western Reserve University) in Cleveland, Ohio. There he had his first experiences working on industrial OR projects.

A project with M&M Candies introduced him to advertising problems, and a project with Cummins Engine, Inc. introduced him to the issues of conflict between a manufacturer and its independent distributors. At M&M, the president had deliberately stopped all advertising after a long period of operating at a high level to see what would happen. The Case team analyzed the resulting sales over time. At first, sales changed very little and then started into a serious decline. With their response analysis and further data, the team calibrated a model which led it to recommend a new policy for buying TV spot advertising. At Cummins, top management was dismayed that, when the company provided extra sales support to its independent distributors, the latter rather quickly reduced their own. John devised a graphical profit analysis showing that such behavior by the distributors was entirely rational. With his accumulation of such real-world experiences, he developed and taught a graduate course—OR in Marketing—that may have been the first such marketing science course.

Although not a problem in marketing, John also worked on the traveling salesman problem (TSP), whose solution procedures for finding the minimum TSP route were computationally challenging. In addressing the problem, he and his coauthors introduced and popularized the term branch and bound as a technique for solving the TSP, as well as other combinatorial optimization problems (Little et al. 1963, 972). For a while, the authors held "the indoor record for size of problem solved, 40 cities on the fastest machine at MIT at the time, an IBM 7090" (Little 2008). It turns out, over four decades later, that the TSP is relevant to modern marketing science. With automatic tracking of supermarket shopping carts and recorded checkout data, researchers are using optimal traveling salesman solutions as reference routes for studying how shoppers actually move around in stores.

John's research at Case (and later work) was clearly influenced by his mentor Philip Morse, who began his career as a theoretical physicist but, in the course of his OR activities, became both an experimentalist and a theoretician. John was guided by the definition of OR promulgated by Morse and George Kimball in their book, *Methods of Operations Research* (Morse and Kimball 1951, 1) "Operations Research is a scientific method of providing executive departments with a quantitative basis for decisions regarding operations under their control." John, in his retrospective review, "Philip M. Morse and the beginnings" (Little 2002) notes: "The definition leaves room for the tremendous development of methodology that we have witnessed in the past 50 years, but it keeps our feet on the ground with the requirement for data, models, and decisions. I like that, and I am sure it is what Morse intended" (Little 2002, 148).

Being true to this paradigm, John, in a study of a real-world problem, would often meet with managers to learn how they perceive the problem. Then he would formulate tentative hypotheses about underlying processes that, if understood, might permit improvement in operations. After that, he would look for relevant data and/or design and execute a plan for collecting it. Whenever feasible, he would do this with students. Often, iteration was required between the observed and modeled worlds until the model was right for the job at hand. This style marked his professional behavior in such disparate subfields of OR as hydroelectric systems, traffic signal synchronization, the process of managing, and eventually marketing.

An exception to the paradigm is the paper underlying Little's Law. This paper, written while John was at Case, established him in OR history. He published the first general proof of the famous queueing formula, $L = \lambda W$ (Little 1961a). Assuming steady state operation, the formula says that the average number of customers in a queueing system over time (L) equals the average arrival rate of customers to the system (λ) multiplied by the average time that each customer spends waiting in the system (W). A customer can be anything from a consumer waiting for a teller in a bank, to an aircraft waiting to land, to a packet of data waiting to be processed in a computer network. Little's Law allows an analyst to obtain all three of these fundamental performance measures of a queue by calculating (or measuring) only two of them. This is useful because the analytic methods used to calculate L and W are usually quite different and, often, one is much easier to carry out than the other.

John taught a queueing course at Case and, among other sources, used Morse's pioneering book on queueing theory for OR applications, *Queues, Inventory and Maintenance* (Morse 1958). In one of John's lectures, he pointed out Morse's observation that the curious formula $L = \lambda W$ always seemed to apply to queues whose operational behavior Morse had solved the long, hard probabilistic way, that is, by making specific assumptions about arrival processes, service time distributions, and queue disciplines (Morse 1958, 75). John went on to sketch a figure on the blackboard, similar to the one that appears in his paper (Little 1961a, 385). He used the figure to give a heuristic argument why the formula should hold in great generality for steady state queues. In discussion after class, one of the students wondered how difficult it would be to prove the general case. John obligingly answered, "It shouldn't be too hard." "Then you should do it!" was the response (Little and Graves 2008, 99).

The discussion stuck in John's mind and he started to think about how he might turn his heuristic proof into a formal one. He bought and read some books on general stochastic processes. In those years, he and Betty would pile themselves, their children, their summer clothes, and a stack of books for John's research projects into their Ford Falcon station wagon and head for Nantucket (Island), Massachusetts to spend the summer. This particular year (1960) he took his new books and worked on $L = \lambda W$ as one of his major projects. The books did not have any magic formulas, but they gave him important ideas—the outcome was his paper, "A proof of the queuing formula $L = \lambda W$" (Little 1961a). John decided then and there not to make a career out of being a measure-theoretic stochastic process mathematician—he has never regretted it. Little's Law has entered OR folklore. At an ORSA conference in New Orleans, T-shirts were sold to raise money for ORSA. A best seller was the one that proclaimed: "It may be Little, but it's the Law."

> WHAT'S IN A NAME? THAT WHICH WE CALL A ROSE . . . .
>
> There has been much curiosity about what John's D. C. middle initials stand for—he has been asked often. An extreme case of curiosity occurred when an OR teacher in Oklahoma City challenged his OR class to find out the exact middle names of the person after whom Little's Law is named—the first student who did so would get a reward. This resulted in John receiving much email from Oklahoma City. Out of curiosity, John decided to pursue the challenge and searched the Web for the answer. He was surprised to find how difficult the task was. He was able to do it, but admits he was aided by knowing that the names are in two places on the MIT Web site.
>
> Many people have guessed that D. C. came from direct current based on John's early work on hydroelectricity, which, by the way would be A. C. (alternating current). Others have guessed District of Columbia from John's contributions to OR in the U.S. It is not DC Comics, although many people consider John a super-hero, nor DC Shoes for skateboards—John jogs. It is not Dominican College or the Dublin Core, nor is it D. C. United, the Department of Corrections, desert combat, or digital camera.
>
> The answer is Dutton Conant. His father, who was John Dutton Little, did not want John to be called junior and so added another middle name. His father was close to his grandmother whose maiden name was Conant. Thus, John became John D. C. Little. John has noted that, although there are many people named John Little, he has never found another who was John D. C. Little. He finds this helpful in searching for himself in Web documents (Little 2008).

Return to MIT, Evolution of the Science of Marketing

In 1962, John interviewed for a faculty position at MIT in the School of Industrial Management, now the MIT Sloan School. He had been a tenured associate professor at Case and, without any qualms, he accepted MIT's offer as an untenured associate professor. John viewed the scope of OR broadly and was attracted back to MIT by the promise of new problems and new research directions. MIT was an excellent base of operations with good colleagues and great students—he has never left!

John, perhaps frustrated by Boston drivers, who are alleged to be the worst in the U.S., first continued his work on traffic flow and traffic signal control, a problem he had begun at Case—traffic delays due to additional time to travel over a route as a result of traffic and traffic lights (Little 1961b). He had also worked with a master's student, John Morgan, who programmed the synchronization of traffic signals on a two-way street on the Case computer. This was the first time the problem had ever been approached in this manner. Previously, it had been done graphically by hand. It is trivial to synchronize the signals on a one way street so that cars traveling at an average speed can traverse the length of the street without stopping. The problem becomes combinatorial and quite difficult on a two-way street when it is desired to have the cars in both directions be able to do the same. The fraction of the signal cycle time for which cars in both directions can travel without stopping is known as the bandwidth of the street (Morgan and Little, 1964). Finding the maximum bandwidth (MAXBAND) is a challenging optimization problem.

At MIT, John extended this work to complete street networks, seeking to maximize a linear combination of the bandwidths of the various arteries in the network (Little 1966a). The methodology was based on mixed-integer linear programming. He was joined by colleagues and research assistants and supported by the Federal Highway Administration to produce the software package, MAXBAND—it was distributed to municipalities so they could optimize their street systems (Gartner et al. 1981; Little and Cohen 1982). This stream of research defined a new state of the art in the field of synchronizing traffic signals on arteries and networks.

In a quite different direction, John, now in a business school, had the vision to perceive marketing as source of interesting and relevant unexplored opportunities for OR and management science (MS). As an example, the effectiveness of a company's advertising is likely to vary over time. No matter how good the response function used to calculate an optimal advertising rate at one point in time, it is likely to drift to something different. What to do?—run an experiment to re-measure effectiveness and update the advertising response function. For example, take five medium-sized markets and set them at higher than the currently presumed optimal advertising rate and set another five markets lower. The resulting measurement can be used to reset the advertising response function and obtain a more profitable advertising rate for use nationally. But the 10 experimental markets are being deliberately operated differently from the perceived best rate, thereby incurring a calculable cost. The adaptive system optimization, however, takes the next step by setting the number of experimental markets so as to maximize total system profit, including the cost of the experiment. John thus became the first scholar to develop adaptive control methods for the field of marketing. He was particularly pleased that his model could be applied readily (Little 1966b). For John, it was not enough to develop a nice mathematical solution—he wanted somebody to use it. He also published a generalized version of the mathematics in terms of optimal adaptive control (Little 1977).

During this period, John became increasingly interested in advertising budgeting and media selection. He and Leonard Lodish, one of his Ph.D. students, developed an on-line, computer-based system for selecting and scheduling advertising media. They described it as a media planning calculus and named it MEDIAC (Little and Lodish 1969). MEDIAC replaced heuristic analyses with the optimization of a measure more closely related to sales and profits.

In 1968, while he was conducting an MIT summer session on OR in marketing, John was approached by attendees from Nabisco (formerly the National Biscuit Company) who asked him to develop a model to set advertising spending levels for Oreo cookies. John realized that Nabisco had some explicit, hard data, but that other key data were buried in managers' heads. Managers with experience in this area had implicit knowledge of how sales would respond to advertising. The challenge was how to unlock that information in a manner that could augment rather than replace the hard data. From this challenge, John developed the concept of a decision calculus as described in "Models and managers: The concept of a decision calculus" (Little 1970).

This revolutionary paper starts out with the sentence: "The big problem with management science models is that managers practically never use them" (Little 1970, B466). But it was not a negative paper—John wanted to improve matters. First, it broke with standard practice for empirical models that all constants be estimated at once on a single data set. If data were not available on, say, advertising, then advertising could not be in the model. John's paper took the view of those who wanted to apply management science models and set forth guidelines that would be critical to implementation. John defined a decision calculus as "a model-based set of procedures for processing data and judgments to assist a manager in his decision making," and proposed that models, to be useful to managers, should be "simple, robust, easy to control, adaptive, complete on important issues, and easy to communicate with" (Little 1970, B-469, B483). John's insight was that for managers to use a model, they must understand the model well enough that they could control it. His theme was: "I claim that the model builder should try to design his models to be given away. In other words, as much as possible, the models should become the property of the manager, not the technical people"

(Little 1970, B-483). The decision calculus paper was cited as one of the ten most influential papers published in the first 50 years of the journal *Management Science.*

John demonstrated the relevance of the decision calculus by applying it to the complex problem of selecting the entire marketing mix. He espoused eclectic calibration. Some submodels, like manufacturer advertising and its effect on brand share, are almost sure to include time lags and be dynamic. Others, like seasonality and trend, may be straightforward and standard. Still others, like coupons, premiums, and production capacity constraints might be handled by simple indices based on data analysis or the product manager's prior experience. The resulting model—ADBUDG (Advertising Budget)—is given in Little (1970). This model was later expanded into BRANDAID, which is a more complete on-line marketing-mix model that provides AID for the BRAND manager by permitting the evaluation of new strategies with respect to price, advertising, promotion, and related variables (Little 1975a, b). The latter paper describes a case study for a well-established brand of packaged goods sold through grocery stores.

THE MARKETING DATA EXPLOSION

John, in his paper, "Aggregate advertising models: The state of the art," (Little 1979), summarized and critiqued the previous decade's modeling knowledge and advances in modeling advertising phenomena. After posing a set of modeling questions, he reviewed the published empirical data and studies that bore on them. He then listed five phenomena that a dynamic model of advertising response should, at a minimum, be able to incorporate: assist annual budget setting, geographic allocation of funds, allocation over time, and incorporate media and copy effects.

One of John's concerns was that available data to test and calibrate such models were aggregate in nature, for example, historical time series at a national or market level. He observed (Little 1979, 629): "Although many models have been built, they frequently contradict each other and considerable doubt exists as to which models best represent advertising processes. . . . Future work must join better models with more powerful calibration methods." Central to this objective was the need for accurate data at the point-of-sale (local) level. John noted that such a data revolution was on its way (Little 1979, 663).

Products had begun to be labeled with computer readable Universal Product Codes (UPC), supermarket scanners were being installed that could read such codes, and computer technology was becoming distributed—no longer just mainframes—to collect, organize, and analyze those data. The marketing field would soon be inundated with data.

To address the new issues in using such data, John and Peter Guadagni, one of his master students, built a disaggregated model—a logit model of brand choice calibrated on scanner data—that predicted actions at the level of the individual consumer making individual purchases (Guadagni and Little 1983). A novel aspect of the model was that it included what John termed a loyalty variable; an exponentially smoothed history of past purchases treated as 0–1 variables and, thus, a measure of the customer's past propensity to purchase the product, weighted most for recent purchases. This paper is one of the most cited papers in *Marketing Science* and has been republished as one of that journal's eight classic papers. The logit model has been improved, reanalyzed, expanded, kicked, and modified. New phenomena have been added and new data have been analyzed. But the basic structure (and the power of the loyalty variable) remain. An entire generation of marketing science academics and students have been influenced by the original and extended UPC logit models.

The logit models are powerful, but could be intimidating to mangers who, according to John's decision calculus theme, should be able to understand the model well enough that they could control it. Managers wanted answers in a form they could digest. More importantly, computer technology had gotten to the point where the logit models could work behind the scenes to create automated reports in the form that managers could use. This thinking led to a decision support (expert) system termed CoverStory that was developed for Ocean Spray Cranberries, a fruit-processing cooperative. Ocean Spray tracks sales and assesses the effectiveness of its marketing program using large data bases collected through bar-code scanners in supermarkets (Schmitz et al. 1990).

For a brand manager, CoverStory rapidly and automatically computes and summarizes a large amount of output generated by the system's models. The output—structured as a memorandum to the manager—includes a single page of charts and a series of descriptive lines customized for the markets in which the brand competes, showing performance vis-à-vis competitors brands. The number of brands, individual products, and regions make it infeasible to do such an analysis manually. Modern computers, artificial intelligence, and advanced marketing science models form the CoverStory system—they are combined in such a way as to have a direct, positive impact on marketing practices, as well as managerial efficiency and effectiveness. John had come a long way from Whirlwind.

EDUCATOR, LEADER, ENTREPRENEUR, AND SERVICE

John is an innovative and devoted educator across all programs of the university—undergraduate, MBA, and Ph.D. Since 1990, he has been chair of the Undergraduate Program Committee of the MIT Sloan School. As such, he is involved in policy matters, but he always has a group of undergraduate advisees. MIT undergraduates who major at MIT Sloan receive an S.B. degree in Management Science, which John calls an MIT-style business degree (Little 2008).

From the time he developed a course on OR in marketing at Case, John has been interested in teaching master students how to solve marketing science problems. At MIT, he developed MIT's first course in marketing models—a course that was a staple fixture in the marketing group until marketing models ultimately invaded almost all marketing courses. In the 1970s, he pioneered a new specialty program at the MIT Sloan School called Fast Track. John would read all the files for admitted students and, if they had very strong quantitative skills, he would invite them to join the Fast Track program. He found that the students thrived in the challenging advanced courses in mathematical programming, information technology, and statistics. John has served MIT in many capacities. At MIT, he was director of the Operations Research Center from 1969 to 1975, succeeding Philip Morse. For the MIT Sloan School, he headed the marketing group and eventually the Management Science Area (MSA) from 1972 to 1982. During this period, he was instrumental in making

> AT HOME WITH J.D.C.L.
>
> John often invited students to his house for social functions that usually included squid tasting. A special time was Thanksgiving Day when John would invite foreign students and their families for dinner at his home in Lincoln. John also made it a practice to invite new faculty to Nantucket during the summer to enjoy the island and be exposed to New England culture. A stay in John's little cabin and fishing for bluefish off Miacomet Rip have provided particularly vivid memories for many. John loves seafood and claims that "anything from the sea was good to eat until proven otherwise." Sea urchin roe pizza is a Little specialty (Little 2008).

the MSA cohesive and interdisciplinary. In 1982, John was asked to work his magic again—the Behavioral and Policy Science Area (BPS) at MIT Sloan was formed after a major reorganization. It was not a cohesive group and, surprisingly, did not include anyone who might be labeled either an operations researcher or management scientist. It was primarily a collection of faculty from the less quantitative fields of organizational studies, research and development management, human relations, and strategy. John led the group for 6 years and his legacy was the establishment of a sound foundation for the area and a potent BPS faculty. In 1989, John was appointed an MIT Institute Professor—a special rank and honor reserved for a very few faculty at the university. In this capacity, John has undertaken some sensitive and important MIT-wide projects. He reports directly to the Provost.

JOHN'S FAMILY

John's wife and former fellow physics graduate student, Betty, was an impressive scientist in her own right. For her Ph.D. thesis, she studied the dynamic behavior of domain walls in barium titanate—she finished her thesis before John finished his and published a paper on it in the *Physical Review* (E. Little 1955). She did not pursue a full-time career, being the one who agreed to stay at home as they raised their children: John N. (Jack), Sarah A., Thomas D. C. (Tom), and Ruel D. Betty became a teacher's aide during the time her children were in public school, and later, in 1985 at age 58, having become interested in Nantucket history and its native-American archaeology, received an M.A. in Anthropology, with concentrations in archaeology and geology, from the University of Massachusetts-Amherst. Betty continued her archaeological research and writings for many more years. After a 2-year battle with cancer, she died in 2003.

Their children could not escape their parents' scientific, engineering, and entrepreneurial influences. Jack Little graduated from MIT in electrical engineering and received an M.S. in Electrical Engineering from Stanford University. In 1984, he co-founded MathWorks, a leading developer of technical computing software for engineers and scientists in industry, government, and education. Sarah Little graduated from Stanford in physics and then joined the MIT-Woods Hole Oceanographic Institute Ph.D. program, graduating in geophysics with a thesis that involved making dives in the deep-ocean submersible *Alvin* and collecting data on hydrothermal vents in the Pacific. Tom Little graduated from Rensselaer Polytechnic Institute in biological engineering, and earned a Ph.D. in computer engineering from Syracuse University. He is now a professor in the Department of Computer Engineering, Boston University School of Engineering. With a former student he co-founded a web software and consulting firm, Molecular, Inc., which they have since sold. Ruel Little has a B.A. in physics from Johns Hopkins University and an S.M. in mechanical engineering from MIT. After working for many years for solar energy companies, he helped found GreenRay, a solar energy startup that is developing labor saving technology that simplifies construction and installation of solar modules for delivering electricity directly into home appliances and lighting.

As a grandfather, John answers to eight grandchildren.

John's interest in modeling real-world problems led him, in 1967, to co-found Management Decision Systems, Inc. (MDS), a company whose objective was to create and commercialize marketing models and marketing decision-support software. MDS grew to over 200 employees. In 1985, MDS merged with Information Resources, Inc. (IRI) with John serving on the IRI board until 2003. John continued his entrepreneurial activities by investing and serving on the board of a start-up company, InSite Marketing Technology. InSite provided a new class of e-business applications that identify the buying style of the customer and the selling style of the company, and dynamically integrate them to help the customer through the buying process. In 2000, InSite merged with the Kana Corporation, a multichannel customer service software company that integrates telephone, email, Web chat, and collaboration channels with knowledge management capabilities in a unified application.

John D. C. Little's family, Nantucket Island, summer 2002

(John, Betty, and their children in bold face)

Back row (left to right): **JOHN, RUEL**, Sara, Max, **TOM**, Nancy, **JACK**

Front row (left to right): **BETTY**, Kathy, Avery, Isaac, **SARAH**, Cora, Doug, Emily, Dyson, Erica

The families of John and Betty's children are: **JACK** Little and Nancy Wittenberg/Erica and Emily Little; **SARAH** Little and Doug Hersh/Cora and Isaac Hersh; **TOM** Little and Sara Brown/Max and Stephanie Little (Stephanie not born until 2003); **RUEL** and Kathy Little/Dyson and Avery Little.

John served as president of ORSA in 1979 and president of TIMS in 1984–1985. During his ORSA term of office, he and Frank Bass, then president of TIMS, persuaded the two societies to found the joint journal *Marketing Science.* When the two societies merged in 1995 to form INFORMS, he was elected its first president (he chaired the committee whose efforts led to the merger).

HONORS AND AWARDS

John has been recognized for his innovative and seminal research in marketing by the Paul D. Converse Award, a lifetime achievement award given by the American Marketing Association (AMA) (1992); the AMA Charles Parlin Award for contributions to the practice of marketing research (1979), and MIT's Buck Weaver Award for outstanding contributions to marketing (2003). He was elected to the National Academy of Engineering for outstanding contributions to operational systems engineering, including research, education, applications in industry, and leadership (1989). He has received the ORSA's George E. Kimball Medal for recognition of distinguished service to the society and profession of OR (1987), and the Distinguished Service Medal from TIMS. He is a member of the International Federation of Operational Research Societies' (IFORS) Operational Research Hall of Fame (Larson 2004), and a fellow of INFORMS and of the INFORMS Society of Marketing Science (ISMS). John has received honorary degrees from the University of London; University of Liège, Belgium; and Facultés Universitaires Catholiques de

JOHN D.C. LITTLE AWARD

John presenting plaques to the winners of the 2007 John D. C. Little Award at the 2008 Marketing Science Conference in Vancouver. Shown (left to right) are John, P. K. Kannan, and Brian T. Ratchford (co-author, Lan Luo, was not able to attend).

Mons, Belgium. He has been honored by having a most prestigious annual award in marketing science named for him—the John D. C. Little Award—given annually by ISMS for the best marketing paper published in an INFORMS journals (including *Marketing Science* and *Management Science*).

ACKNOWLEDGMENTS

The authors, with appreciation, wish to note the following sources of material: "IFORS' operational research hall of fame: John D. C. Little," Larson (2004); and an unpublished presentation by John at his 80th birthday celebration. We also thank John Little for adding many historical facts and insights to our summary of his life and work.

REFERENCES

Bellman RE (1957) Dynamic programming. Princeton University Press, Princeton, NJ

Gartner N, Kelson M, Little JDC (1981) MAXBAND: a program for setting signals on arteries and triangular networks. Transport Res Rec 795:40–46

Guadagni P, Little JDC (1983) A logit model of brand choice calibrated on scanner data. Market Sci 2(3):203–238

Larson R (2004) IFORS' operational research hall of fame: John D. C. Little. Int Trans Oper Res 11(3):361–364

Little E (1955) Dynamic behavior of domain walls in barium titanate. Phys Rev 98(4):978–984

Little JDC (1955) Use of storage water in a hydroelectric system. Oper Res 3(2):187–197

Little JDC (1961a) A proof for the queuing formula: $L = \lambda W$. Oper Res 9(3):383–401

Little JDC (1961b) Approximate expected delays for several maneuvers by a driver in Poisson traffic. Oper Res 9(1):39–52

Little JDC (1966a) The synchronization of traffic signals by mixed-integer linear programming. Oper Res 14(4):568–594

Little JDC (1966b) A model of adaptive control of promotional spending. Oper Res 14(6):1075–1097

Little JDC (1970) Managers and models: the concept of a decision calculus. Manage Sci 16(8):B466–485

Little JDC (1975a) BRANDAID: a marketing-mix model. Part 1: structure. Oper Res 23(4):628–655

Little JDC (1975b) BRANDAID: part 2: implementation, calibration, and case study. Oper Res 23(4):656–673

Little JDC (1977) Optimal adaptive control: a multivariate model for marketing applications. IEEE Transact Automat Contr 22(2):187–195

Little JDC (1979) Aggregate advertising models: the state of the art. Oper Res 27(4):629–667

Little JDC (2002) Philip M. Morse and the beginnings. Oper Res 50(1):146–149

Little JDC (2007) Life as the first OR doctoral student – and other prehistoric tales. In: Larson I (ed) The operations research center at MIT. INFORMS Topics in Operations Research Series, Hanover, MD

Little JDC (2008) Personal communication

Little JDC, Cohen S (1982) The MAXBAND program for arterial signal timing plans. Public Roads 46(2):61–65

Little JDC, Graves S (2008) Little's law. In: Chhajed D, Lowe T (eds) Building intuition: insights from basic operations management models and principles. Springer, New York, NY, pp 81–100

Little JDC, Lodish L (1969) A media planning calculus. Oper Res 17(1):1–35

Little JDC, Murty K, Sweeney D, Karel C (1963) An algorithm for the traveling salesman problem. Oper Res 11(6):972–989

Morgan J, Little JDC (1964) Synchronizing traffic signals for maximal bandwidth. Oper Res 12(6):896–912

Morse P (1958) Queues, inventories and maintenance. Wiley, New York, NY

Morse P, Kimball G (1951) Methods of operations research. Wiley, New York, NY

Schmitz J, Armstrong G, Little JDC (1990) CoverStory—automated news finding in marketing. Interfaces 20(6):29–38

37

Evelyn Martin Lansdowne Beale

John A. Tomlin

Evelyn Martin Lansdowne Beale, always known as Martin to his friends and colleagues, was a giant of the operations research (OR) profession, especially in the U.K., and an outstanding contributor to all aspects of mathematical programming (MP). He not only made major contributions to theory and algorithms, but to the development of practical mathematical-programming computer systems. His pioneering work on developing algorithms for real-world problems, and overseeing their implementation in large-scale commercial software systems, made a major impact on the practice of OR at the time and left a lasting imprint.

He was a founding member and chairman of the Mathematical Programming Society. Martin was elected a Fellow of the Royal Society and served on its Council. He was awarded the Silver Medal of the Operational Research Society of Great Britain. He is also remembered as an outstanding contributor to the field of applied statistics, a wonderful colleague, and an inspiring teacher.

Early Life

Martin Beale was born on September 8, 1928, at Stanwell Moor in Middlesex, the elder child of Evelyn Stewart Lansdowne Beale and Muriel Rebecca (Slade). His father was a senior physicist and research engineer at the research station of the Anglo-Persian Oil Company at Sunbury, but his work took the family to Persia for the year 1932, about 1 year after the

A.A. Assad, S.I. Gass (eds.), *Profiles in Operations Research*, International Series in Operations Research & Management Science 147, DOI 10.1007/978-1-4419-6281-2_37,

birth of Martin's brother Julian. On their return, the family moved to Chelsea, keeping a house in Middlesex. Martin then suffered the first of many bouts of ill-health that interrupted his early formal education. It was found that he had contracted malaria in Persia, so he did not start at the Montessori School in London until 1934, attending it for 2 years. His father left the oil company to found a private consultancy firm for industry with R. Denman and worked brilliantly on a wide range of engineering problems.

Martin showed an early aptitude for mathematics at school but, at age 8, also studied Latin with his mother, anticipating his entry into the more conventional educational system. He was enrolled at St. Aubyn's preparatory school (Rottingdean, Sussex) in 1937, but he was at home throughout 1939–1940 due to another occurrence of ill health. This home was now above Treyarnon Bay; the Beales had become captivated by this part of Cornwall and had built Windhover House there. Martin studied at home with Julian and some other children during that year, and this small school continued throughout the war, although Martin returned to St. Aubyn's in 1940. He then aimed for a scholarship to Winchester in 1941, but ill health struck again; measles prevented him from traveling for the examination. After a further year at St. Aubyn's, his mathematics master, the late E. Webber, who coached him for the scholarship, said to his wife: "I cannot teach any more mathematics to Martin Beale, I have taught him all I know" (Dantzig and Tomlin, 1987, 117–118). Martin was at Winchester from 1942 to 1946, having gained the second scholarship in 1942. Although he was the joint winner of the school's Richardson prize for mathematics, he was advised not to try to go from Winchester to Trinity College, Cambridge (as his father did), because the competition might be too strong. Characteristically, Martin rejected this advice and in due course went to Trinity with a scholarship in 1946.

While at Cambridge, Martin's interest in mathematics seems to have been nearly overshadowed by his absorption in bridge. He (and his brother Julian) played for Cambridge against Oxford, and Martin became fascinated with a new bidding system about which he was planning to write a book. Interviewing Beale, his highly respected prospective employer, Steven Vajda of the Admiralty Research Laboratory (ARL), asked him if he "would like to be told about the work he would be expected to do there. However, he did not; nothing was allowed to distract him from his writing a book on Bridge Bidding" (Dantzig and Tomlin 1987, 118). Despite this, after Martin graduated in 1949 and gained his diploma in mathematical statistics in 1950, he joined ARL working under Vajda. The bridge book never materialized.

AT THE ADMIRALTY RESEARCH LABORATORY

The particular projects that Martin worked on for ARL do not seem to be widely known. They apparently included work on direction finding as well as other unpublished work. What has become widely known is that Steven Vajda introduced Martin to the field of MP. He soon outstripped his mentor. As Vajda wrote: "I am sometimes praised for having introduced Martin to linear programming [LP]. I did and I am glad of it. But there is no merit in having done it. I have introduced LP to others as well, but they did not nurture the seed the same way as Martin has done" (Dantzig and Tomlin 1987, 118).

Martin's first research paper, "An alternative method for linear programming" (Beale 1954), described an independently discovered dual simplex method for solving LP problems. This was soon followed by one of his best-known early papers "On minimizing a convex function subject to linear inequalities" (Beale 1955). This important paper describes not only what came to be known as the Beale-Tucker representation of a linear program and one of the first quadratic programming algorithms, but also one of the first treatments of stochastic programming. It is interesting to observe that this latter work was independently carried out in parallel with a very similar approach by George Dantzig (1955), which of course Martin acknowledged. Martin went on to publish many papers on quadratic, nonlinear, and stochastic programming. Further papers on cycling in the dual simplex method, the convex-cost transportation problem, and quadratic programming (Beale 1959) as well as some others on applications and statistical subjects appeared while Martin was at ARL.

On July 1, 1953, Martin married Violette Elizabeth (Betty) Anne Lewis at St Mary's Church, Hampton. They had met in time to celebrate his 22nd birthday together; she was a scientific assistant in the ARL Mathematics Group (he had spent his 21st birthday playing bridge). Betty soon became part of the Beale family, visiting Windhover several times before their marriage and enjoying cycling, walking, and playing bridge with Martin. Their first home was a small terrace house at Strawberry Hill, Twickenham, within cycling distance of ARL, where Betty continued working until a few weeks before the birth of their first son, Nicholas, in February 1955. Their daughter, Rachel, was born in May 1957, to be followed by a second son, Marcus.

The Beale family spent most of 1958 in the U.S., the majority of the time at Princeton, where Martin is reported to have explained to Ralph Gomory that integer programming was "impossible" (Gomory 2002, 79). It was one of the few times that Martin would be quite wrong, and he would spend much time working in that field later in his career. On this trip, he also traveled to California and visited the RAND Corporation. George Dantzig noted: "I don't recall meeting him in person until 1958 when he, in his unassuming way, dropped into my office at RAND Corporation in Santa Monica, California. Still in his twenties, Beale had already written three important papers on linear programming. I was struck by his youthful enthusiasm, a characteristic that stayed with him always. Martin, I believe, made a special detour to the West Coast on his way from England to Princeton to drop in, but he never said so. I suspect he was curious to see what Fulkerson, Johnson, Shapley, Bellman, Wolfe and I were up to" (Dantzig and Tomlin 1987, 118).

The Beales returned to England, and Martin remained with ARL until 1961. During this time, he continued to display some of the eccentricities which were an integral part of his character. Ken Bowen (1986, 8–9) wrote: "I remember, and they will never forget, his performance for a U.S.A.–Canadian group of analysts in Room 39, in the Admiralty. With an array of tabulations and diagrams on a long table, he started by kneeling on a chair and finished on the table, wandering about the data on hands and knees." However, as Vajda tells us, "He was highly appreciated at ARL. When members had to be assessed, it might have been asked: 'Is (s)he as good as Beale?' Few were" (Dantzig and Tomlin 1987, 119).

CONSULTING AT C-E-I-R, LTD.

A.S. (Sandy) Douglas and the distinguished statistician Maurice (later Sir Maurice) Kendall founded the Corporation for Economic and Industrial Research (CEIR), U.K., in 1960 as a subsidiary of the Washington DC consulting services company C-E-I-R, Inc. Martin joined the young company as employee number 29 in 1961. Outstanding as his work at the Admiralty had been, CEIR provided the framework in which Martin really made his mark as the great pioneer of real-world MP. He consulted on all manner of applications, pursued his research on computational MP, and saw it applied. Through his work with the LP/90/94 computer-based software team (and, hence, with CEIR in the U.S.), he established many of the procedures and practices used in MP computer systems to this day. Eli Hellerman (of the U.S. company) recalled: "His insight into how algorithms could be implemented on a computer was phenomenal. He was the dynamo behind the extensions to the LP/90/94 System in the areas of separable programming, mixed integer, Dantzig-Wolfe decomposition, and a host of other practical algorithms" (Dantzig and Tomlin 1987, 119).

As Philip Wolfe observed:

> ... in his career Martin had a unique role in the field. Along with basic research, there is difficult and important work to be done in applying research concepts to the real world. For us who work in universities and research centers, the prospect that our research will be applied in practice is a great stimulus; and a few of us participate at times in practical developments. On the application side there are individuals of great ability in industry who deserve much credit for their work and leadership in the use of mathematical programming in real-world applications. In the primary ranks in the field it was only Martin who truly and completely bridged both: scholarly research and practical applications. It was only he who both directed the mathematical programming work of an important consulting and service firm and at the same time contributed outstanding basic research (Dantzig and Tomlin 1987, 119).

In the early to mid-1960s, Martin was the first to make three powerful extensions of LP into practical tools. These were nonlinear programming (via separable programming), the Dantzig-Wolfe decomposition method, and integer programming (via branch and bound). All of these were implemented under Martin's direction as extensions of the CEIR LP/90/94 system, then the most powerful of its day, but whose limitations were severe by today's standards. Foremost among these were the limit of 1024 constraints (including simple upper bounds!).

Philip Hughes (1988), in his memoir about Martin, described how the standard LP/90/94 system was inadequate for solution of an oil-field model, which was both large—thus requiring decomposition—and involved unavoidable nonlinearities, which could only be handled by separable programming. It turned out to be difficult to make decomposition work in practice, but with the implementation of several new features, Martin and his colleagues (in 1965) succeeded in solving this problem (Beale et al. 1965a).

Separable programming was to be a major field of development and publication by Martin, and various refinements, such as automatic interpolation, were to be features of the improved LP/90/94. He had quite definite ideas on the way it should be done. Eli Hellerman remembered:

> ... I was assigned the task of writing the User's Manual for the CEIR Extended LP/90 System. One of the new algorithms on the system was Separable Programming and I tried to find a simple and instructive model to illustrate its use... . When Beale reviewed the first draft, he immediately communicated to me that (my example) was not a proper model and indeed it was unsatisfactory. I wrote back to ask him whether there was anything incorrect in the model and would he please send me a simpler model if he had one.
>
> He responded by repeating the charges he made in the first communication. I again wrote to him for enlightenment. It looked very much like an impasse until one day, quite unexpectedly, Martin strode into my office (he had just arrived from London), put down his briefcase and without even saying "hello" marched up to the blackboard and said, "Now, about this Separable thing, here is how it should be done" (Dantzig and Tomlin 1987, 120–121).

During this same period, Martin oversaw the implementation of the first commercially successful mixed integer programming (MIP) code (Beale and Small 1965). Integer programming turned out to be possible, but the preferred method in practice proved to be branch and bound, not cutting planes, and this was implemented in the LP/90/94 system; the method was a little different from that described by Land and Doig (1960) using a binary tree and depth first (last in, first out, or LIFO) search. This code was able to solve substantial MIP problems with a few dozen integer variables—a significant practical advance.

> MARTIN ON PROGRAMMING
>
> 1. Understand the environment you are modeling.
> 2. Logically and clearly name and order variables.
> 3. Present back to the client his input data clearly laid out.
> 4. Produce some results quickly.
> 5. Do not get slavishly tied to a software package.
> 6. Always keep sight of the numbers.
>
> (Hughes 1988, 7)

Martin's contributions were by no means limited to algorithm development and system implementation. He was perhaps even more active in model formulation and consulting on client problems. He developed a systematic method of approaching new applications of MP and devoted considerable attention to the apparently mundane tasks of matrix generation and report writing—essential features of applied work. LP/90/94 used SHARE standard input where the variables and constraints had six-character names, which could be used to encode the meaning of these entities. Since problems had grown to the size where hand-preparation of this input was out of the question, matrix generators were required to generate the input to the LP code. Martin was emphatic about the need to write these matrix generators (in Fortran) in a systematic way with naming conventions, not just for the LP variables and constraints, but for the arrays and indices in the code itself. His paper presented at the 1967 Princeton Mathematical Programming Symposium (Beale 1970a) dealt with some of these issues at length and was unique in those proceedings for doing so.

Also in 1967, Martin's best-known statistical paper "The discarding of variables in multi-variate analysis" (with M. G. Kendall and D. W. Mann) was published. About 20% of Martin's published work was in statistics, and, thus, mostly beyond the scope of this profile, but this and several other papers were in the related field of multiple regression. He maintained a parallel career as a respected statistician throughout his life.

Martin's research and related developments in MP through 1967 (including quadratic programming, separable programming, decomposition, and partitioning) are given in a survey article "Numerical methods" (Beale 1967, 133–206), Chapter VII in the book *Nonlinear Programming*, edited by J. Abadie (1967).

His eccentricities continued, and even flourished, in the CEIR environment. Sandy Douglas (1986, 120) recalled: "He never worried what people might think of him. If it improved his thought processes he would lie down and raise a leg or climb on the table or stuff his handkerchief in his mouth. (His colleague) Peter Windley did once comment to me that he 'wished Martin would take his handkerchief out of his mouth when dropping one of his pearls of wisdom'—but it was said in amusement rather than annoyance. Indeed no one minded what he did, since he always came up with useful and perceptive remarks, from whatever position and through whatever obstructions they were delivered!"

Consulting at Scicon, Ltd.

In the mid 1960s, British Petroleum acquired CEIR (U.K.) with the parent U.S. company acquired by Control Data Corporation in 1968. Not wishing to have its name associated with any particular hardware vendor, the U.K. company changed its name to Scientific Control Systems, Ltd., eventually abbreviated to Scicon, Ltd. This had little immediate impact on Martin's work, but did coincide with the phasing out of the IBM 7090/94 computer and the Extended LP/90/94 software system. The ultimate capabilities of LP/90/94 are nicely summed up in Martin's first book, *Mathematical Programming in Practice* (1968).

The new system, begun from scratch when Martin called his programmer into his office to write the first subroutine, was to be named UMPIRE (Universal Mathematical Programming System Incorporating Refinements and Extensions) and would include many new ideas not in LP/90/94 (and omit some that were). It was written by a team of four or five people which included at various times the present author, Gautam Mitra, and crucially, John Forrest, who had recently returned from a year studying in Berkeley under Dantzig (on a NATO scholarship arranged by Martin). It was during this exciting period that Martin made the remark that he "would rather work with today's algorithms on yesterday's computers than on today's computers with yesterday's algorithms" (Tomlin 1989, 159), an indication of the importance he placed on algorithmic innovation and improvement.

A notable feature of UMPIRE was the inclusion not only of simple upper bounds, but generalized upper bounds or GUB (Dantzig and Van Slyke 1967). Martin was very enthusiastic about GUB, since it provided a way of extending the solvable size of many models, and main memory was still very limited, even on the new UNIVAC 1108, for which UMPIRE was designed. GUB also received considerable attention in some of the many survey papers he was asked to write (Beale 1970b). Many of these papers were presented at conferences and symposia, where Martin was always to be seen in his uniform of suit, tie, and black shiny shoes and socks. Even in warm sunny locations, such as the south of France, he was always to be seen in this uniform, though it is reported that he was once seen on a French nude beach wearing only the socks and shiny black shoes. One can only imagine.

GUB influenced the UMPIRE system in other ways. It made the concept of sets of variables attractive, which in turn led to special ordered sets (Beale and Tomlin 1970) which enabled non-convex programming without resort to integer variables (Forrest and Tomlin 2007). This was to be a rich source of new developments in non-convex and nonlinear programming throughout the rest of Beale's career.

MARTIN ON MODELING

"It is useful to distinguish between established and new mathematical programming models. An established model is run from time to time with updated data as part of some operational decision-making routine. The purpose is then to suggest a specific course of action to management, and the suggestion will usually be accepted. A *new* model may also be used in this way but is more often employed to gain greater understanding of the situation. It may be run under a variety of assumptions that lead to different conclusions, and the model itself will not suggest which set of assumptions is most appropriate.

"During the model development and data-gathering phase we must therefore be prepared to make many optimization calculations which can be shown to management to see if they are sensible. If what the model recommends is not considered sensible, we have to find out why it is not acceptable. Neither the analyst or the manager should accept the recommendations from the model unless they can be explained qualitatively as the natural consequences of the physical and economic assumptions. We can paraphrase this by saying that the results should only be accepted *if they are obvious*. The reader may think that the model is then of no real use! This, however, is not so, because many things are obvious once someone has pointed them out, when they were not at all obvious beforehand" (Beale 1988, 72).

UMPIRE was also the first MP system to use the Gaussian, rather than Gauss-Jordan form of the product form inverse. The first step in this process was Martin's observation that a block factorization could be used to avoid fill-in of a substantial part of the (sparse) basis, followed by the realization that this could be even further improved by using Gaussian triangular factorization. This would eventually lead to the first practical method of updating triangular factors, with considerable improvements in speed.

Quadratic programming was not implemented in UMPIRE, and its separable programming facilities were less elaborate than those of LP/90/94, but this was offset by the implementation of a new conjugate gradient method for approximation programming to handle large-scale, general nonlinear programs (Beale 1974). An important outcome of this work was a new way of viewing nonlinearities in a model. Whereas previous treatments had considered each constraint as consisting of linear terms plus a function of nonlinear variables, Martin considered the whole problem as if it were a linear program where certain of the matrix coefficients, and the right-hand-sides, were not constants but themselves functions of the nonlinear variables. This approach often drastically reduced the number of variables which had to be considered as nonlinear and was to prove useful in later, more general, nonlinear programming systems.

During the same time period, the first matrix generator generator (MGG) was developed at Scicon. Martin had been strongly in favor of using a general purpose language, such as Fortran, for matrix generation, but MGG, developed primarily by our colleague Peter Alsbury, overcame his objections, and he was to use it in many modeling projects. There were at least two good reasons for this. First, MGG generated an intermediate Fortran program (the MG, or matrix generator), which the developer was free to specialize and modify, if desired. Second, the mathematical, equation-like, model description, later adopted (without acknowledgment) by such systems as GAMS, AMPL, and others, greatly speeded up and simplified the modeling process. Once convinced of the value of this approach, Martin jettisoned his previous stance and seems to have never written a Fortran matrix generator again. This was typical of Martin Beale. He had very firm views on the way things should be done, but if a colleague could come up with, and justify, a new and better idea, he would readily abandon his own and, what is more, publicly give credit where it was due.

In its turn, UMPIRE and the UNIVAC 1108 would also become obsolete, and a new Mathematical Programming System (MPS), to be named SCICONIC, was again written under Martin's direction, with John Forrest the primary developer. A later version, intended to be machine independent, was known as SCICONIC VM and was substantially rewritten and extended by R.C. (Bob) Daniel.

SCICONIC did not include GUB, but did carry on with the further development of special ordered sets, first, through Martin's new technique of pseudo-shadow-prices to guide the tree search (Beale and Forrest 1976), interpolation to handle continuous non-convexities, and later linked ordered sets to handle non-separable non-convex non-linearities. This continued to be a major theme of his work and is perhaps best represented by one of his last survey papers "Integer programming" (Beale 1985a).

Martin continued his development of SCICONIC and its applications with a new generation of young colleagues, who, like earlier ones, benefited from Martin's propensity for teaching. This benefit also extended to his students at Imperial College, London, where he regularly taught on Mondays from 1967 on. He was sometimes jokingly known as the Monday Professor.

Other Professional Interests

In addition to his demands at Scicon and time spent with his family, it is remarkable that Martin found time for a multitude of other interests. His parallel career as a statistician has already been noted, and he was active in the Royal Statistical Society and the Institute of Statisticians, as well as Treasurer of the Statistical Dinner Club for a good many years.

Martin Beale was also a key figure in the foundation and direction of the Mathematical Programming Society. He had been one of the organizers of the 1964 London International Symposium on Mathematical Programming, the first such meeting held outside the U.S. Subsequently, finding a small surplus of funds in their treasury, the organizers designated it the International Programming Fund and chose a small international committee to hold it. This was the first step in the identification of MP as a professional specialty and eventually led to the formation of the Mathematical Programming Society in 1972.

Martin was the second chairman of the newly formed society, serving from 1974 to 1976. Prior to that, he was asked to join the board of senior editors of the Society's new journal, *Mathematical Programming*. He served on the Council of the Society from 1982 to 1985 and otherwise on several of its committees.

HONORS AND AWARDS

There is no doubt that the honor of which Martin was most proud was his election as a Fellow of the Royal Society in 1979. As an ardent conservative, royalist, and pillar of the establishment, as well as a distinguished scientist, he can only have been delighted.

In a 1975 letter supporting Martin's nomination to the Royal Society, George Dantzig wrote:

> I rate Beale among the top two or three people in the world in the field of Operations Research. He is also well known among statisticians. He has done outstanding theoretical work on algorithms for solving linear and nonlinear programming systems. On the practical side, he has played (and continues to play) a key role in the development of software packages. These are sophisticated mathematical and computational tools that have made it possible to model and optimally solve important complex large-scale planning problems in national planning and industry.
>
> In my opinion Martin Beale has one of the finest minds in England (and the world) today (Dantzig and Tomlin 1987, 125).

His citation read in part: "... distinguished for his applications of mathematical and statistical techniques to industrial problems, and for his contributions to the theory of mathematical programming."

Martin Beale's honors did not stop there. In 1980, the Operational Research Society of Great Britain awarded him the Silver Medal, its highest honor. In 1984, he was elected to the Council of the Royal Society.

TOWARD THE END

Martin became seriously ill in the 1980s, but was able to attend the Boston Mathematical Programming Symposium in 1985. Bob Hattersley recalled: "... (in that) year, as always, his interests ranged across the spectrum of linear, mixed integer and non-linear programming; from more accurate methods for inversion and updating the inverse, to cut generation and new approaches to estimation for special ordered sets" (Dantzig and Tomlin 1987, 121–123). He returned to a lifelong interest—stochastic programming—and his last paper (Beale et al. 1986) dealt in a novel way with the difficult problem of multi-time-period stochastic programming.

After treatment in 1984, Beale insisted he was on the road to recovery. He declared himself "too busy to die," and as late as November 1985, he wrote that he was "on the mend and very confident that all will be well" (Beale 1985b). Alas, this was not to be the case.

Martin was active with his students and colleagues until the 11th hour. His devoted secretary, Bev Peberdy, and several of his protégés—Robert Ashford, Bob Daniel, Bob Hattersley, and Robert Watson—were at his bedside listening while Martin talked about his ideas until close to the very end, which came at his home in Cornwall on December 23, 1985 (Daniel 1985). He was 57 years old. His wife Betty, in the year that followed, received hundreds of letters from friends around the world.

Betty and Martin, Windhover House, Cornwall, 1983.

A magnificent Martin Beale Memorial Symposium was held at the Royal Society's premises in Carlton House Terrace, London, in the summer of 1987, with proceedings edited by Michael Powell (1988). Martin's final book, *Introduction to Optimization* (Beale 1988), was posthumously edited by Lynne Mackley.

ACKNOWLEDGMENTS

In preparing this chapter, I have relied on much of the same material that Martin's friends and colleagues provided for the memorial article that G. B. Dantzig and I wrote for publication in *Mathematical Programming* in 1987, where they are acknowledged in detail. Some of the material has been further paraphrased or condensed. I am also extremely grateful to M. J. D. Powell for permission to draw freely on his biographical memoir of

Martin Beale (Powell 1987). That memoir is also invaluable for its complete bibliography of Martin Beale's published works. Finally I wish to acknowledge my own indebtedness to Martin, the finest mentor imaginable.

REFERENCES

Abadie J (ed) (1967) Nonlinear programming. North-Holland, Amsterdam

Beale EML (1954) An alternative method for linear programming. Proc Camb Philos Soc 50(Part 4):513–523

Beale EML (1955) On minimizing a convex function subject to linear inequalities. J R Stat Soc (B) 17(2):173–184

Beale EML (1959) On quadratic programming. Nav Res Logistics Q 6(3):227–243

Beale EML (1967) Numerical methods. In: Abadie J (ed) Nonlinear programming. North-Holland, Amsterdam, pp 133–206

Beale EML (1968) Mathematical programming in practice. Pitman's, London, and Wiley, New York, NY

Beale EML (1970a) Matrix generators and output analyzers. In: Kuhn HW (ed) Proceedings of the princeton symposium on mathematical programming, Princeton University Press, Princeton, NJ, pp 25–36

Beale EML (1970b) Advanced algorithmic features for general mathematical programming systems. In: Abadie J (ed) Integer and nonlinear programming. North Holland, Amsterdam, pp 119–137

Beale EML (1974) A conjugate gradient method of approximation programming. In: Cottle R, Krarup J (eds) Optimization methods for resource allocation. English Universities Press, London, pp 261–277

Beale EML (1985a) Integer programming. In: Schittkowski K (ed) Computational mathematical programming. NATO ASI Series F: Computer and System Sciences, 15, Springer, Berlin, pp 1–24

Beale EML (1985b) Private communication, November 11

Beale EML (1988) Introduction to optimization. Wiley, Chichester

Beale EML, Dantzig G, Watson RD (1986) A first-order approach to a class of multi-time-period stochastic programs. Math Program Study 27:103–117

Beale EML, Forrest J (1976) Global optimization using special ordered sets. Math Program 10(1):52–69

Beale EML, Hughes P, Small R (1965a) Experience in using a decomposition program. Comput J 8:13–18

Beale EML, Small R (1965b) Mixed integer programming by a branch and bound technique. In: Kalenich WA (ed) Proceedings of the IFIP congress 1965, MacMillan/Spartan Press, London/Washington, DC, pp 450–451

Beale EML, Tomlin J (1970) Special facilities in a general mathematical programming system for non-convex problems using ordered sets of variables. In: Lawrence J (ed) Proceedings of the fifth international conference on operational research. Tavistock, London, pp 447–454

Bowen K (1986) Professor E. M. L. Beale – personal tribute. O.R. Newsletter, February 8–9

Daniel R (1985) Obituary of E. M. L. Beale. The Times, December 28

Dantzig G (1955) Linear programming under uncertainty. Manage Sci 1(3–4):197–206

Dantzig G, Tomlin J (1987) E .M. L. Beale, FRS: friend and colleague. Math Program 38(2):117–131

Dantzig G, Van Slyke R (1967) Generalized upper bounding techniques. J Comput Syst Sci 1:213–226

Douglas AS (1986) Obituary: Professor Evelyn Martin Lansdowne Beale, FRS, FIMA. I.M.A. Bulletin 22:120–122

Forrest J, Tomlin J (2007) Branch and bound, integer, and non-integer programming. Ann Oper Res 149(1):81–87

Gomory R (2002) Early integer programming. Oper Res 50(1):78–81

Hughes P (1988) Martin Beale: a personal memory. Math Program 42(1):5–9

Land A, Doig A (1960) An automatic method for solving discrete programming problems. Econometrica 28(3):497–520

Powell M (1987) Evelyn Martin Lansdowne Beale. Biogr Mem Fellows R Soc 33(Dec):22–45

Powell M (ed) (1988) Mathematical models and their solutions: contributions to the Martin Beale Memorial Symposium. Math Program 42(1):1–202

Tomlin J (1989) The influences of algorithmic and hardware developments on computational mathematical programming. E. M. L. Beale Memorial Lecture at the 13th International Symposium on Mathematical Programming, Tokyo. In: Iri M, Tanabe K (eds) Mathematical programming: recent developments and applications. Kluwer, Dordrecht, pp 159–175

38

Ralph E. Gomory

Ellis L. Johnson

Observers of operations research(OR) in the 1950s were faced with a curious phenomenon: While OR was established in some departments of industrial engineering and a few schools of business, it continued to be ignored by mainstream, scientifically oriented departments, especially mathematics departments. This seemed strange as OR rests heavily on aspects of both theoretical and applied mathematics. This explains how, in those days, Ralph Gomory, a student with both a B.A. (Williams College, 1950) and Ph.D. (Princeton University, 1954) in mathematics, could leave the academic environment without ever hearing of OR. Soon after, however, he was introduced to OR while on naval duty with the Office of Naval Research (ONR). He then worked for IBM's Research Division, rising to the positions of Director of Research and Senior Vice-President for Science and Technology.

Ralph is recognized for several fundamental contributions to OR, especially in the development of theoretical and computational advances in integer programming (IP) and the development of an efficient way of solving the general cutting-stock problem. For the latter work, he received the 1963 Operations Research Society of America's (ORSA) Lanchester Prize (joint with Paul Gilmore). His total contribution to OR was recognized by his being awarded the John von Neumann Theory Prize from ORSA and The Institute of Management Sciences (1984). He received the U.S. National Medal of Science in 1988 for his scientific contributions and management of research organizations. He is a member of the National Academy of Sciences and the National Academy of Engineering.

A.A. Assad, S.I. Gass (eds.), *Profiles in Operations Research*, International Series in Operations Research & Management Science 147, DOI 10.1007/978-1-4419-6281-2_38,

Transformation I: From Physics to Mathematics

Ralph Gomory was born on May 7, 1929 in New York City and grew up in Brooklyn Heights, a middle-class neighborhood of that city's borough of Brooklyn that overlooks Manhattan from its location along the East River. His father, Andrew Gomory, was born in Szeged, Hungary, and worked in international banking. His mother Marian (Schellenberg) was born in Brooklyn; he has an older brother Paul. Despite growing up in the Great Depression and World War II (WWII), Ralph enjoyed a comfortable childhood and attended the private Woodward School in Brooklyn through the ninth grade. The Woodward School instilled in Ralph a lifelong love of learning. His summers were spent on the Long Island shore, and sailing became a favorite pastime. At the age of 14, he entered the George School, a Quaker boarding school in Bucks County, Pennsylvania, where he attended high school and excelled in physics. Soccer and tennis were his main sports, and he played both for his high school and college teams.

His love of physics came from its explaining observable natural phenomena. His understanding of mathematics, however, lagged a bit behind. It was not until late in high school that its usefulness began to dawn on him with the reading of the book *What is Mathematics?* (Courant and Robbins 1941). About the same time, he saw that trigonometry could be used to help understand the angle of tacking his sailboat. Despite this increased appreciation, his aptitude test scores in mathematics did not match his otherwise uniformly high scores; he was advised against majoring in physics in college because of his less than glowing scores in mathematics, despite excellent grades in both subjects.

Ralph entered Williams College in 1946 and did major in physics. As his classes moved into atomic physics, he became less interested. At the same time, a mathematics professor, Donald Richmond, helped to stimulate in him a greater interest and understanding of mathematics.

Ralph came to believe that mathematics could model and help explain a wide variety of phenomena. His passion for studying physics was replaced by a similar passion for mathematics. His senior thesis was based on the paper "The heartbeat considered as a relaxation oscillation, and an electrical model of the heart" (B. van der Pol and J. van der Mark 1928) and reflected his new understanding of the power, breadth, and applicability of mathematical models. After graduating in 1950 with a B.A. in mathematics, Ralph spent the summer working with Don Richmond and they produced a paper based on Ralph's thesis (Gomory and Richmond 1951).

After the summer at Williams College, he spent a year studying mathematics at the University of Cambridge, England. This opportunity came by way of an invitation from the Provost of King's College London, a family friend. During that year, Solomon Lefschetz, head of the Princeton University mathematics department, invited Ralph to come to Princeton to study for his Ph.D. In those days, a simple invitation and acceptance could constitute the admission process—Ralph arrived at Princeton in the fall of 1951. He completed his degree in 3 years, writing a dissertation on nonlinear differential equations—"Critical points at infinity and forced oscillations"—with Lefschetz as his advisor. Ralph published three papers from or related to the dissertation (Gomory 1955, 1956; Gomory and Haas 1955).

After receiving his Ph.D., Ralph volunteered and served as an officer in the U.S. Navy from 1954 to 1957. He had been in his early teens during WWII and was, therefore, too young to join the military. Nevertheless, he was left with a lifelong love of country and sense of duty. Volunteering for military service came as a natural act for him.

TRANSFORMATION II: FROM MATHEMATICS TO OPERATIONS RESEARCH

After a 4-month course at the Navy's Officer Candidate School in Newport, Rhode Island, Ralph's main tour of duty was in Washington, DC, where he was assigned to the Physics Branch of the ONR in early 1955. He soon discovered the Operations Research Group (ORG) down the hall and began spending his spare time with its research staff. Ralph's attraction to OR was a continuation of the appeal of being able to develop models of a wide variety of phenomena and the use of mathematics to understand the phenomena better—OR being an area in which such models help resolve a wide variety of decision-making situations. This exposure was a prime reason why Ralph transferred his scientific interests to OR. By way of preparation, he took an evening course in OR from Alan Goldman, a mathematician who had acquired a background in OR, a rare occurrence at that time. Alan had recently received his Ph.D. from Princeton; the course introduced Ralph to the rather new topic of linear programming (LP). Thus, Ralph's career objective changed from working in academia to becoming a practitioner of OR.

His plan, however, was interrupted by an attractive offer from Princeton University. He was appointed Higgins Lecturer in mathematics and then Assistant Professor in mathematics at Princeton University, where he remained from 1957 to 1959. The Navy kept him on as a consultant and Ralph made monthly trips to Washington to work on Navy problems. The ORG staff complained about LP solutions prescribing fractional numbers of such variables as aircraft carriers. Ralph set out to combine Diophantine equations with LP so as to get rid of such answers. When that proved difficult, he started his ground-breaking work on cutting-plane methods. He began by looking at LP solutions and LP tableaus and realized that he could find optimum solutions by various means. For a maximizing IP problem, the problem is first relaxed by assuming that all variables are continuous, thus yielding a standard LP maximization problem. A key observation was that if the objective had integer coefficients and the maximum value of the LP solution was fractional, then the objective function of the IP solution could not be greater than the rounded-down value of the continuous LP problem. Ralph moved to looking at other rows of the tableau for further opportunities to do a similar rounding. Studying the integer and fractional

> FRACTIONAL CUTS AND TINGLING TOES
>
> "Late in the afternoon of the eighth day of this [working on solving integer problems] I had run out of ideas. Yet I still believed that, if I had to, in one way or another, I would always be able to get at an integer answer to any particular numerical example. At that point I said to myself, suppose you really had to solve some particular problem and get the answer by any means, what would be the first thing that you would do? The immediate answer was that as a first step I would solve the linear programming (maximization) problem and, if the answer turned out to be 7¼, then I would at least know that the integer maximum could not be more than 7. No sooner had I made this obvious remark to myself than I felt a sudden tingling in two of my left toes, and with great excitement realized that I had just done something different, and something that was not a part of classical Diophantine analysis. How exactly had I managed to conclude, almost without thought, that, if the LP answer was 7¼, the integer answer was at most 7?" (Gomory 2002, 78).

parts of the tableau's coefficients, including the right-hand sides of the equations, led to the formalization of his now-famous fractional cutting-plane method for solving IP problems. Ralph showed how to systematically add new inequalities (cuts) to the LP problem that cut away non-integer solutions, and, most important, he gave a finiteness proof of the cutting-plane algorithm (Gomory 1958, 1960, 1963).

The first public announcement of Ralph's IP research, which he termed the method of integer forms, was a short description given in the *Bulletin of the American Mathematical Society* (Gomory 1958). Ralph's first computer program of the algorithm was written for Princeton's Burroughs E101, a small-scale pinboard-based computer, on which he was able to solve problems with few (four to five) variables. Based on his work, Ralph was invited to spend the summer of 1958 at the RAND Corporation, a U.S. Air Force-sponsored research organization in Santa Monica, California. There, Ralph programmed the algorithm using FORTRAN, a rather new programming language; he was able to solve 10–15 variable problems on an IBM 704 computer. He soon discovered that although most small problems solved quickly, some ran on and on. At first he thought this was due to a bug in his program. But, as time went on, he discovered it to be a characteristic of some IP problems.

From the late 1950s to the early 1960s, RAND was an important center for the development of methodologies, computer programs, and applications of OR. Its permanent staff included Richard Bellman, George Dantzig, Ray Fulkerson, Harry Markowitz, William Orchard-Hays, Herbert Scarf, Lloyd Shapley, and Philip Wolfe. An important personal outcome of Ralph's brief summer visit was his becoming friends with this cadre of outstanding researchers, especially with Phil Wolfe, who was particularly helpful in making Ralph's summer productive. A broader outcome of his summer visit was that RAND became the center of a new and immensely important OR area: computation of integer solutions to LP programs.

Back at Princeton, Ralph collaborated with economics Professor William (Bill) Baumol on the first paper to link integer programs and pricing, that is, "dual prices and their relationship to marginal yields of scarce indivisible resources and their efficient allocation" (Gomory and Baumol 1960, 521). Their collaboration would be renewed years later after both had, in some sense, retired. By a strange coincidence, their pricing article was published in the same issue of the journal *Econometrica* that also contained a paper by Land and Doig (1960) that described an alternative approach for solving IP problems; their work initiated the development of the very computational effective methods of branch and bound.

> ON SOLVING IP PROBLEMS
>
> On his return to Princeton, Ralph became acquainted with Professor Albert W. Tucker, the new chairman of the mathematics department and leader of a group working in game theory and mathematical programming. The group included Harold Kuhn, David Gale, and Martin Beale, who was visiting from England. When Martin, who was in charge of organizing seminars for the group, asked Ralph to speak, and Ralph responded that he "would be glad to give a talk on solving linear programs in integers," Martin exclaimed "but that's impossible" (Gomory 2002, 79).

Research and the Organization Man: IBM Research and the Sloan Foundation

In the summer of 1959, Ralph left Princeton and joined IBM's Research Division, Yorktown Heights, New York. [He had been recruited by Herman H. Goldstine, the founding director of IBM's Mathematical Sciences Department, and Emanuel R. Piore, the first director of IBM Research (Goldstine 1987).] At IBM, Ralph's research interests and accomplishments blossomed. He published more than 20 papers in the period 1960–1966 with nine co-authors that included Michel Balinski and IBM research associates Paul Gilmore, T.C. Hu, and Alan Hoffman. These papers made strong contributions to the resolution of some of the most important problems in OR and mathematics: knapsack, traveling salesman, cutting stock, network flows (Gomory and Hu 1961, 1962), linear algebra, machine sequencing, lot sizes and inventories, and, of course, LP (Gomory and Balinski 1963) and IP (Gomory and Hoffman 1963). In 1964, Ralph was named an IBM Fellow (Fellows are granted a wide sphere of independence in the pursuit of their research); he became Director of IBM's Mathematical Sciences Department in the following year.

During the 1960s, Ralph and Paul Gilmore collaborated in several important application areas, in particular the cutting-stock problem, that is, the cutting of material to meet given order and size requirements (Gilmore and Gomory 1961, 1963). To understand the related manufacturing, scheduling, and cutting processes, they visited steel mills, glass plants, and paper mills, finally concentrating on the trim-loss problem of paper mills. The problem comes about when a paper mill must set cutting blades to slice a standard-sized large roll of paper into a mix of smaller sized rolls needed to fulfill customer orders. Since a combination (a setting of the blades) of smaller rolls may not use all of the larger roll, the objective is to select combinations that will produce the rolls on order, while minimizing the total left over parts of the larger rolls, the trim loss. The situation can be formulated as an IP problem, a complexity that is further compounded because of the very large number of possible combinations—the integer decision variables specify how many times each combination should be used to produce the specified set of ordered rolls.

By ignoring integer restrictions, a mill's trim-loss problem can be modeled as a standard LP problem. But, because of its large size, stating the problem explicitly in the usual row and column (matrix) format is by itself difficult and computationally complex. Ralph and Paul developed a novel column-generation procedure that identifies a possible combination to use by solving a side problem in the form of a knapsack problem; the solutions of successive knapsack problems are introduced one at a time into the LP solution until optimality (or a stopping rule) is reached. The resulting non-integer values of the combinations can be readily converted to integers by simple rounding, usually without much of an increase in the trim-loss value. [The first LP approach to the basic trim-loss problem for the cutting of rolls of newsprint is due to Paull (1956).]

In the late 1960s, Ralph introduced the group problem that stems from a relaxation of the integer conditions of an IP problem, studied corner polyhedra associated with the group problem (Gomory 1965, 1967, 1969), and developed the asymptotic theory and subadditive functions of IP (Gomory and Johnson 1973; Johnson 1974). This line of research was perhaps his most satisfying mathematical work and is still an active area. As Ralph noted, "this resulted in the theory of asymptotic integer programming and the discovery of the corner polyhedra, which I consider to be my best work in the field of integer programming" (Gomory 2002, 80).

Ralph served as Director of IBM Research's Mathematical Sciences Department in 1965–1967 and 1968–1970 during an important period of its growth and evolution. This period saw the beginning of Shmuel Winograd's work on complexity of computations and the design of efficient algorithms, especially fast algorithms for arithmetic, and the fractal geometry research of Benoit Mandelbrot, who founded the field.

KNOCKING ON DOORS

While Ralph was Director of Mathematical Sciences, IBM, in 1969, went through the process of unbundling applications software that was provided free as part of the purchase price of a computer. Ralph had the idea that the Mathematical Sciences Department might be helpful to some of the applications software development groups. (During this time I was a member of the research staff and had the opportunity to work with Ralph). He asked me—young, tall, and geeky—to go out and meet with the major cross-industry applications groups. In most cases, I was received politely by a typically bemused manager who listened to who we were and what we do. The only real connection that led to collaboration was with the IBM Mathematical Programming System development group, and Phil Wolfe (then a member of our group) had already made that connection. What was striking to me was that Ralph already had the vision that the Research Division could be much more vital and useful to IBM, and he was fearless, which gave me courage to carry out what was otherwise a very unnatural task for me.

INTERNATIONAL SYMPOSIUM ON MATHEMATICAL PROGRAMMING, ATLANTA, GEORGIA, 2000

Ralph Gomory and Ellis Johnson

Ralph's first IBM corporate assignment was to the Corporate Technical Committee in early 1970. This appointment led to another phase of his career, his becoming the top technical executive in IBM. It also led to a greatly diminished research output. However, he later succeeded in returning to a more active involvement in OR and economics. [He contributed greatly to my paper that generalized subadditive functions on the line to the multi-dimensional case (Johnson 1974). In early 1971, however, he insisted on removing himself as an author because he was so far removed from active participation in the research. It was with regret that I did so. Later, Ralph returned to this area.]

Ralph was made IBM's Director of Research in 1970 with line responsibility for the Research Division. He held that position until 1986, becoming an IBM vice-president in 1973 and a senior vice-president in 1985. In 1986, he was appointed IBM senior vice-president for science and technology. He retired from IBM in 1989 at the mandatory retirement age of 60 (for executives at his level) and took on the job of President of the Alfred P. Sloan Foundation.

The Sloan Foundation was founded in 1934 by Alfred P. Sloan, Jr., the president of General Motors, to support original research and education in science, technology, engineering, mathematics, and economic performance, plus the study of national issues. At Sloan, Ralph led efforts to protect the population from the threat of bio-terrorism—this initiative was begun 2 years before the 9/11 attacks. He also initiated college-level online learning in 1991, when there was no public access to the Internet—the Foundation has nurtured the growth of online learning through all the succeeding years. Ralph retired from the Foundation at the end of 2007, but remains active in research and service. He is currently a research professor at New York University's Leonard N. Stern School of Business. He also blogs for the Web-based *Huffington Post.*

In 1998, I invited Ralph to visit the Georgia Institute of Technology as an Edenfeld Executive-in-Residence (I had joined the School of Industrial and Systems Engineering in 1995). Ralph, as President of the Sloan Foundation, board member of several corporations, a member of the U.S. President's Council of Advisors on Science and Technology (PCAST), and a retired IBM Senior Vice-President, was certainly well qualified as an executive, but my motive for inviting him was to reinitiate research collaboration. There are two sides to Ralph: one in research that I cannot match, but do understand well, and one as a leader in the corporate and governmental world that I can only look upon with wonder. In his visit, we did profit from his experiences through a lecture he gave on the process for a corporate research effort to produce positive returns (Gomory 1987). More interesting to me, in a seminar lecture, he introduced the idea of using a shooting experiment (random selection) of the group polyhedra of a relaxation of an IP problem. A student of mine, Lisa Evans Miller, worked on this approach and it became an important part of her dissertation research. Later, another student, Brady Hunsaker, jointly advised by Craig Tovey and myself, carried on this work as part of his dissertation. My joint research objective was successful: Ralph and I collaborated on three papers that appeared in a special volume of *Mathematical Programming* (Aráoz et al. 2003; Gomory and Johnson 2003; Gomory et al. 2003).

THE REAL-WORLD(?) OF THE ECONOMIST

A major research area for Ralph during his tenure at Sloan began when he encountered and was deeply perplexed by economists' view that any global trade was good for both countries. He first learned of this view while on a PCAST trip to meet with CEOs of computer chip manufacturing companies in California. The economists on PCAST seemed to feel that losing production of computer chips could be made up by producing more potato chips (Gomory 1993). Ralph was genuinely perplexed by this point of view and met with his economist friends, Herb Scarf and William Baumol, to see if they had an explanation. Thus, he learned about the Ricardo model of production and trade. This model assumes that unit production cost increases as production level increases, perhaps a model developed for agricultural production but not suited for, say, automobile manufacturing. Ralph and Bill began exploring a model that assumed a fixed charge for entry (building a factory) and a linear cost thereafter. They found that there were many equilibria and there were definitely winners and losers. In later work, they found flaws in the usual conclusions even when the fixed charge is removed. This work has profound implications for governmental and corporate policy and flies in the face of conventional wisdom. In its combination of good mathematics, applications to the real world, broad implications, and taking on the conventional wisdom, his work with Bill exemplifies Ralph at his best: brilliant, courageous, and tenacious. Because of the difficulty of publishing journal articles that challenge the accepted orthodoxy, much of this work only appears in their book *Global Trade and Conflicting National Interests* (Gomory and Baumol 2001):

"In this book Ralph Gomory and William Baumol adapt classical trade models to the modern world economy. Trade today is dominated by manufactured goods, rapidly moving technology, and huge firms that benefit from economies of scale. This is very different from the largely agricultural world in which the classical theories originated. Gomory and Baumol show that the new and significant conflicts resulting from international trade are inherent in modern economies.

"Today improvement in one country's productive capabilities is often attainable only at the expense of another country's general welfare. The authors describe why and when this is so and why, in a modern free-trade environment, a country might have a vital stake in the competitive strength of its industries" (MIT Press 2009).

Honors and Awards

Ralph has served in many capacities in academic, industrial, and governmental organizations. He is a member of the National Academy of Sciences, the National Academy of Engineering, and the American Philosophical Society—he was elected to serve on all three councils. He has been awarded eight honorary degrees and many prizes: 1963 Lanchester Prize of the ORSA, joint with Paul Gilmore, for their cutting-stock paper (Gilmore and Gomory 1963); 1984 John von Neumann Theory Prize awarded by ORSA and the Institute of Management Sciences; Harry Goode Memorial Award of the American Federation of Information

THE PERIPATETIC SPEAKER

(Left to right) Richard Larson, Harlan Crowder, Ralph Gomory, and Richard Rosenthal. Ralph was the distinguished IFORS lecturer at the joint International CORS and INFORMS meeting, Banff, Canada, May 18, 2004.

(Left to right) Richard Cottle, Margaret Wright, Ralph Gomory, and Saul Gass. Ralph was a speaker at the George B. Dantzig Memorial Lectures at the Washington, DC INFORMS national meeting, October 14, 2008.

Processing Societies (1984); IEEE Engineering Leadership Recognition Award (1988); Arthur M. Bueche Award of the National Academy of Engineering (1993); Heinz Award for Technology, the Economy and Employment (1998); Princeton University's Madison Medal Award to outstanding alumni (1999); and the Sheffield Fellowship Award of the Yale University Faculty of Engineering (2000). Ralph received the National Medal of Science, awarded by President Reagan on July 15, 1988. The citation read: "For his scientific contributions to the mathematics of discrete optimization and its far-reaching influence on information processing; for bringing to a leading position one of industry's most significant research establishments; and for his contributions to public and private scientific enterprise" (National Science Foundation 1988). He was the 1989 Omega Rho OR Honor Society Distinguished Lecturer at the New York INFORMS national meeting, and the International Federation of Operational Research Societies (IFORS) distinguished lecturer at the joint Canadian Operational Research Society (CORS) and INFORMS meeting, Banff, Canada, May 2004. He was made a member of the IFORS' Hall of Fame in 2005 (Johnson 2005) and received the CORS Harold Larnder Prize in 2006.

REFERENCES

Aráoz J, Evans L, Gomory RE, Johnson E (2003) Cyclic group and knapsack facets. Math Program Ser B 96(2):377–408

Courant R, Robbins H (1941) What is mathematics? Oxford University Press, New York, NY

Gilmore P, Gomory RE (1961) A linear programming approach to the cutting stock problem—part I. Oper Res 9(6):849–859

Gilmore P, Gomory RE (1963) A linear programming approach to the cutting stock problem—part II. Oper Res 11(6):863–888

Goldstine H (1987) Reflections on the early days of the department. IBM J Res Dev 31(2):154–157

Gomory RE (1955) Trajectories tending to a critical point in 3-Space. Ann Math Sec Ser 61(1):140–153

Gomory RE (1956) Critical points at infinity and forced oscillation. In: Lefschetz S (ed) Contributions to the theory of nonlinear oscillations, Annals of Mathematics Studies, vol. 3. No. 36, Princeton University Press, Princeton, NJ, pp 85–126

Gomory RE (1958) Outline of an algorithm for integer solutions to linear programs. Bull Am Math Soc 64(5):275–278

Gomory RE (1960) Solving linear programming problems in integers. Proc Symp Appl Math 10(July):211–215

Gomory RE (1963) An algorithm for integer solutions to linear programs. In: Graves R, Wolfe P (eds) Recent advances in mathematical programming. McGraw-Hill, New York, NY, pp 269–302

Gomory RE (1965) On the relation between integer and noninteger solutions to linear programs. Proc Natl Acad Sci 53:260–263

Gomory RE (1967) Faces of an integer polyhedron. Proc Natl Acad Sci 57(1):16–18

Gomory RE (1969) Some polyhedra related to combinatorial problems. Linear Algebra Appl 2(4):451–558

Gomory RE (1987) Science in industry. IBM J Res Dev 31(2):151–153

Gomory RE (1993) Let them eat chips. Bridge 23(4):13–16

Gomory RE (2002) Early integer programming. Oper Res 50(1):78–81

Gomory RE, Balinski M (1963) A mutual-primal-dual simplex method. In: Graves R, Wolfe P (eds) Recent advances in mathematical programming (Proceedings of the 1962 Symposium for Mathematical Programming), McGraw-Hill, New York, NY, pp 17–26

Gomory RE, Baumol W (1960) Integer programming and pricing. Econometrica 28(3):521–550

Gomory RE, Baumol W (2001) Global trade and conflicting national interests. MIT Press, Cambridge, MA

Gomory RE, Haas F (1955) A study of trajectories which tend to a limit cycle in three-space. Ann Math 62(1):152–161

Gomory RE, Hoffman A (1963) On the convergence of an integer-programming process. Nav Res Logistics Q 10(2):121–123

Gomory RE, Hu T (1961) Multi-terminal network flows. J Soc Ind Appl Math 9(4):551–570

Gomory RE, Hu T (1962) An application of generalized linear programming to network flows. J Soc Indust Appl Math 10(2):260–283

Gomory RE, Johnson E (1973) The group problems and subadditive functions. In: Hu T, Robinson R (eds) Mathematical programming. Academic, New York, NY, pp 157–184

Gomory RE, Johnson E (2003) T-space and cutting planes. Math Program Ser B 96(2):341–375

Gomory RE, Johnson E, Evans L (2003) Corner polyhedra and their connection with cutting planes. Math Program Ser B 96(2):321–339

Gomory RE, Richmond D (1951) Boundaries for the limit cycle of Van der Pol's equation. Q Appl Math 9(2):205–209

Johnson E (1974) On the group problem for mixed integer programming. In: Balinski ML (ed) Math Program Study 2:137–179

Johnson E (2005) IFORS' operational research hall of fame: Ralph E. Gomory. Int Trans Oper Res 12(5):539–543

Land A, Doig A (1960) An automatic method for solving discrete programming problems. Econometrica 28(3):497–520

MIT Press (2009) http://mitpress.mit.edu/catalog/item/default.asp?ttype=2&tid=3682. Accessed 25 Jan 2009

National Science Foundation (1988) The President's National Medal of Science: Recipient details. http://www.nsf.gov/od/nms/recip_details.cfm?recip_id=144. Accessed 27 Jan 2009

Paull A (1956) A key to optimum newsprint production. Pulp Paper Mag Can 57(4):145–150

Van der Pol B, van der Mark J (1928) The heartbeat considered as a relaxation oscillation, and an electrical model of the heart. Lond Edin Dublin Phil Mag J 7(6):763–765

39

Alfred Blumstein

Daniel S. Nagin

In the 1950s, when operations research (OR) analysts were making the first inroads in applying OR to non-military settings, few would have envisaged public policy to emerge as a fertile area of application for the scientific-based methods of OR. In the 1960s, however, the value of OR thinking in the analysis of public policy issues was demonstrated with considerable success. A major exemplar of this success was the seminal work of Alfred Blumstein in his role as the Director of the Science and Technology Task Force of President Johnson's Commission on Law Enforcement and Administration of Justice in 1966–1967. Al, who was first introduced to OR by working on air defense and air traffic control problems, became an internationally recognized pioneer, missionary, and driving force in the application of OR to public policy, crime, and criminal justice.

Al is University Professor and the J. Erik Jonsson Professor of Urban Systems and Operations Research at Carnegie Mellon University (CMU). He was president of the Operations Research Society of America (ORSA) in 1977, president of The Institute of Management Sciences (TIMS) in 1987–1988, and president (1996) of the Institute for Operations Research and the Management Sciences (INFORMS), one of only two individuals who have served as president of the three organizations. Al was awarded the ORSA/TIMS Kimball Medal for service to the profession and the society and the ORSA President's Award for service to society. Al received the Stockholm Prize in Criminology, an award that recognizes achievements in criminological research or in the practical implementation of research to combat crime and promote human rights. He is a member of the National Academy of Engineering.

A.A. Assad, S.I. Gass (eds.), *Profiles in Operations Research*, International Series in Operations Research & Management Science 147, DOI 10.1007/978-1-4419-6281-2_39,

In the Beginning

Al Blumstein was born in New York City on June 3, 1930. He attended New York City public schools and graduated from the Bronx High School of Science. His parents divorced when he was very young; he was an only child. Having grown up in modest circumstances, Al set a high priority on developing skills that would enable him to earn a living. But what skills and in which profession? Al saw only two options: engineering and accounting. While he knew nothing about engineering, he had older cousins who were accountants. But, that field held little interest. By default, he decided to major in engineering. He enrolled in a new, 5-year engineering physics program at Cornell. The program was intended to provide more theory than traditional engineering and more practical emphasis on applications than was characteristic of physics programs. Al graduated in 1951 and, although most of his classmates went on to do graduate work in physics, Al set his sights on something different. What had caught his attention was an article in *Fortune* magazine that discussed the exciting "young applied science" of OR (Solow 1951, 105). That article led him to his best job offer, an operations analyst with the Cornell Aeronautical Laboratory (CAL) in Buffalo, New York, a position he held from 1951 to 1961. In his first assignment, he joined a group that was studying fleet air defense under contract to the Office of Naval Research (ONR). It was a great learning opportunity. One of his colleagues was Earl Isaac, later a founder of Fair-Isaac, one of the nation's leading risk/credit assessment organizations.

A subsequent, important opportunity was his assignment to CAL's ONR project (in Washington, DC) to study the operations of the Combat Information Center for a carrier task force. When the study was completed, Al briefed the Chief of Naval Operations on the study's recommendations, which were then implemented. Due to the success of this effort, upon his return to Buffalo, Al was promoted to lead a project developing new designs for carrier task forces.

OR, AIR TRAFFIC CONTROL, AND COUNTERINSURGENCY

It was about this time, when his interest in OR had crystallized, that Al attended the first national meeting of ORSA that was held at the headquarters of the National Bureau of Standards in northwest Washington, DC. The date was November 17–18, 1952. Ever since that initial meeting, Al has been a regular attendee at ORSA meetings. At one such meeting, about 1955, Al met Andy Schultz, an instructor of statistics he had at Cornell. Andy was now chair of Cornell's Department of Industrial Engineering. He informed Al that the department was establishing a Ph.D. program in OR and encouraged Al to enroll. Since CAL offered a graduate fellowship at Cornell to its employees, Al applied for and received one, and enrolled in 1956. After 2 years of study, he returned to Buffalo to do his dissertation on air traffic control.

Al's interest in air traffic control began during the time (1951–1954) he was studying for his M.A. degree in the statistics department, University of Buffalo. His thesis dealt with the simulation of the dynamics of an aircraft on a ground-controlled approach. It was notable as the simulation was written in machine language, an astonishing challenge which is also testimony to the diligence of youth. That led to his first journal paper, "A Monte Carlo analysis of the ground controlled approach system" (Blumstein 1957).

The publication proved timely—1958 was witness to considerable turmoil in the U.S. air traffic control system. In that year, the U.S. experienced the first collision between aircraft being directed by the Civil Aviation Administration's (CAA) air traffic control system; there was a growing recognition that the CAA was not keeping up with the growing demands of air traffic. In response, the Airways Modernization Board was created, which led to the formation of the Federal Aviation Authority (FAA). The FAA started to solicit and fund research in air traffic control. One of Al's early CAL projects in this area was to investigate the factors contributing to a collision (near Edwards Air Force Base, California) between a test aircraft and a commercial airliner. In discussions with Sy Deitchman, a senior colleague at CAL, who was also interested in air traffic control issues, it became clear to Al that a critical problem limiting the capacity of the system was the ability of airports to absorb the increasing traffic demand. Al made this the subject of his Ph.D. dissertation.

Because landing aircraft would naturally have priority over departing aircraft, Al first framed the problem in terms of landing capacity. There was, in fact, an extant literature that addressed this issue using queueing theory. Al realized, however, that, despite their elegance, the published queueing models were too limiting with respect to the reality of a mix of aircraft with heterogeneous speeds. He also recognized another important constraint: the requirement that a three-mile separation between aircraft on landing approach had to be integrated into the analysis. Thus, the real problem was satisfying the two constraints of no more than one aircraft on the runway and the maintaining of the three-mile separation on the glide path to the runway. At that time, there was a movement in the FAA to build high-speed turn-offs on runways to increase runway capacity, largely based on addressing the runway occupancy time constraint. Al's results showed that this constraint was not the dominant one. His work was a major factor in the decision to basically cancel the high-speed turn-off program—his research results were published in the paper "On the landing capacity of a runway" (Blumstein 1959).

Al received his Ph.D. from Cornell University in 1960. During a mid-semester break, Al married Dolores Reguera in Boston on January 26, 1958. They have three children, Lisa (1959), Ellen (1960), and Diane (1961), and four grandchildren.

Al became enthralled with the new dynamism of President John F. Kennedy's administration and, in particular, with the changes that Robert McNamara was bringing to the Department of Defense (DoD). Washington was beckoning Al. His former colleague, Sy Deitchman, had already moved to the newly formed Institute for Defense Analyses (IDA), a think tank for the office of Secretary of Defense. IDA's mission has always been an impressive one: "... to provide objective analyses of national security issues, particularly those requiring scientific and technical expertise, and conduct related research on other national challenges" (IDA 2009). Al moved to IDA in 1961.

His early work at IDA was concerned with what was then called remote area limited war, later to be called counterinsurgency. The concept of remote area limited war was stimulated by Khrushchev's famous observation about the Korean War: limited war is a stalemate, but that wars of national liberation could be a vehicle for communist victory. In reaction to that boast, IDA organized a counterinsurgency research team led by Luis Alvarez (who received the 1968 Nobel Prize in physics), with team members Al and Daniel Ellsberg (the team's young Turks). Al joined them in his first IDA assignment. The team visited Far Eastern countries to learn about developing insurgency problems and the steps being taken to address them. That effort gave rise to a number of recommendations for a DoD research and development counterinsurgency research program. Al continued to study the problems of defending villages against attack, and in the fall of 1962, he spent 3 months in South Vietnam.

THE CRIME COMMISSION

In 1966, Al was contacted by the President Johnson's Commission on Law Enforcement and Administration of Justice. The Commission had been organized with task forces on police, courts, and corrections, and now wanted to establish a task force on science and technology (S&T) to investigate how S&T could be applied to combating crime. Al was asked to join its staff. At first, he was reluctant and suggested some others as possible candidates. The Commission staff persisted, and 3 months later asked Al to join the task force as its director. Al made it clear that he knew nothing about crime or the administration of justice. They in return made it clear that that was okay, as they knew nothing about S&T. That sealed the deal. Al was able to recruit some excellent people who, among others, comprised the regular staff: Dick Larson, who had just graduated MIT with a bachelor's degree in electrical engineering; Ron Christensen, a Ph.D. student in nuclear physics at Berkeley and who had a law degree from Harvard; Saul Gass, a well-known OR consultant then employed by IBM; Sue Johnson, an independent consultant in systems analysis (who had two children and was expecting a third!); and IDA colleagues Joe Navarro and Jean Taylor. This staff was supported by a number of consultants who undertook special studies, including Thomas Schelling, the recipient of the 2005 Nobel Prize in Economics. Because IDA was solely committed to the DoD, it required a joint agreement between Attorney General Nicholas Katzenbach and Defense Secretary Robert McNamara to establish the task force.

Al's work on the Crime Commission led to his major stream of research on crime and criminal careers. This research marks his most important contributions to science and public policy. Nearly all of this work was done after 1969 when he moved to his first permanent academic position at the newly formed School of Urban and Public Affairs (SUPA) at CMU. Later, in 1986, Al became the fourth Dean of SUPA (William W. Cooper was SUPA's first Dean.) Among Al's notable accomplishments was the renaming of SUPA in 1992 as the H. John Heinz III School Public Policy and Management in honor of the late U.S. Senator. Al held the Dean's position until 1993, after which he returned to the faculty to continue his by then internationally recognized research on crime and criminal justice. To understand the impact of Al's research program, we quote from his 2005 INFORMS Morse Lecture, "An OR missionary's visits to the criminal justice system":

> The missionary role involves bringing OR perspectives to the "heathens" in a particular domain, those who haven't yet adopted quantification, modeling, system perspectives, and planning that characterize the hallmark of OR. In the early days, of course, the dominant role of OR people was that of missionary to the military. Phil Morse, after whom the lectureship is named, was the epitome of an OR missionary. An excellent physicist, he was recruited into the military groups to study and find ways to improve military operations. He didn't start out with much knowledge about military operations, but his group explored a range of operations by bringing his commitment to measurement and observation, and his emphasis on appropriate measures of effectiveness to the study of those operations. By bringing more subtle questioning perspectives, they picked the right approach and thereby made major improvement in military operations, and OR groups continue as a major aspect of military planning and analysis.
>
> The history of OR has seen many of our colleagues functioning as missionaries: Bob Hermann and Les Edie were early in transportation, John Little in marketing, Dick Larson and Jan Chaiken in police patrol analysis, Bill Pierskalla in health care, Ed Kaplan and Jon Caulkins in drug control, Tom Cook in yield management for the transportation industry, and more recently, Larry Wein in homeland security.
>
> My own experience as a missionary to the criminal justice system (CJS) began when I was recruited to be the director of the Task Force on Science and Technology for the President's Crime Commission in 1965-67 (Blumstein 2007, 14).

Al brought his missionary's zeal not only to his work on the Crime Commission, but also to all his subsequent work on crime and criminal justice.

THE CRIMINAL JUSTICE SYSTEM AND BEYOND

The most important tangible products of the Commission's work were its report *The Challenge of Crime in a Free Society* (Government Printing Office 1967a) and the IDA report to the Commission, *Task Force Report: Science and Technology* (Government Printing Office 1967b). Of especial importance was the latter report's chapter, "Analysis of crime and the overall criminal justice system," in which Al and his associates described, for the first time, a systemic view of the total criminal justice system (CJS)—police, prosecution, courts, corrections [the system is also described in Blumstein and Larson (1969)]. This work was the basis of a seminal interactive computer-based analysis program—the Justice System Interactive Model (JUSSIM) (Blumstein et al. 1971; Cohen et al. 1973).

The impact of this work, however, transcends specific tangible products. Forty years ago, the CJS was not conceived as an interconnected system. Indeed, the term criminal justice system was not yet conventional nomenclature, and the idea that the CJS could be modeled in mathematical terms had not been imagined. Prior to the 1970s, only the most rudimentary statistical tools were used in criminological analysis; mathematical models were unheard of. Today, the use of standard and advanced statistical methods and mathematical models is routine. Al and his research helped lead the vanguard of young scholars—together they have fundamentally transformed research in the CJS (Blumstein 2002).

Al's tireless efforts to explain and promote JUSSIM also had important impacts on criminal justice policy. JUSSIM made it possible, for the first time, to analyze how actions in one part of the CJS affected other parts of the system. For example, it provides an analytic tool for analyzing how changes in sentencing laws would affect court case loads and, ultimately, prison populations. In this way, Al was a pioneer in bringing the tools of policy analysis to the formulation of criminal justice policy. In his later roles as Chairman of Pennsylvania Commission on Crime and Delinquency (1979–1990) and Member of the Pennsylvania Commission on Sentencing

THE CRIMINAL JUSTICE SYSTEM ANALYSIS FLOWCHART

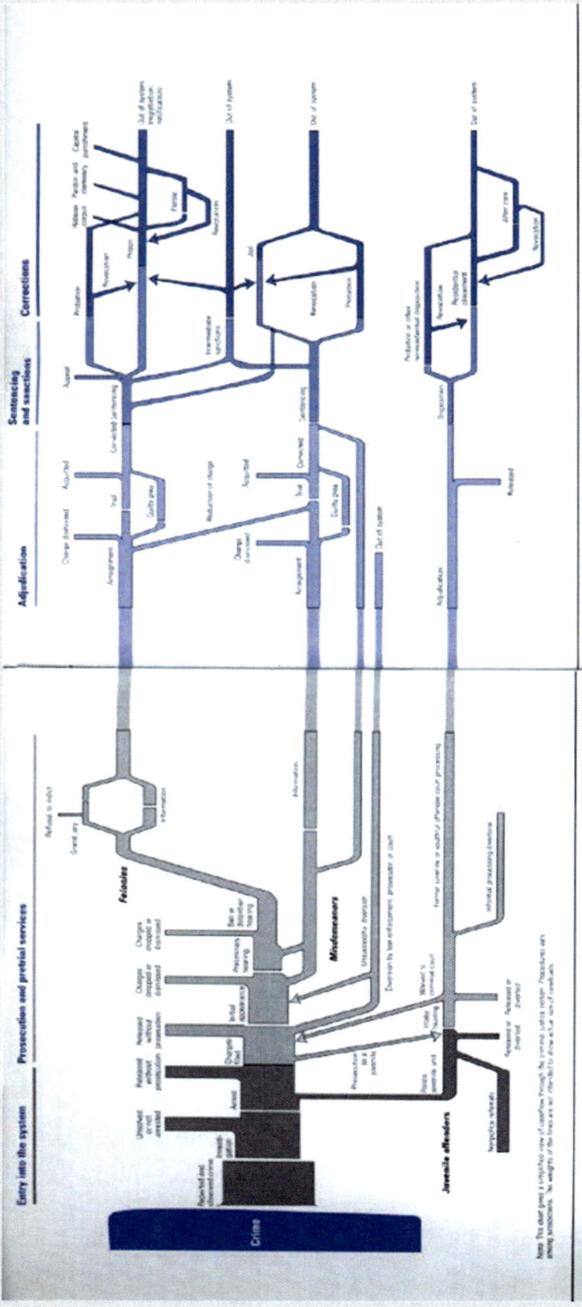

(Bureau of Justice Statistics, U.S. Department of Justice Programs. Source: *The Challenge of Crime in a Free Society*, President's Commission on Law Enforcement and Administration of Justice, 1967a, revised.)

(1987–1997), he insisted that the staff of both these Commissions consider the effects of changes in sentencing policy on costs and prison population. More generally, this linking of scholarly research and policy analysis is a hallmark of Al's multi-faceted research career.

NOT BY THE NUMBERS

"Today, a mathematician helping police to solve crime—like in the hit TV series *Numb3rs*—seems ordinary. But, in 1966, finding a scientist within the criminal justice system was rare. By some stroke of luck, Al Blumstein, a Ph.D. in operations research, was chosen to be the director of the first national-level criminal justice Science and Technology Task Force. Al's systemic view of the interactions between the courts, police, and corrections has proven to be a seminal and lasting contribution. This came about not by theoretical musing in the office, but by Al's scientific philosophy: learning and assimilating everything he could of the system . . . short of getting arrested, prosecuted, and tried" (Saul I. Gass in Ritter 2007, 17).

In the late 1970s, Al turned his attention to formulating what would come to be called the criminal career paradigm. This paradigm generated still another intellectual revolution in criminology. The seed for the criminal career paradigm was first articulated in the report "Deterrence and incapacitation: Estimating the effects of criminal sanctions on crime rates" (Blumstein et al. 1978). A subsequent report, "Criminal careers and 'career criminals'," fully develops the concept (Blumstein et al. 1986).

The criminal career paradigm separates an offending individual's career into three distinct components—(1) the onset of offending, (2) the period of active offending, and (3) the cessation of offending or desistence. These labels for the stages of a criminal career are now standard terminology in both criminological and psychological research. There are now literally hundreds of empirical studies analyzing one or more components of the criminal career. The criminal career paradigm helped to further the use of longitudinal research on

CARNEGIE MELLON CRIME FIGHTERS

Dan Nagin Al Blumstein Jackie Cohen

the developmental origins of crime and violence. In turn, such research spawned the field of developmental criminology, now among the dominant intellectual forces in contemporary criminology. The criminal career paradigm has also deeply affected scholarship outside of criminology, most particularly, in research on the developmental origins of crime and violence in developmental psychology and psychopathology.

Al has been deeply committed to exploring the policy implications of his scholarship on the criminal career paradigm. He has studied the implications of the paradigm in estimating the incapacitative effect of prison, that is, the number of crimes averted by physically separating offenders from the larger population through incarceration. This policy analysis research focused on such issues as the feasibility of identifying particularly high-rate offenders for the purposes of selective incapacitation, and the degree to which very punitive sentencing policies (three-strike laws) have been counterproductive, as they can result in the incarceration of individuals long after their criminal careers would have otherwise ended.

> BIG AL
>
> "He is 5'9" yet the NIJ [National Institute of Justice] staff referred to Blumstein, the dean of criminologists, as 'Big Al.' When he grudgingly welcomed me to the NIJ directorship, he said it was a strange irony, indeed, when the Nation's most important crime research portfolio is handed over to a police officer without a Ph.D. or a long list of juried publications. Big Al's sarcastic welcome was tongue-in-cheek—*he* has an engineering background and is not a traditional social scientist by education or training. That was 25 years ago. Al continues to challenge policy-makers and researchers to be more serious about understanding the causes and correlates of crime in America. With a magic marker and an overhead projector, Al can be very persuasive in using data analysis to demonstrate how misinformed, wrong, and dangerous the conventional wisdom about crime rates can be" (James K. Stewart in Ritter 2007, 16–17).

Over the years, Al has directed his attention to two linked issues: the steady increase in U.S. prison populations that began in the 1970s and accompanying long-term trends and cycles in crime rates. He has been a leader in energizing criminologists to study the forces behind crime trends, as well as crime cycles, such as the rise and fall of homicide rates during the early 1990s. Research on these issues, published in the widely cited volume, *The Crime Drop in America*, co-edited by Blumstein and Wallman (2000), identified the factors contributing to a 40% drop in homicide and robbery in the U.S. between 1993 and 2000.

Of strong interest to Al has been the impact of the Federal Government's harsh penalties for drug violations, in general, and crack cocaine, in particular, on the steady rise in the rate of incarceration. In an important paper, Blumstein and Beck (1999) showed that the war on drugs has indeed been a major factor in the increase in prison population over the past 20 years. This finding was the dominant theme of Al's 1992 American Society of Criminology (ASC) presidential address, "Making rationality relevant," that argued that the rapid increase of incarceration for drug offenses was futile because the resilient drug market would recruit replacements (Blumstein 1993a). His later work then demonstrated the unintended consequences of that incarceration—the replacements were predominantly young people who were armed with handguns, and these young people were the dominant cause of the major rise in homicide in the U.S. between 1985 and 1993 (Blumstein 1995).

A long-standing concern to Al has been the potential impact of criminal justice policy on minorities. It led to his chairing the National Research Council panel that produced the report "Research on sentencing: The search for reform" (Blumstein et al. 1983). His later work "On the racial disproportionality of U.S. prison populations" (Blumstein 1982) has been cited as the classic landmark study in trying to assess the factors contributing to that disproportionality, an area he revisited in Blumstein (1993b).

Since 1996, Al has been Director of the National Consortium of Violence Research (NCOVR). NCOVR is supported by the largest social science grant ever awarded by the U.S. National Science Foundation. The mission of NCOVR is to conduct both basic and policy-related research on the causes and prevention of violence. It also has an important capacity-building function in training the next generation of scholars to carry on this important research mission. That this grant was awarded to a team headed by Alfred Blumstein attests to his stature in the field.

Awards and Honors

Al is a University Professor and the J. Erik Jonsson Professor of Urban Systems and Operations Research and former Dean (from 1986 to 1993) at the H. John Heinz III School of Public Policy and Management of CMU. He is also director of the NCOVR.

Al was president of ORSA in 1977; he was awarded its Kimball Medal for service to the profession and the society in 1985 and its President's Award in 1993 for service to society. He was president of TIMS in 1987–1988 and president of INFORMS in 1996.

ONE MILLION SEK!

Queen Sylvia, Al Blumstein, Terrie Moffitt (Courtesy Laura Dugan)

He is a fellow of the American Association for the Advancement of Science, INFORMS, and the ASC. He was the 1987 recipient of the ASC's Sutherland Award for Research Contributions. In 1998, he was presented with the Wolfgang Award for Distinguished Achievement in Criminology and was elected to the National Academy of Engineering. On June 5, 2007, Al shared the Stockholm Prize in Criminology with Terrie E. Moffitt, King's College London. The Stockholm Prize in Criminology recognizes achievements in criminological research or in the practical implementation of research to combat crime and promote human rights. They were recognized for their longitudinal research into the factors that lead people to commit crime. They shared the prize of one million SEK.

REFERENCES

Blumstein A (1957) A Monte Carlo analysis of the ground controlled approach system. Oper Res 5(3):397–408

Blumstein A (1959) The landing capacity of a runway. Oper Res 7(6):752–763

Blumstein A (1982) On the racial disproportionality of U.S. prison populations. J Crim Law Criminol 73(3):1259–1281

Blumstein A (1993a) Making rationality relevant: the American Society of Criminology Presidential Address. Criminology 31(1):1–16

Blumstein A (1993b) Racial disproportionality of U.S. prison populations revisited. Univ Colo Law Rev 64(3):743–760

Blumstein A (1995) Youth violence, guns, and the illicit-drug industry. J Crim Law Criminol 86(4):10–36

Blumstein A (2002) Crime modeling. Oper Res 50(1):16–24

Blumstein A (2007) An OR missionary's visits to the criminal justice system. Oper Res 55(1):14–23

Blumstein A, Beck A (1999) Population growth in U.S. prisons, 1980–1996. In: Tonry M, Petersilia J (eds) Prisons: crime and justice. University of Chicago Press, Chicago, IL, pp 17–61

Blumstein A, Belkin J, Glass W (1971) JUSSIM, an interactive computer program for analysis of criminal justice systems. Report of the urban systems institute, school of urban and public affairs. Carnegie Mellon University, Pittsburgh, PA

Blumstein A, Cohen J, Martin S, Tonry M (eds) (1983) Research on sentencing: the search for reform (2 volumes). Report of the NRC Panel on Research in Sentencing, National Academy Press, Washington, DC

Blumstein A, Cohen J, Nagin D (1978) Deterrence and incapacitation: estimating the effects of criminal sanctions on crime rates. Report of the Panel on Research and Deterrent and Incapacitative Effects, National Academy of Sciences, Washington, DC

Blumstein A, Cohen J, Visher C (eds) (1986) Criminal careers and "Career Criminals." Appendix B, Report of the National Academy of Sciences Panel on Research on Criminal Careers, National Academy Press, Washington, DC

Blumstein A, Larson R (1969) Models of a total criminal justice system. Oper Res 17(2):199–232

Blumstein A, Wallman J (2000) The crime drop in America. Cambridge University Press, Cambridge, England

Cohen J, Fields K, Letter M, Stafford R, Walker C (1973) Implementation of the JUSSIM model in a criminal justice planning agency. J Res Crime Delinquency 10(2):117–131

Government Printing Office (1967a) The Challenge of Crime in a Free Society. President's Commission on Law Enforcement and Administration of Justice, Washington, DC

Government Printing Office (1967b) Task Force Report: Science and Technology, President's Commission on Law Enforcement and Administration of Justice, Washington, DC

IDA (2009) http://www.ida.org/. Accessed 3 Sept 2009

Ritter N (ed) (2007) Al Blumstein: 40 years of contributions to criminal justice. NIJ J (Issue No. 257):14–18

Solow H (1951) Operations research. Fortune (April) 105–107:146, 148

40

Harvey M. Wagner

Shaler Stidham, Jr.

The second wave of researchers and practitioners in operations research (OR)—roughly speaking, those born in the mid-1920s through the mid-1930s—consisted of graduates of academic programs that included some courses in the principles of OR. Typically, they specialized in mathematics or statistics or economics, but were also exposed to OR topics that had filtered into academic programs, especially linear programming (LP), inventory management, and simulation. Harvey Wagner's entry into OR followed this path, augmented by a long-term arrangement as an OR advisor to a major management consulting firm. Harvey then proceeded to make important contributions to LP, inventory theory, and management consulting. He wrote a pioneering text in OR, *Principles of Operations Research with Applications to Managerial Decisions* (1969), for which he received the 1969 Operations Research Society of America (ORSA) Lanchester Prize for the best publication in OR.

He was awarded the Canadian Operations Research Society's Harold Larnder Prize for Distinguished International Achievement in Operations Research, and the Institute for Operations Research and the Management Sciences (INFORMS) Expository Writing Award for publications in OR and management science (MS) that have set an exemplary standard of exposition. Harvey served as president (1973–1974) of The Institute of Management Sciences (TIMS).

A.A. Assad, S.I. Gass (eds.), *Profiles in Operations Research*, International Series in Operations Research & Management Science 147, DOI 10.1007/978-1-4419-6281-2_40,

WEST COAST INFLUENCE: HIGH SCHOOL TO STANFORD

Harvey Wagner was born in San Francisco on November 20, 1931. His father was born in Vienna and emigrated to California while a teenager. His mother was born in San Francisco. Harvey was their only child (and the first in his family to attend college). At the end of the Great Depression, when he was 10 years old, his family moved to Los Angeles. He attended Los Angeles High School, which was within walking distance of his home near Wilshire Boulevard. Harvey was a prize-winning debater in high school—debating was his passion and his primary extracurricular activity. Academically, in his senior year, he received a third-place award in an all-Los Angeles competition in mathematical skills. Harvey always liked high school mathematics, but wondered how a mathematician could make a living. It took a few years before he found some answers. In 1949 he graduated second in his class of 200 students.

When it came time to consider colleges, Harvey resisted the temptation to go to the University of California at Los Angeles—an option followed by many of his high school classmates. He felt that it would be "like high school, but with a long bus ride" (Wagner 2008). He perused brochures from Harvard, Yale, and Stanford, but his parents made it clear that Stanford was about as far away from home they were prepared to let him go.

Harvey began his undergraduate studies at Stanford in 1949. In his freshman year, he was again a prize-winning debater. His debating skills led his parents to hope he would become a lawyer. He was attracted, however, to economics and statistics after taking an introductory statistics course in the department of economics during his sophomore year. The course was taught by Robert Summers. (Summers's brother, Paul Samuelson, and his brother-in-law, Kenneth Arrow, would both later play important roles in Harvey's career.) A summer job in 1951 with a Los Angeles insurance company, and an industrial engineering course in quality control, further whetted his appetite for statistics and gave him some exposure to applications of mathematics in an industrial setting.

It was not long before Harvey recognized that he needed a firmer foundation in the theory of probability. He enrolled in a first course in probability given by the department of economics that was taught by Arrow. Harvey fondly describes Arrow's unique teaching style:

> He would enter the classroom with an air of disarray, open the textbook, identify the topic for the day, and declare "this is what we should learn." Working without notes—while flipping a piece of chalk nonstop between his middle and third fingers without missing—he would get stuck but always found his way out by the end of the class (Wagner 2008).

Arrow became Harvey's academic advisor and mentor during his undergraduate years at Stanford.

In the summer of 1952, Harvey studied at Oxford University under the tutelage of the economist Paul Streeten. He selected the topic of imperfect competition and wrote about the contributions of the Cambridge economist Joan Robinson. When Streeten looked at the paper, he immediately remarked, "I see that you have mastered applied geometry." That was Harvey's awakening about how the economics profession positioned the importance of applied mathematics (Wagner 2008).

In 1953 Harvey received a B.S. from Stanford, with a double major in statistics and economics. He finished his undergraduate studies at Stanford with the highest grade point average in his class. He was not sure about going to graduate school, but decided that he would go if and only if he received a scholarship. Of the two fields, he felt more drawn to statistics and applied for a scholarship in statistics at Stanford. He received an Honors Fellowship and enrolled in the graduate program in statistics, but continued to take courses in economics. One of these was a course in LP, which was Harvey's introduction to academic OR, at a time when OR was just beginning to emerge as a scholarly discipline.

At the end of the 1953–1954 academic year, Harvey received an M.S. in mathematical statistics. His thesis, "A Monte Carlo study of estimates of simultaneous linear structural equations," was supervised by Arrow. Harvey's use of Monte Carlo simulation was rather novel in econometric research; the thesis was later published in *Econometrica* (Wagner 1958a).

An Intern at RAND

Arrow helped Harvey get a summer position in 1953 as an intern in the RAND Corporation's Logistics Department, headed by the statistician-logistician Murray Geisler. But, before he could begin working at RAND, Harvey had to mark time until he received a security clearance. While waiting, he received his first exposure to the discipline to which he would end up devoting his professional life: Alexander Mood (head of RAND's Mathematics Department) gave him a copy of *Methods of Operations Research* by Morse and Kimball (1951), with a suggestion that he read it while he was waiting for the clearance to come through.

Arrow was a consultant to RAND's economics department and visited there during the summers when Harvey was an intern. During those visits, he advised Harvey on his research pursuits, especially inventory issues that would later play a role in his doctoral dissertation (on revolving funds for inventory replenishment), and research published in *Econometrica* and other journals.

It was at RAND in the summer of 1953 that Harvey (briefly) met George Dantzig. The following summer they had more substantive interactions. Harvey asked Dantzig several questions about the simplex method, and Dantzig suggested that Harvey sit and watch him execute the method at his desk using paper and pencil. Dantzig took out a few sheets of quadrille graph paper and a long clear plastic ruler and, doing all the computations by hand, solved a small illustrative example. While at RAND in the summer of 1954, Harvey wrote a paper on using LP to solve dynamic Leontief type models, which was published a few years later in *Management Science* (Wagner 1957).

At RAND, Harvey also learned how to setup and use an IBM Card-Programmed Electronic Calculator (CPC): "an ingenious lashing together of several existent IBM machines—the 402–417 Accounting Machine with the 604 [Electronic Calculating Punch] and the 941 Auxiliary Storage Unit" (Goldstine 1972, 328). Punch cards were used for both input data and instructions. The 604 was programmable via a wired plugboard (control panel) and could execute a program of up to about 60 steps. William Orchard-Hays, who worked with Dantzig and was responsible for developing RAND's LP computer systems, taught Harvey the trick of putting two instructions instead of one on each card. During that summer, Dantzig and Orchard-Hayes solved a 12-equation LP problem by feeding punch cards through a CPC during an all-night run. When Harvey returned to Stanford, he upgraded Stanford's CPC software to RAND's system—it was then used by Harvey to write his master's thesis.

SOJOURN AT CAMBRIDGE

By this time, Harvey knew he wanted to pursue a Ph.D., but not in statistics. Advised and encouraged by Arrow, he applied for and received a Marshall Scholarship to study economics at the University of Cambridge.

Harvey enrolled at Cambridge in the fall of 1954. Arrow had provided him with an introduction to Professor Richard Stone, a leading empirical economist and the main reason that Harvey went to Cambridge. Harvey became a member of King's College, where Stone was a Fellow. When they met, however, Stone informed him that he would not be able to work with any graduate students that year for personal reasons. Nicholas Kaldor, the director of graduate studies in economics, took Stone's place. At his first meeting with Harvey and the other graduate students, Kaldor announced that he "saw no point in their meeting on a regular basis"; Harvey and the others were left pretty much on their own (Wagner 2008).

> BUSY PROFESSORS
>
> A lecture by John Hicks inspired Harvey to try to solve an open problem in utility theory having to do with revealed preferences, a theory introduced by Samuelson and critiqued by Hicks in his lecture. Harvey was successful in solving the problem. He took his paper to Kaldor, who did not read it but suggested that Harvey show it to Hicks. He sent the paper to Hicks, but Hicks would not read it either. As a last resort, Harvey asked Sir Dennis Robertson, a highly regarded economist at Cambridge, to examine the paper. Several months later, after Harvey had returned to California, he received a warm but succinct reply from Robertson: "The diagrams are interesting" (Wagner 2008). The paper was later published in *Econometrica* (Wagner 1956).

FROM CAMBRIDGE TO CAMBRIDGE: GRADUATE STUDIES AT MIT

Harvey returned to RAND in the summer of 1955 and recounted his experiences at Cambridge to Arrow, who then encouraged him to apply to the Ph.D. program in economics at MIT. Harvey did so and entered MIT in the fall of 1955. There, he spent 2 years as a Ph.D. student in economics, an instructor in the Sloan School of Management, and a research assistant for Professor Thomson Whitin, who was revising his book *The Theory of Inventory Management* (Whitin 1953). The collaboration between Wagner and Whitin led eventually to one of the most well-known and often-cited papers in OR/MS: "Dynamic version of the economic lot size model" (Wagner and Whitin 1958a). This paper was honored in 2004 as one of the ten most influential papers ever published in the journal *Management Science* [reprinted in Veinott (1965)]. Harvey began his MIT dissertation research under the direction of Robert Solow. One of Harvey's MIT friends was Robert B. Fetter, an assistant professor at the Sloan School, who later become his colleague at Yale.

HOW MUCH TO ORDER

The Wagner-Whitin model generalizes the classical economic lot size model to allow different (deterministic) demands in each of a finite number of periods. An optimal policy is shown to have a simple form: produce only when inventory reaches zero and then produce exactly enough to satisfy the demand in the next n periods. All that remains is to determine the optimal value of n for each such period, which can be done by enumeration when the total number of periods is small and more generally by a variant of dynamic programming. Thus, a multi-dimensional, continuous-variable problem is reduced to a much simpler discrete-variable problem. The Wagner-Whitin model is arguably the most widely known dynamic inventory model and forms a crucial building block in multi-echelon models in materials requirement planning (MRP) and modern supply-chain analysis.

RETURN TO STANFORD

After 2 years at MIT (and no Ph.D.), Harvey was lured back to Stanford in the fall of 1957 as an assistant professor in the department of industrial engineering, with a complimentary appointment in statistics. In 1959, the dean of engineering told Harvey that Stanford was prepared to promote him, but only on the condition that he get his Ph.D. now! Harvey got the message. He took off a half year, went to RAND, and completed his doctoral research that dealt with the combination of financial controls and inventory models. His dissertation, "Statistical management of inventory systems," was later published in book form under that title (Wagner 1962). His joint advisors were Robert Solow and Kenneth Arrow.

The impetus of Harvey's dissertation research stems from the following. In the early 1950s, Arrow and his RAND colleagues had developed models that held potential for improving U.S. Air Force weapon systems inventory management. The RAND Logistics Department was the home for many studies that investigated the implementation of the new inventory formulas. One practical issue that was raised related to how these new inventory policies would mesh with fiscal controls and annual budgeting for stock replenishment. One facet dealt with whether it would be possible to devise financial incentives that encouraged the adoption of the new policies. The Air Force used what are called revolving funds to budget replenishment—when inventory is depleted, replacement dollars are generated and credited to an aggregate fund—withdrawals from that fund are then used to replenish inventory. The central idea of the thesis was to investigate how stock fund management could be extended to accommodate new replenishment rules, as well as to motivate Air Force personnel to adopt improved logistics procedures.

Soon after Harvey received his Ph.D. from MIT in 1960, Stanford promoted him to associate professor. Since he was not an engineer, Harvey was not comfortable about his first appointment being in an engineering department. Gerald Lieberman, who was completing his Ph.D. dissertation ("Contributions to sampling inspection") when Harvey was an undergraduate at Stanford, had a joint appointment in the departments of industrial engineering and statistics. He was also director of the newly formed interdepartmental program in OR. Harvey accepted a joint appointment, half in the business school and half in OR. Lieberman became a mentor and close confidant during Harvey's years as a faculty member at Stanford.

Harvey's Stanford research activities, joint with colleagues and graduate students, covered many topics and established him as one of OR's creative thinkers: inventory in a complex organization (Wagner 1960); job shop scheduling (Story and Wagner 1962); machine maintenance and scheduling (Giglio and Wagner 1964; Giglio et al. 1964); optimal inventory policies (O'Hagan et al. 1965); production and employment scheduling (Lippman et al. 1967a, 1967b); knapsack and turnpike models (Shapiro and Wagner 1967). Harvey's first doctoral student was Arthur Geoffrion, who graduated in 1965.

A close working relationship flourished between Harvey and Arthur F. (Pete) Veinott, a professor in industrial engineering. Many ideas formulated during their frequent conversations influenced Harvey's writings, especially in his later pioneering OR text, and helped to generate their joint paper on the calculation of (*s*, *S*) inventory policies (Veinott and Wagner 1965). Harvey also published several papers on the theory, application, and teaching of LP (Wagner 1958b, c, d; 1959a, b).

My personal contact with Harvey began when I was a graduate student in OR at Stanford in 1965–1968. In those days, there was a two-quarter introductory course in LP and Harvey taught the second quarter. Other OR courses at Stanford were more mathematical, but Harvey's teaching was always thoughtfully crafted and artfully delivered (maybe due to his talent as a debater), with a careful balance of intuition, application, and mathematics.

The McKinsey Connection

In 1960, Harvey received a telephone call that would have a major impact on his subsequent career. The management consulting firm of McKinsey & Company asked him to be the OR advisor for their San Francisco office. Harvey's friend from MIT, Robert Fetter, had recommended him. David B. Hertz, who was the editor of ORSA's Publications in Operations Research book series, in which Harvey's dissertation was printed, joined McKinsey's New York City office in 1963. David became a good friend and mentor: "David kindly tutored me on the skills that are required to take university research and make it relevant to business" (Wagner 2002, 224).

From 1960 until the present, Harvey has had an exclusive consulting arrangement with McKinsey. He has essentially been on call to help client companies answer strategic questions about a wide range of operational and management information issues. Examples have included statistical analyses of credit scoring, insurance risks, inventory and capacity expansion, and determining the optimal number of beds in a hospital. Harvey estimates that 60% of the problems on which he has consulted have involved applications of statistics, 30% optimization, and 10% other techniques—a breakdown that closely matches that reported in a number of surveys of industry.

Harvey's extensive involvement as part of McKinsey client project teams had three important impacts on his professional development (Wagner 2008). The first was that it became obvious that practical implementation of OR models requires successful linking of the modeling effort to the data retrieval process that supplies the model's parameter values, as well as to test the robustness of the model outputs—an auspicious revelation for Harvey because his academic interests combine modeling and statistics. In the 1960s, it was difficult to interact with corporate information systems, due in good part to limited computing power and the paucity of data appropriate for OR models. But, over time, both of these limitations changed, and, eventually, the issues became how to access huge amounts of data efficiently and build useful OR models in a time frame that matches corporate managers' need for a fast response. Harvey became adept in tree-driven multivariate analyses (data classification) that used Chi-Square Automatic Interaction Detection (CHAID), well in advance of data mining's wide acceptance of these methods (Kass 1980). The second impact took longer to reveal itself. Eventually, it was evident to Harvey that OR modeling could provide tremendous impact in situations where the models are one-shot efforts in contrast to permanent systems. These models are built rapidly, used for the insights that they provide, and then are never used again. The models are disposable. Spreadsheet technology is the key to building successful one-shot models.

The third impact was a deeper understanding about the importance of the transparency of analytic models and techniques. Most business professionals, who are not steeped in analytics, have a relatively limited comfort zone when asked to act on the basis of computational results. When statistical results emerge from a multivariate process, it is often more effective to sacrifice a modicum of analytic accuracy for greater plausibility of the method. Core findings often emerge, even though a more intuitive process is used rather than the best analytic algorithm.

In 1984, Harvey and colleagues received the TIMS College of Practice Franz Edelman Award for Management Science Achievement. The award was for their work at Blue Bell Industries which Harvey took part in as a McKinsey consultant. Blue Bell, one of the world's largest apparel manufacturers, was concerned with its high investment in working capital and concluded that inventory represented the greatest opportunity for reduction in working capital. A new production planning process was designed, tested, and implemented such that:

> Within 21 months Blue Bell reduced its inventory by more than 31 percent, from 371 to 256 million dollars, with no decrease in sales or services by applying management science models. A combination of innovative problem

> solving and enthusiastic management support ensured success. Many of the models are standard, but a new marker [garment cutting pattern] design and selection model makes the systems approach practical. Management paid close attention to systems development and provided resources that enhanced the effectiveness of the project (Edwards et al. 1985, 34–35).

Harvey describes it as a true OR effort in that success required that the problem had to be solved at the company system level.

A Pioneering OR Text

During his Stanford years, Harvey began writing one of the most influential and well-regarded texts in OR and MS, *Principles of Operations Research with Applications to Managerial Decisions* (Wagner 1969). The book was recognized by ORSA as the best publication in OR in 1969.

From the beginning, Harvey designed his text to satisfy several explicit goals. To understand why he felt these were important, it is necessary to review the state of OR in the second half of the 1960s. Both engineering and business schools in top-rated universities were offering OR courses, notably in LP, inventory, queueing, and Monte Carlo simulation. Computers were not easily accessible for classroom teaching, and time-sharing desktop environments had only recently emerged and were restricted to the solution of small-scale models. Popular programming languages for time-sharing computing were Algol and Basic, with GPSS and SIMSCRIPT for simulation. There were few OR research textbooks and all were written by multiple authors. Their tables of contents and lack of cohesive expository style suggested that OR was a hodgepodge of separate topics. Harvey's most important goal was to produce a text by a single author, written in a unified manner. It would demonstrate that the breadth of OR topics could be grasped by a single individual. It would also establish that the core topics are connected, at least from the perspective of a practitioner. By the end of the 1960s, Harvey's McKinsey experience provided insight as to the likely role that various core topics would play as OR continued to be used in commercial settings. Another goal was to write a book where the presentation is rigorous, yet within reach of business students.

Finally, Harvey also set himself the goal of emulating a classic textbook that had deeply impressed him during his graduate economic studies at MIT. Before taking the MIT doctoral qualifying exams, most students read Paul Samuelson's undergraduate text, *Economics: An Introductory Analysis* (Samuelson 1961), including its footnotes. It was a book that integrated the main themes of economic analyses. Samuelson commented that the text was positioned at the "high end of the market" (Wagner 2008). Although *Principles of Operations Research* was designed to fit the needs for "introductory courses for undergraduates or graduates in business, economics, engineering, liberal arts, and public administration curricula," Harvey hoped that it would also serve OR doctoral students as Samuelson's text did for economics students (Wagner 1969, v). [*Principles of Operations Research* was reworked into a reduced version, *Principles of Management Science with Applications to Executive Decisions* (Wagner 1970), geared to students who intend to be managers.]

HOW TO DRIVE AN ALGORITHM

In his chapter that introduces the simplex method, Harvey explains why he is persuaded that executives and managers must understand the principles of OR methodologies. He offers the following analogy:

"In learning to drive a car, it strains the intellect only mildly to learn to speed up, slow down, go forward or backward, turn, and so forth—in short, to master the skill of driving. To qualify as a topnotch driver, however, you need to know more. You should have an idea of how to care for the battery; otherwise you may make the mistake of playing the radio for a long time with the ignition off. If you drive on icy highways, a knowledge of the braking mechanism should help you control your car when its skids. If your engine overheats suddenly, an understanding of the radiator's function may suggest the correct remedy. Summing up, to be a good driver you must know more than how to handle your auto under ideal conditions. You must know enough about the mechanism to be able to skirt danger when possible. On the other hand, you can obviously be an excellent driver without the training of an automobile mechanic.

"By the same token, the manager who resolutely avoids familiarizing himself with the basic mechanism of his operations research application is flirting with trouble. If he really wants to maintain control, he must nurture his insight to the approach. Only a modest effort is required to reach the appropriate level of knowledge; it does not entail your becoming an expert theoretician" (Wagner 1969, 91–92).

Nearly 40 years have passed since the publication of the book. Today, Harvey does not think the goals described above would be practical or relevant any more; at least those goals would not be a selling point for such a text (Wagner 2008). The field has come full circle in that most researchers work in highly specialized subjects, albeit not the same disciplines as 40 years ago. OR applications in business schools are almost completely subsumed within classic business school departments, such as finance, marketing, and operations (or synonyms). Other departments and schools where OR is central have evolved into focused variations of applied mathematics that require deep mathematical skills.

The Move to Yale

In 1967, Harvey joined the faculty of the Department of Administrative Sciences at Yale University. He was recruited by his friend Robert Fetter and economists Martin Shubik and Herbert Scarf (the latter a prior colleague at Stanford). The move to Yale was based on the prospect (and promise) that Yale would become a major player in OR and that Harvey could be an active leader in this effort. At that time, the Department's faculty was divided into two major components, quantitative and organizational behavior.

The two groups had reached a stable equilibrium in which they coexisted in a cordial détente with very little interaction. That equilibrium was perturbed by Yale President Kingman Brewster, who (along with others) felt that Yale needed a business school. Brewster's vision of the school (reported retrospectively) noted that he "saw operations research as part of a 'different' brand of business school, which was to be grounded in organization and management rather than finance—the cornerstone of traditional business schools" (Post 2005). The result was a commitment to establish a School of Organization and Management (SOM), with the existing Department of Administrative Sciences as its core.

Others who soon joined SOM included Matthew Sobel, a 1967 Stanford Ph.D., and well-established operations researchers George Fishman, Eric Denardo, and Ward Whitt. By the mid-1970s, SOM had become a strong center of academic OR.

While at Yale, Harvey continued and extended aspects of his research, joining with several of his Yale Ph.D. students: inventory and production control (Wagner 1974); LP and bounded variables (Duguay et al. 1973); and production scheduling and employment smoothing (Horen et al. 1975). Of particular interest is his paper, "The ABC's of OR" (Wagner 1971), that addressed the Achievements, Beliefs, and Content of OR. Based on his extensive academic and real-world consulting activities, and concentrating on the private sector, Harvey discusses the battle of survival, challenges from the past, and challenges for the future. Although the discussion is somewhat dated, it has, over 30 years later, elements of importance to the future viability of OR, and concludes with high-priority tasks for practitioners, theoreticians, and educators.

NO FREE LUNCH

Shortly after arriving at Yale, Harvey was welcomed and invited to lunch by Tjalling C. Koopmans. In the 1950s, Koopmans contributed significantly to the theory of simultaneous structural equations, and that work was the impetus to Harvey's master's thesis. Toward the end of what was a very cordial lunch, Koopmans said there was something important that he wanted to discuss. He went on to say that he realized Harvey had been at fine research institutions such as MIT and Stanford, but that Yale was different in an important way. He said that teaching Yale undergraduates was an important faculty obligation, and that all faculty members, no matter what their status, were expected to teach and be available to undergrads. What he feared was that Harvey might not be aware of this responsibility. This conversation made a strong and indelible impression on Harvey. What Koopmans stressed was consonant with Harvey's own experiences at Stanford, both as a student and a faculty member. But in the years since that lunch, Harvey came to appreciate why Koopmans might have speculated otherwise.

(Koopmans was awarded the 1975 Nobel Prize in economics for contributions to the theory of optimum allocation of resources.)

While on leave from Yale, Harvey spent the spring semester of 1974 and the following academic year as a visiting professor at Harvard's Graduate School of Business and School of Public Health. One of his motivations was his concern that Yale's SOM needed to pay more attention to its own internal organization and management, and delineate and focus its role among business schools. What he learned at Harvard was that, whatever one might feel (pro or con) about Harvard Business School's emphasis on case studies, it had led to a system of teaching, research, and consulting in which the components enhanced each other. To improve teaching required faculty spending time in the business world observing and consulting, developing and writing cases, publishing related research, and bringing to the classroom case studies of current value. Harvey decided that the SOM (indeed, any business school) needed a similar system—not necessarily based on case studies—to ensure that teaching, research, and consulting would be mutually reinforcing, rather than conflicting.

North Carolina Years and Beyond

In 1976, Harvey had a chance to put these ideas into practice when he moved to the University of North Carolina at Chapel Hill (UNC-CH) as Dean of the School of Business. In pursuit of establishing UNC-CH as a top-ranked business school, he brought about many fundamental changes in the school's educational missions and governance process. But his tenure as dean only lasted 2 years. In retrospect, Harvey had the right ideas, but lacked the people skills to develop faculty support for a set of core goals. In characteristic fashion, he decided to chalk up his time as dean as a learning experience and move on. From that point forward, he has devoted himself to teaching, consulting, and guidance of doctoral students by joint appointments in the School of Business and in the Department of Operations Research. Harvey continued with his main lines of research in LP (Rubin and Wagner 1990; Wagner et al. 1992) and inventory theory (Bayus et al. 2004; Jacobs and Wagner 1989a, b; Wagner 2002).

His consulting activity introduced him to new problem areas, one of which dealt with exploring the link between gender diversity in top management teams and U.S. corporate financial performance (Catalyst 2004): Do companies [Fortune 500] with more women executives have a better financial performance? (Business Week Online 2004). Harvey was responsible for advising on the conceptual design, methodology, and interpretation of data. The study showed:

> The group of companies with the highest representation of women on their top management teams experienced better financial performance than the group of companies with the lowest women's representation. This finding holds for both financial measures analyzed: Return on Equity (ROE), which is 35.1 percent higher, and Total Return to Shareholders (TRS), which is 34.0 percent higher.
>
> Financial performance was also analyzed by industry, and in each of the five industries analyzed, the group of companies with the highest women's representation on their top management teams experienced a higher ROE than the group of companies with the lowest women's representation.
>
> In four out of the five industries analyzed, the group of companies with the highest women's representation on their top management teams experienced

a higher TRS than the group of companies with the lowest women's representation (Catalyst 2004, 2).

THE SPREADING OF SPREADSHEETS

Early on, Harvey recognized the power of spreadsheet modeling in teaching and applying OR. He became a proselytizer in his classes and among his colleagues at UNC-CH and elsewhere. (He continues to teach a course in modeling with spreadsheets to undergraduates.) This has changed his view about teaching introductory OR modeling and has caused him to rethink his view about how an OR textbook should be organized. The standard textbook sequence of topics is LP at the start and simulation near the end. When a spreadsheet is the vehicle for building models, a better sequence is to *start* with deterministic and stochastic simulation models, followed by queueing and inventory replenishment models, nonlinear optimization models, LP, networks, and ending with combinatorial models and integer optimization. This sequence starts with the most common form of managerial spreadsheet models and ends with important optimization models that require the full power and excitement of the add-in options of spreadsheets (Wagner 2008).

A Life in the Arts

Since his days at Stanford, Harvey has been keenly interested in the visual and performing arts, especially contemporary visual art—American painters initially and, more recently, Australian indigenous artists. The dust jackets of his textbooks reproduce works by American artists Frank Stella and Roy Lichtenstein. He has supported and advised the American Dance Festival, held each summer at Duke University. Harvey was Chair of the North Carolina Symphony Society in 1977–1978.

Awards and Honors

In 1980, Harvey received an honorary doctoral degree from the Katholieke Universiteit Leuven in Belgium. He was recipient of the first INFORMS Expository Writing Award for publications in OR and MS that have set an exemplary standard of exposition (1996), and the Harold Larnder Award from the Canadian Operations Research Society given to an individual who has achieved international distinction in OR (1988). He is a fellow of the American Statistical Association (1964), a fellow of INFORMS (2002), and distinguished fellow of the Manufacturing and Service Operations Management Society (2006). His 1969 book, *Principles of Operations Research with Applications to Managerial*

Decisions, received the ORSA Lanchester Prize for the best publication in OR in 1969 and the American Institute of Industrial Engineers Maynard Book of the Year Award in 1970. In 1984, Harvey and colleagues received the TIMS College of Practice Franz Edelman Award for Management Science Achievement for their study of the Blue Bell, Inc. inventory and production system. He was the 1973–1974 President of TIMS.

REFERENCES

Bayus B, Souza G, Wagner HM (2004) Production and inventory under competition. Manage Sci 50(4):537–549

Business Week Online (2004) The bottom line on women at the top. http://www.businessweek.com/bwdaily/dnflash/jan2004/nf20040126_3378_db035.htm. Accessed 24 Dec 2009

Catalyst Inc. (2004) http://www.catalyst.org/file/44/the%20bottom%20line%20connecting%20corporate%20performance%20and%20gender%20diversity.pdf. Accessed 1 Aug 2009

Duguay C, Todd M, Wagner HM (1973) Linear programming with relative bounded variables. Manage Sci 19(7):751–759

Edwards J, Wagner HM, Wood W (1985) Blue Bell trims its inventory. Interfaces 15(1):34–52

Giglio R, Glaser G, Wagner HM (1964) Preventive maintenance scheduling by mathematical programming. Manage Sci 10(2):316–334

Giglio R, Wagner HM (1964) Approximate solutions to the three-machine scheduling problem. Oper Res 12(2):305–324

Goldstine H (1972) The computer from Pascal to von Neumann. Princeton University Press, Princeton, NJ

Horen J, Wagner HM, Yuan J (1975) Optimal multi-product production scheduling and employment smoothing with deterministic demands. Manage Sci 21(11):1250–1262

Jacobs R, Wagner HM (1989a) Reducing inventory systems costs by using robust demand estimators. Manage Sci 35(7):771–787

Jacobs R, Wagner HM (1989b) Lowering inventory systems costs by using regression-derived estimators of demand variability. Decis Sci 20(3):558–574

Kass GV (1980) An exploratory technique for investigating large quantities of categorical data. J Appl Stat 29(2):119–127

Lippman S, Rolfe A, Wagner HM, Yuan J (1967a) A theory of production scheduling and employment smoothing. Manage Sci 14(3):127–158

Lippman S, Rolfe A, Wagner HM, Yuan J (1967b) Algorithms for optimal production scheduling and employment smoothing. Oper Res 15(6):1011–1029

Morse P, Kimball G (1951) Methods of operations research. Wiley, New York, NY
O'Hagan M, Lundh B, Wagner HM (1965) An empirical study of exactly and approximately optimal inventory policies. Manage Sci 11(7):690–723
Post J (2005) Operations research soldiers on. Yale Daily News, February 24
Rubin D, Wagner HM (1990) Shadow prices: tips and traps for managers and instructors. Interfaces 20(4):150–157
Samuelson P (1961) Economics: an introductory analysis, 5th edn. McGraw-Hill, New York, NY
Shapiro J, Wagner HM (1967) A finite renewal algorithm for knapsack and turnpike models. Oper Res 15(2):319–341
Story A, Wagner HM (1963) Computational experience with integer programming for job shop scheduling. In: Muth J, Thompson G (eds) Industrial scheduling. Prentice-Hall, Englewood Cliffs, NJ, pp 207–219
Veinott A (ed) (1965) Mathematical studies in management science. Macmillan, New York, NY
Veinott A, Wagner HM (1965) Computation of optimal (s,S) policies. Manage Sci 11(5):525–552
Wagner HM (1956) An eclectic approach to the pure theory of customer behavior. Econometrica 24 (October):451–466
Wagner HM (1957) A linear programming solution to dynamic Leontief type models. Manage Sci 3(3):234–254
Wagner HM (1958a) A Monte Carlo study of estimates of simultaneous linear structural equations. Econometrica 26(1):117–133
Wagner HM (1958b) The simplex method for beginners. Oper Res 6(2):190–199
Wagner HM (1958c) A practical guide to the dual theorem. Oper Res 6(3):368–384
Wagner HM (1958d) The dual simplex algorithm for bounded variables. Naval Res Logistics Q 5(3):257–261
Wagner HM (1959a) Linear programming techniques for regression analysis. J Am Stat Assoc 54(285):206–212
Wagner HM (1959b) An integer linear programming model for machine scheduling. Nav Res Logistics Q 6(2):131–140
Wagner HM (1960) A model of inventory in a complex organization. In: West Churchman C, Verhulst M (eds) Management sciences, models and techniques, vol 1. Pergamon, New York, NY, pp 489–517
Wagner HM (1962) Statistical management of inventory system. Wiley, New York, NY
Wagner HM (1969) Principles of operations research with applications to managerial decisions. Prentice-Hall, Englewood Cliffs, NJ
Wagner HM (1970) Principles of management science with applications to executive decisions. Prentice-Hall, Englewood Cliffs, NJ
Wagner HM (1971) The ABC's of OR. Oper Res 19(6):1259–1281

Wagner HM (1974) The design of production and inventory systems for multi-facility and multi-warehouse companies. Oper Res 12(6):278–291

Wagner HM (2002) And then there were none. Oper Res 50(1):217–226

Wagner HM (2008) Personal communication

Wagner HM, Vargas V, Kathuria N (1993) The accuracy of linear programming production planning models. In: Sarin R (ed) Perspectives in operations management: essays in honor of Elwood S. Buffa, Kluwer, Boston, MA, pp 359–387

Wagner HM, Whitin TM (1958a) Dynamic version of the economic lot size model. Manage Sci 5(1):89–96

Whitin TM (1953) The theory of inventory management. Princeton University Press, Princeton, NJ

41

Seth Bonder

W. Peter Cherry

The decade after World War II marked the remarkable worldwide diffusion of operations research (OR) from its military origins to business, industry, and governmental operations. But military OR remained an active area and a new generation of OR professionals were called upon to apply their skills to develop and analyze new applications as members of military OR organizations and consultancies. Chief among them, we find the OR military consultant, researcher, academic, and entrepreneur, Seth Bonder.

Seth came to military OR during his doctoral studies at Ohio State University (OSU) and continued with his first academic position at the University of Michigan. He left full-time academia and founded Vector Research, Inc. (VRI), the pre-eminent nongovernmental military OR consultant company that served, at the highest levels, the U.S. Army and the Department of Defense (DoD) for over 30 years. Seth has always coupled his consulting activities with his career in academia by being an astute commentator on OR education and practice, the future of OR, and the role of OR in society.

For the Military Operations Research Society (MORS), Seth served on its board of directors and as vice president and president. He is a fellow of MORS and was awarded its Vance R. Wanner Memorial Award for distinguished service to the profession. He was president (1978) of the Operations Research Society of America (ORSA). He received the Military Applications Society's Steinhardt Memorial Prize for outstanding contributions to the practical application of OR techniques and to the solution of military problems. For his contributions and service to the U.S. Army, he received the Award for Patriotic Civilian Service from the

A.A. Assad, S.I. Gass (eds.), *Profiles in Operations Research*, International Series in Operations Research & Management Science 147, DOI 10.1007/978-1-4419-6281-2_41,

Secretary of the Army. He is a member of the National Academy of Engineering (NAE).

BASKETBALL, FLYING, TANKS, AND OFF TO OR

Seth Bonder was born in the South Bronx of New York City on July 14, 1932. His parents, Al and Minnie Bonder, had emigrated separately from Russia in the early 1900s. They met while working in New York City's garment district. Seth took little interest in education during his youth and did poorly academically (Bonder 2008). (Although they were not acquaintances at that time, both Seth and Colin Powell attended the Morris High School; they would later meet and become friends when Seth consulted for the Army Chief of Staff.) Seth did become an accomplished pool, billiards, and basketball player in the South Bronx streets. The latter skill led him to enroll in the City College of New York (CCNY), whose basketball team had recently (1950) won both the NCAA and NIT tournament titles. But, he left college in the wake of the basketball scandals when previous CCNY players were caught shaving points during games. Dropping out of college, he drove a truck in the city before enlisting in the U.S. Air Force.

Because of the shortage of pilots during the Korean War, the Air Force allowed enlisted personnel to apply for flying school. Seth applied and received a commission and wings through the aviation cadet program. He served in the Air Force for 5 years at various bases in the continental U.S. and overseas. During his service, it became evident that a college degree was necessary if he were to have a meaningful Air Force career. Rather than take extension courses over several years in preparation for college, he left the Air Force in 1956 and enrolled in the University of Maryland, College Park, in 1957. After a rocky start, he excelled academically, started a freshman tutoring service, created and participated in a flying club, and drove a taxicab at night. Seth received his B.S. in mechanical engineering in June 1960. He went to work for Westinghouse Air Arm Division in Baltimore, Maryland, but he found the job less than challenging. He first learned about OR from an Air Force friend, Irv Goodman, who was pursuing a Ph.D. in OR at Johns Hopkins University. When Seth expressed an interest in this new discipline, Goodman submitted applications on Seth's behalf for admission to the graduate schools of Hopkins and OSU.

In 1960, Seth was accepted to OSU's graduate program and received a graduate assistantship as the program's first Systems Fellow. Dan Howland was the Director of the Systems Research Group (SRG) in the Industrial Engineering Department; he became Seth's mentor and Ph.D. dissertation advisor. Howland suggested that Seth work on SRG's government-sponsored research program with Fort Knox investigating problems associated with requirements for the Army's new armored systems. Based on his interactions with Fort Knox personnel, Seth identified two problem areas that he thought were interesting and challenging: the feasibility of the requirements—could a system be built that met the requirements?, and the operational effectiveness of the system—would the resulting system be of value to the operating forces? Howland encouraged Seth to prepare a proposal to the Army to develop methods for addressing these problems. Thus, Seth, as a graduate student, received a large contract (administered by OSU) to run a four-year research program using faculty and graduate students. The program produced models of tank systems performance characteristics for assessing tank development feasibility, as well as the initial structure of an analytic model of combat for assessing tank effectiveness on the battlefield. This work linked Seth into the Army's extensive research and experimentation program on combat operations, and set the course for his future professional career in military OR.

Seth's first major theoretical contribution to military OR originated from the Fort Knox Armor School—he was asked to evaluate alternative designs for the Sheridan tank. He attempted to use a form of the classic Lanchester model of combat for the evaluations, but he soon recognized that the attrition rate coefficients in the tank model's differential equations were not constants, as required by Lanchester theory. Rather, in a battle, these rates vary over time and are random variables. He also recognized that a way of estimating the rates, a priori, did not exist. Using data and information from many sources—Army operational experimentation in Europe, experiments he conducted with Ohio National Guard armored units, measurements taken on the firing ranges at Fort Knox, and tank firing doctrines—Seth developed the first models of tank systems attrition rates as stochastic variables that were functions of the weapon system's performance characteristics that varied with time, its crew, and the threat (Bonder 1967). He used these results in the Lanchester differential equations to analyze and make recommendations regarding the spectrum of Sheridan design alternatives for the Army developers. This initial analytic research, which was incorporated, in part, in his dissertation, has been the basis of much of Seth's subsequent research and analysis activities.

Seth was awarded a Ph.D. in industrial engineering in 1965. His dissertation title was "A generalized Lanchester model to predict weapon performance in dynamic combat." Married, with son Eric and daughter Lisa soon to be born, Seth intended to embark on a career in industry. But, on an interview trip to Chrysler Motors in Detroit, he also visited the University of Michigan, Ann Arbor, where he gave a seminar to the industrial engineering faculty. Impressed by the technical caliber of the seminar discussions with faculty members, in particular Bob Thrall, Ralph Disney, and Herb Galliher, he changed his mind about industry and accepted Michigan's offer of a faculty position.

THE ACADEMIC/RESEARCH LIFE

Seth joined the faculty of the Industrial Engineering Department at the University of Michigan in 1965. He taught courses in mathematical programming, production control, inventory analysis, and decision analysis. In 1967, he developed and taught innovative modeling courses that provided students direct experience with these processes. His approach and course structure evolved into the concept of the Modeling Studio, as taught by Stephen Pollock, who joined the faculty in 1969. Seth gave substantial thought to the nature of OR and the education and development of practitioners, distilling complex and (for students) challenging concepts into clear, easily understood descriptions of the art and science. Based in part on his analysis experience at OSU, his ongoing consulting with Chrysler, and his teaching experiences, Seth authored an important paper that described his view of the deficiencies in OR curricula for prospective practicing OR analysts, and detailed processes that needed to be taught to alleviate these deficiencies (Bonder 1973).

WHAT'S A MODEL?

"*Models* are defined conventionally as the formal premises. At the risk of offending my philosophy-of-science colleagues, I define OR models to include not only the premises, which I shall call *premise models*, but also the deductively or inductively derived solutions or conclusions, which I shall refer to as *conclusion models*. The premise model reflects the analyst's assumptions about the process dynamics, while the conclusion models, which are directly derived from the premises, reflect his mathematical conclusions about them. Thus, in the example of the cashier waiting-line problem, the birth-death equation is the premise model and the derived relation for the mean number of customers in the system is the conclusion model.

"*Model solution* is the activity that leads from the descriptive premises to the conclusions or hypotheses to be verified. It can be experimental replication of a simulation model or a deductive process for mathematical premises. Sometimes the solution procedure can lead to a descriptive solution (like the expected number of people in the line), or it can be a mathematical programming technique leading to a prescriptive solution" (Bonder 1973, 799).

TWO WORLDS

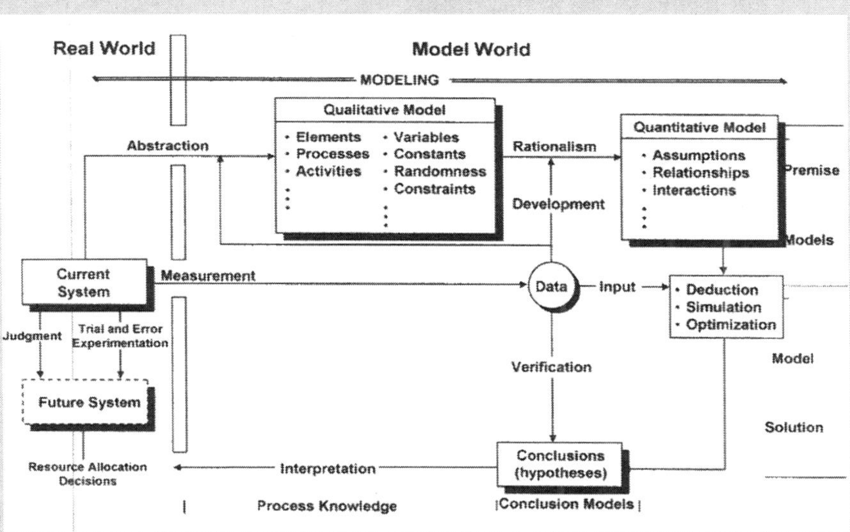

(Adapted from Bonder 1973, 797–799)

In 1966, Seth organized the Systems Research Laboratory (SRL) at Michigan to continue the research he started at OSU and to provide students with modeling and analysis experience. Master's students were mentored by Ph.D. students who, in turn, were mentored by Seth, other faculty, and research staff. A key member of the SRL was Robert (Bob) Farrell, a Michigan graduate with an M.A. in mathematics. Seth and Bob began a lifelong collaboration while working together on SRL projects. Seth obtained support for the SRL from the Office of Naval Research, the Assistant Vice Chief of Staff of the U.S. Army, and other defense organizations to conduct research on analytic models of combat and other related processes.

From 1966 to 1972, the SRL became a leader in research and the development of analytic models of combat processes and combat operations. Using arguments from renewal theory, especially Blackwell's renewal theorem, Bob and Seth demonstrated that a theoretically sound definition of the attrition rate in differential models of combat operations is the reciprocal of the expected time to kill a target (Blackwell 1948). Working with research staff and graduate students, Seth built on his dissertation research to address stochastic models of line-of-sight (LOS), target acquisition, optimal allocation of fires, and the employment of multiple systems, both friendly and threat (Bonder and Farrell 1970). Over a period of 6 years, Seth mentored a series of Ph.D. and master's students that included both civilians and active duty officers, all of whom contributed to the research program under his guidance. The seminal theoretical research performed at the SRL became the basis of subsequent research, model building, and analysis methods used by Seth, the Army, other U.S. defense agencies, and the international defense community throughout the Cold War.

In 1969, one of Seth's SRL sponsors, Lieutenant General William E. (Bill) Depuy [then Assistant Vice Chief of Staff of the U.S. Army and later, as a General, the first Commander of the U.S. Army Training and Doctrine Command (TRADOC)], asked Seth to analyze alternatives for a tank acquisition program that Depuy was debating with David Packard, the Deputy Secretary of Defense. Since the University of Michigan had a moratorium on classified research at that time, Depuy suggested that Seth start an external company that was cleared to conduct classified government work. In 1969, Seth, with Bob Farrell, organized VRI, a firm Seth headed for 32 years. Moonlighting from their University's responsibilities, Seth and Bob completed the study in 3 months, allowing Depuy to counter Packard's objections to the tank development program.

In the course of their work, Seth and Bob created the first hybrid analytic/simulation model of land combat, referred to as the Bonder/IUA (independent unit action) model. This model, and the many related models that evolved from it, represented small units in combat as homogeneous groups and analytically integrated stochastic process submodels of detection, firing, lethality, and vulnerability processes to compute expected attrition over short time intervals during which tactical conditions were constant. Movement and tactical decision making were then simulated on a digital model of terrain (which allowed for explicit calculation of LOS between combatants). This was a major advance over both pure analytic constructs (that lacked sufficient detail to be useful) and pure Monte Carlo simulations extant at that time (that took too long to evaluate a single replication). In addition to computational efficiency, the model provided increased analytic transparency to analyze cause/effect relationships. [The structure and initial development of the 1969 hybrid analytic/simulation model are described in the review paper "Mathematical modeling of military conflict situations" (Bonder 1981). This paper reviews military analytic model structures prior to 1965 and advanced modeling developments from 1965 to 1975.]

Seth joined with John Honig, the Army project officer for the tank acquisition study, to co-author a paper on the Bonder/IUA which they presented at the 1971 MORS Symposium. For this work, they were awarded the MORS Rist Prize that recognizes the best implemented study presented at a MORS symposium (Bonder and Honig 1971). In 1972, the success of this study, and the challenge of the problems facing the Army as it recovered from its Vietnam experience, led Seth to take on an adjunct role at the University of Michigan and apply himself to the full-time management of VRI.

The Consultant/Research Life

In addition to his role as VRI's Chief Executive Officer, Seth also served as its lead analyst and maintained close, technical involvement with all of its projects. (The author, who joined VRI in 1972, recalls intense, technical discussions in Seth's office throughout their collaborative activities at VRI—so intense that the support

Seth Bonder Peter Cherry
(Courtesy Saul I. Gass)

staff requested the doors be closed to avoid frightening the junior staff.) Seth's commitment to excellence, and his creativity in modeling the physics of operational processes, very quickly led to the recognition of VRI as a leader within the military OR consultancy community.

Under Seth's guidance, VRI extended the Bonder/IUA concepts to model company-to-battalion level operations, the Battalion Level Differential Models (BLDM). These models were used extensively in the 1970s and beyond to address a broad spectrum of tactical level issues including design of effective systems, cost/effectiveness of alternative systems, system employment concepts, mix of weapon systems in a unit, and design of unit combat tactics. The models that evolved from the Bonder/IUA/BLDM structure were used to analyze issues of command, control, and communications; intelligence; electronic warfare; combat support (air and ground); air–air warfare; logistics; and tactical nuclear warfare. These models had broad applicability and were employed by many U.S. and international defense analysis organizations.

In the early 1970s, Seth, recognizing the need to address higher level defense issues, initiated a program to develop models of corps level combined arms operations and joint service AirLand campaigns. (Sponsorship was provided only when Seth agreed to a request by the Secretary of Defense and the Central Intelligence Agency to go to Germany for 7 weeks to interrogate Soviet expatriates regarding the extent and quality of Soviet OR activities.) Starting with an initial prototype model, Vector-0, a series of hybrid/analytic simulation Vector models, with increasing functionality and applicability to various conflict regions around the world, were developed from 1971 to 1990. The models were structured to respond to a spectrum of customers' analysis needs and were developed using analytic modeling of operational processes to facilitate analysis of outputs and interpretation of cause/effect relationships. Many of the component models were linked to data from field experiments and tests. In a project that was unusual at the time and only rarely accomplished since, the Vector-2 version of the series was successfully tested against the 1973 Golan Heights campaign prior to performing a Middle East balance of forces study for the Central Intelligence Agency (Bonder 1984; Cherry and Bonder 1984; Farrell et al. 1986).

LESSONS LEARNED

"Do not rely on intuition to determine reasonableness of model results—models and analyses help enhance intuition. Intuition is built on experience with the existing system, not the future one being analyzed.

Conduct extensive parametric analyses on uncertain operational situation variables to understand the impacts of uncertain environmental or threat processes not under your control.

If possible, use multiple models to address major resource decisions.

Involve the client in all aspects of the analysis—it provides immediate buy-in and an advocate for the results. If the analysis is conducted in isolation, it is likely that you will provide a good answer to the wrong question or not have addressed the most innovative alternatives. Both you and the client will learn during the analysis.

Mentoring is critical. It takes 10–15 years to develop a journeyman analyst capable of independent analyses—analyses that produce useful results on time, provide meaningful insights, and provide results that can hold under detailed, technical peer review.

Finally, remember it is the analyst not the model that produces meaningful and useful results. Improve the former before the latter! Too many resources have been devoted to 'model improvement programs' and too little to improving military OR analysts" (Bonder 2002, 32).

Application of Vector-2 in work performed for the Supreme Headquarters Allied Powers Europe led VRI to create statistical models of corps level campaigns and a technique called MACRO that was used to screen force equipment and force mix alternatives initially in the NATO/Warsaw Pact context. MACRO was also used to prepare for operations in Southwest Asia, including Operation Desert Storm.

With the conclusion of the Cold War and the occurrence of the Desert Storm War, Seth turned his attention to the profound changes in the global security environment. He created new military operations knowledge and OR concepts to address the challenges facing the U.S. defense establishment in the 1990s. In particular, he introduced the notion of versatility planning that suggested that, in the new uncertain security environment, U.S. forces should be designed to maximize their versatility to operate in many regions rather than maximizing their effectiveness in one or two situations. The versatility concept was applied in designing the Rapid Reaction Force for the Supreme Allied Commander-Europe in 1992 (Bonder 1993, 1994).

In related work, Seth created two additional analytical constructs with particular relevance to the force projection aspect of the new global coalition of forces security environment (Bonder 1993). The first of these is the window of risk, the period of time from initiation of U.S. deployment during which the coalition forces cannot defend successfully. Analyses using this metric underscored the nonlinear importance of early-arriving forces in defending successfully and reducing casualties, one of the principles underlying the Army's Future Combat Systems Program. The second concept is the war-fighting casualty-capability curve. It depicts estimated casualties and the probability of winning a campaign as a function of the force-exchange ratio (the ratio of the percentage of enemy losses to the percentage of coalition losses). Analysis using these two constructs strongly suggested that early commitment of overmatching forces minimizes casualties, maximizes the probability of winning quickly, and provides force robustness against future uncertainties. These results provided formative evidence for the Powell Doctrine (governmental and military conditions that must be met with respect to decisions to employ U.S. forces).

The development and application of such concepts all arose from studies that Seth directed in response to requests by senior leaders in the Army, the DoD, and the Commanders in Chief of the Unified Commands. Seth was thus able to bring OR ideas and methods to the highest levels of national security leadership and policy decision makers in the late 1980s and 1990s. Maintaining close relationships with these leaders was a characteristic of Seth's approach to and success in applying OR to national security issues, a characteristic quite unique in the ever-changing assignments in which most consultants find themselves. Seth is an exponent of the main theme of what OR is all about—providing executives with a quantitative basis for decisions regarding the operations under their control, in a relationship based on mutual respect.

The national security sector was not the only focus of VRI. In 1972, based on a proposal to the National Institutes of Health, VRI conducted an analysis of the national nursing supply and demand. That analysis was the starting point for a significant line of business for VRI: first in the Military Health Services System and later in national health care issues. Seth obtained sponsorship for and developed the structure of the first enterprise-level health care delivery model, a hybrid/analytic simulation model—Healthcare Complex Model (HCM)—that integrates the resources (physicians, laboratories, equipment), patient flows, and activities (diagnosis, treatment, monitoring) in a health care delivery complex consisting of 2–3 major medical centers, 5–10 hospitals, and 10–20 clinics. These efforts led to an initial prototype of the health care complex associated with telemedicine deployment in Bosnia (Bonder and Zajtchuk 1997). During the period 1997–2001, Seth supervised analyses of a number of strategic health care delivery issues using HCM, including successfully testing the model against services provided by the Madigan Army Medical Center Complex.

From its beginning in 1969 as a two-person operation, VRI grew under Seth's leadership to an organization of over 400 professionals. The increase in its staff was due to internal expansion in VRI's key areas of national security, health care delivery, and financial institutions. In 1999, Seth, recognized the limits of internal growth and prepared a draft strategy that called for VRI to grow by acquisitions. Unfortunately, Seth was diagnosed with kidney cancer in October of that year. He then abandoned this growth strategy and, instead, prepared the company for sale. In 2001, he sold VRI to the Altarum Institute, a nonprofit research company specializing in health systems research and solutions. Rather than fade away into retirement, Seth declared 2001 as the beginning of his age of freedom—freedom not to manage anymore, freedom not to market anymore, and freedom to work on only on those projects in which he could make a significant difference to our national security and to society. He has remained fully occupied.

STAYING INVOLVED

From 2001 to 2008, Seth worked on a number of NAE national security projects. These included ways to improve the DoD's systems acquisition and operational testing processes for the Undersecretary of Defense (Acquisition), and a study for the Air Force Chief of Staff to determine requirements for the next generation stealth interdiction aircraft. At the request of the President of the Institute of Medicine (IOM) of the National Academies, Seth was the only OR member—the only non-psychiatrist—on an IOM study to determine how to reengineer the U.S. mental health care delivery system. He provided the system structure for the study; influenced many of the recommendations, including a critical one on integrating physical and mental health providers in joint primary care practices; and developed a strategy for implementing the recommendations.

One of Seth's consulting engagements deals with defeating the improvised explosive devices (IED) used by insurgents and terrorists in Iraq and Afghanistan. In 2006, he was appointed a Senior Advisor and

Mentor to the DoD's Joint Improvised Explosive Device Defeat Organization (JIEDDO). His technical efforts have been devoted to development of a set of OR methods for analyzing a spectrum of IED issues. He provides advice on on-going OR projects and mentors JIEDDO contractors and OR analysts in Iraq and Afghanistan.

In 2008, he served on an IOM project to determine how OR approaches could be used to improve the delivery of health care for patients with traumatic brain injury (TBI). He structured and implemented a workshop involving mental health psychiatrists, psychiatric TBI researchers, and operations researchers into five working groups. These groups developed OR approaches in five main areas: means of detecting TBI, screening program for a population of returning military personnel who were exposed to blast effects in Iraq or Afghanistan, proactive tracking of TBI patients, forecasting TBI demand, and resource allocation methods for planning and operation of a TBI health care delivery system.

Seth serves on the advisory boards of the Comprehensive Cancer Center and the Kellogg Eye Center at the University of Michigan, providing advice on strategy development. He is also on the foundation board of Washtenaw Community College, where he advises on financial matters and strategy developments. He has endowed two undergraduate scholarships at the college for science and mathematics students.

2007 BONDER SCHOLARSHIP AWARDS

Seth with recipients of the 2007 Bonder Scholarship awards at the INFORMS Annual Meeting, Seattle, Washington. Shown (left to right) are Burhaneddin Sandikci, Health Applications Society; Seth Bonder; and Kuo-Hao Chang, Military Applications Society.

Seth has continued his involvement with OR education and the OR profession. He maintains his appointment as an adjunct professor in the Industrial and Operations Engineering Department at the University of Michigan; he also serves as an advisor to engineering and business schools, and mathematics departments. Seth has contributed to the education of OR professionals by aiding in the development of bright OR graduates and encouraging them to pursue careers in applied OR. In 2001, he endowed 2 yearly OR Ph.D.-level scholarships administered by the Institute for Operations Research and the Management Sciences (INFORMS)—one in the applied military OR area and one in the applied health OR area. He also endowed OR Ph.D. scholarships at OSU and the University of Michigan to encourage graduate students to pursue careers in applied OR areas.

For Seth, although it would not be easy to discern, life has not been all OR. He still finds time for fine food, fine wines, and an extreme fitness routine. While maintaining his prowess at billiards, he became a fierce competitor on the tennis court when his children learned the game in the 1970s. His son, Eric, later competed at the collegiate level for Ohio State and his daughter, Lisa, was a highly ranked tennis professional. (Seth has continually claimed that their skill is a result of patrilineal genetics.) After his bout with kidney cancer, he stopped playing tennis and took up golf with equal passion and skill.

Honors and Awards

Seth served as a member of the MORS Board of Directors, its Vice President (1968–1969) and President (1969–1970). He is a fellow of MORS and was awarded its 1986 Vance R. Wanner Memorial Award for distinguished service to the profession. He served as President of ORSA (1978), and received its Kimball Award for distinguished service to the society and to the profession of OR in 1993. He was Vice President of the International Federation of Operations Research Societies (1985–1989). He is a fellow of INFORMS and received its President's Award in 1991 for important contributions to the welfare of society. He received the Military Applications Society's Steinhardt Memorial Prize for outstanding contributions to the practical application of OR techniques and to the solution of military problems (1999). Seth has served as a member or consultant to the U.S. Army Science Board (ASB) since 1973 and served in the same capacities on the U.S. Defense Science Board from 1988 to 2002. For his contributions and service to the Army, he received the Award for Patriotic Civilian Service from the Secretary of the Army (1999). Seth was elected to the NAE in 2000 and has served as Chair of the Industrial, Manufacturing, and Operational Systems Engineering Section of the NAE until 2008.

REFERENCES

Blackwell D (1948) A renewal theorem. Duke Math J 15(1):145–150

Bonder S (1967) The Lanchester attrition-rate coefficient. Oper Res 15(2):221–232

Bonder S (1973) Operations research education: some requirements and deficiencies. Oper Res 21(3):796–809

Bonder S (1981) Mathematical modeling of military conflict situations. In: Gass SI (ed.) Operations research: mathematics and models. Proceedings of symposia in applied mathematics, vol 25. American Mathematical Society, Providence, RI, pp 1–51

Bonder S (1984) Summary of a verification study of VECTOR-2 with the Arab-Israeli War. In: Huber RK (ed.) Systems analysis and modeling in defense: development, trends, and issues. Plenum Press, New York, NY, pp 155–170

Bonder S (1993) Defense planning in the new global security environment. ARMY August, 18–26

Bonder S (1994) Impact of the new global environment on national security planning—challenges to the OR Community. Int Trans Oper Res 1(1):31–39

Bonder S (2002) Army operations research—historical perspectives and lesson learned. Oper Res 50(1):25–34

Bonder S (2008) Personal communication

Bonder S, Farrell R (eds) (1970) Development of models for defense systems planning. Systems Research Laboratory, Department of Industrial Engineering, The University of Michigan, Ann Arbor, MI

Bonder S, Honig J (1971) An analytical model of grand combat: design and application. In: Proceedings of the 27th military operations research symposium, Military Operations Research Society, Alexandria, VI, 73–107

Bonder S, Zajtchuk R (1997) Changing the paradigm for telemedicine development and evaluation: a prospective model-based approach. Socioecon Plann Sci 31(4):257–280

Cherry P, Bonder S (1984) VECTOR-2 land/air campaign model. In: Huber RK (ed.) Systems analysis and modeling in defense: development, trends, and issues. Plenum Press, New York, NY, pp 139–154

Farrell R, Bonder S, Proegler L, Miller G, Thompson D (1986) Capturing expertise: some approaches to modeling command decision making in combat analysis. IEEE Trans Syst, Man Cybern 16(6):766–773

42

Bernard Roy

Denis Bouyssou and
Daniel Vanderpooten

During the 1950s and early 1960s, as the ideas, methods, and applications of operations research (OR) spread around the world, each country tended to adapt OR to fit its professional, academic, and cultural environments. Often, there were a few dedicated persons who led the way and became recognized as the country's OR pioneers. In the case of France, such a person was Bernard Roy. After a first career as a consultant, during which he made major breakthroughs in graph theory and project scheduling, he started a second career as an academic interested in multiple criteria decision making (MCDM). Among his many achievements, he is the developer of the activity-on-node project scheduling technique and of the famous ELECTRE methods for resolving decision problems with multiple criteria. Through his research, teaching, consulting, and service to the community, he has been one of the major promoters of OR in France.

Bernard served as vice president (1974–1976) and president (1976–1978) of the Association Française pour la Cybernétique Économique et Technique (AFCET, the French OR society at that time). Bernard was president of The Federation of European OR Societies (EURO) in 1985–1986, and was awarded the 1992 EURO gold medal, the highest distinction granted by EURO.

A.A. Assad, S.I. Gass (eds.), *Profiles in Operations Research*, International Series in Operations Research & Management Science 147, DOI 10.1007/978-1-4419-6281-2_42,

FAMILY AND CHILDHOOD

Bernard was born on March 15, 1934 in Moulins-sur-Allier, a medium-sized town in the center of France. He is the only child of René Roy (born 1906) and Jeanne Chérasse (born 1913). Both parents completed their studies with the brevet (a diploma given to pupils at the end of the ninth grade). Bernard's maternal grandfather was a railway station manager; his paternal grandfather built and repaired millwheels. Bernard was the first member of his family to pursue an advanced education. René started his career as a bank teller. In 1934, he became an insurance agent at the Compagnie du Nord. With the help of Jeanne, he was responsible for a portfolio of clients. René took part in World War II (WWII) and, after the defeat of France, he was sent to Germany as a war prisoner. He escaped in 1943. After the war, he kept close contacts with several fellow prisoners by inviting them to yearly family gatherings. One of the ex-prisoners was the father of Bernard's future wife, Françoise. Bernard first met her during one of these gatherings.

During WWII, Moulins-sur-Allier was in the occupied part of France, but located quite close to the demarcation line. At that time, the communication between the two parts of France was highly problematic. Both Bernard's mother and aunt would cross the demarcation line to transmit mail between the two zones (sometimes even helping people to cross the line). His aunt was arrested by the Germans, but she was soon liberated following a bureaucratic error. German soldiers, realizing the error, paid frequent visits to the family's home; these visits made a very strong impression on the young Bernard. The war years were very bleak. Fortunately, Jeanne had relatives living in the countryside, so the family had access to food products that were cruelly missing, and Bernard could enjoy peaceful holidays.

In 1940, at the age of 6, Bernard started his formal education at a local school. Soon after, he began experiencing vision problems. Due to the war, it was not easy to have access to an ophthalmologist, but his parents did manage, as best as they could. The first one consulted advised that these problems were somatic. Because things were not getting any better, several other famous specialists were consulted making various diagnoses, such as a compression of the optical nerves. It was not until 1955 that a correct diagnosis was established; Bernard was suffering from a very rare type of retina problem (atypical retinitis pigmentosa). As a result, Bernard gradually lost sight, while keeping a limited peripheral vision. Reading became more and more difficult. Writing also became problematic; after some time, hardly any one could decipher his letters. Bernard kept writing, however, by using the new Reynolds ballpoint pens that just arrived in France. He did so during elementary school (5 years in France) and through his second year of secondary school. Year after year, Jeanne helped him by reading his notes and books.

Bernard started secondary school in 1945 (consisting of 4 years of collège and 3 years of lycée). He soon abandoned writing, taking notes on a mechanical typewriter during classes. He managed to take exams using the typewriter through the two baccalauréats, which meant, at that time, the end of secondary school. Bernard's interest in mathematics was not immediate, but grew during this period. Over time, Bernard had his typewriter customized with some Greek letters added to the keyboard. He started studying English as his vision deteriorated. His father assembled for him a basic bilingual dictionary that used very large letters that Bernard could read. However, his mastering of the language was uncertain and, during the first part of his career, he published mostly in French. (He continues to favor publishing in French.) Bernard passed his second baccalauréat (in the mathématiques élémentaires section) in 1952, with the highest possible mention. At that time, even with his declining peripheral vision, Bernard could walk by himself; he rode his bicycle until the age of 22, with severe falls from time to time. But, it was obvious that his handicap would prevent him from occupying certain professions.

HIGHER STUDIES: THE ROAD TO OR

Bernard wanted to be an engineer (he had built a radio while he was in secondary school). The traditional way to become an engineer in France is not through universities, but through the distinct system of Grandes Écoles in which students are selected on the basis of a competitive exam that could only be taken after 2 years of Classes Préparatoires. Bernard went to Paris for his first year of Classes Préparatoires at the Lycée Chaptal. His results were so high that he was admitted for the second year to one of the most prestigious Classes Préparatoires at the Lycée Louis-le-Grand, usually the first step to the École Polytechnique or the École Normale Supérieure. In class, Bernard was using his relatively quiet typewriter to take notes. But, his physics teacher thought that the noise was intolerable and did not allow him to use the typewriter. Thus, Bernard, not being able to take notes and rather shaken by this decision, left the Lycée Louis-Le-Grand and the Classes Préparatoires system in October 1953. Thus ended his dream of entering the École Normale Supérieure. He immediately decided to enroll in the Université de Paris and study for a degree in mathematics. At that time, the Licence de Mathématiques meant obtaining three certificates: this usually took 3 years (the Licence had to be preceded by a general mathematics certificate that Bernard had passed while he was at the Lycée Chaptal). In the academic year 1953–1954, Bernard completed two of the three certificates (calculus and probability). He was taught by some great mathematicians who became famous: Laurent Schwartz (founder of the theory of distributions and a member of the Bourbaki group), Jacques-Louis Lions (one of the major promoters of applied mathematics in France, a president of the International Mathematical Union, and father of the future Fields medal laureate Pierre-Louis Lions), Gustave Choquet (developer of the theory of capacities), and Robert Fortet (founder of the most important French research group in the theory of probability). Bernard got even with the École Normale Supérieure—he completed his calculus certificate with the highest possible mention, ending up tied with a student from that school.

Bernard as a student at the age of 20.

During the 1953–1954 academic year at Université de Paris, Bernard met Patrice Bertier, a fellow student in mathematics. Patrice suffered poliomyelitis during his youth and was using a wheel chair. He and Bernard became great friends. They spent the year studying together and helping each other. Patrice completed his Licence in June 1954 having passed the three certificates. His plan was to take courses at the Institut d'Études Politiques (IEP) in the next academic year. IEP was a relatively special Grande École, mainly oriented toward economics and political science; it was the usual first step to the highest positions in the French civil service. At that time, the teaching of economics and political science had little to do with mathematical economics and, for someone holding a degree in mathematics, enrolling in IEP was extremely uncommon. But, Patrice was attracted to economics. He persuaded Bernard to join him in this adventure. The only problem was that Bernard had not completed his Licence: he had to obtain his third certificate. Bernard then decided to study for his missing certificate during summer. He finally obtained this certificate (in rational mechanics) in September 1954, thus completing his 3 years of Licence in only 1 year.

Both Bernard and Patrice joined IEP in October 1954. As this was really unusual—mathematics students at IEP—they also enrolled in the Institut de Statistique de l'Université de Paris (ISUP), an interfaculty department that granted diplomas in statistics and probability. IEP was located at rue Saint-Guillaume, west of the Latin Quarter, while ISUP was located at rue Pierre-et-Marie-Curie, near the Jardin du Luxembourg, south of the Latin Quarter. During the years 1954 and 1955, people walking on the Boulevard Saint-Michel would often observe a strange event: Bernard, half blind, pushing the wheel chair of Patrice, as they went back and forth between ISUP and IEP. At ISUP, Bernard had several remarkable teachers:

Bernard (left) and Patrice (right) on holidays in 1969

Georges Darmois, Georges Morlat, Dickran Indjoudjian, Germain Kreweras, René Roy. ISUP was then one of the rare places in France in which applied probability and statistics were taught to highly trained mathematics students. Here, Bernard discovered mathematical statistics and econometrics; applied statistics was not forgotten, although all computations had to be done on electric non-programmable calculators. At IEP he attended the courses of Alfred Sauvy (an economist and demographer who, in 1952, first used the expression Tiers Monde [Third World]), Jean Fourastié (an economist who coined the expression Les Trente Glorieuses [The Glorious Thirty]—the 30 years from 1945 to 1975), Paul Delouvrier (an economist and urban planner), and André Siegfried (a sociologist specialized in electoral studies). This unique combination of mathematics and economics aroused the interest of Bernard for the application of mathematics to the real world.

The years 1954–1955 were exciting times for Bernard. Several people—Georges-Théodule Guilbaud, Germain Kreweras, Jean Abadie, Jean Ville, Pierre Bouzitat, Marc Barbut, Michel Rosensthiel, Jean Mothes, Claude Berge—began giving unofficial lectures and seminars on OR; OR was not part of any course in France. Bernard especially remembers the lectures of Guilbaud. They were attended by huge crowds in the Amphithéâtre Hermite of the prestigious Institut Henri Poincaré. Bernard had found his way to applying mathematics in the real world. He wanted to do OR. The emerging French OR community was beginning to organize itself and, in 1956, the Société Française de Recherche Opérationnelle (SOFRO) was established. [In 1964, SOFRO became AFIRO (Association Française d'Informatique et de Recherche Opérationnelle), after a merger with a society of computer scientists; in 1968, it became AFCET (Association Française pour la Cybernétique Économique et Technique), after a merger with a society of cyberneticians; and in 1998, AFCET split apart with the French OR society becoming ROADEF (Société Française de Recherche Opérationnelle et d'Aide à la Décision) (Roy 2006)].

These were years of intense activity for Bernard. Besides the courses at IEP and ISUP, he also obtained additional certificates in mathematics (mathematical methods of physics, algebra, and number theory). He completed his master's degree at ISUP in 1957 with his first research in OR: a master's thesis on the newsboy problem presented as the baker's problem (problème du boulanger) (Roy 1957). He decided to start a Ph.D. on the same subject, but soon abandoned it in favor of graph theory.

In July 1956, Robert Fortet managed to obtain positions as junior researchers for both Bernard and Patrice at the Centre National de la Recherche Scientifique (CNRS, the national research agency, created in 1939, to promote fundamental research in France). The jobs paid little but offered immense freedom. As CNRS did not have an office at that time, Bernard and Patrice were also recruited as interns at Électricité de France (EDF, the newly nationalized electricity company) under the supervision of Marcel Boiteux, who was in charge of EDF's Service des Études Économiques Générales (he later became CEO of EDF). Bernard completed his master's thesis for ISUP during this period; he benefited from the advice of Marcel Boiteux on how to write a paper. At that time, EDF had no computing facilities. Small linear programming (LP) models were used to plan production between thermal and hydraulic plants. Bernard and Patrice were still interns at that time and, since the problems involved strategic elements, they did not have full access to the data and results; their main role was to devise the general structure of the LP models. These problems, although small sized (around 50 variables), were still too large to be efficiently solved by hand. Marcel Boiteux and Pierre Massé (the vice-CEO of EDF) were sending these problems by ordinary mail to George Dantzig at the RAND Corporation in Santa Monica, California, with the results also returned by mail. At that time, processing time did not reduce to computation time.

CONSULTANT AT SEMA

Bernard married Françoise Jolivet in July 1957. They had six children, Sylvie (1958†), Laurence (1961), Isabelle (1964), Solange (1966), Patrice (1968), and Philippe (1970†), and nine grandchildren. The meager salary from the CNRS was not adequate to support the young couple. Bernard left CNRS when he was recruited by a newly created OR consulting company, the Société d'Études Pratiques de Recherche Opérationnelle (SEPRO). Meanwhile, the Société de Mathématiques Appliquées (SMA) was created as a joint venture between the Banque de Paris et des Pays-Bas (more commonly known as Paribas) and an independent consulting company led by Marcel Loichot. The aim of SMA was to be a consulting company that would promote the use of management science (MS) in French companies. Jacques Lesourne was appointed as CEO. Bernard left SEPRO to join SMA as a consultant in October 1957, together with Patrice Bertier. SMA quickly became SEMA (Société d'Économie et de Mathématiques

Appliquées). After having created several subsidiaries in Europe, SEMA became SEMA (Metra International).

Françoise and Bernard, 1988

SEMA started with around 10 employees and almost no contracts. Bernard's first task, with Patrice, was to translate into French several chapters of the OR text written by Churchman et al. (1957). They also put the final touches to the book by Lesourne (1958), one of the first OR books written and published in French.

Contracts began to arrive in 1958 and Bernard started to work on applied OR problems, mainly from the private sector. He worked on a variety of problems that involved many ideas and techniques: probability and queueing theory (reducing the waiting time at a ferry), data analysis (choosing the name of a new brand of cigarettes), transportation studies (developing a forecasting model for transportation planning), cutting stock (designing cardboard boxes), location (choosing sites for plants), and finance (optimizing cash management). Many of these applications were later published in *METRA*, the future academic journal sponsored by SEMA. Bernard's most important works were concerned with project scheduling and related graph theory problems.

SEMA was growing steadily during this time. In 1962, it acquired a Control Data computer (CDC 6600) for which several LP and integer linear programming (ILP) codes were developed that enabled larger problems to be solved. Before that, all computations were performed by a bureau de calcul employing many persons working on electric calculators.

In between contracts, Bernard worked on his Ph.D. dissertation in graph theory and its application to project scheduling (together with a minor dissertation on abstract algebra). He received his Ph.D. in 1961 (dissertation on "Cheminement et connexité dans les graphes: Application aux problèmes d'ordonnancement" [Roy 1961)]) from the

Université de Paris, under the supervision of Claude Berge [the author of one of the first books on graph theory (Berge 1958)]. That same year, Bernard was offered a position at the Université de Paris in mathematics. OR, at that time, was not part of the mathematics curriculum and the teaching of mathematics was slanted toward pure mathematics—this period was highly influenced by the Bourbaki group. As accepting the position meant returning to pure mathematics, Bernard declined the offer. Taking advantage of SEMA's policy that encouraged its consultants to teach, Bernard did become involved in teaching OR courses at the Centre Inter-armées de Recherche Opérationnelle (a permanent education program in OR for French officers) and, with Claude Berge, taught seminars on graph theory and combinatorial problems.

In 1962, Jacques Lesourne created within SEMA a scientific group called Direction Scientifique, with the objective of helping consultants in applying new scientific and computational techniques. Bernard joined this group as a consultant of consultants. He became its director in 1964. For many years, this high-powered, multidisciplinary group was the site of intense activity; its members included Raphaël Benayoun, Patrice Bertier, Éric Jacquet-Lagrèze, Hubert Le Boulanger, Benjamin Matalon, Jean de Montgolfier, Hervé Raynaud, and Gilbert Sussmann. At the same time, SEMA launched a quarterly journal called *METRA* to popularize the new techniques it promoted (they included OR techniques, but also covered every aspect of MS). Bernard was appointed its editor-in-chief and remained so until the journal ceased publication in 1977. *METRA* published papers written by SEMA consultants and from its European subsidiaries in four languages (French, Spanish, Italian, and English). It is remarkable that the editorial policy of *METRA* was to promote the techniques developed at SEMA. Its methodological advances could appear in the journal after observing a publication lag of about 2 years which SEMA required to protect its competitive advantage.

Although edited by a commercial company, *METRA* had a standard academic way to process papers and had a scientific editorial board that included academics (most notably Stafford Beer and Paul Gillis). In those times, few French libraries had subscriptions to *Management Science*, *Journal of the Operational Research Society*, or *Operations Research*. Thus, *METRA*, together with *RIRO* [*Revue d'Informatique et de Recherche Opérationnelle*, the newly created journal of AFIRO that would later become *RAIRO* (*Revue d'Automatique, d'Informatique et de*

Recherche Opérationnelle)] played an important part in the diffusion of OR techniques in France.

Consulting, therefore, greatly influenced Bernard's view of OR techniques and applications. Most often, the lack of appropriate software, the paucity or poor quality of data, the softness of some constraints, and the presence of multiple conflicting objectives made the quest for an optimal solution illusory. A good solution that could not be proved optimal was often a major breakthrough in practice. These real-world concerns greatly influenced Bernard's approach to his future research.

PROJECT SCHEDULING AND GRAPH THEORY

One of the most famous contributions of Bernard is in the field of project scheduling. In 1958, when working at SEMA, he was faced with the problem of scheduling the construction of new buildings for the headquarters of a large company in Paris. Managing this project, involving several hundreds of tasks and more than one thousand constraints, required a specific methodology. At this occasion, Bernard developed a method called MPM (*Méthode des Potentiels Metra*). MPM was based on what is now known as the activity-on-node (AON) formulation (Roy 1959a, 1962). While its theoretical foundations were being established (in terms of existence and optimality of schedules), this method was applied successfully to several other scheduling problems (production of crankshafts at Mavilor Motors, design of an appropriate cycle for the new Tracoba house-building process). These applications involved potential constraints (i.e., constraint of the form $t_j - t_i \geq a_{ij}$, where t_j is the starting time of task j and a_{ij} is the minimum time between the start of tasks i and j) and more difficult constraints such as disjunctive or cumulative constraints (disjunctive constraints impose that two tasks do not occur simultaneously, and cumulative constraints require that simultaneous tasks do not consume more than a given amount of resources; this typology of constraints was developed by Bernard). MPM was one of the first computer-based software systems for project scheduling: CONCORD (CONception et Coordination de l'ORDonnancement) (Roy and Dibon 1966). [The AON approach was proposed independently in the U.S. under the name Precedence Diagramming Method by Fondahl (1961).] The existence of a large number of difficult constraints, in the context of scheduling the equipment of the steamship liner France (the largest in the world in 1960), eventually led to the development of another technique, description segmentée, designed to quickly spot incompatible constraints in a system of linear inequalities (Roy 1963; Roy and Simonnard 1961).

Simultaneously and independently, methods like PERT or CPM, based on an activity-on-arc (AOA) formulation were developed in the U.S. in the late 1950s (at DuPont de Nemours, RAND Corporation, and the U.S. Navy for the deployment of the POLARIS missile). It is now widely acknowledged that the AON formulation is superior to the AOA formulation, since it is more systematic, without requiring modeling tricks such as dummy arcs, and its ability to readily handle changes or additions to constraints.

Bernard also obtained results on more theoretical aspects of graph theory, related, for example, to optimal paths, connectivity, transitivity, and chromaticity (Roy 1958, 1959b, 1967, 1969b). As discussed by Hansen and de Werra (2002), some of these pioneering results, obtained over 50 years ago, are still the basis of currently published results.

Also well known is the so-called Roy–Warshall's algorithm that computes the transitive closure of a digraph (Roy 1959b; Warshall 1962). This algorithm was discovered independently by Bernard in 1959 and Stephen Warshall in 1962. In the subfield of network flows, the algorithm to determine a minimum cost flow by successive shortest paths is known as Busacker and Gowen's (1961) algorithm in the U.S. and as Roy's algorithm in Europe. Bernard independently developed this approach in the early 1960s and presented it at several conferences (Roy 1970).

Bernard is the author of a remarkable two-volume, 1300-page textbook on graph theory (Roy 1969c, 1970). Even if it is now outdated on some points, it includes an original treatment on many topics that should be of interest to anyone in this field. Bernard organized two summer schools on graph theory and discrete mathematics. The first one, co-organized with Frank Harary, took place in 1966 in Italy with more than 100 participants. The second one was in Versailles, France, in 1974. Both schools gathered most of the major names in the field of graph theory and combinatorial optimization. The proceedings of the second school were published in Roy (1975b). With Patrice Bertier, Bernard was also among the pioneers who developed and formalized branch and bound procedures in the mid-1960s (Bertier and Roy 1965; Roy 1969a).

ELECTRE and Multiple Criteria Decision Aiding (MCDA)

Bernard's research on multiple criteria decision problems was motivated by real-world problems encountered by SEMA clients. This led to the development of the first ELECTRE method, ELECTRE I (Roy 1968), for solving such problems. A media planning problem led to the development of ELECTRE II (Roy and Bertier 1973). At that time (mid-1960s), Bernard was unaware of the parallel developments in the U.S. by Howard Raiffa, Ralph Keeney, and many others. Bernard accepted the invitation of George Dantzig to organize two sessions on MCDM for the 1970 Mathematical Programming Symposium to be held in The Hague (Roy 1971). These sessions were among the first of their kind to be given at such conferences. During this time, Bernard, working early in the mornings, completed his two-volume exposition on graph theory and its applications (Roy 1969c, 1970).

ELECTRE Methods: An Exposition

ELECTRE (Élimination et Choix TRaduisant la Réalité) methods were first developed in the mid-1960s to answer real-world problems brought to Bernard by SEMA consultants, such as the selection of research projects or of investment opportunities. SEMA had developed a technique, called MARSAN (Méthode d'Analyse et de Recherche pour la Sélection des Activités Nouvelles), that was designed to help firms in selecting new activities. To do so, activities were evaluated on a series of 48 dimensions (the word criterion was not used then). They included quantitative as well as qualitative dimensions. Qualitative dimensions were translated on a numeric scale more or less arbitrarily. A weighted sum of all these numbers was computed to measure the attractiveness of these new activities.

It soon became clear that the use of a weighted sum allowed compensation effects that were not desirable: small advantages on several dimensions could compensate for major weaknesses on some others, which was not felt to be desirable. Moreover, the transformation of qualitative dimensions into numbers was playing an important part in the final result.

Bernard devised a method that would deal both with qualitative dimensions without the need for transforming them into quantitative dimensions and that would not tolerate compensation effects that were felt undesirable. This was the birth of ELECTRE I (Benayoun et al. 1966; Roy 1968). Basically, in ELECTRE I, alternatives are compared in pairs using the following reasoning: Alternative *a* will be declared at least as good as alternative *b* if (1) the proposition is supported by a sufficient majority of dimensions (concordance condition), and (2) among the dimensions opposing the proposition, there is none on which the opposition is too strong (non-discordance condition).

Such an at-least-as-good-as relation (soon called an outranking relation) can be built on the basis of purely ordinal considerations. The non-discordance condition prevents undesirable compensation effects from occurring. The application of the concordance condition leads to assigning weights to each dimension. To decide if a majority of dimensions is sufficiently important, the sum of the weights is compared to a threshold called the concordance threshold (note that these weights are quite different from the weights used in a weighted sum; they are never multiplied with scores and are, therefore, independent from the scale used to measure scores). Similarly, the strength of the opposition of dimensions is computed using a veto threshold.

A specific feature of this relation is that it does not have to be transitive (even in its asymmetric part, because of Condorcet-like effects) or complete (some alternatives may remain incomparable). Therefore, deriving a prescription on this basis is not an easy task and calls for the application of specific techniques, called exploitation techniques. They differ on the type of recommendation that is looked for. ELECTRE I has been designed in a choice problem formulation—it aims at recommending a subset of alternatives (as small as possible) that is likely to contain the best alternatives. Technically, viewing the outranking relation on the set of alternatives as a graph, Bernard suggested using the kernel (an independent and dominating subset) of this graph.

ELECTRE II (Roy and Bertier 1973) is a variant of ELECTRE I that is designed to rank order alternatives. It uses two outranking relations instead of one. The ranking is not necessarily complete: it preserves incomparability between alternatives that appear difficult to compare. ELECTRE III (Roy 1978) is a far-reaching generalization of ELECTRE II that uses a fuzzy outranking relation instead of two crisp ones. Furthermore, it refines the preference modeling on each dimension with the introduction of thresholds preventing small differences between scores from being interpreted as a definite advantage. Such thresholds were introduced in a new version of ELECTRE I, called ELECTRE IS. ELECTRE IV (Roy and Hugonnard 1982) is a variant of ELECTRE III designed to deal with situations in which weights are difficult to elicit, given the diversity of opinions. ELECTRE TRI (Roy and Bouyssou 1993) is designed to deal with a sorting problem formulation in which each alternative is assigned to a category pre-defined by norms which, for example, separate good and bad credit files.

All these methods were developed to deal with specific real-world problems. ELECTRE methods have been applied to a large variety of problems in many countries (Figueira et al. 2005; Roy 1991; Roy and Bouyssou 1993).

MCDA: An Original Perspective on OR

Bernard's concept of OR was influenced by two major themes: the starting of his career as a consultant and his later work in MCDM. Their synergistic interaction led him to develop a decision-aiding methodology that is original and rather non-standard in the OR profession (Roy 1975a, 1977, 1985, 1990, 1993). He noticed that the application of OR models and methods were characterized by the adherence to three main assumptions:

1. The quest for rationality implies the use of a unique criterion that should be optimized.

2. Qualitative information and ambiguous data should be avoided as much as possible.

3. Science aims at describing a reality that is mainly independent from the observer. Reference to this outside reality is central to the validation of a scientific model.

Bernard soon became rather skeptical about these three assumptions and proposed a decision-aiding methodology that would dispense with them (Roy 1981).

Indeed, Bernard quickly acknowledged the fact that in many real-world problems, several actors are involved. These several stakeholders have different opinions. Quite often, their opinions are not always completely structured. Also, there may be no real decision maker. Moreover, what is feasible or what is not feasible is often fuzzy (Roy 1988). This undermines the first assumption and calls for the use of multiple criteria. This does not mean that optimizing is useless, but simply that optimality within a model does not guarantee an acceptable solution, let alone an optimal one, in the real world.

Real-world situations abound with qualitative information. Contrary to the second assumption, information is often uncertain, imprecise, and ill-determined. Trying, by all possible means, to convert all that is qualitative into quantitative information is a difficult task and often leads to a result that is seldom meaningful. Spending time to obtain information of better quality is often an inappropriate use of resources and may lead to instrumental bias (recall the drunkard looking for his keys under a street lamp without really knowing where he lost them). In all real-world problems, irreducible uncertainty, imprecision, and inaccurate determination will remain (Roy 1989). Hence, we should reconcile ourselves that we must deal with the available qualitative information, using techniques that allow robust conclusions to result (Roy 1998).

Decision aiding inevitably means working with preferences. When facing a complex problem, it is rare to have the actor(s)'s preferences clearly stated and completely well structured (Roy and Vincke 1984). The analyst must question the actor(s) and, thus, contributes to the shaping of the preferences, as well as describing them. A clear violation of the third assumption. This learning process, which is often a creation process, is an inevitable part of applying OR models (Roy 1987).

Over the years, Bernard has proposed a complete decision-aiding methodology that does not rely on the above three assumptions (Roy 1985; Roy and Bouyssou 1993). This explains why Bernard prefers to speak of MCDA instead of MCDM.

Bernard's most recent research deals with robustness in decision aiding. In many decision contexts, model parameters are often defined approximately due to uncertainty, imprecision, or ill-determination (Roy 1998). Rather than looking for optimal solutions, it is then more appropriate to look for robust solutions that are resisting to vague approximations and areas of ignorance, that is, which behave well for all, or at least most, plausible values of the parameters. Such a perspective, often well received by practitioners, gives rise to many challenging theoretical questions. Bernard's approach to robustness is discussed in Roy (2010).

Professorship

In the late 1960s, following the May 1968 events in France leading to a 1-month general strike, Bernard started wondering about his future career. Jacques Lesourne had announced that he would soon leave SEMA. During this time, Bernard was asked to give a doctoral course on OR at the newly created Université Paris-Dauphine (this experimental university was created in 1968 and occupied the former NATO headquarters in Paris). In 1971, he was appointed associate professor in mathematics (later joining the computer science department). The following year, he was made full professor. He

(Left to right) Daniel Vanderpooten, Bernard Roy, and Denis Bouyssou (2007)

kept his position at SEMA until 1974, progressively reducing his involvement, as SEMA reduced its OR activities; he remained associated with SEMA as a scientific advisor until 1979. One of Bernard's early academic duties was to reshape the MS curriculum within the management program. In 1974, Bernard created a research group called LAMSADE (Laboratoire d'Analyse et Modélisation de Systèmes pour l'Aide à la Décision) which became affiliated with CNRS in 1976. LAMSADE was one of the few research groups in France oriented toward applied OR. Over the years, as LAMSADE kept growing, it expanded its base of interest to include research topics in computer science.

Bernard made sure that the Dauphine OR curriculum included a doctoral program, Méthodes Scientifiques de Gestion, and thus, through the years, he began his supervision of over 50 doctoral students (both authors of this text are his former doctoral students). His research at LAMSADE became more and more oriented toward MCDM, or rather MCDA.

Although Bernard devoted much energy to the development of LAMSADE and served as its director until 1999, he also undertook several important responsibilities within Université Paris Dauphine, including the directorship of a doctoral school. In addition, in 1980, Bernard became scientific advisor of RATP (Régie Autonome des Transports Parisiens; the company that operates all public transports in the Paris region).

Bernard is the author of more than 80 papers in refereed journals and nearly 50 papers in contributed volumes. A selected list of Bernard's publications is available from LAMSADE (2009).

Bernard retired in 2001 with the title of professor emeritus. A Festschrift honoring him was published on the occasion of his retirement (Bouyssou et al. 2002). He remains quite engaged in his scientific and consulting activities.

HONORS AND AWARDS

Bernard has received six honorary doctoral degrees (Vrije Universiteit Brussels, Belgium, 1978; Université de Liège, Belgium, 1978; Université de Fribourg, Switzerland, 1982; Poznan University of Technology, Poland, 1992; Université Laval, Canada, 1998; Technical University of Crete, Greece, 2002). He received the 1992 EURO gold medal, the highest distinction granted by EURO. He holds the gold medal from the MCDM International Society, as well as the Hermès de la Recherche Prix from the Université Laval, Québec, Canada.

Bernard served as vice-president (1974–1976) and president (1976–1978) of AFCET. He was the president of EURO (1985–1986), after having served on the executive committee for several years. In 1975, he founded one of the most active and long-lasting working groups in OR, the EURO working group on MCDA.

THE EURO WORKING GROUP: MULTIPLE CRITERIA DECISION AIDING

EURO is a federation of the national European OR societies. The first EURO conference was held in Brussels in 1975. Bernard created the EURO working group on multiple criteria decision aiding (MCDA). The group, which usually meets twice a year, aims to promote original research on MCDA in Europe. The meetings of the group are not conferences. They are designed to foster discussions and exchanges. The group has around 350 members, from about 30 countries, and meetings usually gather between 50 and 100 persons. The success of the group is attested by the fact that most texts on MCDM now speak of a European school of MCDA (Roy and Vanderpooten 1996). The 69th meeting took place in Brussels, Belgium, April 2–3, 2009. More details on this working group can be found at http://www.inescc.pt/~ewgmcda/index.html (viewed December 24, 2009).

50th anniversary meeting of the MCDA working group convened at Château de Cerisy-La-Salle in 1999. Bernard is in the second row in the middle (wearing the lighter jacket in this row).

References

Benayoun R, Roy B, Sussmann G (1966) ELECTRE: Une méthode pour guider le choix en présence de points de vue multiples. Note de travail 49, SEMA (Metra International), Direction Scientifique)

Berge C (1958) Théorie des Graphes et ses Applications. Dunod, Paris

Bertier P, Roy B (1965) Une procédure de résolution pour une classe de problèmes pouvant avoir un caractère combinatoire. ICC Bull 4:19–28

Bouyssou D, Jacquet-Lagrèze É, Perny P, Slowinski R, Vanderpooten D, Vincke Ph (2002) (eds) Aiding decisions with multiple criteria: essays in honor of Bernard Roy. Kluwer, Boston, MA

Busacker R, Gowen P (1961) A procedure for determining a family of minimal-cost network flow patterns. Operations Research Office Technical Report 15, J. Hopkins University, Baltimore, MD

Churchman C, Ackoff R, Arnoff E (1957) Introduction to operations research. Wiley, New York, NY. French translation: Eléments de recherche opérationnelle. Dunod, Paris, (1961)

Figueira J, Mousseau V, Roy B (2005) ELECTRE methods. In: Figueira J, Greco S, Ehrgott M (eds) Multiple criteria decision analysis: state of the art surveys. Springer, Boston, MA, pp 133–162

Fondahl J (1961) A non-computer approach to the critical path method for the construction industry. Technical report 9, Department of Civil Engineering, Stanford University

Hansen P, de Werra D (2002) Connectivity, transitivity and chromaticity: the pioneering work of Bernard Roy in graph theory. In: Bouyssou D, Jacquet-Lagrèze E, Perny P, Slowinski R, Vanderpooten D, Vincke Ph. (eds) Aiding decisions with multiple criteria: essays in honor of Bernard Roy. Kluwer, Boston, MA, pp 23–42

LAMSADE (2009) http://www.lamsade.dauphine.fr/~roy/roy_publications.htm. Accessed 14 Sept, 2009

Lesourne J (1958) Techniques économiques et gestion industrielle, Dunod, Paris

Roy B (1957) Recherche d'un programme d'approvisionnement ou de production. Revue de Recherche Opérationnelle 1(4): 172–184

Roy B (1958) Sur quelques propriétés des graphes fortement connexes. Comptes rendus de l'Académie des Sciences 247:399–401

Roy B (1959a). Contribution de la théorie des graphes à l'étude de certains problèmes linéaires. Comptes rendus de l'Académie des Sciences 248:2437–2439

Roy B (1959b). Transitivité et connexité. Comptes rendus des séances de l'Académie des Sciences 249(6):216–218

Roy B (1961) Cheminement et connexité dans les graphes—Application aux problèmes d'ordonnancement. Doctorat d'État de Sciences Mathématiques, Faculté des Sciences de Paris

Roy B (1962) Graphes et ordonnancement. Revue Française de Recherche Opérationnelle (25/4e trimestre):323–333

Roy, B (1963) Programmation mathématique et description segmentée. Revue METRA 2(4):523–535

Roy B (1967) Nombre chromatique et plus longs chemins d'un graphe. RIRO 1(5):129–132

Roy B (1968) Classement et choix en présence de points de vue multiples (la méthode ELECTRE). RIRO 2(8):57–75

Roy B (1969a) Procédure d'exploration par séparation et évaluation (PSEP et PSES). RIRO 3(V-1):61–90

Roy B (1969b). Graphe partiel s-connexe extremum. Revue Roumaine de Mathématiques Pures et Appliquées 14(9):1355–1368

Roy B (1969c). Algèbre moderne et théorie des graphes orientées vers les sciences économiques et sociales: Volume 1: Notions et résultats fondamentaux. Dunod, Paris.

Roy B (1970) Algèbre moderne et théorie des graphes orientées vers les sciences économiques et sociales: Volume 2: Applications et problèmes spécifiques, Dunod, Paris

Roy B (1971) Problems and methods with mutiple objective functions, Math Program 1(2):239–266

Roy B (1975a). Vers une méthodologie générale d'aide à la décision. Revue METRA 14(3):459–497

Roy B (ed.) (1975b). Combinatorial programming: methods and applications. D. Reidel, Dordrecht

Roy B (1977) Partial preference analysis and decision-aid: the fuzzy outranking relation concept. In: Bell D, Keeney R, Raiffa H (eds) Conflicting objectives in decisions. Wiley, New York, NY, pp 40–75

Roy B. (1978) ELECTRE III: Un algorithme de classements fondé sur une représentation floue des préférences en présence de critères multiples. Cahiers du Centre d'études de Recherche Opérationnelle 20(1):3–24

Roy B (1981) The optimisation problem formulation: Criticism and overstepping. J Oper Res Soc 32(6):427–436

Roy B (1985) Méthodologie multicritère d'aide à la décision. Economica, Paris. (English translation: Multicriteria methodology for decision analysis. Kluwer Academic Publishers, 1996. Polish and Spanish translations are also available)

Roy B (1987) Meaning and validity of interactive procedures as tools for decision making. Eur J Oper Res 31(3):297–303

Roy B (1988) Des critères multiples en Recherche Opérationnelle: Pourquoi? In: Rand G (ed.) Operational research '87. Elsevier, Amsterdam, pp 829–842

Roy B (1989) Main sources of inaccurate determination, uncertainty and imprecision in decision models. Math Comput Model 12(10/11):1245–1254

Roy B (1990) Decision-aid and decision-making. Eur J Oper Res 45(2–3):324–331

Roy B (1991) The outranking approach and the foundations of ELECTRE methods. Theory Decis 31(1):49–73

Roy B (1993) Decision science or decision-aid science? Eur J Oper Res 66(2):184–203

Roy B (1998) A missing link in OR-DA: robustness analysis. Foundations Comput Decis Sci 23(3):141–160

Roy B (2006) Regard historique sur la place de la recherche opérationnelle et de l'aide à la décision en France. Mathématiques et Sciences Humaines 175:25–40

Roy B (2010) Robustness in operational research and decision aiding: a multi-faceted issue. Eur J Oper Res 200(3):629–638

Roy B, Bertier P (1973) La méthode ELECTRE II—Une application au média-planning. In: Ross M (ed.) OR '72. North-Holland, Amsterdam, 291–302

Roy B, Bouyssou D (1993) Aide multicritère à la décision: Méthodes et cas. Economica, Paris

Roy B, Dibon M (1966) L'ordonnancement par la méthode des potentiels—Le programme CONCORD. Automatisme 2:1–11

Roy B, Hugonnard J-C (1982) Ranking of suburban line extension projects on the Paris metro system by a multicriteria method. Trans Res 16A(4):301–312

Roy B, Simonnard M (1961) Nouvelle méthode permettant d'explorer un ensemble de possibilités et de déterminer un optimum. Revue Française de Recherche Opérationnelle (18/1e trimestre):15–54

Roy B, Vanderpooten D (1996) The European school of MCDA: Emergence, basic features and current works. J Multi Criteria Decis Anal 5(1):22–38

Roy B, Vincke P (1984) Relational systems of preference with one or more pseudo-criteria: some new concepts and results. Manage Sci 30(11):1323–1335

Warshall S (1962) A theorem on Boolean matrices. J ACM 9(1):11–12

43

Ronald A. Howard

James E. Matheson

Ron Howard is recognized as a pioneer in the fields of Markov decision processes and the general field of decision analysis (DA), a term that he introduced. In a series of papers published in the 1960s, Ron laid out the principles of applied decision theory and brought the techniques of practical DA to various audiences, especially operations research (OR) and engineering. He furthered the application of DA by combining his academic teaching objectives with the founding of the first DA management consulting organization.

Ron is a fellow of Institute for Operations Research and the Management Sciences (INFORMS) and the Institute of Electrical and Electronics Engineers and a member of the National Academy of Engineering. He served as president of The Institute of Management Sciences (TIMS) in 1967. He was the 1986 recipient of the Operations Research Society of America (ORSA) Frank P. Ramsey Medal for distinguished contributions in DA.

The Early Years

Ron was born on August 27, 1934 in Brooklyn, New York, the only child of William and Susan Howard. His parents met and became engaged in Belfast, Northern Ireland. William immigrated to the U.S. in 1926. Susan followed in 1927 and they married soon after. Both of Ron's grandfathers worked on the ill-fated Titanic. After it sank on its maiden voyage, his paternal grandfather returned to the shipyard to find metal punchings from the area of the ship hit by the iceberg and engraved and chrome plated them, thereby creating a keepsake that Ron prizes today.

A.A. Assad, S.I. Gass (eds.), *Profiles in Operations Research*, International Series in Operations Research & Management Science 147, DOI 10.1007/978-1-4419-6281-2_43,

During World War II, Ron grew up in Rosedale, Queens, in his parents' candy store. Ron's mother ran the candy store while his father worked in a Grumman aircraft factory. The family lived in one room behind the candy store, "but it was a big room" (Howard 2008). Besides living every child's dream of inexhaustible treats, he also was able to read all of the latest magazines and newspapers. Across the side street, he discovered the Rosedale public library, "where you didn't even have to pay" (Howard 2008). Ron devoured a portion of the collection, routinely reading as many as six books a week that ranged from how to make a battery to opera librettos. Ron was bored with school, Public School 138 Queens, and attended the minimum hours required to avoid being a truant. This experience set him off early on the path of independent learning. Ron realized in grade school that he would have to be responsible for his own education.

In 1946, Ron and his mother made a trip to Ireland, where he was tutored for his sixth grade and taught algebra by the local schoolmaster. On returning, he entered seventh grade in Rockville Center on Long Island. At the end of the school year his report card said, "promoted to high school," which began with the eighth grade. Then he moved to the next town, Lynbrook, where high school began at the ninth grade. When his mother took him to enroll and the administrator said, "welcome to high school," they both kept quiet (Howard 2008). Ron had just skipped the eighth grade.

The high spot of his Lynbrook high-school experience was chemistry and physics, where his teacher, Mr. Carr, encouraged Ron by giving him unlimited and unsupervised access to the laboratory after school. Ron now had a place to explore his scientific interests. Mr. Carr recommended that Ron go to the Massachusetts Institute of Technology (MIT). A 4-year Grumman scholarship enabled Ron to attend MIT, as well as providing 4 years of summer work at Grumman.

Ron entered MIT as an electrical engineering student. While he enjoyed pursuing this technical program, it seemed, after a while, too narrowly focused and lacked the people element. Looking for other options, Ron learned of a program in engineering and economics. By taking the equivalent of another year of course work, he would be able to get a bachelors' degree in engineering economics, as well. Adding this program enabled Ron to take courses in probability by Robert Solow and labor relations by George Schultz (future Nobel laureate and future U.S. Secretary of State, respectively).

By working extra hard, Ron was able to complete this ambitious program in only 4 years. MIT had a policy that any student could take an advanced standing exam in any course for five dollars. By passing this exam, a version of the course final, the student would receive full credit. Through these exams and course overloads, Ron graduated in 1955 with an S.B. in both electrical engineering and engineering economics.

Ron also received practical training through his student government activities. After 2 years of participation in the finance committee, which handled student activity funds, he served as its chairman in his senior year. Here, he learned and practiced accounting, as well as how to run effective business meetings and apply Robert's Rules of Order.

For the Want of a Transmission: An Eventful Transition

Late in his senior year, Ron began looking for a summer job. One day, he saw an MIT bulletin-board posting that an interesting California company would be interviewing in one hour. He hurried to an interview with Servomechanisms, Inc., which ended in an offer to visit the company in Los Angeles. Shortly thereafter, at spring break, Ron took his first domestic airplane trip for the interview. He was amazed to be treated like a member of the company for a few days, and he even worked in a day's visit to Tijuana, Mexico. This trip resulted in an offer for a summer job.

Ron then planned with his roommate and best friend from MIT, Carroll Barlow, to drive to Los Angeles, where Carroll also had a summer job. In early summer, they embarked on the trip in Ron's 1952 English Ford Zephyr. Life is punctuated by breakdowns, and this was one of those times—the car had a serious transmission breakdown in Pittsburgh. Parts had to be ordered from New York City and would take a week to arrive.

Carroll left after a few days to take an alternate job in Boston. Because the YMCA was too expensive, Ron moved into an inexpensive room over a suburban Pittsburgh bar. By the time the car was repaired, Ron decided to return to Massachusetts to take a job he had turned down, but that was still available, working on transistor magnetic power amplifiers for Raytheon Research Laboratory in Waltham.

According to Ron, "Everything would be different if that car hadn't broken down. Everything! It is probably the most important single event in my life" (Howard 2008). Because the breakdown took Ron back to Boston, he spent a lot of time with Polly Hathaway, a young woman he had been dating since his junior year. She was attending Simmons College and was working for the summer at a Boston hospital as a physical therapist, before completing the last half-year of her undergraduate work. They were married at the end of the summer. At Raytheon, Ron met William (Bill) Linvill, an MIT associate professor of electrical engineering, who was to have a great deal of influence on the direction Ron's future would take.

Graduate School

Initially, Ron did not have any ambition to pursue a graduate degree. But, as the time for a 1955–1956 school term decision approached, he applied for and obtained an MIT Schlumberger fellowship. Because MIT did not make a distinction between admission for master or doctoral degree candidates, Ron decided to go for the latter.

Ron continued to pursue electrical engineering and, at the same time, broaden his interests by taking courses in industrial engineering. At the end of his first year of graduate school, he received his S.M. degree in electrical engineering. Also, by the end of that school year, he completed all of the course work, but not the thesis, for the S.M. in industrial management. In addition, Ron acquired a third interest, the new field of OR.

Professor Philip Morse, a physicist who established and led the first OR group in the U.S. during World War II, was head of the new MIT interdepartmental Operations Research Center (ORC). Morse's wartime OR colleague and collaborator George Kimball, a chemical physicist from Columbia University, was also active in the ORC. Morse and Kimball had written the first book on OR, *Methods of Operations Research*, a few years earlier (Morse and Kimball 1951). Ron, with his strong interest in practical applications, was thus attracted to OR. In addition, he asked for and received permission to work at Raytheon for 2 days per week during the school year, where he would continue to associate with Bill Linvill outside the academic setting.

To support Ron's second graduate year, Bill suggested he apply for the prestigious Ramo–Wooldridge fellowship. Ron did not get the fellowship, but due to Bill's efforts, Ron did receive a fellowship supported by IBM. Bill also helped Ron find summer work with the Operations Research Group of Arthur D. Little, Inc. (ADL), a management consulting firm just down the street from MIT on Memorial Drive. There he met John Magee, a mathematician and first member of ADL's OR group (who later became president, chairman, and CEO of ADL), and George Kimball, who was recently appointed ADL's Scientific Advisor. Ron worked closely with George on ADL projects.

> A LESSON IN LIFE
>
> "While I was studying for my master's degree in electrical engineering at MIT, I took all the coursework required to obtain a masters degree in industrial management as well. Most of the courses were offered in a building that Alfred P. Sloan had donated to house the School of Industrial Management. One day, Mr. Sloan came to campus to dedicate the building; thereafter, the school would be known as the Sloan School. As part of the ceremony, he gave a brief talk in the beautiful marble lobby of the Sloan Building. At the conclusion of the talk, the floor was opened to questions from the assembled students. As I remember it, one student asked a question whose answer had a profound effect on my life. The question was, 'Mr. Sloan, your success in business is an inspiration to all of us. Was it worth it in terms of your personal life?' Mr. Sloan's reply was brief: 'No, my children are strangers. Next question.' At that point, I resolved never to make financial or professional success the main goal of my life" (Howard 2008).

In the 1950s, OR was expanding into business and industry from its birth in military applications, with ADL in the forefront. Once more, Ron had a great summer employment experience, followed by 2 days per week employment at ADL during the school year.

Bill Linvill encouraged Ron to apply again for the Ramo–Wooldridge fellowship for the following year, and this time he was successful. The fellowship included the opportunity to work at the Los Angeles offices of Ramo–Wooldridge during the summer of 1957. When he asked his superiors at ADL about the advisability of spending the summer in Los Angeles, rather than at ADL, they supported the plan. They wanted him to make the future choice of continuing with ADL to be based upon knowledge of, rather than ignorance of, alternative careers. Sage advice. Ron packed up his family and flew to Los Angeles with Polly and their young daughter, Kim. There he worked in the nascent OR department with Andrew Vazsonyi, a mathematician and OR enthusiast interested in industrial applications.

This was 1957, when computers were still in their infancy. Ramo–Wooldridge was interested in applying computers to the complex and tedious task of computing trucking rates. Ron spent 2 days with a freight specialist learning how to compute these rates using a five-foot shelf of tariff books. He was surprised when back at Ramo–Wooldridge he was regarded as a trucking rate expert. Told that he was now a "two-day expert," he realized a new concept: the degree of expertise in any aspect of life can be measured by the number of days of study (Howard 2008).

About the time Ron began looking for his doctoral dissertation topic, he went to a lecture by Stanislaus Ulam, a physicist who had worked on the design of the atom bomb. Ulam's lecture on Markov processes piqued Ron's interest in the field. At MIT, Bob Sittler (a student of Bill Linvill) had been applying systems analysis and graphic methods to Markov processes (Sittler 1956). Also, a recent paper by Richard Bellman, the inventor of dynamic programming (Bellman 1957a; Dreyfus 2002), had introduced dynamic programming for addressing decision situations governed by a Markov process (Bellman 1957b).

Ron applied a systems analysis approach and created typical problems to illustrate its power. An important question was how to determine long-run optimal policies. Ron was challenged by the problem of how to directly determine these long-run policies. His policy iteration method of solution became a primary result of his Sc.D. dissertation, "Studies in discrete dynamic programming" (Howard 1958). Later, Bellman (1961, p. 160) commented, "Of these methods of successive approximation, the most important one is due to R. Howard."

For Ron, this was an exciting period of study and research:

> I spent most of my spare time during the summer at Ramo-Wooldridge reading about Bellman's dynamic programming. I found the idea fascinating and wondered how to incorporate it in a research topic. When I returned to MIT and Arthur D. Little, I talked to George Kimball. We explored value iteration on a few states to see how it might work. Thinking back to the notion of Markov processes, I wondered if there was some way to formulate the idea of a system that evolved according to a Markov process, earned rewards

corresponding to a transition, and where both the transition probabilities and rewards for a given state could be chosen from many possible alternatives. For example, if there were 50 states and 50 different alternatives for a state, then there would be 50^{50} possible policies for controlling the system. Which of these would provide the highest average reward?

In my thesis [dissertation], I initially explored simulation and other approaches. Finally I was able to construct a policy iteration algorithm that guaranteed the highest average return for a single chain process. The algorithm was implemented on the MIT IBM 704 and 709 computers at a time when the inputs were prepared on punched cards. When I wrote the book, *Dynamic Programming and Markov Processes* [Howard 1960], a year later, I was able to extend the algorithm to the multiple chain case (Howard 2008).

Ron returned to MIT, wrote his dissertation, and received his Sc.D. in electrical engineering in 1958. Phil Morse was Ron's official dissertation advisor, with George Kimball his de facto advisor. At ADL, the Markov decision process model was used to determine an optimum policy for the distribution of Sears Roebuck catalogs (Howard 2002). Application was limited in the late 1950s, since few organizations had the data necessary to aid in assigning the state transition probabilities. The widespread current availability of digital data acquisition has removed that limitation.

Ron continued working in this area and, about 10 years later, published a two-volume

THE FROG AND THE LILY PAD

Ron's big insight about probabilistic systems came while writing the two volumes of *Dynamic Probabilistic Systems* (Howard 1971). When you think about steady state in a physical system, it is something you could see on an oscilloscope—you can watch the system itself settle down into steady state. But an informational steady state, such as the one arrived at by backwards induction, is only in your head. Ron likes the analogy of a frog jumping from lily pad to lily pad with probabilities that depend only on the pad the frog is currently resting on and the pad the frog might jump to next (the transition matrix). If you don't look at the frog for a long while, your belief on where the frog might currently be located is described by a vector of steady-state probabilities. However, the frog does not know that he is in steady state. He just jumps the same as always. Of course if you peek at the frog, your belief is no longer described by the steady-state probabilities, but it makes no difference to the frog.

set, *Dynamic Probabilistic Systems* (Howard 1971). Having a perverse sense of humor, Ron noted in the foreword that Jim Matheson (the author of this profile) is responsible for any mistakes in these books, rather than Ron, because Ron bought Jim a first-class dinner in New York in return for his proof reading the entire two-volume set!

Professorship and a Fateful Haircut

After graduation, Ron became assistant professor of electrical engineering and assistant professor of industrial management at MIT. This was an exciting time—renowned professors such as Claude Shannon and Norbert Wiener were colleagues. Ron's first teaching assignment was a course in circuit theory. Ernst Guillemin, the grand old man of circuit theory, gave the main lectures. Then, the young professors, like Ron, would teach smaller sections to supplement the lectures. This was the only course that Ron ever taught that he did not develop.

Ron was involved in the Department of Electrical Engineering and the Sloan School of Management, but he was most interested in the ORC, where he became Associate Director. In 1959, Ron and other ORC staff toured a few NATO countries—Portugal, Belgium, and France—delivering lectures on OR. In 1965 they made another tour that included Israel, India, and Japan, with each such visit contributing to international collegiality and professional development, and the spreading of the word about emerging OR advances and professional activities in the U.S.

Besides his MIT duties, Ron was also very busy in external activities. The Ford Foundation had funded an Institute of Basic Mathematics for Application to Business at Harvard, under the direction of the mathematician Howard Raiffa. Ron gave a one-day-a-week course on modern computation for the Institute. He also taught statistical decision theory and Markov

> TO TEACH OR NOT TO TEACH
>
> While at MIT, Ron faced a difficult job decision. One Friday, he was offered a corporate position as Director of Research, at a fantastic salary, for a company he served as a consultant. He had until Monday to make up his mind. Ron was then an assistant professor in a system without tenure track positions, and where only one in seven would be promoted to associate professor without tenure. He was living on one floor of a two-family house, where he had lived as a graduate student, with a wife and two kids to support. How tempting was the offer! But, by Monday, he realized that teaching was his calling; he turned the job down. Ron's decision model has been lost to history.

decision processes in a middle engineering management course for General Electric (GE) Company.

In 1961, Myron Tribus, Dean of Engineering at Dartmouth, sent Ron a manuscript on probability written by professor Edward Jaynes, a physicist at Washington University, St. Louis. Ron took it to read while getting a haircut. At first, he was skeptical about the manuscript's contents, but, by the time his haircut was finished, he came off the barber's chair transformed, mentally, as well as physically. Jaynes's theme was that there is no such thing as an objective probability: a probability reflects an individual person's knowledge (or equivalently ignorance) about some uncertain distinction. This understanding goes back to Laplace and maybe Bayes, yet somehow it was an idea that had been lost over time. Jaynes evolved his manuscript over the rest of his life—it was published 5 years after he died (Jaynes 2003). The manuscript had a profound influence on Ron; it caused him to take a new look at the then state of the art of decision theory in terms of Jaynes's view of probability. As Ron recalled,

> The time that I spent in the operations research group of Arthur D. Little was the heyday of operations research for business in the U.S. Some corporations had formed internal corporate operations research groups reporting to the highest level of executives. But by the mid-1960s this trend had been reversed. Operations research groups moved to lower and lower supporting roles in corporations; executives seldom consulted them on matters of strategic importance. One reason was that the application of operations research appeared to depend on the availability of operational data. Perhaps its name unwittingly constrained its field of practice. One of my mentors, Phil Morse, once said that you could [only] begin to do operations research when you had 30 observations of a process.
>
> I had noticed that the highest level of executive decision-making often dealt with decisions where there was little or no data to support the choice. I resolved to focus my attention on assisting executives with problems that kept them awake at night. The only philosophy that was consistent with this goal was decision theory. Unfortunately, decision theory was currently developed and taught in the realm of coin-tossing, dice, and ball-filled urns. Making it applicable to actual important decisions required supplementing it with systems engineering, and Jaynesian probability (Howard 2008).

California Hiatus: Decision Analysis

In the spring of 1964, Ron was invited by Stanford University's Operations Research Department to give a talk about his recent work on systems analysis of semi-Markov processes (Howard 1964). His MIT colleague and friend Bill Linvill had moved to Stanford and was now head of the Institute in Engineering Economic Systems. At Stanford, Bill was Ph.D. advisor for Jim Matheson and suggested that Jim attend Ron's talk and join Ron for lunch. Over a picnic table, Ron and Jim realized that they had similar backgrounds and views on applying decision theory. Jim's dissertation, "Optimum teaching procedures derived from mathematical learning models," used dynamic programming to optimize Markov processes with indirectly observable states (Matheson 1964), which was closely related to Ron's work. Since Jim was about to graduate with his Ph.D. that summer, Ron asked if he would like an offer from the MIT ORC. But, there was a complication. Bill had invited Ron to come to Stanford for a 1-year visit the following academic year. Without Ron at MIT, it seemed best that Jim return to his former employer, Westinghouse Research Laboratories in Pittsburgh. (Linvill had initiated an internship program for Stanford Ph.D. students with Westinghouse.) Jim's first assignment was to investigate scientific methods for managing Westinghouse's research and development projects. He arranged for Ron to be a consultant; this was the start of their lifelong collaboration.

During his visiting year at Stanford (1964–1965), Ron, besides continuing his research and teaching, became involved in consulting activities, most of which contributed to the development of his mainstream DA research. Also, through his connections with GE, he was asked to give a series of Monday night lectures on decision theory for the staff of its Nuclear Division in San Jose, California. There he was asked to consult on a decision of whether to put a superheater on GE boiling-water reactors systems. The decision was thought to depend primarily on the average lifetime of a new material, but Ron showed that the decision depended more fundamentally on the question of time preference for profit. Ron used this example in his first paper defining DA (Howard 1966, 66–70).

During the 1950s and early 1960s, the major ideas in DA were based on the expected utility model of John von Neumann and Oskar Morgenstern, the subjective expected utility model of Leonard Savage, and the work and methods of Ward Edwards (Shanteau et al. 1999). Ron's view of this field was first given in a paper at the International Federation of Operational Research Societies' fourth international conference on OR, held at MIT in 1965. It was published soon afterwards, a seminal paper in which the field of DA was first defined (Howard 1966, 56):

> Decision analysis is a logical procedure for the balancing of the factors that influence a decision. The procedure incorporates uncertainties, values, and preferences in a basic structure that models the decision. Typically, it includes technical, marketing, competitive, and environmental factors. The essence of the procedure is the construction of a structural model of the decision in a form suitable for computation and manipulation; the realization of this model is often a set of computer programs.

Ron's initial and subsequent DA research papers contributed greatly to the practice of DA in OR and engineering. His work helped bring DA to the forefront of academic management teaching and research, as well as helping to set the ideas of DA within management consulting organizations (Howard 1968).

CALIFORNIA PERMANENT

At the end of his visiting year at Stanford, Ron decided to accept Stanford's offer of a professorship in the new Engineering-Economic Systems Department that was headed by Bill Linvill. A drawback for Ron, coming from MIT, was the lack of consulting activities that were so beneficial to his research and that of his students, especially his association with ADL. Further, Bill Linvill advocated that all Ph.D. students take practical experience internships at companies having qualified mentors. Bill and Ron focused their attention on the nearest think tank, the Stanford Research Institute (SRI). At that time, SRI was closely associated with Stanford University, but there were few faculty connections. They approached SRI with success and the Joint Engineering-Economic Systems Program was established with seed funding from two SRI departments. Soon afterwards, Ron suggested to Jim Matheson that he take a consulting position with SRI, which he did in early 1966. About a year later, SRI created a Decision Analysis Group (DAG), headed by Jim, with Ron as the key academic member. In this role, Ron actively collaborated with Jim and the DAG members in marketing and conducting DA project work.

> A BARGAIN!
> "The overall aim of decision analysis is insight, not numbers. If the decision-maker does not feel that the analysis has captured his knowledge and concerns and that it has produced a course of action he believes in then the decision analyst has failed. But this is rarely the case. In a recent study the decision analyst presented his final conclusions to the entrepreneur who had hired him. At the conclusion of the presentation the decision analyst asked about the amount of written reporting that would be required. The entrepreneur replied, 'I believe the results of the analysis and I am going to act in accordance with the recommendation. Why should I pay more for a report?' " (Howard 1980a, 9).

The DAG was a pioneering DA research and consulting organization that provided a place where faculty and interns could participate in real-world projects. It also employed many Stanford graduates and allowed them and others to continue their interest in this new field. Ron was instrumental in developing early project work for DAG that often included Stanford graduate students. One of the first projects evolved from a talk Ron gave in Mexico, an ambitious project for the Mexican Electrical Commission to determine their policy toward nuclear power generation. The Commission sent four staff members to SRI for a year so they could participate in developing a state-of-the-art electrical system planning model that would help it analyze system expansion decisions, including nuclear generation (Matheson 1970). A project for GE was to develop a planning methodology for exploring Mars. This project created the then world's largest probabilistic modeling and decision tree program and set a new standard for space program planning (Matheson and Roths 1967). Other pioneering DA studies dealt with banking problems and decisions surrounding the sources and uses of funds, sponsored by Herb Ayers, Vice President of OR for the Morgan Guaranty Trust Company, and the analysis of the decision by National Oceanic and Atmospheric Administration to seed hurricanes to mitigate their destructive force. This study was sponsored by Myron Tribus, who had recently resigned as Dean of Engineering at Dartmouth to become Assistant Secretary of Science and Technology for the U.S. Department of Commerce. The report, "The decision to seed hurricanes," recommended hurricane seeding "... if minimizing the expected loss in terms of property damage (and the cost of seeding) is the only criterion..." and "that seeding be permitted on an emergency basis" (Howard et al. 1972, 1196, 1201). Because of the chaos surrounding President Nixon's resignation, Tribus left his position and further investigation of the decision to seed was not pursued. The paper has become one of the most reprinted and least used DA study. Descriptions of many other projects, papers, and analyses from this period are given in Howard and Matheson (1977, 1983).

Out of a body of work by SRI for the Department of Defense "to develop automated decision aids . . . for making policy decisions based on intelligence information" came the invention of influence diagrams in the early 1970s (Howard and Matheson 2005a, 144). "An influence diagram is a way of describing the dependencies among aleatory [chance] variables and decisions" and "is used to visualize the probabilistic dependencies on a decision analysis and to specify the states of information for which independencies can be assumed to exist" (Howard and Matheson 2005b, 130). Ron also applied the name knowledge map to an influence diagram having no decisions. The method's generality and importance as a tool for DA were soon recognized. Influence diagrams have been used in many fields with applications in business strategy, marketing, manufacturing, breast cancer screening, and technology transfer. An influence diagram processor is embedded in Microsoft Windows operating system as an aid to intelligent help (Howard and Matheson 2005a).

In 1981, Ron, Jim Matheson, and others from the SRI DAG formed a new management consulting firm, Strategic Decisions Group (SDG), with Ron as its Chairman. SDG grew quite rapidly, broadening applications of DA in the strategy area and conducting an extensive set of executive education courses. Between the original DAG and SDG, hundreds of decision analysts have been trained, thousands of executives have been familiarized with the subject, and hundreds of projects have been completed.

DECISION ANALYSIS EXPANSION

In the late 1970s, Ron developed an interest in life-and-death decisions, the use of DA for making safety decisions, and, in general, decisions involving a risk of death or serious bodily harm (Howard 1978, 1980b, 1984, 1989, 1999). Ron defined a micromort as a unit of risk measuring a one-in-a-million probability of death (from micro- and mortality), and a microprobability is a one-in-a million chance of some event; thus a micromort is the microprobability of death. He showed that undertaking small risks (say less than 100 micromorts) could be valued by assessing a monetary value per micromort and then multiplying this by the number of micromorts. This same micromort value applies when eliminating small risks. For large risks, however, he distinguished the buying and selling price of one's life. It does not make sense to talk about the value of life without distinguishing between what one would have to be paid to be killed and what would need to be paid to save one's life when faced with certain death. The answer to the first question would usually be infinite in the absence of any benefit to others; the answer to the second question would be limited by the total resources one can command (Howard 1980b).

THE HOBBYIST

Ron's chief hobbies are travel; camping in his giant motor home, where he often retreats to write professional papers; and sailing. Ron rode a motorcycle for 50 years. Ron is also a tool collector, who has more tools than most mechanics, and loves to use them. However, he prefers to repair with the simplest tools, such as his pocket knife. Below are Ron and Jim Matheson repairing a yacht somewhere on the Aegean Sea—with a little hammer.

Jim Matheson Ron Howard

Placing a value on a life-and-death outcomes permits making many safety decisions such as whether to take a beneficial drug with possible deadly side effects. Also, by including value explicitly in any decision problem, the implication of the differences among alternatives in monetary terms is made clear; one can analyze the value of costly actions before they are undertaken, such as acquiring further information. Ron has stated his concerns about the trend to use multiattribute approaches that make it impossible to deal with valuation issues, and to calculate such fundamental quantities as the value of information (Howard 2007).

Based on their success in applying DA to a variety of real-world management applications, Ron and colleagues began to explore how to extend DA to local community concerns, in particular, educating youths in the concepts and use of DA as an aid for decision making in their lives. This led to the formation of the nonprofit Decision Education Foundation (DEF) devoted to elementary and high-school DA education. The mission of the DEF is to improve the lives of young people by teaching them how to make better decisions (Decision Education Foundation 2008).

DEF has a small permanent staff, plus many volunteers who instruct teachers on how to impart decision skills to their students. Surprisingly, DA has been accepted in the English and history curricula, in addition to mathematics and sciences. Teachers are introducing the basic ideas even in kindergarten.

Over many years, Ron has been concerned about how society could benefit from the efficiencies and clear reasoning DA approaches provide. This led him naturally into the study of voluntary social systems, which Ron believes can produce more of what people want than can centralized or coercive systems. He has developed lecture materials and has taught classes in this area over the last three decades that come under the heading of "Designing a Free Society."

Recognizing that the application of DA (as well as other OR methodologies) can be applied and interpreted to support a bad end, Ron took an interest in professional ethics (Howard 1992, 2001). He developed the only ethics course in the Stanford School of Engineering and titled it "The Ethical Analyst." Ron's sage advice to his students is quite direct:

> Ethical difficulties are much easier to avoid than to resolve. You can avoid many ethical difficulties by following three practices: first, declining to be part of organizations that have ethical codes and behavior inconsistent with your own ethical code; second, avoiding participating in ethically objectionable activities; and, finally, treating all people as you would those you care about (Howard 2001, 82).

Ron co-authored a book on ethics with his doctoral student Clint Korver (Howard and Korver 2008). They term their view an engineering approach to ethics, and encourage each reader to develop their personal ethical code based on principles from DA, philosophy, and behavioral research. Their unique approach is summarized by the following:

> Instead of finding "good reasons" to compromise, we will find convincing reasons to take right action. Instead of shaving off pieces of our character with faulty thinking, we will think through our decisions skillfully and live more satisfying lives. When we reach the end of our lives, we will carry that much less burden, feel that much less remorse, and create that much more satisfaction from using ethics to make our life better (Howard and Korver 2008, 154).

Ron and Polly had four children, Kim Howard Saxe, David, John, and Robert, and six grandchildren. Polly died in 1997 after surviving cancer surgery for about a decade—Ron applied DA to help determine her course of treatment. In 2001, Ron married Joyce Mattea, who died suddenly in 2005.

In his 50-year career as a professor at MIT and Stanford, Ron has supervised the doctoral work of about 100 graduate students. They have gone on to distinguished careers in academia, industry, and consulting all over the world. Notwithstanding his professional career interest in decision making, Ron marvels at how much of our lives have been the result of events and encounters that were beyond our control (Howard 2008).

Honors and Awards

Ron is a fellow of INFORMS and the Institute of Electrical and Electronics Engineers and a member of the National Academy of Engineering. He served as president of TIMS in 1967. He was the 1986 recipient of ORSA's Frank P. Ramsey Medal for distinguished contributions in DA. He received the first INFORMS award for the Teaching of Operations Research/Management Science Practice (1998). Ron is an honorary member of the Omega Rho International Honor Society for Operations Research and Management Science; he presented the 1999 Omega Rho Distinguished Plenary Lecture at the INFORMS Cincinnati National Meeting.

References

Bellman R (1957a) Dynamic programming. Princeton University Press, Princeton, NJ

Bellman R (1957b) A Markovian decision process. J Math Mech 6(4):679–684

Bellman R (1961) Adaptive control processes: a guided tour. Princeton University Press, Princeton, NJ

Decision Education Foundation (2008) www.decisioneducation.org. Accessed 26 Dec 2009

Dreyfus S (2002) Richard Bellman on the birth of dynamic programming. Oper Res 50(1):48–51

Howard RA (1958) Studies in discrete dynamic programming. Sc.D. dissertation, MIT, Cambridge, MA

Howard RA (1960) Dynamic programming and Markov processes. Technology Press-Wiley, Cambridge, MA

Howard RA (1964) Systems analysis of semi-Markov processes. Trans IEEE Prof Group Mil Electron 8(2):114–124

Howard RA (1966) Decision analysis: applied decision theory. In: Hertz D, Melese J (eds) Proceedings of the fourth international conference on operational research. Wiley-Interscience, New York, NY, NY, pp 55–71

Howard RA (1968) The foundations of decision analysis. IEEE Trans Syst Sci Cybern 4(3):211–219

Howard RA (1971) Dynamic probabilistic systems (vols I and II). Wiley, New York, NY. Dover Publications reprint 2007

Howard RA (1978) Life and death decision analysis. In: Schechter R (ed.) Proceedings second Lawrence symposium on systems and decision sciences. Berkeley, CA, pp 271–277

Howard RA (1980a) An assessment of decision analysis. Oper Res 28(1):4–27

Howard RA (1980b) On making life and death decisions. In: Schwinger R, Albers W, Jr (eds) Societal risk assessment: how safe is safe enough? Plenum Press, New York, NY, pp 89–113

Howard RA (1984) On fates comparable to death. Manage Sci 30(4):407–422

Howard RS (1989) Micro risks for medical decision analysis. Int J Technol Assess Health Care 5(3):357–370

Howard RA (1992) Business ethics: tell the truth. J Manage Dev 11(4):4–10

Howard RA (1999) Life and death decisions for individuals and couples. In: Shanteau J, Mellers B, Schum D (eds) Decision science and technology: reflections on the contributions of Ward Edwards. Kluwer, Boston, MA, pp 227–254

Howard RA (2001) The ethical OR/MS professional. Interfaces 31(6): 69–82

Howard RA (2002) Comments on the origin and application of Markov decision processes. Oper Res 50(1):100–102

Howard RA (2007) The foundations of decision analysis revisited. In: Edwards W, Miles R, Jr, von Winterfed D. (eds) Advances in decision analysis: from foundations to applications. Cambridge University Press, Cambridge, England, pp 32–56

Howard RA (2008) Personal communication

Howard RA, Korver C (2008) Ethics for the real world: creating a personal code to guide decisions in work and life. Harvard Business Press, Cambridge, MA

Howard RA, Matheson J (eds) (1977) Readings in decision analysis. Decision Analysis Group, SRI International, Menlo Park, CA

Howard RA, Matheson J (eds) (1983) Readings on the principles and applications of decision analysis, vols I and II. Strategic Decisions Group, Menlo Park, CA

Howard RA, Matheson J (2005a) Influence diagram retrospective. Decis Anal 2(3):144–147

Howard RA, Matheson J (2005b) Influence diagrams. Decis Anal 2(3):127–143

Howard RA, Matheson J, North D (1972) The decision to seed hurricanes. Science 176(4040):1191–1202

Jaynes E (2003) Probability theory: the logic of science. Cambridge University Press, Cambridge, England

Matheson J (1964) Optimum teaching procedures derived from mathematical learning models, Ph.D. dissertation, Stanford University, Stanford, CA

Matheson J (1970) Decision analysis practice: examples and insights. In: Lawrence J (ed.) Proceedings of the fifth international conference on operational research, Venice, 1969. Tavistock Publications, London, England, pp. 677–691

Matheson J, Roths W (1967) Decision analysis of space projects: Voyager Mars. In: Howard RA, Matheson J (eds) 1983. Readings on the principles and applications of decision analysis, vols I and II. Strategic Decisions Group, Menlo Park, CA, 445–475

Morse PM, Kimball G (1951) Methods of operations research. Technology Press of MIT, Cambridge, MA; Wiley, New York, NY

Shanteau J, Mellers B, Schum D (eds) (1999) Decision science and technology: reflections on the contributions of Ward Edwards. Kluwer, Boston, MA

Sittler RW (1956) Systems analysis of discrete Markov processes. IRE Trans Circ Theory 3(4):257–266

Appendix

Table 1 Profiles background

Chapter	Name	Authors	Date of Birth	Birthplace	Discipline-Highest Degree	Year Earned
1	Blackett, Patrick*	Maurice W. Kirby and Johnathan Rosenhead	November 18, 1897	London, U.K.	Physics	1921
2	Vajda, Steven*	Jakob Krarup	August 20, 1901	Budapest, Hungary	Mathematics	1925
3	Morse, Philip McCord*	Robert M. Oliver	August 6, 1903	Shreveport, Louisiana	Physics	1929
4	Von Neumann, John*	Saul I. Gass	December 28, 1903	Budapest, Hungary	Mathematics	1926
5	Goodeve, Charles Frederick*	Maurice W. Kirby	February 21, 1904	Neepawa, Canada	Chemistry	1932
6	Tucker, Albert W.*	Saul I. Gass	November 28, 1905	Ontario, Canada	Mathematics	1933
7	Steinhardt, Jacinto *	Saul I. Gass	May 20, 1906	New York City	Biophysics	1934
8	Kimball, George E.*	John F. Magee	July 12, 1906	Chicago, Illinois	Chemistry	1932
9	Johnson, Ellis A.*	Saul I. Gass	September 2, 1906	Quincy, Massachusetts	Electrical Engineering	1947

(*Continued*)

A.A. Assad, S.I. Gass (eds.), *Profiles in Operations Research*, International Series in Operations Research & Management Science 147, DOI 10.1007/978-1-4419-6281-2,

Table 1 (Continued)

Chapter	Name	Authors	Date of Birth	Birthplace	Discipline-Highest Degree	Year Earned
10	Kantorovich, Leonid Vital'Evich*	Saul I. Gass and Johnathan Rosenhead	January 19, 1912	St. Petersburg, Russia	Mathematics	1935
11	Churchman, C. West*	Arjang A. Assad	August 29, 1913	Mt. Airy, Pennsylvania	Philosophy	1938
12	Cooper, William W.	Timothy W. Ruefli and Robert R. Wiggins	July 23, 1914	Birmingham, Alabama	Economics	1938
13	Dantzig, George B.*	Saul I. Gass	November 8, 1914	Portland, Oregon	Mathematics	1946
14	Simon, Herbert A.*	Arjang A. Assad	June 15, 1916	Milwaukee, Wisconsin	Political Science	1942
15	Vazsonyi, Andrew*	Nancy C. Weida	November 4, 1916	Budapest, Hungary	Mathematics	1938
16	Geisler, Murray Aaron*	Perkins C. Pedrick	March 23, 1917	Brooklyn, New York	Statistics	1960
17	Miser, Hugh Jordan*	Michael H. Rothkopf	May 23, 1917	Fayetteville, Arkansas	Mathematics	1946
18	Charnes, Abraham*	Fred Y. Phillips and Lawrence M. Seiford	September 4, 1917	Hopewell, Virginia	Mathematics	1947
19	Kozmetsky, George*	Lawrence Secrest, David V. Gibson and John S. Butler	October 5, 1917	Seattle, Washington	Business	1950
20	Forrester, Jay Wright	David C. Lane and John D. Sterman	July 14, 1918	Climax, Nebraska	Electrical Engineering	1945

21	Ackoff, Russell Lincoln*	Maurice W. Kirby and Jonathan Rosenhead	February 12, 1919	Philadelphia, Pennsylvania	Philosophy	1946
22	Hertz, David Bendel	Edward K. Baker and Robert T. Plant	March 25, 1919	Yoakum, Texas	Engineering	1949
23	Bellman, Richard E.*	Arjang A. Assad	August 26, 1920	Brooklyn, New York	Mathematics	1946
24	Balas, Egon	Graham K. Rand	June 7, 1922	Cluj, Romania	Economics/ Mathematics	1967
25	Arnoff, E. Leonard*	David F. Rogers	October 15, 1922	Cleveland, Ohio	Mathematics	1952
26	Rivett, Berwyn Hugh Patrick*	Graham K. Rand	April 2, 1923	Oswestry, U.K.	Mathematics	1947
27	Raiffa, Howard	Ralph L. Keeney	January 24, 1924	New York City	Mathematics	1951
28	Fulkerson, D. Ray*	Robert G. Bland and James B. Orlin	August 14, 1924	Tamms, Illinois	Mathematics	1951
29	Kuhn, Harold W.	Saul I. Gass and Guillermo Owen	July 29, 1925	Santa Monica, California.	Mathematics	1950
30	Gass, Saul I.	Arjang A. Assad	February 28, 1926	Chelsea, Massachusetts	Engineering Science	1965
31	Saaty, Thomas L.	Luis G. Vargas	July 18, 1926	Mosul, Iraq	Mathematics	1953
32	Beer, Anthony Stafford*	Jonathan Rosenhead	September 25, 1926	London, U.K.	Cybernetics	*
33	Magee, John F.	Saul I. Gass	December 3, 1926	Bangor, Maine	Mathematics/ Economics	1953

(*Continued*)

Table 1 (Continued)

Chapter	Name	Authors	Date of Birth	Birthplace	Discipline-Highest Degree	Year Earned
34	Wolfe, Philip Starr	Alan J. Hoffman	August 11, 1927	San Francisco, California	Mathematics	1954
35	Markowitz, Harry	John B. Guerard, Jr.	August 24, 1927	Chicago, Illinois	Economics	1954
36	Little, John D. C.	John R. Hauser and Glen L. Urban	February 1, 1928	Boston, Massachusetts	Operations Research/ Physics	1955
37	Beale, Evelyn Martin Lansdowne*	John A. Tomlin	September 8, 1928	Middlesex, UK	Mathematical Statistics	1950
38	Gomory, Ralph E.	Ellis L. Johnson	May 7, 1929	New York City	Mathematics	1954
39	Blumstein, Alfred	Daniel S. Nagin	June 3, 1930	New York City	Operations Research	1960
40	Wagner, Harvey M.	Shaler Stidham, Jr.	November 20, 1931	San Francisco, CA	Economics	1960
41	Bonder, Seth	W. Peter Cherry	July 14, 1932	New York City	Industrial Engineering	1965
42	Roy, Bernard	Denis Bouyssou and Daniel Vanderpooten	March 15, 1934	Moulins-sur-Allier, France	Mathematics	1961
43	Howard, Ronald A.	James E. Matheson	August 27, 1934	Brooklyn, New York	Electrical Engineering	1958
	* Deceased					* No college degree

Table 2 Profiles honors and awards

President OR Societies*	**Nobel Prize**	**Lanchester Prize***	**National Scientific Societies***	**U.S. Presidential Medals***
Ackoff – O 1956	Blackett (Physics) 1948	Beer 1966	Balas – AE 2006	Dantzig – S 1975
Arnoff – T 1968-69	Kantorovich (Economics) 1975	Gomory 1963	Bellman – AE 1977, AS 1983	Forrester – T 1989
Beer – ORS 1970-71	Markowitz (Economics) 1990	Morse 1968	Blackett – RS 1933	Goodeve – F 1946
Blumstein – O 1977, T 1987-1988, I 1996	Simon (Economics) 1978	Raiffa 1976	Blumstein – AE 1998	Kozmetsky – T 1993
Bonder – O 1978		Wagner 1969	Bonder – AE 2000	Simon – S 1986
Charnes –T 1960	**Von Neumann Prize***		Dantzig – AS 1971, AE 1985	Von Neumann – F 1956
Churchman – T 1962	Balas 1995	* U.S. ORSA/TIMS/ INFORMS	Forrester – AE 1967	
Cooper – T 1954	Bellman 1976	Awarded since 1954 for best English language Publication in OR.	Goodeve – RS 1940	*Awards are given for science (S) or technology (T) or freedom (F).
Dantzig – T 1966	Charnes 1982		Gomory – AS 1972, AE 1975	
Gass – O 1976	Cooper 1982	**Kimball Medal***	Howard – AE 2005	**Medal of Merit**
Geisler – T 1961	Dantzig 1975	Ackoff 1975	Kimball – AS 1954	Blackett (U.S.) 1946*
Hertz – T 1964, O 1974	Gomory 1984	Blumstein 1985	Little – AE 1989	Blackett (U.K.) 1967**
Howard – T 1967	Kuhn 1980	Bonder 1993	Morse – AS 1955, AE 1985	Morse (U.S.) 1946*
Kimball – O 1964	Markowitz 1989	Gass 1991	Raiffa – AE 2005	Von Neumann (U.S.)1947*
Kozmetsky– T 1958	Simon 1988	Hertz 1981	Saaty – AE 2005	
Little – O 1979, T 1984-1985, I 1995	Tucker 1980	Little 1987	Simon – AS 1967	*Awarded to civilians of the U.S. and its allies for exceptionally meritorious service or courageous acts during WWII.
Magee – O 1966, T 1971-1972	Wolfe 1992	Magee 1978	Von Neumann – AS 1937	
Miser – O1962	* U.S. ORSA/TIMS/ INFORMS	Miser 1975		**Order of Merit awarded by the British sovereign
Morse – O 1952	Awarded since 1975; recognizes scholars	Morse 1974	*Member U.S. National Academy of Engineering (AE)	
		* U.S. ORSA/TIMS/ INFORMS		

Rivett – ORS 1962-63 Steinhardt – O 1954 Roy – A 1976-78 Vazsonyi – T** 1954 Wagner – T 1973-74 *A – AFCET, French OR Society *ORS– U.K. OR Society *I–INFORMS *O–ORSA *T–TIMS **First past-president, appointed.	who have made fundamental theoretical contributions to OR and MS. **Larnder Prize (CORS)*** Charnes 1989 Dantzig 1997 Gomory 2006 Miser 1990 Rivett 1987 Wagner 1988 *Canadian OR Society Awarded since 1986 for achieving international distinction in OR.	Awarded since 1974 for distinguished Service to society and the profession. **INFORMS Impact Prize*** Cooper 2006 Charnes 2006 Saaty 2008 *Awarded since 2004 in alternate years; recognizes contributions that have had a broad impact on OR and MS.	*Member U.S. National Academy of Sciences (AS) *Fellow U.K. Royal Society (RS) **Silver Medal ORS*** Ackoff 1971 Beale 1980 Dantzig 1986 Goodeve 1964 Kantorovich 1986 Morse 1965 Rivett 1968 *The OR Society's most prestigious award.	for distinguished service in the armed forces, science, art, literature.

Authors' Biographies

Arjang A. Assad

Profiles: Richard E. Bellman, C. West Churchman, Saul I. Gass, Herbert A. Simon

Arjang is Professor and Dean at the School of Management of the University at Buffalo, SUNY. He earned his Ph.D. in management science from the MIT Sloan School of Management under the guidance of Thomas L. Magnanti. He received his B.S. in mathematics, his Chemical Engineer's Degree, and his M.S. in operations research, all from MIT. Arjang was a member of the faculty of the Robert H. Smith School of Business, University of Maryland, College Park, from 1978 to 2008, where he also served as Senior Associate Dean, Chairperson of the Decision and Information Technologies Department, and Director of the IBM Total Quality/QUEST Program. He received the 1996 Maryland Association for Higher Education award for innovation and the 2002 Kirwan Undergraduate Education Award, as well as several teaching awards. He chaired the History and Traditions Committee of INFORMS during 1996–2004. He has published over 50 scholarly papers in routing and distribution, operations management, network modeling and optimization, and quality management. His co-authored books include *Vehicle Routing: Methods and Studies*, *Excellence in Management Science Practice*, *An Annotated Timeline of Operations Research: An Informal History*, and *Profiles in Operations Research.*

EDWARD K. BAKER

Profile: David Bendel Hertz (joint with Robert T. Plant)

Ed is Professor and Chair of the Management Science Department in the School of Business at the University of Miami. He received his B.E.S and M.S. from the Johns Hopkins University and a D.B.A in Management Science and Statistics from the University of Maryland. His early work in operations research examined the airline crew scheduling problem of the Federal Express Corporation and he has maintained a continuing interest in transportation problems. He has been the principal investigator of research projects sponsored by the U.S. Department of Transportation, The Urban Mass Transit Administration, and Ryder Dedicated Logistics. He is also affiliated with the Boating Research Center at the marine school at the University of Miami. He has led studies that have included simulations of marine container port operations, evaluation of hurricane evacuation plans, and simulations of marine traffic patterns to identify manatee protection zones in Florida and to evaluate marine reserves proposed for the Florida Keys. His publications have appeared in *Management Science*, *Operations Research*, *IIE Transactions*, and the *European Journal of Operations Research*. He has been active in various professional organizations and is currently President of the South Florida Chapter of INFORMS.

ROBERT G. BLAND

Profile: D. Ray Fulkerson (joint with James B. Orlin)

Bob is Professor, School of Operations Research & Information Engineering (OR&IE) and Center for Applied Mathematics, Cornell University, Ithaca, New York. Bob received his B.S., M.S., and Ph.D. in OR at Cornell. His Ph.D. dissertation was supervised by Ray Fulkerson. He was a research fellow at CORE (Leuven, Belgium), a Sloan Foundation Research Fellow, a visiting professor at the European Institute for Advanced Studies in Management (Brussels) and the University of British Columbia, and an assistant professor of mathematical sciences at SUNY-Binghamton (now Binghamton University). He has served as Director of the School of OR&IE at Cornell, as well as Director of Graduate Studies and Director of Undergraduate Studies. Bob is one of the co-developers of the theory of oriented matroids; his work in that vein was motivated by Fulkerson's work on frames of vector spaces, and generated combinatorial pivot selection rules that prevent cycling in the simplex method.

Denis Bouyssou

Profile: Bernard Roy (joint with Daniel Vanderpooten)

Denis Bouyssou has an MBA from ESSEC (Paris, France). He obtained his *Doctorat* (Ph.D.) in OR in 1984 and his *Habilitation à Diriger des Recherches* in 1990 from the Université Paris Dauphine (Paris, France). He is a Senior Researcher (*Directeur de Recherche*) at the Centre National de la Recherche Scientifique in Paris, France. From 1987 to 2001 he was professor of decision science at the ESSEC Business School, Paris, France. He has held invited research positions at the Université Libre de Bruxelles (Brussels, Belgium) and Université Laval (Québec, Canada). He has been a consultant for various private and public firms in France. Denis is a former president of ROADEF, the French OR Society, and former secretary of EURO (the European association of OR societies). He is the co-editor-in-chief of *4OR*. He is a member of the editorial committee of *Theory and Decision*, advisory editor of the *Journal of Multi-Criteria Decision Analysis*, and on the editorial board of *Risk, Decision and Policy*. Denis is the co-author of four books on Multicriteria Decision Making (MCDM). He has published over 50 articles in various journals and contributed volumes. His main current research interests are centered on Decision Theory (Preference modeling, Decision under Risk and Uncertainty, Social Choice Theory and MCDM).

John Sibley Butler

Profile: George Kozmetsky (with David V. Gibson and Lawrence Secrest)

John is Director of the IC^2 Institute and the Herb Kelleher Center at the University of Texas at Austin. A professor of management and sociology, he had a close relationship with Professor Kozmetsky, who asked him to continue the tradition of innovation and creativity by taking the leadership of the IC^2 Institute. His research areas are organizational science, with special emphasis on military and entrepreneurial organizations. John has also taught in the MBA programs in Mexico, China, and Japan. He has published extensively in professional journals. His books include *All That We Can Be: Black Leadership the Army Way* (with Charles C. Moskos), *Immigrant and Minority Entrepreneurship: The Continuous Rebirth of American Society* (with George Kozmetsky) *Entrepreneurship and Self-Help Among Black-Americans: A Reconsideration of Race and Economics,* and *An American Story: Mexican American Entrepreneurship and Wealth Creation* (with Alfonso Morales and David L. Torres). President George W. Bush appointed John to the William Fulbright Foreign Scholarship Board. He received his undergraduate degree from Louisiana State University in Baton Rouge and the Ph.D. from Northwestern University in Evanston, Illinois.

W. Peter Cherry

Profile: Seth Bonder

Peter Cherry is a consultant living in Ann Arbor, Michigan. Previously, he was a Vice President of Science Applications International Corporation and Chief Analyst for the Lead Systems Integrator of the U.S. Army's Future Combat Systems Program. He received a B.A. (Hon.) in Mathematics from the University of New Brunswick, a M.Sc. in Mathematics from the University of Toronto and M.Sc. and Ph.D. from the University of Michigan. He began his career in OR in the Directorate of Maritime Operational Research, Department of National Defense in Canada, where the director was Dr. George Lindsay, a member of the first OR teams in the U.K. during WWII. He joined Seth Bonder at Michigan in 1968 as a student and member of the Systems Research Laboratory, completing his Ph.D. in 1972 under the direction of Ralph Disney. He then began a career at Vector Research Inc. (VRI), collaborating closely with Seth on defense issues and Military OR at VRI until its sale to Altarum. Peter was a member of the Army Science Board and was elected to the National Academy of Engineering in 2006. He is married to Dr. Carol Loveland-Cherry, a professor in the School of Nursing at the University of Michigan.

Saul I. Gass

Profiles: George B. Dantzig, Harold W. Kuhn (joint with Guillermo Owen), Ellis A. Johnson, Leonid Vital'evich Kantorovich (joint with Jonathan Rosenhead), John Magee, Jacinto Steinhardt, Albert W. Tucker, John von Neumann.

Saul is Professor Emeritus, Robert H. Smith School of Business, University of Maryland, College Park. He received his B.S. in education/mathematics and M.A. in mathematics from Boston University. At the University of California, Berkeley, he received his Ph.D. in engineering science/operations research under the guidance of George B. Dantzig. He was a member of the U.S. Air Force's Project SCOOP, where Dantzig and others developed the basic concepts and early applications of linear programming; manager of IBM's Project Mercury computational and data flow equipment system; senior vice-president of the consulting group, World Systems Laboratories; vice president and director of consulting and research for the Washington area office of Mathematica; and professor and chair of management science and statistics, University of Maryland. He was president of the Operations Research Society of America. Included in his publications are the books *Linear Programming: Methods and Applications*, *An Illustrated Guide to Linear Programming*, and *Decision Making, Models and Applications*. He is co-author of the *Encyclopedia of Operations Research and Management Science*, *An Annotated Timeline of Operations Research: An Informal History*, and *Profiles in Operations Research*.

DAVID V. GIBSON

Profile: George Kozmetsky (joint with John Sibley Butler and Lawrence Secrest)

Dave Gibson is Associate Director and the Nadya Kozmetsky Scott Centennial Fellow, IC^2 (Innovation, Creativity, Capital) Institute, the University of Texas at Austin (www.ic2.utexas.edu). In 1983, he earned a Ph.D. from Stanford University in Sociology with a concentration in organizational behavior and communication theory. Dave was hired at the Graduate School of Business, the University of Texas at Austin by Dean Kozmetsky. He followed Kozmetsky to the IC^2 Institute and over the next 20 plus years benefited from Kozmetsky's inspiration and creative and innovative "think and do" orientation to life. By being with Kozmetsky, Dave was fortunate to meet and work with a broad range of amazing individuals across academic and professional disciplines at home and abroad. His research and publications are grounded in the inspiration of Dr. Kozmetsky, focusing on technology/knowledge transfer and use; cross-cultural communication and management; and the development and impact of regional technology centers worldwide. Dave is author and editor of 11 books. His journal publications and papers have been translated into Mandarin, Japanese, Korean, Russian, Spanish, Italian, French, German, Finnish, and Portuguese. He is chair of the *International Conference on Technology Policy and Innovation.*

JOHN B. GUERARD, JR.

Profile: Harry Markowitz

John is Director of Quantitative Research at McKinley Capital Management, LLC, in Anchorage, Alaska. He earned his A.B. in Economics from Duke University, M.A. in Economics from the University of Virginia, MSIM from the Georgia Institute of Technology, and Ph.D. in Finance from the University of Texas, Austin. John taught at the University of Virginia, Lehigh University, Rutgers University, and as an adjunct faculty member at the International University of Monaco and the University of Pennsylvania. He worked with the DAIS Group at Drexel, Burnham, Lambert; Daiwa Securities Trust Company; Vantage Global Advisors; and served on the Virtual Research team at GlobeFlex Capital. John co-managed a Japanese equity portfolio with Harry Markowitz at Daiwa Securities Trust Company. John has published several monographs, including *Corporate Financial Policy and R&D Management* (Wiley, 2006, second edition), *Quantitative Corporate Finance* (Springer, 2007, with Eli Schwartz) and *The Handbook of Portfolio Construction: Contemporary Applications of Markowitz Techniques* (Springer, 2009). John serves as an associate editor of the *Journal of Investing* and *The International Journal of Forecasting*. He has published research in *The International Journal of Forecasting, Management Science, The Journal of Forecasting, Journal of Investing, and Research in Finance.*

JOHN R. HAUSER

Profile: John D. C. Little (joint with Glen L. Urban)

John Hauser is the Kirin Professor of Marketing and former Head of the Management Science Area at MIT's Sloan School of Management where he teaches new product development, marketing management, competitive marketing strategy, and research methodology. He is the co-author of two textbooks, *Design and Marketing of New Products* and *Essentials of New Product Management*, and a former editor of *Marketing Science*. He has published over 70 scientific papers. Among his awards include the American Marketing Association's (AMA) Converse Award for outstanding contributions to marketing scholarship, and the joint AMA and AMA Foundation's Parlin Award for outstanding contributions to the field of marketing research. He has consulted for a variety of corporations on product development, sales forecasting, marketing research, voice of the customer, defensive strategy, and R&D management. He is a founder and principal at Applied Marketing Science, Inc.; a trustee of the Marketing Science Institute; on the advisory board of Affinnova, Inc.; an inaugural fellow of the INFORMS Society of Marketing Science; a fellow of INFORMS, and serves on many editorial boards. He is a student of John Little and a long-time colleague at MIT.

ALAN J. HOFFMAN

Profile: Philip Starr Wolfe

Alan is an IBM Fellow Emeritus. He served in the U.S. Army (1943–1946); attended Columbia University, A.B. (1947), Ph.D. in mathematics (1950). He was a Member of the Institute for Advanced Study, Princeton (1950–1951); Mathematician, National Bureau of Standards, Washington, D.C. (1951–1956); Scientific Liaison Officer, Office of Naval Research, London (1956–1957); Consultant, Management Consultation Services, General Electric Company, New York (1957–1961); Research Staff Member, T. J. Watson Research Center, IBM, Yorktown Heights (1961–2002); IBM Fellow (1978). He served as an adjunct or visiting professor at many venues; advisor of 15 Ph.D. students. Honors and awards include: member, National Academy of Sciences (1982); fellow of the New York Academy of Sciences (1975), American Academy of Arts and Sciences (1987), and INFORMS (2002). Honorary D. Sc. degree (1988) from the Technion, Israel Institute of Technology; awarded the ORSA/TIMS John von Neumann Prize (1992) jointly with Philip Wolfe. His research interests include the interplay between linear programming, linear algebra, and combinatorics. In linear programming, Alan is best known for constructing the first example of cycling in the simplex method, the first computer comparison of computer solution methods, and the Lipschitz condition for linear inequalities (Hoffman lemma). He is also known as a pioneer in polyhedral combinatorics (circulation theorem, totally unimodular matrices, total dual integrality, lattice polyhedra), and studies in greedy algorithms (Monge sequences). Among the chapter authors, Alan has the distinction of being a co-author with seven of the profilees. Autobiographical notes, pictures, and photocopies of some of his papers are in *Selected Papers of Alan Hoffman*, edited by C. Micchelli, World Scientific Press, New Jersey, 2003.

ELLIS L. JOHNSON

Profile: Ralph Gomory

Ellis Johnson is the Coca-Cola Chaired Professor in the School of Industrial and Systems Engineering, Georgia Institute of Technology. He received a B.A. in mathematics at Georgia Tech and a Ph.D. in OR at the University of California, Berkeley. Before joining Georgia Tech in 1995, he was at IBM's T. J. Watson Research Center for 26 years. There, he founded and managed the Optimization Center from 1982 until 1990, when he was named IBM Corporate Fellow. In 1980–1981, he was at the University of Bonn, Germany, as recipient of the Alexander Von Humboldt Senior Scientist Award. In 1985, he received the Mathematical Programming Society's George Dantzig Award for his research in mathematical programming. He was awarded ORSA's 1983 Lanchester Prize jointly with Harlan Crowder and Manfred Padberg. In 1988, he was elected to the National Academy of Engineering. In 2000, Dr. Johnson was the co-recipient with Manfred Padberg of INFORMS' John Von Neumann Theory Prize. From 1990 to 1995, he began teaching and conducting research at Georgia Tech, where he co-founded and co-directed the Logistics Engineering Center with George Nemhauser. His research interests in logistics include crew scheduling and real-time repair, fleet assignment and routing, distribution planning, network problems, and combinatorial optimization.

RALPH L. KEENEY

Profile: Howard Raiffa

Professor Keeney is a Research Professor of Decision Sciences at the Fuqua School of Business, Duke University. He has a B.S. in Engineering from UCLA, an M.S and an E.E. in Electrical Engineering from MIT, and a

Ph.D. in Operations Research from MIT. Prior to joining the Duke faculty, Professor Keeney was a faculty member in Management and in Engineering at MIT and at the University of Southern California, a Research Scholar at the International Institute for Applied Systems Analysis in Austria, and the founder of the decision and risk analysis group of a large geotechnical and environmental consulting firm. His professional interests concern structuring decisions, creating innovative alternatives, modeling decisions involving multiple objectives, and applying decision analysis to complex corporate and government problems. Professor Keeney was a doctoral student of Professor Raiffa, with whom he is a co-author of two books: *Decisions with Multiple Objectives* (reprinted by Cambridge University Press, 1993), which won the INFORMS Lanchester Prize, and *Smart Choices: A Practical Guide to Making Better Decisions* (Harvard Business School Press, 1999), also with John S. Hammond, which received the Decision Analysis Society Best Publication Award. It has been translated into 16 languages. Dr. Keeney was awarded the Ramsey Medal for Distinguished Contributions in Decision Analysis by the Decision Analysis Society and is a member of the National Academy of Engineering. He was awarded the 2009 INFORMS President's Award for his substantial contributions to methods for analyzing complex decisions faced by government and society, and for promoting valued-focused decision making by individuals and businesses.

MAURICE W. KIRBY

Profiles: Russell Lincoln Ackoff (with Jonathan Rosenhead), Patrick Blackett (with Jonathan Rosenhead), Charles Goodeve

Maurice Kirby is Professor Emeritus of Economic History in the Management School, Lancaster University. He obtained his B.A. in Economics from the University of Newcastle Upon Tyne and his doctorate in economic history from the University of Sheffield. He was elected a fellow of the Royal Historical Society in 1985. Included in his publications are the books *The British Coalmining Industry: A Political and Economic History, The Decline of British Economic Power since 1870, Men of Business and Politics,* and *The Origins of Railway Enterprise.* Latterly, he has published extensively on the history of operations research following a commission from the U.K. Operational Research Society. His monograph, *Operational Research in War and Peace: the British Experience from the 1930s to 1970* was published in 2003.

Jakob Krarup

Profile: Steven Vajda

Jakob is Professor Emeritus, DIKU (Department of Computer Science, University of Copenhagen). He received his M.Sc. in electrical engineering/applied mathematics and his Ph.D. in operations research (OR) from the Technical University of Denmark. His D.Sc. degree in OR was earned at University of Copenhagen. In addition, he was awarded an honorary degree by the Warsaw Business School. Jakob was among the lucky few who experienced the birth of computer science and applied OR in Denmark; he continued with consultancy in Denmark and abroad until he joined DIKU on a full-time basis in 1975 with focus on general OR, algorithmics and combinatorial optimization. A main parallel activity throughout was the promotion of OR, notably within the frameworks of NATO, IFORS, and EURO. Jakob is past vice president of IFORS, past president of EURO, and has served for 28 years as the Danish representative to both institutions. Jakob was but a young Ph.D. student when he met Steven Vajda for the first time. Steven's influence, however, appeared to be crucial to Jakob's later career. Their paths did not cross again before Steven was 91 years old. Strangely enough, as a side effect of the visible result of their ensuing cooperation, Jakob happened to be the only co-author of a paper with Steven—Steven's last publication which appeared 2 years after he passed away.

David C. Lane

Profile: Jay Wright Forrester (joint with John D. Sterman)

David is a Reader in Management Science at the London School of Economics (LSE) and Political Science. He has a B.Sc. in Mathematics from Bristol University, an M.Sc. in Mathematical Modeling and Numerical Analysis from Oxford University, and a Doctorate in Mathematics also from Oxford. He was a consultant in Shell International (where he led the system dynamics group) and a marketing manager in Shell U.K. His research and publications center on system dynamics. His theoretical work focuses on the formalization of certain concepts arising in social theory. His practical interests are varied, including: problems in healthcare management; the dynamics of project management; and the creation of marketing strategies for cinema films. He has also written on the methodology and history of both OR and system dynamics. He serves on the editorial board of the international journal *Systems Research and Behavioral Science* and is a fellow of the ORS. He received an LSE award for outstanding performance and innovation in teaching (2005) and a student-nominated LSE Students' Union Teaching Excellence Award (2008). In 2007 he was awarded the System Dynamics Society's Jay Wright Forrester Award for outstanding contribution to the field of system dynamics.

JOHN F. MAGEE

Profile: George E. Kimball

John Magee received a B.A. in mathematics and economics from Bowdoin College, an M.B.A. from Harvard University and an M.A. from the University of Maine in mathematics, where his thesis topic was "Economic Applications of Linear Programming." He joined Arthur D. Little, Inc., in 1950, as the first full-time member of its pioneering industrial Operations Research Group. He retired from the company as CEO in 1988, and as non-executive chairman in 1998. He was a founding member and fellow of the Operations Research Society of America, served as president of ORSA and of TIMS, received the Kimball Medal from ORSA and was elected a fellow of INFORMS. He is the author of several papers and books on management; his writings on logistics have been translated and published throughout the world. John worked closely with George Kimball for the decade after Kimball joined Arthur D. Little, Inc., first as a consultant and later as a full-time officer, on assignments for clients in logistics, marketing, and information systems, and they remained close until Kimball's death in 1967.

JAMES E. MATHESON

Profile: Ronald A. Howard

Jim is Chairman and Chief Financial Officer of SmartOrg, Inc., and Consulting Professor of Management Science and Engineering at Stanford University. He also serves on the board of directors of the Decision Education Foundation and on the editorial board of the journal *Decision Analysis*. He received a B.S. in electrical engineering from Carnegie-Mellon University and an M.S and Ph.D. in electrical engineering from Stanford

University. Jim began his career in decision analysis by developing methodology to evaluate research and development projects for the Westinghouse Research Laboratories in 1964. In 1966, he joined the Stanford Research Institute where he directed a new department, the Decision Analysis Group, with Ron Howard as a key advisor and colleague. Jim's group pioneered the application of decision analysis to business and government. In 1981, Jim co-founded the Strategic Decisions Group and, in 2000, he co-founded a spin-off to form SmartOrg, a provider of online software to allow customers with little formal training to apply decision analysis to projects and portfolios. In 1995, Jim received the INFORMS Ramsey Medal for Distinguished Contributions in Decision Analysis, the first practitioner to receive this honor. He is co-author with David Matheson of *The Smart Organization.*

DANIEL S. NAGIN

Profile: Alfred Blumstein

Daniel S. Nagin is Teresa and H. John Heinz III University Professor of Public Policy and Statistics in the Heinz College, Carnegie Mellon University. Al Blumstein was the chair of his dissertation committee. He and Al, along with Jacqueline Cohen, co-authored the 1978 National Research Council Report *Deterrence and Incapacitation: Estimating the Effects of Criminal Sanctions on Crime Rates*. Nagin is an elected fellow of the American Society of Criminology and of the American Society for the Advancement of Science. He is the 2006 recipient of the American Society of Criminology's Edwin H Sutherland Award (for research contributions). In 2008, he was selected to be a University Professor. His research focuses on the evolution of criminal and antisocial behaviors over the life course, the deterrent effect of criminal and non-criminal penalties on illegal behaviors, and the development of statistical methods for analyzing longitudinal data

ROBERT M. OLIVER

Profile: Philip McCord Morse

Bob is Professor Emeritus of Operations Research and Engineering Science at the University of California, Berkeley. He received a Fulbright

Scholarship in Mathematics and Physics to the University of London in 1953, and obtained his doctorate in Physics and Operations Research at the Massachusetts Institute of Technology in 1957. He was appointed Director, Management Science Division, Broadview Research Corporation in 1958 and, jointly with Aryeh Samuel, was awarded the1963 Lanchester Prize of the Operations Research Society of America for research in predicting and optimally scheduling U.S. Post Office mail flow and sorting operations. At Berkeley, he taught courses in forecasting, prediction, and decision-making models. He was Chairman of the IEOR Department from 1964 to 1969 and Director of the Operations Research Center from 1972 to 1975. He is the author of numerous papers and three books. He was awarded the University of California Berkeley Citation in 1993. In 1995, he was appointed Chairman of the Board of the Fair Isaac Companies, a leading provider of scoring and decision models for controlling financial risk in retail credit portfolios. He has been a Trustee of the Mathematical Sciences Research Institute and is former Chairman of the Board of the ANSER Corporation and the Homeland Security Institute for the Department of Homeland Security in Arlington, Virginia. He is a fellow of INFORMS and a member of the National Academy of Engineering.

JAMES B. ORLIN

Profile: D. Ray Fulkerson (joint with Robert G. Bland)

Jim Orlin is the Edward Pennell Brooks Professor of Operations Research at the MIT Sloan School of Management, Cambridge, Massachusetts. He served as co-director of the MIT Operations Research Center from 1998 to 2006. He is an expert in network flow algorithms, and has developed faster algorithms for several combinatorial and network optimization problems including: the shortest path problem, the maximum flow problem, the minimum cost flow problem, the generalized max flow problem, the minimum mean cycle problem, and the minimum (global) cut problem. With Ravi Ahuja and Tom Magnanti, he co-authored a graduate-level text, *Network Flows: Theory, Algorithms, and Applications* (1993). This text, which was offered as a tribute to Ford and Fulkerson for their pioneering work in network flows, was the winner of the 1993 Lanchester Prize for the best English language publication in Operations Research. He is also an INFORMS fellow, a past Fulbright Scholar, and winner of the following awards for research publications: The EXPLOR Award (Online Marketing Research), the Koopmans Prize (Military

Operations Research), the IEEE Leonard G. Abraham Prize (research in communications), and the INFORMS Computing Society Prize.

GUILLERMO OWEN

Profile: Harold W. Kuhn (joint with Saul I. Gass)

Guillermo was born on May 4, 1938 in Bogota, Colombia. He first came to the U.S. in 1947 and stayed there until 1951. After finishing high school in Lausanne, Switzerland, he returned to the U.S. in 1954 to attend Fordham University, where he obtained his B.A. degree in 1958. He attended Princeton University from 1958 to 1961, and obtained his Ph.D. degree in mathematics in 1962. In 1961, he joined the faculty at Fordham University as assistant professor of mathematics, and stayed there until 1969. He then became associate professor of mathematical sciences at Rice University and stayed there until 1977. After a few years as professor of mathematics at the University of the Andes (Bogota, Colombia), he became visiting professor of mathematics at the Naval Postgraduate School (NPS) in 1980. In 1983, he was made professor of mathematics at NPS, and has remained on its faculty to this date. He was faculty chairman in 1991–1992, and was chairman of the department of applied mathematics in 1998–1999. Guillermo has worked as consultant to several industrial firms and think tanks. He is a member of three academies of science and has served as associate editor on five scholarly journals.

PERKINS C. PEDRICK

Profile: Murray Aaron Geisler

Perk is Retired President and Trustee Emeritus, Logistics Management Institute (LMI), McLean, Virginia. Now living in Palm Beach Gardens, Florida, with summer residence in Ocean, New Jersey, Perk holds a B.A. in economics from Haverford College and an M.S. in mathematics from the University of Michigan. He joined LMI in 1963, the same year Murray became Head of the Logistics Department of the RAND Corporation. They shared a vision of how operations research could transform the performance of logistics. They also enjoyed a warm friendship in both collaboration and competition, as Murray built Logistics into RAND's largest department and Perk became President of

LMI. When Murray desired to return from management to analysis for the final decade of his career, Perk created a unique position, Senior Logistician, for him at LMI. Before retirement, Perk also served as Deputy Inspector General of the U.S. Foreign Service and President of the International Club of Washington.

FRED Y. PHILLIPS

Profile: Abraham Charnes (joint with Lawrence M. Seiford)

As a high-school student, Fred Phillips went to work for Abraham Charnes at Northwestern University in 1968. The following year, having persuaded George Kozmetsky to extend scholarship assistance, Charnes brought Fred to Texas for undergraduate studies in mathematics. These led to a doctoral dissertation, under Charnes, on extremal principles for Kullback-Leibler statistics, with applications in transportation and marketing. During his years at Market Research Corporation of America (MRCA), Fred continued to work with Charnes, and with Bill Cooper, in their roles as consultants to MRCA. Later, Fred joined the faculty at UT-Austin and served as Associate Director of the Center for Cybernetic Studies from 1989 until Charnes's death. Fred is now a professor at Maastricht School of Management in the Netherlands, at PUCP-Lima in Peru, and at Alliant International University in San Diego. He is senior editor of the journal *Technological Forecasting & Social Change* and a fellow of UT-Austin's IC^2 Institute.

ROBERT T. PLANT

Profile: David Bendel Hertz (joint with Edward K. Baker)

Robert is an associate professor in the Department of Computer Information Systems at the University of Miami, School of Business Administration, and is the Research Director of the Intelligent Computer Systems Research Institute. He obtained his Ph.D. in Computer Science at the University of Liverpool, England. He previously had studied computation at the Programming Research Group, Oxford University, England, and at Wadham College, Oxford. His early research focused on the validation and verification of knowledge based systems (KBS) as well as the application of formal methods to KBS design methodologies. His recent research has focused on strategy and information systems. His publications have been published in *MIT Sloan Management Review*, *Harvard Business Review*, *Journal of MIS*, and *CACM*. Robert is a Chartered Engineer (U.K.), a European Engineer, a Senior Member of the American Institute of Aeronautics and Astronautics, and a fellow of the British Computer Society. Dr. Plant has held visiting professorial roles at Universidad Gabriela Mistral, Santiago de Chile, and Wuhan University, China.

GRAHAM K. RAND

Profiles: Egon Balas, Berwyn Hugh Patrick Rivett

Graham Rand is a senior lecturer in Operational Research at Lancaster University, U.K. He received his B.Sc. in Computational Science and Statistics from Liverpool University. He has been council member and conference chairman of the British Operational Research Society, and was editor of the *Journal of the Operational Research Society* from 1991 to 1996. For the International Federation of Operational Research Societies (IFORS), he was vice-president (1998–2000), editor of both *International Abstracts in Operations Research* (1980–1991) and *International Transactions in Operational Research* (2000–2005), editor of the proceedings of the 1987 Buenos Aires Conference, and chairman of the program committee for the 1990 Athens Conference. For the European Association of Operational Research Societies, he was chairman of the Gold Medal jury in 1995. For INFORMS, he has been involved with the Edelman Prize as verifier, jury member, and as coach to the prize-winning entry from Dutch Railways in 2008. Graham was a founder member of both the International Society for Inventory Research (1983) and the Operations Management Association (1985). He is European director of Omega Rho. His recent research, with several Ph.D. students, has been in aspects of quality management.

DAVID F. ROGERS

Profile: E. Leonard Arnoff

David Rogers received his B.S. in mathematics/general business and M.B.A. in quantitative methods from Murray State University and his Ph.D. in Management (quantitative methods/operations management) from the Krannert Graduate School of Management, Purdue University. He is currently associate professor, Department of Operations and Business Analytics Management, College of Business, University of Cincinnati, where he began as assistant professor when Len Arnoff became Dean of the College. Dean Arnoff encouraged David to get involved with Omega Rho activities and he initiated the University of Cincinnati Chapter, matriculated through the executive committee to the presidency of Omega Rho, and is currently Omega Rho historian and Web site manager. He was recognized by the Association of College Honor Societies in 2000 as one of the five most influential persons for Omega Rho. Dave has been recognized three times for his teaching excellence, including the Michael Dean Excellence in Undergraduate Classroom Education and Learning award in 2006. He has served on the INFORMS Board of Directors and was General Chair for the 1999 INFORMS conference. Dave's research interests are in supply chain management modeling, aggregation and disaggregation techniques, portfolio risk management, and cellular manufacturing. He is the coordinator for the E. Leonard Arnoff Memorial Lecture on the Practice of Management Science.

JONATHAN ROSENHEAD

Profiles: Russell Lincoln Ackoff (joint with Maurice W. Kirby), Stafford Beer, Patrick Blackett (joint with Maurice W. Kirby), Leonid Vital'evich Kantorovich (joint with Saul I. Gass)

Before becoming an academic, Jonathan Rosenhead worked in the steel industry and as a management consultant. From 1967 he was on the staff of the London School of Economics, and became Professor of Operational Research in 1987. He is now Emeritus Professor but still teaching there. For two periods he held positions at the University of Pennsylvania. He was President of the Operational Research Society in 1986–1987, and holds its President's, Goodeve and Beale Medals. His fields of research interest include problem structuring methods (PSMs), health services planning, community operational research, OR in less developed countries, and the history of OR. Two of these involvements—Community OR and Problem Structuring Methods—grew out of the manifesto for his (contested) election as President of ORS. (His jointly edited book, *Rational Analysis for a Problematic World Revisited*, Wiley 2001, is the standard introduction to PSMs.) He believes that academics have social responsibilities as well as technical challenges. His interest in the history of OR grew out of a desire to better understand the mutual links between OR and the society which generated it.

MICHAEL H. ROTHKOPF

Profile: Hugh J. Miser

Mike Rothkopf died on February 18, 2008 while swimming his daily mile at the McCoy Natatorium at Pennsylvania State University. He was 68 years old. Mike was Smeal Chaired Professor of Supply Chain and Information Systems at Penn State's Smeal College of Business. Before joining Penn State, he was a professor at Rutgers University Business School and Center for Operations Research for 19 years. He had submitted a draft of his Hugh Miser profile in January 2008. We were fortunate to have Mike as a friend and colleague, and very thankful to have his profile of his dear friend Hugh. Mike contributed much to the OR profession in terms of his research and service. He received his Ph.D. in OR (minor in economics) from MIT in 1964. His research expertise included energy economics, auctions, models of markets and competitive bidding, and supply chain analysis. He was awarded ORSA's George E. Kimball Medal for distinguished service to the society and the profession (1997). He was vice-president of TIMS (1979–1982; 1992–1994), ORSA council member (1981–1984), and president of INFORMS in 2004. He was area editor of OR Practice for *Operations Research* (1984–1093), and editor-in-chief of *Interfaces* (1993–2000). Mike was a lifetime member of the National Audubon Society, an avid bird watcher who had recently identified his 1000th bird species.

TIMOTHY W. RUEFLI

Profile: William W. Cooper (joint with Robert R. Wiggins)

Tim Ruefli died on June 19, 2010 in Austin, Texas. He was Bill Cooper's Ph.D. student in the late sixties at Carnegie (Andy Stedry and Yuji Ijiri, both Bill's former students were on the committee). As a result of such a great start in his academic career and having Bill as a role model, he ended up as the Daniel B. Stuart Centennial Professor in Applications of Computers to Business in the Information, Risk & Operations Management Department of the McCombs School of Business, and the Frank C. Erwin Jr. Centennial Research Fellow of the IC^2 Institute of the University of Texas at

Austin. He was also a Research Fellow of the Herb Kelleher Center for Entrepreneurship, The University of Texas at Austin. Tim taught at the Carnegie Institute of Technology and the University of British Columbia, as well as at the University of Texas, and was a member of the extended faculty of Escuela de Graduados en Administración y Dirección de Empresas at the Instituto Technológico y de Estudios Superiores de Monterrey, Monterrey, N.L., Mexico. He was a consultant on strategic management and management science to numerous private and public sector organizations. He held a B.A. from Wesleyan University, an M.S. from Carnegie Institute of Technology and a Ph.D. from Carnegie-Mellon University

LAWRENCE SECREST

Profile: George Kozmetsky (joint with John Sibley Butler and David V. Gibson)

Larry Secrest is Managing Partner of Secrest & Co., a multinational consulting firm focusing on technology innovation and management. He studied under Dean George Kozmetsky at the University of Texas Graduate School of Business Administration where he received an interdisciplinary Ph.D. in Management, Business, and Public Affairs. Larry served as Visiting Research Fellow at Keio University, Tokyo, Japan, where he studied technology commercialization in Japanese corporations. He has served on the faculties of the University of Texas at Austin and Arlington. He designed and taught the first government finance and accounting course at the LBJ School, and has taught graduate courses in strategy, entrepreneurship, international business, and management. Larry has served as CEO or senior executive for a number of fast growth companies including Stevens Water, Hydrolab, Scientific Measurement Systems, NeuroSystems, and SMC.

LAWRENCE M. SEIFORD

Profile: Abraham Charnes (joint with Fred Y. Phillips)

Larry is Professor and Chair of Industrial and Operations Engineering at the University of Michigan. Prior to joining the University of Michigan, he was Program Director of the Operations Research and Production Systems programs at the National Science Foundation (1997–2000), and was a member of the faculty at the

University of Massachusetts, the University of Texas at Austin, the University of Kansas, and York University. He completed his Ph.D. in mathematics under the guidance of Abraham Charnes. His current teaching and research interests are primarily in the areas of quality engineering, productivity analysis, process improvement, and performance measurement. In addition, he is recognized as one of the world's experts in the methodology of Data Envelopment Analysis. Larry has written and co-authored four books and over 100 articles in the areas of quality, productivity, operations management, process improvement, decision analysis, and decision support systems. He is past Editor-in-Chief of *OMEGA, the International Journal of Management Science*, and has been or is on the editorial boards of nine scientific journals. He is a fellow of INFORMS, the Institute of Industrial Engineers, and the American Society for Quality. He received the General Electric Outstanding Teaching Award, the CBA Foundation Award for Research Excellence, and has been a Lily Endowment Teaching Fellow. He was awarded the degree Docteur Honoris Causa from the National Ministry of Education of France in a special recognition ceremony at the Université de la Méditerranée, Aix-Marseille II in November, 2000.

JOHN D. STERMAN

Profile: Jay Wright Forrester (joint with David C. Lane)

John Sterman is the Jay W. Forrester Professor of Management at the MIT Sloan School of Management and Director of the MIT System Dynamics Group. He received an A.B. at Dartmouth College and Ph.D. at MIT, where he worked under Prof. Forrester. He is the author of many scholarly and popular articles on the challenges and opportunities facing organizations today, including the award-winning textbook *Business Dynamics.* John's research centers on improving managerial decision making in complex systems, with applications ranging from the dynamics of organizational change to climate change. He has pioneered the development of management flight simulators of corporate and economic systems, simulators now used by corporations and universities around the world. His work on management flight simulators was named one of the top 50 articles published in the first 50 years of the journal *Management Science.* John has twice been awarded the Jay W. Forrester Prize for the best published work in system dynamics, won a 2005 IBM Faculty Award, won the 2001 Accenture Award for the best paper of the year published in the *California Management Review* (with Nelson Repenning), has six times won awards for teaching excellence from the students of MIT and the Sloan School, and was named one of the Sloan School's outstanding faculty by the *Business Week Guide to the Best Business Schools.*

SHALER STIDHAM, JR.

Profile: Harvey M. Wagner

Shaler Stidham, Jr. is currently Professor Emeritus in the Department of Statistics and Operations Research at the University of North Carolina at Chapel Hill. His research areas include queueing theory (optimal design and control, sample-path analysis), Markov decision processes, telecommunications, and revenue management. He received his B.A. in mathematics from Harvard College, M.S. in operations research and computing from Case Institute of Technology, and Ph.D. in operations research from Stanford University. He has held academic positions at Cornell University, N.C. State University, and the University of North Carolina at Chapel Hill, where he was Chair of the Department of Operations Research from 1990 to 1995. Shaler held visiting positions at the University of Aarhus; Stanford University; the Technical University of Denmark; the University of Cambridge (Overseas Fellow, Churchill College); and INRIA, Sophia-Antipolis, France. He was co-program chair of the Applied Probability Conference at UNC-CH in 1988, and the TIMS International Meeting (Osaka, Japan) in 1989. He has also served INFORMS as chair of the Applied Probability Society and a member of the Expository Writing Prize Committee. He is co-author of *Sample-Path Analysis of Queueing Systems* (1999) and author of *Optimal Design of Queueing Systems* (2009). He has served on the N.C. Museum of Art Foundation and the N.C. Symphony Foundation (President, 2003–2007).

JOHN A. TOMLIN

Profile: Evelyn Martin Lansdowne Beale

John received his B.Sc.(Hons) and Ph.D. degrees in mathematics at the University of Adelaide, South Australia. He then joined Scicon, Ltd. (then called C-E-I-R, Ltd.) in London, and worked under Martin Beale on the development of the UMPIRE Mathematical programming System. One of the outcomes of this work was the publication of "Special Facilities in a General Mathematical Programming System for Non-Convex Problems Using Ordered Sets of Variables" by Beale and Tomlin, beginning a series of papers on this subject, by Beale and others, that continued for many years. This and other innovations in UMPIRE influenced all subsequent Mathematical Programming System development. John subsequently joined the Operations Research Department at Stanford University (under George Dantzig), Ketron, Inc., and IBM Research, where he was a long-term member of the Optimization Subroutine Library development team, as well as a consultant on optimization applications. In 2005, he took up his present position at the newly formed Yahoo! Research Labs.

Glen L. Urban

Profile: John D. C. Little (with John R. Hauser)

Glen is a leading educator and researcher specializing in marketing and new product development. He has been a member of the MIT Sloan School of Management faculty since 1966, was Deputy Dean at the school from 1987 to 1991, and Dean from 1993 to 1998. Glen is co-author of seven books and numerous papers. His papers have won several prestigious awards, including two *Journal of Marketing Research* William F. O'Dell awards. He has received the American Marketing Association's Paul D. Converse Award for outstanding contributions to marketing scholarship and the Charles C. Parlin Marketing Research Award. Over the last 10 years, Glen's research has concentrated on trust-based marketing systems and tools for customer advocacy, supported by MIT-sponsored research projects at Intel, GM, BT, and Suruga Bank. His recent research is on Web site design and morphing digital ads to improve their effectiveness on the Web.

Daniel Vanderpooten

Profile: Bernard Roy (joint with Denis Bouyssou)

Daniel Vanderpooten is a professor in computer science and operations research, Université Paris Dauphine, France. He received his Ph.D. under the guidance of Bernard Roy in 1990 from Université Paris Dauphine. He obtained his *Habilitation à Diriger des Recherches* in 1998.

His main fields of interests are in theoretical and practical aspects of OR, more precisely in multiple criteria decision making, combinatorial optimization, and robust optimization. He is head of the decision aiding research team of LAMSADE. He is in charge of the master program MODO in OR at Université Paris Dauphine. Daniel has been involved in various academic and industrial research projects. He is a member of the editorial board of the *Journal of Multi-Criteria Decision Analysis*. He is the co-author of two books, and has published over 50 articles in various refereed international journals and international conferences proceedings.

LUIS G. VARGAS

Profile: Thomas L. Saaty

Luis is professor of operations, decision sciences and artificial intelligence, the Joseph M. Katz Graduate School of Business, University of Pittsburgh. He received his B.S. in mathematics and M.S. (operations research) from the University of Granada, Spain, and his Ph.D. (mathematics) from the University of Sevilla, Spain. He was the recipient of the Juan March Foundation Scholarship, Madrid, Spain, to the University of Pennsylvania, where he studied under the guidance of Thomas Saaty and received his Ph.D. (social systems sciences) from the Wharton School, University of Pennsylvania. He was chairman of the second International Symposium on the Analytic Hierarchy Process held in Pittsburgh August 11–14, 1991 at the Joseph M. Katz Graduate School of Business. Luis has coauthored five books with Thomas Saaty: *The Logic of Priorities: Applications in Business, Energy, Health and Transportation*; *Prediction, Projection, and Forecasting: The Analytic Hierarchy Process*; *Decision Making in Economic, Political, Social and Technological Environments with the AHP*; *Models, Methods, Concepts & Applications of The Analytic Hierarchy Process*; and *Economic, Political, Social & Technological Applications of The Analytic Network Process (ANP): Benefits, Opportunities, Costs and Risks.*

NANCY C. WEIDA

Profile: Andrew Vazsonyi

Nancy is an associate professor of decision sciences in the School of Management at Bucknell University. A dedicated and energetic teacher, she is winner of numerous teaching and advising awards from her students, from Bucknell University, and from the University of Delaware. She earned an A.B. in mathematics and history from Colgate University, an M.S. in operations research/mathematics from the University of Delaware, and a Ph.D. in operations research/business administration with a concentration in operations management from the University of Delaware. Nancy is co-author with Andy Vazsonyi and Ronny Richardson of the book *Operations Analysis Using Microsoft Excel.* She has published and presented in the areas of mathematical modeling and game theory, and loved debating ideas and working on projects with Andy Vazsonyi. She enjoys being active in professional organizations, having served as a vice president for the Decision Sciences Institute, a founding vice president of the Academy of Business Education, and a member of the board of the Northeast Decision Sciences Institute. In addition, she has held numerous other committee memberships and leadership positions in these and other organizations.

ROBERT R. WIGGINS

Profile: William W. Cooper (joint with Timothy W. Ruefli)

Robert Wiggins was Tim Ruefli's Ph.D. student in the early 1990s at Texas, and Bill Cooper was a very active member of his dissertation committee, as was Preston McAfee, a student of Andrew Whinston's at Purdue, who in turn was yet another of Bill Cooper's Ph.D. students. Robert is currently an Associate Professor of Management in the Fogelman College of Business and Economics at the University of Memphis. He has taught at the A.B. Freeman School of Business at Tulane University and the McCombs School of Business at the University of Texas at Austin, as well as the University of Memphis. His consulting clients have included Goldman Sachs, American Express, McKinsey and Company, and many others. Robert was also a contributing editor for *MacUser* magazine (U.S.) and *CDROM* magazine (Japan), and is currently a board member of Ray West Warehouses, Inc., a privately held logistics firm in South Texas. He holds an A.B. from Vassar College, and an MBA and Ph.D. from the University of Texas at Austin.

Acronyms

A/S	Anti-Submarine
AA	Anti-Aircraft
ABM	Aberdeen Bombing Mission
ACDA	(U.S.) Arms Control and Disarmament Agency
Ack-Ack	Anti-Aircraft
ADBUDG	Advertising Budget
ADL	Arthur D. Little, Inc.
AFCET	Association Française pour la Cybernétique Économique et Technique
AFIRO	Association Française d'Informatique et de Recherche Opérationnelle
AHP	Analytic Hierarchy Process
AI	Artificial Intelligence
AMA	American Management Association
AMC	Air Material Command
ANALAA	ORO study on air defense
AOA	Activity on Arc
AON	Activity on Node
AORG	Army Operational Research Group
ARC	Aeronautical Research Committee
ARL	Admiralty Research Laboratory
ASA	American Statistical Society
ASB	Army Science Board
ASC	American Society of Criminology
ASCA	Airplane Stability and Control Analyzer
ASLIB	Association of Special Libraries & Information Bureau

ASTP	Army Specialized Training Program
ASW	Anti-Submarine Warfare
ASWORG	Anti-Submarine Warfare Operations Research Group
ATI	Austin Technology Incubator
BAH	Booz, Allen, and Hamilton
BISRA	British Iron and Steel Research Association
BLDM	Battalion Level Differential Models
Bonder/IUA	Bonder/Independent Unit Action
BRANDAID	Brand Aid (model to help evaluate the marketing mix)
BRL	Ballistics Research Laboratory
C&O	Chesapeake and Ohio (Railroad)
CAA	Civil Aviation Administration's
CACI	California Analysis Centers, Inc
CAL	Cornell Aeronautical Laboratory
Caltech	California Institute of Technology
Cal-Tech	California Institute of Technology
CAPM	Capital Asset Pricing Model
CBE	Council of Biology Editors
CCNY	City College of New York
CCP	Chance Constrained Programming
CEIR	Corporation for Economic and Industrial Research
CEMI	Central Economics and Mathematics Institute
CEO	Chief Executive Officer
CHAID	Chi-Square Automatic Interaction Detection
CIA	Central Intelligence Agency
CIS	Computer Information Systems
CJS	Criminal Justice System
CMU	Carnegie-Mellon University
CNA	Center for Naval Analyses
CNRS	Centre National de la Recherche Scientifique
COCOMO	COnstructive COst MOdel
CONDOR	Committee on the Next Decade in Operation Research
CORFO	Corporación de Fomento de la Producción
CORS	Canadian Operational Research Society
CPC	Card-Programmed Electronic Calculator (IBM)
CPC	Card Programmed Calculator
CPM/PERT	Critical Path Method/Project Evaluation and Review Technique
CS	Computer Science

CSL	Control and Simulation Language
CSSD	Committee for the Scientific Survey of Air Defence
DA	Decision Analysis
DAG	Decision Analysis Group
DEA	Data Envelopment Analysis
DEC	Digital Equipment Corporation
DEF	Decision Education Foundation
D-Memos	Dynamic Modeling Memos
DoD	Department of Defense
DOE	Department of Energy
DP	Dynamic Programming
DSI	Decision Sciences Institute
DSIR	Road Research Department of the Department of Scientific and Industrial Research
DSMC	Defense Systems Management College
DYNAMO	DYNAMic MOdeling
EAS-E	EASY (EAS stands for entity, attribute, and set)
EDF	Électricité de France
EDVAC	Electronic Discrete Variable Computer
ELECTRE	Élimination et Choix TRaduisant la Réalité
ENIAC	Electronic Numerical Integrator and Computer
EURO	The Federation of European OR Societies
EVANAL	ORO study on operations of equipment under various environmental conditions
FAA	Federal Aviation Authority
FIG	Field Investigation Group
FoOp	Friends of Optimization
FP	Fractional Programming
FRS	Fellow of the Royal Society
GAO	General Accounting (Accountability) Office
GCA	Ground Controlled Approach
GE	General Electric Corporation
GEMS	General Electric Manufacturing Simulator
GI Bill	Servicemen's Readjustment Act of 1944
Gosplan	Gosudarstvennyi Komitet po Planirovaniyu (USSR State Planning Commission)
GPRD	Global Portfolio Research Department
GPS	General Problem Solver
GRO	General Research Office

GSIA	Graduate School of Industrial Administration
GSL	Generalized Simulation Language
GUB	Generalized Upper Bounds
HBR	*Harvard Business Review*
HBS	Harvard Business School
HCM	Healthcare Complex Model
HM	Hungarian Method
HMMS	Holt, Modigliani, Muth, and Simon
HQ USAF	Headquarters, U.S. Air Force
IAOR	*International Abstracts in Operations Research*
IAS	Institute for Advanced Study (Princeton)
IC^2	Institute for Constructive Capitalism/Institute for Innovation, Creativity and Capital
ICMA	International City Managers' Association
IDA	Institute for Defense Analyses
IED	Improvised Explosive Devices
IEEE	Institute of Electrical and Electronics Engineers
IEP	Institut d'Études Politiques
IFORS	International Federation of Operational Research Societies
IIASA	International Institute for Applied Systems Analysis
IIT	Illinois Institute of Technology
IMA	Institute for Mathematics and its Applications
INA	Institute for Numerical Analysis
INFORMS	Institute for Operations Research and the Management Sciences
INTERACT	Institute for Interactive Management
IOM	Institute of Medicine
IOR	Institute for Operational Research
IP	Integer Programming/Integer Program
IPC	International Publishing Corporation
IRI	Information Resources, Inc.
ISEP	Institute of Economic Studies and Planning
ISMS	INFORMS Society for Marketing Science
ISUP	Institut de Statistique de l'Université de Paris
ITI	Identification Technologies International
JHU	Johns Hopkins University
JIEDDO	Joint Improvised Explosive Device Defeat Organization
JLRB	Joint Logistics Review Board
JUSSIM	Justice System Interactive Model

K-12	Kindergarten through 12th Grade (high school)
KKT	Karush-Kuhn-Tucker
KT	Kuhn-Tucker
LAMSADE	Laboratoire d'Analyse et Modélisation de Systèmes pour l'Aide à la Décision
LAV	Least Absolute Value (regression)
LBS	London Business School
LIFO	Last In First Out
LMI	Logistics Management Institute
LOS	Line of Sight
LP	Linear Programming/Linear Program
LSE	London School of Economics
LT	Logic Theorist
M.I.T./MIT	Massachusetts Institute of Technology
MAA	Mathematical Association of America
MAID	ORO study on military aid to foreign countries
MARSAN	Méthode d'Analyse et de Recherche pour la Sélection des Activités Nouvelles
MAXBAND	Maximum Bandwidth
MBA	Master of Business Administration
MCDA	Multiple Criteria Decision Aiding
MCDM	Multiple Criteria Decision Making
MDS	Management Decision Systems
MEDIAC	Media Calculus
MG	Matrix Generator
MGG	Matrix Generator Generator
Mintech	Ministry of Technology
MIP	Mixed Integer Programming
MIPO	Mixed Integer Program Optimizer
MIT	Massachusetts Institute of Technology
MITRE	MIT Research and Engineering
MOLP	Multi-Objective Linear Programming
MORS	Military Operations Research Society
MP	Mathematical Programming
MPM	Méthode des Potentiels Metra
MPS	Mathematical Programming Society
MRP	Materials Requirements Planning
MS	Management Science(s)
MSCM	Minimum Software Cost Model

MSD	Mathematical Sciences Department
MSOM	Manufacturing and Service Operations Management Society
MWORG	Mine Warfare Operations Research Group
NAE	National Academy of Engineering
NAS	National Academy of the Sciences
NASA	National Aeronautics and Space Administration
NATO	North Atlantic Treaty Organization
NBS	National Bureau of Standards
NCAA	National Collegiate Basketball Association
NCB	National Coal Board
NCOVR	National Consortium of Violence Research
NIJ	National Institute of Justice
NIST	National Institute of Standards and Technology
NIT	National Invitation Tournament
NLP	Nonlinear Programming
NOL	Naval Ordnance Laboratory
NP	Nondeterministic Polynomial-Time
NPS	Naval Postgraduate School
NRC	National Research Council
NRL	Naval Research Logistics
NRLQ	Naval Research Logistics Quarterly
NSS	Newell-Simon-Shaw
NYC	New York City
O/R	Operations Research
OA	Operations Analysis
OAO	Operations Analysis Office
OBE	Order of the British Empire
OCS	Officer Candidate School
OEG	Operations Evaluation Group
ONR	Office of Naval Research
OR	Operations Research/Operational Research
OR/MS	Operations Research/Management Sciences
ORC	Operations Research Center
ORG	Operations Research Group
ORQ	*Operational Research Quarterly*
ORS	Operational Research Society (U.K.)
ORS	OR Section
ORSA	Operations Research Society of America

ORSSA	Operations Research Society of South Africa
OSU	Ohio State University
PCAST	President's Council of Advisors on Science and Technology
Penn	University of Pennsylvania
PILOT	Planning Investment Levels Over Time
PIN	Personal Identification Number
PON	Program on Negotiation
PORS	Publications in Operations Research Series
PPG	Pittsburgh Plate Glass
PSAC	President's Science Advisory Committee
PSMs	Problem Structuring Methods
QAIS	Quantitative Analysis and Information Systems
R&D	Research and Development
RA	Research Assistant
RADAR	Radio Detection and Ranging
RAE	Research Assessment Exercise
RAE	Royal Aircraft Establishment
RAF	Royal Air Force
RAIRO	*Revue d'Automatique, d'Informatique et de Recherche Opérationnelle*
RATP	Régie Autonome des Transports Parisiens
RDF	Radio-Direction Finding
RIRO	*Revue d'Informatique et de Recherche Opérationnelle*
RNVR	Royal Navy Volunteer Reserve
ROADEF	Recherche Opérationnelle et Aide à la décision (the French OR Society)
ROE	Return on Equity
ROTC	Reserve Officers' Training Corps
RTI	Research Triangle Institute
S&T	Science and technology
S^3	Social Systems Science
SAGE	Semi-Automatic Ground Environment
SAM	Stochastic Analogue Machine
Scicon	Scientific Control Systems
SCOOP	Scientific Computation of Optimal Programs
SDG	Strategic Decisions Group
SEAC	Standards Eastern Automatic Computer
SEMA	Société d'Économie et de Mathématiques Appliquées
SEPRO	Société d'Études Pratiques de Recherche Opérationnelle

SIGMA	Science in General Management
SIMPLE	Simulation of Industrial Management Problems with Lots of Equations
SIU	Southern Illinois University
SLIM	Software LIfecyle Management
SOFRO	Société Française de Recherche Opérationnelle
SOL	Systems Optimization Laboratory
SOM	School of Organization and Management
SPS1	Simulation Programming System One
SRG	Systems Research Group
SRI	Stanford Research Institute
SRL	Systems Research Laboratory
SSL	Space Sciences Laboratories
SUPA	School of Urban and Public Affairs
SVD	Singular Value Decomposition
SWAC	Standards Western Automatic Computer
TBI	Traumatic Brain Injury
TIMS	The Institute of Management Sciences
TRADOC	Training and Doctrine Command
TRS	Total Return to Shareholders
TSP	Traveling Salesman Problem
TVA	Tennessee Valley Authority
UCB	University of California, Berkeley
UCL	University College, London
UCLA	University of California, Los Angeles
UMPIRE	Universal Mathematical Programming System Incorporating Refinements and Extensions
UN	United Nations
UNC-CH	University of North Carolina at Chapel Hill
UPC	Universal Product Code
UT	University of Texas
USAF	U. S. Air Force
VRI	Vector Research Incorporated
WSEG	Weapons Systems Evaluation Group
WWI	World War I
WWII	World War II

NAME INDEX

C

D

E

F

G

H

L

M

N

O

P

Q

R

S

T

Y

Z

SUBJECT INDEX

B

C

D

E

H

I

J

K

L

M

N

O

P

Q

R

S

T

V

W

Y

Z